Mathematical Foundations of Fuzzy Sets

Mathematical Foundations of Fuzzy Sets

Hsien-Chung Wu
Department of Mathematics
National Kaohsiung Normal University
Kaohsiung
Taiwan

WILEY

This edition first published 2023

Registered Offices
John Wiley & Sons, Inc., 111 River Street, Hoboken, NJ 07030, USA
John Wiley & Sons Ltd, The Atrium, Southern Gate, Chichester, West Sussex, PO19 8SQ, UK

For details of our global editorial offices, customer services, and more information about Wiley products visit us at www.wiley.com.

Wiley also publishes its books in a variety of electronic formats and by print-on-demand. Some content that appears in standard print versions of this book may not be available in other formats.

Library of Congress Cataloging-in-Publication Data Applied for:

Hardback ISBN: 9781119981527

Cover Design: Wiley
Cover Image: © oxygen/Getty Images

Set in 9.5/12.5pt STIXTwoText by Straive, Chennai, India
Printed and bound by CPI Group (UK) Ltd, Croydon, CR0 4YY

C9781119981527_040123

Contents

Preface

The concept of fuzzy set, introduced by L.A. Zadeh in 1965, tried to extend classical set theory. It is well known that a classical set corresponds to an indicator function whose values are only taken to be 0 and 1. With the aid of a membership function associated with a fuzzy set, each element in a set is allowed to take any values between 0 and 1, which can be regarded as the degree of membership. This kind of imprecision draws forth a bunch of applications.

This book is intended to present the mathematical foundations of fuzzy sets, which can rigorously be used as a basic tool to study engineering and economics problems in a fuzzy environment. It may also be used as a graduate level textbook. The main prerequisites for most of the material in this book are mathematical analysis including semi-continuities, supremum, convexity, and basic topological concepts of Euclidean space, $\mathbb{R}^n$. This book presents the current state of affairs in set operations of fuzzy sets, arithmetic operations of fuzzy intervals and fuzzification of crisp functions that are frequently adopted to model engineering and economics problems with fuzzy uncertainty. Especially, the concepts of gradual sets and gradual elements have been presented in order to cope with the difficulty for considering elements of fuzzy sets such as considering elements of crisp sets.

- Chapter 1 presents the mathematical tools that are used to study the essence of fuzzy sets. The concepts of supremum and semi-continuity and their properties are frequently invoked to establish the equivalences among the different settings of set operations and arithmetic operations of fuzzy sets.
- Chapter 2 introduces the basic concepts and properties of fuzzy sets such as membership functions and level sets. The fuzzy intervals are categorized as different types based on the different assumptions of membership functions in order to be used for the different purposes of applications.
- Chapter 3 deals with the intersection and union of fuzzy sets including the complement of fuzzy sets. The general settings by considering aggregation functions have been presented to study the intersection and union of fuzzy sets that cover the conventional ones such as using minimum and maximum functions (t-norm and s-norm) for intersection and union, respectively.
- Chapter 4 extends the conventional extension principle to the so-called generalized extension principle by using general aggregation functions instead of using minimum function or t-norm to fuzzify crisp functions. Fuzzifications of real-valued and vector-valued functions are frequently adopted in engineering and economics problems that involve fuzzy data, which means that the real-valued data cannot be exactly collected owing to the fluctuation of an uncertain situation.

- Chapter 5 presents the methodology for generating fuzzy sets from a nested family or non-nested family of subsets of Euclidean space $\mathbb{R}^n$. Especially, generating fuzzy intervals from a nested family or non-nested family of bounded closed intervals is useful for fuzzifying the real-valued data into fuzzy data. Based on a collection of real-valued data, we can generate a fuzzy set that can essentially represent this collection of real-valued data.
- Chapter 6 deals with the fuzzification of crisp functions. Using the extension principle presented in Chapter 4 can fuzzify crisp functions. This chapter studies another methodology to fuzzify crisp functions using the mathematical expression in the well-known decomposition theorem. Their equivalences are also established under some mild assumptions.
- Chapter 7 studies the arithmetic operations of fuzzy sets. The conventional arithmetic operations of fuzzy sets are based on the extension principle presented in Chapter 4. Many other arithmetic operations using the general aggregation functions haven also been studied. The equivalences among these different settings of arithmetic operations are also established in order to demonstrate the consistent usage in applications.
- Chapter 8 gives a comprehensive and accessible study regarding inner product of fuzzy vectors that can be treated as an application using the methodologies presented in Chapter 7. The potential applications of inner product of fuzzy vectors are fuzzy linear programming problems and the engineering problems that are formulated using the form of inner product involving fuzzy data.
- Chapter 9 introduces the concepts of gradual sets and gradual elements that can be used to propose the concept of elements of fuzzy sets such as the concept of elements of crisp sets. Roughly speaking, a fuzzy set can be treated as a collection of gradual elements. In other words, a fuzzy set consists of gradual elements. In this case, the set operations and arithmetic operations of fuzzy sets can be defined as the operations of gradual elements, like the operations of elements of crisp sets. The equivalences with the conventional set operations and arithmetic operations of fuzzy sets are also established under some mild assumptions.
- Chapter 10 deals with the concept of duality of fuzzy sets by considering the lower α-level sets. The conventional α-level sets are treated as upper α-level sets. This chapter considers the lower α-level sets that can be regarded as the dual of upper α-level sets. The well-known extension principle and decomposition theorem are also established based on the lower α-level sets, and are called the dual extension principle and dual decomposition theorem. The so-called dual arithmetics of fuzzy sets are also proposed based on the lower α-level sets, and a duality relation with the conventional arithmetics of fuzzy sets is also established.

Finally, I would like to thank the publisher for their cooperation in the realization of this book.

Hsien-Chung Wu

Department of Mathematics
National Kaohsiung Normal University
Kaohsiung, Taiwan
e-mail 1: hcwu@mail.nknu.edu.tw
e-mail 2: hsien.chung.wu@gmail.com
Web site: https://sites.google.com/view/hsien-chung-wu
April, 2022

1

Mathematical Analysis

We present some materials from mathematical analysis, which will be used throughout this book. More detailed arguments can be found in any mathematical analysis monograph.

1.1 Infimum and Supremum

Let S be a subset of $\mathbb{R}$. The upper and lower bounds of S are defined below.

- We say that u is an **upper bound** of S when there exists a real number u satisfying $x \leq u$ for every $x \in S$. In this case, we also say that S is bounded above by u.
- We say that l is a **lower bound** of S when there exists a real number l satisfying $x \geq l$ for every $x \in S$. In this case, we also say that S is bounded below by l.

The set S is said to be unbounded above when the set S has no upper bound. The set S is said to be unbounded below when the set S has no lower bound. The maximal and minimal elements of S are defined below.

- We say that u^* is a **maximal element** of S when there exists a real number $u^* \in S$ satisfying $x \leq u^*$ for every $x \in S$. In this case, we write $u^* = \max S$.
- We say that l^* is a **minimal element** of S when there exists a real number $l^* \in S$ satisfying $x \geq l^*$ for every $x \in S$. In this case, we write $l^* = \min S$.

Example 1.1.1 We provide some concrete examples.

(i) The set $\mathbb{R}^+ = (0, +\infty)$ is unbounded above. It has no upper bounds and no maximal element. It is bounded below by 0, but it has no minimal element.
(ii) The closed interval $S = [0,1]$ is bounded above by 1 and is bounded below by 0. We also have $\max S = 1$ and $\min S = 0$.
(iii) The half-open interval $S = [0,1)$ is bounded above by 1, but it has no maximal element. However, we have $\min S = 0$.

Although the set $S = [0,1)$ is bounded above by 1, it has no maximal element. This motivates us to introduce the concepts of supremum and infinum.

Mathematical Foundations of Fuzzy Sets, First Edition. Hsien-Chung Wu.

Definition 1.1.2 Let S be a subset of $\mathbb{R}$.

(i) Suppose that S is bounded above. A real number $\bar{u} \in \mathbb{R}$ is called a **least upper bound** or **supremum** of S when the following conditions are satisfied.
- $\bar{u}$ is an upper bound of S.
- If u is any upper bound of S, then $u \geq \bar{u}$.

In this case, we write $\bar{u} = \sup S$. We say that the supremum $\sup S$ is *attained* when $\bar{u} \in S$.

(ii) Suppose that S is bounded below. A real number $\bar{l} \in \mathbb{R}$ is called a **greatest lower bound** or **infimum** of S when the following conditions are satisfied.
- $\bar{l}$ is a lower bound of S.
- If l is any lower bound of S, then $l \leq \bar{l}$.

In this case, we write $\bar{l} = \inf S$. We say that the infimum $\inf S$ is *attained* when $\bar{l} \in S$.

It is clear to see that if the supremum $\sup S$ is attained, then $\max S = \sup S$. Similarly, if the infimum $\inf S$ is attained, then $\min S = \inf S$.

Example 1.1.3 Let $S = [0,1]$. Then, we have

$$\max S = \sup S = 1 \text{ and } \inf S = \min S = 0.$$

If $S = [0,1)$, then $\max S$ does not exists. However, we have $\sup S = 1$.

Proposition 1.1.4 *Let S be a subset of $\mathbb{R}$ with $\bar{u} = \sup S$. Then, given any $s < \bar{u}$, there exists $t \in S$ satisfying $s < t \leq \bar{u}$.*

Proof. We are going to prove it by contradiction. Suppose that we have $t \leq s$ for all $t \in S$. Then s is an upper bound of S. According to the definition of supremum, we also have $s \geq \bar{u}$. This contradiction implies that $s < t$ for some $t \in S$, and the proof is complete. ■

Proposition 1.1.5 *Given any two nonempty subsets A and B of $\mathbb{R}$, we define $C = A + B$ by*

$$C = \{x + y : x \in A \text{ and } y \in B\}.$$

Suppose that the supremum $\sup A$ and $\sup B$ are attained. Then, the supremum $\sup C$ is attained, and we have

$$\sup C = \sup A + \sup B.$$

Proof. We first have

$$\sup A = \max A \text{ and } \sup B = \max B.$$

We write $a = \sup A$ and $b = \sup B$. Given any $z \in C$, there exist $x \in A$ and $y \in B$ satisfying $z = x + y$. Since $x \leq a$ and $y \leq b$, we have $z = x + y \leq a + b$, which says that $a + b$ is an upper bound of C. Therefore, the definition of $c = \sup C$ says that $c \leq a + b$. Next, we want to show that $a + b \leq c$. Given any $\epsilon > 0$, Proposition 1.1.4 says that there exist $x \in A$ and $y \in B$ satisfying $a - \epsilon < x$ and $b - \epsilon < y$. We also see that $x + y \leq c$. Adding these inequalities, we obtain

$$a + b - 2\epsilon < x + y \leq c,$$

which says that $a + b < c + 2\epsilon$. Since ϵ can be any positive real number, we must have $a + b \leq c$. This completes the proof. ■

Proposition 1.1.6 *Let A and B be any two nonempty subsets of $\mathbb{R}$ satisfying $a \leq b$ for any $a \in A$ and $b \in B$. Suppose that the supremum* $\sup B$ *is attained. Then, the supremum* $\sup A$ *is attained and* $\sup A \leq \sup B$.

Proof. It is left as an exercise. ∎

1.2 Limit Inferior and Limit Superior

Let $\{a_n\}_{n=1}^{\infty}$ be a sequence in $\mathbb{R}$. The **limit superior** of $\{a_n\}_{n=1}^{\infty}$ is defined by

$$\limsup_{n\to\infty} a_n = \inf_{n\geq 1} \sup_{k\geq n} a_k,$$

and the **limit inferior** of $\{a_n\}_{n=1}^{\infty}$ is defined by

$$\liminf_{n\to\infty} a_n = -\limsup_{n\to\infty} (-a_n).$$

Moreover, we can see that

$$\liminf_{n\to\infty} a_n = \sup_{n\geq 1} \inf_{k\geq n} a_k.$$

Let

$$b_n = \sup_{k\geq n} a_k \text{ and } c_n = \inf_{k\geq n} a_k \tag{1.1}$$

It is clear to see that $\{b_n\}_{n=1}^{\infty}$ is a decreasing sequence and $\{c_n\}_{n=1}^{\infty}$ is an increasing sequence. In this case, we have

$$\inf_{n\geq 1} b_n = \lim_{n\to\infty} b_n \text{ and } \sup_{n\geq 1} c_n = \lim_{n\to\infty} c_n,$$

which also says that

$$\limsup_{n\to\infty} a_n = \inf_{n\geq 1} b_n = \lim_{n\to\infty} b_n = \lim_{n\to\infty} \sup_{k\geq n} a_k \tag{1.2}$$

and

$$\liminf_{n\to\infty} a_n = \sup_{n\geq 1} c_n = \lim_{n\to\infty} c_n = \lim_{n\to\infty} \inf_{k\geq n} a_k. \tag{1.3}$$

Some useful properties are given below.

Proposition 1.2.1 *Let $\{a_n\}_{n=1}^{\infty}$ be a sequence of real numbers. Then, the following statements hold true.*

(i) *We have*

$$\liminf_{n\to\infty} a_n \leq \limsup_{n\to\infty} a_n.$$

(ii) *We have*

$$\lim_{n\to\infty} a_n = a$$

if and only if

$$\liminf_{n\to\infty} a_n = \limsup_{n\to\infty} a_n = a \text{ with } |a| < +\infty.$$

(iii) *The sequence diverges to $+\infty$ if and only if*

$$\liminf_{n\to\infty} a_n = \limsup_{n\to\infty} a_n = +\infty.$$

(iv) *The sequence diverges to* $-\infty$ *if and only if*

$$\liminf_{n\to\infty} a_n = \limsup_{n\to\infty} a_n = -\infty.$$

(v) *Let* $\{b_n\}_{n=1}^{\infty}$ *be another sequence satisfying* $a_n \le b_n$ *for all n. Then, we have*

$$\liminf_{n\to\infty} a_n \le \liminf_{n\to\infty} b_n \text{ and } \limsup_{n\to\infty} a_n \le \limsup_{n\to\infty} b_n.$$

Proof. To prove part (i), from (1.1), we see that $c_n \le b_n$ for all n. Using (1.2) and (1.3), we obtain

$$\liminf_{n\to\infty} a_n = \lim_{n\to\infty} c_n \le \lim_{n\to\infty} b_n = \limsup_{n\to\infty} a_n.$$

To prove part (ii), suppose that

$$\lim_{n\to\infty} a_n = a.$$

Then, given any $\epsilon > 0$, there exists an integer N satisfying

$$a - \frac{\epsilon}{2} < a_n < a + \frac{\epsilon}{2} \text{ for } n \ge N,$$

which implies

$$a - \frac{\epsilon}{2} \le \inf_{k\ge n} a_k = c_n \text{ and } b_n = \sup_{k\ge n} a_k \le a + \frac{\epsilon}{2} \text{ for } n \ge N.$$

In other words, we have

$$a - \frac{\epsilon}{2} \le c_n \le b_n \le a + \frac{\epsilon}{2} \text{ for } n \ge N,$$

which also implies

$$|c_n - a| \le \frac{\epsilon}{2} < \epsilon \text{ and } |b_n - a| \le \frac{\epsilon}{2} < \epsilon \text{ for } n \ge N.$$

Therefore, we obtain

$$\lim_{n\to\infty} c_n = a = \lim_{n\to\infty} b_n,$$

which implies, by using (1.2) and (1.3),

$$\liminf_{n\to\infty} a_n = \limsup_{n\to\infty} a_n = a.$$

For the converse, from (1.1) again, we see that $c_n \le a_n \le b_n$ for all $n \ge 1$. Since

$$a = \lim_{n\to\infty} \inf_{k\ge n} a_k = \lim_{n\to\infty} c_n \text{ and } a = \lim_{n\to\infty} \sup_{k\ge n} a_k = \lim_{n\to\infty} b_n.$$

Using the pinching theorem, we obtain the desired limit. The remaining proofs are left as exercise, and the proof is complete. ■

Proposition 1.2.2 *Let* $\{a_n\}_{n=1}^{\infty}$ *and* $\{b_n\}_{n=1}^{\infty}$ *be any two sequences in* $\mathbb{R}$. *Then, we have*

$$\limsup_{n\to\infty} (a_n + b_n) \le \limsup_{n\to\infty} a_n + \limsup_{n\to\infty} b_n$$

and

$$\liminf_{n\to\infty} (a_n + b_n) \ge \liminf_{n\to\infty} a_n + \liminf_{n\to\infty} b_n$$

Proof. For $k \geq n$, we have

$$a_k + b_k \leq \sup_{k \geq n} a_k + \sup_{k \geq n} b_k,$$

which says that

$$\sup_{k \geq n} \left(a_k + b_k\right) \leq \sup_{k \geq n} a_k + \sup_{k \geq n} b_k. \tag{1.4}$$

Therefore, we obtain

$$\begin{aligned}
\limsup_{n \to \infty} \left(a_n + b_n\right) &= \inf_{n \geq 1} \sup_{k \geq n} \left(a_k + b_k\right) = \lim_{n \to \infty} \sup_{k \geq n} \left(a_k + b_k\right) \\
&\leq \lim_{n \to \infty} \left[\sup_{k \geq n} a_k + \sup_{k \geq n} b_k\right] \text{ (using (1.4))} \\
&= \lim_{n \to \infty} \sup_{k \geq n} a_k + \lim_{n \to \infty} \sup_{k \geq n} b_k \text{ (since the limits exist)} \\
&= \limsup_{n \to \infty} a_n + \limsup_{n \to \infty} b_n \text{ (using (1.2) and (1.3)).}
\end{aligned}$$

We similarly have

$$\inf_{k \geq n} \left(a_k + b_k\right) \geq \inf_{k \geq n} a_k + \inf_{k \geq n} b_k. \tag{1.5}$$

Therefore, we also obtain

$$\begin{aligned}
\liminf_{n \to \infty} \left(a_n + b_n\right) &= \sup_{n \geq 1} \inf_{k \geq n} \left(a_k + b_k\right) = \lim_{n \to \infty} \inf_{k \geq n} \left(a_k + b_k\right) \\
&\geq \lim_{n \to \infty} \left[\inf_{k \geq n} a_k + \inf_{k \geq n} b_k\right] \text{ (using (1.5))} \\
&= \lim_{n \to \infty} \inf_{k \geq n} a_k + \lim_{n \to \infty} \inf_{k \geq n} b_k \text{ (since the limits exist)} \\
&= \liminf_{n \to \infty} a_n + \liminf_{n \to \infty} b_n \text{ (using (1.2) and (1.3)).}
\end{aligned}$$

This completes the proof. ■

Proposition 1.2.3 *Let $\{A_n\}_{n=1}^{\infty}$ be a sequence of subsets of $\mathbb{R}^m$ satisfying $A_{n+1} \subseteq A_n$ for all n and $\bigcap_{n=1}^{\infty} A_n = A$, and let f be a real-valued function defined on $\mathbb{R}^m$. Then*

$$\lim_{n \to \infty} \sup_{a \in A_n} f(a) = \sup_{a \in A} f(a) \text{ and } \sup_{a \in A_n} f(a) \geq \sup_{a \in A_{n+1}} f(a)$$

and

$$\lim_{n \to \infty} \inf_{a \in A_n} f(a) = \inf_{a \in A} f(a) \text{ and } \inf_{a \in A_n} f(a) \leq \inf_{a \in A_{n+1}} f(a).$$

Proof. Since

$$\inf_{a \in A} f(a) = -\sup_{a \in A} [-f(a)].$$

It suffices to prove the case of the supremum. Let

$$y_n^* = \sup_{a \in A_n} f(a) \text{ and } y^* = \sup_{a \in A} f(a).$$

Since $A_{n+1} \subseteq A_n$ for all n, we have that $\{y_n^*\}_{n=1}^\infty$ is a decreasing sequence of real numbers. We also have $y_n^* \geq y^*$ for all n, which implies

$$\liminf_{n\to\infty} y_n^* \geq y^*. \tag{1.6}$$

Given any $\epsilon > 0$, according to the concept of supremum, there exists $a_n \in A_n$ satisfying

$$y_n^* - \epsilon \leq f(a_n). \tag{1.7}$$

Let $b_n = \inf_{k\geq n} f(a_k)$. We consider the subsequence $\{\bar{a}_m\}_{m=1}^\infty$ defined by $\bar{a}_m = a_{m+n-1}$ in the sense of

$$\{\bar{a}_1, \bar{a}_2, \ldots, \bar{a}_m, \ldots\} = \{a_n, a_{n+1}, \ldots, a_{m+n-1}, \ldots\}.$$

Then $b_n = \inf_{m\geq n} f(\bar{a}_m)$ and $b_n \leq f(\bar{a}_m)$ for all m. Since $A_{n+1} \subseteq A_n$ for all n and $\bigcap_{n=1}^\infty A_n = A$, the "last term" of the sequence $\{\bar{a}_m\}_{m=1}^\infty$ must be in A, a claim that will be proved below. Since $\bar{a}_k \in A_k \subseteq A_n$ for all $k \geq n$, we have the subsequence $\{\bar{a}_k\}_{k=n}^\infty \subseteq A_n$, which also implies

$$\bar{A} \equiv \bigcap_{n=1}^{\infty} \left(\{\bar{a}_k\}_{k=n}^\infty\right) \subseteq \bigcap_{n=1}^{\infty} A_n = A,$$

where $\bar{A}$ can be regarded as the "last term" and $\bar{A} \subseteq \{\bar{a}_m\}_{m=1}^\infty$. Since y^* is the supremum of f on A, it follows that $f(\bar{a}) \leq y^*$ for each $\bar{a} \in \bar{A} \subseteq A$. Since $b_n \leq f(\bar{a}_m)$ for all m, we see that $b_n \leq y^*$ for all n. Therefore, we obtain

$$\liminf_{n\to\infty} f(a_n) = \sup_{n\geq 1} \inf_{k\geq n} f(a_k) = \sup_{n\geq 1} b_n \leq y^*,$$

which implies, by (1.7),

$$\liminf_{n\to\infty} y_n^* - \epsilon \leq \liminf_{n\to\infty} f(a_n) \leq y^*.$$

Since ϵ is any positive number, we obtain

$$\liminf_{n\to\infty} y_n^* \leq y^*. \tag{1.8}$$

Combining (1.6) and (1.8), we obtain

$$\sup_{n\geq 1} \inf_{k\geq n} y_k^* = \liminf_{n\to\infty} y_n^* = y^*.$$

Since $\{y_n^*\}_{n=1}^\infty$ is a decreasing sequence of real numbers, we conclude that

$$\inf_{n\geq 1} y_n^* = \lim_{n\to\infty} y_n^* = \liminf_{n\to\infty} y_n^* = y^*,$$

and the proof is complete. ■

Proposition 1.2.4 *Let $\{A_n\}_{n=1}^\infty$ be a sequence of subsets of $\mathbb{R}^m$ satisfying $A_n \subseteq A_{n+1}$ for all n and $\bigcup_{n=1}^\infty A_n = A$, and let f be a real-valued function defined on $\mathbb{R}^m$. Then*

$$\lim_{n\to\infty} \sup_{a\in A_n} f(a) = \sup_{a\in A} f(a) \textit{ and } \sup_{a\in A_n} f(a) \leq \sup_{a\in A_{n+1}} f(a)$$

and

$$\lim_{n\to\infty} \inf_{a\in A_n} f(a) = \inf_{a\in A} f(a) \textit{ and } \inf_{a\in A_n} f(a) \geq \inf_{a\in A_{n+1}} f(a).$$

Proof. It suffices to prove the case of the supremum. Let

$$y_n^* = \sup_{a\in A_n} f(a) \text{ and } y^* = \sup_{a\in A} f(a).$$

Since $A_n \subseteq A_{n+1}$ for all n, we have that $\{y_n^*\}_{n=1}^\infty$ is an increasing sequence of real numbers. We also have $y_n^* \le y^*$ for all n, which implies

$$\limsup_{n\to\infty} y_n^* \le y^*. \tag{1.9}$$

Given any $\epsilon > 0$, according to the concept of supremum, there exists $a^* \in A$ satisfying $y^* - \epsilon \le f(a^*)$. Since

$$a^* \in A = \bigcup_{n=1}^\infty A_n,$$

we have that $a^* \in A_{n^*}$ for some integer n^*. We construct a sequence $\{a_n\}_{n=1}^\infty$ satisfying $a_n \in A_n$ for all $n < n^*$ and $a_n = a^*$ for all $n \ge n^*$. Since $A_n \subseteq A_{n+1}$ for all n, it follows that $a_n \in A_n$ for all $n \ge n^*$. Therefore, the sequence $\{a_n\}_{n=1}^\infty$ satisfies $a_n \in A_n$ for all n and

$$a^* \in \{a_k\}_{k=n}^\infty \text{ for all } n,$$

which means that a^* is the "last term" of the sequence $\{a_n\}_{n=1}^\infty$. We also have

$$y_n^* \ge f(a_n). \tag{1.10}$$

Let $b_n = \sup_{k\ge n} f(a_k)$. We consider the subsequence $\{\bar{a}_p\}_{p=1}^\infty$ defined by $\bar{a}_p = a_{p+n-1}$ in the sense of

$$\left\{\bar{a}_1, \bar{a}_2, \ldots, \bar{a}_p, \ldots\right\} = \left\{a_n, a_{n+1}, \ldots, a_{p+n-1}, \ldots\right\}.$$

Then $b_n = \sup_{p\ge n} f(\bar{a}_p)$ and $b_n \ge f(\bar{a}_p)$ for all p. Since a^* is the "last term" of the sequence $\{a_n\}_{n=1}^\infty$, it follows that a^* is also the "last term" of the sequence $\{\bar{a}_m\}_{m=1}^\infty$. Therefore, we have $b_n \ge f(a^*) \ge y^* - \epsilon$ for all n, which implies

$$\limsup_{n\to\infty} f(a_n) = \inf_{n\ge1} \sup_{k\ge n} f(a_k) = \inf_{n\ge1} b_n \ge y^* - \epsilon.$$

Since ϵ is any positive number, it follows that

$$\limsup_{n\to\infty} f(a_n) \ge y^*.$$

Using (1.10), we obtain

$$\limsup_{n\to\infty} y_n^* \ge \limsup_{n\to\infty} f(a_n) \ge y^*. \tag{1.11}$$

Combining (1.9) and (1.11), we obtain

$$\inf_{n\ge1} \sup_{k\ge n} y_k^* = \limsup_{n\to\infty} y_n^* = y^*.$$

Since $\{y_n^*\}_{n=1}^\infty$ is an increasing sequence of real numbers, we conclude that

$$\sup_{n\ge1} y_n^* = \lim_{n\to\infty} y_n^* = \limsup_{n\to\infty} y_n^* = y^*,$$

and the proof is complete. ∎

Given any $x = (x^{(1)}, \ldots, x^{(m)})$ and $y = (y^{(1)}, \ldots, y^{(m)})$ in $\mathbb{R}^m$. The Euclidean distance between x and y is defined by

$$\| x - y \| = \sqrt{(x^{(1)} - y^{(1)})^2 + \cdots + (x^{(m)} - y^{(m)})^2}.$$

Given a point $x \in \mathbb{R}^m$, we consider the open ϵ-ball

$$B(x;\epsilon) = \left\{ y \in \mathbb{R}^m : \| x - y \| < \epsilon \right\}. \tag{1.12}$$

The concept of closure based on open balls will be frequently used throughout this book. For the general concept refer to Kelley and Namioka [55]. In this book, we are going to consider the closure of a subset of $\mathbb{R}^m$, which is given below.

Definition 1.2.5 Let A be a subset of $\mathbb{R}^m$. The **closure** of A is denoted and defined by

$$\text{cl}(A) = \left\{ x \in \mathbb{R}^m : A \cap B(x;\epsilon) \neq \emptyset \text{ for any } \epsilon > 0 \right\}.$$

We say that A is a closed subset of $\mathbb{R}^m$ when $A = \text{cl}(A)$.

Remark 1.2.6 Given any $x \in \text{cl}(A)$, there exists a sequence $\{x_n\}_{n=1}^{\infty}$ in A satisfying $\| x_n - x \| \to 0$ as $n \to \infty$. In particular, for $m = 1$, we see that $x_n \to x$ as $n \to \infty$.

Proposition 1.2.7 *Let A be a subset of $\mathbb{R}$, and let f be a continuous function defined on cl(A). Then*

$$\sup_{a\in A} f(a) = \sup_{a\in \text{cl}(A)} f(a) \text{ and } \inf_{a\in A} f(a) = \inf_{a\in \text{cl}(A)} f(a).$$

Proof. It suffices to prove the case of the supremum, since

$$\inf_{a\in A} f(a) = -\sup_{a\in A} [-f(a)].$$

It is obvious that

$$\sup_{a\in A} f(a) \leq \sup_{a\in \text{cl}(A)} f(a).$$

Given any $\epsilon > 0$, according to the concept of supremum, there exists $a^* \in \text{cl}(A)$ satisfying

$$\sup_{a\in \text{cl}(A)} f(a) - \epsilon \leq f(a^*).$$

We also see that there exists a sequence $\{a_n\}_{n=1}^{\infty}$ in A satisfying $a_n \to a^*$. Since f is continuous on cl(A), we also have $f(a_n) \to f(a^*)$ as $n \to \infty$. Therefore, we obtain

$$\sup_{a\in \text{cl}(A)} f(a) - \epsilon \leq f(a^*) = \lim_{n\to\infty} f\left(a_n\right) \leq \lim_{n\to\infty} \left[\sup_{a\in A} f(a)\right] = \sup_{a\in A} f(a).$$

Since ϵ can be any positive number, it follows that

$$\sup_{a\in \text{cl}(A)} f(a) \leq \sup_{a\in A} f(a).$$

This completes the proof. ■

Let S be a subset of $\mathbb{R}$. For $a \in S$ and a sequence $\{a_n\}_{n=1}^{\infty}$ in $\mathbb{R}$, we write $a_n \uparrow a$ to mean that the sequence $\{a_n\}_{n=1}^{\infty}$ is increasing and converges to a. We also write $a_n \downarrow a$ to mean that the sequence $\{a_n\}_{n=1}^{\infty}$ is decreasing and converges to a.

Proposition 1.2.8 *Let A be a subset of $\mathbb{R}$. The following statements hold true.*

(i) *Let f be a right-continuous function defined on cl(A). Given any fixed $r \in \mathbb{R}$, suppose that there exists a sequence $\{a_n\}_{n=1}^{\infty}$ in A satisfying $a_n \downarrow r$ as $n \to \infty$ and $a_n > r$ for all n. Then, we have*

$$\sup_{\{a\in A: a>r\}} f(a) = \sup_{\{a\in A: a\geq r\}} f(a) \text{ and } \inf_{\{a\in A: a>r\}} f(a) = \inf_{\{a\in A: a\geq r\}} f(a).$$

(ii) *Let f be a continuous function defined on cl(A). Given any fixed $r \in \mathbb{R}$, suppose that there exists a sequence $\{a_n\}_{n=1}^{\infty}$ in A satisfying $a_n \to r$ as $n \to \infty$ and $a_n > r$ for all n. Then, we have*

$$\sup_{\{a\in A:a>r\}} f(a) = \sup_{\{a\in A:a\geq r\}} f(a) \text{ and } \inf_{\{a\in A:a>r\}} f(a) = \inf_{\{a\in A:a\geq r\}} f(a).$$

In particular, we can assume $r \in cl(\{a \in A : a > r\})$.

Proof. It suffices to prove the case of the supremum. It is obvious that

$$\sup_{\{a\in A:a>r\}} f(a) \leq \sup_{\{a\in A:a\geq r\}} f(a).$$

To prove part (i), given any $\epsilon > 0$, according to the concept of supremum $\sup_{\{a\in A:a\geq r\}} f(a)$, there exists $a^* \in A$ with $a^* \geq r$ satisfying

$$\sup_{\{a\in A:a\geq r\}} f(a) - \epsilon \leq f(a^*).$$

We consider the following two cases.

- Suppose that $a^* > r$. Then, we have

$$\sup_{\{a\in A:a\geq r\}} f(a) - \epsilon \leq f(a^*) \leq \sup_{\{a\in A:a>r\}} f(a).$$

- Suppose that $a^* = r$. The assumption says that there exists a sequence $\{a_n\}_{n=1}^{\infty}$ in A satisfying $a_n \downarrow a^*$ as $n \to \infty$ and $a_n > r$ for all n. Since f is right-continuous and $a^* \in \text{cl}(A)$, we also have $f(a_n) \to f(a^*)$ as $n \to \infty$. Therefore, we obtain

$$\sup_{\{a\in A:a\geq r\}} f(a) - \epsilon \leq f(a^*) = \lim_{n\to\infty} f\left(a_n\right) \leq \lim_{n\to\infty} \left[\sup_{\{a\in A:a>r\}} f(a)\right] = \sup_{\{a\in A:a>r\}} f(a).$$

  Since ϵ can be any positive number, it follows that

$$\sup_{\{a\in A:a\geq r\}} f(a) \leq \sup_{\{a\in A:a>r\}} f(a).$$

Part (ii) can be similarly obtained, and the proof is complete. ■

Proposition 1.2.9 *Let $\{A_n\}_{n=1}^{\infty}$ and $\{B_n\}_{n=1}^{\infty}$ be two sequences of subsets of $\mathbb{R}$ satisfying*

$$A_{n+1} \subseteq A_n \text{ and } B_{n+1} \subseteq B_n \text{ for all } n$$

and

$$\bigcap_{n=1}^{\infty} A_n = A \text{ and } \bigcap_{n=1}^{\infty} B_n = B.$$

Then, we have

$$\lim_{n\to\infty} \sup_{a\in A_n} \inf_{b\in B_n} (a-b) = \lim_{n\to\infty} \left(\sup_{a\in A_n} a - \sup_{b\in B_n} b\right) = \sup_{a\in A} a - \sup_{b\in B} b = \sup_{a\in A} \inf_{b\in B} (a-b)$$

and

$$\lim_{n\to\infty} \sup_{b\in B_n} \inf_{a\in A_n} (a-b) = \lim_{n\to\infty} \left(\inf_{a\in A_n} a - \inf_{b\in B_n} b\right) = \inf_{a\in A} a - \inf_{b\in B} b = \sup_{b\in B} \inf_{a\in A} (a-b).$$

Proof. It is obvious that

$$\sup_{a\in A_n} \inf_{b\in B_n} (a-b) = \sup_{a\in A_n} a - \sup_{b\in B_n} b \text{ and } \sup_{b\in B_n} \inf_{a\in A_n} (a-b) = \inf_{a\in A_n} a - \inf_{b\in B_n} b.$$

The results follow immediately from Proposition 1.2.3. ■

Proposition 1.2.10 *Let $\{A_n\}_{n=1}^{\infty}$ and $\{B_n\}_{n=1}^{\infty}$ be two sequences of sets in $\mathbb{R}$ satisfying*

$$A_n \subseteq A_{n+1} \text{ and } B_n \subseteq B_{n+1} \text{ for all } n$$

and

$$\bigcup_{n=1}^{\infty} A_n = A \text{ and } \bigcup_{n=1}^{\infty} B_n = B.$$

Then, we have

$$\lim_{n\to\infty} \sup_{a\in A_n, b\in B_n} \inf (a-b) = \lim_{n\to\infty} \left(\sup_{a\in A_n} a - \sup_{b\in B_n} b \right) = \sup_{a\in A} a - \sup_{b\in B} b = \sup_{a\in A} \inf_{b\in B} (a-b)$$

and

$$\lim_{n\to\infty} \sup_{b\in B_n, a\in A_n} \inf (a-b) = \lim_{n\to\infty} \left(\inf_{a\in A_n} a - \inf_{b\in B_n} b \right) = \inf_{a\in A} a - \inf_{b\in B} b = \sup_{b\in B} \inf_{a\in A} (a-b).$$

Proof. The results follow immediately from Proposition 1.2.4. ■

Proposition 1.2.11 *Let f be a real-valued function defined on a subset A of $\mathbb{R}$, and let k be a constant. Then, we have*

$$\sup_{x\in A} \min \{f(x), k\} = \min \left\{ \sup_{x\in A} f(x), k \right\}$$

and

$$\inf_{x\in A} \max \{f(x), k\} = \max \left\{ \inf_{x\in A} f(x), k \right\}.$$

Proof. We have

$$\min \left\{ \sup_{x\in A} f(x), k \right\} = \begin{cases} k, & \text{if there exists } x \in A \text{satisfying } f(x) > k \\ \sup_{x\in A} f(x), & \text{if } f(x) \le k \text{ for all } x \in A. \end{cases}$$

and

$$\sup_{x\in A} \min \{f(x), k\} = \begin{cases} \max \left\{ \sup_{\{x\in A: f(x)>k\}} \min\{f(x), k\}, \sup_{\{x\in A: f(x)\le k\}} \min\{f(x), k\} \right\}, \\ \quad \text{if there exists } x \in A \text{ satisfying } f(x) > k \\ \sup_{x\in A} f(x), \quad \text{if } f(x) \le k \text{ for all } x \in A \end{cases}.$$

$$= \begin{cases} \max \left\{ k, \sup_{\{x\in A: f(x)\le k\}} f(x) \right\}, \\ \quad \text{if there exists } x \in A \text{ satisfying } f(x) > k \\ \sup_{x\in A} f(x), \quad \text{if } f(x) \le k \text{ for all } x \in A \end{cases}.$$

$$= \begin{cases} k, & \text{if there exists } x \in A \text{ satisfying } f(x) > k \\ \sup_{x\in A} f(x), & \text{if } f(x) \le k \text{ for all } x \in A. \end{cases}$$

Another equality can be similarly obtained. This completes the proof. ■

1.3 Semi-Continuity

Let $f : \mathbb{R}^m \to \mathbb{R}$ be a real-valued function defined on $\mathbb{R}^m$. We say that the supremum $\sup_{x\in S} f(x)$ is *attained* when there exists $x^* \in S$ satisfying $f(x) \le f(x^*)$ for all $x \in S$ with $x \neq x^*$. Equivalently, the supremum $\sup_{x\in S} f(x)$ is attained if and only if

$$\sup_{x\in S} f(x) = \max_{x\in S} f(x).$$

Similarly, the infimum $\inf_{x\in S} f(x)$ is *attained* when there exists $x^* \in S$ satisfying $f(x) \ge f(x^*)$ for all $x \in S$ with $x \neq x^*$. Equivalently, the infimum $\inf_{x\in S} f(x)$ is attained if and only if

$$\inf_{x\in S} f(x) = \min_{x\in S} f(x).$$

Let $\mathbf{x} = (x_1, \dots, x_m)$ be an element in $\mathbb{R}^m$. Recall that the Euclidean norm of $\mathbf{x}$ is given by

$$\| \mathbf{x} \| = \sqrt{x_1^2 + x_2^2 + \cdots + x_m^2}.$$

Definition 1.3.1 Let S be a nonempty set in $\mathbb{R}^m$.

- A real-valued function $f : S \to \mathbb{R}$ defined on S is said to be **upper semi-continuous** at $\bar{\mathbf{x}}$ when the following condition is satisfied: for each $\epsilon > 0$, there exists $\delta > 0$ such that $\| \mathbf{x} - \bar{\mathbf{x}} \| < \delta$ implies $f(\mathbf{x}) < f(\bar{\mathbf{x}}) + \epsilon$ for any $\mathbf{x} \in S$.
- A real-valued function f defined on S is said to be **lower semi-continuous** at $\bar{\mathbf{x}}$ when the following condition is satisfied: for each $\epsilon > 0$, there exists $\delta > 0$ such that $\| \mathbf{x} - \bar{\mathbf{x}} \| < \delta$ implies $f(\bar{\mathbf{x}}) < f(\mathbf{x}) + \epsilon$ for any $\mathbf{x} \in S$.

Remark 1.3.2 We have the following interesting observations.

- If f is upper semi-continuou on S, then $-f$ is lower semi-continuous on S.
- If f is lower semi-continuou on S, then $-f$ is upper semi-continuous on S.
- The real-valued function f is continuous on S if and only if it is both lower and upper semi-continuous in S.
- If f is upper semi-continuous on $\mathbb{R}$, then $\{\mathbf{x} : f(\mathbf{x}) \ge \alpha\}$ is a closed subset of $\mathbb{R}^m$ for all α.
- If f is lower semi-continuous on $\mathbb{R}$, then $\{\mathbf{x} : f(\mathbf{x}) \le \alpha\}$ is a closed subset of $\mathbb{R}^m$ for all α.

Proposition 1.3.3 *Let $f : \mathbb{R}^m \to \mathbb{R}$ be a multi-variable real-valued function, and let each real-valued function $g_i : \mathbb{R} \to \mathbb{R}$ be continuous at $x_0 \in \mathbb{R}$ for $i = 1, \dots, n$. Then, the following statements hold true.*

(i) *Suppose that f is lower semi-continuous at $\mathbf{x}_0 \equiv (g_1(x_0), \dots, g_m(x_0))$. Then, the composition function $h(x) = f\left(g_1(x), \dots, g_m(x)\right)$ is lower semi-continuous at x_0.*
(ii) *Suppose that f is upper semi-continuous at $\mathbf{x}_0 \equiv (g_1(x_0), \dots, g_m(x_0))$. Then, the composition function $h(x) = f\left(g_1(x), \dots, g_m(x)\right)$ is upper semi-continuous at x_0.*

Proof. To prove part (i), since f is lower semi-continuous at $\mathbf{x}_0$, given any $\epsilon > 0$, there exists $\delta^* > 0$ such that

$$\| \mathbf{x} - \mathbf{x}_0 \| < \delta^* \text{ implies } f(\mathbf{x}_0) < f(\mathbf{x}) + \epsilon.$$

Since each g_i is continuous at x_0 for $i = 1, \ldots, n$, given $\delta^*/\sqrt{n}$, there exists $\delta_i > 0$ such that

$$|x - x_0| < \delta_i \text{ implies } |g_i(x) - g_i(x_0)| < \frac{\delta^*}{\sqrt{n}} \text{ for } i = 1, \ldots, n. \tag{1.13}$$

Let $\delta = \min\{\delta_1, \ldots, \delta_m\}$. Then $|x - x_0| < \delta$ implies that the inequality (1.13) is satisfied for all $i = 1, \ldots, n$. Let $\mathbf{x} \equiv (g_1(x), \ldots, g_m(x))$. Then

$$\| \mathbf{x} - \mathbf{x}_0 \| = \sqrt{(g_1(x) - g_1(x_0))^2 + \cdots + (g_m(x) - g_m(x_0))^2} < \delta^*,$$

which implies

$$h(x_0) = f(g_1(x_0), \ldots, g_m(x_0)) = f(\mathbf{x}_0) < f(\mathbf{x}) + \epsilon = f(g_1(x), \ldots, g_m(x)) + \epsilon = h(x) + \epsilon,$$

which says that h is lower semi-continuous at x_0. Part (ii) can be similarly obtained. This completes the proof. ■

Proposition 1.3.4 *Let $f : \mathbb{R}^m \to \mathbb{R}$ be a multi-variable real-valued function, and let each real-valued function $g_i : \mathbb{R} \to \mathbb{R}$ be left-continuous at $x_0 \in \mathbb{R}$ for $i = 1, \ldots, n$. Then, the following statements hold true.*

(i) *Assume that the composition function $h(x) = f(g_1(x), \ldots, g_m(x))$ is increasing. If f is lower semi-continuous at $\mathbf{x}_0 \equiv (g_1(x_0), \ldots, g_m(x_0))$, then h is lower semi-continuous at x_0.*

(ii) *Assume that the composition function $h(x) = f(g_1(x), \ldots, g_m(x))$ is decreasing. If f is upper semi-continuous at $\mathbf{x}_0 \equiv (g_1(x_0), \ldots, g_m(x_0))$, then h is upper semi-continuous at x_0.*

Proof. To prove part (i), since f is lower semi-continuous at $\mathbf{x}_0$, given any $\epsilon > 0$, there exists $\delta^* > 0$ such that

$$\| \mathbf{x} - \mathbf{x}_0 \| < \delta^* \text{ implies } f(\mathbf{x}_0) < f(\mathbf{x}) + \epsilon.$$

Since each g_i is left-continuous at x_0 for $i = 1, \ldots, n$, given $\delta^*/\sqrt{n}$, there exists $\delta_i > 0$ such that

$$0 < x_0 - x < \delta_i \text{ implies } |g_i(x) - g_i(x_0)| < \frac{\delta^*}{\sqrt{n}} \text{ for } i = 1, \ldots, n.$$

The argument in the proof of Proposition 1.3.3 is still valid to show that there exists $\delta > 0$ such that

$$0 < x_0 - x < \delta \text{ implies } h(x_0) < h(x) + \epsilon.$$

For $0 < x - x_0 < \delta$, since h is increasing, it follows that

$$h(x_0) \leq h(x) < h(x) + \epsilon.$$

Therefore, we conclude that

$$|x_0 - x| < \delta \text{ implies } h(x_0) < h(x) + \epsilon,$$

which says that h is lower semi-continuous at x_0.

To prove part (ii), we can similarly show that there exists $\delta > 0$ such that

$$0 < x_0 - x < \delta \text{ implies } h(x) < h(x_0) + \epsilon.$$

For $0 < x - x_0 < \delta$, since h is decreasing, it follows that

$$h(x) \leq h(x_0) < h(x_0) + \epsilon,$$

which says that h is upper semi-continuous at x_0. This completes the proof. ∎

Proposition 1.3.5 *We have the following properties.*

(i) *Suppose that the real-valued functions f_1 and f_2 are lower semi-continuous on the closed interval $[a, b]$. Then, the addition $f_1 + f_2$ is also lower semi-continuous on the closed interval $[a, b]$.*
(ii) *Suppose that the real-valued functions g_1 and g_2 are upper semi-continuous on on the closed interval $[a, b]$. Then, the addition $g_1 + g_2$ is also upper semi-continuous on the closed interval $[a, b]$.*

Proof. To prove part (i), given $\epsilon > 0$, there exist $\delta_1, \delta_2 > 0$ such that

$$|x - x_0| < \delta_1 \text{ implies } f_1(x_0) < f_1(x) + \frac{\epsilon}{2},$$

and that

$$|x - x_0| < \delta_2 \text{ implies } f_2(x_0) < f_2(x) + \frac{\epsilon}{2}.$$

Let $\delta = \min\{\delta_1, \delta_2\}$. Then, for $|x - x_0| < \delta$, we have

$$f_1(x_0) + f_2(x_0) < f_1(x) + \frac{\epsilon}{2} + f_2(x) + \frac{\epsilon}{2} = f_1(x) + f_2(x) + \epsilon$$

which shows that $f_1 + f_2$ is lower semi-continuous at x_0.

To prove part (ii), given $\epsilon > 0$, there exist $\delta_1, \delta_2 > 0$ such that

$$|x - x_0| < \delta_1 \text{ implies } g_1(x_0) + \frac{\epsilon}{2} > g_1(x),$$

and that

$$|x - x_0| < \delta_2 \text{ implies } g_2(x_0) + \frac{\epsilon}{2} > g_2(x).$$

Let $\delta = \min\{\delta_1, \delta_2\}$. Then, for $|x - x_0| < \delta$, we have

$$g_1(x_0) + g_2(x_0) + \epsilon = g_1(x_0) + \frac{\epsilon}{2} + g_2(x_0) + \frac{\epsilon}{2} > g_1(x) + g_2(x),$$

which shows that $g_1 + g_2$ is upper semi-continuous at x_0. This completes the proof. ∎

Proposition 1.3.6 *We have the following properties.*

(i) *Suppose that the real-valued functions f_1 and f_2 are lower semi-continuous on the closed interval $[a, b]$. Then, the real-valued functions* $\min\{f_1, f_2\}$ *and* $\max\{f_1, f_2\}$ *are also lower semi-continuous on the closed interval $[a, b]$.*
(ii) *Suppose that the real-valued functions g_1 and g_2 are upper semi-continuous on on the closed interval $[a, b]$. Then, the real-valued functions* $\min\{g_1, g_2\}$ *and* $\max\{g_1, g_2\}$ *are also upper semi-continuous on the closed interval $[a, b]$.*

Proof. To prove part (i), given $\epsilon > 0$, there exist $\delta_1, \delta_2 > 0$ such that

$$|x - x_0| < \delta_1 \text{ implies } f_1(x_0) < f_1(x) + \epsilon,$$

and that

$$|x - x_0| < \delta_2 \text{ implies } f_2(x_0) < f_2(x) + \epsilon.$$

Let $\delta = \min\{\delta_1, \delta_2\}$. Then, for $|x - x_0| < \delta$, we have

$$\min\{f_1(x_0), f_2(x_0)\} < \min\{f_1(x) + \epsilon, f_2(x) + \epsilon\} = \min\{f_1(x), f_2(x)\} + \epsilon$$

and

$$\max\{f_1(x_0), f_2(x_0)\} < \max\{f_1(x) + \epsilon, f_2(x) + \epsilon\} = \max\{f_1(x), f_2(x)\} + \epsilon,$$

which show that $\min\{f_1, f_2\}$ and $\max\{f_1, f_2\}$ are lower semi-continuous at x_0.

To prove part (ii), given $\epsilon > 0$, there exist $\delta_1, \delta_2 > 0$ such that

$$|x - x_0| < \delta_1 \text{ implies } g_1(x_0) + \epsilon > g_1(x),$$

and that

$$|x - x_0| < \delta_2 \text{ implies } g_2(x_0) + \epsilon > g_2(x).$$

Let $\delta = \min\{\delta_1, \delta_2\}$. Then, for $|x - x_0| < \delta$, we have

$$\min\{g_1(x_0), g_2(x_0)\} + \epsilon = \min\{g_1(x_0) + \epsilon, g_2(x_0) + \epsilon\} > \min\{g_1(x), g_2(x)\}$$

and

$$\max\{g_1(x_0), g_2(x_0)\} + \epsilon = \max\{g_1(x_0) + \epsilon, g_2(x_0) + \epsilon\} > \max\{g_1(x), g_2(x)\},$$

which show that $\min\{g_1, g_2\}$ and $\max\{g_1, g_2\}$ are upper semi-continuous at x_0. This completes the proof. ∎

Proposition 1.3.7 *We have the following properties.*

(i) *Suppose that f is increasing on a subset D of $\mathbb{R}$. Then f is left-continuous on D if and only if f is lower semi-continuous on D.*
(ii) *Suppose that g is decreasing on a subset D of $\mathbb{R}$. Then g is left-continuous on D if and only if g is upper semi-continuous on D.*

Proof. To prove part (i), we first assume that f is left-continuous at $x_0 \in D$. Then, given any $\epsilon > 0$, there exists $\delta > 0$ such that $0 < x_0 - x < \delta$ implies $|f(x_0) - f(x)| < \epsilon$, i.e. $f(x_0) < f(x) + \epsilon$. For $x_0 \in D$ with $0 < x - x_0 < \delta$, since f is increasing, we have

$$f(x_0) \leq f(x) < f(x) + \epsilon.$$

Therefore, we conclude that $|x_0 - x| < \delta$ implies $f(x_0) < f(x) + \epsilon$, which shows that f is lower semi-continuous at $x_0 \in D$.

Conversely, we assume that f is lower semi-continuous at $x_0 \in D$. Then, given any $\epsilon > 0$, there exists $\delta > 0$ such that $|x_0 - x| < \delta$ implies $f(x_0) < f(x) + \epsilon$. If $0 < x_0 - x < \delta$ then we immediately have $f(x_0) - f(x) < \epsilon$ by the lower semi-continuity at x_0. Since f is increasing, we also have

$$f(x) \leq f(x_0) < f(x_0) + \epsilon.$$

Therefore, we conclude that $0 < x_0 - x < \delta$ implies $|f(x_0) - f(x)| < \epsilon$, which shows that f is left-continuous at $x_0 \in D$.

To prove part (ii), we first assume that g is left-continuous at $x_0 \in D$. Then, given any $\epsilon > 0$, there exists $\delta > 0$ such that $0 < x_0 - x < \delta$ implies $|g(x_0) - g(x)| < \epsilon$, i.e. $g(x) < g(x_0) + \epsilon$. For $x_0 \in D$ with $0 < x - x_0 < \delta$, since g is decreasing, we have

$$g(x_0) + \epsilon \geq g(x) + \epsilon > g(x).$$

Therefore, we conclude that $|x_0 - x| < \delta$ implies $g(x) < g(x_0) + \epsilon$, which shows that g is upper semi-continuous at $x_0 \in D$.

Conversely, we assume that f is upper semi-continuous at $x_0 \in D$. Then, given any $\epsilon > 0$, there exists $\delta > 0$ such that $|x_0 - x| < \delta$ implies $g(x) < g(x_0) + \epsilon$. If $0 < x_0 - x < \delta$, then we immediately have $g(x) - g(x_0) < \epsilon$ by the upper semi-continuity at x_0. Since g is decreasing, we also have

$$g(x_0) \leq g(x) < g(x) + \epsilon.$$

Therefore, we conclude that $0 < x_0 - x < \delta$ implies $|g(x_0) - g(x)| < \epsilon$, which shows that g is left-continuous at $x_0 \in D$. This completes the proof. ∎

Let A be a subset of $\mathbb{R}^m$. The **characteristic function** or **indicator function** of A is defined by

$$\chi_A(x) = \begin{cases} 1 & \text{for } x \in A \\ 0 & \text{for } x \notin A. \end{cases} \tag{1.14}$$

Proposition 1.3.8 *Let S be a subset of $\mathbb{R}$, and let $\zeta^L : S \to \mathbb{R}$ and $\zeta^U : S \to \mathbb{R}$ be two bounded real-valued functions defined on S satisfying $\zeta^L(\alpha) \leq \zeta^U(\alpha)$ for each $\alpha \in S$. Suppose that ζ^L is lower semi-continuous on S, and that ζ^U is upper semi-continuous on S. Let $M_\alpha = [\zeta^L(\alpha), \zeta^U(\alpha)]$ for $\alpha \in S$ be closed intervals. Then, for any fixed $x \in \mathbb{R}$, the function $\zeta(\alpha) = \alpha \cdot \chi_{M_\alpha}(x)$ is upper semi-continuous on S.*

Proof. For any fixed $\alpha_0 \in S$, we are going to show that, given any $\epsilon > 0$, there exists $\delta > 0$ such that

$$|\alpha - \alpha_0| < \delta \text{ implies } \zeta(\alpha_0) + \epsilon > \zeta(\alpha).$$

We consider the cases of $x \in M_{\alpha_0}$ and $x \notin M_{\alpha_0}$. For $x \in M_{\alpha_0}$, we have $\zeta(\alpha_0) = \alpha_0$. If $|\alpha - \alpha_0| < \delta = \epsilon$, we have $\alpha_0 + \epsilon > \alpha$. We consider the following cases.

- Suppose that $x \notin M_\alpha$. Then, we have $\zeta(\alpha) = 0$. Therefore, we obtain

  $$\zeta(\alpha_0) + \epsilon = \alpha_0 + \epsilon > 0 = \zeta(\alpha).$$

- Suppose that $x \in M_\alpha$. Then, we have $\zeta(\alpha) = \alpha$. Therefore, we obtain

  $$\zeta(\alpha_0) + \epsilon = \alpha_0 + \epsilon > \alpha = \zeta(\alpha).$$

Now, we consider the case of $x \notin M_{\alpha_0}$, i.e. $x < \zeta^L(\alpha_0)$ or $x > \zeta^U(\alpha_0)$. In this case, we have $\zeta(\alpha_0) = 0$.

- For $x < \zeta^L(\alpha_0)$, let $\epsilon = \zeta^L(\alpha_0) - x$. Since ζ^L is lower semi-continuous at α_0, there exists $\delta > 0$ such that $|\alpha - \alpha_0| < \delta$ implies $\zeta^L(\alpha_0) < \zeta^L(\alpha) + \epsilon$. Therefore, we obtain

  $$\zeta^L(\alpha) > \zeta^L(\alpha_0) - \epsilon = \zeta^L(\alpha_0) + x - \zeta^L(\alpha_0) = x.$$

  This also says that $x \notin M_\alpha$, i.e. $\zeta(\alpha) = 0$ for $|\alpha - \alpha_0| < \delta$.

- For $x > \zeta^U(\alpha_0)$, let $\epsilon = x - \zeta^U(\alpha_0)$. Since ζ^U is upper semi-continuous at α_0, there exists $\delta > 0$ such that $|\alpha - \alpha_0| < \delta$ implies $\zeta^U(\alpha) < \zeta^U(\alpha_0) + \epsilon$. Therefore, we obtain

$$\zeta^U(\alpha) < \zeta^U(\alpha_0) + \epsilon = \zeta^U(\alpha_0) + x - \zeta^U(\alpha_0) = x.$$

This also says that $x \notin M_\alpha$, i.e. $\zeta(\alpha) = 0$ for $|\alpha - \alpha_0| < \delta$.

The above two cases conclude that

$$\zeta(\alpha_0) + \epsilon = \epsilon > 0 = \zeta(\alpha)$$

for $|\alpha - \alpha_0| < \delta$. This completes the proof. ■

Proposition 1.3.9 *Let S be a subset of $\mathbb{R}$, and let $\zeta^L : S \to \mathbb{R}$ and $\zeta^U : S \to \mathbb{R}$ be two bounded real-valued functions defined on S satisfying $\zeta^L(\alpha) \le \zeta^U(\alpha)$ for each $\alpha \in S$. Suppose that the following conditions are satisfied.*

- *ζ^L is an increasing function and ζ^U is a decreasing function on S.*
- *ζ^L and ζ^U are left-continuous on S.*

Let $M_\alpha = [\zeta^L(\alpha), \zeta^U(\alpha)]$ for $\alpha \in S$ be closed intervals. Then, for any fixed $x \in \mathbb{R}$, the function $\zeta(\alpha) = \alpha \cdot \chi_{M_\alpha}(x)$ is upper semi-continuous on S.

Proof. The result follows immediately from Propositions 1.3.8 and 1.3.7. ■

Proposition 1.3.10 *Let S be a subset of $\mathbb{R}$. For each $i = 1, \ldots, n$, let $\zeta_i^L : S \to \mathbb{R}$ and $\zeta_i^U : S \to \mathbb{R}$ be bounded real-valued functions defined on S satisfying $\zeta_i^L(\alpha) \le \zeta_i^U(\alpha)$ for each $\alpha \in S$. Suppose that the following conditions are satisfied.*

- *ζ_i^L are increasing function and ζ_i^U are decreasing function on S for $i = 1, \ldots, n$.*
- *ζ_i^L and ζ_i^U are left-continuous on S for $i = 1, \ldots, n$.*

Let $M_\alpha^{(i)} = [\zeta_i^L(\alpha), \zeta_i^U(\alpha)]$ for $\alpha \in S$ and for $i = 1, \ldots, n$ be closed intervals, and let

$$M_\alpha = M_\alpha^{(1)} \times \cdots \times M_\alpha^{(n)} \subset \mathbb{R}^n.$$

Given any fixed $\mathbf{x} = (x_1, \ldots, x_n) \in \mathbb{R}^n$, the function $\zeta(\alpha) = \alpha \cdot \chi_{M_\alpha}(\mathbf{x})$ is upper semi-continuous on S.

Proof. Proposition 1.3.9 says that the functions $\zeta_i(\alpha) = \alpha \cdot \chi_{M_\alpha^{(i)}}(x_i)$ are upper semi-continuous on S for $i = 1, \ldots, n$. For $r \in S$, we define the sets

$$F_r = \{\alpha \in S : \zeta(\alpha) \ge r\} \text{ and } F_r^{(i)} = \left\{\alpha \in S : \zeta_i(\alpha) \ge r\right\} \text{ for } i = 1, \ldots, n.$$

The upper semi-continuity of ζ_i says that $F_r^{(i)}$ is a closed set for $i = 1, \ldots, n$. We want to claim $F_r = \bigcap_{i=1}^n F_r^{(i)}$. Given any $\alpha \in F_r$, it follows that $\mathbf{x} \in M_\alpha$ and $\alpha \ge r$, i.e. $x_i \in M_\alpha^{(i)}$ and $\alpha \ge r$ for $i = 1, \ldots, n$, which also implies $\zeta_i(\alpha) \ge r$ for $i = 1, \ldots, n$. Therefore, we obtain the inclusion $F_r \subseteq \bigcap_{i=1}^n F_r^{(i)}$. On the other hand, suppose that $\alpha \in F_r^{(i)}$ for $i = 1, \ldots, n$. It follows that $x_i \in M_\alpha^{(i)}$ and $\alpha \ge r$ for $i = 1, \ldots, n$, i.e. $\mathbf{x} \in M_\alpha$ and $\alpha \ge r$. Therefore, we obtain the equality $F_r = \bigcap_{i=1}^n F_r^{(i)}$, which also says that F_r is a closed set, since each $F_r^{(i)}$ is a closed set for $i = 1, \ldots, n$. Therefore, we conclude that ζ is indeed upper semi-continuous on S. This completes the proof. ■

We say that S is a disjoint union of intervals in $\mathbb{R}$ when S can be expressed as

$$S = \bigcup_{i=1}^{\infty} I_i$$

satisfying $I_i \cap I_j = \emptyset$ for $i \neq j$, where each I_i is an interval in $\mathbb{R}$.

Proposition 1.3.11 *Let S be a disjoint union of intervals in $\mathbb{R}$, and let $\zeta^L : S \to \mathbb{R}$ and $\zeta^U : S \to \mathbb{R}$ be two bounded real-valued functions defined on S satisfying $\zeta^L(\alpha) \leq \zeta^U(\alpha)$ for each $\alpha \in S$. For $\alpha \in S$, we define the functions*

$$l(\alpha) = \inf_{\{x \in S : x \geq \alpha\}} \zeta^L(x) \text{ and } u(\alpha) = \sup_{\{x \in S : x \geq \alpha\}} \zeta^U(x).$$

Then l and u are left-continuous on S. Moreover, l is lower semi-continuous on S and u is upper semi-continuous on S.

Proof. Given $\alpha \in S$, since S is a disjoint union of intervals, there exists a sequence $\{\alpha_n\}_{n=1}^{\infty}$ in S satisfying $\alpha_n \uparrow \alpha$ as $n \to \infty$, where we allow $\alpha_n = \alpha$ for some n. Let

$$A_n = \{x \in S : x \geq \alpha_n\} \text{ and } A = \{x \in S : x \geq \alpha\}.$$

Then it is obvious that $A_{n+1} \subseteq A_n$ for all n and $A \subseteq \bigcap_{n=1}^{\infty} A_n$. For $x \in \bigcap_{n=1}^{\infty} A_n$, it means $x \in S$ and $x \geq \alpha_n$ for all n. By taking limit, we obtain $x \geq \alpha$, i.e. $x \in A$. This shows that $A = \bigcap_{n=1}^{\infty} A_n$. Using Proposition 1.2.3, we obtain

$$l(\alpha_n) = \inf_{t \in A_n} \zeta^L(x) \to \inf_{t \in A} \zeta^L(x) = l(\alpha) \text{ for } \alpha_n \uparrow \alpha.$$

This says that l is left-continuous on S. We can similarly show that u is left-continuous on S. Since l is decreasing and u is increasing on S, Proposition 1.3.7 says that l is lower semi-continuous on S and u is upper semi-continuous on S. This completes the proof. ∎

Let S be a disjoint union of intervals in $\mathbb{R}$. We write $\partial^L(S)$ to denote the set of all left endpoints of subintervals in S, and write $\partial^R(S)$ to denote the set of all right endpoints of subintervals in S. For any $\alpha \in S \setminus \partial^R(S)$, i.e. $\alpha \in S$ and $\alpha \notin \partial^R(S)$, it is clear to see that there exists a sequence in S satisfying $\alpha_n \downarrow \alpha$ as $n \to \infty$ with $\alpha_n > \alpha$ for all n.

Proposition 1.3.12 *Let S be a disjoint union of closed intervals in $\mathbb{R}$, and let $\zeta^L : S \to \mathbb{R}$ and $\zeta^U : S \to \mathbb{R}$ be two bounded and right-continuous real-valued functions defined on S satisfying $\zeta^L(\alpha) \leq \zeta^U(\alpha)$ for each $\alpha \in S$. Let $M_\alpha = [\zeta^L(\alpha), \zeta^U(\alpha)]$ for $\alpha \in S$ be closed intervals. Then, the functions*

$$l(\alpha) = \inf_{\{x \in S : x \geq \alpha\}} \zeta^L(x) \text{ and } u(\alpha) = \sup_{\{x \in S : x \geq \alpha\}} \zeta^U(x)$$

are continuous on $S \setminus \partial^R(S)$. Moreover, for $\alpha \in S \setminus \partial^R(S)$ and $\alpha_n \downarrow \alpha$ as $n \to \infty$ with $\alpha_n > \alpha$ for all n, we have $l(\alpha_n) \downarrow l(\alpha)$ and $u(\alpha_n) \uparrow u(\alpha)$ as $n \to \infty$.

Proof. According to Proposition 1.3.11, we remain to show that l and u are right-continuous on $S \setminus \partial^R(S)$. We first note that S is a closed set, i.e. $\text{cl}(S) = S$. We are going to use part (i) of Proposition 1.2.8. Given $\alpha \in S \setminus \partial^R(S)$, there exists a sequence $\{\alpha_n\}_{n=1}^{\infty}$ in S satisfying $\alpha_n \downarrow \alpha$ as $n \to \infty$ with $\alpha_n > \alpha$ for all n. Let

$$A_n = \{x \in S : x \geq \alpha_n\} \text{ and } A^* = \{x \in S : x > \alpha\}.$$

It is clear to see that $A_n \subseteq A_{n+1}$ for all n and $\bigcup_{n=1}^{\infty} A_n \subseteq A^*$. For $x \in A^*$, i.e. $x \in S$ and $x > \alpha$, since $\alpha_n \downarrow \alpha$, there exists α_{n^*} satisfying $\alpha \leq \alpha_{n^*} < x$, which says that $x \in \bigcup_{n=1}^{\infty} A_n$. Therefore, we obtain $\bigcup_{n=1}^{\infty} A_n = A^*$. Using Proposition 1.2.4 and part (i) of Proposition 1.2.8, for $\alpha_n \downarrow \alpha$ with $\alpha_n > \alpha$, we have

$$l(\alpha_n) = \inf_{x \in A_n} \zeta^L(x) \to \inf_{x \in A^*} \zeta^L(x) = \inf_{\{x \in S: x \geq \alpha\}} \zeta^L(x) = l(\alpha).$$

Therefore, we conclude that l is continuous on S. We can similarly show that u is continuous on S. Since l is increasing and u is decreasing, we also have $l(\alpha_n) \downarrow l(\alpha)$ and $u(\alpha_n) \uparrow u(\alpha)$ as $n \to \infty$ for $\alpha_n \downarrow \alpha$ as $n \to \infty$ with $\alpha_n > \alpha$ for all n, and the proof is complete. ∎

Proposition 1.3.13 *Let S be a closed subset of $\mathbb{R}$, and let $\zeta^L : S \to \mathbb{R}$ and $\zeta^U : S \to \mathbb{R}$ be two bounded real-valued functions defined on S satisfying $\zeta^L(\alpha) \leq \zeta^U(\alpha)$ for each $\alpha \in S$. Suppose that ζ^L is lower semi-continuous on S, and that ζ^U is upper semi-continuous on S. Let $M_\alpha = [\zeta^L(\alpha), \zeta^U(\alpha)]$ for $\alpha \in S$ be closed intervals. Then, we have*

$$\bigcup_{\{\beta \in S: \beta \geq \alpha\}} M_\beta = \left[\inf_{\{\beta \in S: \beta \geq \alpha\}} \zeta^L(\beta), \sup_{\{\beta \in S: \beta \geq \alpha\}} \zeta^U(\beta) \right]$$
$$= \left[\min_{\{\beta \in S: \beta \geq \alpha\}} \zeta^L(\beta), \max_{\{\beta \in S: \beta \geq \alpha\}} \zeta^U(\beta) \right] \tag{1.15}$$

for any $\alpha \in S$.

Proof. Since S is a closed set, by Proposition 1.4.4 (which will be given below), the semi-continuities say that the imfimum and supremum are attained given by

$$\inf_{\{\beta \in S: \beta \geq \alpha\}} \zeta^L(\beta) = \min_{\{\beta \in S: \beta \geq \alpha\}} \zeta^L(\beta) \text{ and } \sup_{\{\beta \in S: \beta \geq \alpha\}} \zeta^U(\beta) = \max_{\{\beta \in S: \beta \geq \alpha\}} \zeta^U(\beta).$$

For $x \in \bigcup_{\{\beta \in S: \beta \geq \alpha\}} M_\beta$, there exists $\beta_0 \geq \alpha$ satisfying $x \in M_{\beta_0}$, i.e. $\zeta^L(\beta_0) \leq x \leq \zeta^U(\beta_0)$. Then, we have

$$x \geq \zeta^L(\beta_0) \geq \min_{\{\beta \in S: \beta \geq \alpha\}} \zeta^L(\beta) \text{ and } x \leq \zeta^U(\beta_0) \leq \max_{\{\beta \in S: \beta \geq \alpha\}} \zeta^U(\beta);$$

that is,

$$x \in \left[\min_{\{\beta \in S: \beta \geq \alpha\}} \zeta^L(\beta), \max_{\{\beta \in S: \beta \geq \alpha\}} \zeta^U(\beta) \right].$$

To prove the other direction of inclusion, given any x satisfying

$$\min_{\{\beta \in S: \beta \geq \alpha\}} \zeta^L(\beta) \leq x \leq \max_{\{\beta \in S: \beta \geq \alpha\}} \zeta^U(\beta), \tag{1.16}$$

we want to lead to a contradiction by assuming $x \notin M_\beta$ for each $\beta \in S$ with $\beta \geq \alpha$. Under this assumption, since each M_β is a bounded closed interval, it follows that $x < \zeta^L(\beta)$ for each $\beta \in S$ with $\beta \geq \alpha$ or $x > \zeta^U(\beta)$ for each $\beta \in S$ with $\beta \geq \alpha$. Since the infimum and supremum are attained, we obtain

$$x < \min_{\{\beta \in S: \beta \geq \alpha\}} \zeta^L(\beta) = \inf_{\{\beta \in S: \beta \geq \alpha\}} \zeta^L(\beta) \text{ or } x > \max_{\{\beta \in S: \beta \geq \alpha\}} \zeta^U(\beta) = \sup_{\{\beta \in S: \beta \geq \alpha\}} \zeta^U(\beta),$$

which contradicts (1.16). Therefore, there exists $\beta_0 \in S$ with $\beta_0 \geq \alpha$ satisfying $x \in M_{\beta_0}$. This completes the proof. ∎

1.4 Miscellaneous

The convexity of fuzzy sets is usually assumed for applications in order to simplify the discussion. The concept of convex set in $\mathbb{R}^n$ is given below.

Definition 1.4.1 Let A be a subset of $\mathbb{R}^m$. We say that A is **convex** when, given any $x, y \in A$, the convex combination $\lambda x + (1 - \lambda)y$ belongs to A for any $0 < \lambda < 1$.

Definition 1.4.2 Let $f : A \to \mathbb{R}$ be a real-valued function defined on a convex subset A of $\mathbb{R}^m$. The function f is called **quasi-convex** on A when, for each $x, y \in A$, the following inequality is satisfied:

$$f(\lambda x + (1 - \lambda)y) \leq \max\{f(x), f(y)\}$$

for each $0 < \lambda < 1$.

It is well known that f is quasi-convex on A if and only if the set

$$\{x \in A : f(x) \leq \alpha\}$$

is convex for each $\alpha \in \mathbb{R}$.

The function f is called **quasi-concave** on A when $-f$ is quasi-convex on A. More precisely, the real-valued function f is quasi-concave on A if and only if

$$f(\lambda x + (1 - \lambda)y) \geq \min\{f(x), f(y)\}$$

for each $0 < \lambda < 1$. We also see that f is quasi-concave on A if and only if the set

$$\{x \in A : f(x) \geq \alpha\}$$

is convex for each $\alpha \in \mathbb{R}$.

The following well-known results will be used through out this book.

Proposition 1.4.3 (Apostol [3]) *We have the following results*

(i) *Let f be a continuous real-valued function defined on a connected subset S of $\mathbb{R}^m$. Suppose that $f(x^*) < f(x^\circ)$ for some $x^*, x^\circ \in S$. For each y satisfying $f(x^*) < y < f(x^\circ)$, there exists $x \in S$ satisfying $f(x) = y$.*
(ii) *Let f be a continuous real-valued function defined on a bounded closed interval I in $\mathbb{R}$. Suppose that there are two points $x, y \in I$ satisfying $x < y$ and $f(x) \neq f(y)$. Then f takes every value between $f(x)$ and $f(y)$ in the open interval (x, y).*
(iii) *Let $f : \mathbb{R}^p \to \mathbb{R}^q$ be a vector-valued function. Suppose that f is continuous on a closed and bounded subset X of $\mathbb{R}^p$. Then $f(X)$ is a closed and bounded subset of $\mathbb{R}^q$.*
(iv) *Let $f : \mathbb{R}^p \to \mathbb{R}^q$ be a vector-valued function. Suppose that f is continuous on a connected subset X of $\mathbb{R}^p$. Then $f(X)$ is a connected subset of $\mathbb{R}^q$.*

Proposition 1.4.4 (Royden [100]) *Let f be a real-valued function defined on $\mathbb{R}^m$, and let K be a closed and bounded subset of $\mathbb{R}^m$. Then, we have the following properties.*

(i) *Suppose that f is upper semi-continuous. Then, the supremum is attained in the following sense*

$$\sup_{x \in K} f(x) = \max_{x \in K} f(x).$$

(ii) *Suppose that f is lower semi-continuous. Then, the infimum is attained in the following sense*

$$\inf_{x\in K} f(x) = \min_{x\in K} f(x).$$

Theorem 1.4.5 ***(Cantor Intersection Theorem)****. Let $\{Q_1, Q_2, \ldots\}$ be a countable collection of nonempty subsets of a topological space $\mathbb{R}^m$ such that the following conditions are satisfied:*

- *$Q_{k+1} \subseteq Q_k$ for $k = 1,2,\ldots$;*
- *each Q_k is a nonempty bounded and closed subset of $\mathbb{R}^m$ for all k.*

Then, the intersection $\bigcap_{k=1}^{\infty} Q_k$ is nonempty.

Proposition 1.4.6 *Let $\mathfrak{A} : [0,1]^n \to [0,1]$ be a function defined on $[0,1]^n$. Suppose that*

$$\mathfrak{A}\left(\alpha_1, \ldots, \alpha_m\right) \geq \beta \text{ if and only if } \alpha_i \geq \beta \text{ for all } i = 1, \ldots, n.$$

Then, we have

$$\mathfrak{A}\left(\alpha_1, \ldots, \alpha_m\right) = \min\left\{\alpha_1, \ldots, \alpha_m\right\}.$$

Proof. Since $\alpha_i \geq \min\{\alpha_1, \ldots, \alpha_m\}$ for all $i = 1, \ldots, n$, the assumption says that

$$\mathfrak{A}\left(\alpha_1, \ldots, \alpha_m\right) \geq \min\left\{\alpha_1, \ldots, \alpha_m\right\}$$

by taking $\beta = \min\{\alpha_1, \ldots, \alpha_m\}$. On the other hand, suppose that $\mathfrak{A}(\alpha_1, \ldots, \alpha_m) = \beta$, i.e. $\mathfrak{A}(\alpha_1, \ldots, \alpha_m) \geq \beta$, the assumption says that $\alpha_i \geq \beta$ for all $i = 1, \ldots, n$, which implies

$$\min\left\{\alpha_1, \ldots, \alpha_m\right\} \geq \beta = \mathfrak{A}\left(\alpha_1, \ldots, \alpha_m\right).$$

This completes the proof. ■

Proposition 1.4.7 *Let $\mathfrak{A} : [0,1]^n \to [0,1]$ be a function defined on $[0,1]^n$. Suppose that*

$$\mathfrak{A}\left(\alpha_1, \ldots, \alpha_m\right) \leq \beta \text{ if and only if } \alpha_i \leq \beta \text{ for some } i = 1, \ldots, n.$$

Then, we have

$$\mathfrak{A}\left(\alpha_1, \ldots, \alpha_m\right) = \min\left\{\alpha_1, \ldots, \alpha_m\right\}.$$

Proof. Suppose that $\alpha_i > \min\{\alpha_1, \ldots, \alpha_m\}$ for all $i = 1, \ldots, n$. Then $\min\{\alpha_1, \ldots, \alpha_m\} > \min\{\alpha_1, \ldots, \alpha_m\}$. This contradiction says that $\alpha_i \leq \min\{\alpha_1, \ldots, \alpha_m\}$ for some $i = 1, \ldots, n$. Using the assumption, it follows that

$$\mathfrak{A}\left(\alpha_1, \ldots, \alpha_m\right) \leq \min\left\{\alpha_1, \ldots, \alpha_m\right\}$$

by taking $\beta = \min\{\alpha_1, \ldots, \alpha_m\}$. On the other hand, suppose that $\mathfrak{A}(\alpha_1, \ldots, \alpha_m) = \beta$, i.e. $\mathfrak{A}(\alpha_1, \ldots, \alpha_m) \leq \beta$, the assumption says that $\alpha_i \leq \beta$ for some $i = 1, \ldots, n$, which implies

$$\min\left\{\alpha_1, \ldots, \alpha_m\right\} \leq \beta = \mathfrak{A}\left(\alpha_1, \ldots, \alpha_m\right).$$

This completes the proof. ■

Proposition 1.4.8 *Let F and G be two real-valued functions from $\mathbb{R}^m$ into $(0,1]$. Suppose that*

$$\{x : F(x) \geq \alpha\} = \{x : G(x) \geq \alpha\} \tag{1.17}$$

for each $\alpha \in \mathbb{Q} \cap (0,1]$. Then $F(x) = G(x)$ for all $x \in \mathbb{R}^m$; that is, F and G are identical.

Proof. Assume that there exists $x_0 \in \mathbb{R}^m$ satisfying $G(x_0) \neq 0$ and $F(x_0) < G(x_0)$, where $F(x_0)$ can be 0. Using the denseness of $\mathbb{R}$, there exists $\alpha_0 \in \mathbb{Q} \cap (0,1]$ satisfying $F(x_0) < \alpha_0 < G(x_0)$. This says that

$$x_0 \in \left\{x : G(x) \geq \alpha_0\right\} \text{ and } x_0 \notin \left\{x : F(x) \geq \alpha_0\right\},$$

which contradicts (1.17). If we assume that $F(x_0) \neq 0$ and $F(x_0) > G(x_0)$, where $G(x_0)$ can be 0, then we can similarly obtain a contradiction. Therefore, we conclude that

$$F(x_0) = G(x_0) \neq 0.$$

This completes the proof. ■

2

Fuzzy Sets

The main idea of fuzzy sets is to consider the degree of membership. A fuzzy set is described by a membership function that assigns to each member or element a membership degree. Usually, the range of this membership function is from 0 to 1. A degree of 1 represents complete membership to the set, and degree of 0 represents absolutely no membership to the set. A degree between 0 and 1 represents partial membership to the set.

We can define high fever as a temperature higher than 102 °F. Even if most doctors will agree that the threshold is at about 102 °F (39 °C), this does not mean that a patient with a body temperature of 101.9 °F does not have a high fever while another patient with 102 °F does indeed have a high fever. Therefore, instead of using this rigid definition, each body temperature is associated with a certain degree. For example, we show a possible description of high fever using membership degree as follows

$$\begin{array}{lll} \xi(94\,^\circ F) = 0, & \xi(96\,^\circ F) = 0, & \xi(98\,^\circ F) = 0 \\ \xi(100\,^\circ F) = 0.7, & \xi(102\,^\circ F) = 0.95, & \xi(104\,^\circ F) = 0.99 \\ \xi(106\,^\circ F) = 1, & \xi(108\,^\circ F) = 1, & \xi(110\,^\circ F) = 1. \end{array}$$

The degree of membership can also be represented by a continuous function.

2.1 Membership Functions

Let A be a subset of $\mathbb{R}^m$. Each element $x \in \mathbb{R}^m$ can either belong to or not belong to a set A. This kind of set can be defined by the characteristic function

$$\chi_A(x) = \begin{cases} 1 & \text{for } x \in A \\ 0 & \text{for } x \notin A. \end{cases}$$

That is to say, the characteristic function maps elements of $\mathbb{R}^m$ to elements of the set $\{0,1\}$, which is formally expressed by $\chi_A : \mathbb{R}^m \to \{0,1\}$.

Zadeh [162] proposed a concept of so-called **fuzzy set** by extending the range $\{0,1\}$ of the characteristic function to the unit interval $[0,1]$. A fuzzy set $\tilde{A}$ in $\mathbb{R}^m$ is defined to be a set of ordered pairs

$$\tilde{A} = \{(x, \xi_{\tilde{A}}(x)) : x \in \mathbb{R}^m\},$$

where $\xi_{\tilde{A}} : \mathbb{R}^m \to [0,1]$ is called the **membership function** of $\tilde{A}$. The value $\xi_{\tilde{A}}(x)$ is regarded as the degree of membership of x in $\tilde{A}$. In other words, it indicates the degree to

Mathematical Foundations of Fuzzy Sets, First Edition. Hsien-Chung Wu.

which x belongs to $\tilde{A}$. Any subset A of $\mathbb{R}^m$ can also be regarded as a fuzzy set in $\mathbb{R}^m$ by taking the membership function as the characteristic function of A. In this case, we write $\tilde{1}_A \equiv \chi_A$ by regarding A as a fuzzy set in $\mathbb{R}^m$. When A is a singleton $\{a\}$, we also write $\tilde{1}_{\{a\}}$.

Example 2.1.1 Let $\tilde{A}$ be the set of all real numbers considerably larger than 10. Then $\tilde{A}$ can be described as a fuzzy set in $\mathbb{R}$ with membership function defined by

$$\xi_{\tilde{A}}(x) = \begin{cases} 0, & x \leq 10 \\ \dfrac{1}{1+(x-10)^{-2}}, & x > 10. \end{cases}$$

Let $\tilde{B}$ be the set of all real numbers close (but not equal) to 10. Then $\tilde{B}$ can also be described as a fuzzy set in $\mathbb{R}$ with membership function defined by

$$\xi_{\tilde{A}}(x) = \frac{1}{1+(x-10)^2}.$$

In this case, we may also write $\tilde{B} = \widetilde{10}$ to mean fuzzy real number 10. Therefore, any statements that involve fuzziness can always be represented by a membership function.

2.2 α-level Sets

An interesting and important concept related to fuzzy sets is the α-level set. Let $\tilde{A}$ be a fuzzy set in $\mathbb{R}^m$ with membership function $\xi_{\tilde{A}}$. The range of the membership function $\xi_{\tilde{A}}$ is denoted by $\mathcal{R}(\xi_{\tilde{A}})$. Throughout this book, we shall assume that the range $\mathcal{R}(\xi_{\tilde{A}})$ contains 1. However, the range $\mathcal{R}(\xi_{\tilde{A}})$ is not necessarily equal to the whole unit interval [0,1].

For $\alpha \in (0,1]$, the **α-level set** $\tilde{A}_\alpha$ of $\tilde{A}$ is defined by

$$\tilde{A}_\alpha = \left\{x : \xi_{\tilde{A}}(x) \geq \alpha\right\}. \tag{2.1}$$

Since the range $\mathcal{R}(\xi_{\tilde{A}})$ is assumed to contain 1, it follows that the α-level sets $\tilde{A}_\alpha$ are non-empty for all $\alpha \in (0,1]$. Notice that the 0-level set is not defined by (2.1). The 0-level set will be defined in a different way that will be explained afterward.

Given any $\alpha, \beta \in (0,1]$ satisfying $\alpha < \beta$, it is easy to see

$$\tilde{A}_\beta \subseteq \tilde{A}_\alpha. \tag{2.2}$$

The strict inclusion $\tilde{A}_\beta \subset \tilde{A}_\alpha$ can happen.

Proposition 2.2.1 *Let $\tilde{A}$ be a fuzzy set in $\mathbb{R}^m$ with membership function $\xi_{\tilde{A}}$. Suppose that $\alpha \in \mathcal{R}(\xi_{\tilde{A}})$ and $\beta \in [0,1]$ with $\beta > \alpha$. Then $\tilde{A}_\beta \subset \tilde{A}_\alpha$ (where "$\subset$" means $\tilde{A}_\beta \subseteq \tilde{A}_\alpha$ and $\tilde{A}_\beta \neq \tilde{A}_\alpha$).*

Proof. Since $\alpha \in \mathcal{R}(\xi_{\tilde{A}})$, there exists $x \in \mathbb{R}^m$ satisfying $\xi_{\tilde{A}}(x) = \alpha$. Suppose that there exists $\beta_0 \in [0,1]$ with $\beta_0 > \alpha$ satisfying $x \in \tilde{A}_{\beta_0}$. Then, we have $\xi_{\tilde{A}}(x) \geq \beta_0 > \alpha$ which violates $\xi_{\tilde{A}}(x) = \alpha$. In other words, there exists $x \in \tilde{A}_\alpha$ and $x \notin \tilde{A}_\beta$ for all $\beta \in [0,1]$ with $\beta > \alpha$. This completes the proof. ■

Example 2.2.2 The membership function of a fuzzy set $\tilde{A}$ is given by

$$\xi_{\tilde{A}}(x) = \begin{cases} 0.1 + 0.8 \cdot (x-1) & \text{if } 1 \leq x \leq 1.5 \\ 0.2 + 0.8 \cdot (x-1) & \text{if } 1.5 < x < 2 \\ 1 & \text{if } 2 \leq x \leq 3 \\ 0.2 + 0.8 \cdot (4-x) & \text{if } 3 < x < 3.5 \\ 0.1 + 0.8 \cdot (4-x) & \text{if } 3.5 \leq x \leq 4 \\ 0 & \text{otherwise.} \end{cases}$$

It is clear to see

$$\mathcal{R}(\xi_{\tilde{A}}) = \{0\} \cup [0.1, 0.5] \cup (0.6, 1],$$

which also says that $[0,1] \neq \mathcal{R}(\xi_{\tilde{A}})$. We see that $0.55 \notin \mathcal{R}(\xi_{\tilde{A}})$. However, we still have the 0.55-level set $\tilde{A}_{0.55}$ given by

$$\tilde{A}_{0.55} = \left\{ x : \xi_{\tilde{A}}(x) \geq 0.55 \right\}.$$

Although $0 \leq \alpha < 0.1$ is not in the range $\mathcal{R}(\xi_{\tilde{A}})$, we still have

$$\tilde{A}_{\alpha} = \left\{ x : \xi_{\tilde{A}}(x) \geq \alpha \right\} = [1,4] = \tilde{A}_{0.1} \text{ for each } \alpha \in [0, 0.1).$$

Notice that the expression (2.1) does not include the 0-level set. If we allowed the expression (2.1) taking $\alpha = 0$, the 0-level set of $\tilde{A}$ would be the whole m-dimensional Euclidean space $\mathbb{R}^m$. Defined in this way, the 0-level set would not be helpful for real applications. Therefore, we are going to invoke a topological concept to define the 0-level set. The **support** of a fuzzy set $\tilde{A}$ in $\mathbb{R}^m$ is the crisp set defined by

$$\tilde{A}_{0+} = \{x \in \mathbb{R}^m : \xi_{\tilde{A}}(x) > 0\}. \tag{2.3}$$

The 0-level set $\tilde{A}_0$ of $\tilde{A}$ is defined to be the closure of the support $\tilde{A}_{0+}$, i.e.

$$\tilde{A}_0 = \text{cl}(\tilde{A}_{0+}). \tag{2.4}$$

For the concept of closure, refer to Definition 1.2.5.

Proposition 2.2.3 *Let $\tilde{A}$ be a fuzzy set in $\mathbb{R}^m$ with membership function $\xi_{\tilde{A}}$. Then, we have*

$$\tilde{A}_{0+} = \bigcup_{\{\alpha \in \mathcal{R}(\xi_{\tilde{A}}) : \alpha > 0\}} \tilde{A}_{\alpha} = \bigcup_{0 < \alpha \leq 1} \tilde{A}_{\alpha} \tag{2.5}$$

Proof. Given any $x \in \tilde{A}_{0+}$, i.e. $\alpha \equiv \xi_{\tilde{A}}(x) > 0$, we have $x \in \tilde{A}_{\alpha}$. Since $\alpha \in \mathcal{R}(\xi_{\tilde{A}})$, we have the inclusion

$$\tilde{A}_{0+} \subseteq \bigcup_{\{\alpha \in \mathcal{R}(\xi_{\tilde{A}}) : \alpha > 0\}} \tilde{A}_{\alpha}.$$

For proving the other direction of inclusion, given $x \in \tilde{A}_{\alpha}$ for some $0 < \alpha \in \mathcal{R}(\xi_{\tilde{A}})$, we have $\xi_{\tilde{A}}(x) \geq \alpha > 0$, i.e. $x \in \tilde{A}_{0+}$. This proves the desired equality. ∎

Let A be a subset of $\mathbb{R}^m$. Recall the notation $\chi_A = \tilde{1}_A$. Then, we see that $(\tilde{1}_A)_\alpha = A$ for any $\alpha \in (0,1]$. Also, the 0-level set is given by

$$(\tilde{1}_A)_0 = \text{cl}\left(\bigcup_{0<\alpha\leq 1} (\tilde{1}_A)_\alpha\right) = \text{cl}\left(\bigcup_{0<\alpha\leq 1} A\right) = \text{cl}(A). \tag{2.6}$$

Recall that A is a closed subset of $\mathbb{R}^m$ when $A = \text{cl}(A)$. Now, suppose that A is a closed subset of $\mathbb{R}^m$. Then, we have $(\tilde{1}_A)_\alpha = A$ for any $\alpha \in [0,1]$. In particular, for any $r \in \mathbb{R}^m$, since the singleton $\{r\}$ is a closed subset of $\mathbb{R}^m$, it follows that $(\tilde{1}_{\{r\}})_\alpha = \{r\}$ for any $\alpha \in [0,1]$ because of (2.6).

Example 2.2.4 The membership function of a trapezoidal fuzzy interval is given by

$$\xi_{\tilde{A}}(r) = \begin{cases} (r - a^L)/(a_1 - a^L) & \text{if } a^L \leq r \leq a_1 \\ 1 & \text{if } a_1 < r \leq a_2 \\ (a^U - r)/(a^U - a_2) & \text{if } a_2 < r \leq a^U \\ 0 & \text{otherwise,} \end{cases}$$

which is denoted by $\tilde{A} = (a^L, a_1, a_2, a^U)$. It is clear to see $\mathcal{R}(\xi_{\tilde{A}}) = [0,1]$. For $\alpha \in [0,1]$, the α-level set $\tilde{A}_\alpha$ is a closed interval denoted by $\tilde{A}_\alpha = [\tilde{A}_\alpha^L, \tilde{A}_\alpha^U]$, where

$$\tilde{A}_\alpha^L = (1-\alpha)a^L + \alpha a_1 \text{ and } \tilde{A}_\alpha^U = (1-\alpha)a^U + \alpha a_2.$$

In particular, if $a_1 = a_2 \equiv a$, we have the so-called triangular fuzzy interval $\tilde{A} = (a^L, a, a^U)$.

Let A be a convex set in $\mathbb{R}^m$ (refer to Definition 1.4.1). Then, we see that

$$\{x : \chi_A(x) \geq \alpha\} = A$$

are also convex sets in $\mathbb{R}^m$ for $\alpha \in (0,1]$. Therefore, we can extend the above concept to define the convexity of a fuzzy set in $\mathbb{R}^m$ by replacing the characteristic function with the membership function.

Definition 2.2.5 Let $\tilde{A}$ be a fuzzy set in $\mathbb{R}^m$ with membership function $\xi_{\tilde{A}}$. We say that $\tilde{A}$ is **convex** when the α-level sets

$$\tilde{A}_\alpha = \{x : \xi_{\tilde{A}}(x) \geq \alpha\}$$

are convex sets in $\mathbb{R}^m$ for any $\alpha \in (0,1]$.

Definition 2.2.5 does not include the convexity of the 0-level set. The following proposition can guarantee the convexity of the 0-level set.

Proposition 2.2.6 *Let $\tilde{A}$ be a convex fuzzy set in $\mathbb{R}^m$. Then, the 0-level set $\tilde{A}_0$ is a convex set in $\mathbb{R}^m$.*

Proof. From (2.5), the 0-level set $\tilde{A}_0$ can be expressed as follows

$$\tilde{A}_0 = \text{cl}\left(\bigcup_{0<\alpha\leq 1} \tilde{A}_\alpha\right).$$

By Kelley and Namioka [55, p. 35], the closure cl(K) of a convex set K is also a convex set. In order to show that $\tilde{A}_0$ is a convex set in $\mathbb{R}^m$, it suffices to show that

$$K \equiv \bigcup_{0<\alpha\leq 1} \tilde{A}_\alpha$$

is a convex set in $\mathbb{R}^m$. For any $x, y \in K$, there exist $\alpha_1, \alpha_2 \in (0,1]$ satisfying $x \in \tilde{A}_{\alpha_1}$ and $y \in \tilde{A}_{\alpha_2}$. Let $\alpha_3 = \min\{\alpha_1, \alpha_2\} > 0$. Then, from (2.2), we see that $x, y \in \tilde{A}_{\alpha_3}$. Since $\tilde{A}_{\alpha_3}$ is convex, it follows that

$$\lambda x + (1-\lambda)y \in \tilde{A}_{\alpha_3} \subset K$$

for any $\lambda \in (0,1)$. This shows that K is a convex set in $\mathbb{R}^m$, and the proof is complete. ■

Proposition 2.2.7 *Let $\tilde{A}$ be a fuzzy set in $\mathbb{R}^m$ with membership function $\xi_{\tilde{A}}$. Then $\tilde{A}$ is convex if and only if*

$$\xi_{\tilde{A}}\left(\lambda x_1 + (1-\lambda)x_2\right) \geq \min\left\{\xi_{\tilde{A}}(x_1), \xi_{\tilde{A}}(x_2)\right\}$$

for any $\lambda \in [0,1]$. In other words, $\tilde{A}$ is convex if and only if its membership function $\xi_{\tilde{A}}$ is quasi-concave.

Proof. Suppose that $\tilde{A}_\alpha = \{x : \xi_{\tilde{A}}(x) \geq \alpha\}$ are convex sets for $\alpha \in (0,1]$. For any $x_1, x_2 \in X$, suppose that $\min\left\{\xi_{\tilde{A}}(x_1), \xi_{\tilde{A}}(x_2)\right\} = 0$. Then, we are done, since $\xi_{\tilde{A}}$ is a nonnegative real-valued function. Now, we assume that

$$0 \neq \alpha = \min\left\{\xi_{\tilde{A}}(x_1), \xi_{\tilde{A}}(x_2)\right\},$$

i.e. $\xi_{\tilde{A}}(x_1) \geq \alpha$ and $\xi_{\tilde{A}}(x_2) \geq \alpha$. This says that $x_1, x_2 \in \tilde{A}_\alpha$. Since $\tilde{A}_\alpha$ is convex, we have $\lambda x_1 + (1-\lambda)x_2 \in \tilde{A}_\alpha$, i.e.

$$\xi_{\tilde{A}}\left(\lambda x_1 + (1-\lambda)x_2\right) \geq \alpha = \min\left\{\xi_{\tilde{A}}(x_1), \xi_{\tilde{A}}(x_2)\right\}.$$

To prove the converse, for any $x_1, x_2 \in \tilde{A}_\alpha$, we have $\xi_{\tilde{A}}(x_1) \geq \alpha$ and $\xi_{\tilde{A}}(x_2) \geq \alpha$. By the quasi-concavity, we have

$$\xi_{\tilde{A}}\left(\lambda x_1 + (1-\lambda)x_2\right) \geq \min\left\{\xi_{\tilde{A}}(x_1), \xi_{\tilde{A}}(x_2)\right\} \geq \alpha.$$

This shows that $\lambda x_1 + (1-\lambda)x_2 \in \tilde{A}_\alpha$, i.e. $\tilde{A}_\alpha$ is convex. This completes the proof. ■

For $\alpha \in (0,1]$ and a sequence $\{\alpha_n\}_{n=1}^\infty$ in $[0,1]$, recall that $\alpha_n \uparrow \alpha$ means that the sequence $\{\alpha_n\}_{n=1}^\infty$ is increasing and converges to α. For $\alpha \in [0,1)$, recall that $\alpha_n \downarrow \alpha$ means that the sequence $\{\alpha_n\}_{n=1}^\infty$ is decreasing and converges to α.

Proposition 2.2.8 *Let $\tilde{A}$ be a fuzzy set in $\mathbb{R}^m$.*

(i) *Suppose that $\alpha \in [0,1]$ and $\{\alpha_n\}_{n=1}^\infty$ is a sequence in $[0,1]$ satisfying $\alpha_n \uparrow \alpha$. Then*

$$\tilde{A}_\alpha = \bigcap_{n=1}^\infty \tilde{A}_{\alpha_n}.$$

(ii) *The following statements hold true.*
- *Suppose that $\alpha \in [0,1]$ and $\{\alpha_n\}_{n=1}^\infty$ is a sequence in $[0,1]$ satisfying $\alpha_n \downarrow \alpha$. Then*

$$\tilde{A}_{\alpha+} \subseteq \bigcup_{n=1}^\infty \tilde{A}_{\alpha_n} \subseteq \tilde{A}_\alpha.$$

- *Suppose that $\alpha \in [0,1]$ and $\{\alpha_n\}_{n=1}^{\infty}$ is a sequence in $[0,1]$ satisfying $\alpha_n \downarrow \alpha$ and $\alpha_n > \alpha$ for all n. Then*

$$\tilde{A}_{\alpha+} = \bigcup_{n=1}^{\infty} \tilde{A}_{\alpha_n}.$$

Proof. To prove part (i), since $\alpha_n \leq \alpha$ for all n, we have $\tilde{A}_\alpha \subseteq \tilde{A}_{\alpha_n}$ for all n, which implies $\tilde{A}_\alpha \subseteq \bigcap_{n=1}^{\infty} \tilde{A}_{\alpha_n}$. On the other hand, for $x \in \bigcap_{n=1}^{\infty} \tilde{A}_{\alpha_n}$, we have $\xi_{\tilde{A}}(x) \geq \alpha_n$ for all n, which implies

$$\xi_{\tilde{A}}(x) \geq \sup_n \alpha_n = \lim_{n\to\infty} \alpha_n = \alpha.$$

This shows that $x \in \tilde{A}_\alpha$, and we conclude that the equality holds true.

To prove part (ii), given $x \in \tilde{A}_{\alpha+}$, we have $\xi_{\tilde{A}}(x) > \alpha$. Since $\alpha_n \downarrow \alpha$, given any $0 < \epsilon$ satisfying $\epsilon \leq \xi_{\tilde{A}}(x) - \alpha$, there exists an integer N satisfying $0 < \alpha_N - \alpha < \epsilon$, which says that $\xi_{\tilde{A}}(x) \geq \alpha_N$, i.e. $x \in \bigcup_{n=1}^{\infty} \tilde{A}_{\alpha_n}$. Therefore, we obtain the inclusion $\tilde{A}_{\alpha+} \subseteq \bigcup_{n=1}^{\infty} \tilde{A}_{\alpha_n}$. On the other hand, since $\alpha_n \downarrow \alpha$, we consider the following cases.

- Suppose that $\alpha_n \geq \alpha$ for all n. Then $\tilde{A}_{\alpha_n} \subseteq \tilde{A}_\alpha$, which implies $\bigcup_{n=1}^{\infty} \tilde{A}_{\alpha_n} \subseteq \tilde{A}_\alpha$.
- Suppose that $\alpha_n > \alpha$ for all n. We see that $x \in \tilde{A}_{\alpha_n}$ implies $\xi_{\tilde{A}}(x) \geq \alpha_n > \alpha$, which says that $\tilde{A}_{\alpha_n} \subseteq \tilde{A}_{\alpha+}$ for all n. Therefore, we have $\bigcup_{n=1}^{\infty} \tilde{A}_{\alpha_n} \subseteq \tilde{A}_{\alpha+}$.

Therefore, we obtain the desired equalities and inclusions. This completes the proof. ∎

Proposition 2.2.9 *Let $\tilde{A}$ be a fuzzy set in $\mathbb{R}$. Then, for any fixed $x \in \mathbb{R}^m$, the function*

$$\zeta_x(\alpha) = \alpha \cdot \chi_{\tilde{A}_\alpha}(x)$$

is upper semi-continuous on $[0,1]$.

Proof. We need to show that the following set

$$F_r = \{\alpha \in [0,1] : \zeta_x(\alpha) \geq r\} = \left\{\alpha \in [0,1] : \alpha \cdot \chi_{\tilde{A}_\alpha}(x) \geq r\right\}$$

is a closed subset of $\mathbb{R}^m$ for each $r \in \mathbb{R}$. If $r \leq 0$ then $F_r = [0,1]$ is a closed subset of $\mathbb{R}^m$. If $r > 1$ then $F_r = \emptyset$ is also a closed subset of $\mathbb{R}^m$. Therefore, it remains to be shown that F_r is a closed subset of $\mathbb{R}^m$ for each $r \in (0,1]$. For each $\alpha \in \text{cl}(F_r)$, Remark 1.2.6 says that there exists a sequence $\{\alpha_n\}_{n=1}^{\infty}$ in F_r satisfying $\alpha_n \to \alpha$, i.e. $x \in \tilde{A}_{\alpha_n}$ and $\alpha_n \geq r$ for all n, which also implies

$$\alpha = \lim_{n\to\infty} \alpha_n \geq r > 0.$$

Therefore, there exists a subsequence $\{\alpha_{n_k}\}_{k=1}^{\infty}$ of $\{\alpha_n\}_{n=1}^{\infty}$ satisfying $\alpha_{n_k} \downarrow \alpha$ or $\alpha_{n_k} \uparrow \alpha$.

- Suppose that $\alpha_{n_k} \downarrow \alpha$, i.e. $\alpha \leq \alpha_{n_k}$ for all k. Since $\tilde{A}_{\alpha_{n_k}} \subseteq \tilde{A}_\alpha$, it follows that $x \in \tilde{A}_\alpha$.
- Suppose that $\alpha_{n_k} \uparrow \alpha$. Since $x \in \tilde{A}_{\alpha_{n_k}}$ for all k, part (i) of Proposition 2.2.8 says that

$$x \in \bigcap_{k=1}^{\infty} \tilde{A}_{\alpha_{n_k}} = \tilde{A}_\alpha.$$

The above two cases show that $x \in \tilde{A}_\alpha$. Since

$$\alpha = \lim_{k\to\infty} \alpha_{n_k} \geq r,$$

it follows that $\alpha \in F_r$. Therefore, we conclude that $\mathrm{cl}(F_r) \subseteq F_r$, i.e. $\mathrm{cl}(F_r) = F_r$ which says that F_r is a closed subset of U. This completes the proof. ∎

Let $\tilde{A}$ be a fuzzy set in $\mathbb{R}^m$. Then, for $\alpha \in [0,1)$, the **strong α-level set** of $\tilde{A}$ is denoted and defined by

$$\tilde{A}_{\alpha+} = \left\{x : \xi_{\tilde{A}}(x) > \alpha\right\}. \tag{2.7}$$

The family $\mathcal{A} = \{\tilde{A}_\alpha : \alpha \in [0,1]\}$ of α-level sets is nested in the sense of $\tilde{A}_\alpha \subseteq \tilde{A}_\beta$ for $\beta < \alpha$. The nestedness of α-level sets says that

$$\tilde{A}_\alpha = \bigcup_{\alpha \le \beta \le 1} \tilde{A}_\beta = \bigcup_{\{\beta \in \mathcal{R}(\xi_{\tilde{A}}) : \beta \ge \alpha\}} \tilde{A}_\beta. \tag{2.8}$$

Regarding $\tilde{A}_{\alpha+}$, we have the following interesting results.

Proposition 2.2.10 *Let $\tilde{A}$ be a fuzzy set in $\mathbb{R}^m$.*

(i) *Suppose that $\alpha, \beta \in [0,1]$ with $\alpha < \beta$. Then $\tilde{A}_\beta \subseteq \tilde{A}_{\alpha+} \subseteq \tilde{A}_\alpha$.*

(ii) *For $\alpha \in [0,1)$, we have*

$$\tilde{A}_{\alpha+} = \left\{x : \xi_{\tilde{A}}(x) > \alpha\right\} = \bigcup_{\alpha < \beta \le 1} \tilde{A}_\beta = \bigcup_{\{\beta \in \mathcal{R}(\xi_{\tilde{A}}) : \beta > \alpha\}} \tilde{A}_\beta. \tag{2.9}$$

(iii) *Suppose that $\mathcal{R}(\xi_{\tilde{A}}) = [0,1]$. For $\alpha \in (0,1]$, we have*

$$\tilde{A}_\alpha = \bigcap_{0 \le \beta < \alpha} \tilde{A}_\beta = \bigcap_{0 \le \beta \le \alpha} \tilde{A}_\beta.$$

Proof. Part (i) is obvious. To prove part (ii), if $\beta = \xi_{\tilde{A}}(x) > \alpha$, we have $x \in \tilde{A}_\beta$ with $\beta > \alpha$ and $\beta \in \mathcal{R}(\xi_{\tilde{A}})$, which shows the following inclusions

$$\tilde{A}_{\alpha+} \subseteq \bigcup_{\beta \in \mathcal{R}(\xi_{\tilde{A}}) \cap (\alpha,1]} \tilde{A}_\beta \subseteq \bigcup_{\beta \in [0,1] \cap (\alpha,1]} \tilde{A}_\beta = \bigcup_{\alpha < \beta \le 1} \tilde{A}_\beta.$$

On the other hand, for $x \in \bigcup_{\alpha < \beta \le 1} \tilde{A}_\beta$, there exists $1 \ge \beta > \alpha$ satisfying $x \in \tilde{A}_\beta \ne \emptyset$, i.e. $\xi_{\tilde{A}}(x) \ge \beta > \alpha$. Therefore, we obtain the following inclusion

$$\bigcup_{\alpha < \beta \le 1} \tilde{A}_\beta \subseteq \tilde{A}_{\alpha+},$$

which implies the desired equalities.

To prove part (iii), for $\alpha, \beta \in [0,1]$ with $\beta < \alpha$, we have $\tilde{A}_\alpha \subseteq \tilde{A}_\beta$. Therefore, we have the inclusion

$$\tilde{A}_\alpha \subseteq \bigcap_{0 \le \beta < \alpha} \tilde{A}_\beta.$$

On the other hand, for

$$x \in \bigcap_{0 \le \beta < \alpha} \tilde{A}_\beta,$$

given any $\epsilon > 0$, since $\alpha \in (0,1]$, we have $x \in \tilde{A}_{\alpha-\epsilon}$ and $\alpha - \epsilon \in [0,1] = \mathcal{R}(\xi_{\tilde{A}})$. This says that $\xi_{\tilde{A}}(x) \ge \alpha - \epsilon$, which also implies $\xi_{\tilde{A}}(x) \ge \alpha$, since ϵ is an arbitrary positive number (i.e. we can take $\epsilon \to 0$). Therefore, we conclude that $x \in \tilde{A}_\alpha$, which implies the following inclusion

$$\bigcap_{0 \le \beta < \alpha} \tilde{A}_\beta \subseteq \tilde{A}_\alpha.$$

This completes the proof. ∎

For $\alpha \in (0,1]$, we also define

$$\tilde{A}_{\alpha-} = \bigcap_{0 \le \beta < \alpha} \tilde{A}_\beta.$$

Let $\mathbb{Q}$ denote the set of all rational numbers. It is well known that $\mathbb{Q}$ is dense in $\mathbb{R}$. In other words, given any $r \in \mathbb{R}$, there exists $q \in \mathbb{Q}$ such that q can be arbitrarily close to r. More precisely, given any $\epsilon > 0$, there exists $q \in \mathbb{Q}$ that depends on ϵ satisfying $|q - r| < \epsilon$. Then, we have the following interesting results.

Proposition 2.2.11 *Let $\tilde{A}$ be a fuzzy set in $\mathbb{R}^m$.*

(i) *We have the following equalities*

$$\tilde{A}_{\alpha+} = \bigcup_{\alpha < \beta \le 1} \tilde{A}_\beta = \bigcup_{\{\beta \in \mathcal{R}(\xi_{\tilde{A}}) : \beta > \alpha\}} \tilde{A}_\beta \text{ for } \alpha \in [0,1) \tag{2.10}$$

and

$$\bigcap_{\{\beta \in \mathcal{R}(\xi_{\tilde{A}}) : \beta < \alpha\}} \tilde{A}_\beta \subseteq \bigcap_{0 \le \beta < \alpha} \tilde{A}_\beta = \tilde{A}_{\alpha-} \text{ for } \alpha \in (0,1]. \tag{2.11}$$

(ii) *Given any $\alpha \in (0,1]$, let $\mathcal{D}_\alpha = [0,\alpha) \cap \mathbb{Q}$. Then*

$$\tilde{A}_{\alpha-} = \bigcap_{0 \le \beta < \alpha} \tilde{A}_\beta = \bigcap_{\{\beta \in \mathcal{D}_\alpha : \beta < \alpha\}} \tilde{A}_\beta. \tag{2.12}$$

Proof. To prove part (i), the following inclusion is obvious

$$\bigcup_{\{\beta \in \mathcal{R}(\xi_{\tilde{A}}) : \beta > \alpha\}} \tilde{A}_\beta \subseteq \bigcup_{\alpha < \beta \le 1} \tilde{A}_\beta.$$

To prove the other direction of inclusion, for

$$x \in \bigcup_{\alpha < \beta \le 1} \tilde{A}_\beta,$$

there exists $1 \ge \beta > \alpha$ satisfying $x \in \tilde{A}_\beta$. Therefore, we have $\xi_{\tilde{A}}(x) \ge \beta > \alpha$. Let $\beta_0 = \xi_{\tilde{A}}(x)$. Then, we have $x \in \tilde{A}_{\beta_0}$ with $\beta_0 \in \mathcal{R}(\xi_{\tilde{A}})$ and $\beta_0 > \alpha$. This proves the desired equality (2.10). The inclusion in (2.11) is obvious.

To prove part (ii), we first note that $\mathcal{D}_\alpha$ is dense in $[0,\alpha)$. It suffices to show the following inclusion

$$\bigcap_{\{\beta \in \mathcal{D}_\alpha : \beta < \alpha\}} \tilde{A}_\beta \subseteq \bigcap_{0 \le \beta < \alpha} \tilde{A}_\beta.$$

Suppose that $x \in \tilde{A}_\beta$ for all $\beta \in \mathcal{D}_\alpha$ and $\beta < \alpha$. Given any $\gamma \in [0,\alpha)$, since $\mathcal{D}_\alpha$ is dense in $[0,\alpha)$, there exists $\beta \in \mathcal{D}_\alpha$ satisfying $\gamma < \beta < \alpha$. The nestedness $\tilde{A}_\beta \subseteq \tilde{A}_\gamma$ says that $x \in \tilde{A}_\gamma$, i.e. $x \in \bigcap_{\gamma \in [0,\alpha)} \tilde{A}_\gamma$. Then, we obtain the desired equality in (2.12), and the proof is complete. ■

Let $\tilde{A}$ be a fuzzy set in $\mathbb{R}^m$ with membership function $\xi_{\tilde{A}}$. The inverse function of $\xi_{\tilde{A}}$ does not necessarily exist. However, we can consider the inverse image $\xi_{\tilde{A}}^{-1}(S)$ of any subset S of $(0,1]$ defined by

$$\xi_{\tilde{A}}^{-1}(S) = \left\{ x \in \mathbb{R}^m : \xi_{\tilde{A}}(x) \in S \right\}.$$

Given any $\alpha \in \mathcal{R}(\xi_{\tilde{A}})$, we also write $\xi_{\tilde{A}}^{-1}(\alpha)$ to denote the inverse image of singleton $\{\alpha\}$. More precisely, we have

$$\xi_{\tilde{A}}^{-1}(\alpha) = \xi_{\tilde{A}}^{-1}(\{\alpha\}) = \left\{x \in \mathbb{R}^m : \xi_{\tilde{A}}(x) = \alpha\right\}.$$

The set difference $A \backslash B$ is defined as $A \cap B^c$ (i.e. $x \in A \backslash B$ implies $x \in A$ and $x \notin B$).

Proposition 2.2.12 *Let $\tilde{A}$ be a fuzzy set in $\mathbb{R}^m$, and let $D_\alpha \equiv \tilde{A}_\alpha \backslash \tilde{A}_{\alpha+}$ for $\alpha \in \mathcal{R}(\xi_{\tilde{A}})$ with $\alpha > 0$. Then, we have that*

$$D_\alpha = \xi_{\tilde{A}}^{-1}(\alpha) \neq \emptyset$$

for $\alpha \in \mathcal{R}(\xi_{\tilde{A}})$ with $\alpha > 0$, and the support $\tilde{A}_{0+}$ is the disjoint union of sets D_α given by

$$\tilde{A}_{0+} = \bigcup_{\{\alpha \in \mathcal{R}(\xi_{\tilde{A}}) : \alpha > 0\}} D_\alpha = \bigcup_{\{\alpha \in \mathcal{R}(\xi_{\tilde{A}}) : \alpha > 0\}} \left(\tilde{A}_\alpha \Bigg\backslash \bigcup_{\{\beta \in \mathcal{R}(\xi_{\tilde{A}}) : \beta > \alpha\}} \tilde{A}_\beta,\right)$$

where $D_\alpha \cap D_\beta = \emptyset$ for $\alpha \neq \beta$. Note that if $0 \notin \mathcal{R}(\xi_{\tilde{A}})$, then $\tilde{A}_{0+} = \mathbb{R}$.

Proof. Given any $\alpha \in \mathcal{R}(\xi_{\tilde{A}})$ with $\alpha > 0$, we first note that

$$\begin{aligned} \emptyset \neq \xi_{\tilde{A}}^{-1}(\alpha) &= \left\{x \in \mathbb{R}^m : \xi_{\tilde{A}}(x) = \alpha\right\} = \left\{x \in \mathbb{R}^m : \xi_{\tilde{A}}(x) \geq \alpha\right\} \backslash \left\{x \in \mathbb{R}^m : \xi_{\tilde{A}}(x) > \alpha\right\} \\ &= \tilde{A}_\alpha \backslash \tilde{A}_{\alpha+} = D_\alpha = \tilde{A}_\alpha \Bigg\backslash \bigcup_{\{\beta \in \mathcal{R}(\xi_{\tilde{A}}) : \beta > \alpha\}} \tilde{A}_\beta \text{ (using (2.10)).} \end{aligned} \tag{2.13}$$

From (2.13), we obtain

$$\begin{aligned} \tilde{A}_{0+} &= \left\{x \in \mathbb{R}^m : \xi_{\tilde{A}}(x) > 0\right\} = \bigcup_{\{\alpha \in \mathcal{R}(\xi_{\tilde{A}}) : \alpha > 0\}} \xi_{\tilde{A}}^{-1}(\alpha) = \bigcup_{\{\alpha \in \mathcal{R}(\xi_{\tilde{A}}) : \alpha > 0\}} D_\alpha \\ &= \bigcup_{\{\alpha \in \mathcal{R}(\xi_{\tilde{A}}) : \alpha > 0\}} \left(\tilde{A}_\alpha \Bigg\backslash \bigcup_{\{\beta \in \mathcal{R}(\xi_{\tilde{A}}) : \beta > \alpha\}} \tilde{A}_\beta.\right) \end{aligned}$$

For $\alpha \neq \beta$, it is obvious that

$$D_\alpha \cap D_\beta = \left\{x \in \mathbb{R}^m : \xi_{\tilde{A}}(x) = \alpha\right\} \cap \left\{x \in \mathbb{R}^m : \xi_{\tilde{A}}(x) = \beta\right\} = \emptyset.$$

This completes the proof. ■

Let $\tilde{A}$ be a fuzzy set in $\mathbb{R}^m$. The decomposition theorem says that the membership function $\xi_{\tilde{A}}$ can be expressed in terms of its α-level sets $\tilde{A}_\alpha$.

Theorem 2.2.13 **(Decomposition theorem)** *Let $\tilde{A}$ be a fuzzy set in $\mathbb{R}^m$. The membership function $\xi_{\tilde{A}}$ can be expressed as*

$$\begin{aligned} \xi_{\tilde{A}}(x) &= \sup_{\alpha \in \mathcal{R}(\xi_{\tilde{A}})} \alpha \cdot \chi_{\tilde{A}_\alpha}(x) = \max_{\alpha \in \mathcal{R}(\xi_{\tilde{A}})} \alpha \cdot \chi_{\tilde{A}_\alpha}(x) \\ &= \sup_{0 < \alpha \leq 1} \alpha \cdot \chi_{\tilde{A}_\alpha}(x) = \max_{0 < \alpha \leq 1} \alpha \cdot \chi_{\tilde{A}_\alpha}(x). \end{aligned} \tag{2.14}$$

Proof. Given any fixed $x \in \mathbb{R}^m$, let

$$\alpha_0 = \xi_{\tilde{A}}(x) \in \mathcal{R}(\xi_{\tilde{A}}).$$

Suppose that $\alpha_0 = 0$. If $x \in \tilde{A}_\alpha$ for some $\alpha \in (0,1]$, we have $\xi_{\tilde{A}}(x) \geq \alpha > 0$, which contradicts $\xi_{\tilde{A}}(x) = \alpha_0 = 0$. Therefore, we have $x \notin \tilde{A}_\alpha$ for $\alpha \in (0,1]$. Then $\alpha \cdot \chi_{\tilde{A}_\alpha}(x) = 0$ for all $\alpha \in (0,1]$. This says that the equalities in (2.14) are satisfied.

Now, we assume $\alpha_0 > 0$. Then $x \in \tilde{A}_{\alpha_0}$. For $\alpha > \alpha_0$, if $x \in \tilde{A}_\alpha$, then $\xi_{\tilde{A}}(x) \geq \alpha > \alpha_0$, which contradicts $\alpha_0 = \xi_{\tilde{A}}(x)$. Therefore, we have $x \notin \tilde{A}_\alpha$ for $\alpha > \alpha_0$. If $\alpha \leq \alpha_0$, then $x \in \tilde{A}_{\alpha_0} \subseteq \tilde{A}_\alpha$, which says that $x \in \tilde{A}_\alpha$ for $\alpha \leq \alpha_0$. Therefore, we obtain

$$\begin{aligned}\sup_{0<\alpha\leq 1} \alpha \cdot \chi_{\tilde{A}_\alpha}(x) &= \max\left\{ \sup_{0<\alpha\leq\alpha_0} \alpha \cdot \chi_{\tilde{A}_\alpha}(x), \sup_{\alpha_0<\alpha\leq 1} \alpha \cdot \chi_{\tilde{A}_\alpha}(x) \right\} \\ &= \max\left\{ \sup_{0<\alpha\leq\alpha_0} \alpha, 0 \right\} = \alpha_0 = \xi_{\tilde{A}}(x).\end{aligned}$$

The above supremum is also attained. It means that

$$\xi_{\tilde{A}}(x) = \max_{0<\alpha\leq 1} \alpha \cdot \chi_{\tilde{A}_\alpha}(x).$$

The above arguments are still valid when (0,1] is replaced by $\mathcal{R}(\xi_{\tilde{A}})$. Therefore, we obtain the desired equalities, and the proof is complete. ■

Theorem 2.2.14 **(Decomposition theorem in countable sense)** *Let $\tilde{A}$ be a fuzzy set in $\mathbb{R}^m$. Then, the membership function $\xi_{\tilde{A}}$ can be expressed as*

$$\xi_{\tilde{A}}(x) = \sup_{\alpha\in\mathbb{Q}\cap(0,1]} \alpha \cdot \chi_{\tilde{A}_\alpha}(x) = \max_{\alpha\in\mathbb{Q}\cap(0,1]} \alpha \cdot \chi_{\tilde{A}_\alpha}(x).$$

Proof. Using (2.14), we have

$$0 < \alpha_0 \equiv \xi_{\tilde{A}}(x) = \sup_{0<\alpha\leq 1} \alpha \cdot \chi_{\tilde{A}_\alpha}(x) \geq \sup_{\alpha\in\mathbb{Q}\cap(0,1]} \alpha \cdot \chi_{\tilde{A}_\alpha}(x). \tag{2.15}$$

The denseness of $\mathbb{Q}$ in $\mathbb{R}$ says that there exists an increasing sequence $\{\alpha_n\}_{n=1}^\infty$ in $\mathbb{Q} \cap (0,1]$ satisfying $\alpha_n \uparrow \alpha_0$. Since $\alpha_n \leq \alpha_0$, we have $\tilde{A}_{\alpha_0} \subseteq \tilde{A}_{\alpha_n}$ for all n. Since $x \in \tilde{A}_{\alpha_0}$, it follows that $x \in \tilde{A}_{\alpha_n}$ for all n. Let $\Gamma = \{\alpha_n\}_{n=1}^\infty \subset \mathbb{Q} \cap (0,1]$. Then, we have

$$\begin{aligned}\xi_{\tilde{A}}(x) &= \alpha_0 = \lim_n \alpha_n = \sup_n \alpha_n \text{ (since } \{\alpha_n\}_{n=1}^\infty \text{is increasing)} \\ &= \sup_n \alpha_n \cdot \chi_{\tilde{A}_{\alpha_n}}(x) = \sup_{\alpha\in\Gamma} \alpha \cdot \chi_{\tilde{A}_\alpha}(x) \leq \sup_{\alpha\in\mathbb{Q}\cap(0,1]} \alpha \cdot \chi_{\tilde{A}_\alpha}(x).\end{aligned} \tag{2.16}$$

Combining (2.15) and (2.16), we obtain the equality

$$\xi_{\tilde{A}}(x) = \sup_{\alpha\in\mathbb{Q}\cap(0,1]} \alpha \cdot \chi_{\tilde{A}_\alpha}(x).$$

We also see that the above supremum is attained. This completes the proof. ■

Next, we are going to see how we can define when two fuzzy sets are identical. The concept of identical fuzzy sets is an important issue in applications. Recall that the range $\mathcal{R}(\xi_{\tilde{A}})$ of the membership function $\xi_{\tilde{A}}$ of a fuzzy set $\tilde{A}$ is not necessarily equal to the whole unit interval [0,1]. In other words, the membership function $\xi_{\tilde{A}}$ is not always an onto function.

Proposition 2.2.15 *Let $\tilde{A}$ and $\tilde{B}$ be two fuzzy sets in $\mathbb{R}^m$ with membership functions $\xi_{\tilde{A}}$ and $\xi_{\tilde{B}}$, respectively.*

(i) *Suppose that $\xi_{\tilde{A}}(x) = \xi_{\tilde{B}}(x)$ for all $x \in \mathbb{R}^m$. Then $\tilde{A}_\alpha = \tilde{B}_\alpha$ for all $\alpha \in [0,1]$.*
(ii) *Suppose that $\mathcal{R}(\xi_{\tilde{A}}) = \mathcal{R}(\xi_{\tilde{B}}) \equiv \mathcal{R}$ and $\tilde{A}_\alpha = \tilde{B}_\alpha$ for all $\alpha \in \mathcal{R}$ with $\alpha > 0$. Then $\xi_{\tilde{A}} = \xi_{\tilde{B}}$, i.e. $\xi_{\tilde{A}}(x) = \xi_{\tilde{B}}(x)$ for all $x \in \mathbb{R}^m$.*

(iii) *Suppose that* $\mathcal{R}(\xi_{\tilde{A}}) = \mathcal{R}(\xi_{\tilde{B}})$ *and* $\tilde{A}_\alpha = \tilde{B}_\alpha$ *for all* $\alpha \in \mathbb{Q} \cap (0,1]$. *Then* $\xi_{\tilde{A}} = \xi_{\tilde{B}}$, *i.e.* $\xi_{\tilde{A}}(x) = \xi_{\tilde{B}}(x)$ *for all* $x \in \mathbb{R}^m$.

(iv) *Suppose that* $\tilde{A}_\alpha = \tilde{B}_\alpha$ *for all* $\alpha \in (0,1]$. *Then* $\xi_{\tilde{A}} = \xi_{\tilde{B}}$ *and* $\mathcal{R}(\xi_{\tilde{A}}) = \mathcal{R}(\xi_{\tilde{B}})$.

Proof. To prove part (i), it is obvious that $\mathcal{R}(\xi_{\tilde{A}}) = \mathcal{R}(\xi_{\tilde{B}})$. For $\alpha \in (0,1]$, we have

$$\tilde{A}_\alpha = \left\{x \in \mathbb{R}^m : \xi_{\tilde{A}}(x) \geq \alpha\right\} = \left\{x \in \mathbb{R}^m : \xi_{\tilde{B}}(x) \geq \alpha\right\} = \tilde{B}_\alpha.$$

Using Proposition 2.2.3, we have

$$\tilde{A}_0 = \mathrm{cl}\left(\tilde{A}_{0+}\right) = \mathrm{cl}\left(\bigcup_{0<\alpha\leq 1} \tilde{A}_\alpha\right) = \mathrm{cl}\left(\bigcup_{0<\alpha\leq 1} \tilde{B}_\alpha\right) = \mathrm{cl}\left(\tilde{B}_{0+}\right) = \tilde{B}_0,$$

which proves part (i).

To prove part (ii), we first claim

$$\xi_{\tilde{A}}(x) = 0 \text{ if and only if } \xi_{\tilde{B}}(x) = 0. \tag{2.17}$$

Suppose that $\xi_{\tilde{A}}(x) = 0$. Then $x \notin \tilde{A}_\alpha$ for all $\alpha \in \mathcal{R}$ with $\alpha > 0$, which also implies $x \notin \tilde{B}_\alpha$ for all $\alpha \in \mathcal{R}$ with $\alpha > 0$. From Theorem 2.2.13, we have

$$\xi_{\tilde{B}}(x) = \sup_{\alpha\in\mathcal{R}(\xi_{\tilde{B}})} \alpha \cdot \chi_{\tilde{B}_\alpha}(x) = \sup_{\alpha\in\mathcal{R}} \alpha \cdot \chi_{\tilde{B}_\alpha}(x) = 0.$$

Suppose that $\xi_{\tilde{B}}(x) = 0$. We can similarly obtain $\xi_{\tilde{A}}(x) = 0$. For $x \in \mathcal{R}$ with $0 < \alpha = \xi_{\tilde{A}}(x)$, let $\beta = \xi_{\tilde{B}}(x)$. From (2.17), we see that $\beta > 0$. We also have $x \in \tilde{A}_\alpha$ and $x \in \tilde{B}_\beta$. We consider the following two cases.

- For $x \in \tilde{A}_\alpha$, since $\alpha \in \mathcal{R}(\xi_{\tilde{A}}) = \mathcal{R}$, we have $x \in \tilde{B}_\alpha = \tilde{A}_\alpha$, which says that $\xi_{\tilde{B}}(x) \geq \alpha$, i.e. $\beta \geq \alpha$.
- For $x \in \tilde{B}_\beta$, we have $x \in \tilde{A}_\beta$, which says that $\xi_{\tilde{A}}(x) \geq \beta$, i.e. $\alpha \geq \beta$.

Therefore, we obtain $\alpha = \beta$, i.e. $\xi_{\tilde{A}}(x) = \xi_{\tilde{B}}(x)$.

To prove part (iii), using the argument in part (ii), we can similarly obtain

$$\xi_{\tilde{A}}(x) = 0 \text{ if and only if } \xi_{\tilde{B}}(x) = 0. \tag{2.18}$$

For $x \in \mathcal{R}(\xi_{\tilde{A}}) = \mathcal{R}(\xi_{\tilde{B}})$ with $0 < \alpha = \xi_{\tilde{A}}(x)$, let $\beta = \xi_{\tilde{B}}(x)$. From (2.18), we see that $\beta > 0$. We also have $x \in \tilde{A}_\alpha$ and $x \in \tilde{B}_\beta$. By the denseness, there exist two sequences $\{\alpha_n\}_{n=1}^\infty$ and $\{\beta_n\}_{n=1}^\infty$ in $\mathbb{Q} \cap (0,1]$ satisfying $\alpha_n \uparrow \alpha$ and $\beta_n \uparrow \beta$. In this case, using part (ii) of Proposition 2.2.8, we have

$$\tilde{A}_\alpha = \bigcap_{n=1}^\infty \tilde{A}_{\alpha_n} \text{ and } \tilde{B}_\alpha = \bigcap_{n=1}^\infty \tilde{B}_{\alpha_n} \tag{2.19}$$

and

$$\tilde{A}_\beta = \bigcap_{n=1}^\infty \tilde{A}_{\beta_n} \text{ and } \tilde{B}_\beta = \bigcap_{n=1}^\infty \tilde{B}_{\beta_n}. \tag{2.20}$$

Now, we consider the following two cases.

- For $x \in \tilde{A}_\alpha$, from (2.19), we have $x \in \tilde{A}_{\alpha_n} = \tilde{B}_{\alpha_n}$ for all n, since $\alpha_n \in \mathbb{Q} \cap (0,1]$ for all n, which also says that $x \in \tilde{B}_\alpha$ by (2.19) again. Therefore, we have $\xi_{\tilde{B}}(x) \geq \alpha$, i.e. $\beta \geq \alpha$.
- For $x \in \tilde{B}_\beta$, from (2.20), we have $x \in \tilde{B}_{\beta_n} = \tilde{A}_{\beta_n}$ for all n, since $\beta_n \in \mathbb{Q} \cap (0,1]$ for all n, which also says that $x \in \tilde{A}_\beta$ by (2.20) again. Therefore, we have $\xi_{\tilde{A}}(x) \geq \beta$, i.e. $\alpha \geq \beta$.

The above two cases show that $\alpha = \beta$, i.e. $\xi_{\tilde{A}}(x) = \xi_{\tilde{B}}(x)$.

To prove part (iv), if $\xi_{\tilde{A}}(x) = 0$, then $x \notin \tilde{A}_\alpha$ for all $\alpha \in (0,1]$, which implies $x \notin \tilde{B}_\alpha$ for all $\alpha \in (0,1]$, since $\tilde{A}_\alpha = \tilde{B}_\alpha$ for all $\alpha \in (0,1]$. From Theorem 2.2.13, we have

$$\xi_{\tilde{B}}(x) = \sup_{0<\alpha\leq 1} \alpha \cdot \chi_{\tilde{B}_\alpha}(x) = 0.$$

We can similarly prove that $\xi_{\tilde{B}}(x) = 0$ implies $\xi_{\tilde{A}}(x) = 0$, which says that $\xi_{\tilde{A}}(x) = 0$ if and only if $\xi_{\tilde{B}}(x) = 0$. For $0 < \alpha = \xi_{\tilde{A}}(x)$ and $0 < \beta = \xi_{\tilde{B}}(x)$ with $\alpha, \beta \in (0,1]$, i.e. $x \in \tilde{A}_\alpha$ and $x \in \tilde{B}_\beta$. We consider the following two cases.

- For $x \in \tilde{A}_\alpha$, since $\tilde{A}_\alpha = \tilde{B}_\alpha$, i.e. $x \in \tilde{B}_\alpha$, we have $\xi_{\tilde{B}}(x) \geq \alpha$, i.e. $\beta \geq \alpha$.
- For $x \in \tilde{B}_\beta$, since $\tilde{A}_\beta = \tilde{B}_\beta$, i.e. $x \in \tilde{A}_\beta$, we also have $\xi_{\tilde{A}}(x) \geq \beta$, i.e. $\alpha \geq \beta$.

Therefore, we obtain $\alpha = \beta$. This shows that $\xi_{\tilde{A}}(x) = \xi_{\tilde{B}}(x)$. Therefore, we conclude that $\xi_{\tilde{A}} = \xi_{\tilde{B}}$ with $\mathcal{R}(\xi_{\tilde{A}}) = \mathcal{R}(\xi_{\tilde{B}})$. This completes the proof. ■

2.3 Types of Fuzzy Sets

Now, we are going to consider some special structures of fuzzy sets by classifying the family of all fuzzy sets into many different sub-families in which the α-level sets own elegant structure is useful in applications.

Definition 2.3.1 We denote by $\mathfrak{F}(\mathbb{R}^m)$ the set of all fuzzy sets in $\mathbb{R}^m$ such that each $\tilde{A} \in \mathfrak{F}(\mathbb{R}^m)$ satisfies the following conditions.

(a) $\tilde{A}$ is convex according to Definition 2.2.5, i.e. its membership function $\xi_{\tilde{A}}$ is quasi-concave, according to Proposition 2.2.7.
(b) The membership function $\xi_{\tilde{A}}$ is upper semi-continuous.
(c) The 0-level set $\tilde{A}_0$ is a closed and bounded subset of $\mathbb{R}^m$.

Each element of $\mathfrak{F}(\mathbb{R}^m)$ is called a **fuzzy vector**, and each element of $\mathfrak{F}(\mathbb{R})$ is called a **fuzzy interval**. When the 1-level set of a fuzzy interval is a singleton such as $\tilde{A}_1 = \{p\}$, it is also called a **fuzzy number** with core value p.

The fuzzy vector $\tilde{A}$ with core value p can be regarded as a fuzzification of vector $p \in \mathbb{R}^m$. Recall that the membership function of $\tilde{1}_{\{p\}}$ is given by

$$\tilde{1}_{\{p\}}(r) = \begin{cases} 1 & \text{if } r = p \\ 0 & \text{otherwise.} \end{cases} \tag{2.21}$$

It is clear to see that $(\tilde{1}_{\{p\}})_\alpha = \{p\}$ for all $\alpha \in [0,1]$ and $\tilde{1}_{\{p\}} \in \mathfrak{F}(\mathbb{R}^m)$. We also say that $\tilde{1}_{\{p\}}$ is a **crisp vector** with value p.

Proposition 2.3.2 *Suppose that $\tilde{A} \in \mathfrak{F}(\mathbb{R}^m)$. Then, for each $\alpha \in [0,1]$, the α-level set $\tilde{A}_\alpha$ is a closed, bounded and convex set in $\mathbb{R}^m$. In particular, given a fuzzy interval $\tilde{A} \in \mathfrak{F}(\mathbb{R})$, the α-level set $\tilde{A}_\alpha$ of $\tilde{A}$ is a bounded closed interval and is denoted by*

$$\tilde{A}_\alpha = [\tilde{A}_\alpha^L, \tilde{A}_\alpha^U]$$

for $\alpha \in [0,1]$.

Proof. We first note that the 0-level set $\tilde{A}_0$ is a closed subset of $\mathbb{R}^m$ by definition. From Proposition 2.2.6, we also see that $\tilde{A}_0$ is a convex set in $\mathbb{R}^m$. The upper semi-continuity says that $\tilde{A}_\alpha$ is a closed subset of $\mathbb{R}^m$ for each $\alpha \in [0,1]$. Since $\tilde{A}_0$ is a bounded subset of $\mathbb{R}^m$ and each $\tilde{A}_\alpha$ is a subset of $\tilde{A}_0$ for $\alpha \in (0,1]$, it follows that each $\tilde{A}_\alpha$ is a bounded subset of $\mathbb{R}^m$ for $\alpha \in (0,1]$. This completes the proof. ■

For $\tilde{A} \in \mathfrak{F}(\mathbb{R})$, Proposition 2.3.2 says that each α-level set $\tilde{A}_\alpha$ is a bounded closed interval with degree α, which explains the name of fuzzy interval $\tilde{A}$ in $\mathfrak{F}(\mathbb{R})$. Next, we also propose two kinds of fuzzy intervals.

Definition 2.3.3 Let $\tilde{A} \in \mathfrak{F}(\mathbb{R})$ be a fuzzy interval.

- We say that $\tilde{A}$ is a **standard fuzzy interval** when the membership function $\xi_{\tilde{A}}$ is strictly increasing on the closed interval $[\tilde{A}_0^L, \tilde{A}_1^L]$ and strictly decreasing on the closed interval $[\tilde{A}_1^U, \tilde{A}_0^U]$.
- We say that $\tilde{A}$ is a **canonical fuzzy interval** when the endpoint functions $l(\alpha) \equiv \tilde{A}_\alpha^L$ and $u(\alpha) \equiv \tilde{A}_\alpha^U$ are continuous on $[0,1]$.

Proposition 2.3.4 *Let $\tilde{A} \in \mathfrak{F}(\mathbb{R})$ be a fuzzy interval. Then, the endpoint functions $\zeta^L(\alpha) = \tilde{A}_\alpha^L$ and $\zeta^U(\alpha) = \tilde{A}_\alpha^U$ satisfy the following conditions.*

(i) *ζ^L is increasing on $[0,1]$, and ζ^U is decreasing on $[0,1]$.*
(ii) *ζ^L and ζ^U are left-continuous on $(0,1]$.*
(iii) *ζ^L and ζ^U are right-continuous at 0.*
(iv) *ζ^L and ζ^U have the right-limit at any $\alpha_0 \in [0,1)$.*
(v) *ζ^L is lower semi-continuous on $[0,1]$ and ζ^U is upper semi-continuous on $[0,1]$.*
(vi) *Suppose that $cl(\tilde{A}_{\alpha+}) = \tilde{A}_\alpha$ for all $\alpha \in [0,1)$. Then ζ^L and ζ^U are continuous on $(0,1)$, left-continuous at 1, and right-continuous at 0. In other words, ζ^L and ζ^U are continuous on $[0,1]$, i.e. $\tilde{A}$ is a canonical fuzzy interval.*

Proof. Part (i) is obvious. To prove part (ii), given $\alpha \in (0,1]$, there exists an increasing sequence $\{\alpha_n\}_{n=1}^\infty$ in $[0,1]$ satisfying $\alpha_n \uparrow \alpha$ as $n \to \infty$. Part (ii) of Proposition 2.2.8 says that

$$\left[\zeta^L(\alpha), \zeta^U(\alpha)\right] = \left[\tilde{A}_\alpha^L, \tilde{A}_\alpha^U\right] = \tilde{A}_\alpha = \bigcap_{n=1}^\infty \tilde{A}_{\alpha_n} = \bigcap_{n=1}^\infty \left[\tilde{A}_{\alpha_n}^L, \tilde{A}_{\alpha_n}^U\right] = \bigcap_{n=1}^\infty \left[\zeta^L(\alpha_n), \zeta^U(\alpha_n)\right].$$

Let

$$A = \tilde{A}_\alpha = \left[\zeta^L(\alpha), \zeta^U(\alpha)\right] \text{ and } A_n = \tilde{A}_{\alpha_n} = \left[\zeta^L(\alpha_n), \zeta^U(\alpha_n)\right]$$

for all n. Then $\bigcap_{n=1}^\infty A_n = A$. Since $\alpha_{n+1} \leq \alpha_n$ for all n, we have $A_{n+1} \subseteq A_n$ for all n. We also see that

$$\zeta^U(\alpha_n) = \sup_{p \in \tilde{A}_{\alpha_n}} p = \sup_{p \in A_n} p \text{ and } \zeta^L(\alpha_n) = \inf_{p \in \tilde{A}_{\alpha_n}} p = \inf_{p \in A_n} p.$$

Using Proposition 1.2.3, we have

$$\lim_{n\to\infty} \zeta^U(\alpha_n) = \lim_{n\to\infty} \sup_{p \in A_n} p = \sup_{p \in A} p = \zeta^U(\alpha)$$

and

$$\lim_{n\to\infty}\zeta^L(\alpha_n) = \lim_{n\to\infty}\inf_{p\in A_n} p = \inf_{p\in A} p = \zeta^L(\alpha),$$

which show that ζ^L and ζ^U are left-continuous on (0,1].

To prove part (iv), for any $\alpha_0 \in [0,1)$, we consider the decreasing sequence $\{\alpha_n\}_{n=1}^{\infty}$ satisfying $\alpha_n \downarrow \alpha_0$. Part (i) says that $\zeta^L(\alpha_n) \geq \zeta^L(\alpha_0)$ and $\zeta^U(\alpha_n) \leq \zeta^U(\alpha_0)$ for all n. Therefore, we obtain

$$\inf_{n\geq 1} \zeta^L(\alpha_n) \geq \zeta^L(\alpha_0) \text{ and } \sup_{n\geq 1} \zeta^U(\alpha_n) \leq \zeta^U(\alpha_0).$$

Since $\{\alpha_n\}_{n=1}^{\infty}$ is a decreasing sequence and ζ^L is an increasing function, we have

$$\lim_{n\to\infty}\zeta^L(\alpha_n) = \inf_{n\geq 1} \zeta^L(\alpha_n) \geq \zeta^L(\alpha_0) > -\infty.$$

Since ζ^U is a decreasing function, we also have

$$\lim_{n\to\infty}\zeta^U(\alpha_n) = \sup_{n\geq 1} \zeta^U(\alpha_n) \leq \zeta^U(\alpha_0) < +\infty.$$

Therefore ζ^L and ζ^U have the right-limit at any $\alpha_0 \in [0,1)$.

To prove part (v), using Proposition 1.3.7, we see that ζ^L is lower semi-continuous on (0,1] and ζ^U is upper semi-continuous on (0,1] by parts (i) and (ii). For $\alpha_0 = 0$, since ζ^L is increasing, if $|\alpha_0 - \alpha| < \delta$, i.e. $0 \leq \alpha < \delta$, then

$$\zeta^L(\alpha_0) = \zeta^L(0) \leq \zeta^L(\alpha) < \zeta^L(\alpha) + \epsilon,$$

which says that ζ^L is lower semi-continuous at 0. Since ζ^U is decreasing, if $|\alpha_0 - \alpha| < \delta$, i.e. $0 \leq \alpha < \delta$, then

$$\zeta^U(\alpha_0) + \epsilon = \zeta^U(0) + \epsilon \geq \zeta^U(\alpha) + \epsilon > \zeta^U(\alpha),$$

which also says that ζ^U is upper semi-continuous at 0.

To prove part (vi), since the functions $\zeta^L(\alpha) = \tilde{A}_\alpha^L$ and $\zeta^U(\alpha) = \tilde{A}_\alpha^U$ are left-continuous on (0,1] by part (ii), it remains to show that they are right-continuous on [0,1). Given $\alpha \in [0,1)$. We consider the decreasing sequence $\{\alpha_n\}_{n=1}^{\infty}$ in [0,1] satisfying $\alpha_n \downarrow \alpha$ with $\alpha_n > \alpha$ for all n. Part (ii) of Proposition 2.2.8 says that $\tilde{A}_{\alpha+} = \bigcup_{n=1}^{\infty} \tilde{A}_{\alpha_n}$. Let $A_n = \tilde{A}_{\alpha_n}$, $A_+ = \tilde{A}_{\alpha+}$ and $A = \tilde{A}_\alpha$. Then $\bigcup_{n=1}^{\infty} A_n = A_+$. The assumption also says that $\text{cl}(A_+) = A$. Then, we have

$$\begin{aligned}
\lim_{n\to\infty}\zeta^U(\alpha_n) &= \lim_{n\to\infty}\sup_{p\in\tilde{A}_{\alpha_n}} p = \lim_{n\to\infty}\sup_{p\in A_n} p = \sup_{p\in A_+} p \text{ (by Proposition 1.2.4)}\\
&= \sup_{p\in \text{cl}(A_+)} p \text{ (by Proposition 1.2.7)}\\
&= \sup_{p\in A} p = \sup_{p\in\tilde{A}_\alpha} p = \zeta^U(\alpha)
\end{aligned}$$

and

$$\begin{aligned}
\lim_{n\to\infty}\zeta^L(\alpha_n) &= \lim_{n\to\infty}\inf_{p\in\tilde{A}_{\alpha_n}} p = \lim_{n\to\infty}\inf_{p\in A_n} p = \inf_{p\in A_+} p \text{ (by Proposition 1.2.4)}\\
&= \inf_{p\in \text{cl}(A_+)} p \text{ (by Proposition 1.2.7)}\\
&= \inf_{p\in A} p = \inf_{p\in\tilde{A}_\alpha} p = \zeta^L(\alpha),
\end{aligned}$$

which show that ζ^L and ζ^U are right-continuous on [0,1).

To prove part (iii), we assume that $\alpha_n \downarrow 0$ with $\alpha_n > 0$ for all n. Since $\text{cl}(\tilde{A}_{0+}) = \tilde{A}_0$ by the definition of 0-level set, part (vi) says that ζ^L and ζ^U are right-continuous at 0. This completes the proof. ■

Proposition 2.3.5 *Let $\tilde{A}$ be a fuzzy interval such that the membership function $\xi_{\tilde{A}}$ is strictly increasing on $[\tilde{A}_0^L, \tilde{A}_1^L]$ and strictly decreasing on $[\tilde{A}_1^U, \tilde{A}_0^U]$, i.e. $\tilde{A}$ is a standard fuzzy interval. Then, the functions $\zeta^L(\alpha) = \tilde{A}_\alpha^L$ and $\zeta^U(\alpha) = \tilde{A}_\alpha^U$ are continuous on (0,1), left-continuous at 1, and right-continuous at 0. In other words, ζ^L and ζ^U are continuous on [0,1], i.e. $\tilde{A}$ is a canonical fuzzy interval.*

Proof. The strict monotonicity of a membership function says that, given any x in the 0-level set $\tilde{A}_0$, we have

$$x = \begin{cases} \tilde{A}_\beta^L \text{ for some } \beta \in (0,1] & \text{if } \tilde{A}_0^L < x \leq \tilde{A}_1^L \\ \tilde{A}_\beta^U \text{ for some } \beta \in (0,1] & \text{if } \tilde{A}_1^U \leq x < \tilde{A}_0^U, \end{cases} \tag{2.22}$$

where $\beta = \xi_{\tilde{A}}(x)$. From part (ii) of Proposition 2.3.4, we just need to prove the right continuity on [0,1). Therefore, for $\alpha_0 \in [0,1)$, we consider $0 < \alpha - \alpha_0 < \delta$ to prove the right continuity at α_0. Suppose that $\tilde{A}_{\alpha_0}^L = \tilde{A}_1^L$. Then $\tilde{A}_\alpha^L = \tilde{A}_1^L$ for all $\alpha \in [\alpha_0, 1]$, since $\tilde{A}_\alpha^L$ is increasing with respect to α. This says that $|\tilde{A}_\alpha^L - \tilde{A}_{\alpha_0}^L| = 0$ for $0 < \alpha - \alpha_0 < \delta$. Now, we consider the case of $\tilde{A}_{\alpha_0}^L < \tilde{A}_1^L$. Given any $\epsilon > 0$, we also consider the following cases.

- Suppose that $\tilde{A}_{\alpha_0}^L + \epsilon \leq \tilde{A}_1^L$. Since $\tilde{A}_{\alpha_0}^L < \tilde{A}_{\alpha_0}^L + \epsilon$, the denseness says that there exists $x \in \mathbb{R}$ satisfying $\tilde{A}_{\alpha_0}^L < x < \tilde{A}_{\alpha_0}^L + \epsilon \leq \tilde{A}_1^L$.
- Suppose that $\tilde{A}_{\alpha_0}^L + \epsilon > \tilde{A}_1^L$. Since $\tilde{A}_{\alpha_0}^L < \tilde{A}_1^L$, the denseness says that there exists $x \in \mathbb{R}$ satisfying $\tilde{A}_{\alpha_0}^L < x < \tilde{A}_1^L < \tilde{A}_{\alpha_0}^L + \epsilon$.

From (2.22), there exists $\beta \in (0,1]$ satisfying $x = \tilde{A}_\beta^L$. Therefore, the above two cases say that $\tilde{A}_{\alpha_0}^L < \tilde{A}_\beta^L < \tilde{A}_{\alpha_0}^L + \epsilon$. Since $\tilde{A}_\alpha^L$ is increasing with respect to α, we obtain $\alpha_0 < \beta$. Now, we take $\delta = \beta - \alpha_0$. Then, for $0 < \alpha - \alpha_0 < \delta$, we have $\alpha < \beta$, which says that $\tilde{A}_\alpha^L \leq \tilde{A}_\beta^L < \tilde{A}_{\alpha_0}^L + \epsilon$. Therefore, we obtain $|\tilde{A}_\alpha^L - \tilde{A}_{\alpha_0}^L| < \epsilon$ for $0 < \alpha - \alpha_0 < \delta$. This shows that ζ^L is right-continuous on [0,1). We can similarly show that ζ^U is right-continuous on [0,1), and the proof is complete. ■

Proposition 2.3.5 says that if $\tilde{A}$ is a standard fuzzy interval, then it is also a canonical fuzzy interval.

Definition 2.3.6 Let $\tilde{A}$ be a fuzzy set in $\mathbb{R}$ with membership function $\xi_{\tilde{A}}$.

- We say that $\tilde{A}$ is **nonnegative** when $\xi_{\tilde{A}}(r) = 0$ for all $r < 0$.
- We say that $\tilde{A}$ is **nonpositive** when $\xi_{\tilde{A}}(r) = 0$ for all $r > 0$.
- We say that $\tilde{A}$ is **positive** when $\xi_{\tilde{A}}(r) = 0$ for all $r \leq 0$.
- We say that $\tilde{A}$ is **negative** when $\xi_{\tilde{A}}(r) = 0$ for all $r \geq 0$.

Remark 2.3.7 Let $\tilde{A} \in \mathfrak{F}(\mathbb{R})$ be a fuzzy interval. Then, we have the following observations.

- Suppose that $\tilde{A}$ is nonnegative. Then $\tilde{A}_\alpha^L \geq 0$ for all $\alpha \in [0,1]$, which also says that $\tilde{A}_\alpha^U \geq 0$ for all $\alpha \in [0,1]$.

- Suppose that $\tilde{A}$ is nonpositive. Then $\tilde{A}_\alpha^U \leq 0$ for all $\alpha \in [0,1]$, which also says that $\tilde{A}_\alpha^L \leq 0$ for all $\alpha \in [0,1]$.
- Suppose that $\tilde{A}$ is positive. Then $\tilde{A}_\alpha^L > 0$ for all $\alpha \in [0,1]$, which also says that $\tilde{A}_\alpha^U > 0$ for all $\alpha \in [0,1]$.
- Suppose that $\tilde{A}$ is a negative. Then $\tilde{A}_\alpha^U < 0$ for all $\alpha \in [0,1]$, which also says that $\tilde{A}_\alpha^L < 0$ for all $\alpha \in [0,1]$.

Definition 2.3.8 We say that $\tilde{A}$ is an **LR-fuzzy interval** when its membership function has the following form

$$\xi_{\tilde{A}}(x) = \begin{cases} l_{\tilde{A}}(x), & \text{if } a_1 \leq x < a_2 \\ 1, & \text{if } a_2 \leq x \leq a_3 \\ r_{\tilde{A}}(x), & \text{if } a_3 < x \leq a_4 \\ 0, & \text{otherwise} \end{cases}$$

and satisfies the following conditions:

- $l_{\tilde{A}} : [a_1, a_2) \to [0,1)$ is a right-continuous and increasing function on $[a_1, a_2)$;
- $r_{\tilde{A}} : (a_3, a_4] \to [0,1)$ is a left-continuous and decreasing function on $(a_3, a_4]$.

In this case, we write it as $\tilde{A} = (a_1, a_2, a_3, a_4)_{LR}$. We denote by $\mathfrak{F}_{LR}(\mathbb{R})$ the set of all LR-fuzzy intervals.

In what follows, we are going to claim $\mathfrak{F}(\mathbb{R}) = \mathfrak{F}_{LR}(\mathbb{R})$

Proposition 2.3.9 *Let $\tilde{A} = (a_1, a_2, a_3, a_4)_{LR}$ be an LR-fuzzy interval. Then, we have the following properties.*

(i) *For any $\alpha \in (0,1)$, let*

$$x_\alpha = \inf \left\{x : l_{\tilde{A}}(x) \geq \alpha \text{ for } x \in [a_1, a_2)\right\}$$

and

$$y_\alpha = \sup \left\{x : r_{\tilde{A}}(x) \geq \alpha \text{ for } x \in (a_3, a_4]\right\}$$

Then, the α-level set of $\tilde{A}$ is a closed interval given by

$$\tilde{A}_\alpha = \begin{cases} [a_2, a_3], & \text{if } \alpha = 1 \\ \left[x_\alpha, y_\alpha\right], & \text{if } \alpha \in (0,1) \\ [a_1, a_4], & \text{if } \alpha = 0. \end{cases}$$

(ii) *Suppose that $\xi_{\tilde{A}}$ is continuous on $[a_1, a_4]$, that $l_{\tilde{A}}$ is strictly increasing on $[a_1, a_2)$, and that $r_{\tilde{A}}$ is strictly decreasing function on $(a_3, a_4]$. Then, the α-level set of $\tilde{A}$ is a closed interval given by*

$$\tilde{A}_\alpha = \begin{cases} [a_2, a_3], & \text{if } \alpha = 1 \\ \left[l_{\tilde{A}}^{-1}(\alpha), r_{\tilde{A}}^{-1}(\alpha)\right], & \text{if } \alpha \in (0,1) \\ [a_1, a_4], & \text{if } \alpha = 0. \end{cases}$$

Moreover, we have $\mathfrak{F}_{LR}(\mathbb{R}) \subseteq \mathfrak{F}(\mathbb{R})$.

Proof. To prove part (i), it is obvious for the cases of $\alpha = 0$ and $\alpha = 1$. Now, we consider $\alpha \in (0,1)$. Given any $x_0 \in \tilde{A}_\alpha$, we have $\xi_{\tilde{A}}(x_0) \geq \alpha$. If $x_0 < a_2$, we have $l_{\tilde{A}}(x_0) = \xi_{\tilde{A}}(x_0) \geq \alpha$, which implies $x_0 \geq x_\alpha$. If $x_0 > a_3$, we have $r_{\tilde{A}}(x_0) = \xi_{\tilde{A}}(x_0) \geq \alpha$, which also implies $x_0 \leq y_\alpha$. We conclude $x_0 \in [x_\alpha, y_\alpha]$. This shows the inclusion $\tilde{A}_\alpha \subseteq [x_\alpha, y_\alpha]$.

Next, we want to show $x_\alpha, y_\alpha \in \tilde{A}_\alpha$. By the concept of infimum regarding x_α, there exists a decreasing sequence $\{x_n\}_{n=1}^{\infty}$ in the set $\{x : l_{\tilde{A}}(x) \geq \alpha \text{ for } x \in [a_1, a_2)\}$ satisfying $x_n \downarrow x_\alpha$. Since $l_{\tilde{A}}(x_n) \geq \alpha$ and $l_{\tilde{A}}$ is right-continuous, we have

$$\xi_{\tilde{A}}(x_\alpha) = l_{\tilde{A}}(x_\alpha) = \lim_{n\to\infty} l_{\tilde{A}}(x_n) \geq \alpha,$$

which says that $x_\alpha \in \tilde{A}_\alpha$. Similarly, by the concept of supremum regarding y_α, there exists an increasing sequence $\{y_n\}_{n=1}^{\infty}$ in the set $\{x : r_{\tilde{A}}(x) \geq \alpha \text{ for } x \in (a_3, a_4]\}$ satisfying $y_n \uparrow y_\alpha$. Since $r_{\tilde{A}}(y_n) \geq \alpha$ and $r_{\tilde{A}}$ is left-continuous, we have

$$\xi_{\tilde{A}}(y_\alpha) = r_{\tilde{A}}(y_\alpha) = \lim_{n\to\infty} r_{\tilde{A}}(y_n) \geq \alpha,$$

which says that $y_\alpha \in \tilde{A}_\alpha$. Therefore, we obtain $\tilde{A}_\alpha = [x_\alpha, y_\alpha]$.

Since the α-level sets $\tilde{A}_\alpha$ are closed and convex sets in $\mathbb{R}$, it follows that the membership function $\xi_{\tilde{A}}$ is quasi-concave and upper semi-continuous, i.e. $\tilde{A} \in \mathfrak{F}(\mathbb{R})$. Therefore, we obtain the inclusion $\mathfrak{F}_{LR}(\mathbb{R}) \subseteq \mathfrak{F}(\mathbb{R})$.

To prove part (ii), the strict monotonicity says that the inverse functions $l_{\tilde{A}}^{-1}$ and $r_{\tilde{A}}^{-1}$ exist. The desired results can be easily realized from part (i). This completes the proof. ■

We notice that if $l_{\tilde{A}}$ is continuous on $[a_1, a_2)$ and $r_{\tilde{A}}$ is continuous on $(a_3, a_4]$, then it does not necessarily imply that $\xi_{\tilde{A}}$ is continuous on $[a_1, a_4]$, since $\xi_{\tilde{A}}$ may have jumps at a_2 and a_3. In what follows, we are going to present the converse of Proposition 2.3.9.

Theorem 2.3.10 *Let $\tilde{A}$ be a fuzzy interval. Then, there exists $a_1, a_2, a_3, a_4 \in \mathbb{R}$ and the functions $l_{\tilde{A}}$ and $r_{\tilde{A}}$ satisfying the following conditions:*

- $l_{\tilde{A}} : [a_1, a_2) \to [0,1)$ *is a right-continuous and increasing function on* $[a_1, a_2)$*;*
- $r_{\tilde{A}} : (a_3, a_4] \to [0,1)$ *is a left-continuous and decreasing function on* $(a_3, a_4]$*,*

such that its membership function can be described by

$$\xi_{\tilde{A}}(x) = \begin{cases} l_{\tilde{A}}(x) & \text{if } a_1 \leq x < a_2 \\ 1 & \text{if } a_2 \leq x \leq a_3 \\ r_{\tilde{A}}(x) & \text{if } a_3 < x \leq a_4 \\ 0 & \text{otherwise.} \end{cases}$$

In other words, $\tilde{A}$ is an LR-fuzzy interval. Moreover, we have $\mathfrak{F}(\mathbb{R}) = \mathfrak{F}_{LR}(\mathbb{R})$. Suppose that $\tilde{A}$ is a standard fuzzy interval. Then $l_{\tilde{A}}$ is strictly increasing function on $[a_1, a_2)$ and $r_{\tilde{A}}$ is strictly decreasing function on $(a_3, a_4]$.

Proof. Since the 1-level set $\tilde{A}_1 = [\tilde{A}_1^L, \tilde{A}_1^U]$ and the 0-level set $\tilde{A}_0 = [\tilde{A}_0^L, \tilde{A}_0^U]$ are nonempty, we take

$$a_1 = \tilde{A}_0^L, \quad a_2 = \tilde{A}_1^L, \quad a_3 = \tilde{A}_1^U, \text{ and } a_4 = \tilde{A}_0^U.$$

Then, we define

$$l_{\tilde{A}}(x) = \xi_{\tilde{A}}(x) \text{ for } x \in [a_1, a_2)$$

and

$$r_{\tilde{A}}(x) = \xi_{\tilde{A}}(x) \text{ for } x \in (a_3, a_4].$$

For $a_1 \leq x \leq y < a_2$, there exists $\lambda \in (0,1)$ satisfying $y = \lambda x + (1 - \lambda)a_2$. Since the membership function $\xi_{\tilde{A}}$ is quasi-concave, we have

$$\begin{aligned} l_{\tilde{A}}(y) &= \xi_{\tilde{A}}(y) = \xi_{\tilde{A}}(\lambda x + (1 - \lambda)a_2) \\ &\geq \min\left\{\xi_{\tilde{A}}(x), \xi_{\tilde{A}}(a_2)\right\} = \min\left\{\xi_{\tilde{A}}(x), 1\right\} = \xi_{\tilde{A}}(x) = l_{\tilde{A}}(x), \end{aligned}$$

which shows that $l_{\tilde{A}}$ is increasing on $[a_1, a_2)$. Similarly, For $a_3 < y \leq x \leq a_4$, there exists $\lambda \in (0,1)$ satisfying $y = \lambda x + (1 - \lambda)a_3$. We also have

$$\begin{aligned} r_{\tilde{A}}(y) &= \xi_{\tilde{A}}(y) = \xi_{\tilde{A}}(\lambda x + (1 - \lambda)a_3) \\ &\geq \min\left\{\xi_{\tilde{A}}(x), \xi_{\tilde{A}}(a_3)\right\} = \min\left\{\xi_{\tilde{A}}(x), 1\right\} = \xi_{\tilde{A}}(x) = r_{\tilde{A}}(x), \end{aligned}$$

which shows that $r_{\tilde{A}}$ is decreasing on $(a_3, a_4]$.

For $x_0 \in [a_1, a_2)$, the denseness says that there exists a sequence $\{x_n\}_{n=1}^{\infty}$ in $[a_1, a_2)$ satisfying $x_n \downarrow x_0$. Since $l_{\tilde{A}}$ is increasing, we have $l_{\tilde{A}}(x_n) \geq l_{\tilde{A}}(x_0)$ for all n and

$$\inf_n l_{\tilde{A}}(x_n) = \lim_{n\to\infty} l_{\tilde{A}}(x_n) \geq l_{\tilde{A}}(x_0),$$

which says that the limit exists. Now, we write

$$\alpha = \lim_{n\to\infty} l_{\tilde{A}}(x_n) = \inf_n l_{\tilde{A}}(x_n) \text{ and } \alpha_0 = l_{\tilde{A}}(x_0).$$

Then, we see that

$$\xi_{\tilde{A}}(x_n) = l_{\tilde{A}}(x_n) \geq \alpha \geq \alpha_0 \text{ for all } n,$$

which implies $x_n \in \tilde{A}_\alpha$ for all n. Since $\tilde{A}_\alpha$ is a closed set and $x_n \downarrow x_0$, it follows that

$$x_0 \in \text{cl}(\tilde{A}_\alpha) = \tilde{A}_\alpha.$$

Therefore, we obtain

$$\alpha_0 = l_{\tilde{A}}(x_0) = \xi_{\tilde{A}}(x_0) \geq \alpha,$$

which implies $\alpha = \alpha_0$, i.e.

$$\lim_{n\to\infty} l_{\tilde{A}}(x_n) = \alpha = \alpha_0 = l_{\tilde{A}}(x_0).$$

This shows that $l_{\tilde{A}}$ is right-continuous at x_0.

For $y_0 \in (a_3, a_4]$, there exists a sequence $\{y_n\}_{n=1}^{\infty}$ in $(a_3, a_4]$ satisfying $y_n \uparrow y_0$. Since $r_{\tilde{A}}$ is decreasing, we have $r_{\tilde{A}}(y_n) \geq r_{\tilde{A}}(y_0)$. Now, we write

$$\beta = \lim_{n\to\infty} r_{\tilde{A}}(y_n) = \inf_n r_{\tilde{A}}(y_n) \text{ and } \beta_0 = r_{\tilde{A}}(y_0).$$

Then, we see that

$$\xi_{\tilde{A}}(y_n) = r_{\tilde{A}}(y_n) \geq \beta \geq \beta_0,$$

which implies $y_n \in \tilde{A}_\beta$. Since $\tilde{A}_\beta$ is a closed set and $y_n \uparrow y_0$, it follows that $y_0 \in \tilde{A}_\beta$. Therefore, we obtain

$$\beta_0 = r_{\tilde{A}}(y_0) = \xi_{\tilde{A}}(y_0) \geq \beta,$$

which implies $\beta = \beta_0$, i.e.

$$\lim_{n\to\infty} r_{\tilde{A}}(y_n) = \beta = \beta_0 = r_{\tilde{A}}(y_0).$$

This shows that $r_{\tilde{A}}$ is left-continuous at y_0. Therefore, we conclude that $\tilde{A}$ is an LR-fuzzy interval, which shows the inclusion $\mathfrak{F}(\mathbb{R}) \subseteq \mathfrak{F}_{LR}(\mathbb{R})$. From Proposition 2.3.9, we obtain the equality

$$\mathfrak{F}(\mathbb{R}) = \mathfrak{F}_{LR}(\mathbb{R}).$$

Suppose that $\tilde{A}$ is a standard fuzzy interval. Then, it is clear to see that $l_{\tilde{A}}$ is strictly increasing function on $[a_1, a_2)$ and $r_{\tilde{A}}$ is strictly decreasing function on $(a_3, a_4]$. This completes the proof. ∎

3

Set Operations of Fuzzy Sets

We shall study the intersection and union of fuzzy sets. Let $\tilde{A}$ and $\tilde{B}$ be two fuzzy sets in $\mathbb{R}^m$ with membership functions $\xi_{\tilde{A}}$ and $\xi_{\tilde{B}}$, respectively. The usual intersection and union of $\tilde{A}$ and $\tilde{B}$ by referring to Zadeh [162] are defined using the min and max aggregation functions as follows

$$\xi_{\tilde{A}\cap\tilde{B}}(x) = \min\left\{\xi_{\tilde{A}}(x), \xi_{\tilde{B}}(x)\right\} \text{ and } \xi_{\tilde{A}\cup\tilde{B}}(x) = \max\left\{\xi_{\tilde{A}}(x), \xi_{\tilde{B}}(x)\right\}. \tag{3.1}$$

For more detailed properties, refer to Mizumoto and Tanaka [81], Dubois and Prade [28], and Klir and Yuan [60].

We can extend the min and max aggregation functions to the the t-norm t and s-norm s (t-conorm), respectively. Weber [116] and Yager [158] used the t-norm t and s-norm s to propose the intersection and union as follows

$$\xi_{\tilde{A}\cap\tilde{B}}(x) = t\left(\xi_{\tilde{A}}(x), \xi_{\tilde{B}}(x)\right) \text{ and } \xi_{\tilde{A}\cup\tilde{B}}(x) = s\left(\xi_{\tilde{A}}(x), \xi_{\tilde{B}}(x)\right).$$

The t-norm and s-norm satisfy some suitable conditions in which the boundary condition regarding 0 and 1 in the unit interval [0,1] are taken into account. In this chapter, instead of using the t-norm and s-norm, we shall consider the general aggregation functions to define the intersection and union of fuzzy sets.

3.1 Complement of Fuzzy Sets

Let A be a subset of $\mathbb{R}^m$. The complement set of A is denoted by A^c and given by $A^c = \mathbb{R}^m \setminus A$. It is easy to see that the characteristic function of A^c satisfies the following equality

$$\chi_{A^c}(x) = 1 - \chi_A(x) \text{ for all } x \in \mathbb{R}^m. \tag{3.2}$$

Let $\tilde{A}$ be a fuzzy set in $\mathbb{R}^m$. The complement of $\tilde{A}$ is denoted by $\tilde{A}^c$ and, inspired by (3.2), the membership function of $\tilde{A}^c$ is defined by

$$\xi_{\tilde{A}^c}(x) = 1 - \xi_{\tilde{A}}(x) \text{ for all } x \in \mathbb{R}^m. \tag{3.3}$$

We want to extend the concept of complement of fuzzy set as follows.

Let $c : [0,1] \to [0,1]$ be a mapping that transforms the membership degree of fuzzy set $\tilde{A}$ into the membership degree of its complement $\tilde{A}^c$, i.e.

$$c(\xi_{\tilde{A}}(x)) = \xi_{\tilde{A}^c}(x) \text{ for all } x \in \mathbb{R}^m.$$

Mathematical Foundations of Fuzzy Sets, First Edition. Hsien-Chung Wu.

Then, the extension of fuzzy complement is defined below.

Definition 3.1.1 The function $c : [0,1] \to [0,1]$ is called a **fuzzy complement** when the following conditions are satisfied:

- (boundary conditions) $c(0) = 1$ and $c(1) = 0$;
- (decreasing condition) for any $a, b \in [0,1]$, $a < b$ implies $c(a) \geq c(b)$.

We see that equation (3.3) is obtained by taking $c(x) = 1 - x$, and it is easy to show that this function c satisfies the above conditions. We also have the following fuzzy complements.

(i) **Sugeno class**: For $\lambda \in (-1, \infty)$, we take

$$c_\lambda(a) = \frac{1-a}{1+\lambda a}.$$

(ii) **Yager class**: For $w \in (0, \infty)$, we take

$$c_w(a) = (1 - a^w)^{1/w}. \tag{3.4}$$

We can also consider another definition of fuzzy complement as follows.

Definition 3.1.2 The function $c : [0,1] \to [0,1]$ is called a **fuzzy complement** when the following conditions are satisfied:

- $c(0) = 1$;
- $a < b$ implies $c(a) > c(b)$;
- $c(c(a)) = a$.

It is clear to see $c(1) = c(c(0)) = 0$. The fuzzy complement is seldom used in applications.

3.2 Intersection of Fuzzy Sets

Let A and B be two subsets of $\mathbb{R}^m$. Then, we have two corresponding characteristic functions χ_A and χ_B. The intersection $A \cap B$ also has the corresponding characteristic function $\chi_{A\cap B}$. It is obvious that

$$\chi_{A\cap B}(x) = \min\left\{\chi_A(x), \chi_B(x)\right\}. \tag{3.5}$$

Inspired by (3.5), the intersection of fuzzy sets is defined below by replacing the characteristic functions with the membership functions.

Definition 3.2.1 Let $\tilde{A}$ and $\tilde{B}$ be two fuzzy sets in $\mathbb{R}^m$ with membership functions $\xi_{\tilde{A}}$ and $\xi_{\tilde{B}}$, respectively. The membership function of the intersection $\tilde{A} \wedge \tilde{B}$ of $\tilde{A}$ and $\tilde{B}$ is defined by

$$\xi_{\tilde{A}\wedge\tilde{B}}(x) = \min\{\xi_{\tilde{A}}(x), \xi_{\tilde{B}}(x)\} \tag{3.6}$$

for all $x \in \mathbb{R}^m$.

For more than two fuzzy sets, their intersection is defined inductively. Given any three fuzzy sets $\tilde{A}^{(1)}, \tilde{A}^{(2)}$, and $\tilde{A}^{(3)}$ in $\mathbb{R}^m$, we first consider the intersection $\tilde{A} = \tilde{A}^{(1)} \wedge \tilde{A}^{(2)}$ whose membership function is given in (3.6). Now, the intersection of $\tilde{A}^{(1)}, \tilde{A}^{(2)}$, and $\tilde{A}^{(3)}$ can be defined as

$$\tilde{A}^{\circ} \equiv \left(\tilde{A}^{(1)} \wedge \tilde{A}^{(2)}\right) \wedge \tilde{A}^{(3)} = \tilde{A} \wedge \tilde{A}^{(3)}$$

whose membership function is given by

$$\begin{aligned}\xi_{\tilde{A}^{\circ}}(x) &= \min\left\{\xi_{\tilde{A}}(x), \xi_{\tilde{A}^{(3)}}(x)\right\} \\ &= \min\left\{\min\left\{\xi_{\tilde{A}^{(1)}}(x), \xi_{\tilde{A}^{(2)}}(x)\right\}, \xi_{\tilde{A}^{(3)}}(x)\right\} \\ &= \min\left\{\xi_{\tilde{A}^{(1)}}(x), \xi_{\tilde{A}^{(2)}}(x), \xi_{\tilde{A}^{(3)}}(x)\right\}. \end{aligned} \tag{3.7}$$

Suppose that the intersection is defined as

$$\tilde{A}^{(1)} \wedge \left(\tilde{A}^{(2)} \wedge \tilde{A}^{(3)}\right) \text{ or } \tilde{A}^{(2)} \wedge \left(\tilde{A}^{(3)} \wedge \tilde{A}^{(1)}\right) \text{ or } \left(\tilde{A}^{(3)} \wedge \tilde{A}^{(2)}\right) \wedge \tilde{A}^{(1)}$$

or any other permutation. Then, we can similarly show that their membership functions are identical to (3.7). In this case, we can simply write

$$\tilde{A}^{\circ} \equiv \tilde{A}^{(1)} \wedge \tilde{A}^{(2)} \wedge \tilde{A}^{(3)}.$$

Inductively, the membership function of the intersection

$$\tilde{A}^{(1)} \wedge \ldots \wedge \tilde{A}^{(n)}$$

is given by

$$\xi_{\wedge_{i=1}^{n}\tilde{A}^{(i)}}(x) = \min\left\{\xi_{\tilde{A}^{(1)}}(x), \ldots, \xi_{\tilde{A}^{(n)}}(x)\right\}.$$

Proposition 3.2.2 *Let $\tilde{A}^{(1)}, \ldots, \tilde{A}^{(n)}$ be fuzzy sets in $\mathbb{R}^m$. Then, we have*

$$\left(\tilde{A}^{(1)} \wedge \ldots \wedge \tilde{A}^{(n)}\right)_{\alpha} = \tilde{A}^{(1)}_{\alpha} \cap \ldots \cap \tilde{A}^{(n)}_{\alpha} \tag{3.8}$$

for any $\alpha \in (0,1]$.

Proof. Given any $\alpha \in (0,1]$, we have

$$\begin{aligned}\left(\tilde{A}^{(1)} \wedge \ldots \wedge \tilde{A}^{(n)}\right)_{\alpha} &= \left\{x \in \mathbb{R}^m : \xi_{\wedge_{i=1}^{n}\tilde{A}^{(i)}}(x) \geq \alpha\right\} \\ &= \left\{x \in \mathbb{R}^m : \min\left\{\xi_{\tilde{A}^{(1)}}(x), \ldots, \xi_{\tilde{A}^{(n)}}(x)\right\} \geq \alpha\right\} \\ &= \left\{x \in \mathbb{R}^m : \xi^{(i)}_{\tilde{A}}(x) \geq \alpha \text{ for all } i = 1, \ldots, n\right\} \\ &= \left\{x \in \mathbb{R}^m : x \in \tilde{A}^{(i)}_{\alpha} \text{ for all } i = 1, \ldots, n\right\} \\ &= \tilde{A}^{(1)}_{\alpha} \cap \ldots \cap \tilde{A}^{(n)}_{\alpha}.\end{aligned}$$

This completes the proof. ■

We now want to extend the concept of intersection of fuzzy sets by introducing the t-norm.

Definition 3.2.3 The function $t : [0,1] \times [0,1] \to [0,1]$ is called a **t-norm (triangular norm)** when the following conditions are satisfied:

- (boundary condition) $t(a,1) = a$;
- (commutativity) $t(a, b) = t(b, a)$;
- (increasing property) $a_1 \leq a_2$ and $b_1 \leq b_2$ imply $t(a_1, b_1) \leq t(a_2, b_2)$;
- (associativity) $t(t(a, b), c) = t(a, t(b, c))$.

Remark 3.2.4 The third condition says that $t(0,a) \leq t(0,1)$ for any $a \in [0,1]$. From the first condition, we have $t(0,1) = 0$, which also implies $t(0,a) = 0$ for any $a \in [0,1]$.

Definition 3.2.5 Let t be a t-norm that transforms the membership degrees of fuzzy sets $\tilde{A}$ and $\tilde{B}$ into the membership degree of the intersection $\tilde{A} \wedge \tilde{B}$. The membership function of the intersection $\tilde{A} \wedge \tilde{B}$ is defined by

$$\xi_{\tilde{A}\wedge\tilde{B}}(x) = t(\xi_{\tilde{A}}(x), \xi_{\tilde{B}}(x))$$

for all $x \in \mathbb{R}^m$.

We see that the expression (3.6) is obtained by taking $t(a,b) = \min\{a,b\}$, and it is easy to show that this minimum function satisfies the above conditions. Many well-known t-norms are also shown below:

- **Dombi class**: For $\lambda \in (0,\infty)$, we take

$$t_\lambda^{(D)}(a,b) = \frac{1}{1 + [(\frac{1}{a} - 1)^\lambda + (\frac{1}{b} - 1)^\lambda]^{1/\lambda}}.$$

- **Dubois-Prade class**: For $\alpha \in [0,1]$, we take

$$t_\alpha^{(D\text{-}P)}(a,b) = \frac{ab}{\max\{a,b,\alpha\}}.$$

- **Yager class**: For $w \in (0,\infty)$, we take

$$t_w^{(Y)}(a,b) = 1 - \min\{1, ((1-a)^w + (1-b)^w)^{1/w}\}.$$

- **Schweitzer-Sklar class**: For $p \in \mathbb{R}$, we take

$$t_p^{(S\text{-}S)}(a,b) = 1 - [(1-a)^p + (1-b)^p - (1-a)^p(1-b)^p]^{1/p}.$$

- **Frank class**: For $s > 0$, we take

$$t_s^{(F)}(a,b) = \log_s\left(1 + \frac{(s^a - 1)(s^b - 1)}{s - 1}\right).$$

- **Hamacher product**: For $\lambda \geq 0$, we take

$$t_\lambda^{(HP)}(a,b) = \frac{ab}{\lambda + (1-\lambda)(a+b-ab)}.$$

- **Drastic product**: We take

$$t^{(DP)}(a,b) = \begin{cases} a & \text{if } b = 1 \\ b & \text{if } a = 1 \\ 0 & \text{otherwise.} \end{cases}$$

- **Einstein product**: We take

$$t^{(EP)}(a,b) = \frac{ab}{2 - (a+b-ab)}.$$

- **Algebraic product**: We take

$$t^{(AP)}(a,b) = ab.$$

We have the following relationships

$$\lim_{\lambda\to\infty} t_{\lambda}^{(D)}(a,b) = \min\{a,b\}, \quad \lim_{\lambda\to 0} t_{\lambda}^{(D)}(a,b) = t^{(DP)}(a,b),$$
$$t_{1}^{(S\text{-}S)}(a,b) = ab, \quad \lim_{p\downarrow 0} t_{p}^{(S\text{-}S)}(a,b) = t^{(DP)}(a,b), \quad \lim_{p\to\infty} t_{p}^{(S\text{-}S)}(a,b) = \min\{a,b\}$$
$$t_{1}^{(HP)}(a,b) = ab, \quad \lim_{\lambda\to\infty} t_{\lambda}^{(HP)}(a,b) = t^{(DP)}(a,b).$$

Proposition 3.2.6 (Wang [112]) *For any t-norm t, the following inequalities hold*

$$t^{(DP)}(a,b) \leq t(a,b) \leq min\{a,b\}$$

for any $a,b \in [0,1]$.

The $t^{(DP)}$ operation can be considered as the most "pessimistic" t-norm operation. Now, we consider the interesting Dubois-Prade class $t_{\alpha}^{(D\text{-}P)}(a,b)$ for $\alpha \in [0,1]$. Note that

$$t_{\alpha}^{(D\text{-}P)}(a,b) = \begin{cases} ab/\alpha & \text{if } a,b < \alpha \\ \min\{a,b\} & \text{otherwise.} \end{cases}$$

- For $\alpha = 0$, we have $t_{\alpha}^{(D\text{-}P)}(a,b) = \min\{a,b\}$.
- For $\alpha = 1$, we have $t_{\alpha}^{(D\text{-}P)}(a,b) = ab$.
- For $\alpha \in (0,1)$, it is a function between the minimum and the product.

Next, we shall consider the generalized t-norm. Let t be a t-norm. Since t is associative, we can recursively define the function $T_n : [0,1]^n \to [0,1]$ by

$$T_n(\alpha_1,\ldots,\alpha_{n-1},\alpha_n) = t(T_{n-1}(\alpha_1,\ldots,\alpha_{n-1}),\alpha_n). \tag{3.9}$$

The function T_n is called a **generalized t-norm**. Now, using the axioms of a t-norm, for any $\alpha \in [0,1]$, we have

$$t(\alpha,\alpha) \leq t(\alpha,1) = \alpha,$$

which implies

$$T_3(\alpha,\alpha,\alpha) = t(T_2(\alpha,\alpha),\alpha) = t(t(\alpha,\alpha),\alpha) \leq t(\alpha,\alpha) \leq \alpha$$

and

$$T_4(\alpha,\alpha,\alpha,\alpha) = t(T_3(\alpha,\alpha,\alpha),\alpha) \leq t(\alpha,\alpha) \leq \alpha.$$

Inductively, we obtain

$$T_n(\alpha,\ldots,\alpha) \leq \alpha \tag{3.10}$$

for any $\alpha \in [0,1]$.

Remark 3.2.7 We have the following observations.

- Since the t-norm is increasing in the sense that

  $a_1 \leq b_1$ and $a_2 \leq b_2$ imply $t(a_1,a_2) \leq t(b_1,b_2)$,

  we have that T_n is also increasing in the sense that

  $a_i \leq b_i$ for all $i = 1,\ldots,n$ imply $T_n(a_1,\ldots,a_n) \leq T_n(b_1,\ldots,b_n)$.

However, the converse does not necessarily hold true in general; that is, $T_n(a_1,\ldots,a_n) \leq T_n(b_1,\ldots,b_n)$ does not necessarily imply $a_i \leq b_i$ for all $i = 1,\ldots,n$.

- From Remark 3.2.4, we see that $t(0,a) = t(a,0) = 0$ for any $a \in [0,1]$. Therefore, if one of the a_i is zero, then $T_n(a_1,\ldots,a_n) = 0$. In other words, if $T_n(a_1,\ldots,a_n) > 0$, then $a_i > 0$ for all $i = 1,\ldots,n$.
- According to the boundary condition of t-norm, we see that $t(1,1) = 1$, which also implies $T_n(1,\ldots,1) = 1$.

Instead of considering a t-norm for the intersection, we shall consider the general function that is formally defined below.

Definition 3.2.8 Let $\mathfrak{A}^{\cap} : [0,1]^n \to [0,1]$ be a function defined on $[0,1]^n$. Given fuzzy sets $\tilde{A}^{(1)},\ldots,\tilde{A}^{(n)}$ in $\mathbb{R}^m$, the intersection of $\tilde{A}^{(1)},\ldots,\tilde{A}^{(n)}$ is denoted by

$$\tilde{A}^{(1)} \sqcap \ldots \sqcap \tilde{A}^{(n)} = \sqcap_{i=1}^{n} \tilde{A}^{(i)}, \tag{3.11}$$

and its membership function is defined by

$$\xi_{\sqcap_{i=1}^{n} \tilde{A}^{(i)}}(x) = \mathfrak{A}^{\cap}\left(\xi_{\tilde{A}^{(1)}}(x),\ldots,\xi_{\tilde{A}^{(n)}}(x)\right). \tag{3.12}$$

Since the membership function of the intersection depends on the function $\mathfrak{A}^{\cap}$, we sometimes write

$$\tilde{A}^{(1)} \sqcap \ldots \sqcap \tilde{A}^{(n)}(\mathfrak{A}^{\cap}) = \sqcap_{i=1}^{n} \tilde{A}^{(i)}(\mathfrak{A}^{\cap}). \tag{3.13}$$

In particular, we can take the function $\mathfrak{A}^{\cap}$ as the minimum function or the generalized t-norm as given below

$$\mathfrak{A}^{\cap}\left(\alpha_1,\ldots,\alpha_n\right) = \min\left\{\alpha_1,\ldots,\alpha_n\right\} \text{ or } \mathfrak{A}^{\cap}\left(\alpha_1,\ldots,\alpha_n\right) = T_n\left(\alpha_1,\ldots,\alpha_n\right). \tag{3.14}$$

Inspired by (3.8), we propose the following definition.

Definition 3.2.9 Given any fuzzy sets $\tilde{A}^{(1)},\ldots,\tilde{A}^{(n)}$ in $\mathbb{R}^m$, the concepts of compatibility are defined below.

- We say that the function $\mathfrak{A}^{\cap} : [0,1]^n \to [0,1]$ is **⊆-compatible with set intersection** when the following inclusion holds true

$$\left(\tilde{A}^{(1)} \sqcap \ldots \sqcap \tilde{A}^{(n)}\right)_{\alpha} \subseteq \tilde{A}^{(1)}_{\alpha} \cap \ldots \cap \tilde{A}^{(n)}_{\alpha} \tag{3.15}$$

  for each $\alpha \in (0,1]$.
- We say that the function $\mathfrak{A}^{\cap} : [0,1]^n \to [0,1]$ is **⊇-compatible with set intersection** when the following inclusion holds true

$$\left(\tilde{A}^{(1)} \sqcap \ldots \sqcap \tilde{A}^{(n)}\right)_{\alpha} \supseteq \tilde{A}^{(1)}_{\alpha} \cap \ldots \cap \tilde{A}^{(n)}_{\alpha} \tag{3.16}$$

  for each $\alpha \in (0,1]$.

By referring to (3.8), we see that the function taken by

$$\mathfrak{A}^{\cap}\left(\alpha_1,\ldots,\alpha_n\right) = \min\left\{\alpha_1,\ldots,\alpha_n\right\}$$

is both ⊆-compatible and ⊇-compatible with set intersection.

Proposition 3.2.10 *Suppose that the function $\mathfrak{A}^{\cap} : [0,1]^n \to [0,1]$ is both $\subseteq$-compatible and $\supseteq$-compatible with set intersection. Then, we have*

$$\mathfrak{A}^{\cap}(\alpha_1, \dots, \alpha_n) = \min \left\{\alpha_1, \dots, \alpha_n\right\}$$

for $0 < \alpha_i \in \mathcal{R}(\xi_{\tilde{A}^{(i)}})$ and $i = 1, \dots, n$.

Proof. Given any fixed $\tilde{A}^{(1)}, \dots, \tilde{A}^{(n)}$, since the function $\mathfrak{A}^{\cap} : [0,1]^n \to [0,1]$ is assumed to be both $\subseteq$-compatible and $\supseteq$-compatible with set intersection, it means that

$$\left(\tilde{A}^{(1)} \sqcap \dots \sqcap \tilde{A}^{(n)}\right)_\alpha = \tilde{A}^{(1)}_\alpha \cap \dots \cap \tilde{A}^{(n)}_\alpha \text{ for each } \alpha \in (0,1]. \tag{3.17}$$

Now, we have

$$\begin{aligned}\tilde{A}^{(1)}_\alpha \cap \dots \cap \tilde{A}^{(n)}_\alpha &= \left\{x \in \mathbb{R}^m : x \in \tilde{A}^{(i)}_\alpha \text{ for each } i = 1, \dots, n\right\}\\ &= \left\{x \in \mathbb{R}^m : \xi_{\tilde{A}^{(i)}}(x) \geq \alpha \text{ for each } i = 1, \dots, n\right\}.\end{aligned}$$

From (3.12), we also have

$$\begin{aligned}\left(\tilde{A}^{(1)} \sqcap \dots \sqcap \tilde{A}^{(n)}\right)_\alpha &= \left\{x \in \mathbb{R}^m : \xi_{\sqcap_{i=1}^n \tilde{A}^{(i)}}(x) \geq \alpha\right\}\\ &= \left\{x \in \mathbb{R}^m : \mathfrak{A}^{\cap}\left(\xi_{\tilde{A}^{(1)}}(x), \dots, \xi_{\tilde{A}^{(n)}}(x)\right) \geq \alpha\right\}.\end{aligned}$$

From (3.17), for each $\alpha \in (0,1]$, we see that

$$\left\{x \in \mathbb{R}^m : \xi_{\tilde{A}^{(i)}}(x) \geq \alpha \text{ for each } i = 1, \dots, n\right\} = \left\{x \in \mathbb{R}^m : \mathfrak{A}^{\cap}\left(\xi_{\tilde{A}^{(1)}}(x), \dots, \xi_{\tilde{A}^{(n)}}(x)\right) \geq \alpha\right\}.$$

Equivalently, it means that, for each $\alpha \in (0,1]$ and $x \in \mathbb{R}^m$,

$$\mathfrak{A}^{\cap}\left(\xi_{\tilde{A}^{(1)}}(x), \dots, \xi_{\tilde{A}^{(n)}}(x)\right) \geq \alpha \text{ if and only if } \xi_{\tilde{A}^{(i)}}(x) \geq \alpha \text{ for each } i = 1, \dots, n.$$

We write $\alpha_i = \xi_{\tilde{A}^{(i)}}(x)$ for $i = 1, \dots, n$. Then $\alpha_i \in \mathcal{R}(\xi_{\tilde{A}^{(i)}})$ with $\alpha_i > 0$ for $i = 1, \dots, n$ and

$$\mathfrak{A}^{\cap}\left(\alpha_1, \dots, \alpha_n\right) \geq \alpha \text{ if and only if } \alpha_i \geq \alpha \text{ for each } i = 1, \dots, n. \tag{3.18}$$

We want to claim

$$\mathfrak{A}^{\cap}\left(\alpha_1, \dots, \alpha_n\right) = \min\left\{\alpha_1, \dots, \alpha_n\right\}. \tag{3.19}$$

Let $\alpha = \min\{\alpha_1, \dots, \alpha_n\}$. Since $\alpha_i \geq \min\{\alpha_1, \dots, \alpha_n\}$ for all $i = 1, \dots, n$. Using (3.18), it follows that

$$\mathfrak{A}^{\cap}\left(\alpha_1, \dots, \alpha_n\right) \geq \min\left\{\alpha_1, \dots, \alpha_n\right\}$$

by taking $\alpha = \min\{\alpha_1, \dots, \alpha_n\}$. On the other hand, suppose that $\mathfrak{A}^{\cap}(\alpha_1, \dots, \alpha_n) = \alpha > 0$, i.e. $\mathfrak{A}^{\cap}(\alpha_1, \dots, \alpha_n) \geq \alpha$, the expression (3.18) says that $\alpha_i \geq \alpha$ for each $i = 1, \dots, n$, which implies

$$\min\left\{\alpha_1, \dots, \alpha_n\right\} \geq \alpha = \mathfrak{A}^{\cap}\left(\alpha_1, \dots, \alpha_n\right).$$

Therefore, we obtain the equality (3.19). This completes the proof. ∎

Proposition 3.2.11 *We have the following properties.*

(i) *Suppose that the function $\mathfrak{A}^{\cap} : [0,1]^n \to [0,1]$ satisfies the following condition: for any $\alpha \in [0,1]$,*

$$\mathfrak{A}^{\cap}(\alpha_1, \dots, \alpha_n) \geq \alpha \textit{ implies } \alpha_i \geq \alpha \textit{ for each } i = 1, \dots, n.$$

Then $\mathfrak{A}^{\cap}$ is $\subseteq$-compatible with set intersection.

(ii) *Suppose that the function* $\mathfrak{A}^{\cap} : [0,1]^n \to [0,1]$ *satisfies the following condition: for any* $\alpha \in [0,1]$,

$$\alpha_i \geq \alpha \text{ for each } i = 1, \ldots, n \text{ implies } \mathfrak{A}^{\cap}(\alpha_1, \ldots, \alpha_n) \geq \alpha.$$

Then $\mathfrak{A}^{\cap}$ *is* $\supseteq$*-compatible with set intersection.*

(iii) *Suppose that the function* $\mathfrak{A}^{\cap} : [0,1]^n \to [0,1]$ *satisfies the following condition: for any* $\alpha \in [0,1]$,

$$\mathfrak{A}^{\cap}(\alpha_1, \ldots, \alpha_n) \geq \alpha \text{ if and only if } \alpha_i \geq \alpha \text{ for each } i = 1, \ldots, n.$$

Then

$$\mathfrak{A}^{\cap}\left(\alpha_1, \ldots, \alpha_n\right) = \min\left\{\alpha_1, \ldots, \alpha_n\right\}$$

and $\mathfrak{A}^{\cap}$ *is both* $\subseteq$*-compatible and* $\supseteq$*-compatible with set intersection.*

Proof. The desired results can be similarly obtained from the proof of Proposition 3.2.10. ■

Example 3.2.12 Suppose that the function $\mathfrak{A}^{\cap} : [0,1]^n \to [0,1]$ is defined by

$$\mathfrak{A}^{\cap}\left(\alpha_1, \ldots, \alpha_n\right) = \alpha_1\alpha_2 \cdots \alpha_n = \prod_{i=1}^{n} \alpha_i.$$

Assume that

$$\mathfrak{A}^{\cap}\left(\alpha_1, \ldots, \alpha_n\right) = \alpha_1\alpha_2 \cdots \alpha_n \geq \alpha.$$

Since each $\alpha_i \leq 1$ for all $i = 1, \ldots, n$, we must have $\alpha_i \geq \alpha$ for each $i = 1, \ldots, n$. This shows that $\mathfrak{A}^{\cap}$ is $\subseteq$-compatible with set intersection from part (i) of Proposition 3.2.11.

Example 3.2.13 Suppose that the function $\mathfrak{A}^{\cap} : [0,1]^n \to [0,1]$ satisfies the following conditions.

- $\mathfrak{A}^{\cap}$ is increasing in the following sense:

$$\alpha_i \leq \beta_i \text{ for each } i = 1, \ldots, n \text{ imply } \mathfrak{A}^{\cap}\left(\alpha_1, \ldots, \alpha_n\right) \leq \mathfrak{A}^{\cap}\left(\beta_1, \ldots, \beta_n\right). \tag{3.20}$$

- For any $\alpha \in [0,1]$, the inequality $\mathfrak{A}^{\cap}(\alpha, \ldots, \alpha) \geq \alpha$ holds true.

Assume that $\alpha_i \geq \alpha$ for each $i = 1, \ldots, n$. Then, we have

$$\begin{aligned}\mathfrak{A}^{\cap}\left(\alpha_1, \ldots, \alpha_n\right) &\geq \mathfrak{A}^{\cap}\left(\alpha, \ldots, \alpha\right) \text{ (by the first condition)}\\ &\geq \alpha \text{ (by the second condition).}\end{aligned}$$

This shows that $\mathfrak{A}^{\cap}$ is $\supseteq$-compatible with set intersection from part (ii) of Proposition 3.2.11.

Remark 3.2.14 Suppose that the function $\mathfrak{A}^{\cap}$ is taken to be the generalized t-norm T_n. Since t-norm is increasing, it is clear to see that T_n is also increasing in the sense of (3.20). Using Example 3.2.13, if we wish the function T_n to be $\supseteq$-compatible with set intersection, then T_n must satisfy $T_n(\alpha, \ldots, \alpha) \geq \alpha$ for any $\alpha \in [0,1]$. From (3.10), it follows that $T_n(\alpha, \ldots, \alpha) = \alpha$. In other words, if $T_n(\alpha, \ldots, \alpha) \neq \alpha$, then $T_n(\alpha, \ldots, \alpha) < \alpha$. This contradiction says that if $T_n(\alpha, \ldots, \alpha) \neq \alpha$ then T_n cannot be $\supseteq$-compatible with set intersection.

3.3 Union of Fuzzy Sets

Let A and B be two subsets of $\mathbb{R}^m$. Then, we have two corresponding characteristic functions χ_A and χ_B. The union $A \cup B$ also has the corresponding characteristic function $\chi_{A\cup B}$. We can obtain

$$\chi_{A\cup B}(x) = \max\left\{\chi_A(x), \chi_B(x)\right\}. \tag{3.21}$$

Inspired by (3.21), the union of fuzzy sets is defined below.

Definition 3.3.1 Let $\tilde{A}$ and $\tilde{B}$ be two fuzzy sets in $\mathbb{R}^m$ with membership functions $\xi_{\tilde{A}}$ and $\xi_{\tilde{B}}$, respectively. The membership function of $\tilde{A} \vee \tilde{B}$ is defined by

$$\xi_{\tilde{A}\vee\tilde{B}}(x) = \max\{\xi_{\tilde{A}}(x), \xi_{\tilde{B}}(x)\} \tag{3.22}$$

for all $x \in \mathbb{R}^m$.

For more than two fuzzy sets, their union should be defined inductively. Given any three fuzzy sets $\tilde{A}^{(1)}, \tilde{A}^{(2)}$, and $\tilde{A}^{(3)}$ in $\mathbb{R}^m$, we first consider the union $\tilde{A} = \tilde{A}^{(1)} \vee \tilde{A}^{(2)}$ whose membership function is given by (3.22). Now, the union of $\tilde{A}^{(1)}, \tilde{A}^{(2)}$, and $\tilde{A}^{(3)}$ can be defined as

$$\tilde{A}^{\circ} \equiv \left(\tilde{A}^{(1)} \vee \tilde{A}^{(2)}\right) \vee \tilde{A}^{(3)} = \tilde{A} \vee \tilde{A}^{(3)},$$

whose membership function is given by

$$\begin{aligned}\xi_{\tilde{A}^{\circ}}(x) &= \max\left\{\xi_{\tilde{A}}(x), \xi_{\tilde{A}^{(3)}}(x)\right\}\\ &= \max\left\{\max\left\{\xi_{\tilde{A}^{(1)}}(x), \xi_{\tilde{A}^{(2)}}(x)\right\}, \xi_{\tilde{A}^{(3)}}(x)\right\}\\ &= \max\left\{\xi_{\tilde{A}^{(1)}}(x), \xi_{\tilde{A}^{(2)}}(x), \xi_{\tilde{A}^{(3)}}(x)\right\}. \end{aligned} \tag{3.23}$$

Suppose that the union is defined as

$$\tilde{A}^{(1)} \vee \left(\tilde{A}^{(2)} \vee \tilde{A}^{(3)}\right) \text{ or } \tilde{A}^{(2)} \vee \left(\tilde{A}^{(3)} \vee \tilde{A}^{(1)}\right) \text{ or } \left(\tilde{A}^{(3)} \vee \tilde{A}^{(2)}\right) \vee \tilde{A}^{(1)}$$

or any other permutation. Then, we can similarly show that their membership functions are identical to (3.23). In this case, we can simply write

$$\tilde{A}^{\circ} \equiv \tilde{A}^{(1)} \vee \tilde{A}^{(2)} \vee \tilde{A}^{(3)}.$$

Inductively, the membership function of the union

$$\tilde{A}^{(1)} \vee \ldots \vee \tilde{A}^{(n)}$$

is given by

$$\xi_{\vee_{i=1}^{n}\tilde{A}^{(i)}}(x) = \max\left\{\xi_{\tilde{A}^{(1)}}(x), \ldots, \xi_{\tilde{A}^{(n)}}(x)\right\}.$$

Proposition 3.3.2 *Let $\tilde{A}^{(1)}, \ldots, \tilde{A}^{(n)}$ be fuzzy sets in $\mathbb{R}^m$. Then, we have*

$$\left(\tilde{A}^{(1)} \vee \ldots \vee \tilde{A}^{(n)}\right)_{\alpha} = \tilde{A}^{(1)}_{\alpha} \cup \ldots \cup \tilde{A}^{(n)}_{\alpha} \tag{3.24}$$

for any $\alpha \in (0,1]$.

Proof. Given any $\alpha \in (0,1]$, we have

$$\begin{aligned}\left(\tilde{A}^{(1)} \vee \ldots \vee \tilde{A}^{(n)}\right)_\alpha &= \left\{x \in \mathbb{R}^m : \xi_{\vee_{i=1}^n \tilde{A}^{(i)}}(x) \geq \alpha\right\} \\ &= \left\{x \in \mathbb{R}^m : \max\left\{\xi_{\tilde{A}^{(1)}}(x), \ldots, \xi_{\tilde{A}^{(n)}}(x)\right\} \geq \alpha\right\} \\ &= \left\{x \in \mathbb{R}^m : \xi_{\tilde{A}}^{(i)}(x) \geq \alpha \text{ for some } i = 1, \ldots, n\right\} \\ &= \left\{x \in \mathbb{R}^m : x \in \tilde{A}_\alpha^{(i)} \text{ for some } i = 1, \ldots, n\right\} \\ &= \tilde{A}_\alpha^{(1)} \cup \ldots \cup \tilde{A}_\alpha^{(n)}.\end{aligned}$$

This completes the proof. ■

We want to extend the concept of union of fuzzy sets by introducing the s-norm.

Definition 3.3.3 The function $s : [0,1] \times [0,1] \to [0,1]$ is called an **s-norm (triangular co-norm)** when the following conditions are satisfied:

- (boundary condition) $s(a,0) = a$;
- (commutativity) $s(a,b) = s(b,a)$;
- (increasing property) $a_1 \leq a_2$ and $b_1 \leq b_2$ imply $s(a_1,b_1) \leq s(a_2,b_2)$;
- (associativity) $s(s(a,b),c) = s(a,s(b,c))$.

It is clear to see that $s(1,a) = 1$ for any $a \in [0,1]$. We also have the relation between t-norm and s-norm given by

$$t(a,b) = 1 - s(1-a,1-b) \text{ and } s(a,b) = 1 - t(1-a,1-b).$$

Definition 3.3.4 Let s be an s-norm that transforms the membership degrees of fuzzy sets $\tilde{A}$ and $\tilde{B}$ into the membership degree of the union $\tilde{A} \vee \tilde{B}$. The membership function is defined by

$$\xi_{\tilde{A}\vee\tilde{B}}(x) = s(\xi_{\tilde{A}}(x), \xi_{\tilde{B}}(x))$$

for all $x \in \mathbb{R}^m$.

We see that the expression (3.22) is obtained by taking $s(a,b) = \max\{a,b\}$, and it is clear to see that this maximum function satisfies the above conditions. Now, we also present many other s-norms as follows.

- **Dombi class**: For $\lambda \in (0,\infty)$, we take

$$s_\lambda^{(D)}(a,b) = \frac{1}{1 + [(\frac{1}{a} - 1)^{-\lambda} + (\frac{1}{b} - 1)^{-\lambda}]^{-1/\lambda}}. \tag{3.25}$$

- **Dubois-Prade class**: For $\alpha \in [0,1]$, we take

$$s_\alpha^{(D\text{-}P)}(a,b) = \frac{a + b - ab - \min\{a,b,1-\alpha\}}{\max\{1-a,1-b,\alpha\}}.$$

- **Yager class**: For $w \in (0,\infty)$, we take

$$s_w^{(Y)}(a,b) = \min\{1,(a^w + b^w)^{1/w}\}. \tag{3.26}$$

- **Sugeno class**: For $\lambda \geq -1$, we take

$$s_{\lambda}^{(S)}(a,b) = \min\{1, a+b+\lambda ab\}.$$

- **Drastic sum**: We take

$$s^{(DS)}(a,b) = \begin{cases} a & \text{if } b = 0 \\ b & \text{if } a = 0 \\ 1 & \text{otherwise.} \end{cases}$$

- **Einstein sum**: We take

$$s^{(ES)}(a,b) = \frac{a+b}{1+ab}.$$

- **Algebraic sum**: We take

$$s^{(AS)}(a,b) = a+b-ab.$$

Then, we have the following relationships

$$\lim_{\lambda\to\infty} s_{\lambda}^{(D)}(a,b) = \max\{a,b\} \text{ and } \lim_{\lambda\to 0} s_{\lambda}^{(D)}(a,b) = s^{(DS)}(a,b).$$

Proposition 3.3.5 (Wang *[112]*) *For any s-norm s, the following inequality holds true*

$$\max\{a,b\} \leq s(a,b) \leq s^{(DS)}(a,b)$$

for any $a, b \in [0,1]$.

Let s be an s-norm. Since s is associative, we can recursively define the function $S_n : [0,1]^n \to [0,1]$ by

$$S_n(\alpha_1, \ldots, \alpha_{n-1}, \alpha_n) = s(S_{n-1}(\alpha_1, \ldots, \alpha_{n-1}), \alpha_n). \tag{3.27}$$

The function S_n is called a **generalized s-norm** .

Instead of considering an s-norm for the union, we shall consider the general function that is formally defined below.

Definition 3.3.6 Let $\mathfrak{A}^{\cup} : [0,1]^n \to [0,1]$ be a function defined on $[0,1]^n$. Given fuzzy sets $\tilde{A}^{(1)}, \ldots, \tilde{A}^{(n)}$ in $\mathbb{R}^m$, we define a fuzzy set of the union of fuzzy sets $\tilde{A}^{(1)}, \ldots, \tilde{A}^{(n)}$ as

$$\tilde{A}^{(1)} \sqcup \ldots \sqcup \tilde{A}^{(n)} = \sqcup_{i=1}^{n} \tilde{A}^{(i)}, \tag{3.28}$$

whose membership function is given by

$$\xi_{\sqcup_{i=1}^{n}\tilde{A}^{(i)}}(x) = \mathfrak{A}^{\cup}\left(\xi_{\tilde{A}^{(1)}}(x), \cdots, \xi_{\tilde{A}^{(n)}}(x)\right). \tag{3.29}$$

Since the membership function of a union depends on the function $\mathfrak{A}^{\cup}$, we sometimes write

$$\tilde{A}^{(1)} \sqcup \ldots \sqcup \tilde{A}^{(n)}(\mathfrak{A}^{\cup}) = \sqcup_{i=1}^{n} \tilde{A}^{(i)}(\mathfrak{A}^{\cup}). \tag{3.30}$$

In particular, we can take the function $\mathfrak{A}^{\cup}$ as the maximum function or the generalized s-norm as given below

$$\mathfrak{A}^{\cup}\left(\alpha_1, \ldots, \alpha_n\right) = \max\left\{\alpha_1, \ldots, \alpha_n\right\} \text{ and } \mathfrak{A}^{\cup}\left(\alpha_1, \ldots, \alpha_n\right) = S_n\left(\alpha_1, \ldots, \alpha_n\right). \tag{3.31}$$

Inspired by (3.24), we propose the following definition.

Definition 3.3.7 Given any fuzzy sets $\tilde{A}^{(1)}, \ldots, \tilde{A}^{(n)}$ in $\mathbb{R}^m$, the concepts of compatibility are defined below.

- We say that the function $\mathfrak{A}^{\cup} : [0,1]^n \to [0,1]$ is **$\subseteq$-compatible with set union** when the following inclusions hold true

$$\left(\tilde{A}^{(1)} \sqcup \ldots \sqcup \tilde{A}^{(n)}\right)_\alpha \subseteq \tilde{A}^{(1)}_\alpha \cup \ldots \cup \tilde{A}^{(n)}_\alpha \tag{3.32}$$

  for each $\alpha \in (0,1]$.
- We say that the function $\mathfrak{A}^{\cup} : [0,1]^n \to [0,1]$ is **$\supseteq$-compatible with set union** when the following inclusions hold true

$$\left(\tilde{A}^{(1)} \sqcup \ldots \sqcup \tilde{A}^{(n)}\right)_\alpha \supseteq \tilde{A}^{(1)}_\alpha \cup \ldots \cup \tilde{A}^{(n)}_\alpha \tag{3.33}$$

  for each $\alpha \in (0,1]$.

By referring to (3.24), we see that the function given by

$$\mathfrak{A}^{\cup}\left(\alpha_1, \ldots, \alpha_n\right) = \max\left\{\alpha_1, \ldots, \alpha_n\right\}$$

is both $\subseteq$-compatible and $\supseteq$-compatible with set union.

Proposition 3.3.8 *Suppose that the function $\mathfrak{A}^{\cup} : [0,1]^n \to [0,1]$ is both $\subseteq$-compatible and $\supseteq$-compatible with set union. Then, we have*

$$\mathfrak{A}^{\cup}(\alpha_1, \ldots, \alpha_n) = \max\left\{\alpha_1, \ldots, \alpha_n\right\}$$

for $0 < \alpha_i \in \mathcal{R}(\xi_{\tilde{A}^{(i)}})$ and $i = 1, \ldots, n$.

Proof. Given any fixed $\tilde{A}^{(1)}, \ldots, \tilde{A}^{(n)}$, the assumption says that

$$\left(\tilde{A}^{(1)} \sqcup \ldots \sqcup \tilde{A}^{(n)}\right)_\alpha = \tilde{A}^{(1)}_\alpha \cup \ldots \cup \tilde{A}^{(n)}_\alpha \text{ for each } \alpha \in (0,1]. \tag{3.34}$$

Now, we have

$$\begin{aligned}\tilde{A}^{(1)}_\alpha \cup \ldots \cup \tilde{A}^{(n)}_\alpha &= \left\{x \in \mathbb{R}^m : x \in \tilde{A}^{(i)}_\alpha \text{ for some } i = 1, \ldots, n\right\}\\ &= \left\{x \in \mathbb{R}^m : \xi_{\tilde{A}^{(i)}}(x) \geq \alpha \text{ for some } i = 1, \ldots, n\right\}.\end{aligned}$$

From (3.29), we also have

$$\begin{aligned}\left(\tilde{A}^{(1)} \sqcup \ldots \sqcup \tilde{A}^{(n)}\right)_\alpha &= \left\{x \in \mathbb{R}^m : \xi_{\sqcup_{i=1}^n \tilde{A}^{(i)}}(x) \geq \alpha\right\}\\ &= \left\{x \in \mathbb{R}^m : \mathfrak{A}^{\cup}\left(\xi_{\tilde{A}^{(1)}}(x), \ldots, \xi_{\tilde{A}^{(n)}}(x)\right) \geq \alpha\right\}.\end{aligned}$$

From (3.34), for each $\alpha \in (0,1]$, we see that

$$\left\{x \in \mathbb{R}^m : \xi_{\tilde{A}^{(i)}}(x) \geq \alpha \text{ for some } i = 1, \ldots, n\right\} = \left\{x \in \mathbb{R}^m : \mathfrak{A}^{\cup}\left(\xi_{\tilde{A}^{(1)}}(x), \ldots, \xi_{\tilde{A}^{(n)}}(x)\right) \geq \alpha\right\}.$$

Equivalently, it means that, for each $\alpha \in (0,1]$ and $x \in \mathbb{R}^m$,

$$\mathfrak{A}^{\cup}\left(\xi_{\tilde{A}^{(1)}}(x), \ldots, \xi_{\tilde{A}^{(n)}}(x)\right) \geq \alpha \text{ if and only if } \xi_{\tilde{A}^{(i)}}(x) \geq \alpha \text{ for some } i = 1, \ldots, n.$$

We write $\alpha_i = \xi_{\tilde{A}^{(i)}}(x)$ for $i = 1, \ldots, n$. Then $\alpha_i \in \mathcal{R}(\xi_{\tilde{A}^{(i)}})$ with $\alpha_i > 0$ for $i = 1, \ldots, n$ and

$$\mathfrak{A}^{\cup}\left(\alpha_1, \ldots, \alpha_n\right) \geq \alpha \text{ if and only if } \alpha_i \geq \alpha \text{ for some } i = 1, \ldots, n. \tag{3.35}$$

We want to claim

$$\mathfrak{A}^{\cup}\left(\alpha_1,\ldots,\alpha_n\right)=\max\left\{\alpha_1,\ldots,\alpha_n\right\}. \tag{3.36}$$

Suppose that $\alpha_i < \max\{\alpha_1,\ldots,\alpha_n\}$ for all $i = 1,\ldots,n$. Then $\max\{\alpha_1,\ldots,\alpha_n\} < \max\{\alpha_1,\ldots,\alpha_n\}$. This contradiction says that $\alpha_i \geq \max\{\alpha_1,\ldots,\alpha_n\}$ for some $i = 1,\ldots,n$. Let $\alpha = \max\{\alpha_1,\ldots,\alpha_n\}$. Using (3.35), it follows that

$$\mathfrak{A}^{\cup}\left(\alpha_1,\ldots,\alpha_n\right)\geq\max\left\{\alpha_1,\ldots,\alpha_n\right\}$$

by taking $\alpha = \max\{\alpha_1,\ldots,\alpha_n\}$. On the other hand, suppose that $\mathfrak{A}^{\cup}(\alpha_1,\ldots,\alpha_n) = \alpha > 0$, i.e. $\mathfrak{A}^{\cup}(\alpha_1,\ldots,\alpha_n) \geq \alpha$, the expression (3.35) says that $\alpha_i \geq \alpha$ for some $i = 1,\ldots,n$, which implies

$$\max\left\{\alpha_1,\ldots,\alpha_n\right\}\geq\alpha=\mathfrak{A}^{\cup}\left(\alpha_1,\ldots,\alpha_n\right).$$

Therefore, we obtain the equality (3.36). This completes the proof. ■

Proposition 3.3.9 *We have the following properties.*

(i) *Suppose that the function $\mathfrak{A}^{\cup} : [0,1]^n \to [0,1]$ satisfies the following condition: for any $\alpha \in [0,1]$,*

$$\mathfrak{A}^{\cup}(\alpha_1,\ldots,\alpha_n)\geq\alpha \textit{ implies } \alpha_i\geq\alpha \textit{ for some } i=1,\ldots,n.$$

Then $\mathfrak{A}^{\cup}$ is $\subseteq$-compatible with set union.

(ii) *Suppose that the function $\mathfrak{A}^{\cup} : [0,1]^n \to [0,1]$ satisfies the following condition: for any $\alpha \in [0,1]$,*

$$\alpha_i\geq\alpha \textit{ for some } i=1,\ldots,n \textit{ implies } \mathfrak{A}^{\cup}(\alpha_1,\ldots,\alpha_n)\geq\alpha.$$

Then $\mathfrak{A}^{\cup}$ is $\supseteq$-compatible with set union.

(iii) *Suppose that the function $\mathfrak{A}^{\cup} : [0,1]^n \to [0,1]$ satisfies the following condition: for any $\alpha \in [0,1]$,*

$$\mathfrak{A}^{\cup}(\alpha_1,\ldots,\alpha_n)\geq\alpha \textit{ if and only if } \alpha_i\geq\alpha \textit{ for some } i=1,\ldots,n.$$

Then

$$\mathfrak{A}^{\cup}\left(\alpha_1,\ldots,\alpha_n\right)=\max\left\{\alpha_1,\ldots,\alpha_n\right\}$$

and $\mathfrak{A}^{\cup}$ is both $\subseteq$-compatible and $\supseteq$-compatible with set union.

Proof. The desired results can be similarly obtained from the proof of Proposition 3.3.8. ■

From Propositions 3.2.6 and 3.3.5, we see that, for any membership degrees

$$a=\xi_{\tilde{A}}(r) \text{ and } b=\xi_{\tilde{B}}(r)$$

of arbitrary fuzzy sets $\tilde{A}$ and $\tilde{B}$, the membership degrees of their union defined by an s-norm lies in the interval

$$\left[\max\{a,b\},s^{(DS)}(a,b)\right],$$

and the membership degrees of their intersection defined by t-norm lies in the interval

$$\left[t^{(DP)}(a,b),\min\{a,b\}\right].$$

Let us consider the following two operators.

- **Fuzzy and**: The "fuzzy and" operator

$$v_p(a,b) = p \cdot \min\{a,b\} + \frac{(1-p)(a+b)}{2} \quad \text{for } p \in [0,1]$$

covers the range from $\min\{a,b\}$ to $(a+b)/2$.
- **Fuzzy or**: The "fuzzy or" operator

$$v_\gamma(a,b) = \gamma \cdot \max\{a,b\} + \frac{(1-\gamma)(a+b)}{2} \quad \text{for } \gamma \in [0,1]$$

covers the range from $(a+b)/2$ to $\max\{a,b\}$.

The union and intersection operators presented above cannot cover the interval

$$\left[\min\{a,b\}, \max\{a,b\}\right].$$

Therefore, a so-called averaging operator is defined below.

Definition 3.3.10 The operator that covers the following interval

$$\left[\min\{a,b\}, \max\{a,b\}\right] \tag{3.37}$$

is called an **averaging operator**.

By referring to Wang [112] and Driankov et al. [23], we have the following averaging operators proposed in the literature.

- **Max-Min average**: If we take

$$v_\lambda(a,b) = \lambda \max\{a,b\} + (1-\lambda)\min\{a,b\} \text{ for } \lambda \in [0,1],$$

the max-min average operator covers the whole range from $\min\{a,b\}$ to $\max\{a,b\}$ as given in (3.37) when the parameter λ changes from 0 to 1.
- **Generalized mean**: If we take

$$v_\alpha(a,b) = \left(\frac{a^\alpha + b^\alpha}{2}\right)^{1/\alpha} \quad \text{for } \alpha \in \mathbb{R} \text{ and } \alpha \neq 0,$$

the generalized mean operator covers the whole range from $\min\{a,b\}$ to $\max\{a,b\}$ when α changes from $-\infty$ to ∞.

3.4 Inductive and Direct Definitions

For more than two fuzzy sets, the intersection and union based on the general functions can be defined inductively or directly, which will be explained below. Given three fuzzy sets $\tilde{A}^{(1)}$, $\tilde{A}^{(2)}$, and $\tilde{A}^{(3)}$ in $\mathbb{R}^m$, there are two ways to define the intersection, as follows.

- (**Direct Definition**). By considering a function $\mathfrak{A}_3^{\cap} : [0,1]^3 \to [0,1]$, the membership function of the intersection $\tilde{A}^{\dagger} \equiv \tilde{A}^{(1)} \sqcap \tilde{A}^{(2)} \sqcap \tilde{A}^{(3)}$ can be directly defined by

$$\xi_{\tilde{A}^{\dagger}}(x) = \mathfrak{A}_3^{\cap}\left(\xi_{\tilde{A}^{(1)}}(x), \xi_{\tilde{A}^{(2)}}(x), \xi_{\tilde{A}^{(3)}}(x)\right). \tag{3.38}$$

Let us consider the following function

$$\mathfrak{A}_3^{\cap}\left(\alpha_1, \alpha_2, \alpha_3\right) = \min\left\{\alpha_1\alpha_2, \alpha_3\right\}.$$

We see that

$$\mathfrak{A}_3^{\sqcap}\left(\alpha_1,\alpha_2,\alpha_3\right)=\min\left\{\alpha_1\alpha_2,\alpha_3\right\}\neq\min\left\{\alpha_3\alpha_2,\alpha_1\right\}=\mathfrak{A}_3^{\sqcap}\left(\alpha_3,\alpha_2,\alpha_1\right).$$

It means that, in general, we have

$$\begin{aligned}\mathfrak{A}_3^{\sqcap}\left(\xi_{\tilde{A}^{(3)}}(x),\xi_{\tilde{A}^{(2)}}(x),\xi_{\tilde{A}^{(1)}}(x)\right)&\neq\mathfrak{A}_3^{\sqcap}\left(\xi_{\tilde{A}^{(2)}}(x),\xi_{\tilde{A}^{(3)}}(x),\xi_{\tilde{A}^{(1)}}(x)\right)\\&\neq\mathfrak{A}_3^{\sqcap}\left(\xi_{\tilde{A}^{(1)}}(x),\xi_{\tilde{A}^{(3)}}(x),\xi_{\tilde{A}^{(2)}}(x)\right).\end{aligned}$$

Therefore, we see that

$$\tilde{A}^{(3)}\sqcap\tilde{A}^{(2)}\sqcap\tilde{A}^{(1)}\neq\tilde{A}^{(2)}\sqcap\tilde{A}^{(3)}\sqcap\tilde{A}^{(1)}\neq\tilde{A}^{(1)}\sqcap\tilde{A}^{(3)}\sqcap\tilde{A}^{(2)},\tag{3.39}$$

where there are $3!=6$ permutations of $\tilde{A}^{(1)}$, $\tilde{A}^{(2)}$, and $\tilde{A}^{(3)}$. Suppose that the function $\mathfrak{A}_3^{\sqcap}$ is permutable in the sense of

$$\mathfrak{A}_3^{\sqcap}\left(\alpha_1,\alpha_2,\alpha_3\right)=\mathfrak{A}_3^{\sqcap}\left(\alpha_3,\alpha_2,\alpha_1\right)=\cdots=\mathfrak{A}_3^{\sqcap}\left(\alpha_1,\alpha_3,\alpha_2\right)$$

for all 6 permutations of α_1, α_2, and α_3. Then, the intersections in (3.39) are all identical. For example, the function

$$\mathfrak{A}_3^{\sqcap}\left(\alpha_1,\alpha_2,\alpha_3\right)=\min\left\{\alpha_1,\alpha_2,\alpha_3\right\}$$

is permutable for all 6 permutations.

- **(Inductive Definition)**. We first consider the intersection $\tilde{A}=\tilde{A}^{(1)}\sqcap\tilde{A}^{(2)}$ whose membership function is given by

$$\xi_{\tilde{A}}(x)=\xi_{\tilde{A}^{(1)}\sqcap\tilde{A}^{(2)}}(x)=\mathfrak{A}_2^{\sqcap}\left(\xi_{\tilde{A}^{(1)}}(x),\xi_{\tilde{A}^{(2)}}(x)\right).$$

Given a third fuzzy set $\tilde{A}^{(3)}$, the intersection can be defined as

$$\tilde{A}^{L}\equiv\left(\tilde{A}^{(1)}\sqcap\tilde{A}^{(2)}\right)\sqcap\tilde{A}^{(3)}=\tilde{A}\sqcap\tilde{A}^{(3)}$$

or

$$\tilde{A}^{R}\equiv\tilde{A}^{(1)}\sqcap\left(\tilde{A}^{(2)}\sqcap\tilde{A}^{(3)}\right)$$

whose membership functions are given by

$$\xi_{\tilde{A}^L}(x)=\mathfrak{A}_2^{\sqcap}\left(\xi_{\tilde{A}}(x),\xi_{\tilde{A}^{(3)}}(x)\right)=\mathfrak{A}_2^{\sqcap}\left(\mathfrak{A}_2^{\sqcap}\left(\xi_{\tilde{A}^{(1)}}(x),\xi_{\tilde{A}^{(2)}}(x)\right),\xi_{\tilde{A}^{(3)}}(x)\right)\tag{3.40}$$

and

$$\xi_{\tilde{A}^R}(x)=\mathfrak{A}_2^{\sqcap}\left(\xi_{\tilde{A}^{(1)}}(x),\mathfrak{A}_2^{\sqcap}\left(\xi_{\tilde{A}^{(2)}}(x),\xi_{\tilde{A}^{(3)}}(x)\right)\right),\ \text{respectively.}\tag{3.41}$$

In general, we see that $\tilde{A}^L\neq\tilde{A}^R$. Suppose that $\mathfrak{A}_2^{\sqcap}$ is associative in the following sense

$$\mathfrak{A}_2^{\sqcap}\left(\mathfrak{A}_2^{\sqcap}\left(\alpha_1,\alpha_2\right),\alpha_3\right)=\mathfrak{A}_2^{\sqcap}\left(\alpha_1,\mathfrak{A}_2^{\sqcap}\left(\alpha_2,\alpha_3\right)\right).$$

Then $\tilde{A}^L=\tilde{A}^R$. In this case, the intersection can be simply written as $\tilde{A}^{(1)}\sqcap\tilde{A}^{(2)}\sqcap\tilde{A}^{(3)}$.

By referring to (3.38), it is clear to see $\tilde{A}^L\neq\tilde{A}^\dagger$ and $\tilde{A}^R\neq\tilde{A}^\dagger$ in general, since $\tilde{A}^L$ and $\tilde{A}^R$ are obtained by using the function $\mathfrak{A}_2^{\sqcap}$, and $\tilde{A}^\dagger$ is obtained by using the function $\mathfrak{A}_3^{\sqcap}$. Their relations will be further studied in the remainder of this chapter.

Definition 3.4.1 Let $\mathfrak{A}_2:[0,1]^2\rightarrow[0,1]$ be a function. For $n\geq 3$, we define the function $\mathfrak{A}_n:[0,1]^n\rightarrow[0,1]$ based on $\mathfrak{A}_2$ as follows.

- We say that $\mathfrak{A}_n^L$ is **left-generated** by $\mathfrak{A}_2$ when $\mathfrak{A}_n^L$ is inductively defined as follows:

$$\mathfrak{A}_n^L\left(\alpha_1,\ldots,\alpha_n\right)=\mathfrak{A}_2\left(\mathfrak{A}_{n-1}^L\left(\alpha_1,\ldots,\alpha_{n-1}\right),\alpha_n\right). \tag{3.42}$$

- We say that $\mathfrak{A}_n^R$ is **right-generated** by $\mathfrak{A}_2$ when $\mathfrak{A}_n^R$ is inductively defined as follows:

$$\mathfrak{A}_n^R\left(\alpha_1,\ldots,\alpha_n\right)=\mathfrak{A}_2\left(\alpha_1,\mathfrak{A}_{n-1}^R\left(\alpha_2\cdots,\alpha_n\right)\right). \tag{3.43}$$

Assume that the function $\mathfrak{A}_2$ satisfies

$$\mathfrak{A}_2\left(\mathfrak{A}_2\left(\alpha_1,\alpha_2\right),\alpha_3\right)=\mathfrak{A}_2\left(\alpha_1,\mathfrak{A}_2\left(\alpha_2,\alpha_3\right)\right) \tag{3.44}$$

for all α_1, α_2, and $\alpha_3 \in [0,1]$; that is, the function $\mathfrak{A}_2$ is associative. Then, using induction on $n \geq 3$, it is clear to see that (3.42) and (3.43) are equivalent. In this case, we simply say that $\mathfrak{A}_n$ is **generated** by $\mathfrak{A}_2$.

Example 3.4.2 When we take $\mathfrak{A}_2(\alpha_1,\alpha_2)=\min\{\alpha_1,\alpha_2\}$, it is clear to see

$$\mathfrak{A}_n^L\left(\alpha_1,\ldots,\alpha_n\right)=\mathfrak{A}_n^R\left(\alpha_1,\ldots,\alpha_n\right)=\min\left\{\alpha_1,\ldots,\alpha_n\right\}.$$

When we take $\mathfrak{A}_2(\alpha_1,\alpha_2)=\max\{\alpha_1,\alpha_2\}$, it is clear to see

$$\mathfrak{A}_n^L\left(\alpha_1,\ldots,\alpha_n\right)=\mathfrak{A}_n^R\left(\alpha_1,\ldots,\alpha_n\right)=\max\left\{\alpha_1,\ldots,\alpha_n\right\}.$$

When we take $\mathfrak{A}_2(\alpha_1,\alpha_2)=\alpha_1\cdot\alpha_2$, it is clear to see

$$\mathfrak{A}_n^L\left(\alpha_1,\ldots,\alpha_n\right)=\mathfrak{A}_n^R\left(\alpha_1,\ldots,\alpha_n\right)=\alpha_1\cdot\alpha_2\cdots\alpha_n.$$

Example 3.4.3 We take

$$\mathfrak{A}_2\left(\alpha_1,\alpha_2\right)=\frac{\alpha_1\alpha_2}{\max\left\{\alpha_1,\alpha_2,\beta\right\}}\equiv\gamma \text{ for some constant } \beta\in[0,1].$$

Then

$$\begin{aligned}\mathfrak{A}_3^L\left(\alpha_1,\alpha_2,\alpha_3\right)&=\mathfrak{A}_2\left(\mathfrak{A}_2\left(\alpha_1,\alpha_2\right),\alpha_3\right)=\mathfrak{A}_2\left(\gamma,\alpha_3\right)=\frac{\gamma\alpha_3}{\max\left\{\gamma,\alpha_3,\beta\right\}}\\&=\frac{\alpha_1\alpha_2\alpha_3}{\max\left\{\alpha_1,\alpha_2,\beta\right\}\max\left\{\frac{\alpha_1\cdot\alpha_2}{\max\left\{\alpha_1,\alpha_2,\beta\right\}},\alpha_3,\beta\right\}}\\&=\frac{\alpha_1\alpha_2\alpha_3}{\max\left\{\alpha_1\alpha_2,\alpha_3\max\left\{\alpha_1,\alpha_2,\beta\right\},\beta\max\left\{\alpha_1,\alpha_2,\beta\right\}\right\}}\end{aligned}$$

and

$$\begin{aligned}\mathfrak{A}_3^R\left(\alpha_1,\alpha_2,\alpha_3\right)&=\mathfrak{A}_2\left(\alpha_1,\mathfrak{A}_2\left(\alpha_2,\alpha_3\right)\right)\\&=\frac{\alpha_1\alpha_2\alpha_3}{\max\left\{\alpha_2\alpha_3,\alpha_1\max\left\{\alpha_2,\alpha_3,\beta\right\},\beta\max\left\{\alpha_2,\alpha_3,\beta\right\}\right\}},\end{aligned}$$

which shows that $\mathfrak{A}_3^L(\alpha_1,\alpha_2,\alpha_3)\neq\mathfrak{A}_3^R(\alpha_1,\alpha_2,\alpha_3)$.

Let $\tilde{A}^{(1)}$, $\tilde{A}^{(2)}$, and $\tilde{A}^{(3)}$ be three fuzzy sets in $\mathbb{R}^m$. We are going to consider their intersection. Given a function $\mathfrak{A}_2^{\cap}:[0,1]^2\rightarrow[0,1]$, suppose that the function $\mathfrak{A}_3^{\cap,L}$ is left-generated by $\mathfrak{A}_2^{\cap}$, i.e.

$$\mathfrak{A}_3^{\cap,L}\left(\alpha_1,\alpha_2,\alpha_3\right)=\mathfrak{A}_2^{\cap}\left(\mathfrak{A}_2^{\cap}\left(\alpha_1,\alpha_2\right),\alpha_3\right).$$

Based on the function $\mathfrak{A}_2^{\cap}$, we can define the intersection $(\tilde{A}^{(1)}\sqcap\tilde{A}^{(2)})\sqcap\tilde{A}^{(3)}$ as described in (3.40). Based on the function $\mathfrak{A}_3^{\cap,L}$, by referring to the notations presented in (3.13), we can define the membership function of $\tilde{A}^{(1)} \sqcap \tilde{A}^{(2)} \sqcap \tilde{A}^{(3)}(\mathfrak{A}_3^{\cap,L})$ as described in (3.38). It is clear to see

$$\left(\tilde{A}^{(1)} \sqcap \tilde{A}^{(2)}\right) \sqcap \tilde{A}^{(3)} = \tilde{A}^{(1)} \sqcap \tilde{A}^{(2)} \sqcap \tilde{A}^{(3)}(\mathfrak{A}_3^{\cap,L}). \tag{3.45}$$

Suppose that the function $\mathfrak{A}_3^{\cap,R}$ is right-generated by $\mathfrak{A}_2^{\cap}$, i.e.

$$\mathfrak{A}_3^{\cap,R}\left(\alpha_1, \alpha_2, \alpha_3\right) = \mathfrak{A}_2^{\cap}\left(\alpha_1, \mathfrak{A}_2^{\cap}\left(\alpha_2, \alpha_3\right)\right).$$

Based on the function $\mathfrak{A}_2^{\cap}$, we can define the intersection $\tilde{A}^{(1)} \sqcap (\tilde{A}^{(2)} \sqcap \tilde{A}^{(3)})$ as described in (3.41). Based on the function $\mathfrak{A}_3^{\cap,R}$, we can define the membership function of $\tilde{A}^{(1)} \sqcap \tilde{A}^{(2)} \sqcap \tilde{A}^{(3)}(\mathfrak{A}_3^{\cap,R})$ as described in (3.38). It is clear to see

$$\tilde{A}^{(1)} \sqcap \left(\tilde{A}^{(2)} \sqcap \tilde{A}^{(3)}\right) = \tilde{A}^{(1)} \sqcap \tilde{A}^{(2)} \sqcap \tilde{A}^{(3)}(\mathfrak{A}_3^{\cap,R}). \tag{3.46}$$

The expressions in (3.45) and (3.46) are not necessarily identical. Suppose that we further assume that $\mathfrak{A}_2^{\cap}$ is associative in the sense of (3.44). Then, they are all identical, as shown below

$$\begin{aligned}\tilde{A}^{(1)} \sqcap \left(\tilde{A}^{(2)} \sqcap \tilde{A}^{(3)}\right) &= \left(\tilde{A}^{(1)} \sqcap \tilde{A}^{(2)}\right) \sqcap \tilde{A}^{(3)} = \tilde{A}^{(1)} \sqcap \tilde{A}^{(2)} \sqcap \tilde{A}^{(3)}(\mathfrak{A}_3^{\cap,L}) \\ &= \tilde{A}^{(1)} \sqcap \tilde{A}^{(2)} \sqcap \tilde{A}^{(3)}(\mathfrak{A}_3^{\cap,R}).\end{aligned}$$

In this case, we also simply write

$$\tilde{A}^{(1)} \sqcap \left(\tilde{A}^{(2)} \sqcap \tilde{A}^{(3)}\right) = \left(\tilde{A}^{(1)} \sqcap \tilde{A}^{(2)}\right) \sqcap \tilde{A}^{(3)} = \tilde{A}^{(1)} \sqcap \tilde{A}^{(2)} \sqcap \tilde{A}^{(3)} \tag{3.47}$$

without telling the difference between $\mathfrak{A}_3^{\cap,L}$ and $\mathfrak{A}_3^{\cap,R}$.

Suppose that there is another fuzzy set $\tilde{A}^{(4)}$ in $\mathbb{R}^m$. Then, we can consider the functions $\mathfrak{A}_4^{\cap,L}$ and $\mathfrak{A}_4^{\cap,R}$ that are left-generated and right-generated by $\mathfrak{A}_2$, respectively. More precisely, we have

$$\mathfrak{A}_4^{\cap,L}\left(\alpha_1, \alpha_2, \alpha_3, \alpha_4\right) = \mathfrak{A}_2^{\cap}\left(\mathfrak{A}_2^{\cap}\left(\mathfrak{A}_2^{\cap}\left(\alpha_1, \alpha_2\right), \alpha_3\right), \alpha_4\right)$$

and

$$\mathfrak{A}_4^{\cap,R}\left(\alpha_1, \alpha_2, \alpha_3, \alpha_4\right) = \mathfrak{A}_2^{\cap}\left(\alpha_1, \mathfrak{A}_2^{\cap}\left(\alpha_2, \mathfrak{A}_2^{\cap}\left(\alpha_3, \alpha_4\right)\right)\right).$$

Then, we can obtain

$$\left(\left(\tilde{A}^{(1)} \sqcap \tilde{A}^{(2)}\right) \sqcap \tilde{A}^{(3)}\right) \sqcap \tilde{A}^{(4)} = \tilde{A}^{(1)} \sqcap \tilde{A}^{(2)} \sqcap \tilde{A}^{(3)} \sqcap \tilde{A}^{(4)}(\mathfrak{A}_4^{\cap,L}) \tag{3.48}$$

and

$$\tilde{A}^{(1)} \sqcap \left(\tilde{A}^{(2)} \sqcap \left(\tilde{A}^{(3)} \sqcap \tilde{A}^{(4)}\right)\right) = \tilde{A}^{(1)} \sqcap \tilde{A}^{(2)} \sqcap \tilde{A}^{(3)} \sqcap \tilde{A}^{(4)}(\mathfrak{A}_4^{\cap,R}). \tag{3.49}$$

Suppose that we further assume that $\mathfrak{A}_2^{\cap}$ is associative. Then, the expressions in (3.48) and (3.49) are all identical. The above argument can be inductively extended to the case of $\tilde{A}^{(1)}, \ldots, \tilde{A}^{(n)}$.

Proposition 3.4.4 *We have the following properties.*

(i) *Given a function $\mathfrak{A}_2^{\cap}$, let the functions $\mathfrak{A}_n^{\cap,L}$ and $\mathfrak{A}_n^{\cap,R}$ be left-generated and right-generated by $\mathfrak{A}_2^{\cap}$, respectively.*
   - *Suppose that $\mathfrak{A}_2^{\cap}$ is $\subseteq$-compatible with set intersection. Then $\mathfrak{A}_n^{\cap,L}$ and $\mathfrak{A}_n^{\cap,R}$ are also $\subseteq$-compatible with set intersection.*

- *Suppose that $\mathfrak{A}_2^{\cap}$ is $\supseteq$-compatible with set intersection. Then $\mathfrak{A}_n^{\cap,L}$ and $\mathfrak{A}_n^{\cap,R}$ are also $\supseteq$-compatible with set intersection.*

(ii) *Given a function $\mathfrak{A}_2^{\cup}$, let the functions $\mathfrak{A}_n^{\cup,L}$ and $\mathfrak{A}_n^{\cup,R}$ be left-generated and right-generated by $\mathfrak{A}_2^{\cup}$, respectively.*
- *Suppose that $\mathfrak{A}_2^{\cup}$ is $\subseteq$-compatible with set union. Then $\mathfrak{A}_n^{\cup,L}$ and $\mathfrak{A}_n^{\cup,R}$ are also $\subseteq$-compatible with set union.*
- *Suppose that $\mathfrak{A}_2^{\cup}$ is $\supseteq$-compatible with set union. Then $\mathfrak{A}_n^{\cup,L}$ and $\mathfrak{A}_n^{\cup,R}$ are also $\supseteq$-compatible with set union.*

Proof. By induction on n, it suffices to prove the case of $n = 3$. To prove part (i), suppose that $\mathfrak{A}_2^{\cap}$ is $\subseteq$-compatible with set intersection. Then, for $\alpha \in (0,1]$, we have

$$
\begin{aligned}
&\left\{x \in \mathbb{R}^m : \mathfrak{A}_2^{\cap}\left(\xi_{\tilde{A}^{(1)}}(x), \xi_{\tilde{A}^{(2)}}(x)\right) \geq \alpha\right\} \\
&= \left(\tilde{A}^{(1)} \sqcap \tilde{A}^{(2)}\right)_\alpha \text{ (by the definition of } \alpha\text{-level set)} \\
&\subseteq \tilde{A}_\alpha^{(1)} \cap \tilde{A}_\alpha^{(2)} \text{ (by the } \subseteq\text{-compatibility with set inetersection)} \\
&= \left\{x \in \mathbb{R}^m : x \in \tilde{A}_\alpha^{(1)} \text{ and } x \in \tilde{A}_\alpha^{(2)}\right\} = \left\{x \in \mathbb{R}^m : \xi_{\tilde{A}^{(1)}}(x) \geq \alpha \text{ and } \xi_{\tilde{A}^{(2)}}(x) \geq \alpha\right\}.
\end{aligned}
\tag{3.50}
$$

Therefore, we obtain

$$
\begin{aligned}
&\left(\tilde{A}^{(1)} \sqcap \tilde{A}^{(2)} \sqcap \tilde{A}^{(3)}(\mathfrak{A}_3^{\cap,L})\right)_\alpha = \left\{x \in \mathbb{R}^m : \mathfrak{A}_3^{\cap,L}\left(\xi_{\tilde{A}^{(1)}}(x), \xi_{\tilde{A}^{(2)}}(x), \xi_{\tilde{A}^{(3)}}(x)\right) \geq \alpha\right\} \\
&= \left\{x \in \mathbb{R}^m : \mathfrak{A}_2^{\cap}\left(\mathfrak{A}_2^{\cap}\left(\xi_{\tilde{A}^{(1)}}(x), \xi_{\tilde{A}^{(2)}}(x)\right), \xi_{\tilde{A}^{(3)}}(x)\right) \geq \alpha\right\} \\
&\subseteq \left\{x \in \mathbb{R}^m : \mathfrak{A}_2^{\cap}\left(\xi_{\tilde{A}^{(1)}}(x), \xi_{\tilde{A}^{(2)}}(x)\right) \geq \alpha \text{ and } \xi_{\tilde{A}^{(3)}}(x) \geq \alpha\right\} \text{ (using (3.50))} \\
&\subseteq \left\{x \in \mathbb{R}^m : \xi_{\tilde{A}^{(1)}}(x) \geq \alpha \text{ and } \xi_{\tilde{A}^{(2)}}(x) \geq \alpha \text{ and } \xi_{\tilde{A}^{(3)}}(x) \geq \alpha\right\} \text{ (using (3.50) again)} \\
&= \tilde{A}_\alpha^{(1)} \cap \tilde{A}_\alpha^{(2)} \cap \tilde{A}_\alpha^{(3)}
\end{aligned}
$$

and

$$
\begin{aligned}
&\left(\tilde{A}^{(1)} \sqcap \tilde{A}^{(2)} \sqcap \tilde{A}^{(3)}(\mathfrak{A}_3^{\cap,R})\right)_\alpha = \left\{x \in \mathbb{R}^m : \mathfrak{A}_3^{\cap,R}\left(\xi_{\tilde{A}^{(1)}}(x), \xi_{\tilde{A}^{(2)}}(x), \xi_{\tilde{A}^{(3)}}(x)\right) \geq \alpha\right\} \\
&= \left\{x \in \mathbb{R}^m : \mathfrak{A}_2^{\cap}\left(\xi_{\tilde{A}^{(1)}}(x), \mathfrak{A}_2^{\cap}\left(\xi_{\tilde{A}^{(2)}}(x), \xi_{\tilde{A}^{(3)}}(x)\right)\right) \geq \alpha\right\} \\
&\subseteq \left\{x \in \mathbb{R}^m : \xi_{\tilde{A}^{(1)}}(x) \geq \alpha \text{ and } \mathfrak{A}_2^{\cap}\left(\xi_{\tilde{A}^{(2)}}(x), \xi_{\tilde{A}^{(3)}}(x)\right) \geq \alpha\right\} \text{ (using (3.50))} \\
&\subseteq \left\{x \in \mathbb{R}^m : \xi_{\tilde{A}^{(1)}}(x) \geq \alpha \text{ and } \xi_{\tilde{A}^{(2)}}(x) \geq \alpha \text{ and } \xi_{\tilde{A}^{(3)}}(x) \geq \alpha\right\} \text{ (using (3.50) again)} \\
&= \tilde{A}_\alpha^{(1)} \cap \tilde{A}_\alpha^{(2)} \cap \tilde{A}_\alpha^{(3)},
\end{aligned}
$$

which show that $\mathfrak{A}_n^{\cap,L}$ and $\mathfrak{A}_n^{\cap,R}$ are also $\subseteq$-compatible with set intersection. We can similarly show that if $\mathfrak{A}_2^{\cap}$ is $\supseteq$-compatible with set intersection, then $\mathfrak{A}_n^{\cap,L}$ and $\mathfrak{A}_n^{\cap,R}$ are also $\supseteq$-compatible with set intersection.

To prove part (ii), suppose that u_2 is $\subseteq$-compatible with set union. Then, for $\alpha \in (0,1]$, we have

$$
\begin{aligned}
\left\{x \in \mathbb{R}^m : \mathfrak{A}_2^{\cup}\left(\xi_{\tilde{A}^{(1)}}(x), \xi_{\tilde{A}^{(2)}}(x)\right) \geq \alpha\right\} &= \left(\tilde{A}^{(1)} \sqcup \tilde{A}^{(2)}\right)_\alpha \subseteq \tilde{A}_\alpha^{(1)} \cup \tilde{A}_\alpha^{(2)} \\
&= \left\{x \in \mathbb{R}^m : \xi_{\tilde{A}^{(1)}}(x) \geq \alpha \text{ or } \xi_{\tilde{A}^{(2)}}(x) \geq \alpha\right\}.
\end{aligned}
\tag{3.51}
$$

Therefore, we obtain

$$\left(\tilde{A}^{(1)} \sqcup \tilde{A}^{(2)} \sqcup \tilde{A}^{(3)}(u_3^{\cup,L})\right)_\alpha = \left\{x \in \mathbb{R}^m : u_3^{\cup,L}\left(\xi_{\tilde{A}^{(1)}}(x), \xi_{\tilde{A}^{(2)}}(x), \xi_{\tilde{A}^{(3)}}(x)\right) \geq \alpha\right\}$$
$$= \left\{x \in \mathbb{R}^m : \mathfrak{A}_2^{\cup}\left(\mathfrak{A}_2^{\cup}\left(\xi_{\tilde{A}^{(1)}}(x), \xi_{\tilde{A}^{(2)}}(x)\right), \xi_{\tilde{A}^{(3)}}(x)\right) \geq \alpha\right\}$$
$$\subseteq \left\{x \in \mathbb{R}^m : \mathfrak{A}_2^{\cup}\left(\xi_{\tilde{A}^{(1)}}(x), \xi_{\tilde{A}^{(2)}}(x)\right) \geq \alpha \text{ or } \xi_{\tilde{A}^{(3)}}(x) \geq \alpha\right\} \text{ (using (3.51))}$$
$$\subseteq \left\{x \in \mathbb{R}^m : \xi_{\tilde{A}^{(1)}}(x) \geq \alpha \text{ or } \xi_{\tilde{A}^{(2)}}(x) \geq \alpha \text{ or } \xi_{\tilde{A}^{(3)}}(x) \geq \alpha\right\} \text{ (using (3.51) again)}$$
$$= \tilde{A}_\alpha^{(1)} \cup \tilde{A}_\alpha^{(2)} \cup \tilde{A}_\alpha^{(3)}$$

and

$$\left(\tilde{A}^{(1)} \sqcup \tilde{A}^{(2)} \sqcup \tilde{A}^{(3)}(u_3^{\cup,R})\right)_\alpha = \left\{x \in \mathbb{R}^m : u_3^{\cup,R}\left(\xi_{\tilde{A}^{(1)}}(x), \xi_{\tilde{A}^{(2)}}(x), \xi_{\tilde{A}^{(3)}}(x)\right) \geq \alpha\right\}$$
$$= \left\{x \in \mathbb{R}^m : \mathfrak{A}_2^{\cup}\left(\xi_{\tilde{A}^{(1)}}(x), \mathfrak{A}_2^{\cup}\left(\xi_{\tilde{A}^{(2)}}(x), \xi_{\tilde{A}^{(3)}}(x)\right)\right) \geq \alpha\right\}$$
$$\subseteq \left\{x \in \mathbb{R}^m : \xi_{\tilde{A}^{(1)}}(x) \geq \alpha \text{ or } \mathfrak{A}_2^{\cup}\left(\xi_{\tilde{A}^{(2)}}(x), \xi_{\tilde{A}^{(3)}}(x)\right) \geq \alpha\right\} \text{ (using (3.51))}$$
$$\subseteq \left\{x \in \mathbb{R}^m : \xi_{\tilde{A}^{(1)}}(x) \geq \alpha \text{ or } \xi_{\tilde{A}^{(2)}}(x) \geq \alpha \text{ or } \xi_{\tilde{A}^{(3)}}(x) \geq \alpha\right\} \text{ (using (3.51) again)}$$
$$= \tilde{A}_\alpha^{(1)} \cup \tilde{A}_\alpha^{(2)} \cup \tilde{A}_\alpha^{(3)},$$

which show that $\mathfrak{A}_n^{\cap,L}$ and $\mathfrak{A}_n^{\cap,R}$ are also $\subseteq$-compatible with set union. We can similarly show that if $\mathfrak{A}_2^{\cap}$ is $\supseteq$-compatible with set union, then $\mathfrak{A}_n^{\cap,L}$ and $\mathfrak{A}_n^{\cap,R}$ are also $\supseteq$-compatible with set union. This completes the proof. ■

3.5 α-Level Sets of Intersection and Union

Let A and B be two subsets of $\mathbb{R}^m$. Then, we have the following properties

$$\text{cl}(A \cup B) = \text{cl}(A) \cup \text{cl}(B) \text{ and } \text{cl}(A \cap B) \subseteq \text{cl}(A) \cap \text{cl}(B).$$

Therefore, for any subsets $A_1, \ldots, A_n$ of $\mathbb{R}^m$, we have

$$\text{cl}\left(A_1 \cup \ldots \cup A_n\right) = \text{cl}(A_1) \cup \ldots \cup \text{cl}(A_n) \tag{3.52}$$

and

$$\text{cl}\left(A_1 \cap \ldots \cap A_n\right) \subseteq \text{cl}(A_1) \cap \ldots \cap \text{cl}(A_n). \tag{3.53}$$

However, when $A_1, \ldots, A_n$ are bounded intervals in $\mathbb{R}$, we can show that

$$\text{cl}\left(A_1 \cap \ldots \cap A_n\right) = \text{cl}(A_1) \cap \ldots \cap \text{cl}(A_n). \tag{3.54}$$

We remark that the compatibility in Definition 3.3.7 does not consider the 0-level sets. As we mentioned before, the 0-level set is an important case, so the following propositions present the compatibility regarding the 0-level sets.

Proposition 3.5.1 *Let $\tilde{A}^{(1)}, \ldots, \tilde{A}^{(n)}$ be fuzzy sets in $\mathbb{R}^m$. We consider the function $\mathfrak{A}^{\cup} : [0,1]^n \to [0,1]$.*

(i) *Given any $\alpha_i \in [0,1]$ for $i = 1, \ldots, n$, suppose that*

$$\alpha_i > 0 \textit{ for some } i \textit{ imply } \mathfrak{A}^{\cup}(\alpha_1, \ldots, \alpha_n) > 0.$$

Then, we have the following inclusions

$$\left(\tilde{A}^{(1)} \sqcup \ldots \sqcup \tilde{A}^{(n)}\right)_{0+} \supseteq \tilde{A}^{(1)}_{0+} \cup \ldots \cup \tilde{A}^{(n)}_{0+}$$

and

$$\left(\tilde{A}^{(1)} \sqcup \ldots \sqcup \tilde{A}^{(n)}\right)_{0} \supseteq \tilde{A}^{(1)}_{0} \cup \ldots \cup \tilde{A}^{(n)}_{0}.$$

(ii) *Suppose that $\mathfrak{A}^{\cup}$ is $\subseteq$-compatible with set union. Then, we have the following inclusions*

$$\left(\tilde{A}^{(1)} \sqcup \ldots \sqcup \tilde{A}^{(n)}\right)_{0+} \subseteq \tilde{A}^{(1)}_{0+} \cup \ldots \cup \tilde{A}^{(n)}_{0+} \tag{3.55}$$

and

$$\left(\tilde{A}^{(1)} \sqcup \ldots \sqcup \tilde{A}^{(n)}\right)_{0} \subseteq \tilde{A}^{(1)}_{0} \cup \ldots \cup \tilde{A}^{(n)}_{0}.$$

(iii) *Suppose that $\mathfrak{A}^{\cup}$ is $\supseteq$-compatible with set union. Then, we have the following inclusions*

$$\left(\tilde{A}^{(1)} \sqcup \ldots \sqcup \tilde{A}^{(n)}\right)_{0+} \supseteq \tilde{A}^{(1)}_{0+} \cup \ldots \cup \tilde{A}^{(n)}_{0+}$$

and

$$\left(\tilde{A}^{(1)} \sqcup \ldots \sqcup \tilde{A}^{(n)}\right)_{0} \supseteq \tilde{A}^{(1)}_{0} \cup \ldots \cup \tilde{A}^{(n)}_{0}.$$

Proof. To prove part (i), for $x \in \tilde{A}^{(1)}_{0+} \cup \ldots \cup \tilde{A}^{(n)}_{0+}$, i.e. $x \in \tilde{A}^{(i)}_{0+}$ for some i, it says that $\xi_{\tilde{A}^{(i)}}(x) > 0$ for some i. According to (3.29) and the assumption of $\mathfrak{A}^{\cup}$, we see that

$$\xi_{\sqcup_{i=1}^{n}\tilde{A}^{(i)}}(x) = \mathfrak{A}^{\cup}\left(\xi_{\tilde{A}^{(1)}}(x), \cdots, \xi_{\tilde{A}^{(n)}}(x)\right) > 0,$$

i.e. $x \in (\sqcup_{i=1}^{n}\tilde{A}^{(i)})_{0+}$, which shows the inclusion

$$\tilde{A}^{(1)}_{0+} \cup \ldots \cup \tilde{A}^{(n)}_{0+} \subseteq \left(\sqcup_{i=1}^{n}\tilde{A}^{(i)}\right)_{0+}. \tag{3.56}$$

We also have

$$\begin{aligned}
\left(\sqcup_{i=1}^{n}\tilde{A}^{(i)}\right)_{0} &= \mathrm{cl}\left(\left(\sqcup_{i=1}^{n}\tilde{A}^{(i)}\right)_{0+}\right) \supseteq \mathrm{cl}\left(\bigcup_{i=1}^{n}\tilde{A}^{(i)}_{0+}\right) \text{ (using (3.56))} \\
&= \mathrm{cl}\left(\tilde{A}^{(1)}_{0+}\right) \cup \ldots \cup \mathrm{cl}\left(\tilde{A}^{(n)}_{0+}\right) \text{ (using (3.52))} \\
&= \tilde{A}^{(1)}_{0} \cup \ldots \cup \tilde{A}^{(n)}_{0}.
\end{aligned} \tag{3.57}$$

To prove part (ii), we have

$$\begin{aligned}
\left(\sqcup_{i=1}^{n}\tilde{A}^{(i)}\right)_{0+} &= \bigcup_{0<\alpha\leq 1}\left(\sqcup_{i=1}^{n}\tilde{A}^{(i)}\right)_{\alpha} \text{ (using (2.5))} \\
&\subseteq \bigcup_{0<\alpha\leq 1}\left(\bigcup_{i=1}^{n}\tilde{A}^{(i)}_{\alpha}\right) \text{ (using the } \subseteq \text{-compatiblity with set union)} \\
&= \left(\bigcup_{0<\alpha\leq 1}\tilde{A}^{(1)}_{\alpha}\right) \bigcup \cdots \bigcup \left(\bigcup_{0<\alpha\leq 1}\tilde{A}^{(n)}_{\alpha}\right) \\
&= \tilde{A}^{(1)}_{0+} \cup \ldots \cup \tilde{A}^{(n)}_{0+} \text{ (using (2.5))},
\end{aligned}$$

which proves (3.55). By referring to (3.57), we can similarly obtain

$$\left(\sqcup_{i=1}^{n}\tilde{A}^{(i)}\right)_{0} \subseteq \tilde{A}^{(1)}_{0} \cup \ldots \cup \tilde{A}^{(n)}_{0}.$$

Part (iii) can be similarly obtained by using the ⊇-compatibility with set union. This completes the proof. ■

Proposition 3.5.2 *Let $\tilde{A}^{(1)}, \dots, \tilde{A}^{(n)}$ be fuzzy sets in $\mathbb{R}^m$. Then, we have*

$$\left(\tilde{A}^{(1)} \vee \dots \vee \tilde{A}^{(n)}\right)_{0+} = \tilde{A}^{(1)}_{0+} \cup \dots \cup \tilde{A}^{(n)}_{0+}$$

and

$$\left(\tilde{A}^{(1)} \vee \dots \vee \tilde{A}^{(n)}\right)_{\alpha} = \tilde{A}^{(1)}_{\alpha} \cup \dots \cup \tilde{A}^{(n)}_{\alpha} \textit{ for any } \alpha \in [0,1].$$

Proof. We are going to apply Proposition 3.5.1 by taking the function

$$\mathfrak{A}^{\cup}\left(\alpha_1, \dots, \alpha_n\right) = \max\left\{\alpha_1, \dots, \alpha_n\right\}.$$

From (3.24), we see that $\mathfrak{A}^{\cup}$ is both ⊆-compatible and ⊇-compatible with set union and

$$\left(\tilde{A}^{(1)} \vee \dots \vee \tilde{A}^{(n)}\right)_{\alpha} = \tilde{A}^{(1)}_{\alpha} \cup \dots \cup \tilde{A}^{(n)}_{\alpha} \text{ for any } \alpha \in (0,1].$$

The equality for 0-level sets can be obtained from parts (ii) and (iii) of Proposition 3.5.1, and the proof is complete. ■

Proposition 3.5.3 *Let $\tilde{A}^{(1)}, \dots, \tilde{A}^{(n)}$ be fuzzy sets in $\mathbb{R}^m$. We consider the function $\mathfrak{A}^{\cap} : [0,1]^n \to [0,1]$.*

(i) *Given any $\alpha_i \in [0,1]$ for $i = 1, \dots, n$, suppose that*

$$\alpha_i > 0 \textit{ for all } i \textit{ imply } \mathfrak{A}^{\cap}(\alpha_1, \dots, \alpha_n) > 0.$$

Then, we have the following inclusion

$$\left(\tilde{A}^{(1)} \sqcap \dots \sqcap \tilde{A}^{(n)}\right)_{0+} \supseteq \tilde{A}^{(1)}_{0+} \cap \dots \cap \tilde{A}^{(n)}_{0+}.$$

(ii) *Suppose that $\mathfrak{A}^{\cap}$ is ⊆-compatible with set intersection. Then, we have the following inclusions*

$$\left(\tilde{A}^{(1)} \sqcap \dots \sqcap \tilde{A}^{(n)}\right)_{0+} \subseteq \tilde{A}^{(1)}_{0+} \cap \dots \cap \tilde{A}^{(n)}_{0+} \tag{3.58}$$

and

$$\left(\tilde{A}^{(1)} \sqcap \dots \sqcap \tilde{A}^{(n)}\right)_{0} \subseteq \tilde{A}^{(1)}_{0} \cap \dots \cap \tilde{A}^{(n)}_{0}. \tag{3.59}$$

(iii) *Suppose that $\mathfrak{A}^{\cap}$ is ⊇-compatible with set intersection. Then, we have the following inclusion*

$$\left(\tilde{A}^{(1)} \sqcap \dots \sqcap \tilde{A}^{(n)}\right)_{0+} \supseteq \tilde{A}^{(1)}_{0+} \cap \dots \cap \tilde{A}^{(n)}_{0+}.$$

(iv) *Suppose that the function $\mathfrak{A}^{\cap}$ satisfies all conditions in parts* (i) *and* (ii). *We also assume that $\tilde{A}^{(i)}$ are fuzzy sets in $\mathbb{R}$ for $i = 1, \dots, n$ such that their membership functions are quasi-concave. Then, we have*

$$\left(\tilde{A}^{(1)} \sqcap \dots \sqcap \tilde{A}^{(n)}\right)_{0+} = \tilde{A}^{(1)}_{0+} \cap \dots \cap \tilde{A}^{(n)}_{0+} \tag{3.60}$$

and

$$\left(\tilde{A}^{(1)} \sqcap \dots \sqcap \tilde{A}^{(n)}\right)_{0} = \tilde{A}^{(1)}_{0} \cap \dots \cap \tilde{A}^{(n)}_{0}.$$

Proof. To prove part (i), for $x \in \tilde{A}^{(1)}_{0+} \cap \ldots \cap \tilde{A}^{(n)}_{0+}$, i.e. $x \in \tilde{A}^{(i)}_{0+}$ for each i, it says that $\xi_{\tilde{A}^{(i)}}(x) > 0$ for each i. According to (3.12) and the assumption of $\mathfrak{A}^{\cap}$, we see that

$$\xi_{\sqcap_{i=1}^{n}\tilde{A}^{(i)}}(x) = \mathfrak{A}^{\cap}\left(\xi_{\tilde{A}^{(1)}}(x), \ldots, \xi_{\tilde{A}^{(n)}}(x)\right) > 0,$$

i.e. $x \in (\sqcap_{i=1}^{n}\tilde{A}^{(i)})_{0+}$, which shows the inclusion

$$\tilde{A}^{(1)}_{0+} \cap \cdots \cup \tilde{A}^{(n)}_{0+} \subseteq \left(\sqcap_{i=1}^{n}\tilde{A}^{(i)}\right)_{0+}.$$

To prove part (ii), the following inclusion is obvious

$$\bigcup_{0<\alpha\leq 1}\left(\bigcap_{i=1}^{n}\tilde{A}^{(i)}_{\alpha}\right) \subseteq \left(\bigcup_{0<\alpha\leq 1}\tilde{A}^{(1)}_{\alpha}\right) \cap \cdots \cap \left(\bigcup_{0<\alpha\leq 1}\tilde{A}^{(n)}_{\alpha}\right).$$

The other direction of inclusion needs to consider the nested property of the α-level sets. Given any $x \in \tilde{A}^{(i)}_{\alpha_i}$ with $\alpha_i \in (0,1]$ for all i, let $\alpha_0 = \min\{\alpha_1, \ldots, \alpha_n\} > 0$. Then $\tilde{A}^{(i)}_{\alpha_i} \subseteq \tilde{A}^{(i)}_{\alpha_0}$ for all i, i.e. $x \in \bigcap_{i=1}^{n}\tilde{A}^{(i)}_{\alpha_0}$, which shows the other direction of inclusion. Therefore, we have the following equality

$$\bigcup_{\alpha(0,1]}\left(\bigcap_{i=1}^{n}\tilde{A}^{(i)}_{\alpha}\right) = \left(\bigcup_{\alpha(0,1]}\tilde{A}^{(1)}_{\alpha}\right) \cap \cdots \cap \left(\bigcup_{\alpha(0,1]}\tilde{A}^{(n)}_{\alpha}\right). \tag{3.61}$$

Now, we obtain

$$\begin{aligned}
\left(\sqcap_{i=1}^{n}\tilde{A}^{(i)}\right)_{0+} &\subseteq \bigcup_{\alpha(0,1]}\left(\sqcap_{i=1}^{n}\tilde{A}^{(i)}\right)_{\alpha} \text{ (using (2.5))}\\
&\subseteq \bigcup_{\alpha(0,1]}\left(\bigcap_{i=1}^{n}\tilde{A}^{(i)}_{\alpha}\right) \text{ (using the } \subseteq \text{-compatiblity with set intersection))}\\
&= \left(\bigcup_{\alpha(0,1]}\tilde{A}^{(1)}_{\alpha}\right) \cap \cdots \cap \left(\bigcup_{\alpha(0,1]}\tilde{A}^{(n)}_{\alpha}\right) \text{ (using (3.61))}\\
&= \tilde{A}^{(1)}_{0+} \cap \ldots \cap \tilde{A}^{(n)}_{0+} \text{ (using (2.5))},
\end{aligned} \tag{3.62}$$

which proves (3.58). We also have

$$\begin{aligned}
\left(\sqcap_{i=1}^{n}\tilde{A}^{(i)}\right)_{0} &= \text{cl}\left(\left(\sqcap_{i=1}^{n}\tilde{A}^{(i)}\right)_{0+}\right) \subseteq \text{cl}\left(\bigcap_{i=1}^{n}\tilde{A}^{(i)}_{0+}\right) \text{ (using (3.62))}\\
&\subseteq \text{cl}\left(\tilde{A}^{(1)}_{0+}\right) \cap \ldots \cap \text{cl}\left(\tilde{A}^{(n)}_{0+}\right) \text{ (using (3.53))}\\
&= \tilde{A}^{(1)}_{0} \cap \ldots \cap \tilde{A}^{(n)}_{0}.
\end{aligned}$$

Part (iii) can be similarly obtained by using the $\supseteq$-compatibility with set intersection.

To prove part (iv), we first have the equality (3.60) by using parts (i) and (ii). The quasi-concavity of the membership function says that its α-level sets are convex sets. Therefore, the α-level sets $\tilde{A}^{(i)}_{\alpha}$ are convex sets in $\mathbb{R}$ for all i, which also says that $\tilde{A}^{(i)}_{\alpha}$ are intervals in $\mathbb{R}$ for all i. In this case, the supports

$$\tilde{A}^{(i)}_{0+} = \bigcup_{\{\alpha\in\mathcal{R}(\xi_{\tilde{A}^{(i)}}):\alpha>0\}} \tilde{A}^{(i)}_{\alpha}$$

are also intervals in $\mathbb{R}$ for all i. Then, we obtain

$$\left(\sqcap_{i=1}^{n}\tilde{A}^{(i)}\right)_0 = \text{cl}\left(\left(\sqcap_{i=1}^{n}\tilde{A}^{(i)}\right)_{0+}\right) = \text{cl}\left(\bigcap_{i=1}^{n}\tilde{A}_{0+}^{(i)}\right) \text{ (using (3.60))}$$
$$= \text{cl}\left(\tilde{A}_{0+}^{(1)}\right) \cap \ldots \cap \text{cl}\left(\tilde{A}_{0+}^{(n)}\right) \text{ (using (3.54))}$$
$$= \tilde{A}_0^{(1)} \cap \ldots \cap \tilde{A}_0^{(n)}.$$

This completes the proof. ∎

Proposition 3.5.4 *Let $\tilde{A}^{(1)}, \ldots, \tilde{A}^{(n)}$ be fuzzy sets in $\mathbb{R}^m$. Then, we have the following properties.*

(i) *We have*

$$\left(\tilde{A}^{(1)} \wedge \ldots \wedge \tilde{A}^{(n)}\right)_{0+} = \tilde{A}_{0+}^{(1)} \cap \ldots \cap \tilde{A}_{0+}^{(n)}.$$

For any $\alpha \in (0,1]$, we have

$$\left(\tilde{A}^{(1)} \wedge \ldots \wedge \tilde{A}^{(n)}\right)_{\alpha} = \tilde{A}_{\alpha}^{(1)} \cap \ldots \cap \tilde{A}_{\alpha}^{(n)} \tag{3.63}$$

and

$$\left(\tilde{A}^{(1)} \wedge \ldots \wedge \tilde{A}^{(n)}\right)_{0} \subseteq \tilde{A}_{0}^{(1)} \cap \ldots \cap \tilde{A}_{0}^{(n)}.$$

(ii) *Assume that $\tilde{A}^{(i)}$ are fuzzy sets in $\mathbb{R}$ for $i = 1, \ldots, n$ such that their membership functions are quasi-concave. Then, for any $\alpha \in [0,1]$, we have*

$$\left(\tilde{A}^{(1)} \wedge \ldots \wedge \tilde{A}^{(n)}\right)_{\alpha} = \tilde{A}_{\alpha}^{(1)} \cap \ldots \cap \tilde{A}_{\alpha}^{(n)}.$$

Proof. We are going to use Proposition 3.5.3 by taking

$$\mathfrak{A}^{\cap}\left(\alpha_1, \ldots, \alpha_n\right) = \min\left\{\alpha_1, \ldots, \alpha_n\right\}.$$

It is obvious that the condition in part (i) of Proposition 3.5.3 is satisfied. From (3.8), we also see that $\mathfrak{A}^{\cap}$ is both $\subseteq$-compatible and $\supseteq$-compatible with set intersection such that the equality (3.63) is satisfied for any $\alpha \in (0,1]$. This completes the proof. ∎

3.6 Mixed Set Operations

Next, we consider the mixed set operations. Suppose that we have

$$\left(\left(\tilde{A}^{(1)} \sqcap \tilde{A}^{(2)} \sqcap \tilde{A}^{(3)}\right) \sqcup \left(\tilde{A}^{(4)} \sqcap \tilde{A}^{(5)}\right)\right) \sqcap \left(\tilde{A}^{(6)} \sqcup \tilde{A}^{(7)} \sqcup \tilde{A}^{(8)} \sqcup \tilde{A}^{(9)}\right).$$

Then, we have two ways to define the above expression.

We present the first way as follows by independently defining the functions $\mathfrak{A}_2^{\cap}, \mathfrak{A}_3^{\cap}, \mathfrak{A}_2^{\cup}, \mathfrak{A}_4^{\cup}$. Then, we can consider the following operations.

- The intersection $\tilde{B}^{(1)} \equiv \tilde{A}^{(1)} \sqcap \tilde{A}^{(2)} \sqcap \tilde{A}^{(3)}$ uses $\mathfrak{A}_3^{\cap}$.
- The intersection $\tilde{B}^{(2)} \equiv \tilde{A}^{(4)} \sqcap \tilde{A}^{(5)}$ uses $\mathfrak{A}_2^{\cap}$.

- The union $\tilde{B}^{(3)} \equiv \tilde{A}^{(6)} \sqcup \tilde{A}^{(7)} \sqcup \tilde{A}^{(8)} \sqcup \tilde{A}^{(9)}$ uses $\mathfrak{A}_4^{\cup}$.
- The union $\tilde{B}^{(4)} \equiv \tilde{B}^{(1)} \sqcup \tilde{B}^{(2)}$ uses $\mathfrak{A}_2^{\cup}$.

Finally, the expression $\tilde{B}^{(4)} \sqcap \tilde{B}^{(3)}$ uses $\mathfrak{A}_2^{\cap}$. More precisely, the membership function is given by

$$\mathfrak{A}_2^{\cap}\left(\mathfrak{A}_2^{\cup}\left(\mathfrak{A}_3^{\cap}\left(\xi_{\tilde{A}^{(1)}},\xi_{\tilde{A}^{(2)}},\xi_{\tilde{A}^{(3)}}\right),\mathfrak{A}_2^{\cap}\left(\xi_{\tilde{A}^{(4)}},\xi_{\tilde{A}^{(5)}}\right)\right),\mathfrak{A}_4^{\cup}\left(\xi_{\tilde{A}^{(6)}},\xi_{\tilde{A}^{(7)}},\xi_{\tilde{A}^{(8)}},\xi_{\tilde{A}^{(9)}}\right)\right). \tag{3.64}$$

Now, we present the second way as follows. Given aggregation functions $\mathfrak{A}_2^{\cap}$ and $\mathfrak{A}_2^{\cup}$, we can consider the functions $\mathfrak{A}_3^{\cap,L}$, $\mathfrak{A}_3^{\cap,R}$ that are left-generated and right-generated by $\mathfrak{A}_2^{\cap}$, respectively, and the functions $\mathfrak{A}_4^{\cup,L}$, $\mathfrak{A}_4^{\cup,R}$ that are left-generated and right-generated by $\mathfrak{A}_2^{\cup}$, respectively. Then, the membership functions of the expression have the following different combinations.

$$\begin{aligned}
&\mathfrak{A}_2^{\cap}\left(\mathfrak{A}_2^{\cup}\left(\mathfrak{A}_3^{\cap,L}\left(\xi_{\tilde{A}^{(1)}},\xi_{\tilde{A}^{(2)}},\xi_{\tilde{A}^{(3)}}\right),\mathfrak{A}_2^{\cap}\left(\xi_{\tilde{A}^{(4)}},\xi_{\tilde{A}^{(5)}}\right)\right),\mathfrak{A}_4^{\cup,L}\left(\xi_{\tilde{A}^{(6)}},\xi_{\tilde{A}^{(7)}},\xi_{\tilde{A}^{(8)}},\xi_{\tilde{A}^{(9)}}\right)\right)\\
&\mathfrak{A}_2^{\cap}\left(\mathfrak{A}_2^{\cup}\left(\mathfrak{A}_3^{\cap,L}\left(\xi_{\tilde{A}^{(1)}},\xi_{\tilde{A}^{(2)}},\xi_{\tilde{A}^{(3)}}\right),\mathfrak{A}_2^{\cap}\left(\xi_{\tilde{A}^{(4)}},\xi_{\tilde{A}^{(5)}}\right)\right),\mathfrak{A}_4^{\cup,R}\left(\xi_{\tilde{A}^{(6)}},\xi_{\tilde{A}^{(7)}},\xi_{\tilde{A}^{(8)}},\xi_{\tilde{A}^{(9)}}\right)\right)\\
&\mathfrak{A}_2^{\cap}\left(\mathfrak{A}_2^{\cup}\left(\mathfrak{A}_3^{\cap,R}\left(\xi_{\tilde{A}^{(1)}},\xi_{\tilde{A}^{(2)}},\xi_{\tilde{A}^{(3)}}\right),\mathfrak{A}_2^{\cap}\left(\xi_{\tilde{A}^{(4)}},\xi_{\tilde{A}^{(5)}}\right)\right),\mathfrak{A}_4^{\cup,L}\left(\xi_{\tilde{A}^{(6)}},\xi_{\tilde{A}^{(7)}},\xi_{\tilde{A}^{(8)}},\xi_{\tilde{A}^{(9)}}\right)\right)\\
&\mathfrak{A}_2^{\cap}\left(\mathfrak{A}_2^{\cup}\left(\mathfrak{A}_3^{\cap,R}\left(\xi_{\tilde{A}^{(1)}},\xi_{\tilde{A}^{(2)}},\xi_{\tilde{A}^{(3)}}\right),\mathfrak{A}_2^{\cap}\left(\xi_{\tilde{A}^{(4)}},\xi_{\tilde{A}^{(5)}}\right)\right),\mathfrak{A}_4^{\cup,R}\left(\xi_{\tilde{A}^{(6)}},\xi_{\tilde{A}^{(7)}},\xi_{\tilde{A}^{(8)}},\xi_{\tilde{A}^{(9)}}\right)\right).
\end{aligned}$$

Suppose that $\mathfrak{A}_2^{\cap}$ is associative. Then $\mathfrak{A}_3^{\cap,L} = \mathfrak{A}_3^{\cap,R} \equiv \mathfrak{A}_3^{\cap}$ and we have

$$\begin{aligned}
&\mathfrak{A}_2^{\cap}\left(\mathfrak{A}_2^{\cup}\left(\mathfrak{A}_3^{\cap}\left(\xi_{\tilde{A}^{(1)}},\xi_{\tilde{A}^{(2)}},\xi_{\tilde{A}^{(3)}}\right),\mathfrak{A}_2^{\cap}\left(\xi_{\tilde{A}^{(4)}},\xi_{\tilde{A}^{(5)}}\right)\right),\mathfrak{A}_4^{\cup,L}\left(\xi_{\tilde{A}^{(6)}},\xi_{\tilde{A}^{(7)}},\xi_{\tilde{A}^{(8)}},\xi_{\tilde{A}^{(9)}}\right)\right)\\
&\mathfrak{A}_2^{\cap}\left(\mathfrak{A}_2^{\cup}\left(\mathfrak{A}_3^{\cap}\left(\xi_{\tilde{A}^{(1)}},\xi_{\tilde{A}^{(2)}},\xi_{\tilde{A}^{(3)}}\right),\mathfrak{A}_2^{\cap}\left(\xi_{\tilde{A}^{(4)}},\xi_{\tilde{A}^{(5)}}\right)\right),\mathfrak{A}_4^{\cup,R}\left(\xi_{\tilde{A}^{(6)}},\xi_{\tilde{A}^{(7)}},\xi_{\tilde{A}^{(8)}},\xi_{\tilde{A}^{(9)}}\right)\right).
\end{aligned}$$

Suppose that $\mathfrak{A}_2^{\cup}$ is associative. Then $\mathfrak{A}_4^{\cup,L} = \mathfrak{A}_4^{\cup,R} \equiv \mathfrak{A}_4^{\cup}$ and we have

$$\begin{aligned}
&\mathfrak{A}_2^{\cap}\left(\mathfrak{A}_2^{\cup}\left(\mathfrak{A}_3^{\cap,L}\left(\xi_{\tilde{A}^{(1)}},\xi_{\tilde{A}^{(2)}},\xi_{\tilde{A}^{(3)}}\right),\mathfrak{A}_2^{\cap}\left(\xi_{\tilde{A}^{(4)}},\xi_{\tilde{A}^{(5)}}\right)\right),\mathfrak{A}_4^{\cup}\left(\xi_{\tilde{A}^{(6)}},\xi_{\tilde{A}^{(7)}},\xi_{\tilde{A}^{(8)}},\xi_{\tilde{A}^{(9)}}\right)\right)\\
&\mathfrak{A}_2^{\cap}\left(\mathfrak{A}_2^{\cup}\left(\mathfrak{A}_3^{\cap,R}\left(\xi_{\tilde{A}^{(1)}},\xi_{\tilde{A}^{(2)}},\xi_{\tilde{A}^{(3)}}\right),\mathfrak{A}_2^{\cap}\left(\xi_{\tilde{A}^{(4)}},\xi_{\tilde{A}^{(5)}}\right)\right),\mathfrak{A}_4^{\cup}\left(\xi_{\tilde{A}^{(6)}},\xi_{\tilde{A}^{(7)}},\xi_{\tilde{A}^{(8)}},\xi_{\tilde{A}^{(9)}}\right)\right).
\end{aligned}$$

Suppose that $\mathfrak{A}_2^{\cap}$ and $\mathfrak{A}_2^{\cup}$ are associative. Then, we simply have

$$\mathfrak{A}_2^{\cap}\left(\mathfrak{A}_2^{\cup}\left(\mathfrak{A}_3^{\cap}\left(\xi_{\tilde{A}^{(1)}},\xi_{\tilde{A}^{(2)}},\xi_{\tilde{A}^{(3)}}\right),\mathfrak{A}_2^{\cap}\left(\xi_{\tilde{A}^{(4)}},\xi_{\tilde{A}^{(5)}}\right)\right),\mathfrak{A}_4^{\cup}\left(\xi_{\tilde{A}^{(6)}},\xi_{\tilde{A}^{(7)}},\xi_{\tilde{A}^{(8)}},\xi_{\tilde{A}^{(9)}}\right)\right). \tag{3.65}$$

We need to emphasize that the membership functions (3.64) and (3.65) are different. The aggregation functions $\mathfrak{A}_3^{\cap}$ and $\mathfrak{A}_4^{\cup}$ in (3.64) are independently defined, and the functions $\mathfrak{A}_3^{\cap}$ and $\mathfrak{A}_4^{\cup}$ in (3.65) are generated by $\mathfrak{A}_2^{\cap}$ and $\mathfrak{A}_2^{\cup}$, which are also associative.

The above argument can be inductively extended to the case of $\tilde{A}^{(1)}, \dots, \tilde{A}^{(n)}$. We consider the mixed set operations

$$\tilde{A}^{(1)} \Diamond \tilde{A}^{(2)} \Diamond \cdots \Diamond \tilde{A}^{(n)},$$

where $\Diamond \in \{\sqcap, \sqcup\}$ and the parentheses are implicitly assumed. Then, we can consider the functions $\mathfrak{A}_n^{\cap,L}$ and $\mathfrak{A}_n^{\cap,R}$ generated by $\mathfrak{A}_2^{\cap}$, and the functions $\mathfrak{A}_n^{\cup,L}$ and $\mathfrak{A}_n^{\cup,R}$ generated by $\mathfrak{A}_2^{\cup}$ for $n \geq 3$. Otherwise, we need to independently define each aggregation function $\mathfrak{A}_n^{\cap}$ for $n \geq 2$ and each aggregation function $\mathfrak{A}_n^{\cup}$ for $n \geq 2$ as described above.

Many other interesting approaches are also presented below. Dubois and Prade [26] proposed the intersection and union as follows

$$\xi_{\tilde{A}\cap\tilde{B}}(x) = \xi_{\tilde{A}}(x) * \xi_{\tilde{B}}(x) \text{ and } \xi_{\tilde{A}\cup\tilde{B}}(x) = \xi_{\tilde{A}}(x) \perp \xi_{\tilde{B}}(x),$$

where $*$ and $\perp$ are two aggregation operations on the unit interval [0,1] satisfying some axioms, e.g. commutativity, associativity, De Morgan laws, identity laws and continuity, where the boundary conditions regarding 0 and 1 in the unit interval [0,1] are taken into account.

Klement [58] considered the axiomatic approach for intersection and union of fuzzy sets. Yager [155] proposed the parametric intersection and union of fuzzy sets in which a parameter $p \geq 1$ was considered. Yager [159] proposed the concepts of non-monotonic intersection and union operations. The above approaches also considered the whole unit interval [0,1]. On the other hand, Tan et al. [109] proposed a different intersection and union of fuzzy sets based on the random set in a probability space, where the whole unit interval [0,1] and the α-level set are simultaneously involved in the discussion. Al-Qudah and Hassan [4] and Hu et al. [48] also studied the intersection and union of so-called complex fuzzy sets.

4

Generalized Extension Principle

The main purpose of the extension principle is to fuzzify the crisp functions into fuzzy functions. This kind of fuzzification has been widely used to study economics and engineering problems under a fuzzy environment. Especially, the arithmetics of fuzzy sets are based on the extension principle.

4.1 Extension Principle Based on the Euclidean Space

The set of all fuzzy sets in $\mathbb{R}^n$ is denoted by $\mathcal{F}(\mathbb{R}^n)$. We consider an onto function $f : \mathbb{R}^n \rightarrow \mathbb{R}$, which is also called a **crisp function**. We are going to fuzzify the function f as a function $\tilde{f} : \mathcal{F}(\mathbb{R}^n) \rightarrow \mathcal{F}(\mathbb{R})$, which is also called a **fuzzy function**. In other words, for any $\tilde{A} \in \mathcal{F}(\mathbb{R}^n)$, we have $\tilde{f}(\tilde{A}) \in \mathcal{F}(\mathbb{R})$. The principle for fuzzifying the crisp functions is called the **extension principle**.

For any set A in $\mathbb{R}^n$, the set $f(A)$ is defined by

$$f(A) = \{y : y = f(x), x \in A\}.$$

Since f is onto, we have that the set $\{x : y = f(x)\}$ is nonempty for any $y \in \mathbb{R}$. It is clear to see

$$\chi_{f(A)}(y) = \sup_{\{x:y=f(x)\}} \chi_A(x). \tag{4.1}$$

Inspired by equation (4.1), we are going to fuzzify the crisp function f as a fuzzy function $\tilde{f}$ by replacing the characteristic function with a membership function. Given any fuzzy set $\tilde{A}$ in $\mathbb{R}^n$, the membership function of $\tilde{f}(\tilde{A})$ is defined by

$$\xi_{\tilde{f}(\tilde{A})}(y) = \sup_{\{x:y=f(x)\}} \xi_{\tilde{A}}(x). \tag{4.2}$$

Theorem 2.2.13 says that the membership function of $\tilde{f}(\tilde{A})$ can be expressed as

$$\xi_{\tilde{f}(\tilde{A})}(y) = \sup_{0<\alpha\leq 1} \alpha \cdot \chi_{(\tilde{f}(\tilde{A}))_\alpha}(y). \tag{4.3}$$

Now, we have a different expression of $\xi_{\tilde{f}(\tilde{A})}$ given below.

Mathematical Foundations of Fuzzy Sets, First Edition. Hsien-Chung Wu.

Proposition 4.1.1 *Let $f : \mathbb{R}^n \to \mathbb{R}$ be an onto crisp function, and let $\tilde{f} : \mathcal{F}(\mathbb{R}^n) \to \mathcal{F}(\mathbb{R})$ be a fuzzy function extended from f via the extension principle defined in* (4.2). *Then, given any $\tilde{A} \in \mathcal{F}(\mathbb{R}^n)$, the membership function of $\tilde{f}(\tilde{A}) \in \mathcal{F}(\mathbb{R})$ can be expressed as*

$$\xi_{\tilde{f}(\tilde{A})}(y) = \sup_{0<\alpha\leq 1} \alpha \cdot \chi_{f(\tilde{A}_\alpha)}(y).$$

Proof. We have

$$\begin{aligned}
\xi_{\tilde{f}(\tilde{A})}(y) &= \sup_{\{x:y=f(x)\}} \xi_{\tilde{A}}(x) \text{ (by the extension principle in (4.2))} \\
&= \sup_{\{x:y=f(x)\}} \left(\sup_{\alpha\in(0,1]} \alpha \cdot \chi_{\tilde{A}_\alpha}(x) \right) \text{ (by Theorem 2.2.13)} \\
&= \sup_{\{(x,\alpha):y=f(x),\alpha\in(0,1]\}} \alpha \cdot \chi_{\tilde{A}_\alpha}(x).
\end{aligned}$$

On the other hand, by referring to (4.1), since

$$\chi_{f(\tilde{A}_\alpha)}(y) = \sup_{\{x:y=f(x)\}} \chi_{\tilde{A}_\alpha}(x),$$

we have

$$\begin{aligned}
\sup_{\alpha\in(0,1]} \alpha \cdot \chi_{f(\tilde{A}_\alpha)}(y) &= \sup_{\alpha\in(0,1]} \left(\alpha \cdot \sup_{\{x:y=f(x)\}} \chi_{\tilde{A}_\alpha}(x) \right) \\
&= \sup_{\alpha\in(0,1]} \left(\sup_{\{x:y=f(x)\}} \alpha \cdot \chi_{\tilde{A}_\alpha}(x) \right) \text{ (since } \alpha > 0) \\
&= \sup_{\{(x,\alpha):y=f(x),\alpha\in(0,1]\}} \alpha \cdot \chi_{\tilde{A}_\alpha}(x) = \xi_{\tilde{f}(\tilde{A})}(y).
\end{aligned}$$

This completes the proof. ■

From (4.3) and Proposition 4.1.1, we obtain

$$\xi_{\tilde{f}(\tilde{A})}(y) = \sup_{0<\alpha\leq 1} \alpha \cdot \chi_{f(\tilde{A}_\alpha)}(y) = \sup_{0<\alpha\leq 1} \alpha \cdot \chi_{(\tilde{f}(\tilde{A}))_\alpha}(y).$$

However, the equality $(\tilde{f}(\tilde{A}))_\alpha = f(\tilde{A}_\alpha)$ is not always guaranteed, which will be completely investigated in this chapter.

Let $f : \mathbb{R}^2 \to \mathbb{R}$ be an onto and continuous real-valued function, and let $\tilde{A}, \tilde{B}$ be two fuzzy sets in $\mathbb{R}$ such that the membership functions $\xi_{\tilde{A}}$ and $\xi_{\tilde{B}}$ are upper semi-continuous, and the 0-level sets $\tilde{A}_0$ and $\tilde{B}_0$ are closed and bounded subsets of $\mathbb{R}$. Then, Nguyen [85] has shown that

$$\left(\tilde{f}(\tilde{A}, \tilde{B})\right)_\alpha = f\left(\tilde{A}_\alpha, \tilde{B}_\alpha\right) \tag{4.4}$$

for each $\alpha \in (0,1]$.

Since Nguyen defined the 0-level set as

$$\tilde{A}_0 = \{x \in \mathbb{R} : \xi_{\tilde{A}}(x) \geq 0\},$$

i.e. the 0-level set is the whole Euclidean space $\mathbb{R} = \tilde{A}_0$, it says that the equality (4.4) holds true automatically for $\alpha = 0$ in the sense of Nguyen's definition. However, throughout this book, the 0-level set is defined to be

$$\tilde{A}_0 = \text{cl}\left(\tilde{A}_{0+}\right) = \text{cl}\left(\{x \in \mathbb{R} : \xi_{\tilde{A}}(x) > 0\}\right),$$

which is not necessarily equal to the whole Euclidean space $\mathbb{R}$. Therefore, we need some other argument to guarantee the equality (4.4) for $\alpha = 0$, which will be investigated in the subsequent discussion.

We write

$$\mathcal{F}^n(\mathbb{R}) \equiv \mathcal{F}(\mathbb{R}) \times \cdots \times \mathcal{F}(\mathbb{R}).$$

Note that $\mathcal{F}^n(\mathbb{R}) \neq \mathcal{F}(\mathbb{R}^n)$. Now, we consider the following fuzzy function $\tilde{f} : \mathcal{F}^n(\mathbb{R}) \to \mathcal{F}(\mathbb{R})$ extended from $f : \mathbb{R}^n \to \mathbb{R}$ as follows. Given any fuzzy sets $\tilde{A}^{(i)}$ in $\mathbb{R}$ for $i = 1, \ldots, n$, the membership function of $\tilde{f}(\tilde{A}^{(1)}, \tilde{A}^{(2)}, \ldots, \tilde{A}^{(n)})$ is defined by

$$\xi_{\tilde{f}(\tilde{A}^{(1)}, \tilde{A}^{(2)}, \ldots, \tilde{A}^{(n)})}(y) = \sup_{\{(x_1,\ldots,x_n) : y = f(x_1,\ldots,x_n)\}} \min \left\{ \xi_{\tilde{A}^{(1)}}(x_1), \ldots, \xi_{\tilde{A}^{(n)}}(x_n) \right\}, \tag{4.5}$$

where $\tilde{f}(\tilde{A}^{(1)}, \ldots, \tilde{A}^{(n)})$ is a fuzzy set in $\mathbb{R}$. This mechanism is called an **extension principle**.

Theorem 4.1.2 *Let $f : \mathbb{R}^n \to \mathbb{R}$ be an onto real-valued function defined on $\mathbb{R}^n$, and let $\tilde{f} : \mathcal{F}^n(\mathbb{R}) \to \mathcal{F}(\mathbb{R})$ be a fuzzy function extended from f via the extension principle defined in (4.5). Given any fuzzy sets $\tilde{A}^{(i)}$ in $\mathbb{R}$ for $i = 1, \ldots, n$, suppose that the membership functions $\xi_{\tilde{A}^{(i)}}$ are upper semi-continuous for $i = 1, \ldots, n$. Then, we have the following properties.*

(i) *Suppose that*

$$\{(x_1, \ldots, x_n) : y = f(x_1, \ldots, x_n)\}$$

are closed and bounded subsets of $\mathbb{R}^n$ for all y in the range of f. Then, the α-level set of $\tilde{f}(\tilde{A}^{(1)}, \ldots, \tilde{A}^{(n)})$ is given by

$$\left(\tilde{f}\left(\tilde{A}^{(1)}, \ldots, \tilde{A}^{(n)}\right)\right)_\alpha = f\left(\tilde{A}^{(1)}_\alpha, \ldots, \tilde{A}^{(n)}_\alpha\right) \text{ for } \alpha \in (0,1],$$

and the 0-level set is given by

$$\left(\tilde{f}\left(\tilde{A}^{(1)}, \ldots, \tilde{A}^{(n)}\right)\right)_0 = cl\left(f\left(\tilde{A}^{(1)}_{0+}, \ldots, \tilde{A}^{(n)}_{0+}\right)\right) \subseteq cl\left(f\left(\tilde{A}^{(1)}_0, \ldots, \tilde{A}^{(n)}_0\right)\right). \tag{4.6}$$

(ii) *Suppose that the 0-level sets $\tilde{A}^{(i)}_0$ of $\tilde{A}^{(i)}$ are bounded subsets of $\mathbb{R}$ for all $i = 1, \ldots, n$, and that*

$$\{(x_1, \ldots, x_n) : y = f(x_1, \ldots, x_n)\}$$

are closed subsets of $\mathbb{R}^n$ for all y in the range of f. Then, the α-level set of $\tilde{f}(\tilde{A}^{(1)}, \ldots, \tilde{A}^{(n)})$ is given by

$$\left(\tilde{f}\left(\tilde{A}^{(1)}, \ldots, \tilde{A}^{(n)}\right)\right)_\alpha = f\left(\tilde{A}^{(1)}_\alpha, \ldots, \tilde{A}^{(n)}_\alpha\right) \text{ for } \alpha \in (0,1],$$

and the 0-level set is given by

$$\left(\tilde{f}\left(\tilde{A}^{(1)}, \ldots, \tilde{A}^{(n)}\right)\right)_0 = cl\left(f\left(\tilde{A}^{(1)}_{0+}, \ldots, \tilde{A}^{(n)}_{0+}\right)\right) \subseteq cl\left(f\left(\tilde{A}^{(1)}_0, \ldots, \tilde{A}^{(n)}_0\right)\right).$$

(iii) *Suppose that f is continuous, and that the 0-level sets $\tilde{A}^{(i)}_0$ of $\tilde{A}^{(i)}$ are bounded subsets of $\mathbb{R}$ for all $i = 1, \ldots, n$. Then, the α-level set of $\tilde{f}(\tilde{A}^{(1)}, \ldots, \tilde{A}^{(n)})$ is given by*

$$\left(\tilde{f}\left(\tilde{A}^{(1)}, \ldots, \tilde{A}^{(n)}\right)\right)_\alpha = f\left(\tilde{A}^{(1)}_\alpha, \ldots, \tilde{A}^{(n)}_\alpha\right) \text{ for } \alpha \in [0,1],$$

where the 0-level sets are taken into account.

Proof. To prove part (i), we first consider the α-level sets for $\alpha \in (0,1]$. For $y \in f(\tilde{A}^{(1)}_\alpha, \dots, \tilde{A}^{(n)}_\alpha)$, there exists $(x_1, \dots, x_n) \in \mathbb{R}^n$ satisfying $y = f(x_1, \dots, x_n)$ and $x_i \in \tilde{A}^{(i)}_\alpha$ for all i, which implies $\xi_{\tilde{A}^{(i)}}(x_i) \geq \alpha$ for all i, i.e.

$$\min_{1 \leq i \leq n} \left\{ \xi_{\tilde{A}^{(i)}}(x_i) \right\} \geq \alpha \text{ for all } i.$$

Therefore, we have

$$\xi_{\tilde{f}(\tilde{A}^{(1)}, \dots, \tilde{A}^{(n)})}(y) = \sup_{\{(x_1, \dots, x_n): y = f(x_1, \dots, x_n)\}} \min_{1 \leq i \leq n} \left\{ \xi_{\tilde{A}^{(i)}}(x_i) \right\} \geq \alpha,$$

which says that $y \in (\tilde{f}(\tilde{A}^{(1)}, \dots, \tilde{A}^{(n)}))_\alpha$. Therefore, we obtain the inclusion

$$f\left(\tilde{A}^{(1)}_\alpha, \dots, \tilde{A}^{(n)}_\alpha\right) \subseteq \left(\tilde{f}\left(\tilde{A}^{(1)}, \dots, \tilde{A}^{(n)}\right)\right)_\alpha.$$

For proving the other direction of inclusion, suppose that $y \in (\tilde{f}(\tilde{A}^{(1)}, \dots, \tilde{A}^{(n)}))_\alpha$. Then, we have

$$\sup_{\{(x_1, \dots, x_n): y = f(x_1, \dots, x_n)\}} \min_{1 \leq i \leq n} \left\{ \xi_{\tilde{A}^{(i)}}(x_i) \right\} \geq \alpha. \tag{4.7}$$

Since each $\xi_{\tilde{A}^{(i)}}$ is upper semi-continuous on $\mathbb{R}$ for $i = 1, \dots, n$, we have that

$$\tilde{A}^{(i)}_\alpha = \{x : \xi_{\tilde{A}^{(i)}}(x) \geq \alpha\}$$

are closed subsets of $\mathbb{R}$ for $i = 1, \dots, n$. We also see that

$$C_{i\alpha} = \underbrace{\mathbb{R} \times \cdots \times \mathbb{R}}_{i-1} \times \tilde{A}^{(i)}_\alpha \times \underbrace{\mathbb{R} \times \cdots \times \mathbb{R}}_{n-i}$$

are closed subsets of $\mathbb{R}^n$ for $i = 1, \dots, n$. Therefore,

$$\begin{aligned} \left\{ (x_1, \dots, x_n) : \min_{1 \leq i \leq n} \left\{ \xi_{\tilde{A}^{(i)}}(x_i) \right\} \geq \alpha \right\} &= \left\{ (x_1, \dots, x_n) : \xi_{\tilde{A}^{(i)}}(x_i) \geq \alpha \text{ for all } i = 1, \dots, n \right\} \\ &= \tilde{A}^{(1)}_\alpha \times \cdots \times \tilde{A}^{(n)}_\alpha = \bigcap_{i=1}^n C_{i\alpha} \end{aligned}$$

is a closed subset of $\mathbb{R}^n$, which says that the function

$$g\left(x_1, \dots, x_n\right) = \min_{1 \leq i \leq n} \left\{ \xi_{\tilde{A}^{(i)}}(x_i) \right\}$$

is upper semi-continuous on $\mathbb{R}^n$. Since

$$\{(x_1, \dots, x_n) : y = f(x_1, \dots, x_n)\}$$

are closed and bounded subsets of $\mathbb{R}^n$ by the assumption, Proposition 1.4.4 says that the following supremum is attained

$$\max_{\{(x_1, \dots, x_n): y = f(x_1, \dots, x_n)\}} \min_{1 \leq i \leq n} \left\{ \xi_{\tilde{A}^{(i)}}(x_i) \right\} = \sup_{\{(x_1, \dots, x_n): y = f(x_1, \dots, x_n)\}} \min_{1 \leq i \leq n} \left\{ \xi_{\tilde{A}^{(i)}}(x_i) \right\}. \tag{4.8}$$

Using (4.7), we obtain

$$\max_{\{(x_1, \dots, x_n): y = f(x_1, \dots, x_n)\}} \min_{1 \leq i \leq n} \left\{ \xi_{\tilde{A}^{(i)}}(x_i) \right\} \geq \alpha. \tag{4.9}$$

Therefore, there exists $(x_1, \dots, x_n)$ satisfying

$$y = f(x_1, \dots, x_n) \text{ and } \min_{1 \leq i \leq n} \left\{ \xi_{\tilde{A}^{(i)}}(x_i) \right\} \geq \alpha.$$

This says that $\xi_{\tilde{A}^{(i)}}(x_i) \geq \alpha$ for all $i = 1, \ldots, n$; that is, $x_i \in \tilde{A}_\alpha^{(i)}$ for all $i = 1, \ldots, n$. Therefore, we obtain $y \in f(\tilde{A}_\alpha^{(1)}, \ldots, \tilde{A}_\alpha^{(n)})$, which shows the inclusion

$$\left(\tilde{f}\left(\tilde{A}^{(1)}, \ldots, \tilde{A}^{(n)}\right)\right)_\alpha \subseteq f\left(\tilde{A}_\alpha^{(1)}, \ldots, \tilde{A}_\alpha^{(n)}\right).$$

Therefore, we obtain the desired equality.

Now, we consider the 0-level sets. For notational convenience, we write

$$X_0 = f\left(\tilde{A}_0^{(1)}, \ldots, \tilde{A}_0^{(n)}\right), X_{0+} = f\left(\tilde{A}_{0+}^{(1)}, \ldots, \tilde{A}_{0+}^{(n)}\right), \text{and} Y = \left\{r : \xi_{\tilde{f}(\tilde{A}^{(1)},\ldots,\tilde{A}^{(n)})}(y) > 0\right\}.$$

Then, from (2.4), we see that

$$\left(\tilde{f}\left(\tilde{A}^{(1)}, \ldots, \tilde{A}^{(n)}\right)\right)_0 = \text{cl}(Y). \tag{4.10}$$

By referring to (4.8), we have

$$\begin{aligned} 0 < \xi_{\tilde{f}(\tilde{A}^{(1)},\ldots,\tilde{A}^{(n)})}(y) &= \sup_{\{(x_1,\ldots,x_n):y=f(x_1,\ldots,x_n)\}} \min_{1\leq i\leq n} \left\{\xi_{\tilde{A}^{(i)}}(x_i)\right\} \\ &= \max_{\{(x_1,\ldots,x_n):y=f(x_1,\ldots,x_n)\}} \min_{1\leq i\leq n} \left\{\xi_{\tilde{A}^{(i)}}(x_i)\right\}, \end{aligned}$$

where $y \in Y$. Therefore, there exists $(x_1, \ldots, x_n)$ satisfying $y = f(x_1, \ldots, x_n)$ and $\xi_{\tilde{A}^{(i)}}(x_i) > 0$, i.e. $x_i \in \tilde{A}_{0+}^{(i)} \subseteq \tilde{A}_0^{(i)}$ for all $i = 1, \ldots, n$. This also says that

$$y = f(x_1, \ldots, x_n) \in X_{0+} \subseteq X_0.$$

Therefore, we obtain the inclusions

$$Y \subseteq X_{0+} \subseteq X_0,$$

which implies

$$\text{cl}(Y) \subseteq \text{cl}(X_{0+}) \subseteq \text{cl}(X_0).$$

On the other hand, for any $y \in \text{cl}(X_{0+})$, according to the concept of closure, there exists a sequence $\{y_m\}_{m=1}^\infty$ in X_{0+} satisfying $y_m \to y$ as $m \to \infty$. For $y_m \in X_{0+}$, there exists $(x_1^{(m)}, \ldots, x_p^{(m)})$ satisfying

$$y_m = f\left((x_1^{(m)}, \ldots, x_p^{(m)})\right) \text{ and } x_i^{(m)} \in \tilde{A}_{0+}^{(i)} \text{ for all } i = 1, \ldots, n,$$

i.e. $\xi_{\tilde{A}^{(i)}}(x_i^{(m)}) > 0$ for all $i = 1, \ldots, n$, which implies

$$\min_{1\leq i\leq n} \left\{\xi_{\tilde{A}^{(i)}}\left(x_i^{(m)}\right)\right\} > 0.$$

Therefore, we obtain

$$\xi_{\tilde{f}(\tilde{A}^{(1)},\ldots,\tilde{A}^{(n)})}(y_m) = \sup_{\{(x_1,\ldots,x_n):y_m=f(x_1,\ldots,x_n)\}} \min_{1\leq i\leq n} \left\{\xi_{\tilde{A}^{(i)}}(x_i)\right\} \geq \min_{1\leq i\leq n} \left\{\xi_{\tilde{A}^{(i)}}\left(x_i^{(m)}\right)\right\} > 0,$$

i.e. $y_m \in Y$, which says that $y \in \text{cl}(Y)$, since $y_m \to y$ as $m \to \infty$. This shows the inclusion $\text{cl}(X_{0+}) \subseteq \text{cl}(Y)$. Therefore, we obtain the equality $\text{cl}(X_{0+}) = \text{cl}(Y)$, which proves (4.6).

To prove part (ii), since

$$\min_{1\leq i\leq n} \left\{\xi_{\tilde{A}^{(i)}}(x_i)\right\} = 0$$

outside of the set $\tilde{A}_0^{(1)}, \ldots, \tilde{A}_0^{(n)}$, using (4.7), we have

$$\begin{aligned}
\alpha &\leq \sup_{\{(x_1,\ldots,x_n):y=f(x_1,\ldots,x_n)\}} \min_{1\leq i\leq n} \left\{\xi_{\tilde{A}^{(i)}}(x_i)\right\}\\
&= \sup_{\{(x_1,\ldots,x_n):y=f(x_1,\ldots,x_n) \text{ and } (x_1,\ldots,x_n)\in\tilde{A}_0^{(1)}\times\cdots\times\tilde{A}_0^{(n)}\}} \min_{1\leq i\leq n} \left\{\xi_{\tilde{A}^{(i)}}(x_i)\right\}\\
&= \sup_{\{(x_1,\ldots,x_n):(x_1,\ldots,x_n)\in f^{-1}(\{y\})\cap\tilde{A}_0^{(1)}\times\cdots\times\tilde{A}_0^{(n)}\}} \min_{1\leq i\leq n} \left\{\xi_{\tilde{A}^{(i)}}(x_i)\right\}, \qquad (4.11)
\end{aligned}$$

where $f^{-1}(\{y\})$ denotes the inverse image of the singleton $\{y\}$. Since $\tilde{A}_0^{(i)}$ are bounded subsets of $\mathbb{R}$ for all $i = 1, \ldots, n$, it follows that $\tilde{A}_0^{(1)} \times \cdots \times \tilde{A}_0^{(n)}$ is a bounded subset of $\mathbb{R}^n$. Since $\tilde{A}_0^{(i)}$ are also closed subsets of $\mathbb{R}$ for all $i = 1, \ldots, n$ by the definition of 0-level set, it follows that $\tilde{A}_0^{(1)} \times \cdots \times \tilde{A}_0^{(n)}$ is a closed and bounded subset of $\mathbb{R}^n$. Since the inverse image

$$f^{-1}(\{y\}) = \{(x_1, \ldots, x_n) : y = f(x_1, \ldots, x_n)\}$$

is a closed subset of $\mathbb{R}^n$ by the assumption, it says that

$$f^{-1}(\{y\}) \cap \tilde{A}_0^{(1)} \times \cdots \times \tilde{A}_0^{(n)}$$

is a closed and bounded subset of $\mathbb{R}^n$. Since the function

$$g(x_1, \ldots, x_n) = \min_{1\leq i\leq n} \left\{\xi_{\tilde{A}^{(i)}}(x_i)\right\}$$

is upper semi-continuous on $\mathbb{R}^n$ from the arguments of part (i), Proposition 1.4.4 says that the supremum in (4.11) is attained. By referring to (4.9), the remaining proof follows from the same arguments of part (i).

To prove part (iii), since the singleton $\{y\}$ is a closed subset of $\mathbb{R}$, it follows that $f^{-1}(\{y\})$ is also a closed subset of $\mathbb{R}^n$ by the continuity of f. Therefore, we can have the results from part (ii). Now, it remains to consider the 0-level sets. We want to claim $\text{cl}(Y) = X_0$. Since f is continuous and $\tilde{A}_0^{(i)}$ are closed and bounded for all $i = 1, \ldots, n$, it follows that the set X_0 is closed and bounded by part (iii) of Proposition 1.4.3. From the arguments of part (i), we obtain

$$\text{cl}(Y) \subseteq \text{cl}(X_0) = X_0.$$

To prove the other direction of inclusion, for any $y \in X_0$, there exist

$$x_i \in \tilde{A}_0^{(i)} = \text{cl}(\tilde{A}_{0+}^{(i)}) \text{ for } i = 1, \ldots, n$$

satisfying $y = f(x_1, \ldots, x_n)$. According to the concept of closure, there exist sequences $\{x_i^{(m)}\}_{m=1}^{\infty}$ in $\tilde{A}_0^{(i)}$ satisfying $x_i^{(m)} \to x_i$ as $m \to \infty$, where $\xi_{\tilde{A}^{(i)}}(x_i^{(m)}) > 0$ for all $i = 1, \ldots, n$. Let

$$y_m = f\left(x_1^{(m)}, \ldots, x_n^{(m)}\right).$$

Since f is continuous, we have $y_m \to y$ as $m \to \infty$. We also have

$$\begin{aligned}
\xi_{\tilde{f}(\tilde{A}^{(1)},\ldots,\tilde{A}^{(n)})}(y_m) &= \sup_{\{(x_1,\ldots,x_n):y_m=f(x_1,\ldots,x_n)\}} \min\left\{\xi_{\tilde{A}^{(1)}}(x_1), \ldots, \xi_{\tilde{A}^{(n)}}(x_n)\right\}\\
&\geq \min\left\{\xi_{\tilde{A}^{(1)}}(x_1^{(m)}), \ldots, \xi_{\tilde{A}^{(n)}}(x_n^{(m)})\right\} > 0.
\end{aligned}$$

This says that $\{y_m\}_{m=1}^{\infty}$ is a sequence in Y. Therefore, we obtain $y \in \text{cl}(Y)$, since $y_m \to y$ as $m \to \infty$. This shows the inclusion $X_0 \subseteq \text{cl}(Y)$. Using (4.10), we obtain

$$f\left(\tilde{A}_0^{(1)}, \ldots, \tilde{A}_0^{(n)}\right) = \left(\tilde{f}\left(\tilde{A}^{(1)}, \ldots, \tilde{A}^{(n)}\right)\right)_0.$$

This completes the proof. ∎

Theorem 4.1.3 *Let $f : \mathbb{R}^n \to \mathbb{R}$ be a continuous and onto real-valued function, and let $\tilde{f} : \mathcal{F}^n(\mathbb{R}) \to \mathcal{F}(\mathbb{R})$ be a fuzzy function extended from f via the extension principle defined in (4.5). Suppose that $\tilde{A}^{(i)}$ are fuzzy intervals for $i = 1, \ldots, n$. Then $\tilde{f}(\tilde{A}^{(1)}, \ldots, \tilde{A}^{(n)})$ is also a fuzzy interval and its α-level set is given by*

$$\left(\tilde{f}\left(\tilde{A}^{(1)}, \ldots, \tilde{A}^{(n)}\right)\right)_\alpha = f\left(\tilde{A}^{(1)}_\alpha, \ldots, \tilde{A}^{(n)}_\alpha\right)$$
$$= \left\{f(x_1, \ldots, x_n) : x_i \in \tilde{A}^{(i)}_\alpha \quad \text{for } i = 1, \ldots, n\right\}$$
$$= \left\{f(x_1, \ldots, x_n) : (\tilde{A}^{(i)})^L_\alpha \leq x_i \leq (\tilde{A}^{(i)})^U_\alpha \quad \text{for } i = 1, \ldots, n\right\}$$

for any $\alpha \in [0,1]$, where the 0-level sets are taken into account.

Proof. We need to check the conditions in Definition 2.3.1. Since $\tilde{A}^{(i)}_\alpha$ are closed intervals for all $i = 1, \ldots, n$ and all $\alpha \in [0,1]$, by Theorem 4.1.2, we obtain

$$\left\{y \in \mathbb{R} : \xi_{\tilde{f}(\tilde{A}^{(1)}, \ldots, \tilde{A}^{(n)})}(y) \geq \alpha\right\} = \left(\tilde{f}(\tilde{A}^{(1)}, \ldots, \tilde{A}^{(n)})\right)_\alpha$$
$$= \left\{f(x_1, \ldots, x_n) : x_i \in \tilde{A}^{(i)}_\alpha \quad \text{for } i = 1, \ldots, n\right\} \tag{4.12}$$

for each $\alpha \in [0,1]$. Since f is continuous, it follows that the set in (4.12) is a closed interval by parts (iii) and (iv) of Proposition 1.4.3, i.e. a convex set in $\mathbb{R}$ for each $\alpha \in [0,1]$, which shows that the membership function $\xi_{\tilde{f}(\tilde{A}^{(1)}, \ldots, \tilde{A}^{(n)})}$ is upper semi-continuous and quasi-concave. This completes the proof. ■

Example 4.1.4 Let $\tilde{A}$ and $\tilde{B}$ be two fuzzy intervals. Using the extension principle, the membership function of the maximum $\widetilde{\max}\{\tilde{A}, \tilde{B}\}$ of $\tilde{A}$ and $\tilde{B}$ is given by

$$\xi_{\widetilde{\max}\{\tilde{A}, \tilde{B}\}}(z) = \sup_{\{(x,y) : \max\{x,y\} = z\}} \min\{\xi_{\tilde{A}}(x), \xi_{\tilde{B}}(y)\}.$$

From Theorem 4.1.3, we have

$$\left(\widetilde{\max}\{\tilde{A}, \tilde{B}\}\right)_\alpha = \max\left\{\tilde{A}_\alpha, \tilde{B}_\alpha\right\}$$
$$= \left\{\max\{x, y\} : x \in \tilde{A}_\alpha \text{ and } y \in \tilde{B}_\alpha\right\}$$
$$= \left\{\max\{x, y\} : \tilde{A}^L_\alpha \leq x \leq \tilde{A}^U_\alpha \text{ and } \tilde{B}^L_\alpha \leq y \leq \tilde{B}^U_\alpha\right\}$$
$$= \left[\max\left\{\tilde{A}^L_\alpha, \tilde{B}^L_\alpha\right\}, \max\left\{\tilde{A}^U_\alpha, \tilde{B}^U_\alpha\right\}\right]$$

for each $\alpha \in [0,1]$.

4.2 Extension Principle Based on the Product Spaces

Now, we consider the following onto crisp function

$$f : \mathbb{R}^{n_1} \times \cdots \times \mathbb{R}^{n_p} \to \mathbb{R}^n,$$

where $n_1, \ldots, p_p$ are positive integers. The purpose is to fuzzify the crisp function f as a fuzzy function

$$\tilde{f} : \mathcal{F}(\mathbb{R}^{n_1}) \times \cdots \times \mathcal{F}(\mathbb{R}^{n_p}) \to \mathcal{F}(\mathbb{R}^n).$$

Given any subsets A_i of $\mathbb{R}^{n_i}$ for $i = 1, \ldots, p$, we have

$$f(A_1, \ldots, A_p) = \left\{y : y = f(x_1, \ldots, x_p) \text{ for } x_i \in A_i \text{ and } i = 1, \ldots, n\right\}.$$

Since f is onto, we have that the set

$$\{(x_1,\ldots,x_p) : y = f(x_1,\ldots,x_p)\}$$

is nonempty for any $y \in \mathbb{R}^n$. Regarding the characteristic functions, it is easy to show

$$\chi_{f(A_1,\ldots,A_p)}(y) = \sup_{\{(x_1,\ldots,x_p):y=f(x_1,\ldots,x_p)\}} \min\left\{\chi_{A_1}(x_1),\ldots,\chi_{A_p}(x_p)\right\}. \tag{4.13}$$

Inspired by equation (4.13), we are going to fuzzify the crisp function f as a fuzzy function $\tilde{f}$ by replacing the characteristic functions with membership functions.

Given any fuzzy sets $\tilde{A}^{(i)}$ in $\mathbb{R}^{n_i}$ for $i = 1,\ldots,p$, we have that $\tilde{f}(\tilde{A}^{(1)},\tilde{A}^{(2)},\ldots,\tilde{A}^{(p)})$ is a fuzzy set in $\mathbb{R}^n$. The membership function of $\tilde{f}(\tilde{A}^{(1)},\tilde{A}^{(2)},\ldots,\tilde{A}^{(p)})$ is defined by

$$\xi_{\tilde{f}(\tilde{A}^{(1)},\tilde{A}^{(2)},\ldots,\tilde{A}^{(p)})}(y) = \sup_{\{(x_1,\ldots,x_p):y=f(x_1,\ldots,x_p)\}} \min\left\{\xi_{\tilde{A}^{(1)}}(x_1),\ldots,\xi_{\tilde{A}^{(p)}}(x_p)\right\}. \tag{4.14}$$

Theorem 2.2.13 says that

$$\begin{aligned} \xi_{\tilde{f}(\tilde{A}^{(1)},\tilde{A}^{(2)},\ldots,\tilde{A}^{(p)})}(y) &= \sup_{\alpha\in[0,1]} \alpha \cdot \chi_{(\tilde{f}(\tilde{A}^{(1)},\tilde{A}^{(2)},\ldots,\tilde{A}^{(p)}))_\alpha}(y) \\ &= \sup_{0<\alpha\leq 1} \alpha \cdot \chi_{(\tilde{f}(\tilde{A}^{(1)},\tilde{A}^{(2)},\ldots,\tilde{A}^{(p)}))_\alpha}(y). \end{aligned} \tag{4.15}$$

Next, we are going to present a different expression given by

$$\xi_{\tilde{f}(\tilde{A}^{(1)},\ldots,\tilde{A}^{(p)})}(y) = \sup_{\alpha\in[0,1]} \alpha \cdot \chi_{f(\tilde{A}^{(1)}_\alpha,\ldots,\tilde{A}^{(p)}_\alpha)}(y) = \sup_{0<\alpha\leq 1} \alpha \cdot \chi_{f(\tilde{A}^{(1)}_\alpha,\ldots,\tilde{A}^{(p)}_\alpha)}(y).$$

We first present some useful lemmas.

Lemma 4.2.1 *Let $\tilde{A}$ be a fuzzy set in $\mathbb{R}^n$. Given any fixed $x \in \mathbb{R}^n$, let*

$$\alpha^* = \sup_{\alpha\in[0,1]} \alpha \cdot \chi_{\tilde{A}_\alpha}(x).$$

Then $x \in \tilde{A}_\alpha$ for $\alpha \in [0,1]$ with $\alpha < \alpha^*$, and $x \notin \tilde{A}_\alpha$ for $\alpha \in [0,1]$ with $\alpha > \alpha^*$.

Proof. Suppose that there exists $\alpha_1 \in [0,1]$ satisfying $\alpha_1 < \alpha^*$ and $x \notin \tilde{A}_{\alpha_1}$. Then $x \notin \tilde{A}_\alpha$ for all $\alpha \in [0,1]$ with $\alpha > \alpha_1$, since $\tilde{A}_\alpha \subseteq \tilde{A}_\beta$ for $\alpha, \beta \in [0,1]$ with $\beta < \alpha$. This says that

$$\sup_{\alpha\in[0,1]} \alpha \cdot \chi_{\tilde{A}_\alpha}(x) \leq \alpha_1 < \alpha^*,$$

which is a contradiction. Therefore, we conclude $x \in \tilde{A}_\alpha$ for $\alpha \in [0,1]$ with $\alpha < \alpha^*$.

On the other hand, suppose that there exists $\alpha_2 \in [0,1]$ satisfying $\alpha_2 > \alpha^*$ and $x \in \tilde{A}_{\alpha_2}$. Then, we have

$$\sup_{\alpha\in[0,1]} \alpha \cdot \chi_{\tilde{A}_\alpha}(x) \geq \alpha_2 > \alpha^*,$$

which is also a contradiction. Therefore, we conclude $x \notin \tilde{A}_\alpha$ for $\alpha \in [0,1]$ with $\alpha > \alpha^*$. This completes the proof. ∎

Lemma 4.2.2 *Let $\tilde{A}^{(i)}$ be fuzzy sets in $\mathbb{R}^{n_i}$ for $i = 1,\ldots,p$. Then, we have*

$$\min_{1\leq i\leq p}\left\{\sup_{\alpha\in[0,1]} \alpha \cdot \chi_{\tilde{A}^{(i)}_\alpha}(x_i)\right\} = \sup_{\alpha\in[0,1]} \min_{1\leq i\leq p}\left\{\alpha \cdot \chi_{\tilde{A}^{(i)}_\alpha}(x_i)\right\}.$$

Proof. Let

$$\alpha_i^* = \sup_{\alpha\in[0,1]} \alpha \cdot \chi_{\tilde{A}_\alpha^{(i)}}(x_i) \text{ for } i = 1,\dots,p$$

and let

$$\alpha^* = \min_{1\le i\le p} \alpha_i^*.$$

For $\alpha \in [0,1]$ with $\alpha > \alpha^*$, i.e. $\alpha \in [0,1]$ with $\alpha > \alpha_i^*$ for some $i = 1,\dots,p$, Lemma 4.2.1 says that $x_i \notin \tilde{A}_\alpha^{(i)}$ for some $i = 1,\dots,p$. Therefore, we obtain

$$\min_{1\le i\le p}\left\{\alpha \cdot \chi_{\tilde{A}_\alpha^{(i)}}(x_i)\right\} = 0.$$

For $\alpha \in [0,1]$ with $\alpha < \alpha^*$, i.e. $\alpha \in [0,1]$ with $\alpha < \alpha_i^*$ for all $i = 1,\dots,p$, Lemma 4.2.1 says that $x_i \in \tilde{A}_\alpha^{(i)}$ for all $i = 1,\dots,p$. Therefore, we obtain

$$\min_{1\le i\le p}\left\{\alpha \cdot \chi_{\tilde{A}_\alpha^{(i)}}(x_i)\right\} = \alpha,$$

which also shows that

$$\begin{aligned}
&\sup_{\alpha\in[0,1]}\min_{1\le i\le p}\left\{\alpha \cdot \chi_{\tilde{A}_\alpha^{(i)}}(x_i)\right\}\\
&= \max\left\{\sup_{\{\alpha\in[0,1]:\alpha>\alpha^*\}}\min_{1\le i\le p}\left\{\alpha \cdot \chi_{\tilde{A}_\alpha^{(i)}}(x_i)\right\}, \sup_{\{\alpha\in[0,1]:\alpha<\alpha^*\}}\min_{1\le i\le p}\left\{\alpha \cdot \chi_{\tilde{A}_\alpha^{(i)}}(x_i)\right\},\right.\\
&\qquad\left.\min_{1\le i\le p}\left\{\alpha^* \cdot \chi_{\tilde{A}_{\alpha^*}^{(i)}}(x_i)\right\}\right\}\\
&= \max\left\{0, \sup_{\{\alpha\in[0,1]:\alpha<\alpha^*\}}\alpha, \min_{1\le i\le p}\left\{\alpha^* \cdot \chi_{\tilde{A}_{\alpha^*}^{(i)}}(x_i)\right\}\right\}\\
&= \max\left\{\alpha^*, \min_{1\le i\le p}\left\{\alpha^* \cdot \chi_{\tilde{A}_{\alpha^*}^{(i)}}(x_i)\right\}\right\} = \alpha^*.
\end{aligned}$$

This completes the proof. ■

Theorem 4.2.3 (Representation Theorem) *Let $f : \mathbb{R}^{n_1} \times \cdots \times \mathbb{R}^{n_p} \to \mathbb{R}^n$ be an onto crisp function, and let $\tilde{f} : \mathcal{F}(\mathbb{R}^{n_1}) \times \cdots \times \mathcal{F}(\mathbb{R}^{n_p}) \to \mathcal{F}(\mathbb{R}^n)$ be a fuzzy function extended from f via the extension principle defined in (4.14). Let $\tilde{A}^{(i)}$ be fuzzy sets in $\mathbb{R}^{n_i}$ for $i = 1,\dots,p$. Then, the membership function of fuzzy set $\tilde{f}(\tilde{A}^{(1)},\dots,\tilde{A}^{(p)})$ in $\mathbb{R}^n$ can be expressed as*

$$\xi_{\tilde{f}(\tilde{A}^{(1)},\dots,\tilde{A}^{(p)})}(y) = \sup_{\alpha\in[0,1]} \alpha \cdot \chi_{f(\tilde{A}_\alpha^{(1)},\dots,\tilde{A}_\alpha^{(p)})}(y) = \sup_{0<\alpha\le 1} \alpha \cdot \chi_{f(\tilde{A}_\alpha^{(1)},\dots,\tilde{A}_\alpha^{(p)})}(y).$$

Proof. We have

$$\begin{aligned}
\xi_{\tilde{f}(\tilde{A}^{(1)},\dots,\tilde{A}^{(p)})}(y) &= \sup_{\{(x_1,\dots,x_p):y=f(x_1,\dots,x_p)\}}\min_{1\le i\le p}\left\{\xi_{\tilde{A}^{(i)}}(x_i)\right\} \text{ (using (4.14))}\\
&= \sup_{\{(x_1,\dots,x_p):y=f(x_1,\dots,x_p)\}}\min_{1\le i\le p}\left\{\sup_{\alpha\in[0,1]}\alpha \cdot \chi_{\tilde{A}_\alpha^{(i)}}(x_i)\right\} \text{ (using Theorem 2.2.13).}
\end{aligned} \tag{4.16}$$

On the other hand, from (4.13), we see that

$$\chi_{f(\tilde{A}_\alpha^{(1)},\dots,\tilde{A}_\alpha^{(p)})}(y) = \sup_{\{(x_1,\dots,x_p):y=f(x_1,\dots,x_p)\}}\min_{1\le i\le p}\left\{\chi_{\tilde{A}_\alpha^{(i)}}(x_i)\right\}.$$

Therefore, we have

$$\begin{aligned}\sup_{\alpha\in[0,1]} \alpha\cdot\chi_{f(\tilde{A}_\alpha^{(1)},\ldots,\tilde{A}_\alpha^{(p)})}(x_p) &= \sup_{\alpha\in[0,1]}\alpha\cdot\left(\sup_{\{(x_1,\ldots,x_p):y=f(x_1,\ldots,x_p)\}}\min_{1\le i\le p}\left\{\chi_{\tilde{A}_\alpha^{(i)}}(x_i)\right\}\right)\\ &= \sup_{\alpha\in[0,1]}\sup_{\{(x_1,\ldots,x_p):y=f(x_1,\ldots,x_p)\}}\min_{1\le i\le p}\left\{\alpha\cdot\chi_{\tilde{A}_\alpha^{(i)}}(x_i)\right\}\\ &= \sup_{\{(x_1,\ldots,x_p):y=f(x_1,\ldots,x_p)\}}\sup_{\alpha\in[0,1]}\min_{1\le i\le p}\left\{\alpha\cdot\chi_{\tilde{A}_\alpha^{(i)}}(x_i)\right\}. \end{aligned}\tag{4.17}$$

Using Lemma 4.2.2, the result follows from (4.16) and (4.17) immediately. ■

From 4.15 and Proposition 4.2.3, we see that

$$\xi_{\tilde{f}(\tilde{A}^{(1)},\ldots,\tilde{A}^{(p)})}(y) = \sup_{0<\alpha\le 1}\alpha\cdot\chi_{f(\tilde{A}_\alpha^{(1)},\ldots,\tilde{A}_\alpha^{(p)})}(y) = \sup_{0<\alpha\le 1}\alpha\cdot\chi_{(\tilde{f}(\tilde{A}^{(1)},\ldots,\tilde{A}^{(p)}))_\alpha}(y).$$

However, the equality

$$f\left(\tilde{A}_\alpha^{(1)},\ldots,\tilde{A}_\alpha^{(p)}\right) = \left(\tilde{f}(\tilde{A}^{(1)},\ldots,\tilde{A}^{(p)})\right)_\alpha$$

is not always guaranteed, which will be completely investigated in this chapter. First of all, we have the following interesting inclusion.

Proposition 4.2.4 *Let $f:\mathbb{R}^{n_1}\times\cdots\times\mathbb{R}^{n_p}\to\mathbb{R}^n$ be an onto crisp function, and let $\tilde{f}:\mathcal{F}(\mathbb{R}^{n_1})\times\cdots\times\mathcal{F}(\mathbb{R}^{n_p})\to\mathcal{F}(\mathbb{R}^n)$ be a fuzzy function extended from f via the extension principle defined in (4.14). Then, we have the inclusion*

$$f\left(\tilde{A}_\alpha^{(1)},\ldots,\tilde{A}_\alpha^{(p)}\right) \subseteq \left(\tilde{f}(\tilde{A}^{(1)},\ldots,\tilde{A}^{(p)})\right)_\alpha$$

for each $\alpha\in(0,1]$.

Proof. For $y\in f(\tilde{A}_\alpha^{(1)},\ldots,\tilde{A}_\alpha^{(p)})$, there exists $(x_1,\ldots,x_p)$ satisfying $y=f(x_1,\ldots,x_p)$ and $x_i\in\tilde{A}_\alpha^{(i)}$ for all $i=1,\ldots,p$, which implies $\xi_{\tilde{A}^{(i)}}(x_i)\ge\alpha$ for all $i=1,\ldots,p$, i.e.

$$\min_{1\le i\le p}\left\{\xi_{\tilde{A}^{(i)}}(x_i)\right\}\ge\alpha.$$

Therefore, we have

$$\xi_{\tilde{f}(\tilde{A}^{(1)},\ldots,\tilde{A}^{(p)})}(y) = \sup_{\{(x_1,\ldots,x_p):y=f(x_1,\ldots,x_p)\}}\min_{1\le i\le p}\left\{\xi_{\tilde{A}^{(i)}}(x_i)\right\}\ge\alpha,$$

which says that $y\in(\tilde{f}(\tilde{A}^{(1)},\ldots,\tilde{A}^{(p)}))_\alpha$. This completes the proof. ■

We write

$$\prod_{i=1}^p\mathbb{R}^{n_i}=\mathbb{R}^{n_1}\times\cdots\times\mathbb{R}^{n_p}\text{ and }\prod_{i=1}^p\mathcal{F}(\mathbb{R}^{n_i})=\mathcal{F}(\mathbb{R}^{n_1})\times\cdots\times\mathcal{F}(\mathbb{R}^{n_p}).$$

For convenience, we also write

$$\mathbf{x}=(x_1,\ldots,x_p)\in\mathbb{R}^{n_1}\times\cdots\times\mathbb{R}^{n_p}$$

and

$$\tilde{\mathbf{A}}=\left(\tilde{A}^{(1)},\ldots,\tilde{A}^{(p)}\right)\in\mathcal{F}(\mathbb{R}^{n_1})\times\cdots\times\mathcal{F}(\mathbb{R}^{n_p})$$

and

$$\tilde{\mathbf{A}}_\alpha = \left(\tilde{A}^{(1)}_\alpha, \ldots, \tilde{A}^{(p)}_\alpha\right) \text{ and } \tilde{\mathbf{A}}_{0+} = \left(\tilde{A}^{(1)}_{0+}, \ldots, \tilde{A}^{(p)}_{0+}\right).$$

Next, we present some interesting results.

Theorem 4.2.5 *Let $f : \prod_{i=1}^p \mathbb{R}^{n_i} \to \mathbb{R}^n$ be an onto crisp function, and let $\tilde{f} : \prod_{i=1}^p \mathcal{F}(\mathbb{R}^{n_i}) \to \mathcal{F}(\mathbb{R}^n)$ be a fuzzy function extended from f via the extension principle defined in (4.14). Given any fuzzy sets $\tilde{A}^{(i)}$ in $\mathbb{R}^{n_i}$ such that their membership functions $\xi_{\tilde{A}^{(i)}}$ are upper semi-continuous for $i = 1, \ldots, p$, we have the following properties.*

(i) *Suppose that $\{\mathbf{x} : y = f(\mathbf{x})\}$ are closed and bounded subsets of $\prod_{i=1}^p \mathbb{R}^{n_i}$ for all y in the range of f. Then, the α-level set of $\tilde{f}(\tilde{\mathbf{A}})$ is given by*

$$\left(\tilde{f}(\tilde{\mathbf{A}})\right)_\alpha = f\left(\tilde{\mathbf{A}}_\alpha\right) \text{ for } \alpha \in (0,1]$$

and the 0-level set is given by

$$\left(\tilde{f}(\tilde{\mathbf{A}})\right)_0 = cl\left(f\left(\tilde{\mathbf{A}}_{0+}\right)\right) \subseteq cl\left(f\left(\tilde{\mathbf{A}}_0\right)\right).$$

(ii) *Suppose that the 0-level sets $\tilde{A}^{(i)}_0$ of $\tilde{A}^{(i)}$ are closed and bounded subsets of $\mathbb{R}^{n_i}$ for $i = 1, \ldots, p$, and that $\{\mathbf{x} : y = f(\mathbf{x})\}$ are closed subsets of $\prod_{i=1}^p \mathbb{R}^{n_i}$ for all y in the range of f. Then, the α-level set of $\tilde{f}(\tilde{\mathbf{A}})$ is given by*

$$\left(\tilde{f}(\tilde{\mathbf{A}})\right)_\alpha = f\left(\tilde{\mathbf{A}}_\alpha\right) \text{ for } \alpha \in (0,1]$$

and the 0-level set is given by

$$\left(\tilde{f}(\tilde{\mathbf{A}})\right)_0 = cl\left(f\left(\tilde{\mathbf{A}}_{0+}\right)\right) \subseteq cl\left(f\left(\tilde{\mathbf{A}}_0\right)\right).$$

(iii) *Suppose that the 0-level sets $\tilde{A}^{(i)}_0$ of $\tilde{A}^{(i)}$ are closed and bounded subsets of $\mathbb{R}^{n_i}$ for $i = 1, \ldots, p$, and that the function $f : \prod_{i=1}^p \mathbb{R}^{n_i} \to \mathbb{R}^n$ is continuous. Then, the α-level set of $\tilde{f}(\tilde{\mathbf{A}})$ is given by*

$$\left(\tilde{f}(\tilde{\mathbf{A}})\right)_\alpha = f\left(\tilde{\mathbf{A}}_\alpha\right) \text{ for } \alpha \in [0,1],$$

where the 0-level sets are taken into account.

Proof. To prove part (i), we consider the α-level sets for $\alpha \in (0,1]$. From Proposition 4.2.4, we have the inclusion $f(\tilde{\mathbf{A}}_\alpha) \subseteq (\tilde{f}(\tilde{\mathbf{A}}))_\alpha$. For proving the other direction of inclusion, suppose that $y \in (\tilde{f}(\tilde{\mathbf{A}}))_\alpha$. Then, we have

$$\sup_{\{\mathbf{x}:y=f(\mathbf{x})\}} \min_{1\le i\le p} \left\{\xi_{\tilde{A}^{(i)}}(x_i)\right\} = \xi_{\tilde{f}(\tilde{\mathbf{A}})}(y) \ge \alpha. \tag{4.18}$$

Since $\xi_{\tilde{A}^{(i)}}$ are upper semi-continuous on $\mathbb{R}^{n_i}$ for $i = 1, \ldots, p$, it follows that

$$\tilde{A}^{(i)}_\alpha = \{x : \xi_{\tilde{A}^{(i)}}(x) \ge \alpha\}$$

are closed subsets of $\mathbb{R}^{n_i}$ for $i = 1, \ldots, p$. Since

$$\begin{aligned}\left\{\mathbf{x} : \min_{1\le i\le p} \left\{\xi_{\tilde{A}^{(i)}}(x_i)\right\} \ge \alpha\right\} &= \left\{\mathbf{x} : \xi_{\tilde{A}^{(i)}}(x_i) \ge \alpha \text{ for all } i = 1, \ldots, p\right\} \\ &= \tilde{A}^{(1)}_\alpha \times \cdots \times \tilde{A}^{(p)}_\alpha = \bigcap_{i=1}^p C_{i\alpha}\end{aligned}$$

is a closed subset of $\prod_{i=1}^{p} \mathbb{R}^{n_i}$, it shows that the following function

$$g(x_1, \ldots, x_p) = \min_{1 \le i \le p} \left\{ \xi_{\tilde{A}^{(i)}}(x_i) \right\}$$

is upper semi-continuous on $\prod_{i=1}^{p} \mathbb{R}^{n_i}$. Since $\{\mathbf{x} : y = f(\mathbf{x})\}$ is a closed and bounded subset of $\prod_{i=1}^{p} \mathbb{R}^{n_i}$, Proposition 1.4.4 says that the following supremum is attained

$$\max_{\{\mathbf{x}:y=f(\mathbf{x})\}} \min_{1 \le i \le p} \left\{ \xi_{\tilde{A}^{(i)}}(x_i) \right\} = \sup_{\{\mathbf{x}:y=f(\mathbf{x})\}} \min_{1 \le i \le p} \left\{ \xi_{\tilde{A}^{(i)}}(x_i) \right\}. \tag{4.19}$$

Using (4.18), we have

$$\max_{\{\mathbf{x}:y=f(\mathbf{x})\}} \min_{1 \le i \le p} \left\{ \xi_{\tilde{A}^{(i)}}(x_i) \right\} = \sup_{\{\mathbf{x}:y=f(\mathbf{x})\}} \min_{1 \le i \le p} \left\{ \xi_{\tilde{A}^{(i)}}(x_i) \right\} \ge \alpha. \tag{4.20}$$

Therefore, there exists $\mathbf{x}$ satisfying

$$y = f(\mathbf{x}) \text{ and } \min_{1 \le i \le p} \left\{ \xi_{\tilde{A}^{(i)}}(x_i) \right\} \ge \alpha.$$

This says that $\xi_{\tilde{A}^{(i)}}(x_i) \ge \alpha$ for all $i = 1, \ldots, p$; that is, $x_i \in \tilde{A}_{\alpha}^{(i)}$ for all $i = 1, \ldots, p$. Therefore, we obtain $y \in f(\tilde{\mathbf{A}}_{\alpha})$, which shows the inclusion $(\tilde{f}(\tilde{\mathbf{A}}))_{\alpha} \subseteq f(\tilde{\mathbf{A}}_{\alpha})$.

Now, we consider the 0-level sets. For notational convenience, we write

$$X_0 = f\left(\tilde{\mathbf{A}}_0\right), X_{0+} = f\left(\tilde{\mathbf{A}}_{0+}\right), \text{ and } Y = \left\{ r : \xi_{\tilde{f}(\tilde{A}^{(1)}, \ldots, \tilde{A}^{(p)})}(y) > 0 \right\}.$$

Then, we see that $(\tilde{f}(\tilde{\mathbf{A}}))_0 = \text{cl}(Y)$. Given $y \in Y$, by referring to (4.19), we have

$$0 < \xi_{\tilde{f}(\tilde{\mathbf{A}})}(y) = \sup_{\{\mathbf{x}:y=f(\mathbf{x})\}} \min_{1 \le i \le p} \left\{ \xi_{\tilde{A}^{(i)}}(x_i) \right\} = \max_{\{\mathbf{x}:y=f(\mathbf{x})\}} \min_{1 \le i \le p} \left\{ \xi_{\tilde{A}^{(i)}}(x_i) \right\}.$$

Therefore, there exists $\mathbf{x}$ satisfying $y = f(\mathbf{x})$ and $\xi_{\tilde{A}^{(i)}}(x_i) > 0$ for all i, i.e.

$$x_i \in \tilde{A}_{0+}^{(i)} \subseteq \tilde{A}_0^{(i)} \text{ for all } i = 1, \ldots, p.$$

This also says that

$$y = f(\mathbf{x}) \in X_{0+} \subseteq X_0.$$

Therefore, we obtain the following inclusions

$$Y \subseteq X_{0+} \subseteq X_0,$$

which implies

$$\text{cl}(Y) \subseteq \text{cl}(X_{0+}) \subseteq \text{cl}(X_0).$$

To prove the other direction of inclusion, given any $y \in \text{cl}(X_{0+})$ and any neighborhood N of y, according to the concept of closure, we have $N \cap X_{0+} \neq \emptyset$. Let $\hat{y} \in N \cap X_{0+}$, i.e. $\hat{y} \in X_{0+}$. There exists $\hat{\mathbf{x}}$ satisfying $\hat{y} = f(\hat{\mathbf{x}})$ and $\hat{x}_i \in \tilde{A}_{0+}^{(i)}$ for all i, i.e. $\xi_{\tilde{A}^{(i)}}(\hat{x}_i) > 0$ for all i, which implies

$$\min_{1 \le i \le p} \left\{ \xi_{\tilde{A}^{(i)}}(\hat{x}_i) \right\} > 0 \text{ for all } i.$$

Therefore, we also have

$$\xi_{\tilde{f}(\tilde{\mathbf{A}})}(\hat{y}) = \sup_{\{\mathbf{x}:\hat{y}=f(\mathbf{x})\}} \min_{1 \le i \le p} \left\{ \xi_{\tilde{A}^{(i)}}(x_i) \right\} \ge \min_{1 \le i \le p} \left\{ \xi_{\tilde{A}^{(i)}}(\hat{x}_i) \right\} > 0,$$

which says that $\hat{y} \in Y$, i.e. $N \cap Y \neq \emptyset$. In other words, any neighborhood of y contains points of Y, which says that $y \in \text{cl}(Y)$, i.e. $\text{cl}(X_{0+}) \subseteq \text{cl}(Y)$. Therefore, we obtain $\text{cl}(X_{0+}) = \text{cl}(Y)$.

To prove part (ii), since

$$\min_{1\leq i\leq p}\left\{\xi_{\tilde{A}^{(i)}}(x_i)\right\}=0$$

outside of the set $\tilde{\mathbf{A}}_0$, from (4.18), we have

$$\begin{aligned}\alpha &\leq \sup_{\{\mathbf{x}:y=f(\mathbf{x})\}}\min_{1\leq i\leq p}\left\{\xi_{\tilde{A}^{(i)}}(x_i)\right\}\\ &= \sup_{\{\mathbf{x}:y=f(\mathbf{x})\text{ and }\mathbf{x}\in\tilde{\mathbf{A}}_0\}}\min_{1\leq i\leq p}\left\{\xi_{\tilde{A}^{(i)}}(x_i)\right\}\\ &= \sup_{\{\mathbf{x}:\mathbf{x}\in f^{-1}(\{y\})\cap\tilde{\mathbf{A}}_0\}}\min_{1\leq i\leq p}\left\{\xi_{\tilde{A}^{(i)}}(x_i)\right\}. \end{aligned}\tag{4.21}$$

Since $\tilde{A}_0^{(i)}$ are closed and bounded subsets of $\mathbb{R}^{n_i}$ for $i=1,\ldots,p$, it follows that $\tilde{\mathbf{A}}_0$ is a closed and bounded subset of $\prod_{i=1}^{p}\mathbb{R}^{n_i}$ by Tychonoff's theorem. Since the inverse image

$$f^{-1}(\{y\})=\{\mathbf{x}:y=f(\mathbf{x})\}$$

is a closed subset of $\prod_{i=1}^{p}\mathbb{R}^{n_i}$ by the assumption, it says that $f^{-1}(\{y\})\cap\tilde{\mathbf{A}}_0$ is a closed subset of $\prod_{i=1}^{p}\mathbb{R}^{n_i}$. Since $f^{-1}(\{y\})\cap\tilde{\mathbf{A}}_0$ is also a subset of $\tilde{\mathbf{A}}_0$, it follows that $f^{-1}(\{y\})\cap\tilde{\mathbf{A}}_0$ is a closed and bounded subset of $\prod_{i=1}^{p}\mathbb{R}^{n_i}$. Since the function

$$g(x_1,\ldots,x_p)=\min_{1\leq i\leq p}\left\{\xi_{\tilde{A}^{(i)}}(x_i)\right\}$$

is upper semi-continuous on $\prod_{i=1}^{p}\mathbb{R}^{n_i}$ from the proof of part (i), Proposition 1.4.4 says that the supremum in (4.21) is attained. By referring to (4.20), the remaining proof follows from the proof of part (i).

To prove part (iii), since f is continuous, it says that $f^{-1}(\{y\})$ are closed subsets of $\prod_{i=1}^{p}\mathbb{R}^{n_i}$ for all y in the range of f. Therefore, the desired results follow from part (ii). Now, it remains to consider the 0-level sets; that is, we want to show that $\mathrm{cl}(Y)=X_0$. Since f is continuous and $\tilde{\mathbf{A}}_0$ is a closed and bounded set, part (iii) of Proposition 1.4.3 says that X_0 is a closed and bounded subset of $\mathbb{R}^n$. From the proof of part (i), we have obtained

$$\mathrm{cl}(Y)\subseteq\mathrm{cl}(X_0)=X_0.$$

To prove the other direction of inclusion, for any $y\in X_0$, there exist

$$x_i\in\tilde{A}_0^{(i)}=\mathrm{cl}(\tilde{A}_{0+}^{(i)})\text{ for }i=1,\ldots,p\text{ satisfying }y=f(\mathbf{x}).$$

Since f is continuous, given any neighborhood N of y, there exist neighborhoods N_i of x_i for $i=1,\ldots,p$ satisfying

$$f(N_1,\ldots,N_p)\subseteq N.$$

Since $x_i\in\mathrm{cl}(\tilde{A}_{0+}^{(i)})$ for all i, we have $N_i\cap\tilde{A}_{0+}^{(i)}\neq\emptyset$ for all i by the concept of closure. Now, we take $\hat{x}_i\in N_i\cap\tilde{A}_{0+}^{(i)}$ for all $i=1,\ldots,p$ and set $\hat{y}=f(\hat{\mathbf{x}})$. Then, we see that

$$\hat{y}\in f(N_1,\ldots,N_p)\subseteq N.$$

Since $\hat{x}_i\in\tilde{A}_{0+}^{(i)}$, we also have $\xi_{\tilde{A}^{(i)}}(\hat{x}_i)>0$ for all i. Therefore, we obtain

$$\xi_{\tilde{f}(\tilde{\mathbf{A}})}(\hat{y})=\sup_{\{\mathbf{x}:\hat{y}=f(\mathbf{x})\}}\min_{1\leq i\leq p}\left\{\xi_{\tilde{A}^{(i)}}(x_i)\right\}\geq\min_{1\leq i\leq p}\left\{\xi_{\tilde{A}^{(i)}}(\hat{x}_i)\right\}>0.$$

This says that $\hat{y} \in Y$, i.e. $N \cap Y \neq \emptyset$. In other words, any neighborhood of y contains points of Y, which says that $y \in \text{cl}(Y)$, i.e. $X_0 \subseteq \text{cl}(Y)$. Therefore, we obtain

$$(\tilde{f}(\tilde{\mathbf{A}}))_0 = \text{cl}(Y) = X_0 = f(\tilde{\mathbf{A}}_0).$$

This completes the proof. ■

Given $\mathbf{u}^{(1)}, \mathbf{u}^{(2)} \in \prod_{i=1}^{p} \mathbb{R}^{n_i}$, the vector addition and scalar multiplication in the product space $\prod_{i=1}^{p} \mathbb{R}^{n_i}$ are respectively defined by

$$\mathbf{u}^{(1)} + \mathbf{u}^{(2)} = \left(u_1^{(1)}, \dots, u_p^{(1)}\right) + \left(u_1^{(2)}, \dots, u_p^{(2)}\right) = \left(u_1^{(1)} + u_1^{(2)}, \dots, u_p^{(1)} + u_p^{(2)}\right)$$

and

$$\lambda\left(u_1, \dots, u_p\right) = \left(\lambda u_1, \dots, \lambda u_p\right) \text{ for } \lambda \in \mathbb{R}.$$

Let $f : \prod_{i=1}^{p} \mathbb{R}^{n_i} \to \mathbb{R}^n$ be a crisp function. We say that f is **linear** when

$$f\left(\lambda \mathbf{u}^{(1)} + \gamma \mathbf{u}^{(2)}\right) = \lambda f\left(\mathbf{u}^{(1)}\right) + \gamma f\left(\mathbf{u}^{(2)}\right),$$

where $\lambda, \gamma \in \mathbb{R}$. Next, we are going to investigate the extension principle by considering the fuzzy vector in Definition 2.3.1.

Theorem 4.2.6 *Let $f : \prod_{i=1}^{p} \mathbb{R}^{n_i} \to \mathbb{R}^n$ be a linear, continuous, and onto crisp function, and let $\tilde{f} : \prod_{i=1}^{p} \mathcal{F}(\mathbb{R}^{n_i}) \to \mathcal{F}(\mathbb{R}^n)$ be a fuzzy function extended from f via the extension principle defined in (4.14). Given any fuzzy vectors $\tilde{A}^{(i)} \in \mathfrak{F}(\mathbb{R}^{n_i})$ for $i = 1, \dots, n$, we have $\tilde{f}(\tilde{\mathbf{A}}) \in \mathfrak{F}(\mathbb{R}^n)$, which is also a fuzzy vector in $\mathbb{R}^n$ and its α-level set is given by $(\tilde{f}(\tilde{\mathbf{A}}))_\alpha = f(\tilde{\mathbf{A}}_\alpha)$ for any $\alpha \in [0,1]$, where the 0-level sets are taken into account. In other words, the α-level sets $(\tilde{f}(\tilde{\mathbf{A}}))_\alpha$ are convex, closed, and bounded subsets of $\mathbb{R}^n$ for all $\alpha \in [0,1]$.*

Proof. We are going to apply part (iii) of Theorem 4.2.5 to check the conditions in Definition 2.3.1. Since $\tilde{A}^{(i)} \in \mathfrak{F}(\mathbb{R}^{n_i})$ for all $i = 1, \dots, p$, we have that the 0-level sets $\tilde{A}_0^{(i)}$ are closed and bounded subsets of $\mathbb{R}^{n_i}$ for all $i = 1, \dots, p$. Therefore, we can apply part (iii) of Theorem 4.2.5 to obtain

$$(\tilde{f}(\tilde{\mathbf{A}}))_\alpha = f\left(\tilde{\mathbf{A}}_\alpha\right) \text{ for each } \alpha \in [0,1]. \tag{4.22}$$

Since $\tilde{A}_\alpha^{(i)}$ are convex sets in $\mathbb{R}^{n_i}$ for all i, we want to claim that $f(\tilde{\mathbf{A}}_\alpha)$ is also a convex set in $\mathbb{R}^n$. For any $v_1, v_2 \in f(\tilde{\mathbf{A}}_\alpha)$, there exist $u_i^{(1)}, u_i^{(2)} \in \tilde{A}_\alpha^{(i)}$ for $i = 1, \dots, p$ satisfying

$$v_1 = f\left(u_1^{(1)}, \dots, u_p^{(1)}\right) \text{ and } v_2 = f\left(u_1^{(2)}, \dots, u_p^{(2)}\right).$$

By the linearity of f, it follows that

$$\begin{aligned} \lambda v_1 + (1-\lambda)v_2 &= \lambda f\left(u_1^{(1)}, \dots, u_p^{(1)}\right) + (1-\lambda) f\left(u_1^{(2)}, \dots, u_p^{(2)}\right) \\ &= f\left(\lambda u_1^{(1)} + (1-\lambda)u_1^{(2)}, \dots, \lambda u_1^{(p)} + (1-\lambda)u_p^{(2)}\right). \end{aligned}$$

Since

$$\lambda u_1^{(i)} + (1-\lambda)u_i^{(2)} \in \tilde{A}_\alpha^{(i)}$$

by the convexity of $\tilde{A}_\alpha^{(i)}$ for all i, we have

$$\lambda v_1 + (1-\lambda)v_2 \in f(\tilde{\mathbf{A}}_\alpha),$$

which says that $f(\tilde{\mathbf{A}}_\alpha)$ is a convex set in $\mathbb{R}^n$. From (4.22), we also see that $(\tilde{f}(\tilde{\mathbf{A}}))_\alpha$ is a convex set in $\mathbb{R}^n$.

Since $\tilde{A}^{(i)}_\alpha$ are closed and bounded subsets of $\mathbb{R}^{n_i}$ for all i by Proposition 2.3.2, it follows that $\tilde{\mathbf{A}}_\alpha$ is also a closed and bounded subset of $\prod_{i=1}^p \mathbb{R}^{n_i}$ by the Tychonoff's theorem. Since f is continuous, using part (iii) of Proposition 1.4.3 and (4.22), we see that $(\tilde{f}(\tilde{\mathbf{A}}))_\alpha$ is a closed and bounded subset of $\mathbb{R}^n$. From (4.22), we have $(\tilde{f}(\tilde{\mathbf{A}}))_0 = f(\tilde{\mathbf{A}}_0)$. By the continuity of f, Proposition 1.4.3 says that $(\tilde{f}(\tilde{\mathbf{A}}))_0$ is also a closed and bounded subset of $\mathbb{R}^n$. This completes the proof. ∎

The linearity of f given in Theorem 4.2.6 can be replaced by some other assumptions that will be shown below.

Theorem 4.2.7 *Let $f : \prod_{i=1}^p \mathbb{R}^{n_i} \to \mathbb{R}^n$ be a continuous and onto crisp function, and let $\tilde{f} : \prod_{i=1}^p \mathcal{F}(\mathbb{R}^{n_i}) \to \mathcal{F}(\mathbb{R}^n)$ be a fuzzy function extended from f via the extension principle defined in* (4.14)*. Suppose that $f(A_1, \dots, A_p)$ is a convex set in $\mathbb{R}^n$ for any convex sets A_i in $\mathbb{R}^{n_i}$ and for $i = 1, \dots, p$. Given any fuzzy vectors $\tilde{A}^{(i)} \in \mathfrak{F}(\mathbb{R}^{n_i})$ for $i = 1, \dots, n$, we have $\tilde{f}(\tilde{\mathbf{A}}) \in \mathfrak{F}(\mathbb{R}^n)$, which is also a fuzzy vector in $\mathbb{R}^n$ and its α-level set is $(\tilde{f}(\tilde{\mathbf{A}}))_\alpha = f(\tilde{\mathbf{A}}_\alpha)$ for any $\alpha \in [0,1]$, where the 0-level sets are taken into account. In other words, the α-level sets $(\tilde{f}(\tilde{\mathbf{A}}))_\alpha$ are convex, closed, and bounded subsets of $\mathbb{R}^n$ for all $\alpha \in [0,1]$.*

Proof. According to the proof of Theorem 4.2.6, we just need to show that $(\tilde{f}(\tilde{\mathbf{A}}))_\alpha$ is a convex set in $\mathbb{R}^n$ for each $\alpha \in (0,1]$. Since $\tilde{A}^{(i)}_\alpha$ are convex sets in $\mathbb{R}^{n_i}$ for all $\alpha \in (0,1]$ and $i = 1, \dots, p$, the assumption of f says that $f(\tilde{\mathbf{A}}_\alpha)$ is indeed a convex set in $\mathbb{R}^n$ for each $\alpha \in (0,1]$. This also says that $(\tilde{f}(\tilde{\mathbf{A}}))_\alpha$ is a convex set in $\mathbb{R}^n$ for each $\alpha \in (0,1]$ by (4.22). This completes the proof. ∎

Theorem 4.2.8 *Let $f : \prod_{i=1}^p \mathbb{R}^{n_i} \to \mathbb{R}^n$ be an onto crisp function, and let $\tilde{f} : \prod_{i=1}^p \mathcal{F}(\mathbb{R}^{n_i}) \to \mathcal{F}(\mathbb{R}^n)$ be a fuzzy function extended from f via the extension principle defined in* (4.14)*. Given any fuzzy sets $\tilde{A}^{(i)}$ in $\mathbb{R}^{n_i}$ for $i = 1, \dots, p$, we have the following properties.*

(i) *Suppose that the equality $(\tilde{f}(\tilde{\mathbf{A}}))_\alpha = f(\tilde{\mathbf{A}}_\alpha)$ holds true for each $\alpha \in (0,1]$. Then, for each $y \in \mathbb{R}^n$, the following supremum*

$$\sup_{\{\mathbf{x}: y = f(\mathbf{x})\}} \min_{1 \le i \le p} \left\{ \xi_{\tilde{A}^{(i)}}(x_i) \right\} \tag{4.23}$$

is attained; that is, we have

$$\sup_{\{\mathbf{x}: y = f(\mathbf{x})\}} \min_{1 \le i \le p} \left\{ \xi_{\tilde{A}^{(i)}}(x_i) \right\} = \max_{\{\mathbf{x}: y = f(\mathbf{x})\}} \min_{1 \le i \le p} \left\{ \xi_{\tilde{A}^{(i)}}(x_i) \right\}.$$

(ii) *Suppose that the supremum in (4.23) is attained. Then, we have $(\tilde{f}(\tilde{\mathbf{A}}))_\alpha = f(\tilde{\mathbf{A}}_\alpha)$ for each $\alpha \in (0,1]$ and*

$$\left(\tilde{f}(\tilde{\mathbf{A}})\right)_0 = cl\left(f\left(\tilde{\mathbf{A}}_{0+}\right)\right) \subseteq cl\left(f\left(\tilde{\mathbf{A}}_0\right)\right).$$

(iii) *Suppose that the function $f : \prod_{i=1}^p \mathbb{R}^{n_i} \to \mathbb{R}^n$ is continuous. Then, the equality $(\tilde{f}(\tilde{\mathbf{A}}))_\alpha = f(\tilde{\mathbf{A}}_\alpha)$ holds true for each $\alpha \in [0,1]$ if and only if, for each $y \in \mathbb{R}^n$, the supremum in (4.23) is attained, where the 0-level sets are taken into account.*

Proof. To prove part (i), for $y \in \mathbb{R}^n$, let

$$\xi_{\tilde{f}(\tilde{\mathbf{A}})}(y) = \sup_{\{\mathbf{x}:y=f(\mathbf{x})\}} \min\left\{\xi_{\tilde{A}^{(1)}}(x_1), \ldots, \xi_{\tilde{A}^{(p)}}(x_p)\right\} = \alpha_0.$$

Then $y \in (\tilde{f}(\tilde{\mathbf{A}}))_{\alpha_0}$. If $\alpha_0 = 0$, then

$$\min\{\xi_{\tilde{A}^{(1)}}(\hat{x}_1), \ldots, \xi_{\tilde{A}^{(p)}}(\hat{x}_p)\} = 0$$

for any $\hat{x}_i \in \mathbb{R}^n_i$ and for $i = 1, \ldots, p$ with $y = f(\hat{\mathbf{x}})$. This shows that the supremum is attained. Now, we assume $\alpha_0 > 0$. Since $y \in (\tilde{f}(\tilde{\mathbf{A}}))_{\alpha_0} = f(\tilde{\mathbf{A}}_{\alpha_0})$, there exist $\hat{x}_i \in \tilde{A}^{(i)}_{\alpha_0}$ for $i = 1, \ldots, p$ satisfying $y = f(\hat{\mathbf{x}})$ and

$$\min\left\{\xi_{\tilde{A}^{(1)}}(\hat{x}_1), \ldots, \xi_{\tilde{A}^{(p)}}(\hat{x}_p)\right\} \le \sup_{\{\mathbf{x}:y=f(\mathbf{x})\}} \min\left\{\xi_{\tilde{A}^{(1)}}(x_1), \ldots, \xi_{\tilde{A}^{(p)}}(x_p)\right\} = \alpha_0. \tag{4.24}$$

Since $\hat{x}_i \in \tilde{A}^{(i)}_{\alpha_0}$, i.e. $\xi_{\tilde{A}^{(i)}}(\hat{x}_i) \ge \alpha_0$ for all $i = 1, \ldots, p$, it says that

$$\min\left\{\xi_{\tilde{A}^{(1)}}(\hat{x}_1), \ldots, \xi_{\tilde{A}^{(p)}}(\hat{x}_p)\right\} \ge \alpha_0,$$

which implies

$$\min\left\{\xi_{\tilde{A}^{(1)}}(\hat{x}_1), \ldots, \xi_{\tilde{A}^{(p)}}(\hat{x}_p)\right\} = \alpha_0$$

by (4.24). Therefore, the supremum is attained.

To prove part (ii), we assume that the supremum in (4.23) is attained. From Proposition 4.2.4, it remains to prove the inclusion $(\tilde{f}(\tilde{\mathbf{A}}))_\alpha \subseteq f(\tilde{\mathbf{A}}_\alpha)$. For $y \in (\tilde{f}(\tilde{\mathbf{A}}))_\alpha$, we have

$$\alpha \le \xi_{\tilde{f}(\tilde{\mathbf{A}})}(y) = \sup_{\{\mathbf{x}:y=f(\mathbf{x})\}} \min_{1\le i\le p} \left\{\xi_{\tilde{A}^{(i)}}(x_i)\right\} = \max_{\{\mathbf{x}:y=f(\mathbf{x})\}} \min_{1\le i\le p} \left\{\xi_{\tilde{A}^{(i)}}(x_i)\right\}.$$

Therefore, there exist $\hat{\mathbf{x}}$ satisfying $y = f(\hat{\mathbf{x}})$ and

$$\alpha \le \min_{1\le i\le p} \left\{\xi_{\tilde{A}^{(i)}}(\hat{x}_i)\right\},$$

which implies $\xi_{\tilde{A}^{(i)}}(\hat{x}_i) \ge \alpha$ for all $i = 1, \ldots, p$. Therefore, we have $\hat{x}_i \in \tilde{A}^{(i)}_\alpha$ for all $i = 1, \ldots, p$. In other words, we obtain

$$y = f(\hat{\mathbf{x}}) \in f\left(\tilde{\mathbf{A}}_\alpha\right),$$

which implies the equality $(\tilde{f}(\tilde{\mathbf{A}}))_\alpha = f(\tilde{\mathbf{A}}_\alpha)$ for each $\alpha \in (0,1]$. For the case of 0-level sets, since the supremum in (4.23) is assumed to be attained, using the same arguments in the proof of part (i) of Theorem 4.2.5, we can obtain the desired equalities.

To prove part (iii), suppose that the supremum in (4.23) is attained. Then, from part (ii), we have $(\tilde{f}(\tilde{\mathbf{A}}))_\alpha = f(\tilde{\mathbf{A}}_\alpha)$ for each $\alpha \in (0,1]$. Using the same arguments in the proof of part (iii) of Theorem 4.2.5, we can also obtain $(\tilde{f}(\tilde{\mathbf{A}}))_0 = f(\tilde{\mathbf{A}}_0)$. This completes the proof. ■

4.3 Extension Principle Based on the Triangular Norms

We are going to generalize the above results by considering the generalized t-norm T_p defined in (3.9). Let $f : \prod_{i=1}^p \mathbb{R}^{n_i} \to \mathbb{R}^n$ be an onto crisp function. Given any fuzzy sets $\tilde{A}^{(i)}$ in $\mathbb{R}^{n_i}$ for $i = 1, \ldots, p$, the membership function of fuzzy set $\tilde{f}(\tilde{\mathbf{A}})$ in $\mathbb{R}^n$ is defined by

$$\xi_{\tilde{f}(\tilde{\mathbf{A}})}(y) = \sup_{\{\mathbf{x}:y=f(\mathbf{x})\}} T_p\left(\xi_{\tilde{A}^{(1)}}(x_1), \ldots, \xi_{\tilde{A}^{(p)}}(x_p)\right) \tag{4.25}$$

for $y \in \mathbb{R}^n$. Suppose that we take $t(x, y) = \min\{x, y\}$, which is a t-norm. Then, we see that

$$T_p(\mathbf{x}) = \min\{x_1, \dots, x_p\}.$$

This says that the above definition extends the definition given in (4.14).

Proposition 4.3.1 *Let $f : \prod_{i=1}^{p} \mathbb{R}^{n_i} \to \mathbb{R}^n$ be an onto crisp function, and let $\tilde{f} : \prod_{i=1}^{p} \mathcal{F}(\mathbb{R}^{n_i}) \to \mathcal{F}(\mathbb{R}^n)$ be a fuzzy function extended from f via the extension principle defined in (4.25). Given any fuzzy sets $\tilde{A}^{(i)}$ in $\mathbb{R}^{n_i}$ for $i = 1, \dots, p$, we have the following properties.*

(i) *We have the following inclusions*

$$\bigcup_{\{(\alpha_1,\dots,\alpha_p):T_p(\alpha_1,\dots,\alpha_p)>0\}} f\left(\tilde{A}^{(1)}_{\alpha_1}, \dots, \tilde{A}^{(p)}_{\alpha_p}\right) \subseteq f\left(\tilde{\mathbf{A}}_{0+}\right) \subseteq f\left(\tilde{\mathbf{A}}_0\right) \tag{4.26}$$

$$\bigcup_{\{(\alpha_1,\dots,\alpha_p):T_p(\alpha_1,\dots,\alpha_p)>0\}} f\left(\tilde{A}^{(1)}_{\alpha_1}, \dots, \tilde{A}^{(p)}_{\alpha_p}\right) \subseteq \left(\tilde{f}(\tilde{\mathbf{A}})\right)_{0+} \subseteq \left(\tilde{f}(\tilde{\mathbf{A}})\right)_0 \tag{4.27}$$

$$cl\left(\bigcup_{\{(\alpha_1,\dots,\alpha_p):T_p(\alpha_1,\dots,\alpha_p)>0\}} f\left(\tilde{A}^{(1)}_{\alpha_1}, \dots, \tilde{A}^{(p)}_{\alpha_p}\right)\right) \subseteq \left(\tilde{f}(\tilde{\mathbf{A}})\right)_0. \tag{4.28}$$

(ii) *Suppose that $f : \prod_{i=1}^{p} \mathbb{R}^{n_i} \to \mathbb{R}^n$ is a continuous function. Then, we have the following inclusions*

$$cl\left(\bigcup_{\{(\alpha_1,\dots,\alpha_p):T_p(\alpha_1,\dots,\alpha_p)>0\}} f\left(\tilde{A}^{(1)}_{\alpha_1}, \dots, \tilde{A}^{(p)}_{\alpha_p}\right)\right) \subseteq cl\left(f\left(\tilde{\mathbf{A}}_{0+}\right)\right) \subseteq f\left(\tilde{\mathbf{A}}_0\right). \tag{4.29}$$

We further assume that

$$\alpha_i > 0 \text{ for all } i = 1, \dots, p \text{ imply } T_p(\alpha_1, \dots, \alpha_p) > 0. \tag{4.30}$$

Then, we have the following equalities

$$cl\left(\bigcup_{\{(\alpha_1,\dots,\alpha_p):T_p(\alpha_1,\dots,\alpha_p)>0\}} f\left(\tilde{A}^{(1)}_{\alpha_1}, \dots, \tilde{A}^{(p)}_{\alpha_p}\right)\right) = cl\left(f\left(\tilde{\mathbf{A}}_{0+}\right)\right) = f\left(\tilde{\mathbf{A}}_0\right).$$

Proof. To prove part (i), since $\tilde{A}^{(i)}_{0+} \subseteq \tilde{A}^{(i)}_0$ for $i = 1, \dots, p$, it follows that $f(\tilde{\mathbf{A}}_{0+}) \subseteq f(\tilde{\mathbf{A}}_0)$. Since $\tilde{A}^{(i)}_{\alpha_i} \subseteq \tilde{A}^{(i)}_{0+}$ for any $\alpha_i \in (0,1]$ and $i = 1, \dots, p$, we can obtain inclusion (4.26). Now, for

$$y \in \bigcup_{\{(\alpha_1,\dots,\alpha_p):T_p(\alpha_1,\dots,\alpha_p)>0\}} f\left(\tilde{A}^{(1)}_{\alpha_1}, \dots, \tilde{A}^{(p)}_{\alpha_p}\right),$$

there exist $\beta_1, \dots, \beta_p$ satisfying

$$T_p(\beta_1, \dots, \beta_p) > 0 \text{ and } y \in f(\tilde{A}^{(1)}_{\beta_1}, \dots, \tilde{A}^{(p)}_{\beta_p}).$$

In other words, there exist $\hat{x}_i \in \tilde{A}_{\beta_i}$ for $i = 1, \dots, p$ satisfying $y = f(\hat{\mathbf{x}})$, which also says that $\xi_{\tilde{A}^{(i)}}(\hat{x}_i) \geq \beta_i$ for all $i = 1, \dots, p$. Since T_p is increasing by Remark 3.2.7, we have

$$\begin{aligned}
\xi_{\tilde{f}(\tilde{\mathbf{A}})}(y) &= \sup_{\{\mathbf{x}:y=f(\mathbf{x})\}} T_p\left(\xi_{\tilde{A}^{(1)}}(x_1), \dots, \xi_{\tilde{A}^{(p)}}(x_p)\right) \\
&\geq T_p\left(\xi_{\tilde{A}^{(1)}}(\hat{x}_1), \dots, \xi_{\tilde{A}^{(p)}}(\hat{x}_p)\right) \geq T_p(\beta_1, \dots, \beta_p) > 0.
\end{aligned}$$

This shows that $y \in (\tilde{f}(\tilde{\mathbf{A}}))_{0+}$. Therefore, we obtain the inclusion (4.27). The inclusion shown in (4.28) can be obtained by taking the closure on the inclusions shown in (4.27).

To prove part (ii), since the 0-level sets $\tilde{A}_0^{(i)}$ are closed and bounded subsets of $\mathbb{R}^{n_i}$ for all $i = 1, \ldots, p$, using part (iii) of Proposition 1.4.3, the continuity of f says that $f(\tilde{\mathbf{A}}_0)$ is also a closed and bounded subset of $\mathbb{R}^n$, i.e. $\text{cl}(f(\tilde{\mathbf{A}}_0)) = f(\tilde{\mathbf{A}}_0)$. Therefore, by taking the closure from (4.26), we obtain the inclusions (4.29). Now, we write

$$Y = \bigcup_{\{(\alpha_1,\ldots,\alpha_p):T_p(\alpha_1,\ldots,\alpha_p)>0\}} f\left(\tilde{A}_{\alpha_1}^{(1)}, \ldots, \tilde{A}_{\alpha_p}^{(p)}\right).$$

For any $y \in f(\tilde{\mathbf{A}}_0)$, there exist

$$x_i \in \tilde{A}_0^{(i)} = \text{cl}(\tilde{A}_{0+}^{(i)}) \text{ for } i = 1, \ldots, p \text{ satisfying } y = f(\mathbf{x}).$$

Since f is continuous, given any neighborhood N of y, there exist neighborhoods N_i of x_i for $i = 1, \ldots, p$ satisfying

$$f\left(N_1, \ldots, N_p\right) \subseteq N.$$

Since $x_i \in \text{cl}(\tilde{A}_{0+}^{(i)})$ for all $i = 1, \ldots, p$, according to the concept of closure, we have $N_i \cap \tilde{A}_{0+}^{(i)} \neq \emptyset$ for $i = 1, \ldots, p$. Now, we take $\hat{x}_i \in N_i \cap \tilde{A}_{0+}^{(i)}$ for $i = 1, \ldots, p$ and set $\hat{y} = f(\hat{\mathbf{x}})$. Then, we see that

$$\hat{y} \in f(N_1, \ldots, N_p) \subseteq N.$$

Since $\hat{x}_i \in \tilde{A}_{0+}^{(i)}$, we also have $\xi_{\tilde{A}^{(i)}}(\hat{x}_i) > 0$, i.e. $\hat{x}_i \in \tilde{A}_{\alpha_i}^{(i)}$ for some $\alpha_i > 0$, $i = 1, \ldots, p$, which implies $T_p(\alpha_1, \ldots, \alpha_p) > 0$ by the assumption. This says that $\hat{y} \in Y$, i.e. $N \cap Y \neq \emptyset$. In other words, any neighborhood N of y contains points of Y, which says that $y \in \text{cl}(Y)$, i.e.

$$f\left(\tilde{\mathbf{A}}_0\right) \subseteq \text{cl}\left(\bigcup_{\{(\alpha_1,\ldots,\alpha_p):T_p(\alpha_1,\ldots,\alpha_p)>0\}} f\left(\tilde{A}_{\alpha_1}^{(1)}, \ldots, \tilde{A}_{\alpha_p}^{(p)}\right)\right). \tag{4.31}$$

From (4.29) and (4.31), we obtain the desired equalities, and the proof is complete. ∎

Note that the assumption (4.30) in part (iii) of Proposition 4.3.1 is automatically satisfied when we take

$$T_p(\alpha_1, \ldots, \alpha_p) = \min\left\{\alpha_1, \ldots, \alpha_p\right\}.$$

Theorem 4.3.2 *Let $f : \prod_{i=1}^p \mathbb{R}^{n_i} \to \mathbb{R}^n$ be an onto function, and let $\tilde{f} : \prod_{i=1}^p \mathcal{F}(\mathbb{R}^{n_i}) \to \mathcal{F}(\mathbb{R}^n)$ be a fuzzy function extended from f via the extension principle defined in (4.25). Given any fuzzy sets $\tilde{A}^{(i)}$ in $\mathbb{R}^{n_i}$ for $i = 1, \ldots, p$, we have the following properties.*

(i) *Suppose that the following equality*

$$\left(\tilde{f}(\tilde{\mathbf{A}})\right)_\alpha = \bigcup_{\{(\alpha_1,\ldots,\alpha_p):T_p(\alpha_1,\ldots,\alpha_p)\geq\alpha\}} f\left(\tilde{A}_{\alpha_1}^{(1)}, \ldots, \tilde{A}_{\alpha_p}^{(p)}\right) \tag{4.32}$$

holds true for each $\alpha \in (0,1]$. Then, for each $y \in \mathbb{R}^n$, the following supremum

$$\sup_{\{\mathbf{x}:y=f(\mathbf{x})\}} T_p\left(\xi_{\tilde{A}^{(1)}}(x_1), \ldots, \xi_{\tilde{A}^{(p)}}(x_p)\right) \tag{4.33}$$

is attained.

(ii) *Suppose that the supremum in* (4.33) *is attained for each* $y \in \mathbb{R}^n$. *Then, the equality in* (4.32) *holds true for each* $\alpha \in (0,1]$, *and we also have the following equalities*

$$\left(\tilde{f}(\tilde{\mathbf{A}})\right)_{0+} = \bigcup_{\{(\alpha_1,\ldots,\alpha_p):T_p(\alpha_1,\ldots,\alpha_p)>0\}} f\left(\tilde{A}^{(1)}_{\alpha_1},\ldots,\tilde{A}^{(p)}_{\alpha_p}\right) \tag{4.34}$$

and

$$\left(\tilde{f}(\tilde{\mathbf{A}})\right)_{0} = cl\left(\bigcup_{\{(\alpha_1,\ldots,\alpha_p):T_p(\alpha_1,\ldots,\alpha_p)>0\}} f\left(\tilde{A}^{(1)}_{\alpha_1},\ldots,\tilde{A}^{(p)}_{\alpha_p}\right)\right). \tag{4.35}$$

(iii) *Suppose that* $f : \prod_{i=1}^{p} \mathbb{R}^{n_i} \to \mathbb{R}^n$ *is a continuous function, and that the supremum in* (4.33) *is attained for each* $y \in \mathbb{R}^n$. *Then, the equality in* (4.32) *holds true for each* $\alpha \in (0,1]$, *and we have the following inclusions*

$$\left(\tilde{f}(\tilde{\mathbf{A}})\right)_{0} = cl\left(\bigcup_{\{(\alpha_1,\ldots,\alpha_p):T_p(\alpha_1,\ldots,\alpha_p)>0\}} f\left(\tilde{A}^{(1)}_{\alpha_1},\ldots,\tilde{A}^{(p)}_{\alpha_p}\right)\right) \subseteq cl\left(f\left(\tilde{\mathbf{A}}_{0+}\right)\right) \subseteq f\left(\tilde{\mathbf{A}}_{0}\right).$$

We further assume that

$$\alpha_i > 0 \textit{ for all } i = 1,\ldots,p \textit{ imply } T_p(\alpha_1,\ldots,\alpha_p) > 0.$$

Then, we have the following equalities

$$\left(\tilde{f}(\tilde{\mathbf{A}})\right)_{0} = cl\left(\bigcup_{\{(\alpha_1,\ldots,\alpha_p):T_p(\alpha_1,\ldots,\alpha_p)>0\}} f\left(\tilde{A}^{(1)}_{\alpha_1},\ldots,\tilde{A}^{(p)}_{\alpha_p}\right)\right) = cl\left(f\left(\tilde{\mathbf{A}}_{0+}\right)\right) = f\left(\tilde{\mathbf{A}}_{0}\right).$$

Proof. To prove part (i), given $y \in \mathbb{R}^n$, let

$$\xi_{\tilde{f}(\tilde{\mathbf{A}})}(y) = \sup_{\{\mathbf{x}:y=f(\mathbf{x})\}} T_p\left(\xi_{\tilde{A}^{(1)}}(x_1),\ldots,\xi_{\tilde{A}^{(p)}}(x_p)\right) = \alpha_0 > 0, \tag{4.36}$$

which says that $y \in (\tilde{f}(\tilde{\mathbf{A}}))_{\alpha_0}$. Since $\alpha_0 > 0$, from (4.32), we have

$$y \in \left(\tilde{f}(\tilde{\mathbf{A}})\right)_{\alpha_0} = \bigcup_{\{(\alpha_1,\ldots,\alpha_p):T_p(\alpha_1,\ldots,\alpha_p)\geq\alpha_0\}} f\left(\tilde{A}^{(1)}_{\alpha_1},\ldots,\tilde{A}^{(p)}_{\alpha_p}\right).$$

Therefore, there exist $\beta_1,\ldots,\beta_p$ satisfying

$$T_p(\beta_1,\ldots,\beta_p) \geq \alpha_0 \text{ and } y \in f(\tilde{A}^{(1)}_{\beta_1},\ldots,\tilde{A}^{(p)}_{\beta_p}),$$

i.e. there exist $\hat{x}_i \in \tilde{A}^{(i)}_{\beta_i}$ for $i = 1,\ldots,p$ satisfying $y = f(\hat{\mathbf{x}})$. We also see that $\xi_{\tilde{A}^{(i)}}(\hat{x}_i) \geq \beta_i$ for $i = 1,\ldots,p$, which implies

$$T_p\left(\xi_{\tilde{A}^{(1)}}(\hat{x}_1),\ldots,\xi_{\tilde{A}^{(p)}}(\hat{x}_p)\right) \geq T_p(\beta_1,\ldots,\beta_p) \geq \alpha_0,$$

since T_p is increasing by Remark 3.2.7. Therefore, we obtain

$$\begin{aligned}\alpha_0 &= \sup_{\{\mathbf{x}:y=f(\mathbf{x})\}} T_p\left(\xi_{\tilde{A}^{(1)}}(x_1),\ldots,\xi_{\tilde{A}^{(p)}}(x_p)\right)\\ &\geq T_p\left(\xi_{\tilde{A}^{(1)}}(\hat{x}_1),\ldots,\xi_{\tilde{A}^{(p)}}(\hat{x}_p)\right) \geq T_p(\beta_1,\ldots,\beta_p) \geq \alpha_0,\end{aligned}$$

which shows that

$$T_p(\xi_{\tilde{A}^{(1)}}(\hat{x}_1),\ldots,\xi_{\tilde{A}^{(p)}}(\hat{x}_p)) = \alpha_0 > 0,$$

i.e. the supremum in (4.36) is attained. Now, we assume that $\alpha_0 = 0$. Since T_p is nonnegative, using (4.36), we see that

$$T_p(\xi_{\tilde{A}^{(1)}}(\hat{x}_1), \dots, \xi_{\tilde{A}^{(p)}}(\hat{x}_p)) = 0 = \alpha_0$$

for any $\hat{x}_i \in \mathbb{R}^n_i$ and for all $i = 1, \dots, p$ with $y = f(\hat{\mathbf{x}})$. Therefore, we conclude that the supremum in (4.33) is attained.

To prove part (ii), for

$$y \in \bigcup_{\{(\alpha_1,\dots,\alpha_p):T_p(\alpha_1,\dots,\alpha_p)\geq\alpha\}} f(\tilde{A}^{(1)}_{\alpha_1}, \dots, \tilde{A}^{(p)}_{\alpha_p}),$$

there exist $\beta_1, \dots, \beta_p$ satisfying $T_p(\beta_1, \dots, \beta_p) \geq \alpha$ and $y \in f(\tilde{A}^{(1)}_{\beta_1}, \dots, \tilde{A}^{(p)}_{\beta_p})$. In other words, there exist $\hat{x}_i \in \tilde{A}^{(i)}_{\beta_i}$ for $i = 1, \dots, p$ satisfying $y = f(\hat{\mathbf{x}})$, which also says that $\xi_{\tilde{A}^{(i)}}(\hat{x}_i) \geq \beta_i$ for all $i = 1, \dots, p$. Therefore, we have

$$\begin{aligned}
\xi_{\tilde{f}(\tilde{\mathbf{A}})}(y) &= \sup_{\{\mathbf{x}:y=f(\mathbf{x})\}} T_p\left(\xi_{\tilde{A}^{(1)}}(x_1), \dots, \xi_{\tilde{A}^{(p)}}(x_p)\right) \\
&\geq T_p\left(\xi_{\tilde{A}^{(1)}}(\hat{x}_1), \dots, \xi_{\tilde{A}^{(p)}}(\hat{x}_p)\right) \geq T_p(\beta_1, \dots, \beta_p) \geq \alpha > 0,
\end{aligned}$$

since T_p is increasing by Remark 3.2.7. This shows that $y \in (\tilde{f}(\tilde{\mathbf{A}}))_\alpha$. Therefore, we obtain the inclusion

$$\bigcup_{\{(\alpha_1,\dots,\alpha_p):T_p(\alpha_1,\dots,\alpha_p)\geq\alpha\}} f\left(\tilde{A}^{(1)}_{\alpha_1}, \dots, \tilde{A}^{(p)}_{\alpha_p}\right) \subseteq (\tilde{f}(\tilde{\mathbf{A}}))_\alpha.$$

To prove the other direction of inclusion, given $y \in (\tilde{f}(\tilde{\mathbf{A}}))_\alpha$ for $\alpha > 0$, i.e.

$$\sup_{\{\mathbf{x}:y=f(\mathbf{x})\}} T_p\left(\xi_{\tilde{A}^{(1)}}(x_1), \dots, \xi_{\tilde{A}^{(p)}}(x_p)\right) = \xi_{\tilde{f}(\tilde{\mathbf{A}})}(y) \geq \alpha > 0.$$

Since the supremum is assumed to be attained, there exist $\hat{x}_i \in \mathbb{R}^n_i$ for $i = 1, \dots, p$ satisfying $y = f(\hat{\mathbf{x}})$ and

$$T_p\left(\xi_{\tilde{A}^{(1)}}(\hat{x}_1), \dots, \xi_{\tilde{A}^{(p)}}(\hat{x}_p)\right) = \sup_{\{\mathbf{x}:y=f(\mathbf{x})\}} T_p\left(\xi_{\tilde{A}^{(1)}}(x_1), \dots, \xi_{\tilde{A}^{(p)}}(x_p)\right) \geq \alpha > 0.$$

By taking $\beta_i = \xi_{\tilde{A}^{(i)}}(\hat{x}_i)$ for $i = 1, \dots, p$, we have $T_p(\beta_1, \dots, \beta_p) \geq \alpha > 0$, $\hat{x}_i \in \tilde{A}^{(i)}_{\beta_i}$ for $i = 1, \dots, p$, and

$$y = f(\hat{\mathbf{x}}) \in f(\tilde{A}^{(1)}_{\beta_1}, \dots, \tilde{A}^{(p)}_{\beta_p}).$$

This shows that

$$y \in \bigcup_{\{(\alpha_1,\dots,\alpha_p):T_p(\alpha_1,\dots,\alpha_p)\geq\alpha\}} f(\tilde{A}^{(1)}_{\alpha_1}, \dots, \tilde{A}^{(p)}_{\alpha_p}).$$

Therefore, the equality (4.32) holds true for each $\alpha \in (0,1]$.

For any $\alpha > 0$, from (4.32), we have

$$\begin{aligned}
(\tilde{f}(\tilde{\mathbf{A}}))_\alpha &= \bigcup_{\{(\alpha_1,\dots,\alpha_p):T_p(\alpha_1,\dots,\alpha_p)\geq\alpha\}} f\left(\tilde{A}^{(1)}_{\alpha_1}, \dots, \tilde{A}^{(p)}_{\alpha_p}\right) \\
&\subseteq \bigcup_{\{(\alpha_1,\dots,\alpha_p):T_p(\alpha_1,\dots,\alpha_p)>0\}} f\left(\tilde{A}^{(1)}_{\alpha_1}, \dots, \tilde{A}^{(p)}_{\alpha_p}\right),
\end{aligned}$$

which implies

$$\bigcup_{0<\alpha\leq 1} (\tilde{f}(\tilde{\mathbf{A}}))_\alpha \subseteq \bigcup_{\{(\alpha_1,\ldots,\alpha_p):T_p(\alpha_1,\ldots,\alpha_p)>0\}} f\left(\tilde{A}^{(1)}_{\alpha_1},\ldots,\tilde{A}^{(p)}_{\alpha_p}\right) \tag{4.37}$$

From (2.5), we have

$$\left(\tilde{f}(\tilde{\mathbf{A}})\right)_{0+} = \bigcup_{0<\alpha\leq 1} (\tilde{f}(\tilde{\mathbf{A}}))_\alpha.$$

Using (4.37), we obtain the following inclusion

$$\left(\tilde{f}(\tilde{\mathbf{A}})\right)_{0+} \subseteq \bigcup_{\{(\alpha_1,\ldots,\alpha_p):T_p(\alpha_1,\ldots,\alpha_p)>0\}} f\left(\tilde{A}^{(1)}_{\alpha_1},\ldots,\tilde{A}^{(p)}_{\alpha_p}\right).$$

From (4.27), the equality (4.34) holds true. We also have

$$\mathrm{cl}\left(\left(\tilde{f}(\tilde{\mathbf{A}})\right)_{0+}\right) = \left(\tilde{f}(\tilde{\mathbf{A}})\right)_0,$$

which implies equality (4.35) by taking closure on both sides of (4.34). Finally, part (iii) follows immediately from part (ii) of Proposition 4.3.1. This completes the proof. ∎

Proposition 4.3.3 *Let $f : \prod_{i=1}^p \mathbb{R}^{n_i} \to \mathbb{R}^n$ be an onto crisp function, and let $\tilde{f} : \prod_{i=1}^p \mathcal{F}(\mathbb{R}^{n_i}) \to \mathcal{F}(\mathbb{R}^n)$ be a fuzzy function extended from f via the extension principle defined in* (4.25)*. Then, given any fuzzy sets $\tilde{A}^{(i)}$ in $\mathbb{R}^{n_i}$ for $i = 1,\ldots,p$, we have the following equalities*

$$\left\{f(\mathbf{x}) : T_p\left(\xi_{\tilde{A}^{(1)}}(x_1),\ldots,\xi_{\tilde{A}^{(p)}}(x_p)\right) \geq \alpha\right\} = \bigcup_{\{(\alpha_1,\ldots,\alpha_p):T_p(\alpha_1,\ldots,\alpha_p)\geq\alpha\}} f\left(\tilde{A}^{(1)}_{\alpha_1},\ldots,\tilde{A}^{(p)}_{\alpha_p}\right)$$

for each $\alpha \in (0,1]$ and

$$\left\{f(\mathbf{x}) : T_p\left(\xi_{\tilde{A}^{(1)}}(x_1),\ldots,\xi_{\tilde{A}^{(p)}}(x_p)\right) > 0\right\} = \bigcup_{\{(\alpha_1,\ldots,\alpha_p):T_p(\alpha_1,\ldots,\alpha_p)>0\}} f\left(\tilde{A}^{(1)}_{\alpha_1},\ldots,\tilde{A}^{(p)}_{\alpha_p}\right).$$

Proof. For notational convenience, we write

$$\begin{aligned}
Y_\alpha &= \{f(\mathbf{x}) : T_p\left(\xi_{\tilde{A}^{(1)}}(x_1),\ldots,\xi_{\tilde{A}^{(p)}}(x_p)\right) \geq \alpha\}\\
Y_{0+} &= \{f(\mathbf{x}) : T_p\left(\xi_{\tilde{A}^{(1)}}(x_1),\ldots,\xi_{\tilde{A}^{(p)}}(x_p)\right) > 0\}\\
Z_\alpha &= \bigcup_{\{(\alpha_1,\ldots,\alpha_p):T_p(\alpha_1,\ldots,\alpha_p)\geq\alpha\}} f(\tilde{A}^{(1)}_{\alpha_1},\ldots,\tilde{A}^{(p)}_{\alpha_p})\\
Z_{0+} &= \bigcup_{\{(\alpha_1,\ldots,\alpha_p):T_p(\alpha_1,\ldots,\alpha_p)>0\}} f(\tilde{A}^{(1)}_{\alpha_1},\ldots,\tilde{A}^{(p)}_{\alpha_p})
\end{aligned}$$

For $\alpha \in (0,1]$ and $y \in Y_\alpha$, there exists $\hat{\mathbf{x}} = (\hat{x}_1,\ldots,\hat{x}_p)$ satisfying

$$y = f(\hat{\mathbf{x}}) \text{ and } T_p(\xi_{\tilde{A}^{(1)}}(\hat{x}_1),\ldots,\xi_{\tilde{A}^{(p)}}(\hat{x}_p)) \geq \alpha.$$

Let $\beta_i = \xi_{\tilde{A}^{(i)}}(\hat{x}_i)$ for all $i = 1,\ldots,p$. Then, we see that $T_p(\beta_1,\ldots,\beta_p) \geq \alpha$ and $\hat{x}_i \in \tilde{A}^{(i)}_{\beta_i}$ for all $i = 1,\ldots,p$, i.e.

$$y = f(\hat{\mathbf{x}}) \in f(\tilde{A}^{(1)}_{\beta_1},\ldots,\tilde{A}^{(p)}_{\beta_p}).$$

This shows the inclusion $Y_\alpha \subseteq Z_\alpha$. For $y \in Y_{0+}$, there exists $\hat{\mathbf{x}}$ satisfying

$$y = f(\hat{\mathbf{x}}) \text{ and } T_p(\xi_{\tilde{A}^{(1)}}(\hat{x}_1),\ldots,\xi_{\tilde{A}^{(p)}}(\hat{x}_p)) > 0.$$

The above same arguments can imply the inclusion $Y_{0+} \subseteq Z_{0+}$. On the other hand, for $z \in Z_\alpha$, there exist $\beta_1, \ldots, \beta_p$ satisfying

$$z = f(\hat{\mathbf{x}}) \text{ and } T_p(\beta_1, \ldots, \beta_p) \geq \alpha$$

for some $\hat{x}_i \in \tilde{A}^{(i)}_{\beta_i}$, i.e. $\xi_{\tilde{A}^{(i)}}(\hat{x}_i) \geq \beta_i$, for all $i = 1, \ldots, p$. Since T_p is nondereasing by Remark 3.2.7, we have

$$T_p(\xi_{\tilde{A}^{(1)}}(\hat{x}_1), \ldots, \xi_{\tilde{A}^{(p)}}(\hat{x}_p)) \geq T_p(\beta_1, \ldots, \beta_p) \geq \alpha,$$

which says that $z \in Y_\alpha$. Therefore, we obtain $Y_\alpha = Z_\alpha$. For $z \in Z_{0+}$, there exist $\beta_1, \ldots, \beta_p$ satisfying

$$z = f(\hat{\mathbf{x}}) \text{ and } T_p(\beta_1, \ldots, \beta_p) > 0$$

for some $\hat{x}_i \in \tilde{A}^{(i)}_{\beta_i}$, i.e. $\xi_{\tilde{A}^{(i)}}(\hat{x}_i) \geq \beta_i$, for all $i = 1, \ldots, p$. The above same arguments can imply the equality $Y_{0+} = Z_{0+}$, and the proof is complete. ■

Theorem 4.3.4 *Let $f : \prod_{i=1}^p \mathbb{R}^{n_i} \to \mathbb{R}^n$ be a continuous and onto crisp function, and let $\tilde{f} : \prod_{i=1}^p \mathcal{F}(\mathbb{R}^{n_i}) \to \mathcal{F}(\mathbb{R}^n)$ be a fuzzy function extended from f via the extension principle defined in* (4.25). *Suppose that the generalized t-norm $T_p : [0,1]^p \to [0,1]$ is upper semi-continuous. Given any fuzzy sets $\tilde{A}^{(i)}$ in $\mathbb{R}^{n_i}$ for $i = 1, \ldots, p$, we also assume that the membership functions $\xi_{\tilde{A}^{(i)}}$ are upper semi-continuous and that the 0-level sets $\tilde{A}^{(i)}_0$ are bounded subsets of $\mathbb{R}^{n_i}$ for all $i = 1, \ldots, p$. Then, we have the following properties.*

(i) *The membership function $\xi_{\tilde{f}(\tilde{\mathbf{A}})}$ of fuzzy set $\tilde{f}(\tilde{\mathbf{A}})$ in $\mathbb{R}^n$ is upper semi-continuous.*
(ii) *For each $\alpha \in (0,1]$, the following equalities*

$$\begin{aligned}(\tilde{f}(\tilde{\mathbf{A}}))_\alpha &= \bigcup_{\{(\alpha_1,\ldots,\alpha_p):T_p(\alpha_1,\ldots,\alpha_p)\geq\alpha\}} f\left(\tilde{A}^{(1)}_{\alpha_1}, \ldots, \tilde{A}^{(p)}_{\alpha_p}\right)\\ &= \left\{f(\mathbf{x}) : T_p\left(\xi_{\tilde{A}^{(1)}}(x_1), \ldots, \xi_{\tilde{A}^{(p)}}(x_p)\right) \geq \alpha\right\}\end{aligned}$$

hold true.
(iii) *We have the following inclusions*

$$\begin{aligned}(\tilde{f}(\tilde{\mathbf{A}}))_0 &= cl\left(\bigcup_{\{(\alpha_1,\ldots,\alpha_p):T_p(\alpha_1,\ldots,\alpha_p)>0\}} f\left(\tilde{A}^{(1)}_{\alpha_1}, \ldots, \tilde{A}^{(p)}_{\alpha_p}\right)\right)\\ &\subseteq cl\left(f\left(\tilde{\mathbf{A}}\right)\right) \subseteq f\left(\tilde{\mathbf{A}}_0\right).\end{aligned}$$

We further assume that

$$\alpha_i > 0 \text{ for } i = 1, \ldots, p \text{ imply } T_p(\alpha_1, \ldots, \alpha_p) > 0.$$

Then, we also have the following equalities

$$\begin{aligned}(\tilde{f}(\tilde{\mathbf{A}}))_0 &= cl\left(\bigcup_{\{(\alpha_1,\ldots,\alpha_p):T_p(\alpha_1,\ldots,\alpha_p)>0\}} f\left(\tilde{A}^{(1)}_{\alpha_1}, \ldots, \tilde{A}^{(p)}_{\alpha_p}\right)\right)\\ &= cl\left(f\left(\tilde{\mathbf{A}}_{0+}\right)\right) = f\left(\tilde{\mathbf{A}}_0\right),\end{aligned}$$

and the α-level sets $(\tilde{f}(\tilde{\mathbf{A}}))_\alpha$ of $\tilde{f}(\tilde{\mathbf{A}})$ are closed and bounded subsets of $\mathbb{R}^n$ for all $\alpha \in [0,1]$.

Proof. By part (iii) of Theorem 4.3.2, it suffices to show that, for each $y \in \mathbb{R}^n$, the following supremum

$$\sup_{\{\mathbf{x}:y=f(\mathbf{x})\}} T_p\left(\xi_{\tilde{A}^{(1)}}(x_1), \dots, \xi_{\tilde{A}^{(p)}}(x_p)\right)$$

is attained. We define the function

$$\phi : \prod_{i=1}^{p} \mathbb{R}^{n_i} \to [0,1] \text{ by } \mathbf{x} \mapsto T_p\left(\xi_{\tilde{A}^{(1)}}(x_1), \dots, \xi_{\tilde{A}^{(p)}}(x_p)\right).$$

Since $T_p(\xi_{\tilde{A}^{(1)}}(x_1), \dots, \xi_{\tilde{A}^{(p)}}(x_p)) = 0$ for $\mathbf{x}$ that is outside of the set $\tilde{\mathbf{A}}_0$, we have

$$\sup_{\{\mathbf{x}:y=f(\mathbf{x})\}} T_p\left(\xi_{\tilde{A}^{(1)}}(x_1), \dots, \xi_{\tilde{A}^{(p)}}(x_p)\right) = \sup_{\{\mathbf{x}:\mathbf{x}\in f^{-1}(\{y\})\cap\tilde{\mathbf{A}}_0\}} \phi(\mathbf{x}).$$

Since each $\tilde{A}_0^{(i)}$ is a closed subset of $\mathbb{R}^{n_i}$ for $i = 1, \dots, p$ by the definition of 0-level set, it follows that each $\tilde{A}_0^{(i)}$ is a closed and bounded subset of $\mathbb{R}^{n_i}$ for $i = 1, \dots, p$ by the assumption. Therefore, we see that $\tilde{\mathbf{A}}_0$ is a closed and bounded subset of $\prod_{i=1}^p \mathbb{R}^{n_i}$ by Tychonoff's theorem. Since the singleton $\{y\}$ is a closed subset of $\mathbb{R}^n$, we have that $f^{-1}(\{y\})$ is a closed subset of $\prod_{i=1}^p \mathbb{R}^{n_i}$ by the continuity of f. Hence $f^{-1}(\{y\}) \cap \tilde{\mathbf{A}}_0$ is a closed subset of $\prod_{i=1}^p \mathbb{R}^{n_i}$, which also says that $f^{-1}(\{y\}) \cap \tilde{\mathbf{A}}_0$ is a closed and bounded subset of $\prod_{i=1}^p \mathbb{R}^{n_i}$.

Now, we want to show that the function ϕ is upper semi-continuous. For any fixed $\alpha \in (0,1]$, we need to show that the set

$$\{\mathbf{x} : \phi(\mathbf{x}) \geq \alpha\} = \left\{\mathbf{x} : T_p\left(\xi_{\tilde{A}^{(1)}}(x_1), \dots, \xi_{\tilde{A}^{(p)}}(x_p)\right) \geq \alpha\right\} \tag{4.38}$$

is a closed subset of $\prod_{i=1}^p \mathbb{R}^{n_i}$. Let

$$\Gamma = \{(\alpha_1, \dots, \alpha_p) : T_p(\alpha_1, \dots, \alpha_p) \geq \alpha\}. \tag{4.39}$$

Since T_p is upper semi-continuous, we see that Γ is a closed subset of $[0,1]^p$. Since T_p is increasing by Remark 3.2.7, we see that if $(\hat{\alpha}_1, \dots, \hat{\alpha}_p) \in \Gamma$ then

$$T_p\left(\alpha_1, \dots, \alpha_p\right) \geq T_p\left(\hat{\alpha}_1, \dots, \hat{\alpha}_p\right) \geq \alpha$$

for any $\alpha_i \geq \hat{\alpha}_i$ and for all $i = 1, \dots, p$. This says that

$$[\hat{\alpha}_1, 1] \times \cdots \times [\hat{\alpha}_p, 1] \subseteq \Gamma.$$

Therefore, we have

$$\bigcup_{(\hat{\alpha}_1,\dots,\hat{\alpha}_p)\in\Gamma} [\hat{\alpha}_1, 1] \times \cdots \times [\hat{\alpha}_p, 1] \subseteq \Gamma.$$

Since the other direction of inclusion is obvious, it follows that

$$\Gamma = \bigcup_{(\hat{\alpha}_1,\dots,\hat{\alpha}_p)\in\Gamma} [\hat{\alpha}_1, 1] \times \cdots \times [\hat{\alpha}_p, 1]. \tag{4.40}$$

Since Γ is a closed subset of $[0,1]^p$, from (4.40), we also have

$$\Gamma = [\beta_1, 1] \times \cdots \times [\beta_p, 1] \tag{4.41}$$

for some $\beta_i \in [0,1]$ and $i = 1, \dots, p$. Combining (4.41) and (4.39), we see that

$$T_p(\alpha_1, \dots, \alpha_p) \geq \alpha \text{ if and only if } \alpha_i \geq \beta_i \text{ for } i = 1, \dots, p. \tag{4.42}$$

Therefore, from (4.38) and (4.42), we obtain

$$\{\mathbf{x} : \phi(\mathbf{x}) \geq \alpha\} = \left\{\mathbf{x} : \xi_{\tilde{A}^{(i)}}(x_i) \geq \beta_i \text{ for } i = 1, \ldots, p\right\}. \tag{4.43}$$

Since $\xi_{\tilde{A}^{(i)}}$ are upper semi-continuous for all $i = 1, \ldots, p$, we see that each $\tilde{A}^{(i)}_{\beta_i}$ is a closed subset of $\mathbb{R}^{n_i}$ for all $i = 1, \ldots, p$. Therefore, from (4.43), we also see that

$$\{\mathbf{x} : \phi(\mathbf{x}) \geq \alpha\} = \tilde{A}^{(1)}_{\beta_1} \times \cdots \times \tilde{A}^{(p)}_{\beta_p}$$

is a closed subset of $\prod_{i=1}^{p} \mathbb{R}^{n_i}$. This shows that ϕ is indeed an upper semi-continuous function. Therefore, the function ϕ assumes its maximum on the closed and bounded set $f^{-1}(\{y\}) \cap \tilde{\mathbf{A}}_0$ for all $y \in \mathbb{R}^n$ by Proposition 1.4.4. Therefore, parts (ii) and (iii) can be obtained by Proposition 4.3.3 and part (iii) of Theorem 4.3.2. Using (4.38) and part (ii), we see that the α-level sets $(\tilde{f}(\tilde{\mathbf{A}}))_\alpha$ are closed subsets of $\prod_{i=1}^{p} \mathbb{R}^{n_i}$ for each $\alpha \in (0,1]$, which proves part (i). Moreover, by the continuity of f, since each $\tilde{A}^{(i)}_0$ is a closed and bounded subset of $\mathbb{R}^{n_i}$ for $i = 1, \ldots, p$, part (iii) of Proposition 1.4.3 says that $\tilde{f}(\tilde{\mathbf{A}}_0)$ is a closed and bounded subset of $\mathbb{R}^n$. Since $(\tilde{f}(\tilde{\mathbf{A}}))_0 = f(\tilde{\mathbf{A}}_0)$, we have that $(\tilde{f}(\tilde{\mathbf{A}}))_0$ is a closed and bounded subset of $\mathbb{R}^n$. For each (0,1], since the closed set $(\tilde{f}(\tilde{\mathbf{A}}))_\alpha$ is contained in the closed and bounded set $(\tilde{f}(\tilde{\mathbf{A}}))_0$, part (iii) of Proposition 1.4.4 says that $(\tilde{f}(\tilde{\mathbf{A}}))_\alpha$ is also a closed and bounded set, and the proof is complete. ■

Example 4.3.5 Suppose that the assumptions in Theorem 4.3.4 are all satisfied. We consider many kinds of t-norms by applying part (ii) of Theorem 4.3.4.

- Suppose that the t-norm is given by $t(x, y) = xy$. Then, for each $\alpha \in (0,1]$, we have

$$\begin{aligned}(\tilde{f}(\tilde{A}, \tilde{B}))_\alpha &= \bigcup_{\{(\alpha_1,\alpha_2):\alpha_1\cdot\alpha_2\geq\alpha\}} f\left(\tilde{A}_{\alpha_1}, \tilde{B}_{\alpha_2}\right) \\ &= (\tilde{f}(\tilde{A}, \tilde{B}))_\alpha = \left\{f(x, y) : \xi_{\tilde{A}}(x) \cdot \xi_{\tilde{B}}(y) \geq \alpha\right\}.\end{aligned}$$

- Suppose that the t-norm is given by $t(x, y) = \max\{0, x + y - 1\}$. Then, for each $\alpha \in (0,1]$, we have

$$\begin{aligned}(\tilde{f}(\tilde{A}, \tilde{B}))_\alpha &= \bigcup_{\{(\alpha_1,\alpha_2):\alpha_1+\alpha_2-1\geq\alpha\}} f\left(\tilde{A}_{\alpha_1}, \tilde{B}_{\alpha_2}\right) \\ &= (\tilde{f}(\tilde{A}, \tilde{B}))_\alpha = \left\{f(x, y) : \xi_{\tilde{A}}(x) + \xi_{\tilde{B}}(y) - 1 \geq \alpha\right\}.\end{aligned}$$

4.4 Generalized Extension Principle

We can consider a more general function $\mathfrak{A}$ instead of the generalized t-norm T_p to fuzzify a crisp function. Let $\mathfrak{A} : [0,1]^p \to [0,1]$ be a function defined on $[0,1]^p$, which does not assume any extra conditions. Let $f : \prod_{i=1}^{p} \mathbb{R}^{n_i} \to \mathbb{R}^n$ be an onto crisp function. Given any fuzzy sets $\tilde{A}^{(i)}$ in $\mathbb{R}^{n_i}$ for $i = 1, \ldots, p$, the fuzzy set $\tilde{f}(\tilde{\mathbf{A}})$ in $\mathbb{R}^n$ is defined by

$$\xi_{\tilde{f}(\tilde{\mathbf{A}})}(y) = \sup_{\{\mathbf{x}:y=f(\mathbf{x})\}} \mathfrak{A}\left(\xi_{\tilde{A}^{(1)}}(x_1), \ldots, \xi_{\tilde{A}^{(p)}}(x_p)\right) \tag{4.44}$$

for each $y \in \mathbb{R}^n$, which extends the definition in (4.25).

Proposition 4.4.1 *Let $f : \prod_{i=1}^{p} \mathbb{R}^{n_i} \to \mathbb{R}^n$ be an onto crisp function, and let $\tilde{f} : \prod_{i=1}^{p} \mathcal{F}(\mathbb{R}^{n_i}) \to \mathcal{F}(\mathbb{R}^n)$ be a fuzzy function extended from f via the extension principle defined in (4.44). Given any fuzzy sets $\tilde{A}^{(i)}$ in $\mathbb{R}^{n_i}$ for $i = 1, \ldots, p$, we have the following properties.*

(i) *Suppose that*

$$\mathfrak{A}(\alpha_1, \ldots, \alpha_p) > 0 \text{ implies } \alpha_i > 0 \text{ for all } i = 1, \ldots, p.$$

Then, we have the following inclusion

$$\left\{f(\mathbf{x}) : \mathfrak{A}\left(\xi_{\tilde{A}^{(1)}}(x_1), \ldots, \xi_{\tilde{A}^{(p)}}(x_p)\right) > 0\right\} \subseteq f\left(\tilde{\mathbf{A}}_{0+}\right).$$

(ii) *Suppose that*

$$\alpha_i > 0 \text{ for } i = 1, \ldots, p \text{ imply } \mathfrak{A}(\alpha_1, \ldots, \alpha_p) > 0.$$

Then, we have the following inclusion

$$f\left(\tilde{\mathbf{A}}_{0+}\right) \subseteq \left\{f(\mathbf{x}) : \mathfrak{A}\left(\xi_{\tilde{A}^{(1)}}(x_1), \ldots, \xi_{\tilde{A}^{(p)}}(x_p)\right) > 0\right\}. \tag{4.45}$$

Proof. To prove part (i), for $\mathfrak{A}(\xi_{\tilde{A}^{(1)}}(\hat{x}_1), \ldots, \xi_{\tilde{A}^{(p)}}(\hat{x}_p)) > 0$, the assumption says that $\xi_{\tilde{A}^{(i)}}(\hat{x}_i) > 0$ for all i, i.e. $\hat{x}_i \in \tilde{A}^{(i)}_{0+}$ for all i. This shows the desired inclusion.

To prove part (ii), for $y \in f(\tilde{\mathbf{A}}_{0+})$, there exists $\hat{\mathbf{x}}$ satisfying $y = f(\hat{\mathbf{x}})$ and $\hat{x}_i \in \tilde{A}^{(i)}_{0+}$ for all i, i.e. $\xi_{\tilde{A}^{(i)}}(\hat{x}_i) > 0$, for all i. Therefore, we have $\mathfrak{A}(\xi_{\tilde{A}^{(1)}}(\hat{x}_1), \ldots, \xi_{\tilde{A}^{(p)}}(\hat{x}_p)) > 0$. This shows the desired inclusion. ■

Suppose that we take $\mathfrak{A} = T_p$. Then, the assumption says that

$$\mathfrak{A}(\alpha_1, \ldots, \alpha_p) > 0 \text{ implies } \alpha_i > 0 \text{ for all } i = 1, \ldots, p$$

is satisfied automatically by Remark 3.2.7.

Proposition 4.4.2 *Let $f : \prod_{i=1}^{p} \mathbb{R}^{n_i} \to \mathbb{R}^n$ be an onto crisp function, and let $\tilde{f} : \prod_{i=1}^{p} \mathcal{F}(\mathbb{R}^{n_i}) \to \mathcal{F}(\mathbb{R}^n)$ be a fuzzy function extended from f via the extension principle defined in (4.44). Then, given any fuzzy sets $\tilde{A}^{(i)}$ in $\mathbb{R}^{n_i}$ for $i = 1, \ldots, p$, we have the following inclusions*

$$\left\{f(\mathbf{x}) : \mathfrak{A}\left(\xi_{\tilde{A}^{(1)}}(x_1), \ldots, \xi_{\tilde{A}^{(p)}}(x_p)\right) \geq \alpha\right\} \subseteq \bigcup_{\{(\alpha_1, \ldots, \alpha_p) : \mathfrak{A}(\alpha_1, \ldots, \alpha_p) \geq \alpha\}} f\left(\tilde{A}^{(1)}_{\alpha_1}, \ldots, \tilde{A}^{(p)}_{\alpha_p}\right)$$

for each $\alpha \in (0,1]$ and

$$\left\{f(\mathbf{x}) : \mathfrak{A}\left(\xi_{\tilde{A}^{(1)}}(x_1), \ldots, \xi_{\tilde{A}^{(p)}}(x_p)\right) > 0\right\} \subseteq \bigcup_{\{(\alpha_1, \ldots, \alpha_p) : \mathfrak{A}(\alpha_1, \ldots, \alpha_p) > 0\}} f\left(\tilde{A}^{(1)}_{\alpha_1}, \ldots, \tilde{A}^{(p)}_{\alpha_p}\right).$$

We further assume that $\mathfrak{A}$ is increasing in the sense that

$$\alpha_i \geq \beta_i \text{ for } i = 1, \ldots, p \text{ imply } \mathfrak{A}(\alpha_1, \ldots, \alpha_p) \geq \mathfrak{A}(\beta_1, \ldots, \beta_p).$$

Then, we have the following equalities

$$\left\{f(\mathbf{x}) : \mathfrak{A}\left(\xi_{\tilde{A}^{(1)}}(x_1), \ldots, \xi_{\tilde{A}^{(p)}}(x_p)\right) \geq \alpha\right\} = \bigcup_{\{(\alpha_1, \ldots, \alpha_p) : \mathfrak{A}(\alpha_1, \ldots, \alpha_p) \geq \alpha\}} f\left(\tilde{A}^{(1)}_{\alpha_1}, \ldots, \tilde{A}^{(p)}_{\alpha_p}\right)$$

for each $\alpha \in (0,1]$ and

$$\left\{f(\mathbf{x}) : \mathfrak{A}\left(\xi_{\tilde{A}^{(1)}}(x_1), \ldots, \xi_{\tilde{A}^{(p)}}(x_p)\right) > 0\right\} = \bigcup_{\{(\alpha_1, \ldots, \alpha_p) : \mathfrak{A}(\alpha_1, \ldots, \alpha_p) > 0\}} f\left(\tilde{A}^{(1)}_{\alpha_1}, \ldots, \tilde{A}^{(p)}_{\alpha_p}\right).$$

Proof. For notational convenience, we write

$$Y_\alpha = \{f(\mathbf{x}) : \mathfrak{A}\left(\xi_{\tilde{A}^{(1)}}(x_1), \ldots, \xi_{\tilde{A}^{(p)}}(x_p)\right) \geq \alpha\} \tag{4.46}$$

$$Y_{0+} = \{f(\mathbf{x}) : \mathfrak{A}\left(\xi_{\tilde{A}^{(1)}}(x_1), \ldots, \xi_{\tilde{A}^{(p)}}(x_p)\right) > 0\} \tag{4.47}$$

$$Z_\alpha = \bigcup_{\{(\alpha_1,\ldots,\alpha_p):\mathfrak{A}(\alpha_1,\ldots,\alpha_p)\geq\alpha\}} f(\tilde{A}^{(1)}_{\alpha_1}, \ldots, \tilde{A}^{(p)}_{\alpha_p}) \tag{4.48}$$

$$Z_{0+} = \bigcup_{\{(\alpha_1,\ldots,\alpha_p):\mathfrak{A}(\alpha_1,\ldots,\alpha_p)>0\}} f(\tilde{A}^{(1)}_{\alpha_1}, \ldots, \tilde{A}^{(p)}_{\alpha_p}). \tag{4.49}$$

For $\alpha \in (0,1]$ and $y \in Y_\alpha$, there exists $\hat{\mathbf{x}} = (\hat{x}_1, \ldots, \hat{x}_p)$ satisfying

$$\mathfrak{A}(\xi_{\tilde{A}^{(1)}}(\hat{x}_1), \ldots, \xi_{\tilde{A}^{(p)}}(\hat{x}_p)) \geq \alpha \text{ and } y = f(\hat{\mathbf{x}}).$$

Let $\beta_i = \xi_{\tilde{A}^{(i)}}(\hat{x}_i)$ for all $i = 1, \ldots, p$. Then, we see that $\mathfrak{A}(\beta_1, \ldots, \beta_p) \geq \alpha$ and $\hat{x}_i \in \tilde{A}^{(i)}_{\beta_i}$ for all $i = 1, \ldots, p$, i.e.

$$y = f(\hat{\mathbf{x}}) \in f(\tilde{A}^{(1)}_{\beta_1}, \ldots, \tilde{A}^{(p)}_{\beta_p}).$$

This shows the inclusion $Y_\alpha \subseteq Z_\alpha$. For $y \in Y_{0+}$, there exists $\hat{\mathbf{x}}$ satisfying

$$\mathfrak{A}(\xi_{\tilde{A}^{(1)}}(\hat{x}_1), \ldots, \xi_{\tilde{A}^{(p)}}(\hat{x}_p)) > 0 \text{ and } y = f(\hat{\mathbf{x}}).$$

The above same arguments can imply the inclusion $Y_{0+} \subseteq Z_{0+}$. On the other hand, for $z \in Z_\alpha$, there exists $\beta_1, \ldots, \beta_p$ satisfying

$$\mathfrak{A}(\beta_1, \ldots, \beta_p) \geq \alpha \text{ and } z = f(\hat{\mathbf{x}})$$

for some $\hat{x}_i \in \tilde{A}^{(i)}_{\beta_i}$, i.e. $\xi_{\tilde{A}^{(i)}}(\hat{x}_i) \geq \beta_i$, for all $i = 1, \ldots, p$. Since $\mathfrak{A}$ is assumed to be increasing, we have

$$\mathfrak{A}(\xi_{\tilde{A}^{(1)}}(\hat{x}_1), \ldots, \xi_{\tilde{A}^{(p)}}(\hat{x}_p)) \geq \mathfrak{A}(\beta_1, \ldots, \beta_p) \geq \alpha,$$

which says that $z \in Y_\alpha$. Therefore, we obtain $Y_\alpha = Z_\alpha$. For $z \in Z_{0+}$, there exists $\beta_1, \ldots, \beta_p$ satisfying

$$\mathfrak{A}(\beta_1, \ldots, \beta_p) > 0 \text{ and } z = f(\hat{\mathbf{x}})$$

for some $\hat{x}_i \in \tilde{A}^{(i)}_{\beta_i}$, i.e. $\xi_{\tilde{A}^{(i)}}(\hat{x}_i) \geq \beta_i$, for all $i = 1, \ldots, p$. The above same arguments can imply the equality $Y_{0+} = Z_{0+}$, and the proof is complete. ■

Since T_p is increasing by Remark 3.2.7, we have that the equalities shown in Proposition 4.4.2 holds true when we take $\mathfrak{A} = T_p$.

Proposition 4.4.3 *Let $f : \prod_{i=1}^{p} \mathbb{R}^{n_i} \to \mathbb{R}^n$ be an onto crisp function, and let $\tilde{f} : \prod_{i=1}^{p} \mathcal{F}(\mathbb{R}^{n_i}) \to \mathcal{F}(\mathbb{R}^n)$ be a fuzzy function extended from f via the extension principle defined in (4.44). Then, given any fuzzy sets $\tilde{A}^{(i)}$ in $\mathbb{R}^{n_i}$ for $i = 1, \ldots, p$, we have the following properties.*

(i) *For each $\alpha \in (0,1]$, suppose that*

$$\mathfrak{A}(\alpha_1, \ldots, \alpha_p) \geq \alpha \textit{ implies } \alpha_i \geq \alpha \textit{ for all } i = 1, \ldots, p.$$

Then, we have the following inclusions

$$\bigcup_{\{(\alpha_1,\dots,\alpha_p):\mathfrak{A}(\alpha_1,\dots,\alpha_p)\geq\alpha\}} f\left(\tilde{A}^{(1)}_{\alpha_1},\dots,\tilde{A}^{(p)}_{\alpha_p}\right)\subseteq f\left(\tilde{\mathbf{A}}_\alpha\right) \tag{4.50}$$

and

$$\bigcup_{\{(\alpha_1,\dots,\alpha_p):\mathfrak{A}(\alpha_1,\dots,\alpha_p)>0\}} f\left(\tilde{A}^{(1)}_{\alpha_1},\dots,\tilde{A}^{(p)}_{\alpha_p}\right)\subseteq f\left(\tilde{\mathbf{A}}_{0+}\right). \tag{4.51}$$

(ii) *For each* $\alpha\in(0,1]$, *suppose that*

$$\alpha_i\geq\alpha \textit{ for } i=1,\dots,p \textit{ imply } \mathfrak{A}(\alpha_1,\dots,\alpha_p)\geq\alpha.$$

Then, we have the following inclusions

$$f\left(\tilde{\mathbf{A}}_\alpha\right)\subseteq\left\{f(\mathbf{x}):\mathfrak{A}\left(\xi_{\tilde{A}^{(1)}}(x_1),\dots,\xi_{\tilde{A}^{(p)}}(x_p)\right)\geq\alpha\right\} \tag{4.52}$$

and

$$f\left(\tilde{\mathbf{A}}_{0+}\right)\subseteq\left\{f(\mathbf{x}):\mathfrak{A}\left(\xi_{\tilde{A}^{(1)}}(x_1),\dots,\xi_{\tilde{A}^{(p)}}(x_p)\right)>0\right\}. \tag{4.53}$$

Proof. To prove part (i), using the notations adopted in (4.46)-(4.49), for $z\in Z_\alpha$, there exists $\beta_1,\dots,\beta_p$ satisfying

$$\mathfrak{A}(\beta_1,\dots,\beta_p)\geq\alpha \text{ and } z=f(\hat{\mathbf{x}})$$

for some $\hat{x}_i\in\tilde{A}^{(i)}_{\beta_i}$. By the assumption of $\mathfrak{A}$, we have $\beta_i\geq\alpha$ for all i. Using (2.2), we also have $\hat{x}_i\in\tilde{A}^{(i)}_{\beta_i}\subseteq\tilde{A}^{(i)}_\alpha$ for all i. This shows the inclusion (4.50). The assumption of $\mathfrak{A}$ also says that

$$\mathfrak{A}(\alpha_1,\dots,\alpha_p)>0 \text{ implies } \alpha_i>0 \text{ for all } i=1,\dots,p.$$

Now, for $z\in Z_{0+}$, there exists $\beta_1,\dots,\beta_p$ satisfying

$$\mathfrak{A}(\beta_1,\dots,\beta_p)>0 \text{ and } z=f(\hat{\mathbf{x}})$$

for some $\hat{x}_i\in\tilde{A}^{(i)}_{\beta_i}$. Therefore, we have $\beta_i>0$ and $\hat{x}_i\in\tilde{A}^{(i)}_{\beta_i}\subseteq\tilde{A}^{(i)}_{0+}$ for all i. This shows the inclusion (4.51).

To prove part (ii), for $y\in f(\tilde{\mathbf{A}}_\alpha)$, there exists $\hat{\mathbf{x}}$ satisfying $y=f(\hat{\mathbf{x}})$ and $\hat{x}_i\in\tilde{A}^{(i)}_\alpha$, i.e. $\xi_{\tilde{A}^{(i)}}(\hat{x}_i)\geq\alpha$, for all i. By the assumption of $\mathfrak{A}$, we have

$$\mathfrak{A}(\xi_{\tilde{A}^{(1)}}(\hat{x}_1),\dots,\xi_{\tilde{A}^{(p)}}(\hat{x}_p))\geq\alpha.$$

This shows the inclusion (4.52). The assumption of $\mathfrak{A}$ also says that

$$\alpha_i>0 \text{ for } i=1,\dots,p \text{ imply } \mathfrak{A}(\alpha_1,\dots,\alpha_p)>0.$$

For $y\in f(\tilde{\mathbf{A}}_{0+})$, there exists $\hat{\mathbf{x}}$ satisfying $y=f(\hat{\mathbf{x}})$ and $\hat{x}_i\in\tilde{A}^{(i)}_{0+}$, i.e. $\xi_{\tilde{A}^{(i)}}(\hat{x}_i)>0$, for all i. By the assumption of $\mathfrak{A}$, we have

$$\mathfrak{A}(\xi_{\tilde{A}^{(1)}}(\hat{x}_1),\dots,\xi_{\tilde{A}^{(p)}}(\hat{x}_p))>0.$$

This shows the inclusion (4.53), and the proof is complete. ■

By combining Propositions 4.4.2 and 4.4.3, we can obtain many inclusion and equality relationships. We omit the details here.

Proposition 4.4.4 *Let $f : \prod_{i=1}^{p} \mathbb{R}^{n_i} \to \mathbb{R}^n$ be an onto crisp function, and let $\tilde{f} : \prod_{i=1}^{p} \mathcal{F}(\mathbb{R}^{n_i}) \to \mathcal{F}(\mathbb{R}^n)$ be a fuzzy function extended from f via the extension principle defined in (4.44). Then, given any fuzzy sets $\tilde{A}^{(i)}$ in $\mathbb{R}^{n_i}$ for $i = 1, \ldots, p$, we have the inclusion*

$$\left\{f(\mathbf{x}) : \mathfrak{A}\left(\xi_{\tilde{A}^{(1)}}(x_1), \ldots, \xi_{\tilde{A}^{(p)}}(x_p)\right) \geq \alpha\right\} \subseteq \left(\tilde{f}(\tilde{\mathbf{A}})\right)_\alpha$$

for each $\alpha \in (0,1]$.

Proof. For $\alpha \in (0,1]$ and

$$y \in \left\{f(\mathbf{x}) : \mathfrak{A}(\xi_{\tilde{A}^{(1)}}(x_1), \ldots, \xi_{\tilde{A}^{(p)}}(x_p)) \geq \alpha\right\},$$

there exists **x** satisfying

$$y = f(\mathbf{x}) \text{ and } \mathfrak{A}(\xi_{\tilde{A}^{(1)}}(x_1), \ldots, \xi_{\tilde{A}^{(p)}}(x_p)) \geq \alpha.$$

This shows that

$$\xi_{\tilde{f}(\tilde{\mathbf{A}})}(y) = \sup_{\{\mathbf{x}: y=f(\mathbf{x})\}} \mathfrak{A}\left(\xi_{\tilde{A}^{(1)}}(x_1), \ldots, \xi_{\tilde{A}^{(p)}}(x_p)\right) \geq \alpha,$$

which says that $y \in (\tilde{f}(\tilde{\mathbf{A}}))_\alpha$, and the proof is complete. ■

The following results for considering the 0-level sets are useful for further discussions.

Proposition 4.4.5 *Let $f : \prod_{i=1}^{p} \mathbb{R}^{n_i} \to \mathbb{R}^n$ be an onto crisp function, and let $\tilde{f} : \prod_{i=1}^{p} \mathcal{F}(\mathbb{R}^{n_i}) \to \mathcal{F}(\mathbb{R}^n)$ be a fuzzy function extended from f via the extension principle defined in (4.44). Then, given any fuzzy sets $\tilde{A}^{(i)}$ in $\mathbb{R}^{n_i}$ for $i = 1, \ldots, p$, we have the following properties.*

(i) *We have the following inclusions*

$$\left\{f(\mathbf{x}) : \mathfrak{A}\left(\xi_{\tilde{A}^{(1)}}(x_1), \ldots, \xi_{\tilde{A}^{(p)}}(x_p)\right) > 0\right\} \subseteq \left(\tilde{f}(\tilde{\mathbf{A}})\right)_{0+} \subseteq \left(\tilde{f}(\tilde{\mathbf{A}})\right)_0 \tag{4.54}$$

and

$$cl\left(\left\{f(\mathbf{x}) : \mathfrak{A}\left(\xi_{\tilde{A}^{(1)}}(x_1), \ldots, \xi_{\tilde{A}^{(p)}}(x_p)\right) > 0\right\}\right) \subseteq \left(\tilde{f}(\tilde{\mathbf{A}})\right)_0. \tag{4.55}$$

We further assume that

$$\mathfrak{A}(\alpha_1, \ldots, \alpha_p) > 0 \text{ implies } \alpha_i > 0 \text{ for all } i = 1, \ldots, p.$$

Then, we also have the following inclusions

$$\left\{f(\mathbf{x}) : \mathfrak{A}\left(\xi_{\tilde{A}^{(1)}}(x_1), \ldots, \xi_{\tilde{A}^{(p)}}(x_p)\right) > 0\right\} \subseteq f\left(\tilde{\mathbf{A}}_{0+}\right) \subseteq f\left(\tilde{\mathbf{A}}_0\right). \tag{4.56}$$

(ii) *Suppose that $\tilde{A}_0^{(i)}$ are bounded subsets of $\mathbb{R}^{n_i}$ for all $i = 1, \ldots, p$, and that $f : \prod_{i=1}^{p} \mathbb{R}^{n_i} \to \mathbb{R}^n$ is a continuous function. Then, we have the following properties.*
   - *Assume that*

$$\mathfrak{A}(\alpha_1, \ldots, \alpha_p) > 0 \text{ implies } \alpha_i > 0 \text{ for all } i = 1, \ldots, p.$$

   Then, we have the following inclusions

$$cl\left(\left\{f(\mathbf{x}) : \mathfrak{A}\left(\xi_{\tilde{A}^{(1)}}(x_1), \ldots, \xi_{\tilde{A}^{(p)}}(x_p)\right) > 0\right\}\right) \subseteq cl\left(f\left(\tilde{\mathbf{A}}_{0+}\right)\right) \subseteq f\left(\tilde{\mathbf{A}}_0\right). \tag{4.57}$$

- *Assume that*

$$\mathfrak{A}(\alpha_1,\ldots,\alpha_p)>0 \textit{ if and only if } \alpha_i>0 \textit{ for all } i=1,\ldots,p.$$

Then, we have the following equalities

$$cl\left(\left\{f(\mathbf{x}) : \mathfrak{A}\left(\xi_{\tilde{A}^{(1)}}(x_1),\ldots,\xi_{\tilde{A}^{(p)}}(x_p)\right)>0\right\}\right)=cl\left(f\left(\tilde{\mathbf{A}}_{0+}\right)\right)=f\left(\tilde{\mathbf{A}}_0\right).$$

Proof. To prove part (i), for

$$y\in\{f(\mathbf{x}) : \mathfrak{A}(\xi_{\tilde{A}^{(1)}}(x_1),\ldots,\xi_{\tilde{A}^{(p)}}(x_p))>0\},$$

there exists $\hat{\mathbf{x}}$ satisfying

$$y=f(\hat{\mathbf{x}}) \text{ and } \mathfrak{A}(\xi_{\tilde{A}^{(1)}}(\hat{x}_1),\ldots,\xi_{\tilde{A}^{(p)}}(\hat{x}_p))>0.$$

Therefore, we have

$$\begin{aligned}\xi_{\tilde{f}(\tilde{\mathbf{A}})}(y)&=\sup_{\{\mathbf{x}:y=f(\mathbf{x})\}}\mathfrak{A}\left(\xi_{\tilde{A}^{(1)}}(x_1),\ldots,\xi_{\tilde{A}^{(p)}}(x_p)\right)\\&\geq\mathfrak{A}\left(\xi_{\tilde{A}^{(1)}}(\hat{x}_1),\ldots,\xi_{\tilde{A}^{(p)}}(\hat{x}_p)\right)>0.\end{aligned}$$

This shows that $y\in(\tilde{f}(\tilde{\mathbf{A}}))_{0+}$. Therefore, we obtain the inclusion (4.54). Now, we further assume that

$$\mathfrak{A}(\alpha_1,\ldots,\alpha_p)>0 \text{ implies } \alpha_i>0 \text{ for all } i=1,\ldots,p.$$

For $\mathfrak{A}(\xi_{\tilde{A}^{(1)}}(\hat{x}_1),\ldots,\xi_{\tilde{A}^{(p)}}(\hat{x}_p))>0$, we have $\xi_{\tilde{A}^{(1)}}(\hat{x}_i)>0$ for all i, i.e. $\hat{x}_i\in\tilde{A}^{(i)}_{0+}$ for all i. Since $\tilde{A}^{(i)}_{0+}\subseteq\tilde{A}^{(i)}_0$ for all i, we obtain the inclusion (4.56).

To prove part (ii), since $\tilde{A}^{(i)}_0$ are closed subsets of $\mathbb{R}^{n_i}$ for all i by the definition of 0-level sets, it follows that $\tilde{A}^{(i)}_0$ are closed and bounded subsets of $\mathbb{R}^{n_i}$ for all i. Using part (iii) of Proposition 1.4.3, the continuity of f says that $f(\tilde{\mathbf{A}}_0)$ is a closed and bounded subset of $\mathbb{R}^n$, i.e. $\text{cl}(f(\tilde{\mathbf{A}}_0))=f(\tilde{\mathbf{A}}_0)$. Therefore, by taking closure from (4.56), we obtain the inclusions (4.57). To prove the other direction of inclusion, we write

$$Y=\left\{f(\mathbf{x}) : \mathfrak{A}\left(\xi_{\tilde{A}^{(1)}}(x_1),\ldots,\xi_{\tilde{A}^{(p)}}(x_p)\right)>0\right\}.$$

For any $y\in f(\tilde{\mathbf{A}}_0)$, there exist

$$x_i\in\tilde{A}^{(i)}_0=\text{cl}(\tilde{A}^{(i)}_{0+}) \text{ for } i=1,\ldots,p \text{ satisfying } y=f(\mathbf{x}).$$

Since f is continuous, given any neighborhood N of y, there exist neighborhoods N_i of x_i for $i=1,\ldots,p$ satisfying

$$f(N_1,\ldots,N_p)\subseteq N.$$

Since $x_i\in\text{cl}(\tilde{A}^{(i)}_{0+})$ for all i, we have $N_i\cap\tilde{A}^{(i)}_{0+}\neq\emptyset$ for all i. Now, we take $\hat{x}_i\in N_i\cap\tilde{A}^{(i)}_{0+}$ for all i and set $\hat{y}=f(\hat{\mathbf{x}})$. Then, we see that

$$\hat{y}\in f(N_1,\ldots,N_p)\subseteq N.$$

Since $\hat{x}_i\in\tilde{A}^{(i)}_{0+}$, we also have $\xi_{\tilde{A}^{(i)}}(\hat{x}_i)>0$, i.e. $\hat{x}_i\in\tilde{A}^{(i)}_{\alpha_i}$ for some $\alpha_i>0$, which implies $\mathfrak{A}(\alpha_1,\ldots,\alpha_p)>0$ by the assumption. This says that $\hat{y}\in Y$, i.e. $N\cap Y\neq\emptyset$. In other words, any neighborhood N of y contains points of Y, which says that $y\in\text{cl}(Y)$, i.e.

$$f\left(\tilde{\mathbf{A}}_0\right)\subseteq\text{cl}\left(\left\{f(\mathbf{x}) : \mathfrak{A}\left(\xi_{\tilde{A}^{(1)}}(x_1),\ldots,\xi_{\tilde{A}^{(p)}}(x_p)\right)>0\right\}\right). \tag{4.58}$$

From (4.57) and (4.58), we obtain the desired equalities, and the proof is complete. ■

Proposition 4.4.6 *Let $f : \prod_{i=1}^{p} \mathbb{R}^{n_i} \to \mathbb{R}^n$ be an onto crisp function, and let $\tilde{f} : \prod_{i=1}^{p} \mathcal{F}(\mathbb{R}^{n_i}) \to \mathcal{F}(\mathbb{R}^n)$ be a fuzzy function extended from f via the extension principle defined in (4.44). For each $\alpha \in (0,1]$, suppose that*

$$\alpha_i \geq \alpha \text{ for } i = 1, \ldots, p \text{ imply } \mathfrak{A}(\alpha_1, \ldots, \alpha_p) \geq \alpha.$$

Then, given any fuzzy sets $\tilde{A}^{(i)}$ in $\mathbb{R}^{n_i}$ for $i = 1, \ldots, p$, we have the following inclusions

$$\left(\tilde{\mathbf{A}}_\alpha\right) \subseteq \left(\tilde{f}(\tilde{\mathbf{A}})\right)_\alpha \text{ and } f\left(\tilde{\mathbf{A}}_{0+}\right) \subseteq \left(\tilde{f}(\tilde{\mathbf{A}})\right)_{0+}$$

for each $\alpha \in (0,1]$.

Proof. The desired results follow immediately from part (iii) of Proposition 4.4.3, Proposition 4.4.4 and part (i) of Proposition 4.4.5. ■

Theorem 4.4.7 *Let $f : \prod_{i=1}^{p} \mathbb{R}^{n_i} \to \mathbb{R}^n$ be an onto crisp function, and let $\tilde{f} : \prod_{i=1}^{p} \mathcal{F}(\mathbb{R}^{n_i}) \to \mathcal{F}(\mathbb{R}^n)$ be a fuzzy function extended from f via the extension principle defined in (4.44). Then, given any fuzzy sets $\tilde{A}^{(i)}$ in $\mathbb{R}^{n_i}$ for $i = 1, \ldots, p$, we have the following properties.*

(i) *Suppose that the following equality*

$$\left(\tilde{f}(\tilde{\mathbf{A}})\right)_\alpha = \left\{f(\mathbf{x}) : \mathfrak{A}\left(\xi_{\tilde{A}^{(1)}}(x_1), \ldots, \xi_{\tilde{A}^{(p)}}(x_p)\right) \geq \alpha\right\} \tag{4.59}$$

holds true for each $\alpha \in (0,1]$. Then, the following supremum

$$\sup_{\{\mathbf{x}:y=f(\mathbf{x})\}} \mathfrak{A}\left(\xi_{\tilde{A}^{(1)}}(x_1), \ldots, \xi_{\tilde{A}^{(p)}}(x_p)\right) \tag{4.60}$$

is attained for each $y \in \mathbb{R}^n$.

(ii) *Suppose that the supremum given in (4.60) is attained for each $y \in \mathbb{R}^n$. Then, for each $\alpha \in (0,1]$, equality (4.59) holds true. The results regarding the strong 0-level sets are also given below.*
   - *We have*

$$\left(\tilde{f}(\tilde{\mathbf{A}})\right)_{0+} = \left\{f(\mathbf{x}) : \mathfrak{A}\left(\xi_{\tilde{A}^{(1)}}(x_1), \ldots, \xi_{\tilde{A}^{(p)}}(x_p)\right) > 0\right\}. \tag{4.61}$$

   - *We further assume*

$$\mathfrak{A}(\alpha_1, \ldots, \alpha_p) > 0 \text{ if and only if } \alpha_i > 0 \text{ for all } i = 1, \ldots, p.$$

   Then, we have the equality $(\tilde{f}(\tilde{\mathbf{A}}))_{0+} = f(\tilde{\mathbf{A}}_{0+})$.

(iii) *Suppose that the supremum given in (4.60) is attained for each $y \in \mathbb{R}^n$. Then, for each $\alpha \in (0,1]$, equality (4.59) holds true. The results regarding the 0-level sets are also given below.*
   - *We have*

$$\left(\tilde{f}(\tilde{\mathbf{A}})\right)_0 = cl\left(\left\{f(\mathbf{x}) : \mathfrak{A}\left(\xi_{\tilde{A}^{(1)}}(x_1), \ldots, \xi_{\tilde{A}^{(p)}}(x_p)\right) > 0\right\}\right).$$

   - *We further assume*

$$\mathfrak{A}(\alpha_1, \ldots, \alpha_p) > 0 \text{ if and only if } \alpha_i > 0 \text{ for all } i = 1, \ldots, p.$$

   Then, we have the following equality

$$\left(\tilde{f}(\tilde{\mathbf{A}})\right)_0 = cl\left(f\left(\tilde{\mathbf{A}}_{0+}\right)\right) \subseteq cl\left(f\left(\tilde{\mathbf{A}}_0\right)\right).$$

(iv) *Suppose that $\tilde{A}_0^{(i)}$ are bounded subsets of $\mathbb{R}^{n_i}$ for all $i = 1, \ldots, p$, that $f : \prod_{i=1}^{p} \mathbb{R}^{n_i} \to \mathbb{R}^n$ is a continuous function, and that the supremum given in (4.60) is attained for each $y \in \mathbb{R}^n$. Then, for each $\alpha \in (0,1]$, equality (4.59) holds true. We further assume*

$$\mathfrak{A}(\alpha_1, \ldots, \alpha_p) > 0 \textit{ if and only if } \alpha_i > 0 \textit{ for all } i = 1, \ldots, p.$$

Then, we have the following equalities

$$\left(\tilde{f}(\tilde{\mathbf{A}})\right)_0 = cl\left(\left\{f(\mathbf{x}) : \mathfrak{A}\left(\xi_{\tilde{A}^{(1)}}(x_1), \ldots, \xi_{\tilde{A}^{(p)}}(x_p)\right) > 0\right\}\right) = cl\left(f\left(\tilde{\mathbf{A}}_{0+}\right)\right) = f\left(\tilde{\mathbf{A}}_0\right).$$

Proof. To prove part (i), let $y \in \mathbb{R}^n$ satisfying

$$\xi_{\tilde{f}(\tilde{\mathbf{A}})}(y) = \sup_{\{\mathbf{x}:y=f(\mathbf{x})\}} \mathfrak{A}\left(\xi_{\tilde{A}^{(1)}}(x_1), \ldots, \xi_{\tilde{A}^{(p)}}(x_p)\right) = \alpha_0. \tag{4.62}$$

Suppose that $\alpha_0 = 0$. Then

$$\mathfrak{A}\left(\xi_{\tilde{A}^{(1)}}(\hat{x}_1), \ldots, \xi_{\tilde{A}^{(p)}}(\hat{x}_p)\right) = 0$$

for any $\hat{x}_i \in \mathbb{R}_i^p$ and $i = 1, \ldots, p$ satisfying $y = f(\hat{\mathbf{x}})$, since $\mathfrak{A}$ is a nonnegative function. This shows that the supremum is attained. Now, we assume $\alpha_0 > 0$. Then, we see that $y \in (\tilde{f}(\tilde{\mathbf{A}}))_{\alpha_0}$. Since equality (4.59) holds true, there exists $\hat{\mathbf{x}}$ satisfying

$$y = f(\hat{\mathbf{x}}) \text{ and } \mathfrak{A}\left(\xi_{\tilde{A}^{(1)}}(\hat{x}_1), \ldots, \xi_{\tilde{A}^{(p)}}(\hat{x}_p)\right) \geq \alpha_0. \tag{4.63}$$

From (4.62), we also see that

$$\mathfrak{A}\left(\xi_{\tilde{A}^{(1)}}(\hat{x}_1), \ldots, \xi_{\tilde{A}^{(p)}}(\hat{x}_p)\right) \leq \sup_{\{\mathbf{x}:y=f(\mathbf{x})\}} \mathfrak{A}\left(\xi_{\tilde{A}^{(1)}}(x_1), \ldots, \xi_{\tilde{A}^{(p)}}(x_p)\right) = \alpha_0,$$

which implies

$$\mathfrak{A}(\xi_{\tilde{A}^{(1)}}(\hat{x}_1), \ldots, \xi_{\tilde{A}^{(p)}}(\hat{x}_p)) = \alpha_0$$

by (4.63). Therefore, the supremum is attained.

To prove part (ii), for $\alpha \in (0,1]$, Proposition 4.4.4 has obtained the following inclusion

$$\left\{f(\mathbf{x}) : \mathfrak{A}\left(\xi_{\tilde{A}^{(1)}}(x_1), \ldots, \xi_{\tilde{A}^{(p)}}(x_p)\right) \geq \alpha\right\} \subseteq (\tilde{f}(\tilde{\mathbf{A}}))_\alpha.$$

On the other hand, suppose that $y \in (\tilde{f}(\tilde{\mathbf{A}}))_\alpha$. Then, we have

$$\sup_{\{\mathbf{x}:y=f(\mathbf{x})\}} \mathfrak{A}\left(\xi_{\tilde{A}^{(1)}}(x_1), \ldots, \xi_{\tilde{A}^{(p)}}(x_p)\right) = \xi_{\tilde{f}(\tilde{\mathbf{A}})}(y) \geq \alpha.$$

Since the supremum is attained, there exists $\hat{\mathbf{x}}$ satisfying

$$y = f(\hat{\mathbf{x}}) \text{ and } \mathfrak{A}(\xi_{\tilde{A}^{(1)}}(\hat{x}_1), \ldots, \xi_{\tilde{A}^{(p)}}(\hat{x}_p)) \geq \alpha.$$

Therefore, we obtain the inclusion

$$(\tilde{f}(\tilde{\mathbf{A}}))_\alpha \subseteq \left\{f(\mathbf{x}) : \mathfrak{A}\left(\xi_{\tilde{A}^{(1)}}(x_1), \ldots, \xi_{\tilde{A}^{(p)}}(x_p)\right) \geq \alpha\right\}.$$

This proves the equality (4.59). Now, we consider the strong 0-level sets. For convenience, we write

$$\hat{X} = \left\{f(\mathbf{x}) : \mathfrak{A}\left(\xi_{\tilde{A}^{(1)}}(x_1), \ldots, \xi_{\tilde{A}^{(p)}}(x_p)\right) > 0\right\}$$

and

$$Y = \left\{r : \xi_{\tilde{f}(\tilde{\mathbf{A}})}(y) > 0\right\} = \left(\tilde{f}(\tilde{\mathbf{A}})\right)_{0+}.$$

For $y \in Y$, since the following supremum is attained

$$0 < \xi_{\tilde{f}(\tilde{\mathbf{A}})}(y) = \sup_{\{\mathbf{x}: y = f(\mathbf{x})\}} \mathfrak{A}\left(\xi_{\tilde{A}^{(1)}}(x_1), \ldots, \xi_{\tilde{A}^{(p)}}(x_p)\right),$$

there exists $\hat{\mathbf{x}}$ satisfying

$$\mathfrak{A}(\xi_{\tilde{A}^{(1)}}(\hat{x}_1), \ldots, \xi_{\tilde{A}^{(p)}}(\hat{x}_p)) > 0 \text{ and } y = f(\hat{\mathbf{x}}),$$

which implies $y \in \hat{X}$. Therefore, we obtain the inclusion $Y \subseteq \hat{X}$. To prove the other direction of inclusion, for any $y \in \hat{X}$, there exists $\hat{\mathbf{x}}$ satisfying

$$y = f(\hat{\mathbf{x}}) \text{ and } \mathfrak{A}\left(\xi_{\tilde{A}^{(1)}}(\hat{x}_1), \ldots, \xi_{\tilde{A}^{(p)}}(\hat{x}_p)\right) > 0.$$

Therefore, we also have

$$\begin{aligned}\xi_{\tilde{f}(\tilde{\mathbf{A}})}(y) &= \sup_{\{\mathbf{x}: y = f(\mathbf{x})\}} \mathfrak{A}\left(\xi_{\tilde{A}^{(1)}}(x_1), \ldots, \xi_{\tilde{A}^{(p)}}(x_p)\right) \\ &\geq \mathfrak{A}\left(\xi_{\tilde{A}^{(1)}}(\hat{x}_1), \ldots, \xi_{\tilde{A}^{(p)}}(\hat{x}_p)\right) > 0.\end{aligned}$$

This says that $y \in Y$, i.e. $Y = \hat{X}$. Therefore, we obtain the equality given in (4.61). The equality $(\tilde{f}(\tilde{\mathbf{A}}))_{0+} = f(\tilde{\mathbf{A}}_{0+})$ can be obtained by applying Proposition 4.4.1.

To prove part (iii), since $\tilde{A}^{(i)}_{0+} \subseteq \tilde{A}^{(i)}_{0}$ for $i = 1, \ldots, p$, we have $f(\tilde{\mathbf{A}}_{0+}) \subseteq f(\tilde{\mathbf{A}}_0)$. The desired results follow immediately from part (ii) by taking the closure. Finally, part (iv) follows immediately from part (ii) of Proposition 4.4.5 and part (iii) of this theorem. This completes the proof. ∎

Proposition 4.4.8 *Let $f : \prod_{i=1}^{p} \mathbb{R}^{n_i} \to \mathbb{R}^n$ be an onto crisp function, and let $\tilde{f} : \prod_{i=1}^{p} \mathcal{F}(\mathbb{R}^{n_i}) \to \mathcal{F}(\mathbb{R}^n)$ be a fuzzy function extended from f via the extension principle defined in (4.44). Suppose that the supremum given in (4.60) is attained for each $y \in \mathbb{R}^n$. For each $\alpha \in (0,1]$, we also assume that*

$$\mathfrak{A}(\alpha_1, \ldots, \alpha_p) \geq \alpha \textit{ implies } \alpha_i \geq \alpha \textit{ for all } i = 1, \ldots, p.$$

Then, given any fuzzy sets $\tilde{A}^{(i)}$ in $\mathbb{R}^{n_i}$ for $i = 1, \ldots, p$, we have the inclusion

$$(\tilde{f}(\tilde{\mathbf{A}}))_\alpha \subseteq f(\tilde{\mathbf{A}}_\alpha) \textit{ for each } \alpha \in (0,1].$$

Proof. The desired results follow immediately from part (ii) of Theorem 4.4.7, Proposition 4.4.2, and part (i) of Proposition 4.4.3. ∎

Theorem 4.4.9 *Let $f : \prod_{i=1}^{p} \mathbb{R}^{n_i} \to \mathbb{R}^n$ be an onto crisp function, and let $\tilde{f} : \prod_{i=1}^{p} \mathcal{F}(\mathbb{R}^{n_i}) \to \mathcal{F}(\mathbb{R}^n)$ be a fuzzy function extended from f via the extension principle defined in (4.44). Suppose that the supremum given in (4.60) is attained for each $y \in \mathbb{R}^n$. For each $\alpha \in (0,1]$, we also assume*

$$\mathfrak{A}(\alpha_1, \ldots, \alpha_p) \geq \alpha \textit{ if and only if } \alpha_i \geq \alpha \textit{ for all } i = 1, \ldots, p.$$

Then, we have the following equalities

$$\left(\tilde{f}(\tilde{\mathbf{A}})\right)_\alpha = f\left(\tilde{\mathbf{A}}_\alpha\right) \textit{ and } \left(\tilde{f}(\tilde{\mathbf{A}})\right)_{0+} = f\left(\tilde{\mathbf{A}}_{0+}\right) \textit{ for each } \alpha \in (0,1].$$

We further assume that the function $f : \prod_{i=1}^{p} \mathbb{R}^{n_i} \to \mathbb{R}^n$ is continuous. Then, we also have the equality $(\tilde{f}(\tilde{\mathbf{A}}))_0 = f(\tilde{\mathbf{A}}_0)$. As a matter of fact, the function $\mathfrak{A}$ must be a minimum function given by

$$\mathfrak{A}\left(\alpha_1, \ldots, \alpha_p\right) = \min\left\{\alpha_1, \ldots, \alpha_p\right\}.$$

Proof. For $\alpha \in (0,1]$, the results follows immediately from Propositions 4.4.6 and 4.4.8. It is easy to see that if the following assumption holds true

$$\text{for each } \alpha \in (0,1],\ \mathfrak{A}(\alpha_1, \dots, \alpha_p) \geq \alpha \text{ if and only if } \alpha_i \geq \alpha \text{ for all } i = 1, \dots, p,$$

then the following assumption also holds true

$$\mathfrak{A}(\alpha_1, \dots, \alpha_p) > 0 \text{ if and only if } \alpha_i > 0 \text{ for all } i = 1, \dots, p.$$

For the 0-level sets, the result follows immediately from parts (ii) and (iv) of Theorem 4.4.7. Finally, Proposition 1.4.6 says that the aggregation function $\mathfrak{A}$ must be a minimum function, and the proof is complete. ■

Recall that the function $\mathfrak{A}$ is increasing when

$$\alpha_i \geq \beta_i \text{ for all } i = 1, \dots, p \text{ imply } \mathfrak{A}(\alpha_1, \dots, \alpha_p) \geq \mathfrak{A}(\beta_1, \dots, \beta_p).$$

This definition does not necessarily say that

$$\mathfrak{A}(\alpha_1, \dots, \alpha_p) \geq \mathfrak{A}(\beta_1, \dots, \beta_p) \text{ implies } \alpha_i \geq \beta_i \text{ for all } i = 1, \dots, p.$$

We have the following interesting result.

Lemma 4.4.10 *Given any fuzzy sets $\tilde{A}^{(i)}$ in $\mathbb{R}^{n_i}$ for $i = 1, \dots, p$ such that the membership functions $\xi_{\tilde{A}^{(i)}}$ are upper semi-continuous for all $i = 1, \dots, p$, we define the function $\phi : \prod_{i=1}^{p} \mathbb{R}^{n_i} \to [0,1]$ by*

$$\mathbf{x} \mapsto \mathfrak{A}\left(\xi_{\tilde{A}^{(1)}}(x_1), \dots, \xi_{\tilde{A}^{(p)}}(x_p)\right).$$

Suppose that the function $\mathfrak{A} : [0,1]^p \to [0,1]$ is upper semi-continuous and increasing. Then, the function ϕ is also upper semi-continuous.

Proof. It is clear to see that

$$\{\mathbf{x} : \phi(\mathbf{x}) \geq 0\} = \prod_{i=1}^{p} \mathbb{R}^{n_i}$$

is a closed subset of $\prod_{i=1}^{p} \mathbb{R}^{n_i}$. Therefore, for any fixed $\alpha \in (0,1]$, we need to show that the set

$$\{\mathbf{x} : \phi(\mathbf{x}) \geq \alpha\} = \left\{\mathbf{x} : \mathfrak{A}\left(\xi_{\tilde{A}^{(1)}}(x_1), \dots, \xi_{\tilde{A}^{(p)}}(x_p)\right) \geq \alpha\right\} \tag{4.64}$$

is also a closed subset of $\prod_{i=1}^{p} \mathbb{R}^{n_i}$. Let

$$\Gamma_\alpha = \left\{(\alpha_1, \dots, \alpha_p) : \mathfrak{A}(\alpha_1, \dots, \alpha_p) \geq \alpha\right\}.$$

The upper semi-continuity of function $\mathfrak{A}$ says that Γ_α is a closed subset of $[0,1]^p$. The following inclusion

$$\Gamma_\alpha \subseteq \bigcup_{(\alpha_1^\circ, \dots, \alpha_p^\circ) \in \Gamma_\alpha} [\alpha_1^\circ, 1] \times \cdots \times [\alpha_p^\circ, 1]$$

is obvious. Suppose that $(\alpha_1^\circ, \dots, \alpha_p^\circ) \in \Gamma_\alpha$. Given any α_i satisfying $\alpha_i \geq \alpha_i^\circ$ for all $i = 1, \dots, p$, the increasing assumption of $\mathfrak{A}$ says that

$$\mathfrak{A}(\alpha_1, \dots, \alpha_p) \geq \mathfrak{A}(\alpha_1^\circ, \dots, \alpha_p^\circ) \geq \alpha,$$

i.e. $(\alpha_1, \dots, \alpha_p) \in \Gamma_\alpha$, which implies

$$[\alpha_1^\circ, 1] \times \cdots \times [\alpha_p^\circ, 1] \subseteq \Gamma_\alpha.$$

Therefore, we obtain the inclusion

$$\bigcup_{(\alpha_1^\circ,\ldots,\alpha_p^\circ)\in\Gamma_\alpha} [\alpha_1^\circ,1]\times\cdots\times[\alpha_p^\circ,1]\subseteq\Gamma_\alpha,$$

which also says that

$$\Gamma_\alpha = \bigcup_{(\alpha_1^\circ,\ldots,\alpha_p^\circ)\in\Gamma_\alpha} [\alpha_1^\circ,1]\times\cdots\times[\alpha_p^\circ,1]. \tag{4.65}$$

Since Γ_α is a closed subset of $[0,1]^p$, from (4.65), we must have the form

$$\Gamma_\alpha = [\beta_1,1]\times\cdots\times[\beta_p,1]$$

for some $\beta_i \in [0,1]$ for all $i = 1,\ldots,p$, which also says that

$$\mathfrak{A}(\alpha_1,\ldots,\alpha_p) \geq \alpha \text{ implies } \alpha_i \geq \beta_i \text{ for all } i = 1,\ldots,p.$$

Therefore, from (4.64), we obtain

$$\{\mathbf{x} : \phi(\mathbf{x}) \geq \alpha\} = \left\{\mathbf{x} : \xi_{\tilde{A}^{(i)}}(x_i) \geq \beta_i \text{ for } i = 1,\ldots,p\right\}. \tag{4.66}$$

The upper semi-continuity of $\xi_{\tilde{A}^{(i)}}$ says that $\tilde{A}^{(i)}_{\beta_i}$ are closed subsets of $\mathbb{R}^{n_i}$ for $\beta_i \in (0,1]$ and $i = 1,\ldots,p$. Now, we write $A_{\beta_i} = \tilde{A}^{(i)}_{\beta_i}$ for $\beta_i \in (0,1]$ and $A_{i0} = \mathbb{R}^{n_i}$ for $i = 1,\ldots,p$. It is clear to see that each A_{β_i} is a closed subset of $\mathbb{R}^{n_i}$ for $i = 1,\ldots,p$. Using (4.66), we also see that

$$\{\mathbf{x} : \phi(\mathbf{x}) \geq \alpha\} = A_{\beta_1}\times\cdots\times A_{\beta_p}$$

is a closed subset of $\prod_{i=1}^p \mathbb{R}^{n_i}$. This completes the proof. ∎

In order to apply Theorem 4.4.7, we need to check that the supremum given in (4.60) is attained. Now, we provide some sufficient conditions to achieve this purpose.

Proposition 4.4.11 *Let* $f : \prod_{i=1}^p \mathbb{R}^{n_i} \to \mathbb{R}^n$ *be an onto crisp function, and let* $\tilde{f} : \prod_{i=1}^p \mathcal{F}(\mathbb{R}^{n_i}) \to \mathcal{F}(\mathbb{R}^n)$ *be a fuzzy function extended from f via the extension principle defined in (4.44). Given any fuzzy sets* $\tilde{A}^{(i)}$ *in* $\mathbb{R}^{n_i}$ *for* $i = 1,\ldots,p$ *such that the membership functions* $\xi_{\tilde{A}^{(i)}}$ *are upper semi-continuous for all* $i = 1,\ldots,p$*, suppose that the function* $\mathfrak{A} : [0,1]^p \to [0,1]$ *is upper semi-continuous and increasing, and that any one of the following statements holds true:*

(a) *The sets* $\{\mathbf{x} : y = f(\mathbf{x})\}$ *are closed and bounded subsets of* $\prod_{i=1}^p \mathbb{R}^{n_i}$ *for all y in the range of f.*
(b) *The sets* $\{\mathbf{x} : y = f(\mathbf{x})\}$ *are bounded subsets of* $\prod_{i=1}^p \mathbb{R}^{n_i}$ *for all y in the range of f, and the function* $f : \prod_{i=1}^p \mathbb{R}^{n_i} \to \mathbb{R}^n$ *is continuous.*
(c) *The 0-level sets* $\tilde{A}^{(i)}_0$ *of* $\tilde{A}^{(i)}$ *are bounded subsets of* $\mathbb{R}^{n_i}$ *for* $i = 1,\ldots,p$*, and* $\{\mathbf{x} : y = f(\mathbf{x})\}$ *are closed subsets of* $\prod_{i=1}^p \mathbb{R}^{n_i}$ *for all y in the range of f; we further assume that*

$$\text{if any one of } \{\alpha_1,\ldots,\alpha_p\} \text{ is zero, then } \mathfrak{A}(\alpha_1,\ldots,\alpha_p) = 0. \tag{4.67}$$

(d) *The 0-level sets* $\tilde{A}^{(i)}_0$ *of* $\tilde{A}^{(i)}$ *are bounded subsets of* $\mathbb{R}^{n_i}$ *for* $i = 1,\ldots,p$*, and the function* $f : \prod_{i=1}^p \mathbb{R}^{n_i} \to \mathbb{R}^n$ *is continuous; we further assume that*

$$\text{if any one of } \{\alpha_1,\ldots,\alpha_p\} \text{ is zero, then } \mathfrak{A}(\alpha_1,\ldots,\alpha_p) = 0.$$

Then, the supremum given in (4.60) is attained.

Proof. Considering statement (a), Lemma 4.4.10 says that the function ϕ is upper semi-continuous. Therefore, the supremum given in *(4.60)* is attained by Proposition 1.4.4.

Considering statement (b), since f is continuous, it says that the inverse images

$$f^{-1}(\{y\}) = \{\mathbf{x} : y = f(\mathbf{x})\}$$

are closed subsets of $\prod_{i=1}^{p} \mathbb{R}^{n_i}$ for all y in the range of f. Therefore, the desired results follow immediately by referring to the arguments of considering statement (a).

Considering statement (c), since $\mathfrak{A}(\xi_{\tilde{A}^{(1)}}(x_1), \ldots, \xi_{\tilde{A}^{(p)}}(x_p)) = 0$ outside of the set $\tilde{\mathbf{A}}_0$ by the further assumption (4.67) of $\mathfrak{A}$, we have

$$\begin{aligned}&\sup_{\{\mathbf{x}:y=f(\mathbf{x})\}} \mathfrak{A}\left(\xi_{\tilde{A}^{(1)}}(x_1), \ldots, \xi_{\tilde{A}^{(p)}}(x_p)\right)\\ &= \sup_{\{\mathbf{x}:y=f(\mathbf{x}) \text{ and } \mathbf{x}\in\tilde{\mathbf{A}}_0\}} \mathfrak{A}\left(\xi_{\tilde{A}^{(1)}}(x_1), \ldots, \xi_{\tilde{A}^{(p)}}(x_p)\right) = \sup_{\{\mathbf{x}:\mathbf{x}\in f^{-1}(\{y\})\cap\tilde{\mathbf{A}}_0\}} \phi(\mathbf{x}). \qquad (4.68)\end{aligned}$$

Since $\tilde{A}_0^{(i)}$ are closed subsets of $\mathbb{R}^{n_i}$ for $i = 1, \ldots, p$ by the definition of 0-level set, it follows that $\tilde{\mathbf{A}}_0$ is a closed and bounded subset of $\prod_{i=1}^{p} \mathbb{R}^{n_i}$ by Tychonoff's theorem. Since

$$f^{-1}(\{y\}) = \{\mathbf{x} : y = f(\mathbf{x})\}$$

is a closed subset of $\prod_{i=1}^{p} \mathbb{R}^{n_i}$ by the assumption, we have that $f^{-1}(\{y\}) \cap \tilde{\mathbf{A}}_0$ is a closed and bounded subset of $\prod_{i=1}^{p} \mathbb{R}^{n_i}$. Since ϕ is upper semi-continuous by Lemma 4.4.10, using (4.68), Proposition 1.4.4 says that the supremum given in (4.60) is attained.

Considering statement (d), the continuity of f says that the inverse images

$$f^{-1}(\{y\}) = \{\mathbf{x} : y = f(\mathbf{x})\}$$

are closed subsets of $\prod_{i=1}^{p} \mathbb{R}^{n_i}$ for all y in the range of f. Therefore, the desired result follows immediately by referring to statement (c). This completes the proof. ∎

Notice that Proposition 4.4.11 is applicable to Theorems 4.4.7 and 4.4.9. In what follows, we are going to apply Proposition 4.4.11 to obtain more useful results.

Theorem 4.4.12 *Let $f : \prod_{i=1}^{p} \mathbb{R}^{n_i} \to \mathbb{R}^n$ be a continuous and onto crisp function, and let $\tilde{f} : \prod_{i=1}^{p} \mathcal{F}(\mathbb{R}^{n_i}) \to \mathcal{F}(\mathbb{R}^n)$ be a fuzzy function extended from f via the extension principle defined in* (4.44). *Suppose that the function $\mathfrak{A} : [0,1]^p \to [0,1]$ satisfies the following conditions:*

- *$\mathfrak{A}$ is upper semi-continuous and increasing;*
- *if any one of $\{\alpha_1, \ldots, \alpha_p\}$ is zero, then $\mathfrak{A}(\alpha_1, \ldots, \alpha_p) = 0$.*

Given any fuzzy sets $\tilde{A}^{(i)}$ in $\mathbb{R}^{n_i}$ for $i = 1, \ldots, p$, suppose that the membership functions $\xi_{\tilde{A}^{(i)}}$ are upper semi-continuous for all $i = 1, \ldots, p$ and that the 0-level sets $\tilde{A}_0^{(i)}$ are bounded subsets of $\mathbb{R}^{n_i}$ for all $i = 1, \ldots, p$. Then, we have the following results.

(i) *The membership function $\xi_{\tilde{f}(\tilde{\mathbf{A}})}$ of fuzzy set $\tilde{f}(\tilde{\mathbf{A}})$ in $\mathbb{R}^n$ is upper semi-continuous.*
(ii) *For each $\alpha \in (0,1]$, we have the following equalities*

$$\begin{aligned}\left(\tilde{f}(\tilde{\mathbf{A}})\right)_\alpha &= \bigcup_{\{(\alpha_1,\ldots,\alpha_p):\mathfrak{A}(\alpha_1,\ldots,\alpha_p)\geq\alpha\}} f\left(\tilde{A}_{\alpha_1}^{(1)}, \ldots, \tilde{A}_{\alpha_p}^{(p)}\right)\\ &= \left\{f(\mathbf{x}) : \mathfrak{A}\left(\xi_{\tilde{A}^{(1)}}(x_1), \ldots, \xi_{\tilde{A}^{(p)}}(x_p)\right) \geq \alpha\right\}.\end{aligned}$$

Regarding the 0-level sets, we further assume

$$\mathfrak{A}(\alpha_1, \ldots, \alpha_p) > 0 \text{ if and only if } \alpha_i > 0 \text{ for all } i = 1, \ldots, p.$$

Then, we have the following equalities

$$\begin{aligned}(\tilde{f}(\tilde{\mathbf{A}}))_{0+} &= \bigcup_{\{(\alpha_1,\ldots,\alpha_p):\mathfrak{A}(\alpha_1,\ldots,\alpha_p)>0\}} f\left(\tilde{A}^{(1)}_{\alpha_1}, \ldots, \tilde{A}^{(p)}_{\alpha_p}\right)\\ &= \left\{f(\mathbf{x}) : \mathfrak{A}\left(\xi_{\tilde{A}^{(1)}}(x_1), \ldots, \xi_{\tilde{A}^{(p)}}(x_p)\right) > 0\right\} = f\left(\tilde{\mathbf{A}}_{0+}\right)\end{aligned}$$

and

$$\begin{aligned}(\tilde{f}(\tilde{\mathbf{A}}))_{0} &= cl\left(\bigcup_{\{(\alpha_1,\ldots,\alpha_p):\mathfrak{A}(\alpha_1,\ldots,\alpha_p)>0\}} f\left(\tilde{A}^{(1)}_{\alpha_1}, \ldots, \tilde{A}^{(p)}_{\alpha_p}\right)\right)\\ &= cl\left(\left\{f(\mathbf{x}) : \mathfrak{A}\left(\xi_{\tilde{A}^{(1)}}(x_1), \ldots, \xi_{\tilde{A}^{(p)}}(x_p)\right) > 0\right\}\right)\\ &= cl\left(f\left(\tilde{\mathbf{A}}_{0+}\right)\right) = f\left(\tilde{\mathbf{A}}_{0}\right).\end{aligned}$$

Moreover, the α-level sets $(\tilde{f}(\tilde{\mathbf{A}}))_\alpha$ of $\tilde{f}(\tilde{\mathbf{A}})$ are closed and bounded subsets of $\mathbb{R}^n$ for all $\alpha \in [0,1]$.

Proof. From Lemma 4.4.10, we see that

$$\{\mathbf{x} : \phi(\mathbf{x}) \geq \alpha\} = \left\{\mathbf{x} : \mathfrak{A}\left(\xi_{\tilde{A}^{(1)}}(x_1), \ldots, \xi_{\tilde{A}^{(p)}}(x_p)\right) \geq \alpha\right\}$$

is a closed subset of $\prod_{i=1}^{p} \mathbb{R}^{n_i}$ for each $\alpha \in (0,1]$. By referring to statement (d) of Proposition 4.4.11, the supremum given in *(4.60)* is attained. Using part (iv) of Theorem 4.4.7, we see that the α-level sets

$$\left(\tilde{f}(\tilde{\mathbf{A}})\right)_\alpha = \left\{\mathbf{x} : \mathfrak{A}\left(\xi_{\tilde{A}^{(1)}}(x_1), \ldots, \xi_{\tilde{A}^{(p)}}(x_p)\right) \geq \alpha\right\} = \{\mathbf{x} : \phi(\mathbf{x}) \geq \alpha\}$$

are closed subsets of $\prod_{i=1}^{p} \mathbb{R}^{n_i}$ for each $\alpha \in (0,1]$, which proves part (i).

For the α-level sets with $\alpha \in (0,1]$ and the 0-level sets, the equalities shown in part (ii) can also be obtained by Proposition 4.4.2 and Theorem 4.4.7. Moreover, by the continuity of f, since each $\tilde{A}^{(i)}_0$ is a closed and bounded subset of $\mathbb{R}^{n_i}$ for $i = 1, \ldots, p$, part (iii) of Proposition 1.4.4 says that $\tilde{f}(\tilde{\mathbf{A}}_0)$ is a closed and bounded subset of $\mathbb{R}^n$. Since $(\tilde{f}(\tilde{\mathbf{A}}))_0 = f(\tilde{\mathbf{A}}_0)$, it says that $(\tilde{f}(\tilde{\mathbf{A}}))_0$ is a closed and bounded subset of $\mathbb{R}^n$. For each $(0,1]$, since the closed set $(\tilde{f}(\tilde{\mathbf{A}}))_\alpha$ is contained in the bounded set $(\tilde{f}(\tilde{\mathbf{A}}))_0$, it follows that $(\tilde{f}(\tilde{\mathbf{A}}))_\alpha$ is also closed and bounded, and the proof is complete. ∎

Corollary 4.4.13 *Let $f : \prod_{i=1}^{p} \mathbb{R}^{n_i} \to \mathbb{R}^n$ be a continuous and onto crisp function, and let $\tilde{f} : \prod_{i=1}^{p} \mathcal{F}(\mathbb{R}^{n_i}) \to \mathcal{F}(\mathbb{R}^n)$ be a fuzzy function extended from f via the extension principle defined in* (4.44), *where the function $\mathfrak{A}$ is taken to be the minimum function given by*

$$\mathfrak{A}\left(\alpha_1, \ldots, \alpha_p\right) = min\left\{\alpha_1, \ldots, \alpha_p\right\}.$$

Given any fuzzy sets $\tilde{A}^{(i)}$ in $\mathbb{R}^{n_i}$ for $i = 1, \ldots, p$, suppose that the membership functions $\xi_{\tilde{A}^{(i)}}$ are upper semi-continuous for all $i = 1, \ldots, p$, and that the 0-level sets $\tilde{A}^{(i)}_0$ are bounded subsets of $\mathbb{R}^{n_i}$ for all $i = 1, \ldots, p$. Then, we have the following properties.

(i) *The membership function $\xi_{\tilde{f}(\tilde{\mathbf{A}})}$ of fuzzy set $\tilde{f}(\tilde{\mathbf{A}})$ in $\mathbb{R}^n$ is upper semi-continuous.*

(ii) *For each $\alpha \in (0,1]$, we have the following equalities.*

$$\left(\tilde{f}(\tilde{\mathbf{A}})\right)_\alpha = \bigcup_{\{(\alpha_1,\ldots,\alpha_p):\min\{\alpha_1,\ldots,\alpha_p\}\geq\alpha\}} f\left(\tilde{A}^{(1)}_{\alpha_1},\ldots,\tilde{A}^{(p)}_{\alpha_p}\right)$$
$$= \left\{f(\mathbf{x}) : \min\left\{\xi_{\tilde{A}^{(1)}}(x_1),\ldots,\xi_{\tilde{A}^{(p)}}(x_p)\right\} \geq \alpha\right\} = f\left(\tilde{\mathbf{A}}_\alpha\right).$$

For the 0-level sets, we also have

$$\left(\tilde{f}(\tilde{\mathbf{A}})\right)_{0+} = \bigcup_{\{(\alpha_1,\ldots,\alpha_p):\min\{\alpha_1,\ldots,\alpha_p\}>0\}} f\left(\tilde{A}^{(1)}_{\alpha_1},\ldots,\tilde{A}^{(p)}_{\alpha_p}\right)$$
$$= \left\{f(\mathbf{x}) : \min\left\{\xi_{\tilde{A}^{(1)}}(x_1),\ldots,\xi_{\tilde{A}^{(p)}}(x_p)\right\} > 0\right\} = f\left(\tilde{\mathbf{A}}_{0+}\right).$$

and

$$\left(\tilde{f}(\tilde{\mathbf{A}})\right)_0 = cl\left(\bigcup_{\{(\alpha_1,\ldots,\alpha_p):\min\{\alpha_1,\ldots,\alpha_p\}>0\}} f\left(\tilde{A}^{(1)}_{\alpha_1},\ldots,\tilde{A}^{(p)}_{\alpha_p}\right)\right)$$
$$= cl\left(\ \{f(\mathbf{x}) : \min\left\}\xi_{\tilde{A}^{(1)}}(x_1),\ldots,\xi_{\tilde{A}^{(p)}}(x_p)\right\} > 0\right\}\right)$$
$$= cl\left(f\left(\tilde{A}^{(1)}_{0+},\ldots,\tilde{A}^{(p)}_{0+}\right)\right) = f\left(\tilde{\mathbf{A}}_0\right).$$

Moreover, the α-level sets $(\tilde{f}(\tilde{\mathbf{A}}))_\alpha$ of $\tilde{f}(\tilde{\mathbf{A}})$ are closed and bounded subsets of $\mathbb{R}^n$ for all $\alpha \in [0,1]$.

Proof. It is clear to see that the minimum function satisfies all the assumptions of function $\mathfrak{A}$ in Theorem 4.4.12. Therefore, the desired results follow immediately from Theorems 4.4.9 and 4.4.12. ■

Remark 4.4.14 We have the following observations.

- Using Proposition 4.4.2, Theorem 4.3.2 can be obtained by Theorem 4.4.7. In other words, Theorem 4.3.2 can be extended to Theorem 4.4.7 by considering the weaker form of function $\mathfrak{A}$.
- Using Proposition 4.4.2, Theorem 4.3.4 can be obtained by Theorem 4.4.12. In other words, Theorem 4.3.4 can be extended to Theorem 4.4.12 by considering the weaker form of function $\mathfrak{A}$.

Theorem 4.4.15 *Let $f : \prod_{i=1}^p \mathbb{R}^{n_i} \to \mathbb{R}^n$ be a linear, continuous, and onto crisp function, and let $\tilde{f} : \prod_{i=1}^p \mathcal{F}(\mathbb{R}^{n_i}) \to \mathcal{F}(\mathbb{R}^n)$ be a fuzzy function extended from f via the extension principle defined in* (4.44). *Suppose that the function $\mathfrak{A} : [0,1]^p \to [0,1]$ satisfies the following conditions:*

- *$\mathfrak{A}$ is upper semi-continuous and increasing;*
- *if any one of $\{\alpha_1,\ldots,\alpha_p\}$ is zero, then $\mathfrak{A}(\alpha_1,\ldots,\alpha_p) = 0$;*
- *$\mathfrak{A}\left(\min\{a_1,b_1\},\ldots,\min\{a_p,b_p\}\right) \geq \min\left\{\mathfrak{A}(a_1,\ldots,a_p),\mathfrak{A}(b_1,\ldots,b_p)\right\}$;*
- *$\mathfrak{A}(\alpha_1,\ldots,\alpha_p) > 0$ if and only if $\alpha_i > 0$ for all $i = 1,\ldots,p$.*

Then, given any fuzzy vectors $\tilde{A}^{(i)} \in \mathfrak{F}(\mathbb{R}^{n_i})$ for $i = 1,\ldots,p$, we have the following properties.

(i) *We have that $\tilde{f}(\tilde{\mathbf{A}}) \in \mathfrak{F}(\mathbb{R}^n)$ is a fuzzy vector in $\mathbb{R}^n$. The α-level sets $(\tilde{f}(\tilde{\mathbf{A}}))_\alpha$ are convex, closed, and bounded subsets of $\mathbb{R}^n$ for all $\alpha \in [0,1]$.*

(ii) *For each* $\alpha \in (0,1]$, *we have the following equalities*

$$\begin{aligned}(\tilde{f}(\tilde{\mathbf{A}}))_\alpha &= \bigcup_{\{(\alpha_1,\ldots,\alpha_p):\mathfrak{A}(\alpha_1,\ldots,\alpha_p)\geq\alpha\}} f\left(\tilde{A}^{(1)}_{\alpha_1},\ldots,\tilde{A}^{(p)}_{\alpha_p}\right)\\ &= \left\{f(\mathbf{x}) : \mathfrak{A}\left(\xi_{\tilde{A}^{(1)}}(x_1),\ldots,\xi_{\tilde{A}^{(p)}}(x_p)\right) \geq \alpha\right\}.\end{aligned}$$

For the 0*-level sets, we also have*

$$\begin{aligned}(\tilde{f}(\tilde{\mathbf{A}}))_{0+} &= \bigcup_{\{(\alpha_1,\ldots,\alpha_p):\mathfrak{A}(\alpha_1,\ldots,\alpha_p)>0\}} f\left(\tilde{A}^{(1)}_{\alpha_1},\ldots,\tilde{A}^{(p)}_{\alpha_p}\right)\\ &= \left\{f(\mathbf{x}) : \mathfrak{A}\left(\xi_{\tilde{A}^{(1)}}(x_1),\ldots,\xi_{\tilde{A}^{(p)}}(x_p)\right) > 0\right\} = f\left(\tilde{\mathbf{A}}_{0+}\right)\end{aligned}$$

and

$$\begin{aligned}(\tilde{f}(\tilde{\mathbf{A}}))_0 &= cl\left(\bigcup_{\{(\alpha_1,\ldots,\alpha_p):\mathfrak{A}(\alpha_1,\ldots,\alpha_p)>0\}} f\left(\tilde{A}^{(1)}_{\alpha_1},\ldots,\tilde{A}^{(p)}_{\alpha_p}\right)\right)\\ &= cl\left(\left\{f(\mathbf{x}) : \mathfrak{A}\left(\xi_{\tilde{A}^{(1)}}(x_1),\ldots,\xi_{\tilde{A}^{(p)}}(x_p)\right) > 0\right\}\right)\\ &= cl\left(f\left(\tilde{A}^{(1)}_{0+},\ldots,\tilde{A}^{(p)}_{0+}\right)\right) = f\left(\tilde{\mathbf{A}}_0\right).\end{aligned}$$

Proof. We are going to check the conditions in Definition 2.3.1. Part (i) of Theorem 4.4.12 says that the membership function of $\tilde{f}(\tilde{\mathbf{A}})$ is upper semi-continuous. From part (ii) of Theorem 4.4.12, we have

$$(\tilde{f}(\tilde{\mathbf{A}}))_\alpha = \left\{f(\mathbf{x}) : \mathfrak{A}\left(\xi_{\tilde{A}^{(1)}}(x_1),\ldots,\xi_{\tilde{A}^{(p)}}(x_p)\right) \geq \alpha\right\} \tag{4.69}$$

for each $\alpha \in (0,1]$. Let $v_1 = f(\mathbf{x})$ and $v_2 = f(\mathbf{y})$, where

$$\mathfrak{A}\left(\xi_{\tilde{A}^{(1)}}(x_1),\ldots,\xi_{\tilde{A}^{(p)}}(x_p)\right) \geq \alpha \text{ and } \mathfrak{A}\left(\xi_{\tilde{A}^{(1)}}(y_1),\ldots,\xi_{\tilde{A}^{(p)}}(y_p)\right) \geq \alpha.$$

The linearity of f says that

$$\begin{aligned}\lambda v_1 + (1-\lambda)v_2 &= \lambda f(\mathbf{x}) + (1-\lambda) f(y_1,\ldots,y_p)\\ &= f\left(\lambda x_1 + (1-\lambda)y_1,\ldots,\lambda x_p + (1-\lambda)y_p\right).\end{aligned}$$

In order to claim that the α-level set $(\tilde{f}(\tilde{\mathbf{A}}))_\alpha$ is convex, by (4.69), we need to show

$$\mathfrak{A}\left(\xi_{\tilde{A}^{(1)}}(\lambda x_1 + (1-\lambda)y_1),\ldots,\xi_{\tilde{A}^{(p)}}(\lambda x_p + (1-\lambda)y_1\right) \geq \alpha.$$

Now, for any $i = 1,\ldots,p$, since $\tilde{A}^{(i)}$ is convex, by Proposition 2.2.7, we have

$$\xi_{\tilde{A}^{(i)}}(\lambda x_i + (1-\lambda)y_i) \geq \min\left\{\xi_{\tilde{A}^{(i)}}(x_i),\xi_{\tilde{A}^{(i)}}(y_i)\right\}.$$

Using the assumptions of $\mathfrak{A}$, we obtain

$$\begin{aligned}&\mathfrak{A}\left(\xi_{\tilde{A}^{(1)}}(\lambda x_1 + (1-\lambda)y_1),\ldots,\xi_{\tilde{A}^{(p)}}(\lambda x_p + (1-\lambda)y_1\right)\\ &\geq \mathfrak{A}\left(\min\left\{\xi_{\tilde{A}^{(1)}}(x_1),\xi_{\tilde{A}^{(1)}}(y_1)\right\},\ldots,\min\left\{\xi_{\tilde{A}^{(p)}}(x_p),\xi_{\tilde{A}^{(p)}}(y_p)\right\}\right)\\ &\quad\times (\text{increasing assumption})\\ &\geq \min\left\{\mathfrak{A}\left(\xi_{\tilde{A}^{(1)}}(x_1),\ldots,\xi_{\tilde{A}^{(p)}}(x_p)\right),\mathfrak{A}\left(\xi_{\tilde{A}^{(1)}}(y_1),\ldots,\xi_{\tilde{A}^{(p)}}(y_p)\right)\right\} \geq \alpha.\end{aligned}$$

Therefore, we conclude that $\tilde{f}(\tilde{\mathbf{A}})$ is a fuzzy vector in $\mathbb{R}^n$. From part (ii) of Theorem 4.4.12 and Proposition 2.3.2, we also see that the α-level sets $(\tilde{f}(\tilde{\mathbf{A}}))_\alpha$ are convex, closed, and bounded subsets of $\mathbb{R}^n$ for all $\alpha \in [0,1]$. This completes the proof. ■

Remark 4.4.16 Suppose that the following assumption holds true:

$$\text{if any one of } \{\alpha_1, \ldots, \alpha_p\} \text{ is zero, then } \mathfrak{A}(\alpha_1, \ldots, \alpha_p) = 0.$$

Then, we see that the following assumption also holds true:

$$\mathfrak{A}(\alpha_1, \ldots, \alpha_p) > 0 \text{ implies } \alpha_i > 0 \text{ for all } i = 1, \ldots, p.$$

Therefore, we can reduce the assumptions of Theorem 4.4.15.

Remark 4.4.17 We have the following observations.

- Suppose that we take $\mathfrak{A} = T_p$. Then $\mathfrak{A}(1, \ldots, 1) = 1$ is automatically satisfied by Remark 3.2.7.
- Suppose that $\mathfrak{A}$ is increasing. Since $\min\{a_i, b_i\} \leq a_i$ for all $i = 1, \ldots, p$, we see that
  $$\mathfrak{A}\left(\min\{a_1, b_1\}, \ldots, \min\{a_p, b_p\}\right) \leq \mathfrak{A}(a_1, \ldots, a_p).$$
  Similarly, since $\min\{a_i, b_i\} \leq b_i$ for all $i = 1, \ldots, p$, we also have
  $$\mathfrak{A}\left(\min\{a_1, b_1\}, \ldots, \min\{a_p, b_p\}\right) \leq \mathfrak{A}(b_1, \ldots, b_p).$$
  Therefore, we obtain
  $$\mathfrak{A}\left(\min\{a_1, b_1\}, \ldots, \min\{a_p, b_p\}\right) \leq \min\left\{\mathfrak{A}(a_1, \ldots, a_p), \mathfrak{A}(b_1, \ldots, b_p)\right\}.$$
  Under the assumptions of $\mathfrak{A}$ given in Theorem 4.4.15, we have the equality
  $$\mathfrak{A}\left(\min\{a_1, b_1\}, \ldots, \min\{a_p, b_p\}\right) = \min\left\{\mathfrak{A}(a_1, \ldots, a_p), \mathfrak{A}(b_1, \ldots, b_p)\right\}.$$
- Suppose that $\mathfrak{A}$ is increasing and $\mathfrak{A}(\alpha, \ldots, \alpha) = \alpha$ for each $\alpha \in (0,1]$. Then, we see that $\alpha_i \geq \alpha$ for $i = 1, \ldots, p$ imply $\mathfrak{A}(\alpha_1, \ldots, \alpha_p) \geq \mathfrak{A}(\alpha, \ldots, \alpha) = \alpha$.

Suppose that we consider the real-valued function $f : \prod_{i=1}^p \mathbb{R}^{n_i} \to \mathbb{R}$. Then, many assumptions are not needed, which will be presented below.

Theorem 4.4.18 *Let $f : \prod_{i=1}^p \mathbb{R}^{n_i} \to \mathbb{R}$ be a continuous and onto crisp real-valued function, and let $\tilde{f} : \prod_{i=1}^p \mathcal{F}(\mathbb{R}^{n_i}) \to \mathcal{F}(\mathbb{R})$ be a fuzzy function extended from f via the extension principle defined in (4.44), where the function $\mathfrak{A}$ is taken to be the minimum function given by*

$$\mathfrak{A}\left(\alpha_1, \ldots, \alpha_p\right) = \min\left\{\alpha_1, \ldots, \alpha_p\right\}.$$

Then, given any fuzzy vectors $\tilde{A}^{(i)} \in \mathfrak{F}(\mathbb{R}^{n_i})$ for $i = 1, \ldots, p$, we have that $\tilde{f}(\tilde{\mathbf{A}})$ is a fuzzy interval and its α-level set is a bounded closed interval in $\mathbb{R}$ given by $(\tilde{f}(\tilde{\mathbf{A}}))_\alpha = f(\tilde{\mathbf{A}}_\alpha)$ for each $\alpha \in [0,1]$, where the 0-level sets are taken into account.

Proof. We are going to check the conditions in Definition 2.3.1 by following the similar arguments of Theorem 4.4.15. Using Corollary 4.4.13, we have

$$\left\{y : \xi_{\tilde{f}(\tilde{\mathbf{A}})}(y) \geq \alpha\right\} = \left(\tilde{f}(\tilde{\mathbf{A}})\right)_\alpha = f\left(\tilde{\mathbf{A}}_\alpha\right) \tag{4.70}$$

for each $\alpha \in [0,1]$. Since f is continuous, and $\tilde{A}_\alpha^{(i)}$ are closed and bounded subsets of $\mathbb{R}^{n_i}$ for all $i = 1, \ldots, p$ and $\alpha \in [0,1]$, by Tychonoff's theorem and part (iii) of Proposition 1.4.3, we see that the set presented in (4.70) is a closed and bounded subset of $\mathbb{R}$. Since $\tilde{A}_\alpha^{(i)}$ are also convex sets in $\mathbb{R}^{n_i}$, i.e. connected subsets of $\mathbb{R}^{n_i}$ for all $i = 1, \ldots, p$ and $\alpha \in [0,1]$, by part (iv) of Proposition 1.4.3, we conclude that the set presented in (4.70) is a bounded closed interval, i.e. a convex set in $\mathbb{R}$ for each $\alpha \in [0,1]$. This completes the proof. ■

Example 4.4.19 Let $\tilde{A}^{(1)}, \ldots, \tilde{A}^{(p)}$ be fuzzy intervals. According to the extension principle, the membership function of the maximum $\widetilde{\max}\{\tilde{A}^{(1)}, \ldots, \tilde{A}^{(p)}\}$ is defined by

$$\xi_{\widetilde{\max}\{\tilde{A}^{(1)},\ldots,\tilde{A}^{(p)}\}}(z) = \sup_{\{z:\max\{\mathbf{x}\}=z\}} \min\left(\xi_{\tilde{A}^{(1)}}(x_1), \ldots, \xi_{\tilde{A}^{(p)}}(x_p)\right),$$

Suppose that we take $f : \mathbb{R}^p \to \mathbb{R}$ by

$$f(x_1, \ldots, x_p) = \max\left\{x_1, \ldots, x_p\right\}.$$

Then, we see that the function f is onto and continuous. From Theorem 4.4.18, for each $\alpha \in [0,1]$, we have

$$\begin{aligned}
\left(\widetilde{\max}\{\tilde{A}^{(1)}, \ldots, \tilde{A}^{(p)}\}\right)_\alpha &= (\tilde{f}(\tilde{\mathbf{A}}))_\alpha = f\left(\tilde{\mathbf{A}}_\alpha\right)\\
&= f\left(\tilde{A}^{(1)}, \ldots, \tilde{A}^{(p)}\right) = \left\{f(x_1, \ldots, x_p) : x_i \in \tilde{A}^{(i)} \text{ for } i = 1, \ldots, p\right\}\\
&= \left\{\max\left\{x_1, \ldots, x_p\right\} : (\tilde{A}^{(i)})^L_\alpha \le x_i \le (\tilde{A}^{(i)})^U_\alpha \text{ for } i = 1, \ldots, p\right\}\\
&= \left[\max\left\{(\tilde{A}^{(1)})^L_\alpha, \ldots, (\tilde{A}^{(p)})^L_\alpha\right\}, \max\left\{(\tilde{A}^{(1)})^L_\alpha, \ldots, (\tilde{A}^{(p)})^L_\alpha\right\}\right],
\end{aligned}$$

where the 0-level sets are taken into account.

5

Generating Fuzzy Sets

In economics and engineering problems, when the fuzziness is taken into account, the observed data sometimes are fuzzified to be fuzzy numbers. This kind of fuzzification may result in many different types of fuzzy numbers depending on the methodology adopted by the decision makers, where the subjectivity via the viewpoint of decision makers may be biased and not an accurate representation of reality. In this chapter, we propose a general methodology that can get rid of the subjectivity by directly generating the related fuzzy sets based on the observed data without involving the possible biased viewpoint of decision makers. The main idea is based on the solid and nested families of sets.

Suppose that we want to measure the water level in the summer season. Owing to the fluctuation of the water level, we cannot simply say that the water level is now 10 meters. We should say that the water level is around 10 meters. Therefore, the reasonable way to model the water level is to treat it as a fuzzy interval or fuzzy number. Under this consideration, the water level should be taken as a fuzzy number $\widetilde{10}$. The problem is that the determination of the membership function $\xi_{\widetilde{10}}$ of $\widetilde{10}$ is subjective depending on the viewpoint of decision makers. In other words, there are infinite ways to set up the membership functions. It may happen that the different membership functions can result in different final results. Therefore, the best way is to follow a mechanical procedure to set up the membership functions, which is the main purpose of this chapter. We briefly address this mechanical procedure as follows.

We assume that there are 100 days in summer. Suppose that the engineers can measure the water level two times each day. In other words, the engineers can obtain a bounded closed interval each day by setting the lower and upper bounds of this interval as the low and high water levels, respectively. More precisely, we consider 100 values of α in the unit interval [0,1]. Then, we can obtain a bounded closed interval $[m_\alpha^L, m_\alpha^U]$, where m_α^L denotes the low water level and m_α^U denotes the high water level in the $(100 \cdot \alpha)$th day. For example, the value $\alpha = 0.08$ means the 8th day of this summer. In this case, we obtain a family of closed intervals given by

$$\mathcal{M} = \left\{ M_\alpha : M_\alpha = \left[m_\alpha^L, m_\alpha^U \right] \text{ for } \alpha = 1, \ldots, 100 \right\}.$$

This family cannot be nested in the sense of $M_\alpha \subseteq M_\beta$ for $\alpha > \beta$. However, we can rearrange this family $\mathcal{M}$ as a nested family $\mathcal{M}^{(\eta)}$ given by

$$\mathcal{M}^{(\eta)} = \left\{ M_\alpha^{(\eta)} : M_\alpha^{(\eta)} = M_{\eta(\alpha)} \right\}, \tag{5.1}$$

Mathematical Foundations of Fuzzy Sets, First Edition. Hsien-Chung Wu.

for some suitable function $\eta : (0,1] \to (0,1]$ such that we have the nestedness $M_\alpha^{(\eta)} \subseteq M_\beta^{(\eta)}$ for $\alpha > \beta$. The main purpose of this chapter is to generate a fuzzy set $\tilde{A}$ such that its α-level set $\tilde{A}_\alpha$ is identical to the closed interval $M_\alpha^{(\eta)}$. In this case, the fuzzy set $\tilde{A}$ can be used to describe the water level in summer in which the biased subjectivity raised by the decision makers can be avoided. In other words, this mechanical procedure is independent of the decision makers.

5.1 Families of Sets

Let $\mathcal{M} = \{M_\alpha : \alpha \in (0,1]\}$ be a family of nonempty subsets of $\mathbb{R}^m$, and let $\eta : (0,1] \to (0,1]$ be a function defined on (0,1]. The range of η is denoted by $\mathcal{R}(\eta)$. Then, we can obtain a new family given by

$$\mathcal{M}^{(\eta)} = \left\{M_{\eta(\alpha)} : \alpha \in (0,1]\right\}.$$

For convenience, we write $M_\alpha^{(\eta)} \equiv M_{\eta(\alpha)}$ for $\alpha \in (0,1]$. We have the following observations.

- Suppose that η is the identity function id. Then $\mathcal{M}^{(id)} = \mathcal{M}$.
- Suppose that η is an injective (one-to-one) function. Then, for each $\alpha \in \mathcal{R}(\eta)$, there exists $\beta \in (0,1]$ satisfying $\eta(\beta) = \alpha$. In this case, we also write $\beta = \eta^{-1}(\alpha)$. Therefore, when the function η is injective, we have

$$M_\alpha = M_{\eta^{-1}(\alpha)}^{(\eta)} \text{ for } \alpha \in \mathcal{R}(\eta). \tag{5.2}$$

Example 5.1.1 Let a_1, a_2, b_1, b_2 be real numbers satisfying $b_1 < a_1 < a_2 < b_2$. For $\alpha \in [0,1]$, we define

$$m_\alpha^L = \alpha b_1 + (1-\alpha)a_1 \text{ and } m_\alpha^U = \alpha b_2 + (1-\alpha)a_2.$$

Then, we see that $m_\alpha^L \leq m_\alpha^U$. We define M_α to be a closed interval in $\mathbb{R}$ by $M_\alpha = [m_\alpha^L, m_\alpha^U]$ for $\alpha \in (0,1]$. We define a function η by $\eta(\alpha) = 1 - \alpha$. Then $\mathcal{R}(\eta) = (0,1]$. We also define

$$\bar{m}_\alpha^L = m_{1-\alpha}^L \text{ and } \bar{m}_\alpha^U = m_{1-\alpha}^U \text{ for } \alpha \in (0,1].$$

Then, for $\alpha \in (0,1]$, we see that

$$\bar{m}_\alpha^L = (1-\alpha)b_1 + \alpha a_1 \text{ and } \bar{m}_\alpha^U = (1-\alpha)b_2 + \alpha a_2.$$

Therefore, we have

$$M_\alpha^{(\eta)} = M_{\eta(\alpha)} = M_{1-\alpha} = \left[m_{1-\alpha}^L, m_{1-\alpha}^U\right] = \left[\bar{m}_\alpha^L, \bar{m}_\alpha^U\right] \text{ for } \alpha \in (0,1].$$

It is clear to see

$$M_\alpha^{(\eta)} \subset M_\beta^{(\eta)} \text{ with } M_\alpha^{(\eta)} \neq M_\beta^{(\eta)} \text{ for } \alpha > \beta. \tag{5.3}$$

We also have

$$\bigcup_{0<\alpha\leq 1} M_\alpha^{(\eta)} = \bigcup_{0<\alpha\leq 1} \left[\bar{m}_\alpha^L, \bar{m}_\alpha^U\right] = \left(\bar{m}_0^L, \bar{m}_0^U\right) = (b_1, b_2)$$

by the continuities of $\bar{m}_\alpha^L$ and $\bar{m}_\alpha^U$ with respect to α.

For $\alpha \in (0,1)$, we write

$$A_{\alpha}^{(\eta)} = \bigcup_{\alpha \leq \beta \leq 1} M_{\beta}^{(\eta)} \text{ and } A_{\alpha+}^{(\eta)} = \bigcup_{\alpha < \beta \leq 1} M_{\beta}^{(\eta)}. \tag{5.4}$$

Inspired by Proposition 2.2.12, we propose the following definition.

Definition 5.1.2 Let η be a function from $(0,1]$ into $(0,1]$. Let $\mathcal{M} = \{M_{\alpha} : \alpha \in (0,1]\}$ be a family of nonempty subsets of $\mathbb{R}^m$. We say that $\mathcal{M}$ is a **solid family with respect to** η when

$$D_{\alpha}^{(\eta)} \equiv A_{\alpha}^{(\eta)} \backslash A_{\alpha+}^{(\eta)} \neq \emptyset$$

for $\alpha \in (0,1)$.

Example 5.1.3 Continued from Example 5.1.1, we want to show

$$A_{\alpha+}^{(\eta)} = \bigcup_{\alpha < \beta \leq 1} M_{\beta}^{(\eta)} = \bigcup_{\alpha < \beta \leq 1} \left[\bar{m}_{\beta}^{L}, \bar{m}_{\beta}^{U}\right] = \left(\bar{m}_{\alpha}^{L}, \bar{m}_{\alpha}^{U}\right).$$

If $x \in A_{\alpha+}^{(\eta)}$, then $x \in [\bar{m}_{\beta_0}^{L}, \bar{m}_{\beta_0}^{U}]$ for some $\beta_0 \in (0,1]$ with $\beta_0 > \alpha$. From (5.3), we have

$$\bar{m}_{\alpha}^{L} < \bar{m}_{\beta_0}^{L} \leq x \leq \bar{m}_{\beta_0}^{U} < \bar{m}_{\alpha}^{U},$$

which says that $x \in (\bar{m}_{\alpha}^{L}, \bar{m}_{\alpha}^{U})$. Therefore, we obtain the inclusion $A_{\alpha+}^{(\eta)} \subseteq (\bar{m}_{\alpha}^{L}, \bar{m}_{\alpha}^{U})$. On the other hand, since $\bar{m}_{\alpha}^{L}$ and $\bar{m}_{\alpha}^{U}$ are continuous with respect to α, using the strict monotonicity of $\bar{m}_{\alpha}^{L}$ and $\bar{m}_{\alpha}^{U}$ with respect to α, for any $x \in (\bar{m}_{\alpha}^{L}, \bar{m}_{\alpha}^{U})$, there exists $\beta_1 \in (0,1]$ with $\beta_1 > \alpha$ satisfying

$$x \geq \bar{m}_{\beta_1}^{L} > \bar{m}_{\alpha}^{L} \text{ and } x \leq \bar{m}_{\beta_1}^{U} < \bar{m}_{\alpha}^{U},$$

which says that $x \in [\bar{m}_{\beta_1}^{L}, \bar{m}_{\beta_1}^{U}]$, i.e. $x \in A_{\alpha+}^{(\eta)}$. Therefore, we obtain the desired equality. From (5.3), we can also obtain

$$A_{\alpha}^{(\eta)} = \bigcup_{\alpha \leq \beta \leq 1} M_{\beta}^{(\eta)} = \bigcup_{\alpha \leq \beta \leq 1} \left[\bar{m}_{\beta}^{L}, \bar{m}_{\beta}^{U}\right] = \left[\bar{m}_{\alpha}^{L}, \bar{m}_{\alpha}^{U}\right],$$

which implies

$$D_{\alpha}^{(\eta)} = A_{\alpha}^{(\eta)} \backslash A_{\alpha+}^{(\eta)} = \left[\bar{m}_{\alpha}^{L}, \bar{m}_{\alpha}^{U}\right] \setminus \left(\bar{m}_{\alpha}^{L}, \bar{m}_{\alpha}^{U}\right) = \left\{\bar{m}_{\alpha}^{L}, \bar{m}_{\alpha}^{U}\right\} \neq \emptyset.$$

This shows that $\mathcal{M}$ is a solid family with respect to η.

Proposition 5.1.4 *Let η be a function from $(0,1]$ into $(0,1]$, and let $\mathcal{M} = \{M_{\alpha} : \alpha \in (0,1]\}$ be a family of nonempty subsets of $\mathbb{R}^m$ such that it is a solid family with respect to η. Then, the following statements hold true.*

(i) *We have $D_{\alpha}^{(\eta)} \cap D_{\beta}^{(\eta)} = \emptyset$ for $\alpha, \beta \in (0,1)$ with $\alpha \neq \beta$.*

(ii) *Given any $\alpha \in (0,1)$, we have $M_{\beta}^{(\eta)} \cap D_{\alpha}^{(\eta)} = \emptyset$ for $\beta \in (0,1]$ with $\beta > \alpha$. In particular, we have $M_{1}^{(\eta)} \cap D_{\alpha}^{(\eta)} = \emptyset$ for all $\alpha \in (0,1)$.*

(iii) *Suppose that $M^* \subseteq M_{\alpha}^{(\eta)}$ for all $\alpha \in (0,1]$. Then $M^* \cap D_{\alpha}^{(\eta)} = \emptyset$ for $\alpha \in (0,1)$.*

Proof. To prove part (i), given $\alpha, \beta \in (0,1)$, it suffices to assume $\alpha < \beta$. Then, we have

$$D_{\beta}^{(\eta)} \subseteq A_{\beta}^{(\eta)} \subseteq A_{\alpha+}^{(\eta)} \subseteq A_{\alpha}^{(\eta)}.$$

Since $D_{\alpha}^{(\eta)} = A_{\alpha}^{(\eta)} \backslash A_{\alpha+}^{(\eta)}$, it follows that $D_{\alpha}^{(\eta)} \cap D_{\beta}^{(\eta)} = \emptyset$ for $\alpha < \beta$.

To prove part (ii), suppose that $x \in D_\alpha^{(\eta)}$ for $\alpha \in (0,1)$. By the definition of $D_\alpha^{(\eta)}$, we see that $x \notin A_{\alpha+}^{(\eta)}$, i.e. $x \notin M_\beta^{(\eta)}$ for all $\beta \in (0,1]$ with $\beta > \alpha$. Therefore, we obtain the desired result.

To prove part (iii), suppose that $x \in D_\alpha^{(\eta)}$ for $\alpha \in (0,1)$. It says that $x \notin A_{\alpha+}^{(\eta)}$, i.e. $x \notin M_\beta^{(\eta)}$ for all $\beta \in (0,1]$ with $\beta > \alpha$. The assumption says that $M^* \subseteq M_\beta^{(\eta)}$, which implies $x \notin M^*$. Therefore, we obtain $M^* \cap D_\alpha^{(\eta)} = \emptyset$. This completes the proof. ∎

Proposition 5.1.5 *Let η be a function from $(0,1]$ into $(0,1]$, and let $\mathcal{M} = \{M_\alpha : \alpha \in (0,1]\}$ be a family of nonempty subsets of $\mathbb{R}^m$ such that it is a solid family with respect to η. Let T be a subset of $(0,1)$. Given any $\alpha \in (0,1]$, we have*

$$M_\alpha^{(\eta)} \bigcap \left(\bigcup_{\beta \in T} D_\beta^{(\eta)} \right) = M_\alpha^{(\eta)} \bigcap \left(\bigcup_{\{\beta \in T : \beta \geq \alpha\}} D_\beta^{(\eta)} \right).$$

Proof. Given any fixed $x \in M_\alpha^{(\eta)}$ and $x \in D_\beta^{(\eta)}$ for some $\beta \in T$, we have

$$x \in D_\beta^{(\eta)} = A_\beta^{(\eta)} \setminus A_{\beta+}^{(\eta)} = \bigcup_{\beta \leq \gamma \leq 1} M_\gamma^{(\eta)} \setminus \bigcup_{\beta < \gamma \leq 1} M_\gamma^{(\eta)}. \tag{5.5}$$

For $\alpha > \beta$, the equality (5.5) says that $x \notin M_\alpha^{(\eta)}$, which contradicts $x \in M_\alpha^{(\eta)}$. Therefore, we must have $\alpha \leq \beta$, which implies the following inclusion

$$M_\alpha^{(\eta)} \bigcap \left(\bigcup_{\beta \in T} D_\beta^{(\eta)} \right) \subseteq M_\alpha^{(\eta)} \bigcap \left(\bigcup_{\{\beta \in T : \beta \geq \alpha\}} D_\beta^{(\eta)} \right).$$

The other direction of inclusion is obvious. This completes the proof. ∎

5.2 Nested Families

In what follows, we shall consider the concepts of nested families and study the interesting properties of nested families, which will be used to generate the fuzzy sets in the subsequent discussion.

Definition 5.2.1 Let $\mathcal{M} = \{M_\alpha : \alpha \in (0,1]\}$ be a family of nonempty subsets of $\mathbb{R}^m$.

- We say that $\mathcal{M}$ is a **nested family** when $M_\alpha \subseteq M_\beta$ for $\alpha, \beta \in (0,1]$ with $\beta < \alpha$.
- We say that $\mathcal{M}$ is a **strictly nested family** when $M_\alpha \subset M_\beta$ with $M_\alpha \neq M_\beta$ for $\alpha, \beta \in (0,1]$ satisfying $\beta < \alpha$.

Example 5.2.2 Given any real numbers a_1, a_2, b_1, b_2, let

$$m_\alpha^L = \min\left\{(1-\alpha)b_1 + \alpha a_1, (1-\alpha)b_2 + \alpha a_2\right\}$$

and

$$m_\alpha^U = \max\left\{(1-\alpha)b_1 + \alpha a_1, (1-\alpha)b_2 + \alpha a_2\right\}.$$

We define M_α to be a closed interval in $\mathbb{R}$ by $M_\alpha = [m_\alpha^L, m_\alpha^U]$ for $\alpha \in (0,1]$. Then $\mathcal{M}$ is not a nested family. Suppose that $b_1 < a_1 < a_2 < b_2$. Then, we see that

$$\mathcal{M} = \{M_\alpha : \alpha \in (0,1]\} = \left\{[(1-\alpha)b_1 + \alpha a_1, (1-\alpha)b_2 + \alpha a_2] : \alpha \in (0,1]\right\}$$

is a strictly nested family.

Proposition 5.2.3 *Let $\mathcal{M} = \{M_\alpha : \alpha \in (0,1]\}$ be a nested family of subsets of $\mathbb{R}^m$. Suppose that $M_1 \subseteq \bigcap_{n=1}^\infty M_{\alpha_n}$ when $\alpha_n \uparrow 1$ with $\alpha_n \in (0,1]$ for all n. Then $M_1 \subseteq M_\alpha$ for all $\alpha \in (0,1)$.*

Proof. We see that there exists a sequence $\{\alpha_n\}_{n=1}^\infty$ in $(0,1]$ satisfying $\alpha_n \uparrow 1$. Now, given any $\alpha \in (0,1)$, there exists α_N in $\{\alpha_n\}_{n=1}^\infty$ satisfying $\alpha < \alpha_N \leq 1$. By the assumption, we have

$$M_1 \subseteq \bigcap_{n=1}^{\infty} M_{\alpha_n} \subseteq M_{\alpha_N} \subseteq M_\alpha.$$

This completes the proof. ∎

Definition 5.2.4 Let $\mathcal{M} = \{M_\alpha : \alpha \in (0,1]\}$ be a family of nonempty subsets of $\mathbb{R}^m$, and let η be a function from $(0,1]$ into $(0,1]$.

- We say that $\mathcal{M}$ is a **nested family with respect to** η when $M_\alpha^{(\eta)} \subseteq M_\beta^{(\eta)}$ for $\alpha, \beta \in (0,1]$ with $\beta < \alpha$.
- We say that $\mathcal{M}$ is a **strictly nested family with respect to** η when $M_\alpha^{(\eta)} \subset M_\beta^{(\eta)}$ with $M_\alpha^{(\eta)} \neq M_\beta^{(\eta)}$ for $\alpha, \beta \in (0,1]$ satisfying $\beta < \alpha$.

Example 5.1.1 shows that a non-nested family $\mathcal{M} = \{M_\alpha : \alpha \in (0,1]\}$ can be rearranged as a nested family $\mathcal{M}^{(\eta)} = \{M_\alpha^{(\eta)} : \alpha \in (0,1]\}$ by taking $\eta(\alpha) = 1 - \alpha$.

Proposition 5.2.5 *Let η be a function from $(0,1]$ into $(0,1]$, and let $\mathcal{M} = \{M_\alpha : \alpha \in (0,1]\}$ be a family of nonempty subsets of $\mathbb{R}^m$ such that the following conditions are satisfied:*

- *$\mathcal{M}$ is a nested family with respect to η;*
- *$\bigcap_{n=1}^\infty M_{\alpha_n}^{(\eta)} \subseteq M_\alpha^{(\eta)}$ when $\alpha_n \uparrow \alpha$ with $\alpha, \alpha_n \in (0,1]$ for all n.*

Given any fixed $x \in \mathbb{R}^m$ and $r \in [0,1]$, the following set

$$F_r = \left\{\alpha \in (0,1] : \alpha \cdot \chi_{M_\alpha^{(\eta)}}(x) \geq r\right\}$$

is a closed set in the sense of $cl(F_r) = F_r$; that is, the function

$$\zeta_x(\alpha) = \alpha \cdot \chi_{M_\alpha^{(\eta)}}(x)$$

is upper semi-continuous on $[0,1]$.

Proof. Given $\alpha \in \text{cl}(F_r)$, we want to show that $\alpha \in F_r$. The concept of closure says that there exists a sequence $\{\alpha_n\}_{n=1}^\infty$ in F_r satisfying $\alpha_n \to \alpha$ with $\alpha_n \neq \alpha$ for all n. Since $\alpha_n \in F_r$, we have $\alpha_n \geq r$ and $x \in M_{\alpha_n}^{(\eta)}$ for all n. By taking the limit, we also have $\alpha \geq r$ and $x \in \bigcap_{n=1}^\infty M_{\alpha_n}^{(\eta)}$. We consider the following two cases.

- For $\alpha_n \uparrow \alpha$, since $x \in \bigcap_{n=1}^\infty M_{\alpha_n}^{(\eta)} \subseteq M_\alpha^{(\eta)}$ by the second condition, we obtain

  $$\alpha \cdot \chi_{M_\alpha^{(\eta)}}(x) = \alpha \geq r,$$

  which says that $\alpha \in F_r$.
- For $\alpha_n \downarrow \alpha$, since $\alpha < \alpha_n$ for all n and $\mathcal{M}$ is a nested family with respect to η, we have $M_{\alpha_n}^{(\eta)} \subseteq M_\alpha^{(\eta)}$ for all n, which says that $x \in \bigcap_{n=1}^\infty M_{\alpha_n}^{(\eta)} \subseteq M_\alpha^{(\eta)}$. The above argument is still valid to obtain $\alpha \in F_r$.

Therefore, we conclude that $\text{cl}(F_r) = F_r$, i.e. the set F_r is indeed a closed set. This completes the proof. ■

Proposition 5.2.6 *Let η be a function from $(0,1]$ into $(0,1]$, and let $\mathcal{M} = \{M_\alpha : \alpha \in (0,1]\}$ be a family of nonempty subsets of $\mathbb{R}^m$ such that it is a nested family with respect to η.*

(i) *Given any $\alpha \in (0,1]$, we have*

$$M_\alpha^{(\eta)} = \bigcap_{0<\beta\leq\alpha} M_\beta^{(\eta)}.$$

(ii) *Given $\alpha \in (0,1]$, suppose that $\alpha_n \uparrow \alpha$ with $\alpha_n \in (0,1]$ and $\alpha_n < \alpha$ for all n. Then*

$$\bigcap_{0<\beta\leq\alpha} M_\beta^{(\eta)} \subseteq \bigcap_{0<\beta<\alpha} M_\beta^{(\eta)} = \bigcap_{n=1}^{\infty} M_{\alpha_n}^{(\eta)}.$$

(iii) *Given $\alpha \in (0,1]$, suppose that $\bigcap_{n=1}^{\infty} M_{\alpha_n}^{(\eta)} \subseteq M_\alpha^{(\eta)}$ for $\alpha_n \uparrow \alpha$ with $\alpha_n \in (0,1]$ and $\alpha_n < \alpha$ for all n. Then*

$$\bigcap_{0<\beta\leq\alpha} M_\beta^{(\eta)} = \bigcap_{0<\beta<\alpha} M_\beta^{(\eta)} = \bigcap_{n=1}^{\infty} M_{\alpha_n}^{(\eta)} = M_\alpha^{(\eta)}.$$

Proof. To prove part (i), since $\alpha > 0$ and $M_\alpha^{(\eta)} \subseteq M_\beta^{(\eta)}$ for any $\beta \in (0,1]$ with $\beta < \alpha$, we immediately have

$$\bigcap_{0<\beta\leq\alpha} M_\beta^{(\eta)} = M_\alpha^{(\eta)}.$$

To prove part (ii), since $\alpha_n < \alpha$ for all n, it is obvious that

$$\bigcap_{0<\beta<\alpha} M_\beta^{(\eta)} \subseteq \bigcap_{n=1}^{\infty} M_{\alpha_n}^{(\eta)}.$$

To prove the other direction of inclusion, given any $\beta < \alpha$ for $\beta \in (0,1]$ and $x \in \bigcap_{n=1}^{\infty} M_{\alpha_n}^{(\eta)}$, i.e. $x \in M_{\alpha_n}^{(\eta)}$ for all n, since $\alpha_n \uparrow \alpha$, there exists α_N satisfying $\beta < \alpha_N \leq \alpha$. Therefore, we obtain $x \in M_{\alpha_N}^{(\eta)} \subseteq M_\beta^{(\eta)}$. Since β can be any number with $\beta < \alpha$ and $\beta \in (0,1]$, this shows that

$$x \in \bigcap_{0<\beta<\alpha} M_\beta^{(\eta)},$$

which implies the inclusion

$$\bigcap_{n=1}^{\infty} M_{\alpha_n}^{(\eta)} \subseteq \bigcap_{0<\beta<\alpha} M_\beta^{(\eta)}.$$

Finally, part (iii) follows from parts (i) and (ii) immediately. This completes the proof. ■

Proposition 5.2.7 *Let η be a function from $(0,1]$ into $(0,1]$, and let $\mathcal{M} = \{M_\alpha : \alpha \in (0,1]\}$ be a family of nonempty subsets of $\mathbb{R}^m$ such that it is a solid and nested family with respect to η. Given any $\alpha \in (0,1)$, we have*

$$M_1^{(\eta)} \cup \left(\bigcup_{\alpha\leq\beta<1} D_\beta^{(\eta)} \right) \subseteq M_\alpha^{(\eta)}. \tag{5.6}$$

Proof. Since $\mathcal{M}$ is a solid and nested family with respect to η, we first have $M_1^{(\eta)} \subseteq M_\alpha^{(\eta)}$. Given any $\alpha \in (0,1)$, we have

$$\emptyset \neq D_\alpha^{(\eta)} \subseteq A_\alpha^{(\eta)} = M_\alpha^{(\eta)} \text{ and } \emptyset \neq D_\beta^{(\eta)} \subseteq M_\beta^{(\eta)} \subseteq M_\alpha^{(\eta)}$$

for $\beta \in (0,1)$ with $\alpha < \beta$. Therefore, we obtain the inclusion (5.6), and the proof is complete. ∎

Proposition 5.2.8 *Let η be a function from $(0,1]$ into $(0,1]$, and let $\mathcal{M} = \{M_\alpha : \alpha \in (0,1]\}$ be a family of nonempty subsets of $\mathbb{R}^m$ such that $\mathcal{M}$ is a solid and nested family with respect to η. Then*

$$\bigcup_{0<\beta\leq 1} M_\beta^{(\eta)} = M_1^{(\eta)} \bigcup \left(\bigcup_{0<\beta<1} D_\beta^{(\eta)} \right) \tag{5.7}$$

if and only if

$$M_\alpha^{(\eta)} = M_1^{(\eta)} \bigcup \left(\bigcup_{\alpha\leq\beta<1} D_\beta^{(\eta)} \right) \text{ for } \alpha \in (0,1). \tag{5.8}$$

Proof. To prove the sufficiency, using (5.6), we have the following inclusion

$$M_1^{(\eta)} \bigcup \left(\bigcup_{\alpha\leq\beta<1} D_\beta^{(\eta)} \right) \subseteq M_\alpha^{(\eta)}.$$

For proving the other direction of inclusion, we have

$$\begin{aligned}
M_\alpha^{(\eta)} &= M_\alpha^{(\eta)} \bigcap \left(\bigcup_{0<\beta\leq 1} M_\beta^{(\eta)} \right) \\
&= M_\alpha^{(\eta)} \bigcap \left(M_1^{(\eta)} \bigcup \left(\bigcup_{0<\beta<1} D_\beta^{(\eta)} \right) \right) \text{ (by assumption (5.7))} \\
&= M_\alpha^{(\eta)} \bigcap \left(M_1^{(\eta)} \bigcup \left(\bigcup_{\alpha\leq\beta<1} D_\beta^{(\eta)} \right) \right)
\end{aligned}$$

(using Proposition 5.1.5 by taking $T = (0,1)$),

which implies

$$M_\alpha^{(\eta)} \subseteq M_1^{(\eta)} \bigcup \left(\bigcup_{\alpha\leq\beta<1} D_\beta^{(\eta)} \right).$$

For proving the necessity, we have

$$\begin{aligned}
\bigcup_{0<\alpha\leq 1} M_\alpha^{(\eta)} &= M_1^{(\eta)} \bigcup \left(\bigcup_{0<\alpha<1} M_\alpha^{(\eta)} \right) \\
&= M_1^{(\eta)} \bigcup \left(\bigcup_{0<\alpha<1} \left(M_1^{(\eta)} \bigcup \left(\bigcup_{\alpha\leq\beta<1} D_\beta^{(\eta)} \right) \right) \right) \text{ (by assumption (5.8)} \\
&= M_1^{(\eta)} \bigcup \left(\bigcup_{0<\alpha<1} D_\alpha^{(\eta)} \right).
\end{aligned}$$

This completes the proof. ∎

Proposition 5.2.9 *Let η be a function from* (0,1] *into* (0,1]*, and let* $\mathcal{M} = \{M_\alpha : \alpha \in (0,1]\}$ *be a family of nonempty subsets of* $\mathbb{R}^m$ *such that* $\mathcal{M}$ *is a solid and nested family with respect to* η*. For any* $\alpha \in (0,1)$*, we have the strict inclusion* $M_1^{(\eta)} \subset M_\alpha^{(\eta)}$.

Proof. Since $\mathcal{M}$ is a nested family with respect to η, for any $\alpha \in (0,1)$, we have the inclusion $M_1^{(\eta)} \subseteq M_\alpha^{(\eta)}$. Suppose that there exists $\beta_0 \in (0,1)$ satisfying $M_1^{(\eta)} = M_{\beta_0}^{(\eta)}$. Then, for any $\beta \in (0,1)$ with $\beta > \beta_0$, since

$$M_1^{(\eta)} \subseteq M_\beta^{(\eta)} \subseteq M_{\beta_0}^{(\eta)} = M_1^{(\eta)},$$

it follows that

$$M_1^{(\eta)} = M_\beta^{(\eta)} = M_{\beta_0}^{(\eta)}$$

for any $\beta \in (0,1)$ with $\beta > \beta_0$, which implies

$$A_{\beta_0+}^{(\eta)} = \bigcup_{\beta_0<\beta\leq 1} M_\beta^{(\eta)} = M_1^{(\eta)} = M_{\beta_0}^{(\eta)}.$$

Since $\mathcal{M}$ is a nested family with respect to η, we also have

$$A_{\beta_0}^{(\eta)} = \bigcup_{\beta_0\leq\beta\leq 1} M_\beta^{(\eta)} = M_{\beta_0}^{(\eta)}.$$

Therefore, we obtain

$$D_{\beta_0}^{(\eta)} = A_{\beta_0}^{(\eta)} \backslash A_{\beta_0+}^{(\eta)} = \emptyset,$$

which contradicts $D_\alpha^{(\eta)} \neq \emptyset$ for any $\alpha \in (0,1)$. This completes the proof. ∎

Proposition 5.2.10 *Let η be a function from* (0,1] *into* (0,1]*, and let* $\mathcal{M} = \{M_\alpha : \alpha \in (0,1]\}$ *be a family of nonempty subsets of* $\mathbb{R}^m$ *such that* $\mathcal{M}$ *is a solid and nested family with respect to* η*. Suppose that the following equality holds true*

$$M_\alpha^{(\eta)} = M_1^{(\eta)} \bigcup \left(\bigcup_{\alpha\leq\beta<1} D_\beta^{(\eta)} \right). \tag{5.9}$$

Then, we have

$$\bigcap_{n=1}^{\infty} M_{\alpha_n}^{(\eta)} = M_1^{(\eta)}$$

for $\alpha_n \uparrow 1$ *with* $\alpha_n \in (0,1)$ *for all* n.

Proof. According to Proposition 5.2.9, we define

$$N_\alpha = M_\alpha^{(\eta)} \backslash M_1^{(\eta)} \neq \emptyset \text{ for } \alpha \in (0,1). \tag{5.10}$$

Then, we immediately have $N_\alpha \subseteq N_\beta$ for $\beta < \alpha$. Suppose that $\bigcap_{n=1}^{\infty} N_{\alpha_n} \neq \emptyset$ for $\alpha_n \uparrow 1$ with $\alpha_n \in (0,1)$ for all n. We are going to lead to a contradiction. For $x \in N_{\alpha_n}$ for all n, given any $\alpha \in (0,1)$, since $\alpha_n \uparrow 1$ with $\alpha_n \in (0,1)$ for all n, there exists α_{m_1} satisfying $\alpha_{m_1} > \alpha$, which says that $x \in N_{\alpha_{m_1}} \subseteq N_\alpha$, i.e. $x \in N_\alpha$. Using (5.10) and the assumption (5.9), we see that

$x \in D^{(\eta)}_{\beta_1}$ for some $\beta_1 \in (0,1)$ with $\beta_1 \geq \alpha$. Given any $\beta_1 \in (0,1)$, there exists $\beta_2 \in (0,1)$ satisfying $\beta_2 > \beta_1$. We consider the following cases.

- Since $M^{(\eta)}_1 \cap D^{(\eta)}_{\beta_1} = \emptyset$ for $\beta_1 \in (0,1)$ by part (ii) of Proposition 5.1.4, it follows that $x \notin M^{(\eta)}_1$.
- For any $\beta \in (0,1)$ with $\beta \geq \beta_2$, since $\beta_2 > \beta_1$, it follows that $\beta \neq \beta_1$. Since $D^{(\eta)}_{\beta_1} \cap D^{(\eta)}_{\beta} = \emptyset$ by part (i) of Proposition 5.1.4, we obtain $x \notin D^{(\eta)}_{\beta}$.

Therefore, by the assumption (5.9), the above two cases imply that

$$x \notin M^{(\eta)}_1 \bigcup \left(\bigcup_{\beta_2 \leq \beta < 1} D^{(\eta)}_{\beta} \right) = M^{(\eta)}_{\beta_2}.$$

Since $N_{\beta_2} \subseteq M^{(\eta)}_{\beta_2}$, it follows that $x \notin N_{\beta_2}$. Since $\alpha_n \uparrow 1$ with $\alpha_n \in (0,1)$ for all n, there exists α_{m_2} satisfying $\alpha_{m_2} \in (0,1)$ and $\alpha_{m_2} > \beta_2$, which says that $N_{\alpha_{m_2}} \subseteq N_{\beta_2}$. Therefore, we obtain $x \notin N_{\alpha_{m_2}}$, which contradicts $x \in N_{\alpha_n}$ for all n. This shows that $\bigcap_{n=1}^{\infty} N_{\alpha_n} = \emptyset$. It is easy to see that

$$\bigcap_{n=1}^{\infty} \left(M^{(\eta)}_{\alpha_n} \backslash M^{(\eta)}_1 \right) = \bigcap_{n=1}^{\infty} N_{\alpha_n} = \emptyset \text{ if and only if } \bigcap_{n=1}^{\infty} M^{(\eta)}_{\alpha_n} = M^{(\eta)}_1$$

when $\alpha_n \uparrow 1$ with $\alpha_n \in (0,1)$ for all n. This completes the proof. ■

The converse of Proposition 5.2.10 can be obtained by providing some mild assumptions. We first present some useful inclusions

Proposition 5.2.11 *Let η be a function from $(0,1]$ into $(0,1]$, and let $\mathcal{M} = \{M_\alpha : \alpha \in (0,1]\}$ be a family of nonempty subsets of $\mathbb{R}^m$ such that $\mathcal{M}$ is a solid and nested family with respect to η. For any $\alpha \in (0,1)$, we have the strict inclusion $M^{(\eta)}_1 \subsetneqq A^{(\eta)}_{\alpha+}$.*

Proof. For any $\alpha \in (0,1)$, since $M^{(\eta)}_1 \subseteq M^{(\eta)}_\gamma$ for all $0 < \gamma \leq 1$, we have

$$M^{(\eta)}_1 \subseteq \bigcup_{\alpha < \gamma \leq 1} M^{(\eta)}_\gamma = A^{(\eta)}_{\alpha+}. \tag{5.11}$$

We want to show $M^{(\eta)}_1 \neq A^{(\eta)}_{\alpha+}$. Suppose that $M^{(\eta)}_1 = A^{(\eta)}_{\alpha+}$. We are going to lead to a contradiction. There exists $\gamma \in (0,1]$ satisfying $\alpha < \gamma \leq 1$. Therefore, we obtain

$$M^{(\eta)}_1 \subseteq M^{(\eta)}_\gamma \subseteq \bigcup_{\alpha < \beta \leq 1} M^{(\eta)}_\beta = A^{(\eta)}_{\alpha+} = M^{(\eta)}_1,$$

i.e. $M^{(\eta)}_1 = M^{(\eta)}_\gamma$. Since $\mathcal{M}$ is a nested family with respect to η, we also have

$$A^{(\eta)}_\gamma = \bigcup_{\gamma \leq \beta \leq 1} M^{(\eta)}_\beta = M^{(\eta)}_\gamma.$$

Since $M^{(\eta)}_1 \subseteq A^{(\eta)}_{\gamma+}$ from (5.11), we obtain

$$D^{(\eta)}_\gamma = A^{(\eta)}_\gamma \backslash A^{(\eta)}_{\gamma+} = M^{(\eta)}_\gamma \backslash A^{(\eta)}_{\gamma+} \subseteq M^{(\eta)}_\gamma \backslash M^{(\eta)}_1 = \emptyset,$$

which contradicts $D^{(\eta)}_\alpha \neq \emptyset$ for $\alpha \in (0,1)$. Therefore, we have the desired strict inclusion, and the proof is complete. ■

Theorem 5.2.12 *Let η be a function from $(0,1]$ into $(0,1]$, and let $\mathcal{M} = \{M_\alpha : \alpha \in (0,1]\}$ be a family of nonempty subsets of $\mathbb{R}^m$ such that the following conditions are satisfied:*

- *$\mathcal{M}$ is a solid and nested family with respect to η;*
- *$\bigcap_{n=1}^{\infty} M_{\alpha_n}^{(\eta)} \subseteq M_\alpha^{(\eta)}$ when $\alpha_n \uparrow \alpha$ for $\alpha < 1$.*

For any $\alpha \in (0,1)$, we have

$$M_\alpha^{(\eta)} = M_1^{(\eta)} \bigcup \left(\bigcup_{\alpha \leq \beta < 1} D_\beta^{(\eta)} \right) \tag{5.12}$$

if and only if

$$\bigcap_{n=1}^{\infty} M_{\alpha_n}^{(\eta)} = M_1^{(\eta)}$$

for $\alpha_n \uparrow 1$ with $\alpha_n \in (0,1)$ for all n.

Proof. The sufficiency follows from Proposition 5.2.10 immediately. To prove the necessity, assume that $\bigcap_{n=1}^{\infty} M_{\alpha_n}^{(\eta)} = M_1^{(\eta)}$ for $\alpha_n \uparrow 1$ with $\alpha_n \in (0,1)$ for all n. From Proposition 5.2.7, we have the following inclusion

$$M_1^{(\eta)} \bigcup \left(\bigcup_{\alpha \leq \beta < 1} D_\beta^{(\eta)} \right) \subseteq M_\alpha^{(\eta)}.$$

To prove the other direction of inclusion, for any $\alpha \in (0,1)$, according to Proposition 5.2.9, we define

$$N_\alpha = M_\alpha^{(\eta)} \backslash M_1^{(\eta)} \neq \emptyset. \tag{5.13}$$

We want to show

$$\left(\bigcup_{\alpha < \beta \leq 1} M_\beta^{(\eta)} \right) \Big\backslash M_1^{(\eta)} = \bigcup_{\alpha < \beta < 1} N_\beta. \tag{5.14}$$

Suppose that $x \in M_\beta^{(\eta)}$ for some β satisfying $\alpha < \beta < 1$ and $x \notin M_1^{(\eta)}$. Then $x \in N_\beta$ for some β satisfying $\alpha < \beta < 1$. Therefore, we obtain the inclusion

$$\left(\bigcup_{\alpha < \beta \leq 1} M_\beta^{(\eta)} \right) \Big\backslash M_1^{(\eta)} \subseteq \bigcup_{\alpha < \beta < 1} N_\beta.$$

To prove the other direction of inclusion, if $y \in N_\beta$ for some β satisfying $\alpha < \beta < 1$, then $y \in M_\beta^{(\eta)}$ for some β satisfying $\alpha < \beta < 1$ and $y \notin M_1^{(\eta)}$. This shows the equality (5.14). From (5.13), it is easy to we see that, for $\alpha_n \uparrow 1$ with $\alpha_n \in (0,1)$, we have

$$\bigcap_{n=1}^{\infty} M_{\alpha_n}^{(\eta)} = M_1^{(\eta)} \text{ if and only if } \bigcap_{n=1}^{\infty} N_{\alpha_n} = \emptyset. \tag{5.15}$$

Now, we are in a position to prove the following inclusion

$$M_\alpha^{(\eta)} \subseteq M_1^{(\eta)} \bigcup \left(\bigcup_{\alpha \leq \beta < 1} D_\beta^{(\eta)} \right).$$

For $\alpha \in (0,1)$ and $x \in N_\alpha = M_\alpha^{(\eta)} \backslash M_1^{(\eta)}$, suppose that $x \notin D_\alpha^{(\eta)}$ and $x \notin D_\beta^{(\eta)}$ for all β satisfying $\alpha < \beta < 1$. We shall lead to a contradiction. Since $x \in N_\alpha$ and $x \notin D_\alpha^{(\eta)} = A_\alpha^{(\eta)} \backslash A_{\alpha+}^{(\eta)}$ by the hypothesis, it follows that $x \in A_{\alpha+}^{(\eta)}$. Using Proposition 5.2.11, i.e. $M_1^{(\eta)} \subsetneqq A_{\alpha+}^{(\eta)}$, we also have $x \in A_{\alpha+}^{(\eta)} \backslash M_1^{(\eta)}$. Now, from (5.14), we obtain

$$x \in A_{\alpha+}^{(\eta)} \backslash M_1^{(\eta)} = \left(\bigcup_{\alpha<\beta\leq 1} M_\beta^{(\eta)} \right) \Big\backslash M_1^{(\eta)} = \bigcup_{\alpha<\beta<1} N_\beta,$$

which says that there exists β_1 satisfying $1 > \beta_1 > \alpha$ and $x \in N_{\beta_1}$. Since $x \notin D_{\beta_1}^{(\eta)}$ by the hypothesis, we can similarly show that there exists β_2 satisfying $1 > \beta_2 > \beta_1$ and $x \in N_{\beta_2}$. Also, since $x \notin D_{\beta_2}^{(\eta)}$ by the hypothesis, there exists β_3 satisfying $1 > \beta_3 > \beta_2$ and $x \in N_{\beta_3}$. Continuing this process and argument, we can construct a strictly increasing sequence $\{\beta_n\}_{n=1}^\infty$ contained in the open interval $(\alpha, 1)$ satisfying $x \in N_{\beta_n}$ for all n, i.e. $x \in \bigcap_{n=1}^\infty N_{\beta_n} \neq \emptyset$. We shall claim $1 = \sup_n \beta_n$. Suppose that it is not true, i.e. $\alpha < \sup_n \beta_n = \beta^* < 1$. Then, we have $\bigcap_{n=1}^\infty M_{\beta_n}^{(\eta)} \subseteq M_{\beta^*}^{(\eta)}$ by the second condition. Therefore, we obtain

$$x \in \bigcap_{n=1}^{\infty} N_{\beta_n} = \bigcap_{n=1}^{\infty} \left(M_{\beta_n}^{(\eta)} \backslash M_1^{(\eta)} \right) = \left(\bigcap_{n=1}^{\infty} M_{\beta_n}^{(\eta)} \right) \Big\backslash M_1^{(\eta)} \subseteq M_{\beta^*}^{(\eta)} \backslash M_1^{(\eta)} = N_{\beta^*},$$

which says that $x \in N_{\beta^*}$ and $x \notin D_{\beta^*}^{(\eta)}$. Using the previous arguments for the procedure of constructing the strictly increasing sequence, we can similarly show that there exists β° satisfying $1 > \beta^\circ > \beta^*$ and $x \in N_{\beta^\circ}$. This says that β° must be in the sequence, and that

$$\sup_n \beta_n \geq \beta^\circ > \beta^*,$$

which contradicts $\sup_n \beta_n = \beta^*$. Therefore, we must have $1 = \sup_n \beta_n$, i.e. $\beta_n \uparrow 1$ and $\bigcap_{n=1}^\infty N_{\beta_n} \neq \emptyset$, which also contradicts the assumption $\bigcap_{n=1}^\infty N_{\beta_n} = \emptyset$ given in (5.15). Therefore, we conclude that $x \in D_\alpha^{(\eta)}$ or $x \in D_\beta^{(\eta)}$ for some β satisfying $1 > \beta > \alpha$, which proves the equality (5.12). This completes the proof. ∎

5.3 Generating Fuzzy Sets from Nested Families

Now, we are in a position to generate fuzzy sets from solid and nested families. In applications, the data may be formulated as the solid and nested families by referring to (5.1) and Example 5.1.3.

Theorem 5.3.1 *Let η be a function from (0,1] into (0,1], and let $\mathcal{M} = \{M_\alpha : \alpha \in (0,1]\}$ be a family of nonempty subsets of $\mathbb{R}^m$ such that $\mathcal{M}$ is a solid and nested family with respect to η. Let κ be an increasing and bijective function from (0,1] into (0,1]. Assume that*

$$\bigcup_{0<\alpha\leq 1} M_\alpha^{(\eta)} = M_1^{(\eta)} \bigcup \left(\bigcup_{0<\alpha<1} D_\alpha^{(\eta)} \right). \tag{5.16}$$

- *If $\bigcup_{0<\alpha\leq 1} M_\alpha^{(\eta)} = \mathbb{R}^m$, we define a fuzzy set $\tilde{A}^\star$ in $\mathbb{R}^m$ with membership function given by*

$$\xi_{\tilde{A}^\star}(x) = \begin{cases} \kappa(\alpha), & \text{if } x \in D_\alpha^{(\eta)} \text{ for } \alpha \in (0,1) \\ 1, & \text{if } x \in M_1^{(\eta)}. \end{cases}$$

This also says that $\xi_{\tilde{A}^\star}(x) > 0$ for $x \in \mathbb{R}^m$.

- *If $\bigcup_{0<\alpha\leq 1} M_\alpha^{(\eta)} \neq \mathbb{R}^m$, we define a fuzzy set $\tilde{A}^\circ$ in $\mathbb{R}^m$ with membership function given by*

$$\xi_{\tilde{A}^\circ}(x) = \begin{cases} \kappa(\alpha), & \text{if } x \in D_\alpha^{(\eta)} \text{ for } \alpha \in (0,1) \\ 1, & \text{if } x \in M_1^{(\eta)} \\ 0, & \text{if } x \notin \bigcup\limits_{0<\alpha\leq 1} M_\alpha^{(\eta)}. \end{cases} \tag{5.17}$$

This also says that $\xi_{\tilde{A}^\circ}(x) > 0$ for $x \in \bigcup_{0<\alpha\leq 1} M_\alpha^{(\eta)}$.

Then, we have the following properties:
- $\mathcal{R}(\xi_{\tilde{A}^\star}) = (0,1]$ *and* $\mathcal{R}(\xi_{\tilde{A}^\circ}) = [0,1]$;
- $\tilde{A}^\star_{\kappa(\alpha)} = \tilde{A}^\circ_{\kappa(\alpha)} = M_\alpha^{(\eta)} = M_{\eta(\alpha)}$ *and* $\tilde{A}^\star_\alpha = \tilde{A}^\circ_\alpha = M^{(\eta)}_{\kappa^{-1}(\alpha)} = M_{\eta(\kappa^{-1}(\alpha))}$ *for each* $\alpha \in (0,1]$; *if* η *is injective, we have*

$$\tilde{A}^*_{\kappa(\eta^{-1}(\alpha))} = \tilde{A}^\circ_{\kappa(\eta^{-1}(\alpha))} = M_\alpha$$

for each $\alpha \in \mathcal{R}(\eta)$.

We also have

$$\tilde{A}^\star_0 = cl(\tilde{A}^\star_{0+}) = \mathbb{R}^m = \bigcup_{0<\alpha\leq 1} M_\alpha^{(\eta)} \text{ and } \tilde{A}^\circ_0 = cl(\tilde{A}^\circ_{0+}) = cl\left(\bigcup_{0<\alpha\leq 1} M_\alpha^{(\eta)}\right).$$

Proof. From Proposition 5.1.4, we see that the membership functions $\xi_{\tilde{A}^\star}$ and $\xi_{\tilde{A}^\circ}$ are well defined. Next, we want to show

$$\tilde{A}^\star_{\kappa(\alpha)} = \tilde{A}^\circ_{\kappa(\alpha)} = M_\alpha^{(\eta)} \text{ for each } \alpha \in (0,1].$$

It suffices to show $\tilde{A}^\circ_{\kappa(\alpha)} = M_\alpha^{(\eta)}$ for each $\alpha \in (0,1]$. We first claim that

$$\xi^{-1}_{\tilde{A}^\circ}([\kappa(\alpha), 1]) = \bigcup_{\{\gamma\in\mathcal{R}(\xi_{\tilde{A}^\circ}):\gamma\geq\kappa(\alpha)\}} \xi^{-1}_{\tilde{A}^\circ}(\gamma) \text{ for each } \alpha \in (0,1], \tag{5.18}$$

where $\xi^{-1}_{\tilde{A}^\circ}([\kappa(\alpha), 1])$ denotes the inverse image of the closed interval $[\kappa(\alpha), 1]$ and $\xi^{-1}_{\tilde{A}^\circ}(\gamma)$ denotes the inverse image of singleton $\{\gamma\}$ for function $\xi_{\tilde{A}^\circ}$. Given $x \in \xi^{-1}_{\tilde{A}^\circ}([\kappa(\alpha), 1])$, we have $\xi_{\tilde{A}^\circ}(x) \geq \kappa(\alpha)$. Let

$$\gamma = \xi_{\tilde{A}^\circ}(x) \in \mathcal{R}(\xi_{\tilde{A}^\circ}).$$

Then, we have $\gamma \geq \kappa(\alpha)$ and $x \in \xi^{-1}_{\tilde{A}^\circ}(\gamma)$, which says that

$$x \in \bigcup_{\{\gamma\in\mathcal{R}(\xi_{\tilde{A}^\circ}):\gamma\geq\kappa(\alpha)\}} \xi^{-1}_{\tilde{A}^\circ}(\gamma).$$

On the other hand, given any

$$x \in \bigcup_{\{\gamma\in\mathcal{R}(\xi_{\tilde{A}^\circ}):\gamma\geq\kappa(\alpha)\}} \xi^{-1}_{\tilde{A}^\circ}(\gamma),$$

there exists $\gamma \in \mathcal{R}(\xi_{\tilde{A}^\circ})$ with $\gamma \geq \kappa(\alpha)$ satisfying $x \in \xi^{-1}_{\tilde{A}^\circ}(\gamma)$, which also says that $\xi_{\tilde{A}^\circ}(x) = \gamma \geq \kappa(\alpha)$. Therefore, it follows that $x \in \xi^{-1}_{\tilde{A}^\circ}([\kappa(\alpha), 1])$, which proves the equality (5.18).

Since κ is an increasing and bijective function on (0,1], it follows that $\mathcal{R}(\xi_{\tilde{A}^\circ}) = [0,1]$ and $\kappa(\alpha) > 0$ for $\alpha \in (0,1]$, the equality (5.18) can be rewritten as

$$\xi^{-1}_{\tilde{A}^\circ}([\kappa(\alpha), 1]) = \bigcup_{\kappa(\alpha)\leq\gamma\leq 1} \xi^{-1}_{\tilde{A}^\circ}(\gamma) \tag{5.19}$$

We also see that $\kappa(1) = 1$. Therefore, for $\alpha \in (0,1)$, we have

$$\begin{aligned}\bigcup_{\kappa(\alpha)\leq\gamma<1} \xi^{-1}_{\tilde{A}^\circ}(\gamma) &= \bigcup_{\{\beta\in(0,1]:\kappa(\alpha)\leq\kappa(\beta)<1\}} \xi^{-1}_{\tilde{A}^\circ}(\kappa(\beta)) \text{ (since } \kappa \text{is bijective on } (0,1])\\ &= \bigcup_{\{\beta\in(0,1]:\alpha\leq\beta<1\}} \xi^{-1}_{\tilde{A}^\circ}(\kappa(\beta)) \text{ (since } \kappa \text{is increasing and bijective on } (0,1]). \end{aligned} \tag{5.20}$$

Now, for each $\alpha \in (0,1)$, we obtain

$$\begin{aligned}\tilde{A}^\circ_{\kappa(\alpha)} &= \xi^{-1}_{\tilde{A}^\circ}([\kappa(\alpha), 1]) = \bigcup_{\kappa(\alpha)\leq\gamma\leq 1} \xi^{-1}_{\tilde{A}^\circ}(\gamma) \text{ (by (5.19))}\\ &= \xi^{-1}_{\tilde{A}^\circ}(1) \bigcup \left(\bigcup_{\kappa(\alpha)\leq\gamma<1} \xi^{-1}_{\tilde{A}^\circ}(\gamma) \right)\\ &= \xi^{-1}_{\tilde{A}^\circ}(1) \bigcup \left(\bigcup_{\{\beta\in(0,1]:\alpha\leq\beta<1\}} \xi^{-1}_{\tilde{A}^\circ}(\kappa(\beta)) \right) \text{ (by (5.20)}\\ &= M_1^{(\eta)} \bigcup \left(\bigcup_{\{\beta\in(0,1]:\alpha\leq\beta<1\}} D_\beta^{(\eta)} \right) \text{ (by definition of } \xi_{\tilde{A}^\circ} \text{in (5.17))}\\ &= M_\alpha^{(\eta)} \text{ (by Proposition 5.2.8 using assumption (5.16)).}\end{aligned} \tag{5.21}$$

We also have

$$\begin{aligned}\tilde{A}^\circ_{\kappa(1)} &= \tilde{A}^\circ_1 = \xi^{-1}_{\tilde{A}^\circ}([1,1]) = \left\{x \in \mathbb{R}^m : \xi_{\tilde{A}^\circ}(x) \geq 1\right\}\\ &= \left\{x \in \mathbb{R}^m : \xi_{\tilde{A}^\circ}(x) = 1\right\} = \xi^{-1}_{\tilde{A}^\circ}(1) = M_1^{(\eta)}.\end{aligned}$$

and

$$\begin{aligned}\tilde{A}^\circ_{0+} &= \bigcup_{0<\alpha\leq 1} \tilde{A}^\circ_\alpha = \bigcup_{0<\alpha\leq 1} \tilde{A}^\circ_{\kappa(\alpha)} \quad \text{(since } \kappa \text{is bijective on } (0,1])\\ &= \bigcup_{0<\alpha\leq 1} M_\alpha^{(\eta)} \text{ (by (5.105)).}\end{aligned}$$

We can similarly obtain

$$\tilde{A}^\star_{0+} = \bigcup_{0<\alpha\leq 1} M_\alpha^{(\eta)},$$

Suppose that η is injective. Then $M_\alpha = M^{(\eta)}_{\eta^{-1}(\alpha)}$ for $\alpha \in \mathcal{R}(\eta)$ by (5.2). Therefore, we can obtain $\tilde{A}^\circ_{\kappa(\eta^{-1}(\alpha))} = M_\alpha$. This completes the proof. ■

In particular, the function κ in Theorem 5.3.1 can be taken to be the identity function, given by

$$\kappa(\alpha) = \alpha \text{ for } \alpha \in (0,1].$$

In this case, we can generate a fuzzy set $\tilde{A}^\star$ or $\tilde{A}^\circ$ such that their α-level sets are given by

$$\tilde{A}^\star_\alpha = \tilde{A}^\circ_\alpha = M_\alpha^{(\eta)} = M_{\eta(\alpha)} \text{ for } \alpha \in (0,1].$$

Example 5.3.2 Continued from Example 5.1.3, we can obtain

$$M_1 \bigcup \left(\bigcup_{0<\alpha\leq 1} D_\alpha^{(\eta)} \right) = M_1 \bigcup \left(\bigcup_{0<\alpha\leq 1} \{\bar{m}_\alpha^L, \bar{m}_\alpha^U\} \right)$$
$$= (\bar{m}_0^L, \bar{m}_0^U) = (b_1, b_2) = \bigcup_{0<\alpha\leq 1} M_\alpha^{(\eta)} \neq \mathbb{R}.$$

Therefore, according to Theorem 5.3.1, we can define a fuzzy set $\tilde{A}^\circ$ in $\mathbb{R}$ with membership function given by

$$\xi_{\tilde{A}^\circ}(x) = \begin{cases} \kappa(\alpha), & \text{if } x \in D_\alpha^{(\eta)} = \{\bar{m}_\alpha^L, \bar{m}_\alpha^U\} \text{ for } 0 < \alpha < 1 \\ 1, & \text{if } x \in M_1 \\ 0, & \text{if } x \notin \bigcup_{0<\alpha\leq 1} M_\alpha^{(\eta)} \end{cases}$$
$$= \begin{cases} \kappa(\alpha), & \text{if } x = (1-\alpha)b_1 + \alpha a_1 \text{or } (1-\alpha)b_2 + \alpha a_2 \text{ for } 0 < \alpha < 1 \\ 1, & \text{if } x \in [a_1, a_2] \\ 0, & \text{if } x \leq b_1 \text{or } x \geq b_2. \end{cases}$$

For each $\alpha \in (0,1]$, we have

$$\tilde{A}_{\kappa(\alpha)}^\circ = M_\alpha^{(\eta)} = M_{\eta(\alpha)} = M_{1-\alpha} = [\bar{m}_\alpha^L, \bar{m}_\alpha^U] = [(1-\alpha)b_1 + \alpha a_1, (1-\alpha)b_2 + \alpha a_2]$$

and the α-level set of $\tilde{A}^\circ$ is given by

$$\tilde{A}_\alpha^\circ = M_{\kappa^{-1}(\alpha)}^{(\eta)} = M_{\eta(\kappa^{-1}(\alpha))} = M_{1-\kappa^{-1}(\alpha)} = \left[\bar{m}_{\kappa^{-1}(\alpha)}^L, \bar{m}_{\kappa^{-1}(\alpha)}^U\right]$$
$$= \left[\left(1-\kappa^{-1}(\alpha)\right) b_1 + \kappa^{-1}(\alpha)a_1, \left(1-\kappa^{-1}(\alpha)\right) b_2 + \kappa^{-1}(\alpha)a_2\right].$$

Moreover, the 0-level set is given by $\tilde{A}_0^\circ = [b_1, b_2]$.

Theorem 5.3.3 *Let η be a function from (0,1] into (0,1], and let $\mathcal{M} = \{M_\alpha : \alpha \in (0,1]\}$ be a family of nonempty subsets of $\mathbb{R}^m$ such that the following conditions are satisfied:*

- *$\mathcal{M}$ is a solid and nested family with respect to η;*
- *$\bigcap_{n=1}^\infty M_{\alpha_n}^{(\eta)} \subseteq M_\alpha^{(\eta)}$ when $\alpha_n \uparrow \alpha$ for $\alpha < 1$.*

Let κ be an increasing and bijective function on (0,1]. Assume that

$$\bigcap_{n=1}^\infty M_{\alpha_n}^{(\eta)} \subseteq M_1^{(\eta)} \text{ when } \alpha_n \uparrow 1 \text{ with } \alpha_n \in (0,1) \text{ for all } n.$$

- *If $\bigcup_{0<\alpha\leq 1} M_\alpha^{(\eta)} = \mathbb{R}^m$, we define a fuzzy set $\tilde{A}^\star$ in $\mathbb{R}^m$ with membership function given by*

$$\xi_{\tilde{A}^\star}(x) = \begin{cases} \kappa(\alpha), & \text{if } x \in D_\alpha^{(\eta)} \text{ for } \alpha \in (0,1) \\ 1, & \text{if } x \in M_1^{(\eta)}. \end{cases}$$

This says that $\xi_{\tilde{A}^\star}(x) > 0$ for $x \in \mathbb{R}^m$.

- *If $\bigcup_{0<\alpha\leq 1} M_\alpha^{(\eta)} \neq \mathbb{R}^m$, we define a fuzzy set $\tilde{A}^\circ$ in $\mathbb{R}^m$ with membership function given by*

$$\xi_{\tilde{A}^\circ}(x) = \begin{cases} \kappa(\alpha), & \text{if } x \in D_\alpha^{(\eta)} \text{ for } \alpha \in (0,1) \\ 1, & \text{if } x \in M_1^{(\eta)} \\ 0, & \text{if } x \notin \bigcup_{0<\alpha\leq 1} M_\alpha^{(\eta)}. \end{cases}$$

This says that $\xi_{\tilde{A}^\circ}(x) > 0$ for $x \in \bigcup_{0<\alpha\leq 1} M_\alpha^{(\eta)}$.

Then, we have the following properties:

- $\mathcal{R}(\xi_{\tilde{A}^\star}) = (0,1]$ *and* $\mathcal{R}(\xi_{\tilde{A}^\circ}) = [0,1]$;
- $\tilde{A}^\star_{\kappa(\alpha)} = \tilde{A}^\circ_{\kappa(\alpha)} = M^{(\eta)}_\alpha = M_{\eta(\alpha)}$ *and* $\tilde{A}^\star_\alpha = \tilde{A}^\circ_\alpha = M^{(\eta)}_{\kappa^{-1}(\alpha)} = M_{\eta(\kappa^{-1}(\alpha))}$ *for each* $\alpha \in (0,1)$; *if* η *is injective, we have*

$$\tilde{A}^*_{\kappa(\eta^{-1}(\alpha))} = \tilde{A}^\circ_{\kappa(\eta^{-1}(\alpha))} = M_\alpha$$

for each $\alpha \in \mathcal{R}(\eta)$.

We also have

$$\tilde{A}^\star_0 = cl(\tilde{A}^\star_{0+}) = \mathbb{R}^m = \bigcup_{0<\alpha\leq 1} M^{(\eta)}_\alpha \text{ and } \tilde{A}^\circ_0 = cl(\tilde{A}^\circ_{0+}) = cl\left(\bigcup_{0<\alpha\leq 1} M^{(\eta)}_\alpha\right).$$

Proof. Using the nestedness, we see that

$$M^{(\eta)}_1 \subseteq \bigcap_{n=1}^\infty M^{(\eta)}_{\alpha_n} \text{ when } \alpha_n \uparrow 1 \text{ with } \alpha_n \in (0,1) \text{ for all } n.$$

Therefore, combining it with the assumption, we obtain

$$M^{(\eta)}_1 = \bigcap_{n=1}^\infty M^{(\eta)}_{\alpha_n} \text{ when } \alpha_n \uparrow 1 \text{ with } \alpha_n \in (0,1) \text{ for all } n.$$

By Theorem 5.2.12 and Proposition 5.2.8, we see that the assumption of Theorem 5.3.1 is satisfied. Therefore, the results follow immediately from Theorem 5.3.1. This completes the proof. ∎

5.4 Generating Fuzzy Sets Based on the Expression in the Decomposition Theorem

We are going to present many methods to construct fuzzy sets in $\mathbb{R}^m$ based on the expression in the Decomposition Theorem 2.2.13.

5.4.1 The Ordinary Situation

Let $\mathcal{M} = \{M_\alpha : \alpha \in [0,1]\}$ be a family of nonempty sets in $\mathbb{R}^m$. We can generate a fuzzy set $\tilde{A}^\perp$ in $\mathbb{R}^m$ with membership function defined by

$$\xi_{\tilde{A}^\perp}(x) = \sup_{0<\alpha\leq 1} \alpha \cdot \chi_{M_\alpha}(x). \tag{5.22}$$

For $x \in \bigcup_{0<\alpha\leq 1} M_\alpha$, we have $x \in M_{\alpha_0}$ for some $\alpha_0 \in (0,1]$, which says that

$$\xi_{\tilde{A}^\perp}(x) = \sup_{0<\alpha\leq 1} \alpha \cdot \chi_{M_\alpha}(x) \geq \alpha_0 > 0.$$

For $x \notin \bigcup_{0<\alpha\leq 1} M_\alpha$, we have $x \notin M_\alpha$ for all $\alpha \in (0,1]$. By the definition of $\xi_{\tilde{A}^\perp}$ in (5.22), we have $\xi_{\tilde{A}^\perp}(x) = 0$. Therefore, we obtain

$$\xi_{\tilde{A}^\perp}(x) = \begin{cases} \sup_{0<\alpha\leq 1} \alpha \cdot \chi_{M_\alpha}(x) > 0, & \text{if } x \in \bigcup_{0<\alpha\leq 1} M_\alpha \\ 0, & \text{otherwise.} \end{cases} \tag{5.23}$$

Proposition 5.4.1 *We have the following equality*

$$\tilde{A}^{\perp}_{0+} = \bigcup_{0<\alpha\leq 1} M_\alpha. \tag{5.24}$$

Proof. For $x \notin \bigcup_{0<\alpha\leq 1} M_\alpha$, we have $\xi_{\tilde{A}^{\perp}}(x) = 0$, i.e. $x \notin \tilde{A}^{\perp}_\alpha$ for all $\alpha \in (0,1]$. Therefore, we have the inclusions

$$\tilde{A}^{\perp}_\alpha \subseteq \bigcup_{0<\alpha\leq 1} M_\alpha \text{ for all } \alpha \in (0,1],$$

which implies

$$\tilde{A}^{\perp}_{0+} \subseteq \bigcup_{0<\alpha\leq 1} M_\alpha$$

by (2.5). The inclusion

$$\bigcup_{0<\alpha\leq 1} M_\alpha \subseteq \tilde{A}^{\perp}_{0+}$$

is obvious by (5.23). This completes the proof. ■

Given any fixed $\alpha \in (0,1]$, we write

$$A_\alpha = \bigcup_{\alpha\leq\beta\leq 1} M_\beta,$$

and given any fixed $\alpha \in (0,1)$, we write

$$A_{\alpha+} = \bigcup_{\alpha<\beta\leq 1} M_\beta.$$

It is clear to see $A_1 = M_1$.

Proposition 5.4.2 *Given any fixed $\alpha \in (0,1)$, we have*

$$\tilde{A}^{\perp}_{\alpha+} \subseteq A_\alpha \subseteq \tilde{A}^{\perp}_\alpha \text{ and } A_{\alpha+} = \tilde{A}^{\perp}_{\alpha+}.$$

Proof. For $x \in A_\alpha$, it means that $x \in M_\beta$ for some β satisfying $\alpha \leq \beta \leq 1$. From (5.22), it follows that

$$\xi_{\tilde{A}^{\perp}}(x) = \sup_{0<\gamma\leq 1} \gamma \cdot \chi_{M_\gamma}(x) \geq \beta \geq \alpha,$$

which shows the inclusion $A_\alpha \subseteq \tilde{A}^{\perp}_\alpha$. We can similarly obtain the inclusion $A_{\alpha+} \subseteq \tilde{A}^{\perp}_{\alpha+}$. Suppose that $x \notin A_\alpha$. It means that $x \notin M_\beta$ for all β satisfying $\alpha \leq \beta \leq 1$. Therefore, we obtain

$$\xi_{\tilde{A}^{\perp}}(x) = \sup_{0<\beta\leq 1} \beta \cdot \chi_{M_\beta}(x) = \sup_{0<\beta<\alpha} \beta \cdot \chi_{M_\beta}(x) \leq \alpha,$$

i.e. $x \notin \tilde{A}^{\perp}_{\alpha+}$, which shows the inclusion $\tilde{A}^{\perp}_{\alpha+} \subseteq A_\alpha$. We can similarly obtain the inclusion $\tilde{A}^{\perp}_{\alpha+} \subseteq A_{\alpha+}$, and the proof is complete. ■

Proposition 5.4.3 *Let $\{M_\alpha : \alpha \in [0,1]\}$ be a family of nonempty subsets of $\mathbb{R}^m$. We consider the membership function $\xi_{\tilde{A}^{\perp}}$ defined in (5.22). Suppose that $A_\alpha \backslash A_{\alpha+} \neq \emptyset$ for all $\alpha \in (0,1)$. Then, we have $\mathcal{R}(\xi_{\tilde{A}^{\perp}}) = [0,1]$.*

Proof. It is clear to see $0 \in \mathcal{R}(\xi_{\tilde{A}^{\perp}})$. Since $M_1 \neq \emptyset$, we also see that $1 \in \mathcal{R}(\xi_{\tilde{A}^{\perp}})$. Now, given any fixed $\alpha \in (0,1)$ and any $x \in D_\alpha$, we see that $x \in A_\alpha$ and $x \notin A_{\alpha+}$. Using Proposition 5.4.2, we have $x \in \tilde{A}_\alpha^{\perp}$ and $x \notin \tilde{A}_{\alpha+}^{\perp}$, which say that $\xi_{\tilde{A}^{\perp}}(x) \geq \alpha$ and $\xi_{\tilde{A}^{\perp}}(x) \leq \alpha$. Therefore, we obtain $\xi_{\tilde{A}^{\perp}}(x) = \alpha$, which also says that $(0,1) \subseteq \mathcal{R}(\xi_{\tilde{A}^{\perp}})$, and the proof is complete. ■

Theorem 5.4.4 (Negoita, Puri and Ralescu [84, 92]) *Let $\{M_\alpha : \alpha \in [0,1]\}$ be a family of nonempty subsets of $\mathbb{R}^m$ such that the following conditions are satisfied:*

- $M_0 = \mathbb{R}^m$;
- $M_\alpha \subseteq M_\beta$ *for* $\alpha, \beta \in [0,1]$ *with* $\beta < \alpha$;
- $\bigcap_{n=1}^{\infty} M_{\alpha_n} \subseteq M_\alpha$ *when* $\alpha_n \uparrow \alpha$ *with* $\alpha > 0$.

We can generate a fuzzy set $\tilde{A}^*$ with membership function defined by

$$\xi_{\tilde{A}^*}(x) = \sup_{\alpha \in [0,1]} \alpha \cdot \chi_{M_\alpha}(x) = \sup_{0<\alpha\leq 1} \alpha \cdot \chi_{M_\alpha}(x).$$

Then, we have

$$\tilde{A}_\alpha^* = \{x \in \mathbb{R}^m : \xi_{\tilde{A}^*}(x) \geq \alpha\} = M_\alpha$$

for every $\alpha \in [0,1]$.

The assumption of $M_0 = \mathbb{R}^m$ in Theorem 5.4.4 can be omitted. In what follows, we can consider $M_0 \neq \mathbb{R}^m$. We first provide some useful properties.

Proposition 5.4.5 *Let $\mathcal{M} = \{M_\alpha : \alpha \in [0,1]\}$ be a family of nonempty subsets of $\mathbb{R}^m$ satisfying $M_\beta \subseteq M_\alpha$ for $\alpha, \beta \in (0,1]$ with $\alpha < \beta$. We consider the membership function $\xi_{\tilde{A}^{\perp}}$ defined in (5.22). Then, the following statements hold true.*

(i) *Given any $\alpha \in (0,1]$, we have*

$$M_\alpha \subseteq \tilde{A}_\alpha^{\perp} \tag{5.25}$$

and

$$\tilde{A}_\alpha^{\perp} = \bigcap_{0<\beta<\alpha} M_\beta \subseteq \bigcap_{0<\beta<\alpha} \tilde{A}_\beta^{\perp}. \tag{5.26}$$

(ii) *Given any $\alpha \in (0,1]$, suppose that $\bigcap_{n=1}^{\infty} M_{\alpha_n} \subseteq M_\alpha$ when $\alpha_n \uparrow \alpha$. Then, we have*

$$\tilde{A}_\alpha^{\perp} = M_\alpha = \bigcap_{0<\beta\leq\alpha} M_\beta = \bigcap_{0<\beta<\alpha} M_\beta. \tag{5.27}$$

Proof. To prove part (i), the inclusion (5.25) is obvious by applying the nestedness to Proposition 5.4.2, where

$$A_\alpha = \bigcup_{\alpha\leq\beta\leq 1} M_\beta = M_\alpha$$

by the nestedness. To prove (5.26), given any $x \in \tilde{A}_\alpha^{\perp}$ and β satisfying $0 < \beta < \alpha$, we want to show that $x \in M_\beta$. Since $\xi_{\tilde{A}^{\perp}}(x) \geq \alpha$, we consider the following cases.

- Assume that $\xi_{\tilde{A}^{\perp}}(x) > \alpha$. Let $\epsilon = \xi_{\tilde{A}^{\perp}}(x) - \alpha > 0$. From (5.22), the concept of supremum says that there exists $\alpha_0 \in (0,1]$ satisfying $x \in M_{\alpha_0}$ and $\xi_{\tilde{A}^{\perp}}(x) - \epsilon < \alpha_0$, which implies $\alpha < \alpha_0$. This also says that $x \in M_\beta$, since $M_{\alpha_0} \subseteq M_\alpha \subseteq M_\beta$ by the nestedness.

- Assume that $\xi_{\tilde{A}^{\perp}}(x) = \alpha$. Let $\epsilon = \alpha - \beta > 0$. The concept of supremum says that there exists $\alpha_0 \in (0,1]$ satisfying $x \in M_{\alpha_0}$ and $\alpha - \epsilon < \alpha_0$, which implies $\beta < \alpha_0$. This also says that $x \in M_\beta$ by the nestedness $M_{\alpha_0} \subseteq M_\beta$.

  Since β can be any number satisfying $0 < \beta < \alpha$, we conclude that

$$x \in \bigcap_{0<\beta<\alpha} M_\beta.$$

By combining the inclusion (5.25), we obtain the inclusion (5.26).

For proving the other direction of inclusion, suppose that $x \in M_\beta$ for all β satisfying $0 < \beta < \alpha$. By the definition of $\xi_{\tilde{A}^{\perp}}$ in (5.22), we also see that $\xi_{\tilde{A}^{\perp}}(x) \geq \beta$ for all β satisfying $0 < \beta < \alpha$. We want to show that $\xi_{\tilde{A}^{\perp}}(x) \geq \alpha$. Suppose that $\xi_{\tilde{A}^{\perp}}(x) < \alpha$. There exists $\overline{\beta}$ satisfying $\xi_{\tilde{A}^{\perp}}(x) < \overline{\beta} < \alpha$, which contradicts $\xi_{\tilde{A}^{\perp}}(x) \geq \beta$ for all β satisfying $0 < \beta < \alpha$. Therefore, we obtain $x \in \tilde{A}_\alpha^{\perp}$, which proves the equality (5.26) by applying the inclusion (5.26).

To prove part (ii), given any $x \in \tilde{A}_\alpha^{\perp}$, we want to show that $x \in M_\alpha$. Since $\xi_{\tilde{A}^{\perp}}(x) \geq \alpha$, we consider the following cases.

- Assume that $\xi_{\tilde{A}^{\perp}}(x) > \alpha$. Let $\epsilon = \xi_{\tilde{A}^{\perp}}(x) - \alpha > 0$. The supremum in (5.22) says that there exists $\alpha_0 \in (0,1]$ satisfying $x \in M_{\alpha_0}$ and $\xi_{\tilde{A}^{\perp}}(x) - \epsilon < \alpha_0$, which implies $\alpha < \alpha_0$. This also says that $x \in M_\alpha$, since $M_{\alpha_0} \subseteq M_\alpha$ by the nestedness.
- Assume that $\xi_{\tilde{A}^{\perp}}(x) = \alpha$. The supremum in (5.22) says that there exists a sequence $\{\alpha_n\}_{n=1}^{\infty}$ in $(0,1]$ satisfying $x \in M_{\alpha_n}$ for all n and $\alpha_n \uparrow \alpha$, which implies

$$x \in \bigcap_{n=1}^{\infty} M_{\alpha_n} \subseteq M_\alpha$$

  by the assumption, and concludes that $x \in M_\alpha$.

Therefore, we obtain the inclusion $\tilde{A}_\alpha^{\perp} \subseteq M_\alpha$. The equalities (5.27) follow immediately from (5.25) and the nestedness. This completes the proof. ■

Proposition 5.4.6 *Let $\{M_\alpha : \alpha \in [0,1]\}$ be a family of nonempty subsets of $\mathbb{R}^m$ such that the following conditions are satisfied:*

- *$M_\alpha \subseteq M_\beta$ for $\alpha, \beta \in (0,1]$ with $\beta < \alpha$;*
- *$\bigcap_{n=1}^{\infty} M_{\alpha_n} \subseteq M_\alpha$ when $\alpha_n \uparrow \alpha$.*

Consider the membership function $\xi_{\tilde{A}^{\perp}}$ defined in (5.22). Then, we have

$$\tilde{A}_\alpha^{\perp} = M_\alpha \text{ for every } \alpha \in (0,1] \tag{5.28}$$

and

$$\tilde{A}_{0+}^{\perp} = \bigcup_{0<\alpha\leq 1} M_\alpha. \tag{5.29}$$

We further assume that each M_α is a convex set in $\mathbb{R}^m$ for $\alpha \in (0,1]$. Then, the membership function $\xi_{\tilde{A}^{\perp}}$ is quasi-concave.

Proof. The equality (5.28) follows from part (ii) of Proposition 5.4.5. Also, the equality (5.29) follows from (5.24) immediately. It remains to show that the membership function $\xi_{\tilde{A}^{\perp}}$ is quasi-concave. For any $x_1, x_2 \in \mathbb{R}^m$, let

$$\alpha = \min\{\xi_{\tilde{A}^{\perp}}(x_1), \xi_{\tilde{A}^{\perp}}(x_2)\}.$$

If $\alpha = 0$, it is clear to see that

$$\xi_{\tilde{A}^{\perp}}(kx_1 + (1-k)x_2) \geq 0 = \min\left\{\xi_{\tilde{A}^{\perp}}(x_1), \xi_{\tilde{A}^{\perp}}(x_2)\right\}.$$

For $\alpha > 0$, we have $\xi_{\tilde{A}^{\perp}}(x_1) \geq \alpha$ and $\xi_{\tilde{A}^{\perp}}(x_2) \geq \alpha$, i.e. $x_1, x_2 \in \tilde{A}^{\perp}_{\alpha}$. Since $\tilde{A}^{\perp}_{\alpha} = M_\alpha$ is convex for $\alpha \in (0,1]$, we also have $kx_1 + (1-k)x_2 \in \tilde{A}^{\perp}_{\alpha}$ for any $k \in (0,1)$. Therefore, we obtain.

$$\xi_{\tilde{A}^{\perp}}(kx_1 + (1-k)x_2) \geq \alpha = \min\left\{\xi_{\tilde{A}^{\perp}}(x_1), \xi_{\tilde{A}^{\perp}}(x_2)\right\}.$$

This completes the proof. ∎

Theorem 5.4.7 *Let $\{M_\alpha : \alpha \in [0,1]\}$ be a family of nonempty subsets of $\mathbb{R}^m$ such that the following conditions are satisfied:*

- *$M_\alpha \subseteq M_\beta$ for $\alpha, \beta \in (0,1]$ with $\beta < \alpha$;*
- *$\bigcap_{n=1}^{\infty} M_{\alpha_n} \subseteq M_\alpha$ when $\alpha_n \uparrow \alpha$.*

Then, we can generate a fuzzy set $\tilde{A}^{\perp}$ in $\mathbb{R}^m$ with membership function defined by

$$\xi_{\tilde{A}^{\perp}}(x) = \sup_{0<\alpha\leq 1} \alpha \cdot \chi_{M_\alpha}(x) \tag{5.30}$$

satisfying

$$\tilde{A}^{\perp}_{\alpha} = M_\alpha \text{ for } \alpha \in (0,1].$$

We also have

$$\tilde{A}^{\perp}_{0+} = \bigcup_{0<\alpha\leq 1} M_\alpha$$

and

$$\tilde{A}^{\perp}_{0} = cl(\tilde{A}^{\perp}_{0+}) = cl\left(\bigcup_{0<\alpha\leq 1} \tilde{A}^{\perp}_{\alpha}\right) = cl\left(\bigcup_{0<\alpha\leq 1} M_\alpha\right). \tag{5.31}$$

Suppose that $A_\alpha \backslash A_{\alpha+} \neq \emptyset$ for all $\alpha \in (0,1)$. Then, we have $\mathcal{R}(\xi_{\tilde{A}^{\perp}}) = [0,1]$. We further assume that the following conditions are satisfied.

- *M_α are closed and convex sets in $\mathbb{R}^m$ for $\alpha \in (0,1]$.*
- *The union $\bigcup_{0<\alpha\leq 1} M_\alpha$ is a bounded subset of $\mathbb{R}^m$.*

Then $\tilde{A}^{\perp} \in \mathfrak{F}(\mathbb{R}^m)$ is a fuzzy vector.

Proof. The results regarding the α-level sets can follow from Proposition 5.4.6, and the results regarding the range $\mathcal{R}(\xi_{\tilde{A}^{\perp}})$ of membership function can follow from Proposition 5.4.3. Finally, we want to claim that $\tilde{A}^{\perp} \in \mathfrak{F}(\mathbb{R}^m)$ by checking all the conditions in Definition 2.3.1.

- Since each M_α is assumed to be convex set in $\mathbb{R}^m$ for $\alpha \in (0,1]$, Proposition 5.4.6 says that the membership function $\xi_{\tilde{A}^{\perp}}$ is quasi-concave.
- Since each M_α is assumed to be closed subset of $\mathbb{R}^m$ for $\alpha \in (0,1]$, it follows that

  $$\left\{x \in \mathbb{R}^m : \xi_{\tilde{A}^{\perp}}(x) \geq \alpha\right\} = \tilde{A}^{\perp}_{\alpha} = M_\alpha$$

  is a closed subset of $\mathbb{R}^m$ for $\alpha \in (0,1]$, which also says that the membership function $\xi_{\tilde{A}^{\perp}}$ is upper semi-continuous.
- From (5.31), it follows that the 0-level set $\tilde{A}^{\perp}_{0}$ is bounded in $\mathbb{R}^m$.

Therefore, we conclude that $\tilde{A}^{\perp} \in \mathfrak{F}(\mathbb{R}^m)$, and the proof is complete. ∎

The above results assume that $\{M_\alpha : \alpha \in [0,1]\}$ is a nested family in the sense of $M_\beta \subseteq M_\alpha$ for $\alpha < \beta$. Next, the family $\{M_\alpha : \alpha \in [0,1]\}$ will not be assumed to be nested.

Proposition 5.4.8 *Let $\{M_\alpha : \alpha \in [0,1]\}$ be a family of nonempty subsets of $\mathbb{R}^m$. Suppose that, for any fixed $x \in \mathbb{R}^m$, the function $\zeta_x(\alpha) = \alpha \cdot \chi_{M_\alpha}(x)$ is upper semi-continuous on $[0,1]$. Consider the membership function $\xi_{\tilde{A}^\perp}$ defined in (5.22). Then*

$$\tilde{A}_\alpha^\perp = \bigcup_{\alpha\le\beta\le1} M_\beta \text{ for every } \alpha \in (0,1] \tag{5.32}$$

and

$$\tilde{A}_{0+}^\perp = \bigcup_{0<\alpha\le1} M_\alpha. \tag{5.33}$$

Proof. For $\alpha \in (0,1]$, suppose that $x \in \tilde{A}_\alpha^\perp$ and $x \notin M_\beta$ for all β satisfying $\alpha \le \beta \le 1$. Then $\beta \cdot \chi_{M_\beta}(x) < \alpha$ for all $0 < \beta \le 1$. Since ζ_x is upper semi-continuous on $[0,1]$ by the assumption, Proposition 1.4.4 says that the following supremum

$$\sup_{0\le\beta\le1} \zeta_x(\beta) = \max_{0\le\beta\le1} \zeta_x(\beta)$$

is attained. This says that

$$0 < \xi_{\tilde{A}^\perp}(x) = \sup_{0\le\beta\le1} \zeta_x(\beta) = \max_{0\le\beta\le1} \zeta_x(\beta) = \beta^* \cdot \chi_{M_{\beta^*}}(x) < \alpha$$

for some $\beta^* \in [0,1]$. Since $\beta^* \neq 0$, it follows that $0 < \beta^* \le 1$, which violates $x \in \tilde{A}_\alpha^\perp$. Therefore, there exists β_0 satisfying $\alpha \le \beta_0 \le 1$ and $x \in M_{\beta_0}$, which shows the following inclusion

$$\tilde{A}_\alpha^\perp \subseteq \bigcup_{\alpha\le\beta\le1} M_\beta.$$

The following inclusion

$$\bigcup_{\alpha\le\beta\le1} M_\beta \subseteq \left\{ x \in \mathbb{R}^m : \sup_{0<\beta\le1} \beta \cdot \chi_{M_\beta}(x) \ge \alpha \right\} = \{x \in \mathbb{R}^m : \xi_{\tilde{A}^\perp}(x) \ge \alpha\} = \tilde{A}_\alpha^\perp$$

is obvious, which proves the equality (5.32). Finally, the equality (5.33) follows immediately from (5.24), and the proof is complete. ∎

Theorem 5.4.9 **(Non-Nested Family).** *Let $\{M_\alpha : \alpha \in [0,1]\}$ be a family of nonempty subsets of $\mathbb{R}^m$. Suppose that, for any fixed $x \in \mathbb{R}^m$, the function $\zeta_x(\alpha) = \alpha \cdot \chi_{M_\alpha}(x)$ is upper semi-continuous on $[0,1]$. Then, we can generate a fuzzy set $\tilde{A}^\perp$ in $\mathbb{R}^m$ with membership function defined by*

$$\xi_{\tilde{A}^\perp}(x) = \sup_{0<\alpha\le1} \alpha \cdot \chi_{M_\alpha}(x) \tag{5.34}$$

satisfying

$$\tilde{A}_\alpha^\perp = \bigcup_{\alpha\le\beta\le1} M_\beta \text{ for } \alpha \in (0,1].$$

We also have

$$\tilde{A}_{0+}^\perp = \bigcup_{0<\alpha\le1} M_\alpha$$

and

$$\tilde{A}_0^{\perp} = cl(\tilde{A}_{0+}^{\perp}) = cl\left(\bigcup_{0<\alpha\leq 1} M_\alpha\right).$$

Suppose that $A_\alpha \setminus A_{\alpha+} \neq \emptyset$ *for all* $\alpha \in (0,1)$. *Then, we have* $\mathcal{R}(\xi_{\tilde{A}^{\perp}}) = [0,1]$.

Proof. Using Proposition 5.4.8 and a similar argument to the proof of Theorem 5.4.7, the proof is complete. ∎

5.4.2 Based on One Function

We consider another approach for generating fuzzy sets in $\mathbb{R}^m$. Let $\{M_\alpha : \alpha \in [0,1]\}$ be a family of nonempty subsets of $\mathbb{R}^m$, and let $\lambda : [0,1] \to [0,1]$ be a real-valued function defined on $[0,1]$. We can generate a fuzzy set $\tilde{A}^\diamond$ in $\mathbb{R}^m$ with membership function defined by

$$\xi_{\tilde{A}^\diamond}(x) = \sup_{\alpha\in[0,1]} \lambda(\alpha) \cdot \chi_{M_\alpha}(x). \tag{5.35}$$

Let

$$S^\bullet = \{\alpha \in [0,1] : \lambda(\alpha) > 0\}. \tag{5.36}$$

Then, we see that

$$\xi_{\tilde{A}^\diamond}(x) = \sup_{\alpha\in[0,1]} \lambda(\alpha) \cdot \chi_{M_\alpha}(x) = \sup_{\alpha\in S^\bullet} \lambda(\alpha) \cdot \chi_{M_\alpha}(x). \tag{5.37}$$

We see that (5.22) is a special case of (5.37) by taking $\lambda(\alpha) = \alpha$.

For $x \in \bigcup_{\alpha\in S^\bullet} M_\alpha$, we have $x \in M_{\alpha_0}$ for some $\alpha_0 \in S^\bullet$, which says that

$$\xi_{\tilde{A}^\diamond}(x) = \sup_{\alpha\in S^\bullet} \lambda(\alpha) \cdot \chi_{M_\alpha}(x) \geq \lambda(\alpha_0) > 0.$$

For $x \notin \bigcup_{\alpha\in S^\bullet} M_\alpha$, we have $x \notin M_\alpha$ for all $\alpha \in S^\bullet$. By the definition of $\xi_{\tilde{A}^\diamond}$ in (5.37), we have $\xi_{\tilde{A}^\diamond}(x) = 0$. Therefore, we obtain

$$\xi_{\tilde{A}^\diamond}(x) = \begin{cases} \sup\limits_{\alpha\in S^\bullet} \lambda(\alpha) \cdot \chi_{M_\alpha}(x) > 0, & \text{if } x \in \bigcup\limits_{\alpha\in S^\bullet} M_\alpha \\ 0, & \text{otherwise.} \end{cases} \tag{5.38}$$

Proposition 5.4.10 *We have the following equality*

$$\tilde{A}_{0+}^\diamond = \bigcup_{\alpha\in S^\bullet} M_\alpha. \tag{5.39}$$

Proof. For $x \notin \bigcup_{\alpha\in S^\bullet} M_\alpha$, we have $\xi_{\tilde{A}^\diamond}(x) = 0$, i.e. $x \notin \tilde{A}_\alpha^\diamond$ for all $\alpha \in (0,1]$. Therefore, we have

$$\tilde{A}_\alpha^\diamond \subseteq \bigcup_{\alpha\in S^\bullet} M_\alpha \text{ for all } \alpha \in (0,1].$$

Using (2.5), we can obtain the inclusion

$$\tilde{A}_{0+}^\diamond \subseteq \bigcup_{\alpha\in S^\bullet} M_\alpha.$$

The inclusion

$$\bigcup_{\alpha \in S^{\bullet}} M_{\alpha} \subseteq \tilde{A}^{\diamond}_{0+}$$

is obvious by (5.38). This completes the proof. ∎

Proposition 5.4.11 *Suppose that $\mathcal{M} = \{M_{\alpha} : \alpha \in [0,1]\}$ is a nested family in the sense of $M_{\alpha} \subseteq M_{\beta}$ for $\alpha, \beta \in S^{\bullet}$ with $\lambda(\beta) < \lambda(\alpha)$. We also assume that any one of the following conditions is satisfied.*

- *The infimum $\inf_{\alpha \in S^{\bullet}} \lambda(\alpha)$ is attained.*
- *$S^{\bullet}$ is a closed and bounded set, and λ is lower semi-continuous on $S^{\bullet}$.*

Then, we have

$$\tilde{A}^{\diamond}_{0+} = \bigcup_{\alpha \in S^{\bullet}} M_{\alpha} = M_{\overline{\alpha}},$$

where

$$\overline{\alpha} = \inf_{\alpha \in S^{\bullet}} \lambda(\alpha).$$

Proof. Suppose that the infimum $\inf_{\alpha \in S^{\bullet}} \lambda(\alpha)$ is attained; that is, there exists $\overline{\alpha} \in S^{\bullet}$ satisfying

$$\lambda(\overline{\alpha}) = \min_{\alpha \in S^{\bullet}} \lambda(\alpha) = \inf_{\alpha \in S^{\bullet}} \lambda(\alpha).$$

In this case, we have $M_{\alpha} \subseteq M_{\overline{\alpha}}$ for all $\alpha \in S^{\bullet}$, which says that $\bigcup_{\alpha \in S^{\bullet}} M_{\alpha} = M_{\overline{\alpha}}$.

Suppose that $S^{\bullet}$ is a closed and bounded set, and that λ is lower semi-continuous on $S^{\bullet}$. Then, the infimum $\inf_{\alpha \in S^{\bullet}} \lambda(\alpha)$ is automatically attained. This completes the proof. ∎

Given any fixed $\alpha \in [0,1)$, we write

$$B^{(\lambda)}_{\alpha} = \bigcup_{\{\beta \in S^{\bullet} : \lambda(\beta) \geq \lambda(\alpha)\}} M_{\beta} \tag{5.40}$$

and

$$B^{(\lambda)}_{\alpha+} = \begin{cases} \emptyset, & \text{if } \lambda(\beta) \leq \lambda(\alpha) \text{ for all } \beta \in S^{\bullet} \\ \displaystyle\bigcup_{\{\beta \in S^{\bullet} : \lambda(\beta) > \lambda(\alpha)\}} M_{\beta}, & \text{otherwise.} \end{cases} \tag{5.41}$$

Proposition 5.4.12 *Given any fixed $\alpha \in [0,1)$, we have*

$$\tilde{A}^{\diamond}_{\lambda(\alpha)+} \subseteq B^{(\lambda)}_{\alpha} \subseteq \tilde{A}^{\diamond}_{\lambda(\alpha)} \text{ and } B^{(\lambda)}_{\alpha+} = \tilde{A}^{\diamond}_{\lambda(\alpha)+}.$$

Proof. For $x \in B^{(\lambda)}_{\alpha}$, it means that $x \in M_{\beta}$ for some $\beta \in S^{\bullet}$ satisfying $\lambda(\beta) \geq \lambda(\alpha)$. From (5.37), it follows that

$$\xi_{\tilde{A}^{\diamond}}(x) = \sup_{\gamma \in S^{\bullet}} \lambda(\gamma) \cdot \chi_{M_{\gamma}}(x) \geq \lambda(\beta) \geq \lambda(\alpha),$$

which shows the inclusion $B_\alpha^{(\lambda)} \subseteq \tilde{A}^\diamond_{\lambda(\alpha)}$. We can similarly obtain the inclusion $B_\alpha^{(\lambda)} \subseteq \tilde{A}^\diamond_{\lambda(\alpha)+}$. Suppose that $x \notin B_\alpha^{(\lambda)}$. It means that $x \notin M_\beta$ for all $\beta \in S^\bullet$ satisfying $\lambda(\beta) \geq \lambda(\alpha)$. Therefore, we obtain

$$\xi_{\tilde{A}^\diamond}(x) = \sup_{\beta \in S^\bullet} \lambda(\beta) \cdot \chi_{M_\beta}(x) = \sup_{\{\beta \in S^\bullet : \lambda(\beta) < \lambda(\alpha)\}} \lambda(\beta) \cdot \chi_{M_\beta}(x) \leq \lambda(\alpha),$$

i.e. $x \notin \tilde{A}^\diamond_{\lambda(\alpha)+}$, which shows the inclusion $\tilde{A}^\diamond_{\lambda(\alpha)+} \subseteq B_\alpha^{(\lambda)}$. If $B_{\alpha+}^{(\lambda)} \neq \emptyset$, we can similarly obtain the following inclusion $\tilde{A}^\diamond_{\lambda(\alpha)+} \subseteq B_{\alpha+}^{(\lambda)}$. If $B_{\alpha+}^{(\lambda)} = \emptyset$, we have $\lambda(\beta) \leq \lambda(\alpha)$ for all $\beta \in S^\bullet$, which says that $\xi_{\tilde{A}^\diamond}(x) \leq \lambda(\alpha)$ for any $x \in \mathbb{R}^m$ by (5.37), i.e.

$$\{x \in \mathbb{R}^m : \xi_{\tilde{A}^\diamond}(x) > \lambda(\alpha)\} = \emptyset.$$

This completes the proof. ■

The range $\mathcal{R}(\xi_{\tilde{A}^\diamond})$ of function $\xi_{\tilde{A}^\diamond}$ is an important issue. Therefore, the following results will be used in the subsequent discussion.

Proposition 5.4.13 *Let $\{M_\alpha : \alpha \in [0,1]\}$ be a family of nonempty subsets of $\mathbb{R}^m$, and let λ be a function from $[0,1]$ into $[0,1]$. We consider the membership function $\xi_{\tilde{A}^\diamond}$ defined in (5.37). Then, the following statements hold true.*

(i) *Given any fixed $x \in \bigcup_{\alpha \in S^\bullet} M_\alpha$, we have the following properties.*
- *Suppose that the supremum $\xi_{\tilde{A}^\diamond}(x)$ in (5.37) is attained. Then, we have $\xi_{\tilde{A}^\diamond}(x) \in \mathcal{R}(\lambda)$.*
- *Suppose that the supremum $\xi_{\tilde{A}^\diamond}(x)$ in (5.37) is not attained. Then, we have*

$$\xi_{\tilde{A}^\diamond}(x) \in cl(\mathcal{R}(\lambda)) \backslash \mathcal{R}(\lambda).$$

(ii) *We have $0 \in \mathcal{R}(\xi_{\tilde{A}^\diamond}) \subseteq cl(\mathcal{R}(\lambda)) \cup \{0\}$.*

(iii) *Suppose that $B_\alpha^{(\lambda)} \backslash B_{\alpha+}^{(\lambda)} \neq \emptyset$ for all $\alpha \in [0,1) \cap S^\bullet$. Then, we have*

$$\mathcal{R}(\lambda) \cup \{0\} \subseteq \mathcal{R}(\xi_{\tilde{A}^\diamond}) \subseteq cl(\mathcal{R}(\lambda)) \cup \{0\} = cl(\mathcal{R}(\lambda) \cup \{0\})$$

and

$$cl(\mathcal{R}(\xi_{\tilde{A}^\diamond})) = cl(\mathcal{R}(\lambda) \cup \{0\}).$$

In particular, we have the following properties.
- *Suppose that the supremum $\xi_{\tilde{A}^\diamond}(x)$ in (5.37) is attained for all $x \in \bigcup_{\alpha \in S^\bullet} M_\alpha$. Then, we have*

$$\mathcal{R}(\xi_{\tilde{A}^\diamond}) = \mathcal{R}(\lambda) \cup \{0\}.$$

- *Suppose that $\mathcal{R}(\lambda) \cup \{0\}$ is a closed subset of $[0,1]$. Then, we have*

$$\mathcal{R}(\xi_{\tilde{A}^\diamond}) = \mathcal{R}(\lambda) \cup \{0\}.$$

Proof. To prove part (i), if the supremum in (5.37) is attained, it means that there exists $\alpha_0 \in S^\bullet$ satisfying

$$\xi_{\tilde{A}^\diamond}(x) = \lambda(\alpha_0) \in \mathcal{R}(\lambda) \text{ and } x \in M_{\alpha_0}.$$

Suppose that the supremum in (5.37) is not attained, i.e. $\xi_{\tilde{A}^\perp}(x) \notin \mathcal{R}(\lambda)$. According to the concept of supremum, there exists a sequence $\{\alpha_n\}_{n=1}^\infty$ in $[0,1]$ satisfying $x \in M_{\alpha_n}$ for all n and $\lambda(\alpha_n) \to \xi_{\tilde{A}^\diamond}(x)$ as $n \to \infty$, which says that $\xi_{\tilde{A}^\diamond}(x) \in \mathrm{cl}(\mathcal{R}(\lambda))$. It is obvious that $0 \in \mathcal{R}(\xi_{\tilde{A}^\diamond})$ from (5.38). Therefore, part (ii) can be realized from part (i).

To prove part (iii), let $E_\alpha^{(\lambda)} = B_\alpha^{(\lambda)} \backslash B_{\alpha+}^{(\lambda)}$. Then $E_\alpha^{(\lambda)} \neq \emptyset$ for all $\alpha \in [0,1) \cap S^\bullet$. Given any $\gamma \in \mathcal{R}(\lambda)\backslash\{0\}$, there exists $\alpha \in [0,1] \cap S^\bullet$ satisfying $\lambda(\alpha) = \gamma > 0$. Given any $x \in E_\alpha^{(\lambda)}$, we consider the following cases.

- Suppose that $B_{\alpha+}^{(\lambda)} = \emptyset$. Then $\lambda(\beta) \leq \lambda(\alpha)$ for all $\beta \in [0,1]$, i.e. $\xi_{\tilde{A}^\diamond}(x) \leq \lambda(\alpha)$ for any $x \in \mathbb{R}^m$ by (5.37). For any $x \in E_\alpha^{(\lambda)}$, we also have $x \in B_\alpha^{(\lambda)}$. Using Proposition 5.4.12, we obtain $x \in \tilde{A}^\diamond_{\lambda(\alpha)}$, i.e. $\xi_{\tilde{A}^\diamond}(x) \geq \lambda(\alpha)$, which implies $\xi_{\tilde{A}^\diamond}(x) = \lambda(\alpha) = \gamma$, i.e. $\gamma \in \mathcal{R}(\xi_{\tilde{A}^\diamond})$.
- Suppose that $B_{\alpha+}^{(\lambda)} \neq \emptyset$. For any $x \in E_\alpha^{(\lambda)}$, we see that $x \in B_\alpha^{(\lambda)}$ and $x \notin B_{\alpha+}^{(\lambda)}$. Using Proposition 5.4.12, we have $x \in \tilde{A}^\diamond_{\lambda(\alpha)}$ and $x \notin \tilde{A}^\diamond_{\lambda(\alpha)+}$, i.e. $\xi_{\tilde{A}^\diamond}(x) \geq \lambda(\alpha)$ and $\xi_{\tilde{A}^\diamond}(x) \leq \lambda(\alpha)$, which implies $\xi_{\tilde{A}^\diamond}(x) = \lambda(\alpha) = \gamma$.

Therefore, the above two cases prove the inclusion $\mathcal{R}(\lambda)\backslash\{0\} \subseteq \mathcal{R}(\xi_{\tilde{A}^\diamond})$. Using parts (i) and (ii), the proof is complete. ∎

Proposition 5.4.14 *Let $\{M_\alpha : \alpha \in [0,1]\}$ be a family of nonempty subsets of $\mathbb{R}^m$ satisfying $M_\beta \subseteq M_\alpha$ for $\alpha, \beta \in S^\bullet$ with $\lambda(\alpha) < \lambda(\beta)$, and let λ be a function from $[0,1]$ into $[0,1]$. We consider the membership function $\xi_{\tilde{A}^\diamond}$ defined in (5.37). Then, the following statements hold true.*

(i) *Given any $\alpha \in S^\bullet$, we have*

$$M_\alpha \subseteq \tilde{A}^\diamond_{\lambda(\alpha)} \tag{5.42}$$

and

$$\tilde{A}^\diamond_{\lambda(\alpha)} \subseteq \bigcap_{\{\beta \in S^\bullet : \lambda(\beta) < \lambda(\alpha)\}} M_\beta \subseteq \bigcap_{\{\beta \in S^\bullet : \lambda(\beta) < \lambda(\alpha)\}} \tilde{A}^\diamond_{\lambda(\beta)}. \tag{5.43}$$

(ii) *Suppose that λ is continuous. Given any $\alpha \in S^\bullet$, we have*

$$\tilde{A}^\diamond_{\lambda(\alpha)} = \bigcap_{\{\beta \in S^\bullet : \lambda(\beta) < \lambda(\alpha)\}} M_\beta. \tag{5.44}$$

(iii) *Given any $\alpha \in S^\bullet$, suppose that $\bigcap_{n=1}^\infty M_{\alpha_n} \subseteq M_\alpha$ when $\lambda(\alpha_n) \uparrow \lambda(\alpha)$ with $\alpha_n \in S^\bullet$ for all n. Then, we have*

$$\tilde{A}^\diamond_{\lambda(\alpha)} = M_\alpha = \bigcap_{\{\beta \in S^\bullet : \lambda(\beta) \leq \lambda(\alpha)\}} M_\beta. \tag{5.45}$$

(iv) *Given any $\alpha \in S^\bullet$, suppose that $\bigcap_{n=1}^\infty M_{\alpha_n} \subseteq M_\alpha$ when $\lambda(\alpha_n) \uparrow \lambda(\alpha)$ with $\alpha_n \in S^\bullet$ for all n. We also assume that λ is continuous. Then, we have*

$$\tilde{A}^\diamond_{\lambda(\alpha)} = M_\alpha = \bigcap_{\{\beta \in S^\bullet : \lambda(\beta) \leq \lambda(\alpha)\}} M_\beta = \bigcap_{\{\beta \in S^\bullet : \lambda(\beta) < \lambda(\alpha)\}} M_\beta.$$

Proof. To prove part (i), for any $\alpha \in S^\bullet$, since $M_\beta \subseteq M_\alpha$ for $\alpha, \beta \in S^\bullet$ with $\lambda(\alpha) < \lambda(\beta)$, it follows that $B_\alpha^{(\lambda)} = M_\alpha$. Then, the inclusion (5.42) is obvious by using Proposition 5.4.12. To prove the inclusions (5.43), given any $x \in \tilde{A}^\diamond_{\lambda(\alpha)}$ and $\beta \in S^\bullet$ satisfying $\lambda(\beta) < \lambda(\alpha)$, we want to show that $x \in M_\beta$. Since $\xi_{\tilde{A}^\diamond}(x) \geq \lambda(\alpha)$, we consider the following cases.

- Assume that $\xi_{\tilde{A}^\diamond}(x) > \lambda(\alpha)$. Let

$$\epsilon = \xi_{\tilde{A}^\diamond}(x) - \lambda(\alpha) > 0.$$

From (5.37), the concept of supremum says that there exists $\alpha_0 \in S^\bullet$ satisfying $x \in M_{\alpha_0}$ and

$$\xi_{\tilde{A}^\circ}(x) - \epsilon < \lambda(\alpha_0),$$

which implies $\lambda(\alpha) < \lambda(\alpha_0)$. This also says that $x \in M_\beta$, since

$$M_{\alpha_0} \subseteq M_\alpha \subseteq M_\beta$$

by the nestedness.

- Assume that $\xi_{\tilde{A}^\circ}(x) = \lambda(\alpha)$. Let

$$\epsilon = \lambda(\alpha) - \lambda(\beta) > 0.$$

The concept of supremum says that there exists $\alpha_0 \in S^\bullet$ satisfying $x \in M_{\alpha_0}$ and

$$\lambda(\alpha) - \epsilon < \lambda(\alpha_0),$$

which implies $\lambda(\beta) < \lambda(\alpha_0)$. This also says that $x \in M_\beta$ by the nestedness $M_{\alpha_0} \subseteq M_\beta$.

Since $\lambda(\beta)$ can be any number $\beta \in S^\bullet$ satisfying $\lambda(\beta) < \lambda(\alpha)$, we conclude that

$$x \in \bigcap_{\{\beta \in S^\bullet : \lambda(\beta) < \lambda(\alpha)\}} M_\beta.$$

By combining the inclusion (5.42), we obtain the inclusions (5.43).

To prove part (ii), suppose that $x \in M_\beta$ for all $\beta \in S^\bullet$ satisfying $\lambda(\beta) < \lambda(\alpha)$. By the definition of $\xi_{\tilde{A}^\circ}$ in (5.37), we also see that $\xi_{\tilde{A}^\circ}(x) \geq \lambda(\beta)$ for all $\beta \in S^\bullet$ satisfying $\lambda(\beta) < \lambda(\alpha)$. We want to show that $\xi_{\tilde{A}^\circ}(x) \geq \lambda(\alpha)$. Suppose that $\xi_{\tilde{A}^\circ}(x) < \lambda(\alpha)$. Then, there exists $c \in (0,1)$ satisfying

$$\lambda(\beta_0) \leq \xi_{\tilde{A}^\circ}(x) < c < \lambda(\alpha)$$

for some $\beta_0 \in S^\bullet$ satisfying $\lambda(\beta_0) < \lambda(\alpha)$. Since λ is continuous and $\lambda(\beta_0) < c < \lambda(\alpha)$, Proposition 1.4.3 says that there exists $\overline{\beta} \in [\beta_0, \alpha]$ satisfying $c = \lambda(\overline{\beta})$. This shows that $\xi_{\tilde{A}^\circ}(x) < \lambda(\overline{\beta})$, which contradicts $\xi_{\tilde{A}^\circ}(x) \geq \lambda(\beta)$ for all $\beta \in S^\bullet$ satisfying $\lambda(\beta) < \lambda(\alpha)$. Therefore, we must have $\xi_{\tilde{A}^\circ}(x) \geq \lambda(\alpha)$, i.e. $x \in \tilde{A}^\diamond_{\lambda(\alpha)}$, which proves the equality (5.44) by applying the inclusion (5.43).

To prove part (iii), given any $x \in \tilde{A}^\diamond_{\lambda(\alpha)}$, we want to show that $x \in M_\alpha$. Since $\xi_{\tilde{A}^\circ}(x) \geq \lambda(\alpha)$, we consider the following cases.

- Assume that $\xi_{\tilde{A}^\circ}(x) > \lambda(\alpha)$. Let

$$\epsilon = \xi_{\tilde{A}^\circ}(x) - \lambda(\alpha) > 0.$$

The supremum in (5.37) says that there exists $\alpha_0 \in S^\bullet$ satisfying $x \in M_{\alpha_0}$ and

$$\xi_{\tilde{A}^\circ}(x) - \epsilon < \lambda(\alpha_0),$$

which implies $\lambda(\alpha) < \lambda(\alpha_0)$. This also says that $x \in M_\alpha$, since $M_{\alpha_0} \subseteq M_\alpha$ by the nestedness.

- Assume that $\xi_{\tilde{A}^\circ}(x) = \lambda(\alpha)$. The supremum in (5.37) says that there exists a sequence $\{\alpha_n\}_{n=1}^\infty$ in $S^\bullet$ satisfying $x \in M_{\alpha_n}$ for all n and $\lambda(\alpha_n) \uparrow \lambda(\alpha)$, which implies

$$x \in \bigcap_{n=1}^\infty M_{\alpha_n} \subseteq M_\alpha$$

by the assumption, and concludes that $x \in M_\alpha$.

Therefore, we obtain the inclusion $\tilde{A}^{\diamond}_{\lambda(\alpha)} \subseteq M_\alpha$. The equalities (5.45) follow immediately from (5.42) and the nestedness. Finally, part (iv) follows from parts (ii) and (iii). This completes the proof ■

Proposition 5.4.14 presents the representation $\tilde{A}^{\diamond}_{\lambda(\alpha)}$ for $\lambda(\alpha) \in \mathcal{R}(\lambda)$. If $\gamma \notin \mathcal{R}(\lambda)$, the representation $\tilde{A}^{\diamond}_{\gamma}$ is unknown. The following proposition will provide the representation of $\tilde{A}^{\diamond}_{\gamma}$ in terms of sequence $\{\tilde{A}^{\diamond}_{\lambda(\alpha_n)}\}_{n=1}^{\infty}$ without considering the nestedness in which the representation $\tilde{A}^{\diamond}_{\lambda(\alpha_n)}$ can be realized from Proposition 5.4.14.

Proposition 5.4.15 *Let $\mathcal{M} = \{M_\alpha : \alpha \in [0,1]\}$ be a family of nonempty subsets of $\mathbb{R}^m$, and let λ be a function from [0,1] into [0,1]. Consider the membership function $\xi_{\tilde{A}^{\diamond}}$ defined in (5.37). Then, we have the following properties.*

(i) *For $\gamma \notin \mathcal{R}(\lambda)$, suppose that $\lambda(\alpha_n) \uparrow \gamma$. Then, we have*

$$\tilde{A}^{\diamond}_{\gamma} = \bigcap_{n=1}^{\infty} \tilde{A}^{\diamond}_{\lambda(\alpha_n)}. \tag{5.46}$$

(ii) *For $\gamma \notin \mathcal{R}(\lambda)$, suppose that $\lambda(\beta_m) \downarrow \gamma$. Then, we have*

$$\tilde{A}^{\diamond}_{\gamma+} \subseteq \bigcup_{m=1}^{\infty} \tilde{A}^{\diamond}_{\lambda(\beta_m)+} \subseteq \bigcup_{m=1}^{\infty} \tilde{A}^{\diamond}_{\lambda(\beta_m)} \subseteq \tilde{A}^{\diamond}_{\gamma}.$$

Proof. To prove part (i), the following inclusion is obvious

$$\tilde{A}^{\diamond}_{\gamma} = \left\{x \in \mathbb{R}^m : \xi_{\tilde{A}^{\diamond}}(x) \geq \gamma\right\} \subseteq \bigcap_{n=1}^{\infty} \left\{x \in \mathbb{R}^m : \xi_{\tilde{A}^{\diamond}}(x) \geq \lambda(\alpha_n)\right\} = \bigcap_{n=1}^{\infty} \tilde{A}^{\diamond}_{\lambda(\alpha_n)}.$$

On the other hand, if $\xi_{\tilde{A}^{\diamond}}(x) \geq \lambda(\alpha_n)$ for all n, we have $\xi_{\tilde{A}^{\diamond}}(x) \geq \gamma$ by taking the limit, which shows the other direction of inclusion. Therefore, we obtain the desired equality (5.46).

To prove part (ii), since $\lambda(\beta_m) \downarrow \gamma$, the following inclusion is obvious

$$\tilde{A}^{\diamond}_{\gamma} = \left\{x \in \mathbb{R}^m : \xi_{\tilde{A}^{\diamond}}(x) \geq \gamma\right\} \supseteq \bigcup_{m=1}^{\infty} \left\{x \in \mathbb{R}^m : \xi_{\tilde{A}^{\diamond}}(x) \geq \lambda(\beta_m)\right\} = \bigcup_{m=1}^{\infty} \tilde{A}^{\diamond}_{\lambda(\beta_m)}.$$

Now, for $\xi_{\tilde{A}^{\diamond}}(x) > \gamma$, since $\lambda(\beta_m) \downarrow \gamma$, there exists

$$\lambda(\beta_M) \in \{\lambda(\beta_m)\}_{m=1}^{\infty} \text{ satisfying } \gamma \leq \lambda(\beta_M) < \xi_{\tilde{A}^{\diamond}}(x),$$

which shows the following inclusion

$$\tilde{A}^{\diamond}_{\gamma+} = \left\{x \in \mathbb{R}^m : \xi_{\tilde{A}^{\diamond}}(x) > \gamma\right\} \subseteq \bigcup_{m=1}^{\infty} \left\{x \in \mathbb{R}^m : \xi_{\tilde{A}^{\diamond}}(x) > \lambda(\beta_m)\right\} = \bigcup_{m=1}^{\infty} \tilde{A}^{\diamond}_{\lambda(\beta_m)+}.$$

This completes the proof. ■

Remark 5.4.16 The existence of the sequences $\{\lambda(\alpha_n)\}_{n=1}^{\infty}$ and $\{\lambda(\beta_m)\}_{m=1}^{\infty}$ satisfying $\lambda(\alpha_n) \uparrow \gamma$ and $\lambda(\beta_m) \downarrow \gamma$, respectively, can be realized by considering the denseness of range $\mathcal{R}(\lambda)$. Suppose that the range $\mathcal{R}(\lambda)$ is dense in [0,1]. Given any $r \in (0,1)\backslash\mathcal{R}(\lambda)$, there exists two sequences $\{\lambda(\alpha_n)\}_{n=1}^{\infty}$ and $\{\lambda(\beta_m)\}_{m=1}^{\infty}$ in $\mathcal{R}(\lambda)$ satisfying $\lambda(\alpha_n) \uparrow \gamma$ and $\lambda(\beta_m) \downarrow \gamma$. For $\gamma = 1$, we can only consider the case of $\lambda(\alpha_n) \uparrow \gamma$.

Proposition 5.4.17 *Let λ be a function from* [0,1] *into* [0,1], *and let* $\{M_\alpha : \alpha \in [0,1]\}$ *be a family of nonempty subsets of* $\mathbb{R}^m$ *such that the following conditions are satisfied:*

- $M_\beta \subseteq M_\alpha$ *for* $\alpha, \beta \in S^\bullet$ *with* $\lambda(\alpha) < \lambda(\beta)$;
- $\bigcap_{n=1}^{\infty} M_{\alpha_n} \subseteq M_\alpha$ *when* $\lambda(\alpha_n) \uparrow \lambda(\alpha)$ *with* $\alpha_n, \alpha \in S^\bullet$ *for all* n.

Consider the membership function $\xi_{\tilde{A}^\diamond}$ *defined in* (5.37). *Then, we have*

$$\tilde{A}^\diamond_{\lambda(\alpha)} = M_\alpha \text{ for } \alpha \in S^\bullet \tag{5.47}$$

and

$$\tilde{A}^\diamond_{0+} = \bigcup_{\alpha \in S^\bullet} M_\alpha. \tag{5.48}$$

We also have

$$\tilde{A}^\diamond_\gamma = \begin{cases} M_\alpha, & \text{if } \gamma = \lambda(\alpha) \in \mathcal{R}(\lambda) \\ \bigcap_{n=1}^{\infty} M_{\alpha_n}, & \text{if } \gamma \notin \mathcal{R}(\lambda) \text{ and } \lambda(\alpha_n) \uparrow \gamma \text{ for } \alpha_n \in S^\bullet \text{ for all } n. \end{cases} \tag{5.49}$$

Proof. The equality (5.47) follows from part (iii) of Proposition 5.4.14. Also, The equality (5.48) follows from (5.39) immediately. Using part (i) of Proposition 5.4.15 and (5.47), we obtain (5.49). This completes the proof. ■

Example 5.4.18 We present two different functions regarding λ.

(a) Let $\lambda(\alpha) = 1 - \alpha$ be defined on [0,1]. Then

$$S^\bullet = \{\alpha \in [0,1] : \lambda(\alpha) > 0\} = \{\alpha \in [0,1] : 1 - \alpha > 0\} = [0,1).$$

- The first condition of Proposition 5.4.17 says that $\lambda(\alpha) > \lambda(\beta)$ implies $M_\alpha \subseteq M_\beta$, i.e. $\alpha < \beta$ implies $M_\alpha \subseteq M_\beta$.
- The second condition of Proposition 5.4.17 says that $\lambda(\alpha_n) \uparrow \lambda(\alpha)$ implies $\bigcap_{n=1}^{\infty} M_{\alpha_n} \subseteq M_\alpha$, i.e. $\alpha_n \downarrow \alpha$ implies $\bigcap_{n=1}^{\infty} M_{\alpha_n} \subseteq M_\alpha$.

In this case, we have $\tilde{A}^\diamond_{1-\alpha} = M_\alpha$ for all $\alpha \in [0,1)$. Equivalently, we also have $\tilde{A}^\diamond_\alpha = M_{1-\alpha}$ for all $\alpha \in (0,1]$.

(b) Let $\lambda(\alpha) = \alpha^2$ be defined on [0,1]. Then

$$S^\bullet = \{\alpha \in [0,1] : \lambda(\alpha) > 0\} = \left\{\alpha \in [0,1] : \alpha^2 > 0\right\} = (0,1].$$

- The first condition of Proposition 5.4.17 says that $\alpha^2 < \beta^2$ implies $M_\beta \subseteq M_\alpha$, which also says that $\alpha < \beta$ implies $M_\beta \subseteq M_\alpha$, since $\alpha < \beta$ if and only if $\alpha^2 < \beta^2$ for positive numbers α and β.
- The second condition of Proposition 5.4.17 says that $\alpha_n^2 \uparrow \alpha^2$ implies $\bigcap_{n=1}^{\infty} M_{\alpha_n} \subseteq M_\alpha$, i.e. $\alpha_n \uparrow \alpha$ implies $\bigcap_{n=1}^{\infty} M_{\alpha_n} \subseteq M_\alpha$, since $\alpha_n \uparrow \alpha$ if and only if $\alpha_n^2 \uparrow \alpha^2$ for positive numbers α and α_n for $n = 1,2,\ldots$.

In this case, we have $\tilde{A}^\diamond_{\alpha^2} = M_\alpha$ for all $\alpha \in (0,1]$.

Theorem 5.4.19 **(Based on One Function)**. *Let λ be a function from* [0,1] *into* [0,1], *and let* $\{M_\alpha : \alpha \in [0,1]\}$ *be a family of nonempty subsets of* $\mathbb{R}^m$ *such that the following conditions are satisfied:*

- $M_\alpha \subseteq M_\beta$ *for* $\alpha, \beta \in S^\bullet$ *with* $\lambda(\beta) < \lambda(\alpha)$;
- $\bigcap_{n=1}^{\infty} M_{\alpha_n} \subseteq M_\alpha$ *when* $\lambda(\alpha_n) \uparrow \lambda(\alpha)$ *with* $\alpha_n, \alpha \in S^\bullet$ *for all* n.

Then, we can generate a fuzzy set $\tilde{A}^{\diamond}$ in $\mathbb{R}^m$ with membership function defined by

$$\xi_{\tilde{A}^{\diamond}}(x) = \sup_{\alpha \in S^{\bullet}} \lambda(\alpha) \cdot \chi_{M_\alpha}(x) \tag{5.50}$$

satisfying $\tilde{A}^{\diamond}_{\lambda(\alpha)} = M_\alpha$ for $\alpha \in S^{\bullet}$. We also have

$$\tilde{A}^{\diamond}_{0+} = \bigcup_{\alpha \in S^{\bullet}} M_\alpha \text{ and } \tilde{A}^{\diamond}_{0} = cl(\tilde{A}^{\diamond}_{0+}) = cl\left(\bigcup_{\alpha \in S^{\bullet}} \tilde{A}^{\diamond}_{\alpha}\right) = cl\left(\bigcup_{\alpha \in S^{\bullet}} M_\alpha\right).$$

For $\gamma \notin \mathcal{R}(\lambda)$ satisfying $\lambda(\alpha_n) \uparrow \gamma$, we have

$$\tilde{A}^{\diamond}_{\gamma} = \bigcap_{n=1}^{\infty} M_{\alpha_n} = \bigcap_{n=1}^{\infty} \tilde{A}^{\diamond}_{\lambda(\alpha_n)}.$$

On the other hand, the range $\mathcal{R}(\xi_{\tilde{A}^{\diamond}})$ of $\xi_{\tilde{A}^{\diamond}}$ satisfies the following properties.

(i) *Given any fixed $x \in \bigcup_{\alpha \in S^{\bullet}} M_\alpha$, we have the following properties.*
  - *Suppose that the supremum $\xi_{\tilde{A}^{\diamond}}(x)$ in (5.50) is attained. Then, we have $\xi_{\tilde{A}^{\diamond}}(x) \in \mathcal{R}(\lambda)$.*
  - *Suppose that the supremum $\xi_{\tilde{A}^{\diamond}}(x)$ in (5.50) is not attained. Then, we have*

$$\xi_{\tilde{A}^{\diamond}}(x) \in cl(\mathcal{R}(\lambda)) \setminus \mathcal{R}(\lambda).$$

(ii) *We have $0 \in \mathcal{R}(\xi_{\tilde{A}^{\diamond}}) \subseteq cl(\mathcal{R}(\lambda)) \cup \{0\}$.*

(iii) *Suppose that $B^{(\lambda)}_{\alpha} \setminus B^{(\lambda)}_{\alpha+} \neq \emptyset$ for all $\alpha \in [0,1) \cap S^{\bullet}$. Then, we have*

$$\mathcal{R}(\lambda) \cup \{0\} \subseteq \mathcal{R}(\xi_{\tilde{A}^{\diamond}}) \subseteq cl(\mathcal{R}(\lambda)) \cup \{0\} = cl(\mathcal{R}(\lambda) \cup \{0\}) \tag{5.51}$$

and

$$cl(\mathcal{R}(\xi_{\tilde{A}^{\diamond}})) = cl(\mathcal{R}(\lambda) \cup \{0\}).$$

In particular, we have the following properties.
  - *Suppose that the supremum $\xi_{\tilde{A}^{\diamond}}(x)$ in (5.50) is attained for all $x \in \bigcup_{\alpha \in S^{\bullet}} M_\alpha$. Then, we have*

$$\mathcal{R}(\xi_{\tilde{A}^{\diamond}}) = \mathcal{R}(\lambda) \cup \{0\}.$$

  - *Suppose that $\mathcal{R}(\lambda) \cup \{0\}$ is a closed subset of $[0,1]$. Then, we have*

$$\mathcal{R}(\xi_{\tilde{A}^{\diamond}}) = \mathcal{R}(\lambda) \cup \{0\}.$$

Proof. The results follow immediately from Proposition 5.4.17. The results regarding the range $\mathcal{R}(\xi_{\tilde{A}^{\diamond}})$ can refer to Proposition 5.4.13. ■

Example 5.4.20 Given any real numbers a_1, a_2, b_1, b_2 satisfying $b_1 < a_1 < a_2 < b_2$, let

$$m^L_\alpha = (1-\alpha)b_1 + \alpha a_1 \text{ and } m^U_\alpha = (1-\alpha)b_2 + \alpha a_2.$$

We define M_α to be a closed interval in $\mathbb{R}$ by $M_\alpha = [m^L_\alpha, m^U_\alpha]$ for $\alpha \in [0,1]$. Let λ be a function from $[0,1]$ into $[0,1]$ defined by $\lambda(\alpha) = \alpha^2$. Then, we have $S^{\bullet} = S = (0,1]$.

- It is clear to see that $M_\alpha \subseteq M_\beta$ for $\alpha, \beta \in S^{\bullet}$ with $\lambda(\beta) < \lambda(\alpha)$, i.e. $\beta^2 < \alpha^2$.
- Now, we want to show $\bigcap_{n=1}^{\infty} M_{\alpha_n} \subseteq M_\alpha$ for $\lambda(\alpha_n) \uparrow \lambda(\alpha)$ with $\alpha_n, \alpha \in S^{\bullet}$ for all n; that is, $\alpha_n^2 \uparrow \alpha^2$ with $\alpha_n, \alpha \in S^{\bullet}$ for all n. Equivalently, we want to show $\bigcap_{n=1}^{\infty} M_{\sqrt{\alpha_n}} \subseteq M_{\sqrt{\alpha}}$ for $\alpha_n \uparrow \alpha$ with $\alpha_n, \alpha \in S^{\bullet}$ for all n. Given any $x \in M_{\sqrt{\alpha_n}}$ for all n, we have $m^L_{\sqrt{\alpha_n}} \leq x \leq m^U_{\sqrt{\alpha_n}}$ for all n. Since $m^L_{\sqrt{\alpha_n}}$ and $m^U_{\sqrt{\alpha_n}}$ are continuous with respect to α on $S^{\bullet}$, by taking the limit as $n \to \infty$, we obtain $m^L_{\sqrt{\alpha}} \leq x \leq m^U_{\sqrt{\alpha}}$, i.e. $x \in M_{\sqrt{\alpha}}$, which shows the desired inclusion.

Then, we can generate a fuzzy set $\tilde{A}^{\diamond}$ in $\mathbb{R}$ with membership function given by

$$\xi_{\tilde{A}^{\diamond}}(x) = \sup_{\alpha\in S^{\bullet}} \lambda(\alpha)\cdot \chi_{M_\alpha}(x) = \sup_{0<\alpha\leq 1} \alpha^2\cdot \chi_{M_\alpha}(x).$$

Using Theorem 5.4.19, we have

$$\tilde{A}^{\diamond}_{\lambda(\alpha)} = \tilde{A}^{\diamond}_{\alpha^2} = M_\alpha \text{ for } 0<\alpha\leq 1.$$

We also have

$$\tilde{A}^{\diamond}_{0+} = \bigcup_{\alpha\in S^{\bullet}} M_\alpha = \bigcup_{0<\alpha\leq 1} M_\alpha = \left(m_0^L, m_0^U\right) = \left(b_1, b_2\right)$$

and

$$\tilde{A}^{\diamond}_{0} = \left[b_1, b_2\right].$$

Given any fixed $\alpha\in S^{\bullet} = (0,1]$, we have

$$\begin{aligned} B_\alpha^{(\lambda)} &= \bigcup_{\{\beta\in S^{\bullet}:\lambda(\beta)\geq\lambda(\alpha)\}} M_\beta = \bigcup_{\{\beta\in(0,1]:\beta^2\geq\alpha^2\}} M_\beta = \bigcup_{\{\beta\in(0,1]:\beta\geq\alpha\}} M_\beta \\ &= \bigcup_{\alpha\leq\beta\leq 1} M_\beta = \bigcup_{\alpha\leq\beta\leq 1}\left[m_\beta^L, m_\beta^U\right] = \left[m_\alpha^L, m_\alpha^U\right] \end{aligned}$$

and

$$\begin{aligned} B_{\alpha+}^{(\lambda)} &= \begin{cases} \emptyset, & \text{if } \lambda(\beta)\leq\lambda(\alpha) \text{ for all } \beta\in S^{\bullet} \\ \displaystyle\bigcup_{\{\beta\in S^{\bullet}:\lambda(\beta)>\lambda(\alpha)\}} M_\beta, & \text{otherwise} \end{cases} \\ &= \begin{cases} \emptyset, & \text{if } \beta^2\leq\alpha^2 \text{ for all } \beta\in(0,1] \\ \displaystyle\bigcup_{\{\beta\in(0,1]:\beta^2>\alpha^2\}} M_\beta, & \text{otherwise} \end{cases} \\ &= \begin{cases} \emptyset, & \text{if } \beta\leq\alpha \text{ for all } \beta\in(0,1] \\ \displaystyle\bigcup_{\alpha<\beta\leq 1} M_\beta, & \text{otherwise} \end{cases} \\ &= \bigcup_{\alpha<\beta\leq 1}\left[m_\beta^L, m_\beta^U\right] = \left(m_\alpha^L, m_\alpha^U\right). \end{aligned}$$

Therefore, we have

$$B_\alpha^{(\lambda)}\setminus B_\alpha^{(\lambda+)} = \left[m_\alpha^L, m_\alpha^U\right]\setminus\left(m_\alpha^L, m_\alpha^U\right) = \left\{m_\alpha^L, m_\alpha^U\right\}\neq\emptyset.$$

Since $\mathcal{R}(\lambda) = [0,1]$, using (5.51), it follows that $\mathcal{R}(\xi_{\tilde{A}^{\diamond}}) = [0,1]$.

The above results assume that $\{M_\alpha : \alpha\in[0,1]\}$ is a nested family in the sense of $M_\beta\subseteq M_\alpha$ for $\lambda(\alpha)<\lambda(\beta)$. Next, the family $\{M_\alpha : \alpha\in[0,1]\}$ will not be assumed to be nested in that sense.

Proposition 5.4.21 *Let λ be a function from $[0,1]$ into $[0,1]$, and let $\{M_\alpha : \alpha\in[0,1]\}$ be a family of nonempty subsets of $\mathbb{R}^m$. Suppose that, for any fixed $x\in\mathbb{R}^m$, the following function*

$$\zeta_x(\alpha) = \lambda(\alpha)\cdot\chi_{M_\alpha}(x)$$

is upper semi-continuous on [0,1]. *Consider the membership function* $\xi_{\tilde{A}^\diamond}$ *defined in* (5.37). *Then, we have*

$$\tilde{A}^\diamond_{\lambda(\alpha)} = \bigcup_{\{\beta\in S^\bullet : \lambda(\alpha)\le\lambda(\beta)\}} M_\beta \text{ for every } \alpha \in S^\bullet \tag{5.52}$$

and

$$\tilde{A}^\diamond_{0+} = \bigcup_{\alpha\in S^\bullet} M_\alpha. \tag{5.53}$$

If $\gamma \notin \mathcal{R}(\lambda)$ *satisfies* $0 < \lambda(\alpha_n) \uparrow \gamma$, *we have*

$$\tilde{A}^\diamond_\gamma = \bigcap_{n=1}^{\infty} \bigcup_{\{\beta\in S^\bullet : \lambda(\alpha_n)\le\lambda(\beta)\}} M_\beta. \tag{5.54}$$

Proof. Given any fixed $\alpha \in S^\bullet$, we have $0 < \lambda(\alpha)$. Suppose that $x \in \tilde{A}^\diamond_{\lambda(\alpha)}$ and

$$x \notin M_\beta \text{ for all } \beta \in S^\bullet \text{ satisfying } \lambda(\alpha) \le \lambda(\beta). \tag{5.55}$$

Then, we have

$$\lambda(\beta) \cdot \chi_{M_\beta}(x) < \lambda(\alpha) \text{ for all } \beta \in S^\bullet. \tag{5.56}$$

Since ζ_x is upper semi-continuous on [0,1] by the assumption, Proposition 1.4.4 says that the following supremum

$$\sup_{0\le\beta\le1} \zeta_x(\beta) = \max_{0\le\beta\le1} \zeta_x(\beta)$$

is attained. This says that

$$\begin{aligned} 0 < \lambda(\alpha) \le \xi_{\tilde{A}^\diamond}(x) &= \sup_{0\le\beta\le1} \zeta_x(\beta) = \max_{0\le\beta\le1} \zeta_x(\beta) \\ &= \max_{0\le\beta\le1} \lambda(\beta) \cdot \chi_{M_\beta}(x) = \lambda(\beta^*) \cdot \chi_{M_{\beta^*}}(x), \end{aligned}$$

where $\lambda(\beta^*) \neq 0$, i.e. $\beta^* \in S^\bullet$. Using (5.56), we have

$$\lambda(\alpha) \le \xi_{\tilde{A}^\diamond}(x) = \lambda(\beta^*) \cdot \chi_{M_{\beta^*}}(x) < \lambda(\alpha).$$

This contradiction says that there exists $\beta_0 \in S^\bullet$ satisfying $\lambda(\beta_0) \ge \lambda(\alpha)$ and $x \in M_{\beta_0}$ by referring to (5.55), which shows the following inclusion

$$\tilde{A}^\diamond_{\lambda(\alpha)} \subseteq \bigcup_{\{\beta\in S^\bullet : \lambda(\alpha)\le\lambda(\beta)\}} M_\beta.$$

The following inclusion

$$\begin{aligned} \bigcup_{\{\beta\in S^\bullet : \lambda(\alpha)\le\lambda(\beta)\}} M_\beta &\subseteq \left\{ x \in \mathbb{R}^m : \sup_{\beta\in S^\bullet} \lambda(\beta) \cdot \chi_{M_\beta}(x) \ge \lambda(\alpha) \right\} \\ &= \{x \in \mathbb{R}^m : \xi_{\tilde{A}^\diamond}(x) \ge \lambda(\alpha)\} = \tilde{A}^\diamond_{\lambda(\alpha)} \end{aligned}$$

is obvious, which shows the equality (5.52). The equality (5.53) follows immediately from (5.39). Using part (i) of Proposition 5.4.15, for $\gamma \notin \mathcal{R}(\lambda)$ satisfying $0 < \lambda(\alpha_n) \uparrow \gamma$, we have

$$\tilde{A}^\diamond_\gamma = \bigcap_{n=1}^{\infty} \tilde{A}^\diamond_{\lambda(\alpha_n)}.$$

Using (5.52), we obtain (5.54), and the proof is complete. ■

Theorem 5.4.22 **(Based on One Function and a Non-Nested Family).** *Let λ be a function from $[0,1]$ into $[0,1]$, and let $\{M_\alpha : \alpha \in [0,1]\}$ be a family of nonempty subsets of $\mathbb{R}^m$. Suppose that, for any fixed $x \in \mathbb{R}^m$, the following function*

$$\zeta_x(\alpha) = \lambda(\alpha) \cdot \chi_{M_\alpha}(x)$$

is upper semi-continuous on $[0,1]$. Then, we can generate a fuzzy set $\tilde{A}^\diamond$ in $\mathbb{R}^m$ with membership function defined by

$$\xi_{\tilde{A}^\diamond}(x) = \sup_{\alpha \in S^\bullet} \lambda(\alpha) \cdot \chi_{M_\alpha}(x) \tag{5.57}$$

satisfying

$$\tilde{A}^\diamond_{\lambda(\alpha)} = \bigcup_{\{\beta \in S^\bullet : \lambda(\alpha) \le \lambda(\beta)\}} M_\beta \text{ for } \alpha \in S^\bullet.$$

We also have

$$\tilde{A}^\diamond_{0+} = \bigcup_{\alpha \in S^\bullet} M_\alpha$$

and

$$\tilde{A}^\diamond_0 = cl(\tilde{A}^\diamond_{0+}) = cl\left(\bigcup_{\alpha \in S^\bullet} M_\alpha\right).$$

For $\gamma \notin \mathcal{R}(\lambda)$ satisfying $\lambda(\alpha_n) \uparrow \gamma$, we have

$$\tilde{A}^\diamond_\gamma = \bigcap_{n=1}^{\infty} \tilde{A}^\diamond_{\lambda(\alpha_n)} = \bigcap_{n=1}^{\infty} \bigcup_{\{\beta \in S^\bullet : \lambda(\alpha_n) \le \lambda(\beta)\}} M_\beta.$$

On the other hand, the range $\mathcal{R}(\xi_{\tilde{A}^\diamond})$ of $\xi_{\tilde{A}^\diamond}$ satisfies the following properties.

(i) *Given any fixed $x \in \bigcup_{\alpha \in S^\bullet} M_\alpha$, we have the following properties.*
- *Suppose that the supremum $\xi_{\tilde{A}^\diamond}(x)$ in (5.57) is attained. Then, we have $\xi_{\tilde{A}^\diamond}(x) \in \mathcal{R}(\lambda)$.*
- *Suppose that the supremum $\xi_{\tilde{A}^\diamond}(x)$ in (5.57) is not attained. Then, we have*

$$\xi_{\tilde{A}^\diamond}(x) \in cl(\mathcal{R}(\lambda)) \setminus \mathcal{R}(\lambda).$$

(ii) *We have $0 \in \mathcal{R}(\xi_{\tilde{A}^\diamond}) \subseteq cl(\mathcal{R}(\lambda)) \cup \{0\}$.*

(iii) *Suppose that $B^{(\lambda)}_\alpha \setminus B^{(\lambda+)}_\alpha$ for all $\alpha \in [0,1) \cap S^\bullet$. Then, we have*

$$\mathcal{R}(\lambda) \cup \{0\} \subseteq \mathcal{R}(\xi_{\tilde{A}^\diamond}) \subseteq cl(\mathcal{R}(\lambda)) \cup \{0\} = cl(\mathcal{R}(\lambda) \cup \{0\})$$

and

$$cl(\mathcal{R}(\xi_{\tilde{A}^\diamond})) = cl(\mathcal{R}(\lambda) \cup \{0\}).$$

In particular, we have the following properties.
- *Suppose that the supremum $\xi_{\tilde{A}^\diamond}(x)$ in (5.57) is attained for all $x \in \bigcup_{\alpha \in S^\bullet} M_\alpha$. Then, we have*

$$\mathcal{R}(\xi_{\tilde{A}^\diamond}) = \mathcal{R}(\lambda) \cup \{0\}.$$

- *Suppose that $\mathcal{R}(\lambda) \cup \{0\}$ is a closed subset of $[0,1]$. Then,we have*

$$\mathcal{R}(\xi_{\tilde{A}^\diamond}) = \mathcal{R}(\lambda) \cup \{0\}.$$

Proof. Using Proposition 5.4.21 and a similar argument to the proof of Theorem 5.4.19, the proof is complete. ∎

5.4.3 Based on Two Functions

A more generalization of (5.37) is defined below. Given another function $\eta : [0,1] \to [0,1]$, we can generate a fuzzy set $\tilde{A}^{\top}$ in $\mathbb{R}^m$ with membership function defined by

$$\xi_{\tilde{A}^{\top}}(x) = \sup_{\alpha \in [0,1]} \lambda(\alpha) \cdot \chi_{M_\alpha^{(\eta)}}(x). \tag{5.58}$$

Then, we also have

$$\xi_{\tilde{A}^{\top}}(x) = \sup_{\alpha \in [0,1]} \lambda(\alpha) \cdot \chi_{M_\alpha^{(\eta)}}(x) = \sup_{\alpha \in S^\bullet} \lambda(\alpha) \cdot \chi_{M_\alpha^{(\eta)}}(x). \tag{5.59}$$

We also see that (5.37) is a special case of (5.59) by taking $\eta(\alpha) = \alpha$. We consider two cases as follows.

- Suppose that $x \in \bigcup_{\alpha \in S^\bullet} M_\alpha^{(\eta)}$. Then $x \in M_{\alpha_0}^{(\eta)}$ for some $\alpha_0 \in S^\bullet$, which says that
$$\xi_{\tilde{A}^{\top}}(x) = \sup_{\alpha \in S^\bullet} \lambda(\alpha) \cdot \chi_{M_\alpha^{(\eta)}}(x) \geq \lambda(\alpha_0) > 0.$$
- Suppose that $x \notin \bigcup_{\alpha \in S^\bullet} M_\alpha^{(\eta)}$. Then $x \notin M_\alpha^{(\eta)}$ for all $\alpha \in S^\bullet$, by definition of $\xi_{\tilde{A}^{\top}}$ in (5.59), we have $\xi_{\tilde{A}^{\top}}(x) = 0$.

Therefore, we have

$$\xi_{\tilde{A}^{\top}}(x) = \begin{cases} \sup_{\alpha \in S^\bullet} \lambda(\alpha) \cdot \chi_{M_\alpha^{(\eta)}}(x) > 0, & \text{if } x \in \bigcup_{\alpha \in S^\bullet} M_\alpha^{(\eta)} \\ 0, & \text{otherwise.} \end{cases} \tag{5.60}$$

Proposition 5.4.23 *We have the following equality*

$$\tilde{A}_{0+}^{\top} = \bigcup_{\alpha \in S^\bullet} M_\alpha^{(\eta)}. \tag{5.61}$$

Proof. Suppose that $x \notin \bigcup_{\alpha \in S^\bullet} M_\alpha^{(\eta)}$. Then $\xi_{\tilde{A}^{\top}}(x) = 0$ by referring to (5.60), i.e. $x \notin \tilde{A}_\alpha^{\top}$ for all $\alpha \in (0,1]$. Therefore, we have

$$\tilde{A}_\alpha^{\top} \subseteq \bigcup_{\alpha \in S^\bullet} M_\alpha^{(\eta)} \text{ for all } \alpha \in (0,1].$$

Using (2.5), we can obtain the inclusion

$$\tilde{A}_{0+}^{\top} \subseteq \bigcup_{\alpha \in S^\bullet} M_\alpha^{(\eta)}.$$

The inclusion

$$\bigcup_{\alpha \in S^\bullet} M_\alpha^{(\eta)} \subseteq \tilde{A}_{0+}^{\top}$$

is obvious by referring to (5.60). This completes the proof. ■

Remark 5.4.24 Assume that $\mathcal{M}$ is a nested family in the sense of $M_\alpha^{(\eta)} \subseteq M_\beta^{(\eta)}$ for $\alpha, \beta \in S^\bullet$ with $\lambda(\beta) < \lambda(\alpha)$. Then, we can have similar results to those given in Proposition 5.4.11.

Given any fixed $\alpha \in [0,1)$, we write

$$B_\alpha^{(\eta,\lambda)} = \bigcup_{\{\beta \in S^\bullet : \lambda(\beta) \geq \lambda(\alpha)\}} M_\beta^{(\eta)} \tag{5.62}$$

and

$$B_{\alpha+}^{(\eta,\lambda)} = \begin{cases} \emptyset, & \text{if } \lambda(\beta) \leq \lambda(\alpha) \text{ for all } \beta \in S^{\bullet} \\ \bigcup_{\{\beta \in S^{\bullet} : \lambda(\beta) > \lambda(\alpha)\}} M_{\beta}^{(\eta)}, & \text{otherwise.} \end{cases} \tag{5.63}$$

The relationships between $B_{\alpha}^{(\eta,\lambda)}$ and $\tilde{A}_{\lambda(\alpha)}^{\top}$ are given below.

Proposition 5.4.25 *Given any fixed $\alpha \in [0,1)$, we have*

$$\tilde{A}_{\lambda(\alpha)+}^{\top} \subseteq B_{\alpha}^{(\eta,\lambda)} \subseteq \tilde{A}_{\lambda(\alpha)}^{\top} \text{ and } B_{\alpha+}^{(\eta,\lambda)} = \tilde{A}_{\lambda(\alpha)+}^{\top}.$$

Proof. For $x \in B_{\alpha}^{(\eta,\lambda)}$, it means that $x \in M_{\beta}^{(\eta)}$ for some $\beta \in S^{\bullet}$ with $\lambda(\beta) \geq \lambda(\alpha)$. From (5.59), it follows that

$$\xi_{\tilde{A}^{\top}}(x) = \sup_{\gamma \in S^{\bullet}} \lambda(\gamma) \cdot \chi_{M_{\gamma}}(x) \geq \lambda(\beta) \geq \lambda(\alpha),$$

which shows the inclusion $B_{\alpha}^{(\eta,\lambda)} \subseteq \tilde{A}_{\lambda(\alpha)}^{\top}$. We can similarly obtain the inclusion $B_{\alpha+}^{(\eta,\lambda)} \subseteq \tilde{A}_{\lambda(\alpha)+}^{\top}$. Suppose that $x \notin B_{\alpha}^{(\eta,\lambda)}$. It means that $x \notin M_{\beta}^{(\eta)}$ for all $\beta \in S^{\bullet}$ with $\lambda(\beta) \geq \lambda(\alpha)$. Therefore, we obtain

$$\xi_{\tilde{A}^{\top}}(x) = \sup_{\beta \in S^{\bullet}} \lambda(\beta) \cdot \chi_{M_{\beta}^{(\eta)}}(x) = \sup_{\{\beta \in S^{\bullet} : \lambda(\beta) < \lambda(\alpha)\}} \lambda(\beta) \cdot \chi_{M_{\beta}^{(\eta)}}(x) \leq \lambda(\alpha),$$

i.e. $x \notin \tilde{A}_{\lambda(\alpha)+}^{\top}$, which shows the inclusion $\tilde{A}_{\lambda(\alpha)+}^{\top} \subseteq B_{\alpha}^{(\eta,\lambda)}$. If $B_{\alpha+}^{(\eta,\lambda)} \neq \emptyset$, we can similarly obtain the inclusion $\tilde{A}_{\lambda(\alpha)+}^{\top} \subseteq B_{\alpha+}^{(\eta,\lambda)}$. If $B_{\alpha+}^{(\eta,\lambda)} = \emptyset$, we have $\lambda(\beta) \leq \lambda(\alpha)$ for all $\beta \in S^{\bullet}$, which says that $\xi_{\tilde{A}^{\top}}(x) \leq \lambda(\alpha)$ for any $x \in \mathbb{R}^m$ by (5.59), i.e.

$$\{x \in \mathbb{R}^m : \xi_{\tilde{A}^{\top}}(x) > \lambda(\alpha)\} = \emptyset.$$

This completes the proof. ■

The range $\mathcal{R}(\xi_{\tilde{A}^{\top}})$ of function $\xi_{\tilde{A}^{\top}}$ is an important issue. Therefore, the following results will be used in the subsequent discussion.

Proposition 5.4.26 *Let $\{M_{\alpha} : \alpha \in [0,1]\}$ be a family of nonempty subsets of $\mathbb{R}^m$, and let η and λ be functions from $[0,1]$ into $[0,1]$. We consider the membership function $\xi_{\tilde{A}^{\top}}$ defined in (5.59). Then, the following statements hold true.*

(i) *Given any fixed $x \in \bigcup_{\alpha \in S^{\bullet}} M_{\alpha}^{(\eta)}$, we have the following properties.*
   - *Suppose that the supremum $\xi_{\tilde{A}^{\top}}(x)$ in (5.59) is attained. Then, we have*

     $$\xi_{\tilde{A}^{\top}}(x) \in \mathcal{R}(\lambda).$$

   - *Suppose that the supremum $\xi_{\tilde{A}^{\top}}(x)$ in (5.59) is not attained. Then, we have*

     $$\xi_{\tilde{A}^{\top}}(x) \in cl(\mathcal{R}(\lambda)) \setminus \mathcal{R}(\lambda).$$

(ii) *We have $0 \in \mathcal{R}(\xi_{\tilde{A}^{\top}}) \subseteq cl(\mathcal{R}(\lambda)) \cup \{0\}$.*

(iii) *Suppose that $B_{\alpha}^{(\eta,\lambda)} \setminus B_{\alpha+}^{(\eta,\lambda)} \neq \emptyset$ for all $\alpha \in [0,1) \cap S^{\bullet}$. Then*

$$\mathcal{R}(\lambda) \cup \{0\} \subseteq \mathcal{R}(\xi_{\tilde{A}^{\top}}) \subseteq cl(\mathcal{R}(\lambda)) \cup \{0\} = cl(\mathcal{R}(\lambda) \cup \{0\})$$

and

$$cl(\mathcal{R}(\xi_{\tilde{A}^\top})) = cl(\mathcal{R}(\lambda) \cup \{0\}).$$

In particular, we have the following properties.

- *Suppose that the supremum* $\xi_{\tilde{A}^\top}(x)$ *in (5.59) is attained for all* $x \in \bigcup_{\alpha \in S^\bullet} M_\alpha^{(\eta)}$. *Then, we have*

$$\mathcal{R}(\xi_{\tilde{A}^\top}) = \mathcal{R}(\lambda) \cup \{0\}.$$

- *Suppose that* $\mathcal{R}(\xi_{\tilde{A}^\top}) = \mathcal{R}(\lambda) \cup \{0\}$ *is a closed subset of* $[0,1]$. *Then, we have*

$$\mathcal{R}(\xi_{\tilde{A}^\top}) = \mathcal{R}(\lambda) \cup \{0\}.$$

Proof. To prove part (i), if the supremum in (5.59) is attained, it means that there exists $\alpha_0 \in S^\bullet$ satisfying

$$\xi_{\tilde{A}^\top}(x) = \lambda(\alpha_0) \in \mathcal{R}(\lambda) \text{ and } x \in M_{\alpha_0}^{(\eta)}.$$

Suppose that the supremum in (5.59) is not attained, i.e. $\xi_{\tilde{A}^\top}(x) \notin \mathcal{R}(\lambda)$. According to the concept of supremum, there exists a sequence $\{\alpha_n\}_{n=1}^\infty$ in $[0,1]$ satisfying $x \in M_{\alpha_n}^{(\eta)}$ for all n and $\lambda(\alpha_n) \to \xi_{\tilde{A}^\top}(x)$ as $n \to \infty$, which says that $\xi_{\tilde{A}^\top}(x) \in \text{cl}(\mathcal{R}(\lambda))$. It is obvious that $0 \in \mathcal{R}(\xi_{\tilde{A}^\top})$ from (5.60). Therefore, part (ii) can be realized from part (i).

To prove part (iii), we write

$$E_\alpha^{(\eta,\lambda)} \equiv B_\alpha^{(\eta,\lambda)} \setminus B_{\alpha+}^{(\eta,\lambda)}.$$

Then $E_\alpha^{(\eta,\lambda)} \neq \emptyset$ for $\alpha \in [0,1) \cap S^\bullet$. Given any $\gamma \in \mathcal{R}(\lambda) \setminus \{0\}$, there exists $\alpha \in [0,1) \cap S^\bullet$ satisfying $\lambda(\alpha) = \gamma > 0$. We consider the following cases.

- Suppose that $B_{\alpha+}^{(\eta,\lambda)} = \emptyset$. Then $\lambda(\beta) \leq \lambda(\alpha)$ for all $\beta \in [0,1]$, i.e. $\xi_{\tilde{A}^\top}(x) \leq \lambda(\alpha)$ for any $x \in \mathbb{R}^m$ by (5.59). For any $x \in E_\alpha^{(\eta,\lambda)}$, we also have $x \in B_\alpha^{(\eta,\lambda)}$. Using Proposition 5.4.25, we obtain $x \in \tilde{A}_{\lambda(\alpha)}^\top$, i.e. $\xi_{\tilde{A}^\top}(x) \geq \lambda(\alpha)$, which implies $\xi_{\tilde{A}^\top}(x) = \lambda(\alpha) = \gamma$, i.e. $\gamma \in \mathcal{R}(\xi_{\tilde{A}^\top})$.
- Suppose that $B_{\alpha+}^{(\eta,\lambda)} \neq \emptyset$. For any $x \in E_\alpha^{(\eta,\lambda)}$, we see that $x \in B_\alpha^{(\eta,\lambda)}$ and $x \notin B_{\alpha+}^{(\eta,\lambda)}$. Using Proposition 5.4.25, we have $x \in \tilde{A}_{\lambda(\alpha)}^\top$ and $x \notin \tilde{A}_{\lambda(\alpha)+}^\top$, i.e. $\xi_{\tilde{A}^\top}(x) \geq \lambda(\alpha)$ and $\xi_{\tilde{A}^\top}(x) \leq \lambda(\alpha)$, which implies $\xi_{\tilde{A}^\top}(x) = \lambda(\alpha) = \gamma$, i.e. $\gamma \in \mathcal{R}(\xi_{\tilde{A}^\top})$.

Therefore, the above two cases prove the inclusion

$$\mathcal{R}(\lambda) \setminus \{0\} \subseteq \mathcal{R}(\xi_{\tilde{A}^\top}).$$

Using parts (i) and (ii), the proof is complete. ∎

Proposition 5.4.27 *Let* η *and* λ *be functions from* $[0,1]$ *into* $[0,1]$, *and let* $\mathcal{M} = \{M_\alpha : \alpha \in [0,1]\}$ *be a family of nonempty subsets of* $\mathbb{R}^m$ *satisfying* $M_\beta^{(\eta)} \subseteq M_\alpha^{(\eta)}$ *for* $\alpha, \beta \in S^\bullet$ *with* $\lambda(\alpha) < \lambda(\beta)$. *We consider the membership function* $\xi_{\tilde{A}^\top}$ *defined in (5.59). Then, the following statements hold true.*

(i) *Given any* $\alpha \in S^\bullet$, *we have*

$$M_\alpha^{(\eta)} \subseteq \tilde{A}_{\lambda(\alpha)}^\top \tag{5.64}$$

and

$$\tilde{A}_{\lambda(\alpha)}^\top \subseteq \bigcap_{\{\beta \in S^\bullet : \lambda(\beta) < \lambda(\alpha)\}} M_\beta^{(\eta)} \subseteq \bigcap_{\{\beta \in S^\bullet : \lambda(\beta) < \lambda(\alpha)\}} \tilde{A}_{\lambda(\beta)}^\top. \tag{5.65}$$

(ii) *Suppose that λ is continuous. Given any $\alpha \in S^\bullet$, we have*

$$\tilde{A}^{\top}_{\lambda(\alpha)} = \bigcap_{\{\beta \in S^\bullet : \lambda(\beta) < \lambda(\alpha)\}} M^{(\eta)}_\beta. \tag{5.66}$$

(iii) *Given any $\alpha \in S^\bullet$, suppose that $\bigcap_{n=1}^{\infty} M^{(\eta)}_{\alpha_n} \subseteq M^{(\eta)}_\alpha$ when $\lambda(\alpha_n) \uparrow \lambda(\alpha)$ with $\alpha_n \in S^\bullet$ for all n. Then, we have*

$$\tilde{A}^{\top}_{\lambda(\alpha)} = M^{(\eta)}_\alpha = \bigcap_{\{\beta \in S^\bullet : \lambda(\beta) \leq \lambda(\alpha)\}} M^{(\eta)}_\beta. \tag{5.67}$$

(iv) *Given any $\alpha \in S^\bullet$, suppose that $\bigcap_{n=1}^{\infty} M^{(\eta)}_{\alpha_n} \subseteq M^{(\eta)}_\alpha$ when $\lambda(\alpha_n) \uparrow \lambda(\alpha)$ with $\alpha_n \in S^\bullet$ for all n. We also assume that λ is continuous. Then, we have*

$$\tilde{A}^{\top}_{\lambda(\alpha)} = M^{(\eta)}_\alpha = \bigcap_{\{\beta \in S^\bullet : \lambda(\beta) \leq \lambda(\alpha)\}} M^{(\eta)}_\beta = \bigcap_{\{\beta \in S^\bullet : \lambda(\beta) < \lambda(\alpha)\}} M^{(\eta)}_\beta.$$

Proof. To prove part (i), since $M^{(\eta)}_\beta \subseteq M^{(\eta)}_\alpha$ for $\alpha, \beta \in S^\bullet$ with $\lambda(\alpha) < \lambda(\beta)$, the inclusion (5.64) is obvious by using Proposition 5.4.25, where $B^{(\eta,\lambda)}_\alpha = M^{(\eta)}_\alpha$. To prove the inclusions (5.65), given any $x \in \tilde{A}^{\top}_{\lambda(\alpha)}$ and $\beta \in S^\bullet$ with $\lambda(\beta) < \lambda(\alpha)$, we want to show that $x \in M^{(\eta)}_\beta$. Since $\xi_{\tilde{A}^\top}(x) \geq \lambda(\alpha)$, we consider the following cases.

- Assume that $\xi_{\tilde{A}^\top}(x) > \lambda(\alpha)$. Let

  $$\epsilon = \xi_{\tilde{A}^\top}(x) - \lambda(\alpha) > 0.$$

  From (5.59), the concept of supremum says that there exists $\alpha_0 \in S^\bullet$ satisfying $x \in M^{(\eta)}_{\alpha_0}$ and

  $$\xi_{\tilde{A}^\top}(x) - \epsilon < \lambda(\alpha_0),$$

  which implies $\lambda(\alpha) < \lambda(\alpha_0)$. This also says that $x \in M^{(\eta)}_\beta$, since

  $$M^{(\eta)}_{\alpha_0} \subseteq M^{(\eta)}_\alpha \subseteq M^{(\eta)}_\beta$$

  by the nestedness.
- Assume that $\xi_{\tilde{A}^\top}(x) = \lambda(\alpha)$. Let

  $$\epsilon = \lambda(\alpha) - \lambda(\beta) > 0.$$

  The concept of supremum says that there exists $\alpha_0 \in S^\bullet$ satisfying $x \in M^{(\eta)}_{\alpha_0}$ and

  $$\lambda(\alpha) - \epsilon < \lambda(\alpha_0),$$

  which implies $\lambda(\beta) < \lambda(\alpha_0)$. This also says that $x \in M^{(\eta)}_\beta$ by the nestedness $M^{(\eta)}_{\alpha_0} \subseteq M^{(\eta)}_\beta$.

Since $\lambda(\beta)$ can be any number for $\beta \in S^\bullet$ with $\lambda(\beta) < \lambda(\alpha)$, we conclude that

$$x \in \bigcap_{\{\beta \in S^\bullet : \lambda(\beta) < \lambda(\alpha)\}} M^{(\eta)}_\beta.$$

By combining the inclusion (5.64), we obtain the inclusions (5.65).

To prove part (ii), suppose that $x \in M^{(\eta)}_\beta$ for all $\beta \in S^\bullet$ with $\lambda(\beta) < \lambda(\alpha)$. By the definition of $\xi_{\tilde{A}^\top}$ in (5.59), we also see that $\xi_{\tilde{A}^\top}(x) \geq \lambda(\beta)$ for all $\beta \in S^\bullet$ with $\lambda(\beta) < \lambda(\alpha)$. We want to show that $\xi_{\tilde{A}^\top}(x) \geq \lambda(\alpha)$. Suppose that $\xi_{\tilde{A}^\top}(x) < \lambda(\alpha)$. Then, there exists $c \in (0,1)$ satisfying

$$\lambda(\beta_0) \leq \xi_{\tilde{A}^\top}(x) < c < \lambda(\alpha)$$

for some $\beta_0 \in S^\bullet$ with $\lambda(\beta_0) < \lambda(\alpha)$. Since λ is continuous and $\lambda(\beta_0) < c < \lambda(\alpha)$, Proposition 1.4.3 says that there exists $\overline{\beta} \in [\beta_0, \alpha]$ satisfying $c = \lambda(\overline{\beta})$. This shows that $\xi_{\tilde{A}^\top}(x) < \lambda(\overline{\beta})$, which contradicts $\xi_{\tilde{A}^\top}(x) \geq \lambda(\beta)$ for all $\beta \in S^\bullet$ with $\lambda(\beta) < \lambda(\alpha)$. Therefore, we obtain $x \in \tilde{A}^\top_{\lambda(\alpha)}$, which proves the equality (5.66) by applying the inclusion (5.65).

To prove part (iii), given any $x \in \tilde{A}^\top_{\lambda(\alpha)}$, we want to show that $x \in M^{(\eta)}_\alpha$. Since $\xi_{\tilde{A}^\top}(x) \geq \lambda(\alpha)$, we consider the following cases.

- Assume that $\xi_{\tilde{A}^\top}(x) > \lambda(\alpha)$. Let

$$\epsilon = \xi_{\tilde{A}^\top}(x) - \lambda(\alpha) > 0.$$

The supremum in (5.59) says that there exists $\alpha_0 \in S^\bullet$ satisfying $x \in M^{(\eta)}_{\alpha_0}$ and

$$\xi_{\tilde{A}^\top}(x) - \epsilon < \lambda(\alpha_0),$$

which implies $\lambda(\alpha) < \lambda(\alpha_0)$. This also says that $x \in M^{(\eta)}_\alpha$, since $M^{(\eta)}_{\alpha_0} \subseteq M^{(\eta)}_\alpha$ by the nestedness.

- Assume that $\xi_{\tilde{A}^\top}(x) = \lambda(\alpha)$. The supremum in (5.59) says that there exists a sequence $\{\alpha_n\}_{n=1}^\infty$ in $S^\bullet$ satisfying $x \in M^{(\eta)}_{\alpha_n}$ for all n and $\lambda(\alpha_n) \uparrow \lambda(\alpha)$, which implies

$$x \in \bigcap_{n=1}^\infty M^{(\eta)}_{\alpha_n} \subseteq M^{(\eta)}_\alpha$$

by the assumption, and we conclude that $x \in M^{(\eta)}_\alpha$.

Therefore, we obtain the inclusion $\tilde{A}^\top_{\lambda(\alpha)} \subseteq M^{(\eta)}_\alpha$. The equalities (5.67) follow immediately from (5.64) and the nestedness. Finally, part (iv) follows from parts (ii) and (iii). This completes the proof. ■

Proposition 5.4.27 presents the representation $\tilde{A}^\top_{\lambda(\alpha)}$ for $\lambda(\alpha) \in \mathcal{R}(\lambda)$. If $\gamma \notin \mathcal{R}(\lambda)$, the representation $\tilde{A}^\top_\gamma$ is unknown. The following proposition will provide the representation of $\tilde{A}^\top_\gamma$ in terms of the sequence $\{\tilde{A}^\top_{\lambda(\alpha_n)}\}_{n=1}^\infty$ without considering the nestedness in which the representation $\tilde{A}^\top_{\lambda(\alpha_n)}$ can be realized from Proposition 5.4.27.

Proposition 5.4.28 *Let η and λ be functions from* [0,1] *into* [0,1], *and let $\mathcal{M} = \{M_\alpha : \alpha \in [0,1]\}$ be a family of nonempty subsets of $\mathbb{R}^m$. Consider the membership function $\xi_{\tilde{A}^\top}$ defined in* (5.59). *Then, we have the following properties.*

(i) *For $\gamma \notin \mathcal{R}(\lambda)$, suppose that $\lambda(\alpha_n) \uparrow \gamma$ with $\alpha_n \in S^\bullet$ for all n. Then, we have*

$$\tilde{A}^\top_\gamma = \bigcap_{n=1}^\infty \tilde{A}^\top_{\lambda(\alpha_n)}. \tag{5.68}$$

(ii) *For $\gamma \notin \mathcal{R}(\lambda)$, suppose that $\lambda(\beta_m) \downarrow \gamma$ with $\beta_m \in S^\bullet$ for all m. Then, we have*

$$\tilde{A}^\top_{\gamma+} \subseteq \bigcup_{m=1}^\infty \tilde{A}^\top_{\lambda(\beta_m)+} \subseteq \bigcup_{m=1}^\infty \tilde{A}^\top_{\lambda(\beta_m)} \subseteq \tilde{A}^\top_\gamma.$$

Proof. To prove part (i), the following inclusion is obvious

$$\tilde{A}^\top_\gamma = \{x \in \mathbb{R}^m : \xi_{\tilde{A}^\top}(x) \geq \gamma\} \subseteq \bigcap_{n=1}^\infty \{x \in \mathbb{R}^m : \xi_{\tilde{A}^\top}(x) \geq \lambda(\alpha_n)\} = \bigcap_{n=1}^\infty \tilde{A}^\top_{\lambda(\alpha_n)}.$$

On the other hand, if $\xi_{\tilde{A}^{\top}}(x) \geq \lambda(\alpha_n)$ for all n, we have $\xi_{\tilde{A}^{\top}}(x) \geq \gamma$ by taking the limit, which shows the other direction of inclusion. Therefore, we obtain the desired equality (5.68).

To prove part (ii), since $\lambda(\beta_m) \downarrow \gamma$, the following inclusion is obvious

$$\tilde{A}_{\gamma}^{\top} = \left\{x \in \mathbb{R}^m : \xi_{\tilde{A}^{\top}}(x) \geq \gamma\right\} \supseteq \bigcup_{m=1}^{\infty} \left\{x \in \mathbb{R}^m : \xi_{\tilde{A}^{\top}}(x) \geq \lambda(\beta_m)\right\} = \bigcup_{m=1}^{\infty} \tilde{A}_{\lambda(\beta_m)}^{\top}.$$

Now, for $\xi_{\tilde{A}^{\top}}(x) > \gamma$, since $\lambda(\beta_m) \downarrow \gamma$, there exists $\lambda(\beta_M) \in \{\lambda(\beta_m)\}_{m=1}^{\infty}$ satisfying $r \leq \lambda(\beta_M) < \xi_{\tilde{A}^{\top}}(x)$, which shows the following inclusion

$$\tilde{A}_{\gamma+}^{\top} = \left\{x \in \mathbb{R}^m : \xi_{\tilde{A}^{\top}}(x) > \gamma\right\} \subseteq \bigcup_{m=1}^{\infty} \left\{x \in \mathbb{R}^m : \xi_{\tilde{A}^{\top}}(x) > \lambda(\beta_m)\right\} = \bigcup_{m=1}^{\infty} \tilde{A}_{\lambda(\beta_m)+}^{\top}.$$

This completes the proof. ■

Proposition 5.4.29 *Let η and λ be functions from* [0,1] *into* [0,1], *and let* $\{M_\alpha : \alpha \in [0,1]\}$ *be a family of nonempty subsets of* $\mathbb{R}^m$ *such that the following conditions are satisfied:*

- $M_\beta^{(\eta)} \subseteq M_\alpha^{(\eta)}$ *for* $\alpha, \beta \in S^\bullet$ *with* $\lambda(\alpha) < \lambda(\beta)$;
- $\bigcap_{n=1}^{\infty} M_{\alpha_n}^{(\eta)} \subseteq M_\alpha^{(\eta)}$ *when* $\lambda(\alpha_n) \uparrow \lambda(\alpha)$ *with* $\alpha_n, \alpha \in S^\bullet$ *for all* n.

Consider the membership function $\xi_{\tilde{A}^{\top}}$ *defined in* (5.59). *Then, we have*

$$\tilde{A}_{\lambda(\alpha)}^{\top} = M_\alpha^{(\eta)} \text{ for } \alpha \in S^\bullet \tag{5.69}$$

and

$$\tilde{A}_{0+}^{\top} = \bigcup_{\alpha \in S^\bullet} M_\alpha^{(\eta)}. \tag{5.70}$$

We also have

$$\begin{aligned}\tilde{A}_{\gamma}^{\top} &= \{x \in \mathbb{R}^m : \xi_{\tilde{A}^{\top}}(x) \geq \gamma\} \\ &= \begin{cases} M_\alpha^{(\eta)}, & \text{if } \gamma = \lambda(\alpha) \in \mathcal{R}(\lambda) \\ \displaystyle\bigcap_{n=1}^{\infty} M_{\alpha_n}^{(\eta)}, & \text{if } \gamma \notin \mathcal{R}(\lambda) \text{ and } \lambda(\alpha_n) \uparrow \gamma \text{ for } \alpha_n \in S^\bullet \text{ for all } n. \end{cases}\end{aligned} \tag{5.71}$$

Proof. Equality (5.69) follows from part (iii) of Proposition 5.4.27. Also, Equality (5.70) follows immediately from (5.61). Using part (i) of Proposition 5.4.28 and (5.69), we obtain (5.71). This completes the proof. ■

Theorem 5.4.30 **(Based on Two Functions).** *Let η and λ be two functions from* [0,1] *into* [0,1], *and let* $\mathcal{M} = \{M_\alpha : \alpha \in [0,1]\}$ *be a family of nonempty subsets of* $\mathbb{R}^m$ *such that the following conditions are satisfied:*

- $M_\beta^{(\eta)} \subseteq M_\alpha^{(\eta)}$ *for* $\alpha, \beta \in S^\bullet$ *with* $\lambda(\alpha) < \lambda(\beta)$;
- $\bigcap_{n=1}^{\infty} M_{\alpha_n}^{(\eta)} \subseteq M_\alpha^{(\eta)}$ *when* $\lambda(\alpha_n) \uparrow \lambda(\alpha)$ *with* $\alpha_n, \alpha \in S^\bullet$ *for all* n.

Then, we can generate a fuzzy set $\tilde{A}^{\top}$ *in* $\mathbb{R}^m$ *with membership function defined by*

$$\xi_{\tilde{A}^{\top}}(x) = \sup_{\alpha \in S^\bullet} \lambda(\alpha) \cdot \chi_{M_\alpha^{(\eta)}}(x) \tag{5.72}$$

satisfying $\tilde{A}^{\top}_{\lambda(\alpha)} = M^{(\eta)}_{\alpha}$ *for* $\alpha \in S^{\bullet}$. *We also have*

$$\tilde{A}^{\top}_{0+} = \bigcup_{\alpha \in S^{\bullet}} M^{(\eta)}_{\alpha}$$

and

$$\tilde{A}^{\top}_{0} = cl(\tilde{A}^{\top}_{0+}) = cl\left(\bigcup_{\alpha \in S^{\bullet}} \tilde{A}^{\top}_{\alpha}\right) = cl\left(\bigcup_{\alpha \in S^{\bullet}} M^{(\eta)}_{\alpha}\right).$$

For $\gamma \notin \mathcal{R}(\lambda)$ *satisfying* $\lambda(\alpha_n) \uparrow \gamma$ *and* $\alpha_n \in S^{\bullet}$ *for all n, we have*

$$\tilde{A}^{\top}_{\gamma} = \bigcap_{n=1}^{\infty} M^{(\eta)}_{\alpha_n} = \bigcap_{n=1}^{\infty} \tilde{A}^{\top}_{\lambda(\alpha_n)}.$$

On the other hand, the range $\mathcal{R}(\xi_{\tilde{A}^{\top}})$ *of* $\xi_{\tilde{A}^{\top}}$ *satisfies the following properties.*

(i) *Given any fixed* $x \in \bigcup_{\alpha \in S^{\bullet}} M^{(\eta)}_{\alpha}$, *we have the following properties.*
- *Suppose that the supremum* $\xi_{\tilde{A}^{\top}}(x)$ *in (5.72) is attained. Then, we have*

$$\xi_{\tilde{A}^{\top}}(x) \in \mathcal{R}(\lambda).$$

- *Suppose that the supremum* $\xi_{\tilde{A}^{\top}}(x)$ *in (5.72) is not attained. Then, we have*

$$\xi_{\tilde{A}^{\top}}(x) \in cl(\mathcal{R}(\lambda)) \setminus \mathcal{R}(\lambda).$$

(ii) *We have* $0 \in \mathcal{R}(\xi_{\tilde{A}^{\top}}) \subseteq cl(\mathcal{R}(\lambda)) \cup \{0\}$.

(iii) *Suppose that* $B^{(\eta,\lambda)}_{\alpha} \setminus B^{(\eta,\lambda)}_{\alpha+} \neq \emptyset$ *for all* $\alpha \in [0,1) \cap S^{\bullet}$. *Then, we have*

$$\mathcal{R}(\lambda) \cup \{0\} \subseteq \mathcal{R}(\xi_{\tilde{A}^{\top}}) \subseteq cl(\mathcal{R}(\lambda)) \cup \{0\} = cl(\mathcal{R}(\lambda) \cup \{0\}) \tag{5.73}$$

and

$$cl(\mathcal{R}(\xi_{\tilde{A}^{\top}})) = cl(\mathcal{R}(\lambda) \cup \{0\}).$$

In particular, we have the following properties.
- *Suppose that the supremum* $\xi_{\tilde{A}^{\top}}(x)$ *in (5.72) is attained for all* $x \in \bigcup_{\alpha \in S^{\bullet}} M^{(\eta)}_{\alpha}$. *Then, we have*

$$\mathcal{R}(\xi_{\tilde{A}^{\top}}) = \mathcal{R}(\lambda) \cup \{0\}.$$

- *Suppose that* $\mathcal{R}(\lambda) \cup \{0\}$ *is a closed subset of* $[0,1]$. *Then, we have*

$$\mathcal{R}(\xi_{\tilde{A}^{\top}}) = \mathcal{R}(\lambda) \cup \{0\}.$$

Proof. The results regarding the α-level sets follow immediately from Proposition 5.4.29. For the results regarding the range $\mathcal{R}(\xi_{\tilde{A}^{\circ}})$, refer to Proposition 5.4.26. ■

Example 5.4.31 Given any real numbers a_1, a_2, b_1, b_2 satisfying $b_1 < a_1 < a_2 < b_2$, for $\alpha \in [0,1]$, we define

$$m^L_{\alpha} = \alpha b_1 + (1-\alpha)a_1 \text{ and } m^U_{\alpha} = \alpha b_2 + (1-\alpha)a_2.$$

We also define M_{α} to be a closed interval in $\mathbb{R}$ by $M_{\alpha} = [m^L_{\alpha}, m^U_{\alpha}]$ for $\alpha \in [0,1]$. Let η be a function from $[0,1]$ into $(0,1]$ defined by $\eta(\alpha) = 1 - \alpha$, and let

$$\bar{m}^L_{\alpha} = m^L_{1-\alpha} \text{ and } \bar{m}^U_{\alpha} = m^U_{1-\alpha} \text{ for } \alpha \in [0,1].$$

Then, we see that

$$\bar{m}_\alpha^L = (1-\alpha)b_1 + \alpha a_1 \text{ and } \bar{m}_\alpha^U = (1-\alpha)b_2 + \alpha a_2.$$

Therefore, we have

$$M_\alpha^{(\eta)} = \left[\bar{m}_\alpha^L, \bar{m}_\alpha^U\right] = \left[(1-\alpha)b_1 + \alpha a_1, (1-\alpha)b_2 + \alpha a_2\right].$$

Let λ be a function from [0,1] into (0,1] defined by $\lambda(\alpha) = \alpha^2$. Then, we have $S^\bullet = (0,1]$.

- It is clear to see that $M_\alpha^{(\eta)} \subseteq M_\beta^{(\eta)}$ for $\alpha, \beta \in S^\bullet$ with $\lambda(\beta) < \lambda(\alpha)$, i.e. $\beta^2 < \alpha^2$.
- Now, we want to show $\bigcap_{n=1}^\infty M_{\alpha_n}^{(\eta)} \subseteq M_\alpha^{(\eta)}$ for $\lambda(\alpha_n) \uparrow \lambda(\alpha)$ with $\alpha_n, \alpha \in S^\bullet$ for all n; that is, $\alpha_n^2 \uparrow \alpha^2$ with $\alpha_n, \alpha \in S^\bullet$ for all n. Equivalently, we want to show $\bigcap_{n=1}^\infty M_{\sqrt{\alpha_n}}^{(\eta)} \subseteq M_{\sqrt{\alpha}}^{(\eta)}$ for $\alpha_n \uparrow \alpha$ with $\alpha_n, \alpha \in S^\bullet$ for all n. Given any $x \in M_{\sqrt{\alpha_n}}^{(\eta)}$ for all n, we have $\bar{m}_{\sqrt{\alpha_n}}^L \leq x \leq \bar{m}_{\sqrt{\alpha_n}}^U$ for all n. Since $\bar{m}_{\sqrt{\alpha_n}}^L$ and $\bar{m}_{\sqrt{\alpha_n}}^U$ are continuous with respect to α on $S^\bullet$, by taking the limit as $n \to \infty$, we obtain $\bar{m}_{\sqrt{\alpha}}^L \leq x \leq \bar{m}_{\sqrt{\alpha}}^U$, i.e. $x \in M_{\sqrt{\alpha}}^{(\eta)}$, which shows the desired inclusion.

Then, we can generate a fuzzy set $\tilde{A}^\top$ in $\mathbb{R}$ with membership function given by

$$\xi_{\tilde{A}^\top}(x) = \sup_{\alpha \in S^\bullet} \lambda(\alpha) \cdot \chi_{M_\alpha^{(\eta)}}(x) = \sup_{0<\alpha\leq 1} \alpha^2 \cdot \chi_{M_\alpha^{(\eta)}}(x).$$

Using Theorem 5.4.30, we have

$$\tilde{A}_{\lambda(\alpha)}^\top = \tilde{A}_{\alpha^2}^\top = M_\alpha^{(\eta)} \text{ for } 0 < \alpha \leq 1.$$

We also have

$$\tilde{A}_{0+}^\top = \bigcup_{\alpha \in S^\bullet} M_\alpha^{(\eta)} = \bigcup_{0<\alpha\leq 1} M_\alpha^{(\eta)} = \left(\bar{m}_0^L, \bar{m}_0^U\right) = \left(b_1, b_2\right)$$

and

$$\tilde{A}_0^\top = \left[b_1, b_2\right].$$

Given any fixed $\alpha \in S^\bullet = (0,1]$, we have

$$\begin{aligned} B_\alpha^{(\eta,\lambda)} &= \bigcup_{\{\beta \in S^\bullet : \lambda(\beta) \geq \lambda(\alpha)\}} M_\beta^{(\eta)} = \bigcup_{\{\beta \in (0,1] : \beta^2 \geq \alpha^2\}} M_\beta^{(\eta)} = \bigcup_{\{\beta \in (0,1] : \beta \geq \alpha\}} M_\beta^{(\eta)} \\ &= \bigcup_{\alpha \leq \beta \leq 1} M_\beta^{(\eta)} = \bigcup_{\alpha \leq \beta \leq 1} \left[\bar{m}_\beta^L, \bar{m}_\beta^U\right] = \left[\bar{m}_\alpha^L, \bar{m}_\alpha^U\right] \end{aligned}$$

and

$$\begin{aligned} B_{\alpha+}^{(\eta,\lambda)} &= \begin{cases} \emptyset, & \text{if } \lambda(\beta) \leq \lambda(\alpha) \text{ for all } \beta \in S^\bullet \\ \bigcup_{\{\beta \in S^\bullet : \lambda(\beta) > \lambda(\alpha)\}} M_\beta^{(\eta)}, & \text{otherwise} \end{cases} \\ &= \begin{cases} \emptyset, & \text{if } \beta^2 \leq \alpha^2 \text{ for all } \beta \in (0,1] \\ \bigcup_{\{\beta \in (0,1] : \beta^2 > \alpha^2\}} M_\beta^{(\eta)}, & \text{otherwise} \end{cases} \\ &= \begin{cases} \emptyset, & \text{if } \beta \leq \alpha \text{ for all } \beta \in (0,1] \\ \bigcup_{\alpha < \beta \leq 1} M_\beta^{(\eta)}, & \text{otherwise} \end{cases} \\ &= \bigcup_{\alpha < \beta \leq 1} \left[\bar{m}_\beta^L, \bar{m}_\beta^U\right] = \left(\bar{m}_\alpha^L, \bar{m}_\alpha^U\right). \end{aligned}$$

Therefore, we have

$$B_\alpha^{(\eta,\lambda)} \backslash B_{\alpha+}^{(\eta,\lambda)} = \left[\bar{m}_\alpha^L, \bar{m}_\alpha^U\right] \setminus \left(\bar{m}_\alpha^L, \bar{m}_\alpha^U\right) = \left\{\bar{m}_\alpha^L, \bar{m}_\alpha^U\right\} \neq \emptyset.$$

Since $\mathcal{R}(\lambda) = [0,1]$, using (5.73), it follows that $\mathcal{R}(\xi_{\tilde{A}^{\top}}) = [0,1]$.

The above results assume that $\{M_\alpha : \alpha \in [0,1]\}$ is a nested family in the sense of $M_\beta^{(\eta)} \subseteq M_\alpha^{(\eta)}$ for $\lambda(\alpha) < \lambda(\beta)$. Next, the family $\{M_\alpha : \alpha \in [0,1]\}$ will not be assumed to be nested in that sense.

Proposition 5.4.32 *Let λ and η be functions from $[0,1]$ into $[0,1]$, and let $\{M_\alpha^{(\eta)} : \alpha \in [0,1]\}$ be a family of nonempty subsets of $\mathbb{R}^m$. Suppose that, for any fixed $x \in \mathbb{R}^m$, the following function*

$$\zeta_x(\alpha) = \lambda(\alpha) \cdot \chi_{M_\alpha^{(\eta)}}(x)$$

is upper semi-continuous on $[0,1]$. Consider the function $\xi_{\tilde{A}^{\top}}$ defined in (5.59). Then

$$\tilde{A}_{\lambda(\alpha)}^{\top} = \bigcup_{\{\beta \in S^\bullet : \lambda(\alpha) \leq \lambda(\beta)\}} M_\beta^{(\eta)} \text{ for every } \alpha \in S^\bullet \tag{5.74}$$

and

$$\tilde{A}_{0+}^{\top} = \bigcup_{\alpha \in S^\bullet} M_\alpha^{(\eta)}. \tag{5.75}$$

If $\gamma \notin \mathcal{R}(\lambda)$ satisfies $0 < \lambda(\alpha_n) \uparrow \gamma$ and $\alpha_n \in S^\bullet$ for all n, we have

$$\tilde{A}_\gamma^{\top} = \bigcap_{n=1}^{\infty} \bigcup_{\{\beta \in S^\bullet : \lambda(\alpha_n) \leq \lambda(\beta)\}} M_\beta^{(\eta)}. \tag{5.76}$$

Proof. Given any fixed $\alpha \in S^\bullet$, we have $\lambda(\alpha) > 0$. Suppose that $x \in \tilde{A}_{\lambda(\alpha)}^{\top}$ and

$$x \notin M_\beta^{(\eta)} \text{ for all } \beta \in S^\bullet \text{ with } \lambda(\alpha) \leq \lambda(\beta). \tag{5.77}$$

Then

$$\lambda(\beta) \cdot \chi_{M_\beta^{(\eta)}}(x) < \lambda(\alpha) \text{ for all } \beta \in S^\bullet. \tag{5.78}$$

Since ζ_x is upper semi-continuous on $[0,1]$ by the assumption, Proposition 1.4.4 says that the following supremum

$$\sup_{0 \leq \beta \leq 1} \zeta_x(\beta) = \max_{0 \leq \beta \leq 1} \zeta_x(\beta)$$

is attained. This says that

$$0 < \lambda(\alpha) \leq \xi_{\tilde{A}^{\top}}(x) = \sup_{0 \leq \beta \leq 1} \zeta_x(\beta) = \max_{0 \leq \beta \leq 1} \zeta_x(\beta) = \max_{0 \leq \beta \leq 1} \lambda(\beta) \cdot \chi_{M_\beta^{(\eta)}}(x) = \lambda(\beta^*) \cdot \chi_{M_{\beta^*}}(x),$$

where $\lambda(\beta^*) \neq 0$, i.e. $\beta^* \in S^\bullet$. Therefore, using (5.78), we have

$$\lambda(\alpha) \leq \xi_{\tilde{A}^{\top}}(x) = \lambda(\beta^*) \cdot \chi_{M_{\beta^*}^{(\eta)}}(x) < \lambda(\alpha).$$

This contradiction says that there exists $\beta_0 \in S^\bullet$ satisfying $\lambda(\beta_0) \geq \lambda(\alpha)$ and $x \in M_{\beta_0}^{(\eta)}$ by referring to (5.77), which shows the following inclusion

$$\tilde{A}_{\lambda(\alpha)}^{\top} \subseteq \bigcup_{\{\beta \in S^\bullet : \lambda(\alpha) \leq \lambda(\beta)\}} M_\beta^{(\eta)}.$$

The following inclusion

$$\bigcup_{\{\beta \in S^\bullet : \lambda(\alpha) \leq \lambda(\beta)\}} M_\beta^{(\eta)} \subseteq \left\{ x \in \mathbb{R}^m : \sup_{\beta \in S^\bullet} \lambda(\beta) \cdot \chi_{M_\beta^{(\eta)}}(x) \geq \lambda(\alpha) \right\}$$

$$= \{x \in \mathbb{R}^m : \xi_{\tilde{A}^\top}(x) \geq \lambda(\alpha)\} = \tilde{A}^\top_{\lambda(\alpha)}$$

is obvious. The equality (5.75) follows from (5.61). Using part (i) of Proposition 5.4.28, for $\gamma \notin \mathcal{R}(\lambda)$ satisfying $0 < \lambda(\alpha_n) \uparrow \gamma$ and $\alpha_n \in S^\bullet$ for all n, we have

$$\tilde{A}^\top_\gamma = \bigcap_{n=1}^{\infty} \tilde{A}^\top_{\lambda(\alpha_n)}.$$

Using (5.74), we obtain (5.76), and the proof is complete. ∎

Theorem 5.4.33 (Based on Two Functions and a Non-Nested Family). *Let λ and η be functions from $[0,1]$ into $[0,1]$, and let $\{M_\alpha^{(\eta)} : \alpha \in [0,1]\}$ be a family of nonempty subsets of $\mathbb{R}^m$. Suppose that, for any fixed $x \in \mathbb{R}^m$, the following function*

$$\zeta_x(\alpha) = \lambda(\alpha) \cdot \chi_{M_\alpha^{(\eta)}}(x)$$

is upper semi-continuous on $[0,1]$. Then, we can generate a fuzzy set $\tilde{A}^\top$ in $\mathbb{R}^m$ with membership function defined by

$$\xi_{\tilde{A}^\top}(x) = \sup_{\alpha \in S^\bullet} \lambda(\alpha) \cdot \chi_{M_\alpha^{(\eta)}}(x) \tag{5.79}$$

satisfying

$$\tilde{A}^\top_{\lambda(\alpha)} = \bigcup_{\{\beta \in S^\bullet : \lambda(\alpha) \leq \lambda(\beta)\}} M_\beta^{(\eta)} \text{ for } \alpha \in S^\bullet.$$

We also have

$$\tilde{A}^\top_{0+} = \bigcup_{\alpha \in S^\bullet} M_\alpha^{(\eta)}$$

and

$$\tilde{A}^\top_0 = cl(\tilde{A}^\top_{0+}) = cl\left(\bigcup_{\alpha \in S^\bullet} \tilde{A}^\top_\alpha\right) = cl\left(\bigcup_{\alpha \in S^\bullet} M_\alpha^{(\eta)}\right).$$

For $\gamma \notin \mathcal{R}(\lambda)$ satisfying $\lambda(\alpha_n) \uparrow \gamma$ with $\alpha_n \in S^\bullet$ for all n, we have

$$\tilde{A}^\top_\gamma = \bigcap_{n=1}^{\infty} \tilde{A}^\top_{\lambda(\alpha_n)} = \bigcap_{n=1}^{\infty} \bigcup_{\{\beta \in S^\bullet : \lambda(\alpha_n) \leq \lambda(\beta)\}} M_\beta^{(\eta)}.$$

On the other hand, the range $\mathcal{R}(\xi_{\tilde{A}^\top})$ of $\xi_{\tilde{A}^\top}$ satisfies the following properties.

(i) *Given any fixed $x \in \bigcup_{\alpha \in S^\bullet} M_\alpha^{(\eta)}$, we have the following properties.*
   - *Suppose that the supremum $\xi_{\tilde{A}^\top}(x)$ in (5.79) is attained. Then,we have*

     $$\xi_{\tilde{A}^\top}(x) \in \mathcal{R}(\lambda).$$

   - *Suppose that the supremum $\xi_{\tilde{A}^\top}(x)$ in (5.79) is not attained. Then,we have*

     $$\xi_{\tilde{A}^\top}(x) \in cl(\mathcal{R}(\lambda)) \setminus \mathcal{R}(\lambda).$$

(ii) *We have $0 \in \mathcal{R}(\xi_{\tilde{A}^\top}) \subseteq cl(\mathcal{R}(\lambda)) \cup \{0\}$.*

(iii) $B_{\alpha}^{(\eta,\lambda)} \backslash B_{\alpha+}^{(\eta,\lambda)} \neq \emptyset$ *for all* $\alpha \in [0,1) \cap S^{\bullet}$. *Then, we have*

$$\mathcal{R}(\lambda) \cup \{0\} \subseteq \mathcal{R}(\xi_{\tilde{A}^{\top}}) \subseteq cl(\mathcal{R}(\lambda)) \cup \{0\} = cl(\mathcal{R}(\lambda) \cup \{0\})$$

and

$$cl(\mathcal{R}(\xi_{\tilde{A}^{\top}})) = cl(\mathcal{R}(\lambda) \cup \{0\}).$$

In particular, we have the following properties.

- *Suppose that the supremum* $\xi_{\tilde{A}^{\top}}(x)$ *in (5.79) is attained for all* $x \in \bigcup_{\alpha \in S^{\bullet}} M_{\alpha}^{(\eta)}$. *Then, we have*

$$\mathcal{R}(\xi_{\tilde{A}^{\top}}) = \mathcal{R}(\lambda) \cup \{0\}.$$

- *Suppose that* $\mathcal{R}(\lambda) \cup \{0\}$ *is a closed subset of* $[0,1]$. *Then, we have*

$$\mathcal{R}(\xi_{\tilde{A}^{\top}}) = \mathcal{R}(\lambda) \cup \{0\}.$$

Proof. Using Proposition 5.4.32 and a similar argument to the proof of Theorem 5.4.30, the proof is complete. ■

5.5 Generating Fuzzy Intervals

In what follows, we are going to present the methodology for generating fuzzy intervals in $\mathbb{R}$ by referring to Definition 2.3.1. The main issue is to consider the endpoint functions of bounded closed intervals in $\mathbb{R}$.

Theorem 5.5.1 *Let* $\zeta^L : (0,1] \to \mathbb{R}$ *and* $\zeta^U : (0,1] \to \mathbb{R}$ *be two real-valued functions defined on* $(0,1]$ *such that the following conditions are satisfied.*

- ζ^L *and* ζ^U *are bounded on* $(0,1]$.
- $\zeta^L(\alpha) \leq \zeta^U(\alpha)$ *for each* $\alpha \in (0,1]$.
- ζ^L *is an increasing function and* ζ^U *is a decreasing function on* $(0,1]$.
- ζ^L *and* ζ^U *are left-continuous on* $(0,1]$ *(equivalently,* ζ^L *is lower semi-continuous and* ζ^U *is upper semi-continuous on* $(0,1]$*).*

Let $M_{\alpha} = [\zeta^L(\alpha), \zeta^U(\alpha)]$ *be closed intervals in* $\mathbb{R}$ *for* $\alpha \in (0,1]$. *Then, we can generate a fuzzy interval* $\tilde{A}^{\star}$ *in* $\mathbb{R}$ *with membership function defined by*

$$\xi_{\tilde{A}^{\star}}(x) = \sup_{0<\alpha\leq 1} \alpha \cdot \chi_{M_{\alpha}}(x)$$

satisfying $\tilde{A}^{\star}_{\alpha} = M_{\alpha}$ *for* $\alpha \in (0,1]$. *We also have*

$$\tilde{A}^{\star}_{0+} = \bigcup_{0<\alpha\leq 1} M_{\alpha} = \bigcup_{0<\alpha\leq 1} [\zeta^L(\alpha), \zeta^U(\alpha)]$$

and

$$\tilde{A}^{\star}_{0} = cl(\tilde{A}^{\star}_{0+}) = cl\left(\bigcup_{0<\alpha\leq 1} \tilde{A}^{\star}_{\alpha}\right) = cl\left(\bigcup_{0<\alpha\leq 1} M_{\alpha}\right) = cl\left(\bigcup_{0<\alpha\leq 1} [\zeta^L(\alpha), \zeta^U(\alpha)]\right).$$

We further assume that ζ^L is right-continuous and strictly increasing on $(0,1]$, and that ζ^U is right-continuous and strictly decreasing on $(0,1]$. Then, we have

$$A_\alpha \backslash A_{\alpha+} = \left\{\zeta^L(\alpha), \zeta^U(\alpha)\right\} \neq \emptyset \text{ for all } \alpha \in (0,1)$$

and $\mathcal{R}(\xi_{\tilde{A}^\star}) = [0,1]$, where

$$A_\alpha = \bigcup_{\alpha \leq \beta \leq 1} M_\beta \text{ and } A_{\alpha+} = \bigcup_{\alpha < \beta \leq 1} M_\beta.$$

Proof. We first note that Proposition 1.3.7 says that the lower semi-continuity for ζ^L and the upper semi-continuity for ζ^U are equivalent to the left-continuities for ζ^L and ζ^U. By applying Theorem 5.4.7, we just need to check $\bigcap_{n=1}^{\infty} M_{\alpha_n} \subseteq M_\alpha$ when $\alpha_n \uparrow \alpha$. Assume that $x \in M_{\alpha_n}$ for all n. Then

$$\zeta^L(\alpha_n) \leq x \leq \zeta^U(\alpha_n) \text{ for all } n \text{ satisfying } \alpha_n \uparrow \alpha.$$

The left-continuities of ζ^L and ζ^U say that

$$\zeta^L(\alpha) = \lim_{n\to\infty} \zeta^L(\alpha_n) \leq x \leq \lim_{n\to\infty} \zeta^U(\alpha_n) = \zeta^U(\alpha),$$

which also says that $x \in M_\alpha$.

In order to apply Theorem 5.4.7 to obtain the range $\mathcal{R}(\xi_{\tilde{A}^\star})$, we want to check $A_\alpha \backslash A_{\alpha+} \neq \emptyset$. Since $M_\alpha \subseteq M_\beta$ for $\beta < \alpha$, we have

$$A_\alpha = \bigcup_{\alpha \leq \beta \leq 1} M_\beta = \bigcup_{\alpha \leq \beta \leq 1} \left[\zeta^L(\beta), \zeta^U(\beta)\right] = \left[\zeta^L(\alpha), \zeta^U(\alpha)\right].$$

Now, we want to claim

$$A_{\alpha+} = \bigcup_{\alpha < \beta \leq 1} M_\beta = \bigcup_{\alpha < \beta \leq 1} \left[\zeta^L(\beta), \zeta^U(\beta)\right] = \left(\zeta^L(\alpha), \zeta^U(\alpha)\right).$$

For $x \in A_{\alpha+}$, we have $x \in [\zeta^L(\beta_0), \zeta^U(\beta_0)]$ for some β_0 satisfying $\alpha < \beta_0 \leq 1$. Using the strict monotonicity, we have

$$\zeta^L(\alpha) < \zeta^L(\beta_0) \leq x \leq \zeta^U(\beta_0) < \zeta^U(\alpha),$$

which says that $x \in (\zeta^L(\alpha), \zeta^U(\alpha))$. Therefore, we obtain the inclusion

$$A_{\alpha+} \subseteq (\zeta^L(\alpha), \zeta^U(\alpha)).$$

For proving the other direction of inclusion, suppose that

$$\zeta^L(\alpha) < x < \zeta^U(\alpha)).$$

We take a decreasing sequence $\{\beta_n\}_{n=1}^{\infty}$ satisfying $\beta_n \downarrow \alpha$. Since ζ^L is right-continuous and strictly increasing, we have $\zeta^L(\beta_n) \downarrow \zeta^L(\alpha)$. Since ζ^U is right-continuous and strictly decreasing, we have $\zeta^U(\beta_n) \uparrow \zeta^U(\alpha)$. Therefore, there exists β_N satisfying $\alpha < \beta_N$ and

$$\zeta^L(\alpha) < \zeta^L(\beta_N) \leq x \leq \zeta^U(\beta_N) < \zeta^U(\alpha)),$$

which says that $x \in [\zeta^L(\beta_N), \zeta^L(\beta_N)]$, i.e. $x \in A_{\alpha+}$. Therefore, we obtain the desired equality and

$$A_\alpha \backslash A_{\alpha+} = \left[\zeta^L(\alpha), \zeta^U(\alpha)\right] \setminus \left(\zeta^L(\alpha), \zeta^U(\alpha)\right) = \left\{\zeta^L(\alpha), \zeta^U(\alpha)\right\} \neq \emptyset.$$

Since each α-level set $\tilde{A}^\star_\alpha$ is a closed interval in $\mathbb{R}$ for $\alpha \in (0,1]$, i.e. a closed and convex set in $\mathbb{R}$ for $\alpha \in (0,1]$, the closure of $\tilde{A}^\star_\alpha$ says that the membership function $\xi_{\tilde{A}^\star}$ is upper semi-continuous, and the convexity of $\tilde{A}^\star_\alpha$ says that the membership function $\xi_{\tilde{A}^\star}$ is quasi-concave. Therefore $\tilde{A}^\star$ is indeed a fuzzy interval. This completes the proof. ■

Example 5.5.2 Given any real numbers a_1, a_2, b_1, b_2 satisfying $b_1 < a_1 < a_2 < b_2$, we define two real-valued functions ζ^L and ζ^U on (0,1] by

$$\zeta^L(\alpha) = (1-\alpha)b_1 + \alpha a_1 \text{ and } \zeta^U(\alpha) = (1-\alpha)b_2 + \alpha a_2.$$

Then ζ^L and ζ^U satisfy all the assumptions of Theorem 5.5.1. Let

$$M_\alpha = \left[\zeta^L(\alpha), \zeta^U(\alpha)\right] = \left[(1-\alpha)b_1 + \alpha a_1, (1-\alpha)b_2 + \alpha a_2\right]$$

be closed intervals in $\mathbb{R}$ for $\alpha \in (0,1]$. Then, we can generate a fuzzy interval $\tilde{A}^\star$ in $\mathbb{R}$ with membership function defined by

$$\xi_{\tilde{A}^\star}(x) = \sup_{0<\alpha\leq 1} \alpha \cdot \chi_{M_\alpha}(x)$$

satisfying $\tilde{A}^\star_\alpha = M_\alpha$ for $\alpha \in (0,1]$. The support $\tilde{A}^\star_{0+}$ of $\tilde{A}^\star$ is given by

$$\tilde{A}^\star_{0+} = \bigcup_{0<\alpha\leq 1} \left[\zeta^L(\alpha), \zeta^U(\alpha)\right] = \bigcup_{0<\alpha\leq 1} \left[(1-\alpha)b_1 + \alpha a_1, (1-\alpha)b_2 + \alpha a_2\right].$$

Moreover, the range of $\tilde{A}^\star$ is given by $\mathcal{R}(\xi_{\tilde{A}^\star}) = [0,1]$.

We have mentioned that the 0-level set is an important issue in fuzzy sets theory. Notice that the functions ζ^L and ζ^U considered in Theorem 5.5.1 are defined on (0,1]. Next, we are going to consider the functions ζ^L and ζ^U defined on [0,1]. The difference is that the expressions of 0-level sets are different.

Theorem 5.5.3 *Let $\zeta^L : [0,1] \to \mathbb{R}$ and $\zeta^U : [0,1] \to \mathbb{R}$ be two real-valued functions defined on [0,1] such that the following conditions are satisfied.*

- *ζ^L and ζ^U are bounded on [0,1].*
- *$\zeta^L(\alpha) \leq \zeta^U(\alpha)$ for each $\alpha \in [0,1]$.*
- *ζ^L is a strictly increasing function and ζ^U is a strictly decreasing function on [0,1].*
- *ζ^L and ζ^U are left-continuous on (0,1].*
- *ζ^L and ζ^U are right-continuous at 0.*

Let $M_\alpha = [\zeta^L(\alpha), \zeta^U(\alpha)]$ be closed intervals in $\mathbb{R}$ for $\alpha \in (0,1]$. Then, we can generate a fuzzy interval $\tilde{A}^\star$ in $\mathbb{R}$ with membership function defined by

$$\xi_{\tilde{A}^\star}(x) = \sup_{0<\alpha\leq 1} \alpha \cdot \chi_{M_\alpha}(x) \tag{5.80}$$

satisfying

$$\tilde{A}^\star_\alpha = [\zeta^L(\alpha), \zeta^U(\alpha)] \text{ for } \alpha \in [0,1],$$

and the support of $\tilde{A}^\star$ is an open interval given by

$$\tilde{A}^\star_{0+} = \left(\zeta^L(0), \zeta^U(0)\right).$$

We further assume that ζ^L *and* ζ^U *are right-continuous on* [0,1]. *Then, we have*

$$A_\alpha \backslash A_{\alpha+} = \left\{\zeta^L(\alpha), \zeta^U(\alpha)\right\} \neq \emptyset \text{ for all } \alpha \in (0,1)$$

and $\mathcal{R}(\xi_{\tilde{A}^\star}) = [0,1]$, *where*

$$A_\alpha = \bigcup_{\alpha \le \beta \le 1} M_\beta \text{ and } A_{\alpha+} = \bigcup_{\alpha < \beta \le 1} M_\beta.$$

Proof. The desired results can be similarly obtained from the proof of Theorem 5.5.1 by replacing (0,1] with the whole unit interval [0,1]. It remains to show that $\tilde{A}^\star_{0+} = \left(\zeta^L(0), \zeta^U(0)\right)$. Using the strict monotonicity of functions ζ^L and ζ^U, we see that $\zeta^L(\alpha) > \zeta^L(0)$ and $\zeta^U(\alpha) < \zeta^U(0)$ for all $\alpha \in (0,1]$, i.e.

$$\left[\zeta^L(\alpha), \zeta^U(\alpha)\right] \subseteq \left(\zeta^L(0), \zeta^U(0)\right) \text{ for all } \alpha \in (0,1].$$

It follows that

$$\tilde{A}^\star_{0+} = \bigcup_{0<\alpha\le 1} \left[\zeta^L(\alpha), \zeta^U(\alpha)\right] \subseteq \left(\zeta^L(0), \zeta^U(0)\right).$$

For proving the other direction of inclusion, given any $x \in (\zeta^L(0), \zeta^U(0))$, we consider the following cases.

- Given any $\zeta^L(0) < x < \zeta^L(1)$, we take $0 < \epsilon < x - \zeta^L(0)$. Since ζ^L is right-continuous at 0, there exists $\delta > 0$ such that

  $$0 < (\alpha_1 - 0) < \delta \text{ implies } \left|\zeta^L(\alpha_1) - \zeta^L(0)\right| < \epsilon.$$

  Since ζ^L is increasing, it follows that

  $$\zeta^L(0) \le \zeta^L(\alpha_1) < \zeta^L(0) + \epsilon < x.$$

- Given any $\zeta^U(1) < x < \zeta^U(0)$, we take $0 < \epsilon < \zeta^U(0) - x$. Since ζ^U is right-continuous at 0, there exists $\delta > 0$ such that

  $$0 < (\alpha_2 - 0) < \delta \text{ implies } \left|\zeta^U(0) - \zeta^U(\alpha_2)\right| < \epsilon.$$

  Since ζ^U is decreasing, we have $x < \zeta^U(0) - \epsilon < \zeta^U(\alpha_2) \le \zeta^U(0)$.

Therefore, we obtain

$$\begin{cases} \zeta^L(0) \le \zeta^L(\alpha_2) \le \zeta^L(\alpha_1) < x < \zeta^U(\alpha_2) \le \zeta^U(0), & \text{if } \alpha_1 \ge \alpha_2 \\ \zeta^L(0) \le \zeta^L(\alpha_1) < x < \zeta^U(\alpha_2) \le \zeta^U(\alpha_1) \le \zeta^U(0), & \text{if } \alpha_1 < \alpha_2 \end{cases}$$

for $\alpha_1, \alpha_2 \in (0,1]$, which says that there exists $\alpha \in (0,1]$ satisfying $x \in [\zeta^L(\alpha), \zeta^U(\alpha)]$. This completes the proof. ■

The family $\{M_\alpha : \alpha \in (0,1]\}$ of bounded closed intervals considered in Theorems 5.5.1 and 5.5.3 is nested. Next, we are going to consider the non-nested family of bounded closed intervals.

Theorem 5.5.4 *Let* $\zeta^L : [0,1] \to \mathbb{R}$ *and* $\zeta^U : [0,1] \to \mathbb{R}$ *be two real-valued functions defined on* [0,1] *such that the following conditions are satisfied.*

- ζ^L *and* ζ^U *are bounded on* [0,1].
- $\zeta^L(\alpha) \le \zeta^U(\alpha)$ *for each* $\alpha \in [0,1]$.
- ζ^L *is lower semi-continuous and* ζ^U *is upper semi-continuous on* [0,1].

Let $M_\alpha = [\zeta^L(\alpha), \zeta^U(\alpha)]$ be closed intervals in $\mathbb{R}$ for $\alpha \in [0,1]$. Then, we can generate a fuzzy interval $\tilde{A}^\star$ in $\mathbb{R}$ with membership function defined by

$$\xi_{\tilde{A}^\star}(x) = \sup_{0<\alpha\leq 1} \alpha \cdot \chi_{M_\alpha}(x) \tag{5.81}$$

satisfying

$$\begin{aligned}\tilde{A}^\star_\alpha &= \{x : \xi_{\tilde{A}^\star}(x) \geq \alpha\} = \left\{x : \sup_{0<\beta\leq 1} \beta \cdot \chi_{M_\beta}(x) \geq \alpha\right\} \\ &= \left[\min_{\alpha\leq\beta\leq 1} \zeta^L(\beta), \max_{\alpha\leq\beta\leq 1} \zeta^U(\beta)\right] \end{aligned}\tag{5.82}$$

for $\alpha \in (0,1]$ and

$$\tilde{A}^\star_{0+} = \bigcup_{0<\alpha\leq 1} M_\alpha = \bigcup_{0<\alpha\leq 1} \left[\zeta^L(\alpha), \zeta^U(\alpha)\right].$$

We further assume that ζ^L and ζ^U are right-continuous on $[0,1]$. Then, we have

$$A_\alpha \backslash A_{\alpha+} = \left\{\zeta^L(\alpha), \zeta^U(\alpha)\right\} \neq \emptyset \text{ for all } \alpha \in (0,1)$$

and $\mathcal{R}(\xi_{\tilde{A}^\star}) = [0,1]$, where

$$A_\alpha = \bigcup_{\alpha\leq\beta\leq 1} M_\beta \text{ and } A_{\alpha+} = \bigcup_{\alpha<\beta\leq 1} M_\beta.$$

We also have the support

$$\left(\min_{0\leq\beta\leq 1} \zeta^L(\beta), \max_{0\leq\beta\leq 1} \zeta^U(\beta)\right) \subseteq \tilde{A}^\star_{0+} \subseteq \left[\min_{0\leq\beta\leq 1} \zeta^L(\beta), \max_{0\leq\beta\leq 1} \zeta^U(\beta)\right]$$

and the 0-level set

$$\tilde{A}^\star_0 = \left[\min_{0\leq\beta\leq 1} \zeta^L(\beta), \max_{0\leq\beta\leq 1} \zeta^U(\beta)\right].$$

Proof. Since $[0,1]$ is a closed set, we note that the following infimum and supremum

$$\inf_{\alpha\leq\beta\leq 1} \zeta^L(\beta) = \min_{\alpha\leq\beta\leq 1} \zeta^L(\beta) \text{ and } \sup_{\alpha\leq\beta\leq 1} \zeta^U(\beta) = \max_{\alpha\leq\beta\leq 1} \zeta^U(\beta)$$

are attained by Proposition 1.4.4. Given any fixed x, let $\eta(\beta) = \beta \cdot \chi_{M_\beta}(x)$. Proposition 1.3.8 says that the function $\eta(\beta) = \beta \cdot \chi_{M_\beta}(x)$ is upper semi-continuous on the closed set $[0,1]$ for any fixed x. We first prove the equality (5.82). For $\alpha \in (0,1]$, we consider

$$x \in \left\{x : \sup_{0<\beta\leq 1} \beta \cdot \chi_{M_\beta}(x) \geq \alpha\right\}. \tag{5.83}$$

Suppose that $x \notin M_\beta$ for all β satisfying $\alpha \leq \beta \leq 1$, i.e. $\beta \cdot \chi_{M_\beta}(x) = 0$ for all β satisfying $\alpha \leq \beta \leq 1$. Then, we have

$$\beta \cdot \chi_{M_\beta}(x) < \alpha \text{ for all } \beta \in (0,1]. \tag{5.84}$$

Since $[0,1]$ is a closed set and $\eta(\beta) = \beta \cdot \chi_{M_\beta}(x)$ is upper semi-continuous on $[0,1]$, Proposition 1.4.4 says that

$$\sup_{0<\alpha\leq 1} \beta \cdot \chi_{M_\beta}(x) = \sup_{0\leq\alpha\leq 1} \beta \cdot \chi_{M_\beta}(x) = \max_{0\leq\alpha\leq 1} \beta \cdot \chi_{M_\beta}(x) = \beta^* \cdot \chi_{M_{\beta^*}}(x)$$

for some $\beta^* \in [0,1]$. We consider two cases.

- If $\beta^* = 0$, we have

$$\sup_{0<\alpha\leq 1} \beta \cdot \chi_{M_\beta}(x) = 0,$$

which violates (5.83).
- If $\beta^* \neq 0$, i.e. $\beta^* \in (0,1]$, using (5.84), we have

$$\sup_{0<\alpha\leq 1} \beta \cdot \chi_{M_\beta}(x) = \beta^* \cdot \chi_{M_{\beta^*}}(x) < \alpha,$$

which also violates (5.83).

This says that there exists β_0 satisfying $\alpha \leq \beta_0 \leq 1$ and $x \in M_{\beta_0}$, i.e. $\zeta^L(\beta_0) \leq x \leq \zeta^U(\beta_0)$. Then

$$x \geq \zeta^L(\beta_0) \geq \min_{\alpha\leq\beta\leq 1} \zeta^L(\beta) \text{ and } x \leq \zeta^U(\beta_0) \leq \max_{\alpha\leq\beta\leq 1} \zeta^U(\beta);$$

that is,

$$x \in \left[\min_{\alpha\leq\beta\leq 1} \zeta^L(\beta), \max_{\alpha\leq\beta\leq 1} \zeta^U(\beta)\right].$$

Therefore, we obtain the following inclusion

$$\left\{x : \sup_{0<\alpha\leq 1} \beta \cdot \chi_{M_\beta}(x) \geq \alpha\right\} \subseteq \left[\min_{\alpha\leq\beta\leq 1} \zeta^L(\beta), \max_{\alpha\leq\beta\leq 1} \zeta^U(\beta)\right].$$

For proving the other direction of inclusion, we consider

$$x \in \left[\min_{\alpha\leq\beta\leq 1} \zeta^L(\beta), \max_{\alpha\leq\beta\leq 1} \zeta^U(\beta)\right]. \tag{5.85}$$

Suppose that $x \notin M_\beta$ for all β satisfying $\alpha \leq \beta \leq 1$. Then, from Proposition 1.3.13, we have

$$x \notin \bigcup_{\alpha\leq\beta\leq 1} M_\beta = \left[\min_{\alpha\leq\beta\leq 1} \zeta^L(\beta), \max_{\alpha\leq\beta\leq 1} \zeta^U(\beta)\right],$$

which violates (5.85). Therefore, there exists β_0 satisfying $\beta_0 \geq \alpha$ and $x \in M_{\beta_0}$, i.e.

$$\sup_{0<\alpha\leq 1} \beta \cdot \chi_{M_\beta}(x) \geq \beta_0 \geq \alpha,$$

which says that

$$x \in \left\{x : \sup_{0<\alpha\leq 1} \beta \cdot \chi_{M_\beta}(x) \geq \alpha\right\}.$$

Therefore, we obtain the equality (5.82). Using Theorem 5.4.9, we also have

$$\tilde{A}^\star_{0+} = \bigcup_{0<\alpha\leq 1} M_\alpha = \bigcup_{0<\alpha\leq 1} \left[\zeta^L(\alpha), \zeta^U(\alpha)\right]. \tag{5.86}$$

From the proof of Theorem 5.5.1, we can show that $\tilde{A}^\star$ is a fuzzy interval. The right-continuities of ζ^L and ζ^U also say that

$$A_\alpha \backslash A_{\alpha+} = \left\{\zeta^L(\alpha), \zeta^U(\alpha)\right\} \neq \emptyset \text{ for all } \alpha \in (0,1)$$

and $\mathcal{R}(\xi_{\tilde{A}^\star}) = [0,1]$ by the proof of Theorem 5.5.1 again. Let

$$l(\alpha) = \min_{\alpha\leq\beta\leq 1} \zeta^L(\beta) \text{ and } u(\alpha) = \max_{\alpha\leq\beta\leq 1} \zeta^U(\beta).$$

Since $\zeta^L(\beta) \leq \zeta^U(\beta)$ for each $\beta \in [0,1]$, it follows that $l(\alpha) \leq u(\alpha)$ for all $\alpha \in [0,1]$. Since the function l is increasing and the function u is decreasing on $[0,1]$, it follows that

$$\bigcup_{0<\alpha\leq 1} \left[l(\alpha), u(\alpha)\right] \subseteq \left[l(0), u(0)\right].$$

Since $l(\alpha) \leq u(\alpha)$ for all $\alpha \in [0,1]$, it follows that

$$\sup_{0<\alpha\leq 1} l(\alpha) \leq \inf_{0<\alpha\leq 1} u(\alpha).$$

Given any x with $l(0) < x < u(0)$, we are going to show that there exists $\hat{\alpha} \in [0,1]$ satisfying $l(\hat{\alpha}) \leq x \leq u(\hat{\alpha})$. For any fixed $\alpha \in (0,1]$, since

$$l(0) \leq l(\alpha) \leq \sup_{0<\alpha\leq 1} l(\alpha) \leq \inf_{0<\alpha\leq 1} u(\alpha) \leq u(\alpha) \leq u(0), \tag{5.87}$$

we consider the following cases.

- Suppose that

$$\sup_{0<\alpha\leq 1} l(\alpha) \leq x \leq \inf_{0<\alpha\leq 1} u(\alpha).$$

Then, from (5.87), we can take some $\hat{\alpha} \in [0,1]$ satisfying $l(\hat{\alpha}) \leq x \leq u(\hat{\alpha})$.

- Suppose that

$$l(0) < x < \sup_{0<\alpha\leq 1} l(\alpha).$$

There exists a sequence $\{\alpha_n\}_{n=1}^{\infty}$ in $(0,1]$ satisfying $\alpha_n \downarrow 0$. Since ζ^L is right-continuous on $[0,1]$, from Proposition 1.3.12, we have $l(\alpha_n) \downarrow l(0)$. Therefore, there exists α_N satisfying $l(0) \leq l(\alpha_N) < x$. Using (5.87), we have

$$l(\alpha_N) < x < \sup_{0<\alpha\leq 1} l(\alpha) \leq \inf_{0<\alpha\leq 1} u(\alpha) \leq u(\alpha_N).$$

- Suppose that

$$\inf_{0<\alpha\leq 1} u(\alpha) < x < u(0).$$

Since ζ^U is right-continuous on $[0,1]$, for $\alpha_n \downarrow 0$ with $\alpha_n > 0$ for all n, Proposition 1.3.12 says that $u(\alpha_n) \uparrow u(0)$. Therefore, there exists α_N satisfying $x < u(\alpha_N) \leq u(0)$. Using (5.87), we have

$$l(\alpha_N) \leq \sup_{0<\alpha\leq 1} l(\alpha) \leq \inf_{0<\alpha\leq 1} u(\alpha) < x < u(\alpha_N).$$

Therefore, we conclude that there exists $\alpha_N \in (0,1]$ satisfying $x \in [l(\alpha_N), u(\alpha_N)]$, which shows the following inclusions

$$(l(0), u(0)) \subseteq \bigcup_{0<\alpha\leq 1} \left[l(\alpha), u(\alpha)\right] \subseteq \left[l(0), u(0)\right].$$

From (5.86), we obtain

$$\tilde{A}_0^{\star} = \mathrm{cl}(\tilde{A}_{0+}^{\star}) = \mathrm{cl}\left(\bigcup_{0\leq\alpha\leq 1} \left[l(\alpha), u(\alpha)\right]\right) = \left[l(0), u(0)\right] = \left[\min_{0\leq\alpha\leq 1} \zeta^L(\beta), \max_{0<\alpha\leq 1} \zeta^U(\beta)\right].$$

This completes the proof. ■

Example 5.5.5 Given any real numbers a_1, a_2, b_1, b_2 satisfying $b_1 < a_1 < a_2 < b_2$, we define two real-valued functions ζ^L and ζ^U on [0,1] by

$$\zeta^L(\alpha) = \alpha b_1 + (1-\alpha)a_1 \text{ and } \zeta^U(\alpha) = \alpha b_2 + (1-\alpha)a_2.$$

In this case, we see that ζ^L is not increasing and ζ^U is not decreasing on [0,1]. Therefore, we cannot apply Theorem 5.5.1. However, the functions ζ^L and ζ^U satisfy all the assumptions of Theorem 5.5.4. Let

$$M_\alpha = \left[\zeta^L(\alpha), \zeta^U(\alpha)\right] = \left[\alpha b_1 + (1-\alpha)a_1, \alpha b_2 + (1-\alpha)a_2\right]$$

be closed intervals in $\mathbb{R}$ for $\alpha \in (0,1]$. Then, we can generate a fuzzy interval $\tilde{A}^\star$ in $\mathbb{R}$ with membership function defined by

$$\xi_{\tilde{A}^\star}(x) = \sup_{0<\alpha\leq 1} \alpha \cdot \chi_{M_\alpha}(x)$$

satisfying

$$\tilde{A}^\star_\alpha = \left[\min_{\alpha\leq\beta\leq 1} \zeta^L(\beta), \max_{\alpha\leq\beta\leq 1} \zeta^U(\beta)\right] = \left[\min_{0.8\geq\beta\geq\alpha} \zeta^L(\beta), \max_{0.8\geq\beta\geq\alpha} \zeta^U(\beta)\right]$$

for $\alpha \in (0,1]$. Since ζ^L is decreasing and ζ^U is increasing on [0,1], it follows that

$$\tilde{A}^\star_\alpha = \left[\zeta^L(1), \zeta^U(1)\right] \text{ for } \alpha \in (0,1].$$

Since

$$\left[\zeta^L(\alpha), \zeta^U(\alpha)\right] \subseteq \left[\zeta^L(\beta), \zeta^U(\beta)\right] \text{ for } \beta > \alpha,$$

the support $\tilde{A}^\star_{0+}$ of $\tilde{A}^\star$ is given by

$$\tilde{A}^\star_{0+} = \bigcup_{0<\alpha\leq 1} \left[\zeta^L(\alpha), \zeta^U(\alpha)\right] = \left[\zeta^L(1), \zeta^U(1)\right].$$

Given any fixed $\alpha \in (0,1)$, we have

$$A_\alpha = \bigcup_{\alpha\leq\beta\leq 1} M_\beta = \bigcup_{\alpha\leq\beta\leq 1} \left[\zeta^L(\beta), \zeta^U(\beta)\right] = \left[\zeta^L(1), \zeta^U(1)\right]$$

and

$$A_{\alpha+} = \bigcup_{\alpha<\beta\leq 1} M_\beta = \bigcup_{\alpha<\beta\leq 1} \left[\zeta^L(\beta), \zeta^U(\beta)\right] = \left[\zeta^L(1), \zeta^U(1)\right].$$

Therefore, we obtain $A_\alpha \backslash A_{\alpha+} = \emptyset$. As a matter of fact, we have $\mathcal{R}(\xi_{\tilde{A}^\star}) = \{1\}$.

We also have a more general case shown below.

Theorem 5.5.6 *Let* $\eta_1 : [0,1] \to \mathbb{R}$ *and* $\eta_2 : [0,1] \to \mathbb{R}$ *be two real-valued functions defined on* [0,1], *and let*

$$\zeta^L(\alpha) = \min\{\eta_1(\alpha), \eta_2(\alpha)\} \text{ and } \zeta^U(\alpha) = \max\{\eta_1(\alpha), \eta_2(\alpha)\}.$$

We define a closed interval

$$M_\alpha = \left[\zeta^L(\alpha), \zeta^U(\alpha)\right] = \left[\min\{\eta_1(\alpha), \eta_2(\alpha)\}, \max\{\eta_1(\alpha), \eta_2(\alpha)\}\right]$$

in $\mathbb{R}$ for $\alpha \in (0,1]$. Suppose that ζ^L is lower semi-continuous and ζ^U is upper semi-continuous on $[0,1]$. Then, we can generate a fuzzy interval $\tilde{A}^\star$ in $\mathbb{R}$ with membership function defined by

$$\xi_{\tilde{A}^\star}(x) = \sup_{0<\alpha\leq 1} \alpha \cdot \chi_{M_\alpha}(x) \tag{5.88}$$

satisfying

$$\begin{aligned}
\tilde{A}^\star_\alpha &= \{x : \xi_{\tilde{A}^\star}(x) \geq \alpha\} = \left\{x : \sup_{0<\beta\leq 1} \beta \cdot \chi_{M_\beta}(x) \geq \alpha\right\} \\
&= \left[\min_{\alpha\leq\beta\leq 1} \zeta^L(\beta), \max_{\alpha\leq\beta\leq 1} \zeta^U(\beta)\right] \\
&= \left[\min_{\alpha\leq\beta\leq 1} \min\left\{\eta_1(\beta), \eta_2(\beta)\right\}, \max_{\alpha\leq\beta\leq 1} \max\left\{\eta_1(\beta), \eta_2(\beta)\right\}\right] \\
&= \left[\min\left\{\min_{\alpha\leq\beta\leq 1} \eta_1(\beta), \min_{\alpha\leq\beta\leq 1} \eta_2(\beta)\right\}, \max\left\{\max_{\alpha\leq\beta\leq 1} \eta_1(\beta), \max_{\alpha\leq\beta\leq 1} \eta_2(\beta)\right\}\right]
\end{aligned} \tag{5.89}$$

for $\alpha \in (0,1]$ and

$$\tilde{A}^\star_{0+} = \bigcup_{0<\alpha\leq 1} M_\alpha = \bigcup_{0<\alpha\leq 1} \left[\zeta^L(\alpha), \zeta^U(\alpha)\right] = \bigcup_{0<\alpha\leq 1} \left[\min\{\eta_1(\alpha), \eta_2(\alpha)\}, \max\left\{\eta_1(\alpha), \eta_2(\alpha)\right\}\right].$$

We further assume that ζ^L and ζ^U are right-continuous on $[0,1]$. Then, we have

$$A_\alpha \backslash A_{\alpha+} = \left\{\zeta^L(\alpha), \zeta^U(\alpha)\right\} \neq \emptyset \text{ for all } \alpha \in (0,1)$$

and $\mathcal{R}(\xi_{\tilde{A}^\star}) = [0,1]$, where

$$A_\alpha = \bigcup_{\alpha\leq\beta\leq 1} M_\beta \text{ and } A_{\alpha+} = \bigcup_{\alpha<\beta\leq 1} M_\beta.$$

We also have

$$\left(\min_{0\leq\beta\leq 1} \zeta^L(\beta), \max_{0\leq\beta\leq 1} \zeta^U(\beta)\right) \subseteq \tilde{A}^\star_{0+} \subseteq \left[\min_{0\leq\beta\leq 1} \zeta^L(\beta), \max_{0\leq\beta\leq 1} \zeta^U(\beta)\right]$$

and the 0-level set

$$\tilde{A}^\star_0 = \left[\min_{0\leq\beta\leq 1} \zeta^L(\beta), \max_{0\leq\beta\leq 1} \zeta^U(\beta)\right], \tag{5.90}$$

where

$$\min_{0\leq\beta\leq 1} \zeta^L(\beta) = \min\left\{\min_{0\leq\beta\leq 1} \eta_1(\beta), \min_{0\leq\beta\leq 1} \eta_2(\beta)\right\}$$

and

$$\max_{0\leq\beta\leq 1} \zeta^L(\beta) = \max\left\{\max_{0\leq\beta\leq 1} \eta_1(\beta), \max_{0\leq\beta\leq 1} \eta_2(\beta)\right\}.$$

Proof. The desired results can be obtained from Theorem 5.5.4. Therefore, we see that $\tilde{A}^\star$ is a fuzzy interval with

$$\begin{aligned}
\tilde{A}^\star_\alpha &= \left[\min_{\alpha\leq\beta\leq 1} \zeta^L(\beta), \max_{\alpha\leq\beta\leq 1} \zeta^U(\beta)\right] \\
&= \left[\min_{\alpha\leq\beta\leq 1} \min\{\eta_1(\beta), \eta_2(\beta)\}, \max_{\alpha\leq\beta\leq 1} \max\left\{\eta_1(\beta), \eta_2(\beta)\right\}\right]
\end{aligned}$$

for $\alpha \in (0,1]$. Now, we want to claim

$$\min_{\alpha\le\beta\le1} \min\left\{\eta_1(\beta), \eta_2(\beta)\right\} = \min\left\{\min_{\alpha\le\beta\le1} \eta_1(\beta), \min_{\alpha\le\beta\le1} \eta_2(\beta),\right\}. \tag{5.91}$$

Suppose that

$$\min_{\alpha\le\beta\le1} \min\left\{\eta_1(\beta), \eta_2(\beta)\right\} = \min\left\{\eta_1(\beta^*), \eta_2(\beta^*)\right\}$$

for some β^* satisfying $\alpha \le \beta^* \le 1$ by Proposition 1.4.4. Then, we have

$$\min_{\alpha\le\beta\le1} \eta_1(\beta) \le \eta_1(\beta^*) \text{ and } \min_{\alpha\le\beta\le1} \eta_2(\beta) \le \eta_2(\beta^*),$$

which says that

$$\begin{aligned}\min\left\{\min_{\alpha\le\beta\le1} \eta_1(\beta), \min_{\alpha\le\beta\le1} \eta_2(\beta)\right\} &\le \min\left\{\eta_1(\beta^*), \eta_2(\beta^*)\right\}\\ &= \min_{\alpha\le\beta\le1} \min\left\{\eta_1(\beta), \eta_2(\beta)\right\}.\end{aligned}$$

On the other hand, since

$$\min\left\{\eta_1(\beta), \eta_2(\beta)\right\} \le \eta_1(\beta) \text{ and } \min\left\{\eta_1(\beta), \eta_2(\beta)\right\} \le \eta_2(\beta),$$

we have

$$\min_{\alpha\le\beta\le1} \min\left\{\eta_1(\beta), \eta_2(\beta)\right\} \le \min_{\alpha\le\beta\le1} \eta_1(\beta)$$

and

$$\min_{\alpha\le\beta\le1} \min\left\{\eta_1(\beta), \eta_2(\beta)\right\} \le \min_{\alpha\le\beta\le1} \eta_2(\beta),$$

which says that

$$\min_{\alpha\le\beta\le1} \min\left\{\eta_1(\beta), \eta_2(\beta)\right\} \le \min\left\{\min_{\alpha\le\beta\le1} \eta_1(\beta), \min_{\alpha\le\beta\le1} \eta_2(\beta),\right\}.$$

Therefore, we obtain (5.91). We can similarly obtain

$$\max_{\alpha\le\beta\le1} \max\left\{\eta_1(\beta), \eta_2(\beta)\right\} = \max\left\{\max_{\alpha\le\beta\le1} \eta_1(\beta), \max_{\alpha\le\beta\le1} \eta_2(\beta),\right\}.$$

This completes the proof. ■

Example 5.5.7 Given any real numbers a_1, a_2, b_1, b_2, we define two real-valued functions η_1 and η_2 on [0,1] by

$$\eta_1(\alpha) = \alpha^2 a_1 + \alpha b_1 + c_1 \text{ and } \eta_2(\alpha) = \alpha^2 a_2 + \alpha b_2 + c_2.$$

From Theorem 5.5.6, we can generate a fuzzy interval $\tilde{A}^\star$ in $\mathbb{R}$ with membership function defined by

$$\xi_{\tilde{A}^\star}(x) = \sup_{0<\alpha\le1} \alpha \cdot \chi_{M_\alpha}(x)$$

such that the α-level sets $\tilde{A}^\star_\alpha$ are given in the form of (5.89) for $\alpha \in (0,1]$. Let $\alpha_1^* = -b_1/2a_1$. Then, for any fixed $\alpha \in (0,1]$, we have

$$\min_{\alpha\le\beta\le1} \eta_1(\beta) = \min_{\alpha\le\beta\le1} \eta_1(\beta) = \min_{\alpha\le\beta\le1} \left(\beta^2 a_1 + \beta b_1 + c_1\right) = \begin{cases} \eta_1(\alpha) & \text{if } \alpha > \alpha_1^* \\ \eta_1(\alpha_1^*) & \text{if } \alpha \le \alpha_1^* \le 1 \end{cases}$$

and

$$\max_{\alpha\le\beta\le 1}\eta_1(\beta)=\max_{\alpha\le\beta\le 1}\eta_1(\beta)=\max_{\alpha\le\beta\le 1}\left(\beta^2 a_1+\beta b_1+c_1\right)$$
$$=\begin{cases}\eta_1(1) & \text{if } \alpha>\alpha_1^*\\ \max\{\eta_1(\alpha),\eta_1(1)\} & \text{if } \alpha\le\alpha_1^*\le 1.\end{cases}$$

Let $\alpha_2^*=-b_2/2a_2$. We can similarly obtain

$$\min_{\alpha\le\beta\le 1}\eta_2(\beta)=\begin{cases}\eta_2(\alpha) & \text{if } \alpha>\alpha_2^*\\ \eta_2(\alpha_2^*) & \text{if } \alpha\le\alpha_2^*\le 1\end{cases}$$

and

$$\max_{\alpha\le\beta\le 1}\eta_2(\beta)=\begin{cases}\eta_2(1) & \text{if } \alpha>\alpha_2^*\\ \max\{\eta_2(\alpha),\eta_2(1)\} & \text{if } \alpha\le\alpha_2^*\le 1.\end{cases}$$

Therefore, given any specific values of a_i, b_i, c_i for $i=1,2$, we can obtain the α-level sets $\tilde{A}_\alpha^\star$ given in the form of (5.89). For example, we take $a_1=1$, $b_1=-1$, $c_1=1$, $a_2=5$, $b_1=-6$ and $c_2=1$. Then, we have

$$\eta_1(\alpha)=\alpha^2-\alpha+1 \text{ and } \eta_2(\alpha)=5\alpha^2-6\alpha+1.$$

In this case, we obtain $\alpha_1^*=0.5$ and $\alpha_2^*=0.6$. Given $\alpha=0.3$, we have

$$\min_{\alpha\le\beta\le 1}\eta_1(\beta)=\eta_1(\alpha_1^*)=\eta_1(0.5)=0.75$$
$$\min_{\alpha\le\beta\le 1}\eta_2(\beta)=\eta_2(\alpha_2^*)=\eta_2(0.6)=-1$$
$$\max_{\alpha\le\beta\le 1}\eta_1(\beta)=\max\{\eta_1(\alpha),\eta_1(1)\}=\max\{\eta_1(0.3),\eta_1(1)\}=\max\{0.79,1\}=1$$
$$\max_{\alpha\le\beta\le 1}\eta_2(\beta)=\max\{\eta_2(\alpha),\eta_2(1)\}=\max\{\eta_2(0.3),\eta_2(1)\}=\max\{-0.35,0\}=0.$$

According to (5.89), the 0.3-level set is given by

$$\tilde{A}_{0.3}^\star=[\min\{0.75,-1\},\max\{1,0\}]=[-1,1].$$

5.6 Uniqueness of Construction

Given any two fuzzy sets $\tilde{A}$ and $\tilde{B}$ in $\mathbb{R}^m$, we consider the following families

$$\mathcal{A}=\left\{\tilde{A}_\alpha : \alpha\in(0,1]\right\} \text{ and } \mathcal{B}=\left\{\tilde{B}_\alpha : \alpha\in(0,1]\right\}.$$

We define the concept of identical for $\tilde{A}$ and $\tilde{B}$ as follows.

Definition 5.6.1 Let $\tilde{A}$ and $\tilde{B}$ be two fuzzy sets in $\mathbb{R}^m$ with membership functions $\xi_{\tilde{A}}$ and $\xi_{\tilde{B}}$, respectively.

- We say that $\tilde{A}$ and $\tilde{B}$ are **identical** when $\xi_{\tilde{A}}(x)=\xi_{\tilde{B}}(x)$ for all $x\in\mathbb{R}^m$.
- We say that $\tilde{A}$ and $\tilde{B}$ are **permutably identical** on (0,1] when there exists a bijective function $\kappa:(0,1]\to(0,1]$ satisfying $\tilde{B}_\alpha=\tilde{A}_{\kappa(\alpha)}$ for all $\alpha\in(0,1]$ or $\tilde{A}_\alpha=\tilde{B}_{\kappa(\alpha)}$ for all $\alpha\in(0,1]$. In this case, we also write $\tilde{A}\overset{\kappa}{=}\tilde{B}$.

Example 5.6.2 Let $\tilde{A}$ and $\tilde{B}$ be two fuzzy sets in $\mathbb{R}^m$ satisfying $\tilde{B}_\alpha=\tilde{A}_{\alpha^3}$ for all $\alpha\in(0,1]$. By taking $\kappa:(0,1]\to(0,1]$ as $\kappa(\alpha)=\alpha^3$, we see that $\tilde{A}\overset{\kappa}{=}\tilde{B}$.

Proposition 5.6.3 *Let $\tilde{A}$ and $\tilde{B}$ be two fuzzy sets in $\mathbb{R}^m$. Suppose that $\tilde{A} \stackrel{\kappa}{=} \tilde{B}$ in the sense of $\tilde{B}_\alpha = \tilde{A}_{\kappa(\alpha)}$ for all $\alpha \in (0,1]$ or $\tilde{A}_\alpha = \tilde{B}_{\kappa(\alpha)}$ for all $\alpha \in (0,1]$ for some bijective function κ defined on $(0,1]$. We also assume that*

$$\mathcal{A} = \{\tilde{A}_\alpha : \alpha \in (0,1]\} \text{ or } \mathcal{B} = \{\tilde{B}_\alpha : \alpha \in (0,1]\}$$

is a strictly nested family. Then κ is a strictly increasing function on $(0,1]$.

Proof. It suffices to prove the case of $\tilde{B}_\alpha = \tilde{A}_{\kappa(\alpha)}$ for all $\alpha \in (0,1]$. We separately consider the following situation.

- Suppose that $\mathcal{B}$ is a strictly nested family. For $\beta < \alpha$ with $\alpha, \beta \in (0,1]$, we assume $\kappa(\beta) \geq \kappa(\alpha)$ to lead to a contradiction. In this case, we have

  $$\tilde{B}_\beta = \tilde{A}_{\kappa(\beta)} \subseteq \tilde{A}_{\kappa(\alpha)} = \tilde{B}_\alpha,$$

  which contradicts the strictness $\tilde{B}_\alpha \subset \tilde{B}_\beta$. Therefore, we must have $\kappa(\beta) < \kappa(\alpha)$.
- Suppose that $\mathcal{A}$ is a strictly nested family. Since κ is bijective on $(0,1]$, it means that the inverse function κ^{-1} exists. For $\beta < \alpha$ with $\alpha, \beta \in (0,1]$, we assume $\kappa^{-1}(\beta) \geq \kappa^{-1}(\alpha)$ to lead to a contradiction. In this case, we have

  $$\tilde{A}_\beta = \tilde{B}_{\kappa^{-1}(\beta)} \subseteq \tilde{B}_{\kappa^{-1}(\alpha)} = \tilde{A}_\alpha,$$

  which contradicts the strictness $\tilde{A}_\alpha \subset \tilde{A}_\beta$. Therefore, we must have $\kappa^{-1}(\beta) < \kappa^{-1}(\alpha)$, which says that κ^{-1} is strictly increasing on $(0,1]$. This also shows that κ is strictly increasing on $(0,1]$.

This completes the proof. ■

Let $\tilde{A}$ be a fuzzy set in $\mathbb{R}^m$ with membership function $\xi_{\tilde{A}}$. The inverse function $\xi_{\tilde{A}}^{-1}$ may not exist. However, we write $\xi_{\tilde{A}}^{-1}(\alpha)$ to denote the inverse image of α defined by

$$\xi_{\tilde{A}}^{-1}(\alpha) = \left\{x \in \mathbb{R}^m : \xi_{\tilde{A}}(x) = \alpha\right\}$$

for α in the range $\mathcal{R}(\xi_{\tilde{A}})$ of membership function $\xi_{\tilde{A}}$.

Proposition 5.6.4 *Let $\tilde{A}$ and $\tilde{B}$ be two fuzzy sets in $\mathbb{R}^m$. Suppose that $\tilde{A} \stackrel{\kappa}{=} \tilde{B}$ in the sense of $\tilde{B}_\alpha = \tilde{A}_{\kappa(\alpha)}$ for some bijective function κ defined on $(0,1]$ and for all $\alpha \in (0,1]$. We also assume that*

$$\mathcal{A} = \{\tilde{A}_\alpha : \alpha \in (0,1]\} \text{ or } \mathcal{B} = \{\tilde{B}_\alpha : \alpha \in (0,1]\}$$

is a strictly nested family. Then, we have the following properties.

(i) *Suppose that the function κ is strictly increasing on $(0,1]$. For $\alpha \in (0,1)$ and $x \in \xi_{\tilde{B}}^{-1}(\alpha)$, we have*

$$\xi_{\tilde{B}}^{-1}(\alpha) = \xi_{\tilde{A}}^{-1}(\kappa(\alpha)) \text{ and } \xi_{\tilde{A}}(x) = \kappa(\xi_{\tilde{B}}(x)).$$

(ii) *We have*

$$\xi_{\tilde{B}}^{-1}(1) = \tilde{B}_1 = \tilde{A}_1 = \xi_{\tilde{A}}^{-1}(1) \text{ and } \tilde{A}_{0+} = \tilde{B}_{0+}.$$

Proof. From Proposition 5.6.3, we see that the function $\kappa : (0,1] \to (0,1]$ is bijective and strictly increasing. It follows that $\kappa(1) = 1$. Given any $\alpha \in (0,1)$, we have

$$\begin{aligned}\xi_{\tilde{A}}^{-1}(\alpha) &= \left\{x \in \mathbb{R}^m : \xi_{\tilde{A}}(x) = \alpha\right\} = \left\{x \in \mathbb{R}^m : \xi_{\tilde{A}}(x) \geq \alpha\right\} \setminus \left\{x \in \mathbb{R}^m : \xi_{\tilde{A}}(x) > \alpha\right\} \\ &= \tilde{A}_\alpha \backslash \tilde{A}_{\alpha+} = \tilde{A}_\alpha \Big\backslash \bigcup_{\alpha<\beta\leq 1} \tilde{A}_\beta \text{ (using (2.10)).}\end{aligned} \tag{5.92}$$

Therefore, we also have

$$\begin{aligned}\xi_{\tilde{B}}^{-1}(\alpha) &= \tilde{B}_\alpha \Big\backslash \bigcup_{\alpha<\beta\leq 1} \tilde{B}_\beta \\ &= \left(\tilde{A}_{\kappa(\alpha)} \Big\backslash \bigcup_{\alpha<\beta\leq 1} \tilde{A}_{\kappa(\beta)}\right) = \left(\tilde{A}_{\kappa(\alpha)} \Big\backslash \bigcup_{\kappa(\alpha)<\kappa(\beta)\leq 1} \tilde{A}_{\kappa(\beta)}\right) \\ &\quad \text{(since } \kappa \text{is bijective and strictly increasing)} \\ &= \left(\tilde{A}_{\kappa(\alpha)} \Big\backslash \bigcup_{\kappa(\alpha)<\gamma\leq 1} \tilde{A}_\gamma\right) = \xi_{\tilde{A}}^{-1}(\kappa(\alpha)) \text{ (by (5.92)),}\end{aligned}$$

which implies

$$\xi_{\tilde{A}}(\xi_{\tilde{B}}^{-1}(\alpha)) = \{\kappa(\alpha)\}\,.$$

Therefore, we obtain

$$\xi_{\tilde{A}}(x) = \kappa(\xi_{\tilde{B}}(x)) \text{ for every } x \in \xi_{\tilde{B}}^{-1}(\alpha),$$

which proves part (i).

To prove part (ii), we immediately have

$$\xi_{\tilde{B}}^{-1}(1) = \left\{x \in \mathbb{R}^m : \xi_{\tilde{B}}(x) = 1\right\} = \left\{x \in \mathbb{R}^m : \xi_{\tilde{B}}(x) \geq 1\right\} = \tilde{B}_1 \tag{5.93}$$

and

$$\xi_{\tilde{A}}^{-1}(1) = \left\{x \in \mathbb{R}^m : \xi_{\tilde{A}}(x) = 1\right\} = \left\{x \in \mathbb{R}^m : \xi_{\tilde{A}}(x) \geq 1\right\} = \tilde{A}_1. \tag{5.94}$$

It remains to prove $\tilde{A}_1 = \tilde{B}_1$. From Theorem 2.2.13, we have

$$\xi_{\tilde{A}}(x) = \sup_{0<\alpha\leq 1} \alpha \cdot \chi_{\tilde{A}_\alpha}(x) \text{ and } \xi_{\tilde{B}}(x) = \sup_{0<\alpha\leq 1} \alpha \cdot \chi_{\tilde{B}_\alpha}(x).$$

Since κ is a bijective function on (0,1], the inverse function κ^{-1} is also a bijective function on (0,1]. Therefore, we have

$$\xi_{\tilde{A}}(x) = \sup_{0<\alpha\leq 1} \alpha \cdot \chi_{\tilde{A}_\alpha}(x) = \sup_{0<\alpha\leq 1} \kappa(\alpha) \cdot \chi_{\tilde{A}_{\kappa(\alpha)}}(x) \tag{5.95}$$

and

$$\xi_{\tilde{B}}(x) = \sup_{0<\alpha\leq 1} \alpha \cdot \chi_{\tilde{B}_\alpha}(x) = \sup_{0<\alpha\leq 1} \kappa^{-1}(\alpha) \cdot \chi_{\tilde{B}_{\kappa^{-1}(\alpha)}}(x). \tag{5.96}$$

For $x \in \tilde{A}_1$, i.e. $\xi_{\tilde{A}}(x) = 1$, since $\tilde{A}_1 \subseteq \tilde{A}_\alpha$ for all $\alpha \in (0,1]$, we have $x \in \tilde{A}_\alpha$ for all $\alpha \in (0,1]$. Since $\tilde{A}_\alpha = \tilde{B}_{\kappa^{-1}(\alpha)}$, we also have $x \in \tilde{B}_{\kappa^{-1}(\alpha)}$ for all $\alpha \in (0,1]$. Therefore, from (5.96), we have

$$\xi_{\tilde{B}}(x) = \sup_{0<\alpha\leq 1} \kappa^{-1}(\alpha) = \sup\,(0,1] = 1,$$

which also says that $x \in \tilde{B}_1$, i.e. $\tilde{A}_1 \subseteq \tilde{B}_1$. On the other hand, for $x \in \tilde{B}_1$, i.e. $\xi_{\tilde{B}}(x) = 1$, we have $x \in \tilde{B}_\alpha = \tilde{A}_{\kappa(\alpha)}$ for all $\alpha \in (0,1]$, i.e. $x \in \tilde{A}_{\kappa(\alpha)}$ for all $\alpha \in (0,1]$. Therefore, from (5.95), we have

$$\xi_{\tilde{A}}(x) = \sup_{0<\alpha\leq 1} \kappa(\alpha) = \sup\,(0,1] = 1,$$

which also says that $x \in \tilde{A}_1$, i.e. $\tilde{B}_1 \subseteq \tilde{A}_1$. Therefore, we obtain the equality $\tilde{B}_1 = \tilde{A}_1$. From (2.5), since κ is bijective on $(0,1]$, it follows that

$$\tilde{B}_{0+} = \bigcup_{0<\alpha\leq 1} \tilde{B}_\alpha = \bigcup_{0<\alpha\leq 1} \tilde{A}_{\kappa(\alpha)} = \bigcup_{0<\alpha\leq 1} \tilde{A}_\alpha = \tilde{A}_{0+}.$$

This completes the proof. ∎

By referring to Definition 5.6.1, we also propose the following definition.

Definition 5.6.5 Let η be a function from $(0,1]$ into $(0,1]$, and let $\mathcal{M} = \{M_\alpha : \alpha \in (0,1]\}$ be a family of nonempty subsets of $\mathbb{R}^m$ such that it is a nested family with respect to η. Let $\tilde{A}$ be a fuzzy set in $\mathbb{R}^m$.

- We say that $\mathcal{M}$ and $\tilde{A}$ are **identical** with respect to η when $\tilde{A}_\alpha = M_\alpha^{(\eta)}$ for all $\alpha \in (0,1]$. In this case, we also write $\mathcal{M} \overset{\eta}{=} \tilde{A}$.
- We say that $\mathcal{M}$ and $\tilde{A}$ are **permutably identical** with respect to η when there exists a bijective function $\kappa : (0,1] \to (0,1]$ satisfying $\tilde{A}_{\kappa(\alpha)} = M_\alpha^{(\eta)}$ for all $\alpha \in (0,1]$ or $\tilde{A}_\alpha = M_{\kappa(\alpha)}^{(\eta)}$ for all $\alpha \in (0,1]$. In this case, we also write $\mathcal{M} \overset{(\kappa,\eta)}{=} \tilde{A}$.

Remark 5.6.6 For $\mathcal{M} \overset{(\kappa,\eta)}{=} \tilde{A}$, we notice that the function κ depends on $\tilde{A}$. More precisely, we must write $\mathcal{M} \overset{(\kappa_A,\eta)}{=} \tilde{A}$ and $\mathcal{M} \overset{(\kappa_B,\eta)}{=} \tilde{B}$ for some bijective functions κ_A and κ_B on $(0,1]$, where the function η is fixed.

Proposition 5.6.7 *Let η be a function from $(0,1]$ into $(0,1]$, and let $\mathcal{M} = \{M_\alpha : \alpha \in (0,1]\}$ be a family of nonempty subsets of $\mathbb{R}^m$ such that it is a nested family with respect to η. Let $\tilde{A}$ be a fuzzy set in $\mathbb{R}^m$ satisfying $\mathcal{M} \overset{(\kappa,\eta)}{=} \tilde{A}$. Then, we have the following properties.*

(i) *Suppose that $\mathcal{M}$ is a strictly nested family with respect to η. Then κ is a strictly increasing function.*
(ii) *Suppose that $\mathcal{A} = \{\tilde{A}_\alpha : \alpha \in (0,1]\}$ is a strictly nested family. Then κ is a strictly increasing function.*

Proof. We want to prove the case of $\tilde{A}_{\kappa(\alpha)} = M_\alpha^{(\eta)}$ for all $\alpha \in (0,1]$. To prove part (i), given any $\alpha, \beta \in (0,1]$ with $\beta < \alpha$, we assume $\kappa(\beta) \geq \kappa(\alpha)$ to lead to a contradiction. Now, we have

$$M_\beta^{(\eta)} = \tilde{A}_{\kappa(\beta)} \subseteq \tilde{A}_{\kappa(\alpha)} = M_\alpha^{(\eta)},$$

which contradicts the strictness $M_\alpha^{(\eta)} \subset M_\beta^{(\eta)}$. Therefore, we must have $\kappa(\beta) < \kappa(\alpha)$.

To prove part (ii), since κ is bijective on $(0,1]$, it means that the inverse function κ^{-1} exists. Given any $\alpha, \beta \in (0,1]$ with $\beta < \alpha$, we assume $\kappa^{-1}(\beta) \geq \kappa^{-1}(\alpha)$ to lead to a contradiction. Now, we have

$$\tilde{A}_\beta = M_{\kappa^{-1}(\beta)}^{(\eta)} \subseteq M_{\kappa^{-1}(\alpha)}^{(\eta)} = \tilde{A}_\alpha,$$

which contradicts the strictness $\tilde{A}_\alpha \subset \tilde{A}_\beta$. Therefore, we must have $\kappa^{-1}(\beta) < \kappa^{-1}(\alpha)$, which says that κ^{-1} is strictly increasing on (0,1]. This also shows that κ is strictly increasing on (0,1], and the proof is complete. ■

Proposition 5.6.8 *Let η be a function from* (0,1] *into* (0,1], *and let $\mathcal{M} = \{M_\alpha : \alpha \in (0,1]\}$ be a family of nonempty subsets of $\mathbb{R}^m$ such that it is a nested family with respect to η. Let $\tilde{A}$ be a fuzzy set in $\mathbb{R}^m$ satisfying $\mathcal{M} \overset{(\kappa,\eta)}{=} \tilde{A}$. Then*

$$\xi_{\tilde{A}}(x) \begin{cases} > 0, & \text{if } x \in \bigcup_{0<\alpha\le 1} M_\alpha^{(\eta)} \\ = 0, & \text{otherwise.} \end{cases}$$

We also have

$$\tilde{A}_1 \subseteq \bigcup_{0<\alpha\le 1} M_\alpha^{(\eta)}$$

and

$$\tilde{A}_{0+} = \bigcup_{0<\alpha\le 1} \tilde{A}_\alpha = \bigcup_{\alpha\in\mathcal{R}(\xi_{\tilde{A}})\setminus\{0\}} \tilde{A}_\alpha = \bigcup_{0<\alpha\le 1} M_\alpha^{(\eta)}.$$

Proof. From Proposition 5.6.7, we see that κ is an increasing function. Since

$$\tilde{A}_1 \subseteq \tilde{A}_\alpha = M_{\kappa^{-1}(\alpha)}^{(\eta)} \text{ for all } \alpha \in (0,1],$$

it follows that

$$\tilde{A}_1 \subseteq \bigcup_{0<\alpha\le 1} M_{\kappa^{-1}(\alpha)}^{(\eta)} = \bigcup_{0<\alpha\le 1} M_\alpha^{(\eta)} \text{ (since } \kappa \text{ is bijective).} \tag{5.97}$$

From (2.5), we have

$$\tilde{A}_{0+} = \bigcup_{\alpha\in\mathcal{R}(\xi_{\tilde{A}})\setminus\{0\}} \tilde{A}_\alpha = \bigcup_{0<\alpha\le 1} \tilde{A}_\alpha = \bigcup_{0<\alpha\le 1} M_{\kappa^{-1}(\alpha)}^{(\eta)} = \bigcup_{0<\alpha\le 1} M_\alpha^{(\eta)}.$$

This completes the proof. ■

Let $\mathcal{M} = \{M_\alpha : \alpha \in (0,1]\}$ be a family of nonempty subsets of $\mathbb{R}^m$. We consider two families of fuzzy sets in $\mathbb{R}^m$ given by

$$\mathcal{F}^\circ(\mathcal{M}) = \left\{\tilde{A} \in \mathcal{F}(\mathbb{R}^m) : \mathcal{R}(\xi_{\tilde{A}})\setminus\{0\} = (0,1] \text{ and } \tilde{A} \overset{(\kappa,\eta)}{=} \mathcal{M}\right\}$$

and

$$\hat{\mathcal{F}}^\circ(\mathcal{M}) = \left\{\tilde{A} \in \mathcal{F}(\mathbb{R}^m) : \tilde{A} \overset{\eta}{=} \mathcal{M}\right\}.$$

We notice that $\mathcal{R}(\xi_{\tilde{A}})\setminus\{0\} = (0,1]$ means that $\mathcal{R}(\xi_{\tilde{A}}) = (0,1]$ or $\mathcal{R}(\xi_{\tilde{A}}) = [0,1]$. We adopt the notation $|X|$ to denote the *cardinality* of the family X (i.e. the number of elements in X). We are going to show that $|\hat{\mathcal{F}}^\circ(\mathcal{M})| = 1$. The following results will help to understand those families.

Proposition 5.6.9 *Let $\eta : (0,1] \to (0,1]$ be a bijective function, and let $\mathcal{M} = \{M_\alpha : \alpha \in (0,1]\}$ be a family of nonempty subsets of $\mathbb{R}^m$ such that it is a nested family with respect to η. Then, we have the following properties.*

(i) *Given any* $\tilde{A}, \tilde{B} \in \mathcal{F}^{\circ}(\mathcal{M})$, *we have* $\tilde{A} \stackrel{\kappa}{=} \tilde{B}$ *in the sense of* $\tilde{B}_{\kappa(\alpha)} = \tilde{A}_{\alpha}$ *for* $\alpha \in (0,1]$ *and for some bijective function* κ *defined on* $(0,1]$.
(ii) *We have* $|\hat{\mathcal{F}}^{\circ}(\mathcal{M})| = 1$.

Proof. To prove part (i), according to Remark 5.6.6, we have $\tilde{A} \stackrel{(\kappa_A,\eta)}{=} \mathcal{M}$ and $\tilde{B} \stackrel{(\kappa_B,\eta)}{=} \mathcal{M}$ for some bijective functions κ_A and κ_B defined on $(0,1]$ satisfying

$$\tilde{A}_{\kappa_A(\alpha)} = M_{\alpha}^{(\eta)} = M_{\eta^{-1}(\alpha)} \text{ and } \tilde{B}_{\kappa_B(\alpha)} = M_{\alpha}^{(\eta)} = M_{\eta^{-1}(\alpha)} \text{ for } \alpha \in (0,1],$$

which imply

$$\tilde{A}_{\alpha} = M_{\eta^{-1}(\kappa_A^{-1}(\alpha))} \text{ and } \tilde{B}_{\alpha} = M_{\eta^{-1}(\kappa_B^{-1}(\alpha))}. \tag{5.98}$$

We take

$$\beta = \kappa_B(\kappa_A^{-1}(\alpha)) = (\kappa_B \circ \kappa_A^{-1})(\alpha) \equiv \kappa(\alpha). \tag{5.99}$$

For any $\alpha \in (0,1]$, we have

$$\begin{aligned}\tilde{B}_{\kappa(\alpha)} &= \tilde{B}_{\beta} = M_{\eta^{-1}(\kappa_B^{-1}(\beta))} \text{ (using (5.98))} \\ &= M_{\eta^{-1}(\kappa_A^{-1}(\alpha))} \text{ (using (5.99))} \\ &= \tilde{A}_{\alpha} \text{ (using (5.99))},\end{aligned}$$

which says that $\tilde{A} \stackrel{\kappa}{=} \tilde{B}$ by referring to Definition 5.6.1.

To prove part (ii), for $\tilde{A}, \tilde{B} \in \hat{\mathcal{F}}^{\circ}(\mathcal{M})$, the definition says that

$$\tilde{A}_{\alpha} = M_{\alpha}^{(\eta)} = \tilde{B}_{\alpha} \text{ for all } \alpha \in (0,1].$$

Part (iv) of Proposition 2.2.15 says that $\xi_{\tilde{A}}(x) = \xi_{\tilde{B}}(x)$ for all $x \in \mathbb{R}^m$. Therefore, $\tilde{A}$ and $\tilde{B}$ are identical, i.e. $|\hat{\mathcal{F}}^{\circ}(\mathcal{M})| = 1$. This completes the proof. ■

The following results will be useful for discussing the uniqueness.

Proposition 5.6.10 *Let* $\tilde{A}$ *and* $\tilde{B}$ *be two fuzzy sets in* $\mathbb{R}^m$ *such that* $\tilde{A} \stackrel{\kappa}{=} \tilde{B}$ *in the sense of* $\tilde{B}_{\alpha} = \tilde{A}_{\kappa(\alpha)}$ *for some bijective function* κ *defined on* $(0,1]$ *and for all* $\alpha \in (0,1]$. *Suppose that*

$$\mathcal{A} = \{\tilde{A}_{\alpha} : \alpha \in (0,1]\} \text{ or } \mathcal{B} = \{\tilde{B}_{\alpha} : \alpha \in (0,1]\}$$

is a strictly nested family. Then, the following statements hold true.

- κ *is a strictly increasing function on* $(0,1]$.
- *We have* $\xi_{\tilde{A}}(x) = \kappa(\xi_{\tilde{B}}(x))$ *for* $x \in \tilde{A}_{0+}$ *and* $\xi_{\tilde{A}}(x) = \xi_{\tilde{B}}(x) = 0$ *for* $x \notin \tilde{A}_{0+}$.
- *We have* $\tilde{A}_{0+} = \tilde{B}_{0+}$ *and* $\tilde{A}_1 = \tilde{B}_1$.

Proof. From Proposition 5.6.3, we see that κ is strictly increasing on $(0,1]$. Part (ii) of Proposition 5.6.4 says that $\tilde{A}_{0+} = \tilde{B}_{0+}$ and $\tilde{A}_1 = \tilde{B}_1$. If $x \notin \tilde{A}_{0+}$, i.e. $\xi_{\tilde{A}}(x) = 0$, we have $x \notin \tilde{B}_{0+}$, since $\tilde{A}_{0+} = \tilde{B}_{0+}$, which also says that $\xi_{\tilde{B}}(x) = 0$. Now, we have

$$\tilde{A}_{0+} = \tilde{B}_{0+} = \left\{x \in \mathbb{R}^m : \xi_{\tilde{B}}(x) > 0\right\} = \bigcup_{0<\alpha\leq 1} \xi_{\tilde{B}}^{-1}(\alpha) \tag{5.100}$$

$$= \xi_{\tilde{B}}^{-1}(1) \bigcup \left(\bigcup_{0<\alpha<1} \xi_{\tilde{B}}^{-1}(\alpha) \right). \tag{5.101}$$

Suppose that $x \in \tilde{A}_{0+}$. From (5.100), we see that $x \in \xi_{\tilde{B}}^{-1}(\alpha)$ for some $\alpha \in (0,1]$. We consider the following cases.

- Suppose that $x \notin \xi_{\tilde{B}}^{-1}(1)$, i.e. $\alpha = \xi_{\tilde{B}}(x) < 1$. From from part (i) of Proposition 5.6.4, we see that $\xi_{\tilde{A}}(x) = \kappa(\xi_{\tilde{B}}(x))$.
- Suppose that $x \in \xi_{\tilde{B}}^{-1}(1)$, i.e. $\alpha = \xi_{\tilde{B}}(x) = 1$, which also says that $x \in \tilde{B}_1$. Since $\tilde{A}_1 = \tilde{B}_1$, we see that $x \in \tilde{A}_1$, i.e. $\xi_{\tilde{A}}(x) \geq 1$. Therefore, we must have $\xi_{\tilde{A}}(x) = 1$, which also says that

$$\xi_{\tilde{A}}(x) = 1 = \kappa(1) = \kappa(\xi_{\tilde{B}}(x)).$$

Therefore, we conclude that $\xi_{\tilde{A}}(x) = \kappa(\xi_{\tilde{B}}(x))$ for all $x \in \tilde{A}_{0+}$. This completes the proof. ■

Definition 5.6.11 Let $\kappa : (0,1] \to (0,1]$ be a function from $(0,1]$ into itself. Given a fixed fuzzy set $\tilde{A}^\star$ in $\mathbb{R}^m$, we define $\xi_{\tilde{A}} : \mathbb{R}^m \to [0,1]$ by

$$\xi_{\tilde{A}}(x) = \begin{cases} \kappa(\xi_{\tilde{A}^\star}(x)), & \text{if } x \in \tilde{A}_{0+}^\star \\ 0, & \text{if } x \notin \tilde{A}_{0+}^\star. \end{cases} \tag{5.102}$$

In this case, we write $\xi_{\tilde{A}} \doteqdot \kappa\circ\xi_{\tilde{A}^\star}$.

Remark 5.6.12 Suppose that $\kappa_1 \neq \kappa_2$ and $\mathcal{R}(\xi_{\tilde{A}^\star})\setminus\{0\} = (0,1]$. Then, there exists $y_0 \in (0,1]$ satisfying $\kappa_1(y_0) \neq \kappa_2(y_0)$. We are going to claim that the membership functions $\xi_{\tilde{A}^{(1)}} \doteqdot \kappa_1\circ\xi_{\tilde{A}^\star}$ and $\xi_{\tilde{A}^{(2)}} \doteqdot \kappa_2\circ\xi_{\tilde{A}^\star}$ are not identical. Since $y_0 \in \mathcal{R}(\xi_{\tilde{A}^\star})\setminus\{0\}$, there exists $x_0 \in \mathbb{R}^m$ satisfying $\xi_{\tilde{A}^\star}(x_0) = y_0$, which also says that

$$\xi_{\tilde{A}^{(1)}}(x_0) = \kappa_1(\xi_{\tilde{A}^\star}(x_0)) = \kappa_1(y_0) \neq \kappa_2(y_0) = \kappa_2(\xi_{\tilde{A}^\star}(x_0)) = \xi_{\tilde{A}^{(2)}}(x_0).$$

This means $\tilde{A}^{(1)} \neq \tilde{A}^{(2)}$.

The following proposition will show that Definition 5.6.11 is well defined, and will also present some useful results that will be adopted in the following sections.

Proposition 5.6.13 *Let $\kappa : (0,1] \to (0,1]$ be a function from $(0,1]$ into itself. Given a fixed fuzzy set $\tilde{A}^*$ in $\mathbb{R}^m$, if $\xi_{\tilde{A}} \doteqdot \kappa\circ\xi_{\tilde{A}^*}$, we have $\tilde{A}_{0+} = \tilde{A}_{0+}^*$.*

Proof. If $x \notin \tilde{A}_{0+}^*$, we have $\xi_{\tilde{A}}(x) = 0$ by definition, i.e. $x \notin \tilde{A}_{0+}$, which says that $\tilde{A}_{0+} \subseteq \tilde{A}_{0+}^*$. If $x \in \tilde{A}_{0+}^*$, we have $\xi_{\tilde{A}^*}(x) > 0$. Since $\kappa(\alpha) > 0$ for any $\alpha \in (0,1]$, we obtain $\xi_{\tilde{A}}(x) = \kappa(\xi_{\tilde{A}^*}(x)) > 0$, which says that $x \in \tilde{A}_{0+}$, i.e. $\tilde{A}_{0+}^* \subseteq \tilde{A}_{0+}$. This shows the equality $\tilde{A}_{0+} = \tilde{A}_{0+}^*$, and the proof is complete. ■

We define a set $\mathcal{K}$ by

$$\mathcal{K} = \left\{\kappa : \kappa \text{ is a bijective and strictly increasing function on } (0,1]\right\}.$$

Theorem 5.6.14 *Let $\eta : (0,1] \to (0,1]$ be a function from $(0,1]$ into itself, and let $\mathcal{M} = \{M_\alpha : \alpha \in (0,1]\}$ be a family of nonempty subsets of $\mathbb{R}^m$ such that it is a strictly nested family with respect to η. Given a fixed $\tilde{A}^\star \in \mathcal{F}^\circ(\mathcal{M})$, we have*

$$\mathcal{F}^\circ(\mathcal{M}) = \left\{\tilde{A} \in \mathcal{F}(\mathbb{R}^m) : \xi_{\tilde{A}} \doteqdot \kappa\circ\xi_{\tilde{A}^\star} \text{ for } \kappa \in \mathcal{K}\right\} \tag{5.103}$$

and

$$\left|\mathcal{F}^\circ(\mathcal{M})\right| = |\mathcal{K}|. \tag{5.104}$$

Moreover, for any $\tilde{A} \in \mathcal{F}^\circ(\mathcal{M})$, we have $\tilde{A} \stackrel{\kappa}{=} \tilde{A}^\star$ in the sense of $\tilde{A}_\alpha^\star = \tilde{A}_{\kappa(\alpha)}$ for $\kappa \in \mathcal{K}$.

Proof. To prove part (i), since $\tilde{A}^\star \in \mathcal{F}^\circ(\mathcal{M})$, we have $\mathcal{R}(\xi_{\tilde{A}^\star})\backslash\{0\} = (0,1]$ and $\tilde{A}^\star \overset{(\kappa,\eta)}{=} \mathcal{M}$ satisfying

$$\tilde{A}^\star_{\kappa_*(\alpha)} = M^{(\eta)}_\alpha = M_{\eta(\alpha)},\text{ i.e. } \tilde{A}^\star_\alpha = M_{\eta(\kappa_*^{-1}(\alpha))} \tag{5.105}$$

for some bijective function κ_* on (0,1] and for all $\alpha \in (0,1]$. Given any $\tilde{A} \in \mathcal{F}^\circ(\mathcal{M})$, we also have $\mathcal{R}(\xi_{\tilde{A}})\backslash\{0\} = (0,1]$ and $\tilde{A} \overset{\kappa}{=} \mathcal{M}$ satisfying

$$\tilde{A}_{\kappa_A(\alpha)} = M^{(\eta)}_\alpha = M_{\eta(\alpha)} \tag{5.106}$$

for some bijective function κ_A on (0,1] and for all $\alpha \in (0,1]$. We take

$$\beta = \kappa_A(\kappa_*^{-1}(\alpha)) = (\kappa_A \circ \kappa_*^{-1})(\alpha) \equiv \kappa(\alpha). \tag{5.107}$$

Then κ is a bijective function on (0,1]. Part (i) of Proposition 5.6.7 says that κ_* and κ_A are strictly increasing on (0,1], which also says that the function κ is strictly increasing on (0,1]. This shows that $\kappa \in \mathcal{K}$. Now, we also have

$$\begin{aligned}\tilde{A}_{\kappa(\alpha)} &= \tilde{A}_\beta = \tilde{A}_{\kappa_A(\kappa_A^{-1}(\beta))}\\ &= M_{\eta(\kappa_A^{-1}(\beta))} \text{ (using (5.106))}\\ &= M_{\eta(\kappa_*^{-1}(\alpha))} \text{ (using (5.107))}\\ &= \tilde{A}^\star_\alpha \text{ (using (5.105))}\end{aligned} \tag{5.108}$$

for all $\alpha \in (0,1]$. From Proposition 5.6.10, we have $\xi_{\tilde{A}}(x) = \kappa(\xi_{\tilde{A}^\star}(x))$ for all $x \in \tilde{A}_{0+} = \tilde{A}^\star_{0+}$, and $\xi_{\tilde{A}}(x) = \xi_{\tilde{A}^\star}(x) = 0$ for $x \notin \tilde{A}_{0+} = \tilde{A}^\star_{0+}$. This shows $\xi_{\tilde{A}} \doteqdot \kappa \circ \xi_{\tilde{A}^\star}$, which implies the following inclusion

$$\mathcal{F}^\circ(\mathcal{M}) \subseteq \left\{\tilde{A} \in \mathcal{F}(\mathbb{R}^m) : \xi_{\tilde{A}} \doteqdot \kappa \circ \xi_{\tilde{A}^\star} \text{ for } \kappa \in \mathcal{K}\right\}.$$

On the other hand, given any $\kappa \in \mathcal{K}$ with $\xi_{\tilde{A}} \doteqdot \kappa \circ \xi_{\tilde{A}^\star}$, we want to show $\tilde{A} \in \mathcal{F}^\circ(\mathcal{M})$. We first claim $\mathcal{R}(\xi_{\tilde{A}})\backslash\{0\} = (0,1]$. We first see that $\mathcal{R}(\xi_{\tilde{A}^\star})\backslash\{0\} = (0,1]$ by definition. Proposition 5.6.13 says that we have $\tilde{A}_{0+} = \tilde{A}^\star_{0+}$. Given any $\alpha \in (0,1]$, since κ is bijective on (0,1], there exists $\hat{\alpha} \in (0,1]$ satisfying $\kappa(\hat{\alpha}) = \alpha$. Since $\mathcal{R}(\xi_{\tilde{A}^\star})\backslash\{0\} = (0,1]$, there also exists $x \in \tilde{A}^\star_{0+}$ satisfying $\xi_{\tilde{A}^\star}(x) = \hat{\alpha}$. Since $\xi_{\tilde{A}} \doteqdot \kappa \circ \xi_{\tilde{A}^\star}$, we obtain

$$\xi_{\tilde{A}}(x) = \kappa\left(\xi_{\tilde{A}^\star}(x)\right) = \kappa(\hat{\alpha}) = \alpha > 0,$$

which shows the inclusion $(0,1] \subseteq \mathcal{R}(\xi_{\tilde{A}})\backslash\{0\}$ Therefore, we indeed obtain $\mathcal{R}(\xi_{\tilde{A}})\backslash\{0\} = (0,1]$. It remains to show that $\tilde{A} \overset{(\kappa,\eta)}{=} \mathcal{M}$. Now, given any $\alpha \in (0,1]$, we have

$$\begin{aligned}\tilde{A}_\alpha &= \left\{x \in \mathbb{R}^m : \xi_{\tilde{A}}(x) \geq \alpha\right\} = \left\{x \in \tilde{A}_{0+} : \xi_{\tilde{A}}(x) \geq \alpha\right\}\\ &= \left\{x \in \tilde{A}^\star_{0+} : \xi_{\tilde{A}}(x) \geq \alpha\right\} \text{ (since } \tilde{A}_{0+} = \tilde{A}^\star_{0+})\\ &= \left\{x \in \tilde{A}^\star_{0+} : \kappa\left(\xi_{\tilde{A}^\star}(x)\right) \geq \alpha\right\} \text{ (since } \xi_{\tilde{A}} \doteqdot \kappa \circ \xi_{\tilde{A}^\star})\\ &= \left\{x \in \tilde{A}^\star_{0+} : \xi_{\tilde{A}^\star}(x) \geq \kappa^{-1}(\alpha)\right\} \text{ (since } \kappa \text{is increasing and bijective)}\\ &= \left\{x \in \mathbb{R}^m : \xi_{\tilde{A}^\star}(x) \geq \kappa^{-1}(\alpha)\right\}\\ &= \tilde{A}^\star_{\kappa^{-1}(\alpha)}.\end{aligned}$$

Therefore, we obtain $\tilde{A} \overset{\kappa}{=} \tilde{A}^\star$ in the sense of $\tilde{A}^\star_\alpha = \tilde{A}_{\kappa(\alpha)}$ for $\alpha \in (0,1]$. Since $\tilde{A}^\star \overset{(\kappa,\eta)}{=} \mathcal{M}$ and $\tilde{A}^\star_\alpha = \tilde{A}_{\kappa(\alpha)}$ for $\alpha \in (0,1]$, using the arguments for obtaining (5.108), we can similarly obtain

$\tilde{A} \stackrel{(\kappa,\eta)}{=} \mathcal{M}$. Therefore, we conclude that $\tilde{A} \in \mathcal{F}^{\circ}(\mathcal{M})$, which proves the equality (5.103). The equality (5.104) can be realized from Remark 5.6.12. This completes the proof. ■

Theorem 5.6.15 (**Uniqueness**) *Let $\eta : (0,1] \to (0,1]$ be a function from $(0,1]$ into itself, and let $\mathcal{M} = \{M_\alpha : \alpha \in (0,1]\}$ be a family of nonempty subsets of $\mathbb{R}^m$ such that it is a strictly nested family with respect to η. Suppose that $|\mathcal{K}| = 1$. Then $\mathcal{K} = \{i_S\}$ consists of only one identity function i_S on $(0,1]$, and there exists one and only one fuzzy set $\tilde{A}^{\star}$ in $\mathbb{R}^m$ satisfying $\tilde{A}^{\star} \stackrel{\eta}{=} \mathcal{M}$ in the sense of*

$$\tilde{A}^{\star}_{\alpha} = \tilde{A}^{\star}_{i_S(\alpha)} = M^{(\eta)}_{\alpha} = M_{\eta(\alpha)} \text{ for all } \alpha \in (0,1]. \tag{5.109}$$

In other words, we have $\mathcal{F}^{\circ}(\mathcal{M}) = \{\tilde{A}^{\star}\}$. We also have

$$\tilde{A}^{\star}_{0+} = \bigcup_{0<\alpha\leq 1} \tilde{A}^{\star}_{\alpha} = \bigcup_{0<\alpha\leq 1} M_{\eta(\alpha)}. \tag{5.110}$$

Proof. The cardinality $|\mathcal{K}| = 1$ says that $|\mathcal{F}^{\circ}(\mathcal{M})| = 1$ by referring to (5.104). Therefore, from (5.103), we must have

$$\xi_{\tilde{A}} \doteqdot \kappa \circ \xi_{\tilde{A}^{\star}} = \xi_{\tilde{A}^{\star}},$$

which also says that κ must be an identity function on $(0,1]$, i.e. $\kappa = i_S$. In this case, we obtain $\mathcal{F}^{\circ}(\mathcal{M}) = \{\tilde{A}^{\star}\}$ satisfying $\tilde{A}^{\star} \stackrel{\eta}{=} \mathcal{M}$ in the sense of

$$\tilde{A}^{\star}_{\alpha} = \tilde{A}^{\star}_{i_S(\alpha)} = \tilde{A}^{\star}_{\kappa(\alpha)} = M^{(\eta)}_{\alpha}$$

for all $\alpha \in (0,1]$. The equalities in (5.110) follow immediately from (2.5), and the proof is complete. ■

The sufficient conditions to guarantee $|\mathcal{K}| = 1$ are based on concepts from the field of pure mathematics. Therefore, it is left for the readers to establish the sufficient conditions to guarantee $|\mathcal{K}| = 1$. Next, we are going to present the existence and uniqueness.

Theorem 5.6.16 *Let κ be an increasing and bijective function on $(0,1]$. Let $\eta : (0,1] \to (0,1]$ be a function from $(0,1]$ into itself, and let $\mathcal{M} = \{M_\alpha : \alpha \in (0,1]\}$ be a family of nonempty subsets of $\mathbb{R}^m$ such that it is a solid and strictly nested family with respect to η satisfying*

$$\bigcup_{0<\alpha\leq 1} M^{(\eta)}_{\alpha} = \mathbb{R}^m.$$

The existence and uniqueness are presented below.

- (**Existence**) *Suppose that the following equality*

$$\bigcup_{0<\alpha\leq 1} M^{(\eta)}_{\alpha} = M^{(\eta)}_{1} \bigcup \left(\bigcup_{0<\alpha<1} D^{(\eta)}_{\alpha} \right)$$

holds true, or the following two conditions are satisfied:

$$\bigcap_{n=1}^{\infty} M^{(\eta)}_{\alpha_n} \subseteq M^{(\eta)}_{\alpha} \text{ when } \alpha_n \uparrow \alpha \text{ for } \alpha < 1$$

and

$$\bigcap_{n=1}^{\infty} M_{\alpha_n}^{(\eta)} \subseteq M_1^{(\eta)} \text{ when } \alpha_n \uparrow 1 \text{ with } \alpha_n \in (0,1) \text{ for all } n.$$

Then, there exists a fuzzy set $\tilde{A}^{\star}$ *in* $\mathbb{R}^m$ *with membership function given by*

$$\xi_{\tilde{A}^{\star}}(x) = \begin{cases} \kappa(\alpha), & \text{if } x \in D_{\alpha}^{(\eta)} \text{ for } \alpha \in (0,1] \\ 1, & \text{if } x \in M_1^{(\eta)}. \end{cases} \tag{5.111}$$

In other words, we have $\tilde{A}^{\star} \in \mathcal{F}^{\circ}(\mathcal{M})$*. Moreover, we also have the following equalities.*

$$\mathcal{F}^{\circ}(\mathcal{M}) = \left\{\tilde{A} \in \mathcal{F}(\mathbb{R}^m) : \xi_{\tilde{A}} \doteqdot \kappa \circ \xi_{\tilde{A}^{\star}} \text{ for } \kappa \in \mathcal{K}\right\} \text{ and } \left|\mathcal{F}^{\circ}(\mathcal{M})\right| = |\mathcal{K}|.$$

- **(Uniqueness)** *Suppose that* $|\mathcal{K}| = 1$*. Then, we have* $\mathcal{F}^{\circ}(\mathcal{M}) = \{\tilde{A}^{\star}\}$*, where the membership function of* $\tilde{A}^{\star}$ *is given in* (5.111) *such that the equalities of* (5.109) *and* (5.110) *are satisfied.*

Proof. To prove the existence , the membership function $\xi_{\tilde{A}^{\star}}$ of $\tilde{A}^{\star}$ can be immediately realized from Theorems 5.3.1 and 5.3.3. We also have $\mathcal{R}(\xi_{\tilde{A}^{\star}})\backslash\{0\} = (0,1]$ and $\tilde{A}^{\star}_{\kappa(\alpha)} = M_{\alpha}^{(\eta)}$ for each $\alpha \in (0,1]$. This shows that $\tilde{A}^{\star} \in \mathcal{F}^{\circ}(\mathcal{M})$. Therefore, the desired equalities and the uniqueness can be realized from Theorems 5.6.14 and 5.6.15, respectively. This completes the proof. ■

Theorem 5.6.17 *Let* κ *be an increasing and bijective function on* (0,1]*. Let* $\eta : (0,1] \to (0,1]$ *be a function from* (0,1] *into itself, and let* $\mathcal{M} = \{M_{\alpha} : \alpha \in (0,1]\}$ *be a family of nonempty subsets of* $\mathbb{R}^m$ *such that it is a solid and strictly nested family with respect to* η *satisfying*

$$\bigcup_{0<\alpha\leq 1} M_{\alpha}^{(\eta)} \neq \mathbb{R}^m.$$

- **(Existence)** *Suppose that the following equality*

$$\bigcup_{0<\alpha\leq 1} M_{\alpha}^{(\eta)} = M_1^{(\eta)} \bigcup \left(\bigcup_{0<\alpha<1} D_{\alpha}^{(\eta)}\right)$$

holds true, or the following two conditions are satisfied:

$$\bigcap_{n=1}^{\infty} M_{\alpha_n}^{(\eta)} \subseteq M_{\alpha}^{(\eta)} \text{ when } \alpha_n \uparrow \alpha \text{ for } \alpha < 1$$

and

$$\bigcap_{n=1}^{\infty} M_{\alpha_n}^{(\eta)} \subseteq M_1^{(\eta)} \text{ when } \alpha_n \uparrow 1 \text{ with } \alpha_n \in (0,1) \text{ for all } n.$$

Then, there exists a fuzzy set $\tilde{A}^{\star}$ *in* $\mathbb{R}^m$ *with membership function given by*

$$\xi_{\tilde{A}^{\star}}(x) = \begin{cases} \kappa(\alpha), & \text{if } x \in D_{\alpha}^{(\eta)} \text{ for } \alpha \in (0,1] \\ 1, & \text{if } x \in M_1^{(\eta)} \\ 0, & \text{if } x \notin \bigcup_{0<\alpha\leq 1} M_{\alpha}^{(\eta)}. \end{cases} \tag{5.112}$$

In other words, we have $\tilde{A}^\star \in \mathcal{F}^\circ(\mathcal{M})$. *Moreover, we also have the following equalities.*

$$\mathcal{F}^\circ(\mathcal{M}) = \left\{\tilde{A} \in \mathcal{F}(\mathbb{R}^m) : \xi_{\tilde{A}} \doteqdot \kappa \circ \xi_{\tilde{A}^\star} \text{ for } \kappa \in \mathcal{K}\right\} \text{ and } \left|\mathcal{F}^\circ(\mathcal{M})\right| = |\mathcal{K}|.$$

- **(Uniqueness)** *Suppose that* $|\mathcal{K}| = 1$. *Then, we have* $\mathcal{F}^\circ(\mathcal{M}) = \{\tilde{A}^\star\}$, *where the membership function of* $\tilde{A}^\star$ *is given in* (5.112) *such that the equalities of* (5.109) *and* (5.110) *are satisfied.*

Proof. To prove the existence, the membership function $\xi_{\tilde{A}^\star}$ of $\tilde{A}^\star$ can be realized from Theorems 5.3.1 and 5.3.3. We also have $\mathcal{R}(\xi_{\tilde{A}^\star})\setminus\{0\} = (0,1]$ and $\tilde{A}^\star_{\kappa(\alpha)} = M^{(\eta)}_\alpha$ for $\alpha \in (0,1]$. This shows that $\tilde{A}^\star \in \mathcal{F}^\circ(\mathcal{M})$. Therefore, the desired equalities and the uniqueness can be realized from Theorems 5.6.14 and 5.6.15, respectively. This completes the proof. ■

Theorem 5.6.18 *Let* λ *be an increasing and bijective function on* $[0,1]$. *Let* $\eta : (0,1] \to (0,1]$ *be a function from* $(0,1]$ *into itself, and let* $\mathcal{M} = \{M_\alpha : \alpha \in (0,1]\}$ *be a family of nonempty subsets of* $\mathbb{R}^m$ *such that it is a solid and strictly nested family with respect to* η.

- **(Existence)** *There exists a fuzzy set* $\tilde{A}^\star$ *in* $\mathbb{R}^m$ *with membership function given by*

$$\xi_{\tilde{A}^\star}(x) = \sup_{0<\alpha\leq 1} \lambda(\alpha) \cdot \chi_{M^{(\eta)}_\alpha}(x) \tag{5.113}$$

satisfying $\tilde{A}^\star_{\lambda(\alpha)} = M^{(\eta)}_\alpha$ *for* $\alpha \in (0,1]$. *We also have*

$$\mathcal{F}^\circ(\mathcal{M}) = \left\{\tilde{A} \in \mathcal{F}(\mathbb{R}^m) : \xi_{\tilde{A}} \doteq \lambda \circ \xi_{\tilde{A}^\star} \text{ for } \lambda \in \mathcal{K}\right\} \text{ and } \left|\mathcal{F}^\circ(\mathcal{M})\right| = |\mathcal{K}|.$$

- **(Uniqueness)** *Suppose that* $|\mathcal{K}| = 1$. *Then, we have* $\mathcal{F}^\star(\mathcal{M}) = \{\tilde{A}^\star\}$, *where the membership function of* $\tilde{A}^\star$ *is given in* (5.113) *such that the equalities of* (5.109) *and* (5.110) *are satisfied.*

Proof. We are going to use Theorem 5.4.30 to prove the existence. Since λ is an increasing and bijective function on $[0,1]$, it follows that

$$\mathcal{R}(\lambda) = [0,1] \text{ and } S^\bullet = \{\alpha \in [0,1] : \lambda(\alpha) > 0\} = (0,1].$$

Let us recall the notations $B^{(\lambda)}_\alpha$ and $B^{(\lambda)}_{\alpha+}$ in (5.62) and (5.63), respectively, and the notations $A^{(\eta)}_\alpha$ and $A^{(\eta)}_{\alpha+}$ in (5.4). Since λ is an increasing and bijective function on $[0,1]$, for $\alpha \in (0,1)$, it follows that

$$B^{(\eta,\lambda)}_\alpha = \bigcup_{\{\beta\in S^\bullet : \lambda(\beta)\geq\lambda(\alpha)\}} M^{(\eta)}_\beta = \bigcup_{\{\beta\in(0,1] : \beta\geq\alpha\}} M^{(\eta)}_\beta = A^{(\eta)}_\alpha$$

and

$$B^{(\eta,\lambda)}_{\alpha+} = \bigcup_{\{\beta\in S^\bullet : \lambda(\beta)>\lambda(\alpha)\}} M^{(\eta)}_\beta = \bigcup_{\{\beta\in(0,1] : \beta>\alpha\}} M^{(\eta)}_\beta = A^{(\eta)}_{\alpha+}.$$

We also see that $\alpha \in [0,1) \cap S^\bullet$ is equivalent to say $\alpha \in (0,1)$. Since $\mathcal{M}$ is a solid and strictly nested family with respect to η, we have

- $M^{(\eta)}_\beta \subseteq M^{(\eta)}_\alpha$ for $\alpha, \beta \in S^\bullet = (0,1]$ with $\lambda(\alpha) < \lambda(\beta)$;
- $\bigcap_{n=1}^\infty M^{(\eta)}_{\alpha_n} \subseteq M^{(\eta)}_\alpha$ when $\lambda(\alpha_n) \uparrow \lambda(\alpha)$ with $\alpha_n, \alpha \in S^\bullet = (0,1]$ for all n;
- $B^{(\eta,\lambda)}_\alpha \setminus B^{(\eta,\lambda)}_{\alpha+} \neq \emptyset$ for all $\alpha \in [0,1) \cap S^\bullet$.

Therefore, we can apply Theorem 5.4.30 to guarantee the existence of fuzzy set $\tilde{A}^\star$ satisfying

$$\mathcal{R}(\xi_{\tilde{A}^\star})\setminus\{0\} = (0,1]$$

and

$$\tilde{A}^\star_{\lambda(\alpha)} = M_\alpha^{(\eta)} \text{ for all } \alpha \in (0,1].$$

This shows that $\tilde{A}^\star \in \mathcal{F}^\circ(\mathcal{M})$. Therefore, the desired equalities and the uniqueness can be realized from Theorems 5.6.14 and 5.6.15, respectively. This completes the proof. ■

Theorems 5.6.16–5.6.18 present the methodology to generate a fuzzy set from a family

$$\mathcal{M} = \{M_\alpha : \alpha \in (0,1]\}$$

of nonempty subsets of $\mathbb{R}^m$ such that it satisfies some intuitive conditions. As we mentioned above, when the fuzziness is considered in economics and engineering problems, we can form a family $\mathcal{M}$ from the observed data. In this case, we can generate the fuzzy data based on the observed data using the mechanical procedure as shown in Theorems 5.6.16–5.6.18, which can avoid the biased subjectivity caused by the decision makers for setting up the fuzzy data based on the real-valued data that are observed from the fuzzy environment.

6

Fuzzification of Crisp Functions

Using fuzzy set theory to study engineering and economics problems has received much attention for a long time. The main reason is that we may have problems collecting the actual data from the environment because of the uncertainty. When the uncertainty can be regarded as fuzziness, we can consider fuzzy data in engineering and economics problems. In this case, we may need to fuzzify real-valued functions into fuzzy functions.

For example, the cost functions and benefit functions should be considered as fuzzy functions when fuzzy data are involved in the formulated problems. In this chapter, we are going to use the extension principle and the form of mathematical expression in the decomposition theorem to fuzzify crisp functions into fuzzy functions. The fuzzification of real-valued functions and vector-valued functions will be studied separately.

There are two ways to fuzzify a vector-valued function $\mathbf{f} : \mathbb{R}^n \to \mathbb{R}^m$, where $\mathbf{f} = (f_1, \dots, f_m)$ and f_j denotes the j-th component function of $\mathbf{f}$ for $j = 1, \dots, m$. We can directly fuzzify the vector-valued function $\mathbf{f}$ without considering its component functions to obtain a fuzzy function $\tilde{\mathbf{f}}$. Alternatively, we can just fuzzify its component functions f_j to obtain the fuzzy functions $\tilde{f}_j$ for $j = 1, \dots, m$ and obtain a vector-form of fuzzy functions $(\tilde{f}_1, \dots, \tilde{f}_m)$. In general, the fuzzy function $\tilde{\mathbf{f}}$ is not identical with the vector-form of fuzzy functions $(\tilde{f}_1, \dots, \tilde{f}_m)$. Their relationships will be established in this chapter.

Let $\mathbf{f} : \mathbb{R}^n \to \mathbb{R}^m$ be a vector-valued function defined on $\mathbb{R}^m$. Given fuzzy sets $\tilde{A}^{(1)}, \tilde{A}^{(2)}, \dots, \tilde{A}^{(n)}$ in $\mathbb{R}$, we are going to study the fuzzification $\tilde{\mathbf{f}}^{(EP)}(\tilde{A}^{(1)}, \tilde{A}^{(2)}, \dots, \tilde{A}^{(n)})$ using the extension principle, and the fuzzification $\tilde{\mathbf{f}}^{(DT)}(\tilde{A}^{(1)}, \tilde{A}^{(2)}, \dots, \tilde{A}^{(n)})$ using the form of expression in the decomposition theorem.

6.1 Fuzzification Using the Extension Principle

Let $\mathfrak{A} : [0,1]^n \to [0,1]$ be a function defined on $[0,1]^n$. We consider an onto vector-valued function $\mathbf{f} : \mathbb{R}^n \to \mathbb{R}^m$. Then, the following sets

$$\{\mathbf{x} : \mathbf{y} = \mathbf{f}(\mathbf{x})\}$$

are not empty for each $\mathbf{y} \in \mathbb{R}^m$. According to the generalized extension principle, we can fuzzify the vector-valued function $\mathbf{f}$ using the aggregation $\mathfrak{A}$ to obtain a fuzzy function

$$\tilde{\mathbf{f}}^{(EP)} : \mathfrak{F}(\mathbb{R}) \times \cdots \times \mathfrak{F}(\mathbb{R}) \to \mathfrak{F}(\mathbb{R}^m).$$

Mathematical Foundations of Fuzzy Sets, First Edition. Hsien-Chung Wu.

Given any $\tilde{\mathbf{A}} = (\tilde{A}^{(1)}, \ldots, \tilde{A}^{(n)}) \in \mathfrak{F}^n(\mathbb{R})$, the function value $\tilde{\mathbf{f}}^{(EP)}(\tilde{\mathbf{A}})$ is a fuzzy set in $\mathbb{R}^m$ with membership function given by

$$\xi_{\tilde{\mathbf{f}}^{(EP)}(\tilde{\mathbf{A}})}(\mathbf{y}) = \sup_{\{\mathbf{x}:\mathbf{y}=\mathbf{f}(\mathbf{x})\}} \mathfrak{A}\left(\xi_{\tilde{A}^{(1)}}(x_1), \ldots, \xi_{\tilde{A}^{(n)}}(x_n)\right) \tag{6.1}$$

for $\mathbf{y} \in \mathbb{R}^m$. The fuzzification of component function f_j can be similarly obtained using the above form to obtain the fuzzy function $\tilde{f}_j$ with membership function given by

$$\xi_{\tilde{f}_j^{(EP)}(\tilde{\mathbf{A}})}(y) = \sup_{\{\mathbf{x}:y=f_j(\mathbf{x})\}} \mathfrak{A}\left(\xi_{\tilde{A}^{(1)}}(x_1), \ldots, \xi_{\tilde{A}^{(n)}}(x_n)\right)$$

for $y \in \mathbb{R}$ and for $j = 1, \ldots, m$.

Example 6.1.1 We are going to consider the addition of vectors of fuzzy sets in $\mathbb{R}$. Given an onto vector-valued function $\mathbf{f} : \mathbb{R}^{2n} \to \mathbb{R}^m$ defined by

$$\mathbf{f}(\mathbf{a}, \mathbf{b}) = \mathbf{f}\left(a_1, \ldots, a_n, b_1, \ldots, b_n\right) = \left(a_1 + b_1, \ldots, a_n + b_n\right) = \mathbf{a} + \mathbf{b}.$$

Given any two vectors of fuzzy sets in $\mathbb{R}$

$$\tilde{\mathbf{A}} = \left(\tilde{A}^{(1)}, \ldots, \tilde{A}^{(n)}\right) \text{ and } \tilde{\mathbf{B}} = \left(\tilde{B}^{(1)}, \ldots, \tilde{B}^{(n)}\right),$$

we can define the addition of $\tilde{\mathbf{A}}$ and $\tilde{\mathbf{B}}$ to be

$$\tilde{\mathbf{A}} \oplus_{EP} \tilde{\mathbf{B}} = \tilde{\mathbf{f}}^{(EP)}\left(\tilde{\mathbf{A}}, \tilde{\mathbf{B}}\right) \tag{6.2}$$

with membership function given by (6.1). More precisely, the membership function is given by

$$\begin{aligned}
\xi_{\tilde{\mathbf{A}} \oplus_{EP} \tilde{\mathbf{B}}}(\mathbf{y}) &= \xi_{\tilde{\mathbf{f}}^{(EP)}(\tilde{\mathbf{A}}, \tilde{\mathbf{B}})}(\mathbf{y}) \\
&= \sup_{\{(\mathbf{a},\mathbf{b}):\mathbf{y}=\mathbf{f}(\mathbf{a},\mathbf{b})\}} \mathfrak{A}\left(\xi_{\tilde{A}^{(1)}}(a_1), \ldots, \xi_{\tilde{A}^{(n)}}(a_n), \xi_{\tilde{B}^{(1)}}(b_1), \ldots, \xi_{\tilde{B}^{(n)}}(b_n)\right) \\
&= \sup_{\{(\mathbf{a},\mathbf{b}):\mathbf{y}=\mathbf{a}+\mathbf{b}\}} \mathfrak{A}\left(\xi_{\tilde{A}^{(1)}}(a_1), \ldots, \xi_{\tilde{A}^{(n)}}(a_n), \xi_{\tilde{B}^{(1)}}(b_1), \ldots, \xi_{\tilde{B}^{(n)}}(b_n)\right).
\end{aligned}$$

We can similarly define the difference $\tilde{\mathbf{A}} \ominus_{EP} \tilde{\mathbf{B}}$ considering the onto vector-valued function $\mathbf{f} : \mathbb{R}^{2n} \to \mathbb{R}^m$ defined by

$$\mathbf{f}(\mathbf{a}, \mathbf{b}) = \mathbf{f}\left(a_1, \ldots, a_n, b_1, \ldots, b_n\right) = \left(a_1 - b_1, \ldots, a_n - b_n\right) = \mathbf{a} - \mathbf{b},$$

and define the multiplication $\tilde{\mathbf{A}} \otimes_{EP} \tilde{\mathbf{B}}$ of vectors of fuzzy sets considering the onto vector-valued function $\mathbf{f} : \mathbb{R}^{2n} \to \mathbb{R}^m$ defined by

$$\mathbf{f}(\mathbf{a}, \mathbf{b}) = \mathbf{f}\left(a_1, \ldots, a_n, b_1, \ldots, b_n\right) = \left(a_1 \cdot b_1, \ldots, a_n \cdot b_n\right) = \mathbf{a} \times \mathbf{b}.$$

Let $\tilde{A}^{(1)}, \tilde{A}^{(2)}, \ldots, \tilde{A}^{(n)}$ be fuzzy intervals. The α-level sets of $\tilde{A}^{(i)}$ are bounded closed intervals given by

$$\tilde{A}^{(i)}_\alpha = \left[\left(\tilde{A}^{(i)}\right)^L_\alpha, \left(\tilde{A}^{(i)}\right)^U_\alpha\right] \equiv \left[\tilde{A}^L_{i\alpha}, \tilde{A}^U_{i\alpha}\right].$$

Suppose that the following conditions are satisfied.

- The function $\mathbf{f}$ is continuous.
- The function $\mathfrak{A}$ is upper semi-continuous and increasing.

- Any one of $\{\alpha_1,\dots,\alpha_s\}$ is zero implies $\mathfrak{A}(\alpha_1,\dots,\alpha_s)=0$.
- $\mathfrak{A}(\alpha_1,\dots,\alpha_s)>0$ if and only if $\alpha_i>0$ for all $i=1,\dots,n$.

Then, using Theorem 4.4.12, for each $\alpha\in(0,1]$, we have

$$\begin{aligned}\left(\tilde{\mathbf{f}}^{(EP)}(\tilde{\mathbf{A}})\right)_\alpha &= \left\{\mathbf{f}(\mathbf{x}) : \mathfrak{A}\left(\xi_{\tilde{A}^{(1)}}(x_1),\dots,\xi_{\tilde{A}^{(n)}}(x_n)\right)\geq\alpha\right\}\\ &= \left\{\left(f_1(\mathbf{x}),\dots,f_m(\mathbf{x})\right) : \mathfrak{A}\left(\xi_{\tilde{A}^{(1)}}(x_1),\dots,\xi_{\tilde{A}^{(n)}}(x_n)\right)\geq\alpha\right\}\end{aligned}\tag{6.3}$$

and

$$\left(\tilde{\mathbf{f}}^{(EP)}(\tilde{\mathbf{A}})\right)_0=\mathbf{f}\left(\tilde{A}_0^{(1)},\dots,\tilde{A}_0^{(n)}\right).$$

Moreover, for each $\alpha\in[0,1]$, the α-level sets $(\tilde{\mathbf{f}}^{(EP)}(\tilde{\mathbf{A}}))_\alpha$ of $\tilde{\mathbf{f}}^{(EP)}(\tilde{\mathbf{A}})$ are closed and bounded subsets of $\mathbb{R}^m$.

We write $\tilde{\mathbf{A}}_\alpha$ to denote the m-dimensional closed interval given by

$$\tilde{\mathbf{A}}_\alpha=\left[\tilde{A}_{1\alpha}^L,\tilde{A}_{1\alpha}^U\right]\times\left[\tilde{A}_{2\alpha}^L,\tilde{A}_{2\alpha}^U\right]\times\cdots\times\left[\tilde{A}_{n\alpha}^L,\tilde{A}_{n\alpha}^U\right].\tag{6.4}$$

When the function $\mathfrak{A}:[0,1]^n\to[0,1]$ is taken by

$$\mathfrak{A}\left(\alpha_1,\dots,\alpha_n\right)=\min\left\{\alpha_1,\dots,\alpha_n\right\},$$

for each $\alpha\in(0,1]$, from (6.3), we have

$$\begin{aligned}\left(\tilde{\mathbf{f}}^{(EP)}(\tilde{\mathbf{A}})\right)_\alpha &= \left\{\mathbf{f}(\mathbf{x}) : \min\left\{\xi_{\tilde{A}^{(1)}}(x_1),\dots,\xi_{\tilde{A}^{(n)}}(x_n)\right\}\geq\alpha\right\}\\ &= \left\{\mathbf{f}(\mathbf{x}) : \xi_{\tilde{A}^{(i)}}(x_i)\geq\alpha \text{ for each } i=1,\dots,n\right\}\\ &= \left\{\mathbf{f}(\mathbf{x}) : x_i\in\tilde{A}_\alpha^{(i)}\equiv\left[\tilde{A}_{i\alpha}^L,\tilde{A}_{i\alpha}^U\right] \text{ for each } i=1,\dots,n\right\}=\mathbf{f}\left(\tilde{\mathbf{A}}_\alpha\right).\end{aligned}\tag{6.5}$$

Then, from (6.5), for $\alpha\in(0,1]$, we have

$$\left(\tilde{\mathbf{f}}^{(EP)}(\tilde{\mathbf{A}})\right)_\alpha=\mathbf{f}\left(\tilde{\mathbf{A}}_\alpha\right)=f_1\left(\tilde{\mathbf{A}}_\alpha\right)\times f_2\left(\tilde{\mathbf{A}}_\alpha\right)\times\cdots\times f_m\left(\tilde{\mathbf{A}}_\alpha\right).\tag{6.6}$$

Let $\tilde{f}_j^{(EP)}$ be fuzzified from the jth component function f_j using the extension principle. For $\alpha\in(0,1]$, we can similarly obtain

$$\left(\tilde{f}_j^{(EP)}(\tilde{\mathbf{A}})\right)_\alpha=f_j\left(\tilde{\mathbf{A}}_\alpha\right)\ \text{ for } j=1,\dots,m.$$

By referring to (6.6), for $\alpha\in(0,1]$, we obtain

$$\left(\tilde{\mathbf{f}}^{(EP)}(\tilde{\mathbf{A}})\right)_\alpha=\left(\tilde{f}_1^{(EP)}(\tilde{\mathbf{A}})\right)_\alpha\times\cdots\times\left(\tilde{f}_m^{(EP)}(\tilde{\mathbf{A}})\right)_\alpha.$$

Suppose that the vector-valued function $\mathbf{f}=(f_1,\dots,f_m)$ is continuous. Then, the component functions f_j are also continuous for $j=1,\dots,m$. Using (6.4) and Proposition 1.4.3, we see that $f_j(\tilde{\mathbf{A}}_\alpha)$ are bounded closed intervals given by

$$\left(\tilde{f}_j^{(EP)}(\tilde{\mathbf{A}})\right)_\alpha=f_j\left(\tilde{\mathbf{A}}_\alpha\right)=\left[\inf_{\mathbf{x}\in\tilde{\mathbf{A}}_\alpha}f_j(\mathbf{x}),\sup_{\mathbf{x}\in\tilde{\mathbf{A}}_\alpha}f_j(\mathbf{x})\right]=\left[\min_{\mathbf{x}\in\tilde{\mathbf{A}}_\alpha}f_j(\mathbf{x}),\max_{\mathbf{x}\in\tilde{\mathbf{A}}_\alpha}f_j(\mathbf{x})\right].$$

It follows that

$$\left(\tilde{\mathbf{f}}^{(EP)}(\tilde{\mathbf{A}})\right)_\alpha=\left[\min_{\mathbf{x}\in\tilde{\mathbf{A}}_\alpha}f_1(\mathbf{x}),\max_{\mathbf{x}\in\tilde{\mathbf{A}}_\alpha}f_1(\mathbf{x})\right]\times\cdots\times\left[\min_{\mathbf{x}\in\tilde{\mathbf{A}}_\alpha}f_m(\mathbf{x}),\max_{\mathbf{x}\in\tilde{\mathbf{A}}_\alpha}f_m(\mathbf{x})\right].$$

The above results are summarized in the following theorem.

Theorem 6.1.2 *Let $\mathbf{f} : \mathbb{R}^n \to \mathbb{R}^m$ be a continuous vector-valued function defined on $\mathbb{R}^m$, and let $\tilde{A}^{(1)}, \tilde{A}^{(2)}, \dots, \tilde{A}^{(n)}$ be fuzzy intervals. By referring to (6.1), let $\tilde{\mathbf{f}}^{(EP)}(\tilde{\mathbf{A}})$ be obtained based on the function $\mathfrak{A} : [0,1]^n \to [0,1]$ taken by*

$$\mathfrak{A}\left(\alpha_1, \dots, \alpha_n\right) = \min\left\{\alpha_1, \dots, \alpha_n\right\}.$$

Then, for $\alpha \in [0,1]$, we have

$$\left(\tilde{f}_j^{(EP)}(\tilde{\mathbf{A}})\right)_\alpha = \left[\min_{\mathbf{x}\in\tilde{\mathbf{A}}_\alpha} f_j(\mathbf{x}), \max_{\mathbf{x}\in\tilde{\mathbf{A}}_\alpha} f_j(\mathbf{x})\right] \text{ for } j = 1, \dots, m$$

and

$$\begin{aligned}\left(\tilde{\mathbf{f}}^{(EP)}(\tilde{\mathbf{A}})\right)_\alpha &= \left(\tilde{f}_1^{(EP)}(\tilde{\mathbf{A}})\right)_\alpha \times \cdots \times \left(\tilde{f}_m^{(EP)}(\tilde{\mathbf{A}})\right)_\alpha \\ &= \left[\min_{\mathbf{x}\in\tilde{\mathbf{A}}_\alpha} f_1(\mathbf{x}), \max_{\mathbf{x}\in\tilde{\mathbf{A}}_\alpha} f_1(\mathbf{x})\right] \times \cdots \times \left[\min_{\mathbf{x}\in\tilde{\mathbf{A}}_\alpha} f_m(\mathbf{x}), \max_{\mathbf{x}\in\tilde{\mathbf{A}}_\alpha} f_m(\mathbf{x})\right].\end{aligned}$$

Example 6.1.3 Continued from Example 6.1.1 by considering the fuzzy intervals, we want to obtain the α-level sets $(\tilde{\mathbf{A}} \oplus_{EP} \tilde{\mathbf{B}})_\alpha$ of addition $\tilde{\mathbf{A}} \oplus_{EP} \tilde{\mathbf{B}}$ for $\alpha \in [0,1]$. According to Theorem 6.1.2, we have

$$\begin{aligned}\left(\tilde{\mathbf{A}} \oplus_{EP} \tilde{\mathbf{B}}\right)_\alpha = \left(\tilde{\mathbf{f}}^{(EP)}(\tilde{\mathbf{A}}, \tilde{\mathbf{B}})\right)_\alpha &= \left(\tilde{f}_1^{(EP)}(\tilde{\mathbf{A}}, \tilde{\mathbf{B}})\right)_\alpha \times \cdots \times \left(\tilde{f}_n^{(EP)}(\tilde{\mathbf{A}}, \tilde{\mathbf{B}})\right)_\alpha \\ &= \left(\tilde{A}^{(1)} \oplus_{EP} \tilde{B}^{(1)}\right)_\alpha \times \cdots \times \left(\tilde{A}^{(n)} \oplus_{EP} \tilde{B}^{(n)}\right)_\alpha \\ &= \left[\tilde{A}_{1\alpha}^L + \tilde{B}_{1\alpha}^L, \tilde{A}_{1\alpha}^U + \tilde{B}_{1\alpha}^L\right] \times \cdots \times \left[\tilde{A}_{n\alpha}^L + \tilde{B}_{n\alpha}^L, \tilde{A}_{n\alpha}^U + \tilde{B}_{n\alpha}^L\right]\end{aligned}$$

for each $\alpha \in [0,1]$.

6.2 Fuzzification Using the Expression in the Decomposition Theorem

Let $\tilde{A}^{(1)}, \tilde{A}^{(2)}, \dots, \tilde{A}^{(n)}$ be fuzzy sets in $\mathbb{R}$. We consider the family

$$\{M_\alpha \neq \emptyset : 0 < \alpha \leq 1\} \tag{6.7}$$

that consists of subsets of $\mathbb{R}^m$, where M_α depends on $\mathbf{f}$ and $\tilde{A}^{(i)}$ for $i = 1, \dots, n$. For example, we can take

$$M_\alpha = \mathbf{f}\left(\tilde{A}_\alpha^{(1)}, \dots, \tilde{A}_\alpha^{(n)}\right).$$

We are going to consider three different M_α that can be used to generate three different fuzzy functions.

We say that the family $\{M_\alpha : 0 < \alpha \leq 1\}$ is nested when $M_\beta \subseteq M_\alpha$ for $\alpha < \beta$. We shall study the cases of nested family and non-nested family separately by considering three intuitive families.

According to the expression in the Decomposition Theorem 2.2.13, we can induce a fuzzy function $\tilde{\mathbf{f}}^{(DT)} : \mathfrak{F}^n(\mathbb{R}) \to \mathfrak{F}(\mathbb{R}^m)$. Given any $\tilde{\mathbf{A}} \in \mathfrak{F}^n(\mathbb{R})$, the function value $\tilde{\mathbf{f}}^{(DT)}(\tilde{\mathbf{A}})$ is a fuzzy set in $\mathbb{R}^m$ with membership function defined by

$$\xi_{\tilde{\mathbf{f}}^{(DT)}(\tilde{\mathbf{A}})}(\mathbf{y}) = \sup_{0<\alpha\leq 1} \alpha \cdot \chi_{M_\alpha}(\mathbf{y}) \tag{6.8}$$

for $\mathbf{y} \in \mathbb{R}^m$.

6.2.1 Nested Family Using α-Level Sets

Let $\tilde{A}^{(1)}, \tilde{A}^{(2)}, \ldots, \tilde{A}^{(n)}$ be fuzzy intervals. For $\alpha \in (0,1]$, we take

$$M_\alpha = \mathbf{f}\left(\tilde{A}_\alpha^{(1)}, \ldots, \tilde{A}_\alpha^{(n)}\right) = \mathbf{f}\left(\tilde{\mathbf{A}}_\alpha\right) = \left(f_1\left(\tilde{\mathbf{A}}_\alpha\right), \ldots, f_m\left(\tilde{\mathbf{A}}_\alpha\right)\right). \tag{6.9}$$

Then, we see that

$$M_\alpha = \left\{\mathbf{f}(\mathbf{x}) : x_i \in \tilde{A}_\alpha^{(i)} = \left[\tilde{A}_{i\alpha}^L, \tilde{A}_{i\alpha}^U\right] \text{ for each } i = 1, \ldots, n\right\}.$$

It is clear to see that $\{M_\alpha : 0 < \alpha \leq 1\}$ is a nested family in the sense of $M_\beta \subseteq M_\alpha$ for $\alpha < \beta$. We need some useful lemmas.

Lemma 6.2.1 *Let $\{A_m\}_{m=1}^\infty$ be a sequence of subsets of a universal set U satisfying $A_{m+1} \subseteq A_m$ for all m, and let f be a real-valued function defined on U. Then*

$$f\left(\bigcap_{m=1}^\infty A_m\right) = \bigcap_{m=1}^\infty f\left(A_m\right).$$

Proof. Given any

$$y \in f\left(\bigcap_{m=1}^\infty A_m\right),$$

there exists $x \in \bigcap_{m=1}^\infty A_m$ satisfying $y = f(x)$, which says that

$$y = f(x) \in f\left(A_m\right) \text{ for all } m.$$

Therefore, we obtain the following inclusion

$$f\left(\bigcap_{m=1}^\infty A_m\right) \subseteq \bigcap_{m=1}^\infty f\left(A_m\right).$$

For proving the other direction of inclusion, given any $y \in f(A_m)$ for all m, we have $y = f(x_m)$ for $x_m \in A_m$ and for all m, which says that we obtain a sequence $\{x_m\}_{m=1}^\infty$. Let $\bigcap_{m=1}^\infty A_m = A$. Since $A_{m+1} \subseteq A_m$ for all m, the "last term" of the sequence $\{x_m\}_{m=1}^\infty$ must be in A, which will be proved below. Since $x_k \in A_k \subseteq A_m$ for all $k \geq m$, we have the subsequence $\{x_k\}_{k=m}^\infty \subseteq A_m$, which also implies

$$\hat{A} \equiv \bigcap_{m=1}^\infty \left(\{x_k\}_{k=m}^\infty\right) \subseteq \bigcap_{m=1}^\infty A_m = A,$$

where $\hat{A}$ can be regarded as the "last term" and $\hat{A} \subseteq \{x_m\}_{m=1}^\infty$. For $\bar{x} \in \hat{A}$, it follows that $\bar{x} \in A$ and $\bar{x} = x_{m^*}$ for some x_{m^*} in the sequence $\{x_m\}_{m=1}^\infty$. Therefore, we obtain

$$y = f(x_{m^*}) = f(\bar{x}),$$

which says that $y \in f(A)$. This completes the proof. ∎

Lemma 6.2.2 *Suppose that*

$$A^{(i)} = \bigcap_{m=1}^\infty A_m^{(i)} \text{ for } i = 1, \ldots, n.$$

Then, we have

$$A^{(1)} \times \cdots \times A^{(n)} = \bigcap_{m=1}^\infty \left(A_m^{(1)} \times \cdots \times A_m^{(n)}\right).$$

Proof. Given $\mathbf{x} = (x_1, \ldots, x_n)$ with $x_i \in A^{(i)}$ for $i = 1, \ldots, m$, it says that $x_i \in A_m^{(i)}$ for all m, i.e.

$$\mathbf{x} = (x_1, \ldots, x_n) \in A_m^{(1)} \times \cdots \times A_m^{(n)}$$

for all m, which proves the inclusion

$$A^{(1)} \times \cdots \times A^{(n)} \subseteq \bigcap_{m=1}^{\infty} \left(A_m^{(1)} \times \cdots \times A_m^{(n)}\right).$$

On the other hand, given any $\mathbf{x} = (x_1, \ldots, x_n) \in A_m^{(1)} \times \cdots \times A_m^{(n)}$ for all m, it follows that $x_i \in A_m^{(i)}$ for all i and m, i.e. $x_i \in \bigcap_{m=1}^{\infty} A_m^{(i)} = A^{(i)}$ for all i. This completes the proof. ■

Proposition 6.2.3 *Let $\tilde{A}^{(1)}, \tilde{A}^{(2)}, \ldots, \tilde{A}^{(n)}$ be fuzzy intervals, and let M_α be given in (6.9). Suppose that $\mathbf{f}$ is continuous. Then, we have*

$$M_\alpha = \bigcap_{s=1}^{\infty} M_{\alpha_s} \text{ for } \alpha \in (0,1] \text{ and } 0 < \alpha_s \uparrow \alpha \text{ with } \alpha > \alpha_s \text{ for all } s.$$

Proof. From (6.9), we see that

$$M_\alpha = f_1\left(\tilde{\mathbf{A}}_\alpha\right) \times \cdots \times f_m\left(\tilde{\mathbf{A}}_\alpha\right) \equiv M_\alpha^{(1)} \times \cdots \times M_\alpha^{(m)}, \tag{6.10}$$

where

$$f_j\left(\tilde{\mathbf{A}}_\alpha\right) \equiv M_\alpha^{(j)} \text{ for } j = 1, \ldots, m.$$

The continuity of $\mathbf{f}$ says that the component functions f_j are also continuous for $j = 1, \ldots, m$. Since each $\tilde{A}_\alpha^{(i)}$ is a bounded closed interval, the continuity of f_j says that each $f_j(\tilde{\mathbf{A}}_\alpha)$ is also a bounded closed interval given by

$$M_\alpha^{(j)} = f_j\left(\tilde{\mathbf{A}}_\alpha\right) = \left[\min_{\mathbf{x} \in \tilde{\mathbf{A}}_\alpha} f_j(\mathbf{x}), \max_{\mathbf{x} \in \tilde{\mathbf{A}}_\alpha} f_j(\mathbf{x})\right] \text{ for } j = 1, \ldots, m.$$

It is clear to see that $M_\alpha^{(j)} \subseteq M_\beta^{(j)}$ for $\beta < \alpha$. Given any sequence $\{\alpha_s\}_{s=1}^{\infty}$ satisfying $0 < \alpha_s \uparrow \alpha$ with $\alpha > \alpha_s$ for all s, we see that

$$\tilde{A}_\alpha^{(i)} = \bigcap_{s=1}^{\infty} \tilde{A}_{\alpha_s}^{(i)} \text{ for } i = 1, \ldots, n$$

by referring to Proposition 2.2.8, which also implies $\tilde{\mathbf{A}}_\alpha = \bigcap_{s=1}^{\infty} \tilde{\mathbf{A}}_{\alpha_s}$ by Lemma 6.2.2. Therefore, using Lemma 6.2.1, we can obtain

$$f_j\left(\tilde{\mathbf{A}}_\alpha\right) = f_j\left(\bigcap_{s=1}^{\infty} \tilde{\mathbf{A}}_{\alpha_s}\right) = \bigcap_{s=1}^{\infty} f_j\left(\tilde{\mathbf{A}}_{\alpha_s}\right),$$

which implies

$$M_\alpha^{(j)} = \bigcap_{s=1}^{\infty} M_{\alpha_s}^{(j)} \text{ for } j = 1, \ldots, m.$$

Using (6.10) and Lemma 6.2.2, we complete the proof. ■

Let $\tilde{\mathbf{f}}^{(\diamond DT)}$ be fuzzified from $\mathbf{f}$ obtained from (6.8) using the family $\{M_\alpha : 0 < \alpha \leq 1\}$ defined in (6.9). Then, its α-level sets are given below.

Proposition 6.2.4 *Let $\tilde{A}^{(1)}, \tilde{A}^{(2)}, \dots, \tilde{A}^{(n)}$ be fuzzy intervals. Suppose that $\mathbf{f}$ is continuous. Then*

$$\left(\tilde{\mathbf{f}}^{(\diamond DT)}(\tilde{\mathbf{A}})\right)_\alpha = M_\alpha = \mathbf{f}\left(\tilde{A}_\alpha^{(1)}, \dots, \tilde{A}_\alpha^{(n)}\right) = \mathbf{f}\left(\tilde{\mathbf{A}}_\alpha\right) \text{ for } 0 < \alpha \leq 1. \tag{6.11}$$

Proof. Given any $\mathbf{y} \in M_\alpha$, we see that $\xi_{\tilde{\mathbf{f}}^{(\diamond DT)}(\tilde{\mathbf{A}})}(\mathbf{y}) \geq \alpha$ by (6.8). Therefore, we obtain $\mathbf{y} \in (\tilde{\mathbf{f}}^{(\diamond DT)}(\tilde{\mathbf{A}}))_\alpha$, which proves the inclusion $M_\alpha \subseteq (\tilde{\mathbf{f}}^{(\diamond DT)}(\tilde{\mathbf{A}}))_\alpha$. For proving the other direction of inclusion, given any $\mathbf{y} \in (\tilde{\mathbf{f}}^{(\diamond DT)}(\tilde{\mathbf{A}}))_\alpha$, i.e. $\xi_{\tilde{\mathbf{f}}^{(\diamond DT)}(\tilde{\mathbf{A}})}(\mathbf{y}) \geq \alpha$, let $\hat{\alpha} = \xi_{\tilde{\mathbf{f}}^{(\diamond DT)}(\tilde{\mathbf{A}})}(\mathbf{y})$.

- Assume that $\hat{\alpha} > \alpha$. Let $\epsilon = \hat{\alpha} - \alpha > 0$. According to the concept of supremum, there exists α_0 satisfying $\mathbf{y} \in M_{\alpha_0}$ and $\hat{\alpha} - \epsilon < \alpha_0$, which implies $\alpha < \alpha_0$. This also says that $\mathbf{y} \in M_\alpha$, since $M_{\alpha_0} \subseteq M_\alpha$ by the nestedness.
- Assume that $\hat{\alpha} = \alpha$. Let $\{\alpha_s\}_{s=1}^\infty$ be a sequence satisfying $0 < \alpha_s \uparrow \alpha$ and $\alpha > \alpha_s$ for all s. Let $\epsilon_s = \alpha - \alpha_s > 0$. According to the concept of supremum, there exists α_0 satisfying $\mathbf{y} \in M_{\alpha_0}$ and $\hat{\alpha} - \epsilon_s = \alpha - \epsilon_s < \alpha_0$, which implies $\alpha_0 > \alpha_s$. This also says that $\mathbf{y} \in M_{\alpha_s}$ by the nestedness for all s. Therefore, we obtain $\mathbf{y} \in \bigcap_{s=1}^\infty M_{\alpha_s}$. From Proposition 6.2.3, it follows that $\mathbf{y} \in M_\alpha$.

The above two cases imply that $\mathbf{y} \in M_\alpha$, and the proof is complete. ■

Example 6.2.5 Continued from Example 6.1.1, by referring to (6.9), let

$$\begin{aligned}
M_\alpha &= \mathbf{f}\left(\tilde{A}_\alpha^{(1)}, \dots, \tilde{A}_\alpha^{(n)}, \tilde{B}_\alpha^{(1)}, \dots, \tilde{B}_\alpha^{(n)}\right) \\
&= \left\{\mathbf{f}\left(a_1, \dots, a_n, b_1, \dots, b_n\right) : a_i \in \tilde{A}_\alpha^{(i)} \text{ and } b_i \in \tilde{B}_\alpha^{(i)} \text{ for } i = 1, \dots, n\right\} \\
&= \left\{\left(a_1 + b_1, \dots, a_n + b_n\right) : a_i \in \left[\tilde{A}_{i\alpha}^L, \tilde{A}_{i\alpha}^U\right] \text{ and } b_i \in \left[\tilde{B}_{i\alpha}^L, \tilde{B}_{i\alpha}^U\right] \text{ for } i = 1, \dots, n\right\} \\
&= \left[\tilde{A}_{1\alpha}^L + \tilde{B}_{1\alpha}^L, \tilde{A}_{1\alpha}^U + \tilde{B}_{1\alpha}^U\right] \times \cdots \times \left[\tilde{A}_{n\alpha}^L + \tilde{B}_{n\alpha}^L, \tilde{A}_{n\alpha}^U + \tilde{B}_{n\alpha}^U\right].
\end{aligned}$$

We can define the addition of $\tilde{\mathbf{A}}$ and $\tilde{\mathbf{B}}$ to be

$$\tilde{\mathbf{A}} \oplus_{DT}^{\diamond} \tilde{\mathbf{B}} = \tilde{\mathbf{f}}^{(\diamond DT)}\left(\tilde{\mathbf{A}}, \tilde{\mathbf{B}}\right)$$

with membership function given by

$$\xi_{\tilde{\mathbf{A}} \oplus_{DT}^{\diamond} \tilde{\mathbf{B}}}(\mathbf{y}) = \xi_{\tilde{\mathbf{f}}^{(\diamond DT)}(\tilde{\mathbf{A}}, \tilde{\mathbf{B}})}(\mathbf{y}) = \sup_{0<\alpha\leq 1} \alpha \cdot \chi_{M_\alpha}(\mathbf{y}).$$

For $\alpha \in (0,1]$, Proposition 6.2.4 says that

$$\begin{aligned}
\left(\tilde{\mathbf{A}} \oplus_{DT}^{\diamond} \tilde{\mathbf{B}}\right)_\alpha &= \left(\tilde{\mathbf{f}}^{(\diamond DT)}(\tilde{\mathbf{A}}, \tilde{\mathbf{B}})\right)_\alpha = M_\alpha \\
&= \left[\tilde{A}_{1\alpha}^L + \tilde{B}_{1\alpha}^L, \tilde{A}_{1\alpha}^U + \tilde{B}_{1\alpha}^L\right] \times \cdots \times \left[\tilde{A}_{n\alpha}^L + \tilde{B}_{n\alpha}^L, \tilde{A}_{n\alpha}^U + \tilde{B}_{n\alpha}^U\right].
\end{aligned}$$

We can similarly consider the difference $\tilde{\mathbf{A}} \ominus_{DT}^{\diamond} \tilde{\mathbf{B}}$ of vectors of fuzzy intervals by taking

$$M_\alpha = \left[\tilde{A}_{1\alpha}^L - \tilde{B}_{1\alpha}^U, \tilde{A}_{1\alpha}^U - \tilde{B}_{1\alpha}^L\right] \times \cdots \times \left[\tilde{A}_{n\alpha}^L - \tilde{B}_{n\alpha}^U, \tilde{A}_{n\alpha}^U - \tilde{B}_{n\alpha}^L\right].$$

Now, we take

$$M_\alpha^{(j)} = f_j(\tilde{\mathbf{A}}_\alpha) \text{ for } 0 < \alpha \leq 1 \text{ and } j = 1, \dots, m. \tag{6.12}$$

From (6.9), we see that

$$M_\alpha = M_\alpha^{(1)} \times \cdots \times M_\alpha^{(m)} \subset \mathbb{R}^m. \tag{6.13}$$

Let $\tilde{f}_j^{(\diamond DT)}$ be fuzzified from the jth component function f_j using the expression in the decomposition theorem based on the family $\{M_\alpha^{(j)} : 0 < \alpha \leq 1\}$ defined in (6.12). By referring to Proposition 6.2.4, for $\alpha \in (0,1]$, we can similarly obtain

$$\left(\tilde{f}_j^{(\diamond DT)}(\tilde{\mathbf{A}})\right)_\alpha = M_\alpha^{(j)} = f_j(\tilde{\mathbf{A}}_\alpha) \text{ for } j = 1, \ldots, m. \tag{6.14}$$

For $\alpha \in (0,1]$, using (6.11), (6.13) and (6.14), we have

$$\left(\tilde{\mathbf{f}}^{(\diamond DT)}(\tilde{\mathbf{A}})\right)_\alpha = \left(\tilde{f}_1^{(\diamond DT)}(\tilde{\mathbf{A}})\right)_\alpha \times \cdots \times \left(\tilde{f}_m^{(\diamond DT)}(\tilde{\mathbf{A}})\right)_\alpha. \tag{6.15}$$

For the 0-level sets, using (2.4) and (2.5), we have

$$\left(\tilde{f}_j^{(\diamond DT)}(\tilde{\mathbf{A}})\right)_0 = \text{cl}\left(\tilde{f}_j^{(\diamond DT)}(\tilde{\mathbf{A}})\right)_{0+} = \text{cl}\left(\bigcup_{0<\alpha\leq 1}\left(\tilde{f}_j^{(\diamond DT)}(\tilde{\mathbf{A}})\right)_\alpha\right) = \text{cl}\left(\bigcup_{0<\alpha\leq 1} f_j(\tilde{\mathbf{A}}_\alpha)\right)$$

and

$$\begin{aligned}
\left(\tilde{\mathbf{f}}^{(\diamond DT)}(\tilde{\mathbf{A}})\right)_0 &= \text{cl}\left(\tilde{\mathbf{f}}^{(\diamond DT)}(\tilde{\mathbf{A}})\right)_{0+} = \text{cl}\left(\bigcup_{0<\alpha\leq 1}\left(\tilde{\mathbf{f}}^{(\diamond DT)}(\tilde{\mathbf{A}})\right)_\alpha\right)\\
&= \text{cl}\left(\bigcup_{0<\alpha\leq 1}\left(\tilde{f}_1^{(\diamond DT)}(\tilde{\mathbf{A}})\right)_\alpha\right) \times \cdots \times \text{cl}\left(\bigcup_{0<\alpha\leq 1}\left(\tilde{f}_1^{(\diamond DT)}(\tilde{\mathbf{A}})\right)_\alpha\right)\\
&\quad \text{(using (6.15) and the nestedness)}\\
&= \left(\tilde{f}_1^{(\diamond DT)}(\tilde{\mathbf{A}})\right)_0 \times \cdots \times \left(\tilde{f}_m^{(\diamond DT)}(\tilde{\mathbf{A}})\right)_0.
\end{aligned} \tag{6.16}$$

Suppose that $\mathbf{f}$ is continuous. Then, each f_j is continuous for $j = 1, \ldots, m$. Since each $\tilde{A}^{(i)}$ is assumed to be fuzzy interval for $i = 1, \ldots, n$, i.e. the α-level sets $\tilde{A}_\alpha^{(i)}$ are bounded closed intervals for $\alpha \in [0,1]$ and $i = 1, \ldots, n$, it follows that $f_j(\tilde{\mathbf{A}}_\alpha)$ is also a bounded closed interval given by

$$f_j\left(\tilde{\mathbf{A}}_\alpha\right) = \left[\min_{\mathbf{x}\in\tilde{\mathbf{A}}_\alpha} f_j(\mathbf{x}), \max_{\mathbf{x}\in\tilde{\mathbf{A}}_\alpha} f_j(\mathbf{x})\right] \text{ for } 0 < \alpha \leq 1 \text{ and } j = 1, \ldots, m.$$

Therefore, we obtain

$$\begin{aligned}
\left(\tilde{f}_j^{(\diamond DT)}(\tilde{\mathbf{A}})\right)_0 &= \text{cl}\left(\bigcup_{0<\alpha\leq 1} f_j(\tilde{\mathbf{A}}_\alpha)\right)\\
&= \text{cl}\left(\bigcup_{0<\alpha\leq 1}\left[\min_{\mathbf{x}\in\tilde{\mathbf{A}}_\alpha} f_j(\mathbf{x}), \max_{\mathbf{x}\in\tilde{\mathbf{A}}_\alpha} f_j(\mathbf{x})\right]\right) = \left[\min_{\mathbf{x}\in\tilde{\mathbf{A}}_0} f_j(\mathbf{x}), \max_{\mathbf{x}\in\tilde{\mathbf{A}}_0} f_j(\mathbf{x})\right]
\end{aligned}$$

for $j = 1, \ldots, m$. The above results are summarized below.

Theorem 6.2.6 *Let $\mathbf{f} : \mathbb{R}^n \to \mathbb{R}^m$ be a continuous vector-valued function defined on $\mathbb{R}^m$, and let $\tilde{A}^{(1)}, \tilde{A}^{(2)}, \ldots, \tilde{A}^{(n)}$ be fuzzy intervals. Suppose that the families*

$$\{M_\alpha : 0 < \alpha \leq 1\} \text{ and } \{M_\alpha^{(j)} : 0 < \alpha \leq 1\} \text{ for } j = 1, \ldots, m$$

are given by

$$M_\alpha = \mathbf{f}\left(\tilde{A}_\alpha^{(1)}, \ldots, \tilde{A}_\alpha^{(n)}\right) \text{ and } M_\alpha^{(j)} = f_j\left(\tilde{A}_\alpha^{(1)}, \ldots, \tilde{A}_\alpha^{(n)}\right) \text{ for } j = 1, \ldots, m.$$

For $\alpha \in [0,1]$, we have

$$\left(\tilde{\mathbf{f}}^{(\diamond DT)}(\tilde{\mathbf{A}})\right)_\alpha = \mathbf{f}\left(\tilde{A}_\alpha^{(1)}, \ldots, \tilde{A}_\alpha^{(n)}\right) = \left(\tilde{f}_1^{(\diamond DT)}(\tilde{\mathbf{A}})\right)_\alpha \times \cdots \times \left(\tilde{f}_m^{(\diamond DT)}(\tilde{\mathbf{A}})\right)_\alpha \tag{6.17}$$

and

$$\left(\tilde{f}_j^{(\diamond DT)}(\tilde{\mathbf{A}})\right)_\alpha = f_j\left(\tilde{\mathbf{A}}_\alpha\right) = \left[\min_{\mathbf{x}\in\tilde{\mathbf{A}}_\alpha} f_j(\mathbf{x}), \max_{\mathbf{x}\in\tilde{\mathbf{A}}_\alpha} f_j(\mathbf{x})\right] \text{ for } j = 1, \ldots, m.$$

6.2.2 Nested Family Using Endpoints

Let $\tilde{A}^{(1)}, \tilde{A}^{(2)}, \ldots, \tilde{A}^{(n)}$ be fuzzy intervals. Then, the α-level sets of $\tilde{A}^{(i)}$ are given by $\tilde{A}_\alpha^{(i)} = [\tilde{A}_{i\alpha}^L, \tilde{A}_{i\alpha}^U]$ for $i = 1, \ldots, n$. Now, for $\alpha \in [0,1]$, we write

$$\left(\tilde{A}_{1\alpha}^L, \tilde{A}_{2\alpha}^L, \ldots, \tilde{A}_{n\alpha}^L\right) = \tilde{\mathbf{A}}_\alpha^L \in \mathbb{R}^m \text{ and } \left(\tilde{A}_{1\alpha}^U, \tilde{A}_{2\alpha}^U, \ldots, \tilde{A}_{n\alpha}^U\right) = \tilde{\mathbf{A}}_\alpha^U \in \mathbb{R}^m.$$

Then, we define two vector-valued functions $\boldsymbol{\eta}^L : [0,1] \to \mathbb{R}^m$ and $\boldsymbol{\eta}^U : [0,1] \to \mathbb{R}^m$ by

$$\boldsymbol{\eta}^L(\alpha) = \mathbf{f}\left(\tilde{A}_{1\alpha}^L, \tilde{A}_{2\alpha}^L, \ldots, \tilde{A}_{n\alpha}^L\right) = \mathbf{f}\left(\tilde{\mathbf{A}}_\alpha^L\right) = \left(f_1\left(\tilde{\mathbf{A}}_\alpha^L\right), \ldots, f_m\left(\tilde{\mathbf{A}}_\alpha^L\right)\right)$$

and

$$\boldsymbol{\eta}^U(\alpha) = \mathbf{f}\left(\tilde{A}_{1\alpha}^U, \tilde{A}_{2\alpha}^U, \ldots, \tilde{A}_{n\alpha}^U\right) = \mathbf{f}\left(\tilde{\mathbf{A}}_\alpha^U\right) = \left(f_1\left(\tilde{\mathbf{A}}_\alpha^U\right), \ldots, f_m\left(\tilde{\mathbf{A}}_\alpha^U\right)\right).$$

We also write

$$\boldsymbol{\eta}^L(\alpha) = \left(\eta_1^L(\alpha), \ldots, \eta_m^L(\alpha)\right) \text{ and } \boldsymbol{\eta}^U(\alpha) = \left(\eta_1^U(\alpha), \ldots, \eta_m^U(\alpha)\right),$$

where

$$\eta_j^L(\alpha) = f_j\left(\tilde{\mathbf{A}}_\alpha^L\right) \text{ and } \eta_j^U(\alpha) = f_j\left(\tilde{\mathbf{A}}_\alpha^U\right) \text{ for } j = 1, \ldots, m.$$

For $\alpha \in (0,1]$, we consider the following bounded closed intervals

$$M_\alpha^{(j)} = \left[\min\left\{\eta_j^L(\alpha), \eta_j^U(\alpha)\right\}, \max\left\{\eta_j^L(\alpha), \eta_j^U(\alpha)\right\}\right] \text{ for } j = 1, \ldots, m, \tag{6.18}$$

and the family $\{M_\alpha : 0 < \alpha \leq 1\}$ given by

$$M_\alpha = \left(\bigcup_{\alpha\leq\beta\leq 1} M_\beta^{(1)}\right) \times \cdots \times \left(\bigcup_{\alpha\leq\beta\leq 1} M_\beta^{(m)}\right). \tag{6.19}$$

Then $\{M_\alpha : 0 < \alpha \leq 1\}$ is a nested family in the sense of $M_\alpha \subseteq M_\beta$ for $\alpha > \beta$. Let

$$\zeta_j^L(\alpha) = \min\left\{\eta_j^L(\alpha), \eta_j^U(\alpha)\right\} \text{ and } \zeta_j^U(\alpha) = \max\left\{\eta_j^L(\alpha), \eta_j^U(\alpha)\right\}$$

for $j = 1, \ldots, m$. Then $M_\alpha^{(j)} = \left[\zeta_j^L(\alpha), \zeta_j^U(\alpha)\right]$. Suppose that ζ_j^L is lower semi-continuous on $[0,1]$ and ζ_j^U is upper semi-continuous on $[0,1]$. Then, for $\alpha \in (0,1]$, we can obtain

$$\bigcup_{\alpha\leq\beta\leq 1} M_\beta^{(j)} = \left[\inf_{\alpha\leq\beta\leq 1} \zeta_j^L(\beta), \sup_{\alpha\leq\beta\leq 1} \zeta_j^U(\beta)\right] = \left[\min_{\alpha\leq\beta\leq 1} \zeta_j^L(\beta), \max_{\alpha\leq\beta\leq 1} \zeta_j^U(\beta)\right] \equiv N_\alpha^{(j)}. \tag{6.20}$$

It is clear to see that $\{N_\alpha^{(j)} : 0 < \alpha \le 1\}$ is a nested family. For $j = 1, \ldots, m$, we can show that

$$N_\alpha^{(j)} = \bigcap_{s=1}^{\infty} N_{\alpha_s}^{(j)} \text{ for } \alpha \in (0,1] \text{ and } 0 < \alpha_s \uparrow \alpha \text{ with } \alpha > \alpha_s \text{ for all } s. \tag{6.21}$$

From (6.19) and (6.20), we also have

$$M_\alpha = N_\alpha^{(1)} \times \cdots \times N_\alpha^{(m)}. \tag{6.22}$$

Let $\tilde{f}_j^{(\star DT)}$ be fuzzified from the jth component function f_j obtained from (6.8) using the family $\{N_\alpha^{(j)} : 0 < \alpha \le 1\}$ defined in (6.20). For $\alpha \in (0,1]$, using (6.21) and a similar argument to the proof of Proposition 6.2.4, we can obtain

$$\left(\tilde{f}_j^{(\star DT)}(\tilde{\mathbf{A}})\right)_\alpha = N_\alpha^{(j)} = \bigcup_{\alpha \le \beta \le 1} M_\beta^{(j)}. \tag{6.23}$$

Let $\tilde{\mathbf{f}}^{(\star DT)}$ be fuzzified from the vector-valued function $\mathbf{f}$ obtained from (6.8) using the family $\{M_\alpha : 0 < \alpha \le 1\}$ defined in (6.19). For $\alpha \in (0,1]$ and $0 < \alpha_s \uparrow \alpha$ with $\alpha > \alpha_s$ for all s, we also have

$$\begin{aligned}
\bigcap_{s=1}^{\infty} M_{\alpha_s} &= \bigcap_{s=1}^{\infty} \left(N_{\alpha_s}^{(1)} \times \cdots \times N_{\alpha_s}^{(m)}\right) \text{ (using (6.22))} \\
&= \left(\bigcap_{s=1}^{\infty} N_{\alpha_s}^{(1)}\right) \times \cdots \times \left(\bigcap_{s=1}^{\infty} N_{\alpha_s}^{(m)}\right) \\
&= N_\alpha^{(1)} \times \cdots \times N_\alpha^{(m)} = M_\alpha \text{ (using (6.22) and (6.21))},
\end{aligned}$$

which says that

$$M_\alpha = \bigcap_{s=1}^{\infty} M_{\alpha_s} \text{ for } \alpha \in (0,1] \text{ and } 0 < \alpha_s \uparrow \alpha \text{ with } \alpha > \alpha_s \text{ for all } s. \tag{6.24}$$

For $\alpha \in (0,1]$, using (6.23), (6.24), and a similar argument to the proof of Proposition 6.2.4, we can obtain

$$\left(\tilde{\mathbf{f}}^{(\star DT)}(\tilde{\mathbf{A}})\right)_\alpha = M_\alpha = N_\alpha^{(1)} \times \cdots \times N_\alpha^{(m)} = \left(\tilde{f}_1^{(\star DT)}(\tilde{\mathbf{A}})\right)_\alpha \times \cdots \times \left(\tilde{f}_m^{(\star DT)}(\tilde{\mathbf{A}})\right)_\alpha.$$

By referring to (6.16), we can also obtain the 0-level set

$$\left(\tilde{\mathbf{f}}^{(\star DT)}(\tilde{\mathbf{A}})\right)_0 = \left(\tilde{f}_1^{(\star DT)}(\tilde{\mathbf{A}})\right)_0 \times \cdots \times \left(\tilde{f}_m^{(\star DT)}(\tilde{\mathbf{A}})\right)_0.$$

The above results are summarized below.

Theorem 6.2.7 *Let $\mathbf{f} : \mathbb{R}^n \to \mathbb{R}^m$ be a vector-valued function defined on $\mathbb{R}^m$, and let $\tilde{A}^{(1)}, \tilde{A}^{(2)}, \ldots, \tilde{A}^{(n)}$ be fuzzy intervals. Suppose that the family $\{M_\alpha : \alpha \in (0,1]\}$ is given by*

$$M_\alpha = \left(\bigcup_{\alpha \le \beta \le 1} M_\beta^{(1)}\right) \times \cdots \times \left(\bigcup_{\alpha \le \beta \le 1} M_\beta^{(m)}\right) \subset \mathbb{R}^m,$$

where $M_\alpha^{(j)}$ are bounded closed intervals given by

$$M_\alpha^{(j)} = \left[\min\left\{f_j\left(\tilde{\mathbf{A}}_\alpha^L\right), f_j\left(\tilde{\mathbf{A}}_\alpha^U\right)\right\}, \max\left\{f_j\left(\tilde{\mathbf{A}}_\alpha^L\right), f_j\left(\tilde{\mathbf{A}}_\alpha^U\right)\right\}\right] \text{ for } j = 1, \ldots, m.$$

We also assume that the functions $\zeta_j^L : [0,1] \to \mathbb{R}$ *defined by*

$$\zeta_j^L(\alpha) = \min\left\{f_j\left(\tilde{\mathbf{A}}_\alpha^L\right), f_j\left(\tilde{\mathbf{A}}_\alpha^U\right)\right\}$$

are lower semi-continuous on $[0,1]$ *and the functions* $\zeta_j^U : [0,1] \to \mathbb{R}$ *defined by*

$$\zeta_j^U(\alpha) = \max\left\{f_j\left(\tilde{\mathbf{A}}_\alpha^L\right), f_j\left(\tilde{\mathbf{A}}_\alpha^U\right)\right\}$$

are upper semi-continuous on $[0,1]$ *for* $j = 1, \ldots, m$. *Then, for* $\alpha \in [0,1]$, *we have*

$$\left(\tilde{\mathbf{f}}^{(\star DT)}(\tilde{\mathbf{A}})\right)_\alpha = M_\alpha = N_\alpha^{(1)} \times \cdots \times N_\alpha^{(m)} = \left(\tilde{f}_1^{(\star DT)}(\tilde{\mathbf{A}})\right)_\alpha \times \cdots \times \left(\tilde{f}_m^{(\star DT)}(\tilde{\mathbf{A}})\right)_\alpha \tag{6.25}$$

and

$$\left(\tilde{f}_j^{(\star DT)}(\tilde{\mathbf{A}})\right)_\alpha = N_\alpha^{(j)} = \bigcup_{\alpha\le\beta\le 1} M_\beta^{(j)} \text{ for } j = 1, \ldots, m, \tag{6.26}$$

where

$$\begin{aligned} N_\alpha^{(j)} = \bigcup_{\alpha\le\beta\le 1} M_\beta^{(j)} &= \left[\min_{\alpha\le\beta\le 1} \zeta_j^L(\beta), \max_{\alpha\le\beta\le 1} \zeta_j^U(\beta)\right] \\ &= \left[\min_{\alpha\le\beta\le 1} \min\left\{f_j\left(\tilde{\mathbf{A}}_\beta^L\right), f_j\left(\tilde{\mathbf{A}}_\beta^L\right)\right\}, \max_{\alpha\le\beta\le 1} \max\left\{f_j\left(\tilde{\mathbf{A}}_\beta^L\right), f_j\left(\tilde{\mathbf{A}}_\beta^L\right)\right\}\right] \\ &= \left[\min\left\{\min_{\alpha\le\beta\le 1} f_j\left(\tilde{\mathbf{A}}_\beta^L\right), \min_{\alpha\le\beta\le 1} f_j\left(\tilde{\mathbf{A}}_\beta^U\right)\right\},\right. \\ &\quad \left.\times \max\left\{\max_{\{\beta\in[0,1]:\beta\ge\alpha\}} f_j\left(\tilde{\mathbf{A}}_\beta^L\right), \max_{\{\beta\in[0,1]:\beta\ge\alpha\}} f_j\left(\tilde{\mathbf{A}}_\beta^U\right)\right\}\right] \end{aligned} \tag{6.27}$$

are bounded closed intervals for $j = 1, \ldots, m$.

Example 6.2.8 Continued from Example 6.1.1, we see that the component functions f_j of $\mathbf{f}$ are given by

$$f_j\left(a_1, \ldots, a_n, b_1, \ldots, b_n\right) = a_j + b_j \text{ for } j = 1, \ldots, n.$$

We are going to consider M_α in (6.22). First of all, from (6.18), for $j = 1, \ldots, m$, we have

$$\begin{aligned} M_\alpha^{(j)} &= \left[\min\left\{\eta_j^L(\alpha), \eta_j^U(\alpha)\right\}, \max\left\{\eta_j^L(\alpha), \eta_j^U(\alpha)\right\}\right] \\ &= \left[\min\left\{f_j\left(\tilde{\mathbf{A}}_\alpha^L, \tilde{\mathbf{B}}_\alpha^L\right), f_j\left(\tilde{\mathbf{A}}_\alpha^U, \tilde{\mathbf{B}}_\alpha^U\right)\right\}, \max\left\{f_j\left(\tilde{\mathbf{A}}_\alpha^L, \tilde{\mathbf{B}}_\alpha^L\right), f_j\left(\tilde{\mathbf{A}}_\alpha^U\right), \tilde{\mathbf{B}}_\alpha^U\right\}\right] \\ &= \left[\min\left\{\tilde{A}_{j\alpha}^L + \tilde{B}_{j\alpha}^L, \tilde{A}_{j\alpha}^U + \tilde{B}_{j\alpha}^U\right\}, \max\left\{\tilde{A}_{j\alpha}^L + \tilde{B}_{j\alpha}^L, \tilde{A}_{j\alpha}^U + \tilde{B}_{j\alpha}^U\right\}\right] \\ &= \left[\tilde{A}_{j\alpha}^L + \tilde{B}_{j\alpha}^L, \tilde{A}_{j\alpha}^U + \tilde{B}_{j\alpha}^U\right]. \end{aligned}$$

From (6.23), we have

$$N_\alpha^{(j)} = \bigcup_{\alpha\le\beta\le 1} M_\beta^{(j)} = \bigcup_{\alpha\le\beta\le 1} \left[\tilde{A}_{j\beta}^L + \tilde{B}_{j\beta}^L, \tilde{A}_{j\beta}^U + \tilde{B}_{j\beta}^U\right] = \left[\tilde{A}_{j\alpha}^L + \tilde{B}_{j\alpha}^L, \tilde{A}_{j\alpha}^U + \tilde{B}_{j\alpha}^U\right] = M_\alpha^{(j)}.$$

Using (6.22), we obtain

$$M_\alpha = \left[\tilde{A}_{1\alpha}^L + \tilde{B}_{1\alpha}^L, \tilde{A}_{1\alpha}^U + \tilde{B}_{1\alpha}^U\right] \times \cdots \times \left[\tilde{A}_{n\alpha}^L + \tilde{B}_{n\alpha}^L, \tilde{A}_{n\alpha}^U + \tilde{B}_{n\alpha}^U\right].$$

We can define the addition of $\tilde{\mathbf{A}}$ and $\tilde{\mathbf{B}}$ to be

$$\tilde{\mathbf{A}} \oplus_{DT}^{\star} \tilde{\mathbf{B}} = \tilde{\mathbf{f}}^{(\star DT)}\left(\tilde{\mathbf{A}}, \tilde{\mathbf{B}}\right)$$

with membership function given by

$$\xi_{\tilde{\mathbf{A}} \oplus_{DT}^{\star} \tilde{\mathbf{B}}}(\mathbf{y}) = \xi_{\tilde{\mathbf{f}}^{(\star DT)}(\tilde{\mathbf{A}}, \tilde{\mathbf{B}})}(\mathbf{y}) = \sup_{0<\alpha\leq 1} \alpha \cdot \chi_{M_\alpha}(\mathbf{y}).$$

For $\alpha \in (0,1]$, using (6.25) in Theorem 6.2.7, we have

$$\begin{aligned}\left(\tilde{\mathbf{A}} \oplus_{DT}^{\star} \tilde{\mathbf{B}}\right)_\alpha &= \left(\tilde{\mathbf{f}}^{(\star DT)}(\tilde{\mathbf{A}}, \tilde{\mathbf{B}})\right)_\alpha = M_\alpha \\ &= \left[\tilde{A}_{1\alpha}^L + \tilde{B}_{1\alpha}^L, \tilde{A}_{1\alpha}^U + \tilde{B}_{1\alpha}^U\right] \times \cdots \times \left[\tilde{A}_{n\alpha}^L + \tilde{B}_{n\alpha}^L, \tilde{A}_{n\alpha}^U + \tilde{B}_{n\alpha}^U\right].\end{aligned}$$

6.2.3 Non-Nested Family Using Endpoints

Now, we consider the product set

$$M_\alpha = M_\alpha^{(1)} \times \cdots \times M_\alpha^{(m)} \subset \mathbb{R}^m,$$

where $M_\alpha^{(j)}$ are defined in (6.18) for $j = 1, \ldots, m$. In this case, the family $\{M_\alpha : 0 < \alpha \leq 1\}$ is not necessarily nested. Let us recall the concept of canonical fuzzy interval in Definition 2.3.3.

Lemma 6.2.9 *Let* $\mathbf{f} : \mathbb{R}^n \to \mathbb{R}^m$ *be a continuous vector-valued function defined on* $\mathbb{R}^m$, *and let* $\tilde{A}^{(1)}, \tilde{A}^{(2)}, \ldots, \tilde{A}^{(n)}$ *be canonical fuzzy intervals. Suppose that the family* $\{M_\alpha : \alpha \in (0,1]\}$ *is given by*

$$M_\alpha = M_\alpha^{(1)} \times \cdots \times M_\alpha^{(m)} \subset \mathbb{R}^m,$$

where $M_\alpha^{(j)}$ *is a bounded closed interval given by*

$$M_\alpha^{(j)} = \left[\min\left\{f_j\left(\tilde{\mathbf{A}}_\alpha^L\right), f_j\left(\tilde{\mathbf{A}}_\alpha^U\right)\right\}, \max\left\{f_j\left(\tilde{\mathbf{A}}_\alpha^L\right), f_j\left(\tilde{\mathbf{A}}_\alpha^U\right)\right\}\right]$$

for $j = 1, \ldots, m$. *Then, for any fixed* $y \in \mathbb{R}$ *and* $\mathbf{y} \in \mathbb{R}^m$, *the following functions*

$$\phi_y^{(j)}(\alpha) = \alpha \cdot \chi_{M_\alpha^{(j)}}(y) \text{ for } j = 1, \ldots, m$$

and

$$\phi_{\mathbf{y}}(\alpha) = \alpha \cdot \chi_{M_\alpha}(\mathbf{y})$$

are upper semi-continuous on $[0,1]$.

Proof. Since each $\tilde{A}^{(i)}$ is a canonical fuzzy interval for $i = 1, \ldots, n$, the definition says that $\tilde{A}_{i\alpha}^L$ and $\tilde{A}_{i\alpha}^U$ are continuous with respect to α on $[0,1]$ for $i = 1, \ldots, n$. The continuity of $\mathbf{f}$ also says that the following functions

$$\zeta_j^L(\alpha) = \min\left\{f_j\left(\tilde{\mathbf{A}}_\alpha^L\right), f_j\left(\tilde{\mathbf{A}}_\alpha^U\right)\right\} \text{ and } \zeta_j^U(\alpha) = \max\left\{f_j\left(\tilde{\mathbf{A}}_\alpha^L\right), f_j\left(\tilde{\mathbf{A}}_\alpha^U\right)\right\}$$

are continuous on $[0,1]$. Now, we have $M_\alpha^{(j)} = [\zeta_j^L(\alpha), \zeta_j^U(\alpha)]$. For any $\alpha \in (0,1]$ and any convergent sequence $\{\alpha_s\}_{s=1}^\infty$ in $[0,1]$ with $\alpha_s \to \alpha$ as $s \to \infty$, we are going to claim

$$\bigcap_{s=1}^\infty M_{\alpha_s}^{(j)} \subseteq M_\alpha^{(j)} \text{ and } \bigcap_{s=1}^\infty M_{\alpha_s} \subseteq M_\alpha. \tag{6.28}$$

For any $\mathbf{y} \in \bigcap_{s=1}^{\infty} M_{\alpha_s}^{(j)}$, i.e, $\mathbf{y} \in M_{\alpha_s}^{(j)}$ for all s, it means that $\zeta_j^L(\alpha_s) \leq y \leq \zeta_j^U(\alpha_s)$ for all s. By taking $s \to \infty$ and using continuity, we obtain $\zeta_j^L(\alpha) \leq y \leq \zeta_j^U(\alpha)$, which says that $\mathbf{y} \in M_\alpha^{(j)}$. This proves the inclusion $\bigcap_{s=1}^{\infty} M_{\alpha_s}^{(j)} \subseteq M_\alpha^{(j)}$, which also implies the inclusion $\bigcap_{s=1}^{\infty} M_{\alpha_s} \subseteq M_\alpha$.

Given any fixed $\mathbf{y} \in \mathbb{R}^m$, we define the following set

$$F_r = \left\{\alpha \in [0,1] \,:\, \phi_{\mathbf{y}}(\alpha) \geq r\right\}$$

for $r \in [0,1]$. In order to show that the function $\phi_{\mathbf{y}}$ is upper semi-continuous on $[0,1]$, we want to claim that the set F_r is closed for each $r \in [0,1]$. If $r = 0$, then $F_r = [0,1]$ is a closed set. For $r > 0$ and for each $\alpha \in \mathrm{cl}(F_r)$, the concept of closure says that there exists a sequence $\{\alpha_s\}_{s=1}^{\infty}$ in F_r satisfying $\alpha_s \to \alpha$. Therefore, we have $\phi_{\mathbf{y}}(\alpha_s) \geq r$, which says that $\alpha_s \geq r$ and $\mathbf{y} \in M_{\alpha_s}$ for all s. We also have

$$\alpha = \lim_{n\to\infty} \alpha_s \geq r > 0 \text{ and } \mathbf{y} \in \bigcap_{s=1}^{\infty} M_{\alpha_s},$$

which implies $\mathbf{y} \in M_\alpha$ by (6.28). This says that $\alpha \cdot \chi_{M_\alpha}(\mathbf{y}) \geq r$, i.e. $\alpha \in F_r$. Therefore, we obtain the inclusion $\mathrm{cl}(F_r) \subseteq F_r$, which means that F_r is closed; that is, the function $\phi_{\mathbf{y}}$ is upper semi-continuous on $[0,1]$. We can similarly show that the functions $\phi_{\mathbf{y}}^{(j)}$ are upper semi-continuous on $[0,1]$ for $j = 1, \ldots, m$. This completes the proof. ∎

Let $\tilde{f}_j^{(\dagger DT)}$ be fuzzified from f_j obtained from (6.8) using the family $\{M_\alpha^{(j)} : 0 < \alpha \leq 1\}$ given by

$$M_\alpha^{(j)} = \left[\min\left\{f_j\left(\tilde{\mathbf{A}}_\alpha^L\right), f_j\left(\tilde{\mathbf{A}}_\alpha^U\right)\right\}, \max\left\{f_j\left(\tilde{\mathbf{A}}_\alpha^L\right), f_j\left(\tilde{\mathbf{A}}_\alpha^U\right)\right\}\right].$$

Let $\tilde{\mathbf{f}}^{(\dagger DT)}$ be fuzzified from $\mathbf{f}$ obtained from (6.8) using the family $\{M_\alpha : 0 < \alpha \leq 1\}$ given by

$$M_\alpha = M_\alpha^{(1)} \times \cdots \times M_\alpha^{(m)} \subset \mathbb{R}^m.$$

Their α-level sets are presented below.

Theorem 6.2.10 *Let $\mathbf{f} : \mathbb{R}^n \to \mathbb{R}^m$ be a continuous vector-valued function defined on $\mathbb{R}^m$, and let $\tilde{A}^{(1)}, \tilde{A}^{(2)}, \ldots, \tilde{A}^{(n)}$ be canonical fuzzy intervals. Suppose that the family $\{M_\alpha : 0 < \alpha \leq 1\}$ is given by*

$$M_\alpha = M_\alpha^{(1)} \times \cdots \times M_\alpha^{(m)} \subset \mathbb{R}^m,$$

where $M_\alpha^{(j)}$ is a bounded closed interval given by

$$M_\alpha^{(j)} = \left[\min\left\{f_j\left(\tilde{\mathbf{A}}_\alpha^L\right), f_j\left(\tilde{\mathbf{A}}_\alpha^U\right)\right\}, \max\left\{f_j\left(\tilde{\mathbf{A}}_\alpha^L\right), f_j\left(\tilde{\mathbf{A}}_\alpha^U\right)\right\}\right] \tag{6.29}$$

for $j = 1, \ldots, m$.

(i) *For $\alpha \in (0,1]$, we have*

$$\left(\tilde{\mathbf{f}}^{(\dagger DT)}(\tilde{\mathbf{A}})\right)_\alpha = \bigcup_{\alpha\leq\beta\leq 1} M_\beta = \bigcup_{\alpha\leq\beta\leq 1} \left(M_\beta^{(1)} \times \cdots \times M_\beta^{(m)}\right). \tag{6.30}$$

and

$$\left(\tilde{\mathbf{f}}^{(\dagger DT)}(\tilde{\mathbf{A}})\right)_0 = cl\left(\bigcup_{0<\alpha\leq 1} \left(\tilde{\mathbf{f}}^{(\dagger DT)}(\tilde{\mathbf{A}})\right)_\alpha\right).$$

(ii) *For* $\alpha \in [0,1]$*, we have*

$$\left(\tilde{f}_j^{(\dagger DT)}(\tilde{\mathbf{A}})\right)_\alpha = \bigcup_{\alpha\le\beta\le1} M_\beta^{(j)} \tag{6.31}$$

$$= \left[\min_{\alpha\le\beta\le1}\min\left\{f_j\left(\tilde{\mathbf{A}}_\beta^L\right), f_j\left(\tilde{\mathbf{A}}_\beta^L\right)\right\}, \max_{\alpha\le\beta\le1}\max\left\{f_j\left(\tilde{\mathbf{A}}_\beta^L\right), f_j\left(\tilde{\mathbf{A}}_\beta^L\right)\right\}\right]$$

$$= \left[\min\left\{\min_{\alpha\le\beta\le1} f_j\left(\tilde{\mathbf{A}}_\beta^L\right), \min_{\alpha\le\beta\le1} f_j\left(\tilde{\mathbf{A}}_\beta^U\right)\right\}, \max\left\{\max_{\alpha\le\beta\le1} f_j\left(\tilde{\mathbf{A}}_\beta^L\right), \max_{\alpha\le\beta\le1} f_j\left(\tilde{\mathbf{A}}_\beta^U\right)\right\}\right]$$

are bounded closed intervals for $j = 1, \ldots, m$.

Proof. Given any $\alpha \in (0,1]$, suppose that $\mathbf{y} \in (\tilde{\mathbf{f}}^{(\dagger DT)}(\tilde{\mathbf{A}}))_\alpha$ and $\mathbf{y} \notin M_\beta$ for all $\beta \in [0,1]$ with $\beta \ge \alpha$. Then $\beta \cdot \chi_{M_\beta}(\mathbf{y}) < \alpha$ for all $\beta \in [0,1]$. Since $\phi_\mathbf{y}(\beta) = \beta \cdot \chi_{M_\beta}(\mathbf{y})$ is upper semi-continuous on [0,1] by Lemma 6.2.9, the supremum of the function $\phi_\mathbf{y}$ is attained by Proposition 1.4.4. This says that

$$\xi_{\tilde{\mathbf{f}}^{(\dagger DT)}(\tilde{\mathbf{A}})}(\mathbf{y}) = \sup_{\beta\in[0,1]} \phi_\mathbf{y}(\beta) = \sup_{\beta\in[0,1]} \beta \cdot \chi_{M_\beta}(\mathbf{y}) = \max_{\beta\in[0,1]} \beta \cdot \chi_{M_\beta}(\mathbf{y}) = \beta^* \cdot \chi_{M_{\beta^*}}(\mathbf{y}) < \alpha$$

for some $\beta^* \in [0,1]$, which violates $\mathbf{y} \in (\tilde{\mathbf{f}}^{(\dagger DT)}(\tilde{\mathbf{A}}))_\alpha$. Therefore, there exists β_0 satisfying $\beta_0 \ge \alpha$ and $\mathbf{y} \in M_{\beta_0}$, which shows the following inclusion

$$\left(\tilde{\mathbf{f}}^{(\dagger DT)}(\tilde{\mathbf{A}})\right)_\alpha \subseteq \bigcup_{\alpha\le\beta\le1} M_\beta.$$

On the other hand, the following inclusion

$$\bigcup_{\alpha\le\beta\le1} M_\beta \subseteq \left\{\mathbf{y} \in \mathbb{R}^m : \sup_{\beta\in[0,1]} \beta \cdot \chi_{M_\beta}(\mathbf{y}) \ge \alpha\right\}$$

$$= \{\mathbf{y} \in \mathbb{R}^m : \xi_{\tilde{\mathbf{f}}^{(\dagger DT)}(\tilde{\mathbf{A}})}(\mathbf{y}) \ge \alpha\} = (\tilde{\mathbf{f}}^{(\dagger DT)}(\tilde{\mathbf{A}}))_\alpha$$

is obvious. This shows (6.30). We can similarly obtain

$$\left(\tilde{f}_j^{(\dagger DT)}(\tilde{\mathbf{A}})\right)_\alpha = \bigcup_{\alpha\le\beta\le1} M_\beta^{(j)} \text{ for } j = 1, \ldots, m \text{ and } \alpha \in (0,1]. \tag{6.32}$$

Since $\mathbf{f}$ is continuous and $\tilde{A}^{(i)}$ are taken to be canonical fuzzy intervals, it follows that the functions ζ_j^L and ζ_j^U in Theorem 6.2.7 are continuous on [0,1]. By referring to 6.32 and 6.27, we can obtain 6.31. This completes the proof. ■

We notice that

$$\left(\tilde{\mathbf{f}}^{(\dagger DT)}(\tilde{\mathbf{A}})\right)_\alpha \ne \left(\tilde{f}_1^{(\dagger DT)}(\tilde{\mathbf{A}})\right)_\alpha \times \cdots \times \left(\tilde{f}_m^{(\dagger DT)}(\tilde{\mathbf{A}})\right)_\alpha$$

in general for $\alpha \in (0,1]$.

Example 6.2.11 Continued from Example 6.1.1 by considering the canonical fuzzy intervals, we take

$$M_\alpha = M_\alpha^{(1)} \times \cdots \times M_\alpha^{(m)},$$

where $M_\alpha^{(j)}$ are defined in (6.18) for $j = 1, \ldots, m$. From Example 6.2.8, we have obtained

$$M_\alpha^{(j)} = N_\alpha^{(j)} = \left[\tilde{A}_{j\alpha}^L + \tilde{B}_{j\alpha}^L, \tilde{A}_{j\alpha}^U + \tilde{B}_{j\alpha}^U\right].$$

We can define the addition of $\tilde{\mathbf{A}}$ and $\tilde{\mathbf{B}}$ to be

$$\tilde{\mathbf{A}} \oplus_{DT}^{\dagger} \tilde{\mathbf{B}} = \tilde{\mathbf{f}}^{(\dagger DT)}\left(\tilde{\mathbf{A}}, \tilde{\mathbf{B}}\right)$$

with membership function given by (6.8). More precisely, the membership function is given by

$$\xi_{\tilde{\mathbf{A}} \oplus_{DT}^{\dagger} \tilde{\mathbf{B}}}(\mathbf{y}) = \xi_{\tilde{\mathbf{f}}^{(\dagger DT)}(\tilde{\mathbf{A}}, \tilde{\mathbf{B}})}(\mathbf{y}) = \sup_{0<\alpha\leq 1} \alpha \cdot \chi_{M_\alpha}(\mathbf{y}).$$

For $\alpha \in (0,1]$, using (6.30) in Theorem 6.2.10, we have

$$\begin{aligned}
\left(\tilde{\mathbf{A}} \oplus_{DT}^{\dagger} \tilde{\mathbf{B}}\right)_\alpha &= \left(\tilde{\mathbf{f}}^{(\dagger DT)}(\tilde{\mathbf{A}}, \tilde{\mathbf{B}})\right)_\alpha = \bigcup_{\alpha\leq\beta\leq 1} \left(M_\beta^{(1)} \times \cdots \times M_\beta^{(m)}\right) \\
&= \bigcup_{\alpha\leq\beta\leq 1} \left(\left[\tilde{A}_{1\beta}^L + \tilde{B}_{1\beta}^L, \tilde{A}_{1\beta}^U + \tilde{B}_{1\beta}^U\right] \times \cdots \times \left[\tilde{A}_{n\beta}^L + \tilde{B}_{n\beta}^L, \tilde{A}_{n\beta}^U + \tilde{B}_{n\beta}^U\right]\right) \\
&= \left[\tilde{A}_{1\alpha}^L + \tilde{B}_{1\alpha}^L, \tilde{A}_{1\alpha}^U + \tilde{B}_{1\alpha}^U\right] \times \cdots \times \left[\tilde{A}_{n\alpha}^L + \tilde{B}_{n\alpha}^L, \tilde{A}_{n\alpha}^U + \tilde{B}_{n\alpha}^U\right].
\end{aligned}$$

6.3 The Relationships between EP and DT

In what follows, we are going to present the relationships for the fuzzification using the extension principle and the expression in the decomposition theorem by considering their equivalences and fuzziness.

6.3.1 The Equivalences

We first present the equivalences regarding the fuzzification of vector-valued functions $\mathbf{f} : \mathbb{R}^n \to \mathbb{R}^m$ using the extension principle and the expression in the decomposition theorem.

Theorem 6.3.1 *Let $\mathbf{f} : \mathbb{R}^n \to \mathbb{R}^m$ be a continuous vector-valued function defined on $\mathbb{R}^m$, and let $\tilde{A}^{(1)}, \tilde{A}^{(2)}, \ldots, \tilde{A}^{(n)}$ be fuzzy intervals. Suppose that the family $\{M_\alpha : 0 < \alpha \leq 1\}$ is given by*

$$M_\alpha = \mathbf{f}\left(\tilde{A}_\alpha^{(1)}, \ldots, \tilde{A}_\alpha^{(n)}\right),$$

and that the function $\mathfrak{A} : [0,1]^n \to [0,1]$ is given by

$$\mathfrak{A}\left(\alpha_1, \ldots, \alpha_n\right) = \min\left\{\alpha_1, \ldots, \alpha_n\right\}.$$

Let $\tilde{\mathbf{f}}^{(EP)}(\tilde{\mathbf{A}})$ be obtained from $\mathbf{f}$ based on the function $\mathfrak{A}$ by referring to (6.1), and let $\tilde{\mathbf{f}}^{(\diamond DT)}(\tilde{\mathbf{A}})$ be obtained from $\mathbf{f}$ based on the family $\{M_\alpha : 0 < \alpha \leq 1\}$ by referring to (6.8). Then, we have the equivalence

$$\tilde{\mathbf{f}}^{(EP)}(\tilde{\mathbf{A}}) = \tilde{\mathbf{f}}^{(\diamond DT)}(\tilde{\mathbf{A}}).$$

Proof. It is clear to see that the family $\{M_\alpha : 0 < \alpha \leq 1\}$ is nested. For $j = 1, \ldots, m$, we have

$$\begin{aligned}
&\left\{f_j(\mathbf{x}) : \min\left\{\xi_{\tilde{A}^{(1)}}(x_1), \ldots, \xi_{\tilde{A}^{(n)}}(x_n)\right\} \geq \alpha\right\} \\
&\quad = \left\{f_j(\mathbf{x}) : \xi_{\tilde{A}^{(i)}}(x_i) \geq \alpha \text{ for each } i = 1, \ldots, n\right\}
\end{aligned}$$

$$= \left\{ f_j(\mathbf{x}) : x_i \in \tilde{A}_\alpha^{(i)} \text{ for each } i = 1, \ldots, n \right\}$$
$$= f_j\left(\tilde{A}_\alpha^{(1)}, \ldots, \tilde{A}_\alpha^{(n)}\right), \tag{6.33}$$

which implies

$$\begin{aligned}
&\left\{\mathbf{f}(\mathbf{x}) : \min\left\{\xi_{\tilde{A}^{(1)}}(x_1), \ldots, \xi_{\tilde{A}^{(n)}}(x_n)\right\} \geq \alpha\right\} \\
&= \left\{\left(f_1(\mathbf{x}), \ldots, f_n(\mathbf{x})\right) : \min\left\{\xi_{\tilde{A}^{(1)}}(x_1), \ldots, \xi_{\tilde{A}^{(n)}}(x_n)\right\} \geq \alpha\right\} \\
&= \left(f_1\left(\tilde{A}_\alpha^{(1)}, \ldots, \tilde{A}_\alpha^{(n)}\right), \ldots, f_m\left(\tilde{A}_\alpha^{(1)}, \ldots, \tilde{A}_\alpha^{(n)}\right)\right) \text{ (using (6.33))} \\
&= \mathbf{f}\left(\tilde{A}_\alpha^{(1)}, \ldots, \tilde{A}_\alpha^{(n)}\right) = M_\alpha.
\end{aligned} \tag{6.34}$$

Let $\mathcal{R}_{\tilde{\mathbf{A}}}^{(EP)}$ and $\mathcal{R}_{\tilde{\mathbf{A}}}^{(\diamond DT)}$ denote the ranges of membership functions of $\tilde{\mathbf{f}}^{(EP)}(\tilde{\mathbf{A}})$ and $\tilde{\mathbf{f}}^{(\diamond DT)}(\tilde{\mathbf{A}})$, respectively. We first assume that $0 \notin \mathcal{R}_{\tilde{\mathbf{A}}}^{(EP)}$ and $0 \notin \mathcal{R}_{\tilde{\mathbf{A}}}^{(\diamond DT)}$. From (6.1), given any $\mathbf{y} \in \mathbb{R}^m$, we have

$$\begin{aligned}
0 < r = \xi_{\tilde{\mathbf{f}}^{(EP)}(\tilde{\mathbf{A}})}(\mathbf{y}) &= \sup_{\{\mathbf{x}:\mathbf{y}=\mathbf{f}(\mathbf{x})\}} \mathfrak{A}\left(\xi_{\tilde{A}^{(1)}}(x_1), \ldots, \xi_{\tilde{A}^{(n)}}(x_n)\right) \\
&= \sup_{\{\mathbf{x}:\mathbf{y}=\mathbf{f}(\mathbf{x})\}} \min\left\{\xi_{\tilde{A}^{(1)}}(x_1), \ldots, \xi_{\tilde{A}^{(n)}}(x_n)\right\}.
\end{aligned} \tag{6.35}$$

Let m^* be an integer satisfying $r - \frac{1}{m^*} > 0$. Then $r - \frac{1}{m} > 0$ for all $m \geq m^*$. We also see that $r - \frac{1}{m} \in [0,1]$. According to the concept of supremum by referring to (6.35), given any $m \geq m^*$, there exists $(x_1^{(m)}, \ldots, x_n^{(m)}) \in \mathbb{R}^n$ satisfying

$$0 < r - \frac{1}{m} \leq \min\left\{\xi_{\tilde{A}^{(1)}}\left(x_1^{(m)}\right), \ldots, \xi_{\tilde{A}^{(n)}}\left(x_n^{(m)}\right)\right\} \text{ and } \mathbf{y} = f\left(x_1^{(m)}, \ldots, x_n^{(m)}\right).$$

From (6.34), it follows that $\mathbf{y} \in M_{r-\frac{1}{m}}$ for all $m \geq m^*$, since $r - \frac{1}{m} \in [0,1]$ for all $m \geq m^*$. Then, for each $m \geq m^*$, by referring to (6.8), we have

$$\xi_{\tilde{\mathbf{f}}^{(\diamond DT)}(\tilde{\mathbf{A}})}(\mathbf{y}) = \sup_{0<\alpha\leq 1} \alpha \cdot \chi_{M_\alpha}(\mathbf{y}) \geq r - \frac{1}{m}.$$

By taking the limit as $m \to \infty$, we obtain

$$\xi_{\tilde{\mathbf{f}}^{(\diamond DT)}(\tilde{\mathbf{A}})}(\mathbf{y}) \geq r = \xi_{\tilde{\mathbf{f}}^{(EP)}(\tilde{\mathbf{A}})}(\mathbf{y}).$$

On the other hand, since $0 \notin \mathcal{R}_{\tilde{\mathbf{A}}}^{(\diamond DT)}$ is assumed, from (6.8), given any $\mathbf{y} \in \mathbb{R}^m$, we have

$$0 < r = \xi_{\tilde{\mathbf{f}}^{(\diamond DT)}(\tilde{\mathbf{A}})}(\mathbf{y}) = \sup_{0<\alpha\leq 1} \alpha \cdot \chi_{M_\alpha}(\mathbf{y}). \tag{6.36}$$

Let m^* be an integer satisfying $r - \frac{1}{m^*} > 0$. Then $r - \frac{1}{m} > 0$ for all $m \geq m^*$. We also see that $r - \frac{1}{m} \in [0,1]$. According to the concept of supremum by referring to (6.36), given any $m \geq m^*$, there exists $\alpha^{(m)} \in (0,1]$ satisfying

$$r - \frac{1}{m} \leq \alpha^{(m)} \text{ and } \mathbf{y} \in M_{\alpha^{(m)}}.$$

Since $M_{\alpha^{(m)}} \subseteq M_{r-\frac{1}{m}}$ by the nestendess, it follows that $\mathbf{y} \in M_{r-\frac{1}{m}}$ for all $m \geq m^*$. From (6.34), we obtain

$$\mathbf{y} = \mathbf{f}\left(x_1^{(m)}, \ldots, x_n^{(m)}\right) \text{ and } \min\left\{\xi_{\tilde{A}^{(1)}}\left(x_1^{(m)}\right), \ldots, \xi_{\tilde{A}^{(n)}}\left(x_n^{(m)}\right)\right\} \geq r - \frac{1}{m} > 0$$

for some $(x_1^{(m)}, \ldots, x_n^{(m)}) \in \mathbb{R}^n$. Then, from (6.35), for each $m \geq m^*$, we have

$$\begin{aligned}\xi_{\tilde{\mathbf{f}}^{(EP)}(\tilde{\mathbf{A}})}(\mathbf{y}) &= \sup_{\{\mathbf{x}:\mathbf{y}=\mathbf{f}(\mathbf{x})\}} \min\left\{\xi_{\tilde{A}^{(1)}}(x_1), \ldots, \xi_{\tilde{A}^{(n)}}(x_n)\right\} \\ &\geq \min\left\{\xi_{\tilde{A}^{(1)}}\left(x_1^{(m)}\right), \ldots, \xi_{\tilde{A}^{(n)}}\left(x_n^{(m)}\right)\right\} \geq r - \frac{1}{m}.\end{aligned}$$

By taking the limit as $m \to \infty$, we obtain

$$\xi_{\tilde{\mathbf{f}}^{(EP)}(\tilde{\mathbf{A}})}(\mathbf{y}) \geq r = \xi_{\tilde{\mathbf{f}}^{(\diamond DT)}(\tilde{\mathbf{A}})}(\mathbf{y}).$$

Therefore, we conclude that if $\mathbf{y} \in \mathbb{R}^m$ satisfies $\xi_{\tilde{\mathbf{f}}^{(EP)}(\tilde{\mathbf{A}})}(\mathbf{y}) > 0$ and $\xi_{\tilde{\mathbf{f}}^{(\diamond DT)}(\tilde{\mathbf{A}})}(\mathbf{y}) > 0$, then $\xi_{\tilde{\mathbf{f}}^{(EP)}(\tilde{\mathbf{A}})}(\mathbf{y}) = \xi_{\tilde{\mathbf{f}}^{(\diamond DT)}(\tilde{\mathbf{A}})}(\mathbf{y})$.

Now, we assume that $0 \in \mathcal{R}_{\tilde{\mathbf{A}}}^{(EP)}$ or $0 \in \mathcal{R}_{\tilde{\mathbf{A}}}^{(\diamond DT)}$. Suppose that

$$0 = \xi_{\tilde{\mathbf{f}}^{(EP)}(\tilde{\mathbf{A}})}(\mathbf{y}) = \sup_{\{\mathbf{x}:\mathbf{y}=\mathbf{f}(\mathbf{x})\}} \min\left\{\xi_{\tilde{A}^{(1)}}(x_1), \ldots, \xi_{\tilde{A}^{(n)}}(x_n)\right\}.$$

The nonnegativity says that $\min\left\{\xi_{\tilde{A}^{(1)}}(x_1), \ldots, \xi_{\tilde{A}^{(n)}}(x_n)\right\} = 0$ for any $\mathbf{x} \in \mathbb{R}^m$ with $\mathbf{y} = \mathbf{f}(\mathbf{x})$. In other words, there exists $\xi_{\tilde{A}^{(i)}}(x_i) = 0$ with $\mathbf{y} = \mathbf{f}(\mathbf{x})$, i.e. $x_i \notin \tilde{A}_\alpha^{(i)}$ for each $\alpha \in (0,1]$, which also says that $\mathbf{y} \notin \mathbf{f}\left(\tilde{A}_\alpha^{(1)}, \ldots, \tilde{A}_\alpha^{(n)}\right) = M_\alpha$ for each $\alpha \in (0,1]$. It follows that

$$\xi_{\tilde{\mathbf{f}}^{(\diamond DT)}(\tilde{\mathbf{A}})}(\mathbf{y}) = \sup_{0<\alpha\leq 1} \alpha \cdot \chi_{M_\alpha}(\mathbf{y}) = 0 = \xi_{\tilde{\mathbf{f}}^{(EP)}(\tilde{\mathbf{A}})}(\mathbf{y}).$$

On the other hand, suppose that

$$0 = \xi_{\tilde{\mathbf{f}}^{(\diamond DT)}(\tilde{\mathbf{A}})}(\mathbf{y}) = \sup_{0<\alpha\leq 1} \alpha \cdot \chi_{M_\alpha}(\mathbf{y}).$$

It follows that $\mathbf{y} \notin M_\alpha = \mathbf{f}(\tilde{A}_\alpha^{(1)}, \ldots, \tilde{A}_\alpha^{(n)})$ for any $\alpha \in (0,1]$. In other words, there exists $x_i \notin \tilde{A}_\alpha^{(i)}$ with $\mathbf{y} = \mathbf{f}(\mathbf{x})$ for each $\alpha \in (0,1]$, i.e. $\xi_{\tilde{A}^{(i)}}(x_i) = 0$ with $\mathbf{y} = \mathbf{f}(\mathbf{x})$ for each $\alpha \in (0,1]$, which also says that $\min\left\{\xi_{\tilde{A}^{(1)}}(x_1), \ldots, \xi_{\tilde{A}^{(n)}}(x_n)\right\} = 0$ with any $\mathbf{y} = \mathbf{f}(\mathbf{x})$. It follows that

$$\xi_{\tilde{\mathbf{f}}^{(EP)}(\tilde{\mathbf{A}})}(\mathbf{y}) = \sup_{\{\mathbf{x}:\mathbf{y}=\mathbf{f}(\mathbf{x})\}} \min\left\{\xi_{\tilde{A}^{(1)}}(x_1), \ldots, \xi_{\tilde{A}^{(n)}}(x_n)\right\} = 0 = \xi_{\tilde{\mathbf{f}}^{(\diamond DT)}(\tilde{\mathbf{A}})}(\mathbf{y}).$$

Therefore, we conclude that

$$\xi_{\tilde{\mathbf{f}}^{(\diamond DT)}(\tilde{\mathbf{A}})}(\mathbf{y}) = 0 \text{ if and only if } \xi_{\tilde{\mathbf{f}}^{(EP)}(\tilde{\mathbf{A}})}(\mathbf{y}) = 0.$$

We also have $\mathcal{R}_{\tilde{\mathbf{A}}}^{(EP)} = \mathcal{R}_{\tilde{\mathbf{A}}}^{(\diamond DT)}$. This completes the proof. ■

Recall that the α-level sets of $\tilde{\mathbf{f}}^{(EP)}(\tilde{\mathbf{A}})$ and $\tilde{f}_j^{(EP)}(\tilde{\mathbf{A}})$ for $j = 1, \ldots, m$ can be obtained by Theorem 6.1.2, and the α-level sets of $\tilde{\mathbf{f}}^{(\diamond DT)}(\tilde{\mathbf{A}})$ and $\tilde{f}_j^{(\diamond DT)}(\tilde{\mathbf{A}})$ for $j = 1, \ldots, m$ can be obtained by Theorem 6.2.6.

Theorem 6.3.2 *Let $\mathbf{f} : \mathbb{R}^n \to \mathbb{R}^m$ be a continuous vector-valued function defined on $\mathbb{R}^m$, and let $\tilde{A}^{(1)}, \tilde{A}^{(2)}, \ldots, \tilde{A}^{(n)}$ be fuzzy intervals. Suppose that each component function f_j is monotonic for $j = 1, \ldots, m$; that is, each f_j is increasing or decreasing for $j = 1, \ldots, m$. Let $\tilde{\mathbf{f}}^{(\diamond DT)}(\tilde{\mathbf{A}})$ and $\tilde{f}_j^{(\diamond DT)}(\tilde{\mathbf{A}})$ for $j = 1, \ldots, m$ be obtained from Theorem 6.2.6, and let $\tilde{\mathbf{f}}^{(\star DT)}(\tilde{\mathbf{A}})$ and $\tilde{f}_j^{(\star DT)}(\tilde{\mathbf{A}})$ for $j = 1, \ldots, m$ be obtained from Theorem 6.2.7. Then, we have the following properties.*

(i) *We have*

$$\tilde{\mathbf{f}}^{(\diamond DT)}(\tilde{\mathbf{A}}) = \tilde{\mathbf{f}}^{(\star DT)}(\tilde{\mathbf{A}}) \text{ and } \tilde{f}_j^{(\diamond DT)}(\tilde{\mathbf{A}}) = \tilde{f}_j^{(\star DT)}(\tilde{\mathbf{A}}) \text{ for } j = 1, \ldots, m.$$

For $\alpha \in [0,1]$, *we also have*

$$\begin{aligned}\left(\tilde{\mathbf{f}}^{(\diamond DT)}(\tilde{\mathbf{A}})\right)_\alpha &= \left(\tilde{\mathbf{f}}^{(\star DT)}(\tilde{\mathbf{A}})\right)_\alpha = \left(\tilde{f}_1^{(\diamond DT)}(\tilde{\mathbf{A}})\right)_\alpha \times \cdots \times \left(\tilde{f}_m^{(\diamond DT)}(\tilde{\mathbf{A}})\right)_\alpha \\ &= \left(\tilde{f}_1^{(\star DT)}(\tilde{\mathbf{A}})\right)_\alpha \times \cdots \times \left(\tilde{f}_m^{(\star DT)}(\tilde{\mathbf{A}})\right)_\alpha .\end{aligned} \tag{6.37}$$

More precisely, for $j = 1, \ldots, m$, *we have the following equalities.*

- *If* f_j *is increasing, then*

$$\left(\tilde{f}_j^{(\diamond DT)}(\tilde{\mathbf{A}})\right)_\alpha = \left(\tilde{f}_j^{(\star DT)}(\tilde{\mathbf{A}})\right)_\alpha = \left[f_j\left(\tilde{\mathbf{A}}_\alpha^L\right), f_j\left(\tilde{\mathbf{A}}_\alpha^U\right)\right].$$

- *If* f_j *is decreasing, then*

$$\left(\tilde{f}_j^{(\diamond DT)}(\tilde{\mathbf{A}})\right)_\alpha = \left(\tilde{f}_j^{(\star DT)}(\tilde{\mathbf{A}})\right)_\alpha = \left[f_j\left(\tilde{\mathbf{A}}_\alpha^U\right), f_j\left(\tilde{\mathbf{A}}_\alpha^L\right)\right].$$

(ii) *Suppose that* $\tilde{A}^{(1)}, \tilde{A}^{(2)}, \ldots, \tilde{A}^{(n)}$ *are taken to be canonical fuzzy intervals. Let* $\tilde{\mathbf{f}}^{(\dagger DT)}(\tilde{\mathbf{A}})$ *be obtained from part (i) of Theorem 6.2.10. Then*

$$\tilde{\mathbf{f}}^{(\diamond DT)}(\tilde{\mathbf{A}}) = \tilde{\mathbf{f}}^{(\star DT)}(\tilde{\mathbf{A}}) = \tilde{\mathbf{f}}^{(\dagger DT)}(\tilde{\mathbf{A}}).$$

(iii) *Suppose that* $\tilde{A}^{(1)}, \tilde{A}^{(2)}, \ldots, \tilde{A}^{(n)}$ *are taken to be canonical fuzzy intervals. Let* $\tilde{f}_j^{(\dagger DT)}(\tilde{\mathbf{A}})$ *for* $j = 1, \ldots, m$ *be obtained from part (ii) of Theorem 6.2.10. Then*

$$\tilde{f}_j^{(\diamond DT)}(\tilde{\mathbf{A}}) = \tilde{f}_j^{(\star DT)}(\tilde{\mathbf{A}}) = \tilde{f}_j^{(\dagger DT)}(\tilde{\mathbf{A}}) \text{ for } j = 1, \ldots, m.$$

Proof. The continuity of vector-valued function $\mathbf{f}$ says that the component functions f_j are also continuous for $j = 1, \ldots, m$. To prove part (i), we first assume that f_j is increasing. For $\alpha \in [0,1]$, from Theorem 6.2.6, we have

$$\left(\tilde{f}_j^{(\diamond DT)}(\tilde{\mathbf{A}})\right)_\alpha = \left[f_j\left(\tilde{\mathbf{A}}_\alpha^L\right), f_j\left(\tilde{\mathbf{A}}_\alpha^U\right)\right].$$

For $\alpha \in [0,1]$, from Theorem 6.2.7, we also have

$$\left(\tilde{f}_j^{(\star DT)}(\tilde{\mathbf{A}})\right)_\alpha = \bigcup_{\alpha \le \beta \le 1} M_\beta^{(j)} = N_\alpha^{(j)} = \left[f_j\left(\tilde{\mathbf{A}}_\alpha^L\right), f_j\left(\tilde{\mathbf{A}}_\alpha^U\right)\right] = \left(\tilde{f}_j^{(\diamond DT)}(\tilde{\mathbf{A}})\right)_\alpha .$$

If f_j is decreasing, then, for $\alpha \in [0,1]$, we can similarly obtain

$$\left(\tilde{f}_j^{(\diamond DT)}(\tilde{\mathbf{A}})\right)_\alpha = \left(\tilde{f}_j^{(\star DT)}(\tilde{\mathbf{A}})\right)_\alpha = \left[f_j\left(\tilde{\mathbf{A}}_\alpha^U\right), f_j\left(\tilde{\mathbf{A}}_\alpha^L\right)\right].$$

On the other hand, the equalities (6.37) can be obtained from (6.17) and (6.25). The equivalences

$$\tilde{\mathbf{f}}^{(\diamond DT)}(\tilde{\mathbf{A}}) = \tilde{\mathbf{f}}^{(\star DT)}(\tilde{\mathbf{A}}) \text{ and } \tilde{f}_j^{(\diamond DT)}(\tilde{\mathbf{A}}) = \tilde{f}_j^{(\star DT)}(\tilde{\mathbf{A}}) \text{ for } j = 1, \ldots, m$$

follow immediately from Proposition 2.2.15.

To prove part (ii), from (6.29) and (6.30) by referring to part (i) of Theorem 6.2.10, for $\alpha \in [0,1]$, we have

$$M_\alpha^{(j)} = \left\{\begin{array}{ll}\left[f_j\left(\tilde{\mathbf{A}}_\alpha^L\right), f_j\left(\tilde{\mathbf{A}}_\alpha^U\right)\right] & \text{if } f_j \text{ is increasing} \\ \left[f_j\left(\tilde{\mathbf{A}}_\alpha^U\right), f_j\left(\tilde{\mathbf{A}}_\alpha^L\right)\right] & \text{if } f_j \text{ is decreasing}\end{array}\right\} = \left(\tilde{f}_j^{(\diamond DT)}(\tilde{\mathbf{A}})\right)_\alpha = \left(\tilde{f}_j^{(\star DT)}(\tilde{\mathbf{A}})\right)_\alpha$$

and

$$\left(\tilde{\mathbf{f}}^{(\dagger DT)}(\tilde{\mathbf{A}})\right)_\alpha = \bigcup_{\alpha \le \beta \le 1} \left(M_\beta^{(1)} \times \cdots \times M_\beta^{(m)}\right).$$

Since, for each $j = 1, \ldots, m$, the family $\{M_\beta^{(j)} : \beta \in [0,1] \text{ with } \beta \ge \alpha\}$ is nested, and $f_j(\tilde{\mathbf{A}}_\beta^L)$ and $f_j(\tilde{\mathbf{A}}_\beta^U)$ are continuous with respect to β on $[0,1]$, it follows that

$$\begin{aligned}\left(\tilde{\mathbf{f}}^{(\dagger DT)}(\tilde{\mathbf{A}})\right)_\alpha &= \bigcup_{\alpha \le \beta \le 1} \left(M_\beta^{(1)} \times \cdots \times M_\beta^{(m)}\right) = M_\alpha^{(1)} \times \cdots \times M_\alpha^{(m)} \\ &= \left(\tilde{f}_1^{(\diamond DT)}(\tilde{\mathbf{A}})\right)_\alpha \times \cdots \times \left(\tilde{f}_m^{(\diamond DT)}(\tilde{\mathbf{A}})\right)_\alpha \\ &= \left(\tilde{f}_1^{(\star DT)}(\tilde{\mathbf{A}})\right)_\alpha \times \cdots \times \left(\tilde{f}_m^{(\star DT)}(\tilde{\mathbf{A}})\right)_\alpha .\end{aligned}$$

Using Proposition 2.2.15 and part (i), can we obtain the desired equivalences.

To prove part (iii), from part (ii) of Theorem 6.2.10 and the proof of part (ii), we see that, for $\alpha \in [0,1]$,

$$\left(\tilde{f}_j^{(\dagger DT)}(\tilde{\mathbf{A}})\right)_\alpha = \bigcup_{\alpha \le \beta \le 1} M_\beta^{(j)} = M_\alpha^{(j)} \text{ for } j = 1, \ldots, m.$$

This completes the proof. ∎

Example 6.3.3 Continued from Example 6.1.1 by considering the canonical fuzzy intervals, for $\alpha \in (0,1]$, we have obtained

$$\begin{aligned}(\tilde{\mathbf{A}} \oplus_{ET} \tilde{\mathbf{B}})_\alpha &= (\tilde{\mathbf{A}} \oplus_{DT}^{\diamond} \tilde{\mathbf{B}})_\alpha = (\tilde{\mathbf{A}} \oplus_{DT}^{\star} \tilde{\mathbf{B}})_\alpha = \left(\tilde{\mathbf{A}} \oplus_{DT}^{\dagger} \tilde{\mathbf{B}}\right)_\alpha \\ &= \left[\tilde{A}_{1\alpha}^L + \tilde{B}_{1\alpha}^L, \tilde{A}_{1\alpha}^U + \tilde{B}_{1\alpha}^L\right] \times \cdots \times \left[\tilde{A}_{n\alpha}^L + \tilde{B}_{n\alpha}^L, \tilde{A}_{n\alpha}^U + \tilde{B}_{n\alpha}^L\right].\end{aligned}$$

In other words, we have

$$\tilde{\mathbf{A}} \oplus_{ET} \tilde{\mathbf{B}} = \tilde{\mathbf{A}} \oplus_{DT}^{\diamond} \tilde{\mathbf{B}} = \tilde{\mathbf{A}} \oplus_{DT}^{\star} \tilde{\mathbf{B}} = \tilde{\mathbf{A}} \oplus_{DT}^{\dagger} \tilde{\mathbf{B}},$$

which says that the four additions are equivalent. However, the four differences $\tilde{\mathbf{A}} \ominus_{ET} \tilde{\mathbf{B}}$, $\tilde{\mathbf{A}} \ominus_{DT}^{\diamond} \tilde{\mathbf{B}}$, $\tilde{\mathbf{A}} \ominus_{DT}^{\star} \tilde{\mathbf{B}}$, and $\tilde{\mathbf{A}} \ominus_{DT}^{\dagger} \tilde{\mathbf{B}}$ cannot be all equivalent, which can be checked by the reader.

6.3.2 The Fuzziness

Theorem 6.3.2 is based on the monotonicity of component functions f_j for $j = 1, \ldots, m$. In other words, if the component functions are not monotonic, we have difficulty establishing their equivalence. However, we can compare their fuzziness, which will be presented below.

Given any two fuzzy sets $\tilde{A}$ and $\tilde{B}$ in $\mathbb{R}^m$, we can compare their fuzziness based on the α-level sets. The formal definition is given below.

Definition 6.3.4 Let $\tilde{A}$ and $\tilde{B}$ be two fuzzy sets in $\mathbb{R}^m$. We say that $\tilde{A}$ is **fuzzier** than $\tilde{B}$ when $\tilde{B}_\alpha \subseteq \tilde{A}_\alpha$ for all $\alpha \in (0,1]$.

Remark 6.3.5 By referring to (2.5), if $\tilde{A}$ is fuzzier than $\tilde{B}$ then $\tilde{B}_{0+} \subseteq \tilde{A}_{0+}$, which also implies

$$\tilde{B}_0 = \text{cl}(\tilde{B}_{0+}) \subseteq \text{cl}(\tilde{A}_{0+}) = \tilde{A}_0.$$

Therefore $\tilde{A}$ is fuzzier than $\tilde{B}$ if and only if $\tilde{B}_\alpha \subseteq \tilde{A}_\alpha$ for all $\alpha \in [0,1]$.

Given a real number $x \in \mathbb{R}$, when we say that the fuzzy numbers $\tilde{A}$ and $\tilde{B}$ are approximately equal to x, it means that $\xi_{\tilde{A}}(x) = 1 = \xi_{\tilde{B}}(x)$, where $\xi_{\tilde{A}}$ and $\xi_{\tilde{B}}$ denote the membership functions of $\tilde{A}$ and $\tilde{B}$, respectively. When $\tilde{A}$ is fuzzier than $\tilde{B}$, it is natural to pick $\tilde{B}$, which has less fuzziness.

Suppose now that we plan to collect n real number data $a_1, \ldots, a_n \in \mathbb{R}$. Owing to the unexpected situation, we cannot exactly obtain the desired data. Instead, we can just obtain the fuzzy data $\tilde{A}^{(1)}, \ldots, \tilde{A}^{(n)}$ that can be described by some suitable membership functions. Suppose that the multivariable vector-valued function $\mathbf{f}$ is regarded as a "black box." Given the fuzzy input data $\tilde{A}^{(1)}, \ldots, \tilde{A}^{(n)}$ as described above, we have two ways to obtain the output data through the "black box" $\mathbf{f}$. One is based on the extension principle to obtain $\tilde{\mathbf{f}}^{(EP)}(\tilde{\mathbf{A}})$, and another one is based on the expression in the decomposition theorem to obtain $\tilde{\mathbf{f}}^{(\diamond DT)}(\tilde{\mathbf{A}})$. We shall show that $\tilde{\mathbf{f}}^{(EP)}(\tilde{\mathbf{A}})$ is fuzzier than $\tilde{\mathbf{f}}^{(\diamond DT)}(\tilde{\mathbf{A}})$ and $\tilde{\mathbf{f}}^{(\star DT)}(\tilde{\mathbf{A}})$. In other words, we prefer to take $\tilde{\mathbf{f}}^{(\diamond DT)}(\tilde{\mathbf{A}})$ or $\tilde{\mathbf{f}}^{(\star DT)}(\tilde{\mathbf{A}})$, which have less fuzziness.

Lemma 6.3.6 *For any $\alpha \in (0,1]$, we have the following inclusion*

$$\bigcup_{\alpha \le \beta \le 1} \left(M_\beta^{(1)} \times \cdots \times M_\beta^{(m)} \right) \subseteq \left(\bigcup_{\alpha \le \beta \le 1} M_\beta^{(1)} \right) \times \cdots \times \left(\bigcup_{\alpha \le \beta \le 1} M_\beta^{(m)} \right).$$

Proof. The proof is a routine argument and is left for the reader. ■

Theorem 6.3.7 *Let $\mathbf{f} : \mathbb{R}^n \to \mathbb{R}^m$ be a continuous vector-valued function defined on $\mathbb{R}^m$. Then, we have the following properties.*

(i) *Let $\tilde{\mathbf{f}}^{(EP)}(\tilde{\mathbf{A}})$ and $\tilde{\mathbf{f}}^{(\diamond DT)}(\tilde{\mathbf{A}})$ be obtained from Theorems 6.1.2 and 6.2.6, respectively. Then, we have the following inclusion*

$$\left(\tilde{\mathbf{f}}^{(\diamond DT)}(\tilde{\mathbf{A}}) \right)_\alpha \subseteq \left(\tilde{\mathbf{f}}^{(EP)}(\tilde{\mathbf{A}}) \right)_\alpha \text{ for ear ch } \alpha \in (0,1],$$

which says that $\tilde{\mathbf{f}}^{(EP)}(\tilde{\mathbf{A}})$ is fuzzier than $\tilde{\mathbf{f}}^{(\diamond DT)}(\tilde{\mathbf{A}})$.

(ii) *Let $\tilde{\mathbf{f}}^{(EP)}(\tilde{\mathbf{A}})$ and $\tilde{\mathbf{f}}^{(\star DT)}(\tilde{\mathbf{A}})$ be obtained from Theorems 6.1.2 and 6.2.7, respectively. Then, we have the following inclusion*

$$\left(\tilde{\mathbf{f}}^{(\star DT)}(\tilde{\mathbf{A}}) \right)_\alpha \subseteq \left(\tilde{\mathbf{f}}^{(EP)}(\tilde{\mathbf{A}}) \right)_\alpha \text{ for each } \alpha \in (0,1],$$

which says that $\tilde{\mathbf{f}}^{(EP)}(\tilde{\mathbf{A}})$ is fuzzier than $\tilde{\mathbf{f}}^{(\star DT)}(\tilde{\mathbf{A}})$.

(iii) *Let $\tilde{\mathbf{f}}^{(\star DT)}(\tilde{\mathbf{A}})$ and $\tilde{\mathbf{f}}^{(\diamond DT)}(\tilde{\mathbf{A}})$ be obtained from Theorems 6.2.7 and 6.2.6, respectively. Then, we have the following inclusion*

$$\left(\tilde{\mathbf{f}}^{(\star DT)}(\tilde{\mathbf{A}}) \right)_\alpha \subseteq \left(\tilde{\mathbf{f}}^{(\diamond DT)}(\tilde{\mathbf{A}}) \right)_\alpha \text{ for each } \alpha \in (0,1],$$

which says that $\tilde{\mathbf{f}}^{(\diamond DT)}(\tilde{\mathbf{A}})$ is fuzzier than $\tilde{\mathbf{f}}^{(\star DT)}(\tilde{\mathbf{A}})$.

(iv) *Suppose that $\tilde{A}^{(1)}, \tilde{A}^{(2)}, \ldots, \tilde{A}^{(n)}$ are taken to be canonical fuzzy intervals. Let $\tilde{\mathbf{f}}^{(\star DT)}(\tilde{\mathbf{A}})$ and $\tilde{\mathbf{f}}^{(\dagger DT)}(\tilde{\mathbf{A}})$ be obtained from Theorems 6.2.7 and 6.2.10, respectively. Then, we have the following inclusion*

$$\left(\tilde{\mathbf{f}}^{(\dagger DT)}(\tilde{\mathbf{A}}) \right)_\alpha \subseteq \left(\tilde{\mathbf{f}}^{(\star DT)}(\tilde{\mathbf{A}}) \right)_\alpha \text{ for each } \alpha \in (0,1],$$

which says that $\tilde{\mathbf{f}}^{(\star DT)}(\tilde{\mathbf{A}})$ is fuzzier than $\tilde{\mathbf{f}}^{(\dagger DT)}(\tilde{\mathbf{A}})$.

Proof. To prove part (i), using Theorems 6.2.6 and 6.1.2, we obtain the following inclusions

$$\left(\tilde{\mathbf{f}}^{(\diamond DT)}(\tilde{\mathbf{A}})\right)_\alpha \subseteq \left(\tilde{\mathbf{f}}^{(EP)}(\tilde{\mathbf{A}})\right)_\alpha \text{ for each } \alpha \in (0,1],$$

which also says that $\tilde{\mathbf{f}}^{(EP)}(\tilde{\mathbf{A}})$ is fuzzier than $\tilde{\mathbf{f}}^{(\diamond DT)}(\tilde{\mathbf{A}})$.

To prove part (ii), for $j = 1, \ldots, m$ and $\alpha \in (0,1]$, we have

$$\min\left\{f_j\left(\tilde{\mathbf{A}}_\beta^L\right), f_j\left(\tilde{\mathbf{A}}_\beta^U\right)\right\} \leq \min_{\mathbf{x}\in\tilde{\mathbf{A}}_\beta} f_j(\mathbf{x}) \text{ for } \alpha \leq \beta \leq 1 \tag{6.38}$$

and

$$\max\left\{f_j\left(\tilde{\mathbf{A}}_\beta^L\right), f_j\left(\tilde{\mathbf{A}}_\beta^U\right)\right\} \geq \max_{\mathbf{x}\in\tilde{\mathbf{A}}_\beta} f_j(\mathbf{x}) \text{ for } \alpha \leq \beta \leq 1. \tag{6.39}$$

Therefore, we can obtain the following inclusions

$$\begin{aligned}
N_\alpha^{(j)} &= \left[\min_{\alpha\leq\beta\leq 1} \min\left\{f_j\left(\tilde{\mathbf{A}}_\beta^L\right), f_j\left(\tilde{\mathbf{A}}_\beta^U\right)\right\}, \max_{\alpha\leq\beta\leq 1} \max\left\{f_j\left(\tilde{\mathbf{A}}_\beta^L\right), f_j\left(\tilde{\mathbf{A}}_\beta^U\right)\right\}\right] \\
&\subseteq \left[\min_{\alpha\leq\beta\leq 1}\min_{\mathbf{x}\in\tilde{\mathbf{A}}_\beta} f_j(\mathbf{x}), \max_{\alpha\leq\beta\leq 1}\max_{\mathbf{x}\in\tilde{\mathbf{A}}_\beta} f_j(\mathbf{x})\right] \text{ (using (6.38) and (6.39))} \\
&\subseteq \left[\min_{\mathbf{x}\in\tilde{\mathbf{A}}_\alpha} f_j(\mathbf{x}), \max_{\mathbf{x}\in\tilde{\mathbf{A}}_\alpha} f_j(\mathbf{x})\right].
\end{aligned} \tag{6.40}$$

Using Theorems 6.2.7 and 6.1.2, the inclusions (6.40) imply

$$\left(\tilde{\mathbf{f}}^{(\star DT)}(\tilde{\mathbf{A}})\right)_\alpha \subseteq \left(\tilde{\mathbf{f}}^{(EP)}(\tilde{\mathbf{A}})\right)_\alpha \text{ for each } \alpha \in (0,1],$$

which says that $\tilde{\mathbf{f}}^{(EP)}(\tilde{\mathbf{A}})$ is fuzzier than $\tilde{\mathbf{f}}^{(\star DT)}(\tilde{\mathbf{A}})$.

To prove part (iii), using (6.40), Theorems 6.2.7 and 6.2.6 say that

$$\left(\tilde{f}_j^{(\star DT)}(\tilde{\mathbf{A}})\right)_\alpha = N_\alpha^{(j)} \subseteq \left[\min_{\mathbf{x}\in\tilde{\mathbf{A}}_\alpha} f_j(\mathbf{x}), \max_{\mathbf{x}\in\tilde{\mathbf{A}}_\alpha} f_j(\mathbf{x})\right] = \left(\tilde{f}_j^{(\diamond DT)}(\tilde{\mathbf{A}})\right)_\alpha \text{ for } j = 1, \ldots, m$$

and

$$\left(\tilde{\mathbf{f}}^{(\star DT)}(\tilde{\mathbf{A}})\right)_\alpha \subseteq \left(\tilde{\mathbf{f}}^{(\diamond DT)}(\tilde{\mathbf{A}})\right)_\alpha \text{ for each } \alpha \in (0,1],$$

which says that $\tilde{\mathbf{f}}^{(\diamond DT)}(\tilde{\mathbf{A}})$ is fuzzier than $\tilde{\mathbf{f}}^{(\star DT)}(\tilde{\mathbf{A}})$.

To prove part (iv), for $\alpha \in (0,1]$, we have

$$\begin{aligned}
\left(\tilde{\mathbf{f}}^{(\dagger DT)}(\tilde{\mathbf{A}})\right)_\alpha &= \bigcup_{\alpha\leq\beta\leq 1} M_\beta = \bigcup_{\alpha\leq\beta\leq 1} \left(M_\beta^{(1)} \times \cdots \times M_\beta^{(m)}\right) \text{ (using (6.30))} \\
&\subseteq \left(\bigcup_{\alpha\leq\beta\leq 1} M_\beta^{(1)}\right) \times \cdots \times \left(\bigcup_{\alpha\leq\beta\leq 1} M_\beta^{(m)}\right) \text{ (using Lemma 6.3.6)} \\
&= \left(\tilde{f}_1^{(\star DT)}(\tilde{\mathbf{A}})\right)_\alpha \times \cdots \times \left(\tilde{f}_m^{(\star DT)}(\tilde{\mathbf{A}})\right)_\alpha \text{ (using (6.26))} \\
&= \left(\tilde{\mathbf{f}}^{(\star DT)}(\tilde{\mathbf{A}})\right)_\alpha \text{ (using (6.25))},
\end{aligned}$$

which says that $\tilde{\mathbf{f}}^{(\star DT)}(\tilde{\mathbf{A}})$ is fuzzier than $\tilde{\mathbf{f}}^{(\dagger DT)}(\tilde{\mathbf{A}})$. This completes the proof. ∎

Remark 6.3.8 Suppose that $\tilde{A}^{(1)}, \tilde{A}^{(2)}, \dots, \tilde{A}^{(n)}$ are taken to be canonical fuzzy intervals. Theorem 6.3.7 says that we have the following inclusions

$$\left(\tilde{\mathbf{f}}^{(\dagger DT)}(\tilde{\mathbf{A}})\right)_\alpha \subseteq \left(\tilde{\mathbf{f}}^{(\star DT)}(\tilde{\mathbf{A}})\right)_\alpha \subseteq \left(\tilde{\mathbf{f}}^{(\diamond DT)}(\tilde{\mathbf{A}})\right)_\alpha \subseteq \left(\tilde{\mathbf{f}}^{(EP)}(\tilde{\mathbf{A}})\right)_\alpha$$

for $\alpha \in (0,1]$. Therefore, the relationships regarding fuzziness are obvious. We also see that $\tilde{\mathbf{f}}^{(\dagger DT)}(\tilde{\mathbf{A}})$ has less fuzziness. In other words, we prefer to use $\tilde{\mathbf{f}}^{(\dagger DT)}(\tilde{\mathbf{A}})$ in real applications, where the α-level sets of $\tilde{\mathbf{f}}^{(\dagger DT)}(\tilde{\mathbf{A}})$ can be calculated according to Theorem 6.2.10.

One of the simple applications for fuzzifying the vector-valued function is to study the arithmetics of vectors of fuzzy intervals. Given any two vectors of fuzzy intervals

$$\tilde{\mathbf{A}} = \left(\tilde{A}^{(1)}, \dots, \tilde{A}^{(n)}\right) \text{ and } \tilde{\mathbf{B}} = \left(\tilde{B}^{(1)}, \dots, \tilde{B}^{(n)}\right),$$

the arithmetics between $\tilde{\mathbf{A}}$ and $\tilde{\mathbf{B}}$ are given below.

- We consider an onto vector-valued function $\mathbf{f} : \mathbb{R}^{2n} \to \mathbb{R}^m$ defined by

  $$\mathbf{f}(a_1, \dots, a_n, b_1, \dots, b_n) = (a_1 + b_1, \dots, a_n + b_n).$$

  Then, we can define the additions

  $$\tilde{\mathbf{A}} \oplus_{EP} \tilde{\mathbf{B}} = \tilde{\mathbf{f}}^{(EP)}(\tilde{\mathbf{A}}, \tilde{\mathbf{B}})$$

  and

  $$\tilde{\mathbf{A}} \oplus_{DT}^{\diamond} \tilde{\mathbf{B}} = \tilde{\mathbf{f}}^{(\diamond DT)}(\tilde{\mathbf{A}}, \tilde{\mathbf{B}}), \quad \tilde{\mathbf{A}} \oplus_{DT}^{\star} \tilde{\mathbf{B}} = \tilde{\mathbf{f}}^{(\star DT)}(\tilde{\mathbf{A}}, \tilde{\mathbf{B}}), \text{ and } \tilde{\mathbf{A}} \oplus_{DT}^{\dagger} \tilde{\mathbf{B}} = \tilde{\mathbf{f}}^{(\dagger DT)}(\tilde{\mathbf{A}}, \tilde{\mathbf{B}}).$$

  We have also shown that they are equivalent as follows

  $$\tilde{\mathbf{A}} \oplus_{EP} \tilde{\mathbf{B}} = \tilde{\mathbf{A}} \oplus_{DT}^{\diamond} \tilde{\mathbf{B}} = \tilde{\mathbf{A}} \oplus_{DT}^{\star} \tilde{\mathbf{B}} = \tilde{\mathbf{A}} \oplus_{DT}^{\dagger} \tilde{\mathbf{B}}.$$

- We consider an onto vector-valued function $\mathbf{f} : \mathbb{R}^{2n} \to \mathbb{R}^m$ defined by

  $$\mathbf{f}(a_1, \dots, a_n, b_1, \dots, b_n) = (a_1 - b_1, \dots, a_n - b_n).$$

  Then, we can define the differences

  $$\tilde{\mathbf{A}} \ominus_{EP} \tilde{\mathbf{B}} = \tilde{\mathbf{f}}^{(EP)}(\tilde{\mathbf{A}}, \tilde{\mathbf{B}})$$

  and

  $$\tilde{\mathbf{A}} \ominus_{DT}^{\diamond} \tilde{\mathbf{B}} = \tilde{\mathbf{f}}^{(\diamond DT)}(\tilde{\mathbf{A}}, \tilde{\mathbf{B}}), \quad \tilde{\mathbf{A}} \ominus_{DT}^{\star} \tilde{\mathbf{B}} = \tilde{\mathbf{f}}^{(\star DT)}(\tilde{\mathbf{A}}, \tilde{\mathbf{B}}), \text{ and } \tilde{\mathbf{A}} \ominus_{DT}^{\dagger} \tilde{\mathbf{B}} = \tilde{\mathbf{f}}^{(\dagger DT)}(\tilde{\mathbf{A}}, \tilde{\mathbf{B}}).$$

  However, their equivalence cannot be guaranteed. Based on the fuzziness, we prefer to use $\tilde{\mathbf{A}} \ominus_{DT}^{\dagger} \tilde{\mathbf{B}}$.

- We consider an onto vector-valued function $\mathbf{f} : \mathbb{R}^{2n} \to \mathbb{R}^m$ defined by

  $$\mathbf{f}(a_1, \dots, a_n, b_1, \dots, b_n) = (a_1 b_1, \dots, a_n b_n).$$

  Then, we can define the cross products

  $$\tilde{\mathbf{A}} \otimes_{EP} \tilde{\mathbf{B}} = \tilde{\mathbf{f}}^{(EP)}(\tilde{\mathbf{A}}, \tilde{\mathbf{B}})$$

  and

  $$\tilde{\mathbf{A}} \otimes_{DT}^{\diamond} \tilde{\mathbf{B}} = \tilde{\mathbf{f}}^{(\diamond DT)}(\tilde{\mathbf{A}}, \tilde{\mathbf{B}}), \quad \tilde{\mathbf{A}} \otimes_{DT}^{\star} \tilde{\mathbf{B}} = \tilde{\mathbf{f}}^{(\star DT)}(\tilde{\mathbf{A}}, \tilde{\mathbf{B}}), \text{ and } \tilde{\mathbf{A}} \otimes_{DT}^{\dagger} \tilde{\mathbf{B}} = \tilde{\mathbf{f}}^{(\dagger DT)}(\tilde{\mathbf{A}}, \tilde{\mathbf{B}}).$$

  However, their equivalence cannot be guaranteed. Based on the fuzziness, we prefer to use $\tilde{\mathbf{A}} \otimes_{DT}^{\dagger} \tilde{\mathbf{B}}$.

Example 6.3.9 Let $\tilde{A}^{(1)}, \ldots, \tilde{A}^{(n)}$ be canonical fuzzy intervals. We are going to present the membership function of $\max\{\tilde{A}^{(1)}, \ldots, \tilde{A}^{(n)}\}$. In other words, we want to fuzzify the following real-valued function

$$f(\mathbf{x}) = \max\left\{x_1, \ldots, x_n\right\}. \tag{6.41}$$

According to the extension principle, the membership function of $\max\{\tilde{A}^{(1)}, \ldots, \tilde{A}^{(n)}\}$ is defined by

$$\begin{aligned}\xi_{\max\{\tilde{A}^{(1)},\ldots,\tilde{A}^{(n)}\}}(y) &= \sup_{\{z:f(\mathbf{x})=y\}} \min\left\{\xi_{\tilde{A}^{(1)}}(x_1), \ldots, \xi_{\tilde{A}^{(n)}}(x_n)\right\}\\ &= \sup_{\{z:\max\{x_1,\ldots,x_n\}=y\}} \min\left\{\xi_{\tilde{A}^{(1)}}(x_1), \ldots, \xi_{\tilde{A}^{(n)}}(x_n)\right\}.\end{aligned}$$

In this case, we write

$$\tilde{f}^{(EP)}(\tilde{\mathbf{A}}) = (\mathrm{EP})\text{-}\max\{\tilde{A}^{(1)}, \ldots, \tilde{A}^{(n)}\}.$$

Since the function f defined in (6.41) is continuous, we have

$$\begin{aligned}\left(\tilde{f}^{(EP)}(\tilde{\mathbf{A}})\right)_\alpha &= \left((\mathrm{EP})\text{-}\max\{\tilde{A}^{(1)}, \ldots, \tilde{A}^{(n)}\}\right)_\alpha\\ &= \left\{\max\{\mathbf{x}\} : x_i \in \tilde{A}^{(i)} \text{ for } i = 1, \ldots, n\right\}\\ &= \left[\max\left\{\tilde{A}^L_{1\alpha}, \ldots, \tilde{A}^L_{n\alpha}\right\}, \max\left\{\tilde{A}^U_{1\alpha}, \ldots, \tilde{A}^U_{n\alpha}\right\}\right]\end{aligned}$$

for each $\alpha \in [0,1]$.

Now, we present the membership function of $\max\{\tilde{A}^{(1)}, \ldots, \tilde{A}^{(n)}\}$ based on the expression in the decomposition theorem. In this case, we write

$$\tilde{f}^{(DT)}(\tilde{\mathbf{A}}) = (\mathrm{DT})\text{-}\max\{\tilde{A}^{(1)}, \ldots, \tilde{A}^{(n)}\}.$$

Since the function f defined in (6.41) is increasing, Theorems 6.3.1 and 6.3.2 say that we have the equality $\tilde{f}^{(DT)}(\tilde{\mathbf{A}}) = \tilde{f}^{(EP)}(\tilde{\mathbf{A}})$.

Example 6.3.10 Let $\tilde{A}$ be a canonical fuzzy interval. We are going to present the membership function of $\sin\tilde{A}$. In other words, we want to fuzzify the real-valued function $f(x) = \sin x$. According to the extension principle, the membership function of $\sin\tilde{A}$ is defined by

$$\xi_{\sin\tilde{A}}(y) = \sup_{\{z:f(x)=y\}} \xi_{\tilde{A}}(x) = \sup_{\{z:\sin x=y\}} \xi_{\tilde{A}}(x).$$

In this case, we write

$$\tilde{f}^{(EP)}(\tilde{A}) = (\mathrm{EP})\text{-}\sin\tilde{A}.$$

Since the function f defined in (6.41) is continuous, we have

$$\left(\tilde{f}^{(EP)}(\tilde{A})\right)_\alpha = \left((\mathrm{EP})\text{-}\sin\tilde{A}\right)_\alpha = \left\{\sin x : x \in \tilde{A}_\alpha\right\} = \left[\min_{x\in\tilde{A}_\alpha} \sin x, \max_{x\in\tilde{A}_\alpha} \sin x\right]$$

for each $\alpha \in [0,1]$.

Next, we present the membership function of $\sin\tilde{A}$ based on the expression in the decomposition theorem. In this case, we write

$$\tilde{f}^{(DT)}(\tilde{A}) = (\mathrm{DT})\text{-}\sin\tilde{A}.$$

Since $\tilde{A}^L_\alpha$ and $\tilde{A}^U_\alpha$ are continuous on [0,1] with respect to α by Proposition 2.3.4, the real-valued functions

$$\zeta^L(\alpha) = \min\left\{\sin \tilde{A}^L_\alpha, \sin \tilde{A}^U_\alpha\right\} \text{ and } \zeta^U(\alpha) = \max\left\{\sin \tilde{A}^L_\alpha, \sin \tilde{A}^U_\alpha\right\}$$

are also continuous on [0,1]. Let $M_\alpha = [\zeta^L(\alpha), \zeta^U(\alpha)]$. Using Theorem 6.2.10, the α-level set of $\tilde{f}^{(\dagger DT)}(\tilde{A})$ is given by

$$\begin{aligned}\left(\tilde{f}^{(\dagger DT)}(\tilde{A})\right)_\alpha &= \left[\min_{\alpha\le\beta\le 1} \min\left\{\sin \tilde{A}^L_\beta, \sin \tilde{A}^U_\beta\right\}, \max_{\alpha\le\beta\le 1} \max\left\{\sin \tilde{A}^L_\beta, \sin \tilde{A}^U_\beta\right\}\right] \\ &= \left[\min\left\{\min_{\alpha\le\beta\le 1} \sin \tilde{A}^L_\beta, \min_{\alpha\le\beta\le 1} \sin \tilde{A}^U_\beta\right\}, \max\left\{\max_{\alpha\le\beta\le 1} \sin \tilde{A}^L_\beta, \max_{\alpha\le\beta\le 1} \sin \tilde{A}^U_\beta\right\}\right]\end{aligned}$$

for $\alpha \in [0,1]$. We can see that $\tilde{f}^{(\dagger DT)}(\tilde{A})$ is also a fuzzy interval. Remark 6.3.8 also says that the fuzzy interval $\tilde{f}^{(EP)}(\tilde{A})$ is fuzzier than the fuzzy interval $\tilde{f}^{(\dagger DT)}(\tilde{A})$. This also says that we prefer to take

$$\tilde{f}^{(\dagger DT)}(\tilde{A}) = (\dagger\text{DT})\text{- } \sin \tilde{A}.$$

to be the fuzzification of sine function.

6.4 Differentiation of Fuzzy Functions

We consider the differentiation of fuzzy functions that are taking values on $\mathfrak{F}(\mathbb{R})$. Two types of differentiations will be considered. One is defined on an open interval in $\mathbb{R}$, and the other one is obtained by fuzzifying the differentiation of real-valued functions.

6.4.1 Defined on Open Intervals

Let $\tilde{f} : I \to \mathfrak{F}(\mathbb{R})$ be a fuzzy function defined on an open interval I. Therefore, each $\tilde{f}(x)$ is a fuzzy interval for $x \in I$. For $\alpha \in [0,1]$, we have

$$(\tilde{f}(x))_\alpha = \left[(\tilde{f}(x))^L_\alpha, (\tilde{f}(x))^U_\alpha\right].$$

In this case, we can define two real-valued functions as follows:

$$\tilde{f}^L_\alpha(x) \equiv \left(\tilde{f}(x)\right)^L_\alpha \text{ and } \tilde{f}^U_\alpha(x) \equiv \left(\tilde{f}(x)\right)^U_\alpha$$

for $\alpha \in [0,1]$. According to the expression in the decomposition theorem presented in Theorem 2.2.13, the differentiation is defined below.

Definition 6.4.1 Let I be an open interval in $\mathbb{R}$. We say that the fuzzy function $\tilde{f} : I \to \mathfrak{F}(\mathbb{R})$ is **differentiable** at x^* when the real-valued functions $\tilde{f}^L_\alpha$ and $\tilde{f}^U_\alpha$ are differentiable at x^* for all $\alpha \in [0,1]$. The membership function of the **fuzzy derivative** $\tilde{f}'(x^*)$ is defined as follows. Let

$$\zeta^L(\alpha) = \min\left\{\left(\tilde{f}^L_\alpha\right)'(x^*), \left(\tilde{f}^U_\alpha\right)'(x^*)\right\} \text{ and } \zeta^U(\alpha) = \max\left\{\left(\tilde{f}^L_\alpha\right)'(x^*), \left(\tilde{f}^U_\alpha\right)'(x^*)\right\}, \tag{6.42}$$

and let $M_\alpha = [\zeta^L(\alpha), \zeta^U(\alpha)]$. The membership function of $\tilde{f}'(x^*)$ is defined by

$$\xi_{\tilde{f}'(x^*)}(r) = \sup_{0<\alpha\leq 1} \alpha \cdot \chi_{M_\alpha}(r).$$

The fuzzy derivative $\tilde{f}'(x^*)$ is just a fuzzy set in $\mathbb{R}$. In order to make $\tilde{f}'(x^*)$ a fuzzy interval, more assumptions are needed, which are shown below.

Proposition 6.4.2 *Suppose that $\tilde{f} : I \to \mathcal{F}(\mathbb{R})$ is differentiable at x^*, and that the functions ζ^L and ζ^U defined in* (6.42) *are lower semi-continuous and upper semi-continuous on* [0,1], *respectively. Then, the fuzzy derivative $\tilde{f}'(x^*)$ is a fuzzy interval with the α-level set given by*

$$\left(\tilde{f}'(x^*)\right)_\alpha$$

$$= \{r : \xi_{\tilde{f}'(x^*)}(r) \geq \alpha\} = \left\{ r : \sup_{t\in[0,1]} t \cdot \chi_{M_t}(r) \geq \alpha \right\} = \left[\min_{\alpha\leq t\leq 1} \zeta^L(t), \max_{\alpha\leq t\leq 1} \zeta^U(t)\right]$$

$$= \left[\min_{\alpha\leq t\leq 1} \min\left\{ \left(\tilde{f}_t^L\right)'(x^*), \left(\tilde{f}_t^U\right)'(x^*)\right\}, \max_{\alpha\leq t\leq 1} \max\left\{ \left(\tilde{f}_t^L\right)'(x^*), \left(\tilde{f}_t^U\right)'(x^*)\right\}\right]$$

$$= \left[\min\left\{ \min_{\alpha\leq t\leq 1} \left(\tilde{f}_t^L\right)'(x^*), \min_{\alpha\leq t\leq 1} \left(\tilde{f}_t^U\right)'(x^*)\right\}, \max\left\{ \max_{\alpha\leq t\leq 1} \left(\tilde{f}_t^L\right)'(x^*), \max_{\alpha\leq t\leq 1} \left(\tilde{f}_t^U\right)'(x^*)\right\}\right].$$

for $\alpha \in [0,1]$.

Proof. Using Theorem 5.5.6, it follows that the fuzzy derivative $\tilde{f}'(x^*)$ is a fuzzy interval. The α-level sets of $\tilde{f}'(x^*)$ can also be obtained by using (5.89) and (5.90). This completes the proof. ■

6.4.2 Fuzzification of Differentiable Functions Using the Extension Principle

Let $\tilde{f}$ be a fuzzified function obtained from a real-valued function $f : \mathbb{R} \to \mathbb{R}$. We are going to consider the differentiations of fuzzified functions that are obtained by the extension principle.

According to (6.1), let $\tilde{f}^{(EP)} : \mathfrak{F}(\mathbb{R}) \to \mathfrak{F}(\mathbb{R})$ be a fuzzified function obtained by the extension principle from a real-valued function $f : \mathbb{R} \to \mathbb{R}$. Given $\tilde{A} \in \mathfrak{F}(\mathbb{R})$, Theorem 6.1.2 says

$$\left(\tilde{f}^{(EP)}(\tilde{A})\right)_\alpha = \left[\min_{x\in\tilde{A}_\alpha} f(x), \max_{x\in\tilde{A}_\alpha} f(x)\right] \tag{6.43}$$

for $\alpha \in [0,1]$. We are going to define the membership function of the fuzzy derivative $\tilde{f'}^{(EP)}(\tilde{A})$ at $\tilde{A} \in \mathfrak{F}(\mathbb{R})$ based on the extension principle.

Definition 6.4.3 Let $f : \mathbb{R} \to \mathbb{R}$ be a real-valued function, and let $\tilde{A}$ be a fuzzy interval. We say that the **(EP)-fuzzy derivative** $\tilde{f'}^{(EP)}(\tilde{A})$ at $\tilde{A}$ exists when the real-valued function f is differentiable at any point in the 0-level set $\tilde{A}_0$ of $\tilde{A}$. The membership function of (EP)-fuzzy derivative $\tilde{f'}^{(EP)}(\tilde{A})$ is defined by

$$\xi_{\tilde{f'}^{(EP)}(\tilde{A})}(y) = \sup_{\{x\in\tilde{A}_0 : y=f'(x)\}} \xi_{\tilde{A}}(x).$$

The (EP)-fuzzy derivative $\tilde{f'}^{(EP)}(\tilde{A})$ is just a fuzzy set in $\mathbb{R}$. In order to make $\tilde{f'}^{(EP)}(\tilde{A})$ a fuzzy interval, an extra assumption is needed.

Proposition 6.4.4 *Let $f : \mathbb{R} \to \mathbb{R}$ be a real-valued function, and let $\tilde{A}$ be a fuzzy interval. Suppose that the real-valued function f is differentiable at any point in the 0-level set $\tilde{A}_0$ of $\tilde{A}$, and that f' is continuous on $\tilde{A}_0$. Then, the (EP)-fuzzy derivative $\tilde{f'}^{(EP)}(\tilde{A})$ is a fuzzy interval and the α-level sets of $(\tilde{f}^{(EP)})'(\tilde{A})$ are given by*

$$\left(\tilde{f'}^{(EP)}(\tilde{A})\right)_\alpha = f'\left(\tilde{A}_\alpha\right) = \left\{f'(x) : x \in \tilde{A}_\alpha\right\} = \left[\min_{x\in\tilde{A}_\alpha} f'(x), \max_{x\in\tilde{A}_\alpha} f'(x)\right] \tag{6.44}$$

for $\alpha \in [0,1]$.

Proof. The desired results follow immediately from Theorem 4.1.3. ■

We see that the α-level set of (EP)-fuzzy derivative $\tilde{f'}^{(EP)}(\tilde{A})$ presented in (6.44) seems to take the derivative directly from the form of the α-level set of $\tilde{f}^{(EP)}(\tilde{A})$ presented in (6.43).

6.4.3 Fuzzification of Differentiable Functions Using the Expression in the Decomposition Theorem

Let $\tilde{f}$ be a fuzzified function obtained from a real-valued function $f : \mathbb{R} \to \mathbb{R}$. We are going to consider the differentiations of fuzzified functions that are obtained by using (6.8), that is, based on the expression in the decomposition theorem.

Definition 6.4.5 Let $f : \mathbb{R} \to \mathbb{R}$ be a real-valued function, and let $\tilde{A}$ be a fuzzy interval. We say that the **($\diamond$DT)-fuzzy derivative** $\tilde{f'}^{(\diamond DT)}(\tilde{A})$ at $\tilde{A}$ exists when the real-valued function f is differentiable at any point in the 0-level set $\tilde{A}_0$ of $\tilde{A}$. The membership function of $\tilde{f'}^{(\diamond DT)}(\tilde{A})$ is defined by

$$\xi_{\tilde{f'}^{(\diamond DT)}(\tilde{A})}(r) = \sup_{0<\alpha\leq 1} \alpha \cdot \chi_{M_\alpha}(r),$$

where M_α is given by

$$M_\alpha = f'\left(\tilde{A}_\alpha\right) = \left\{f'(x) : x \in \tilde{A}_\alpha = \left[\tilde{A}^L_\alpha, \tilde{A}^U_\alpha\right]\right\}$$

for $\alpha \in (0,1]$.

It is clear to see that $\{M_\alpha : 0 < \alpha \leq 1\}$ is a nested family in the sense of $M_\beta \subseteq M_\alpha$ for $\alpha < \beta$, since $\{\tilde{A}_\alpha : 0 < \alpha \leq 1\}$ is a nested family. The ($\diamond$DT)-fuzzy derivative $\tilde{f'}^{(\diamond DT)}(\tilde{A})$ is just a fuzzy set in $\mathbb{R}$. In order to make $\tilde{f'}^{(\diamond DT)}(\tilde{A})$ a fuzzy interval, we need to assume that the derivative function f' is also continuous.

Proposition 6.4.6 *Let $f : \mathbb{R} \to \mathbb{R}$ be a real-valued function, and let $\tilde{A}$ be a fuzzy interval. Suppose that the real-valued function f is differentiable at any point in the 0-level set $\tilde{A}_0$ of $\tilde{A}$, and that f' is continuous on $\tilde{A}_0$. Then, given a family $\{M_\alpha : 0 < \alpha \leq 1\}$ defined by*

$$M_\alpha = f'\left(\tilde{A}_\alpha\right) = \left\{f'(x) : x \in \tilde{A}_\alpha = \left[\tilde{A}^L_\alpha, \tilde{A}^U_\alpha\right]\right\},$$

the ($\diamond$DT)-fuzzy derivative $\widetilde{f'}^{(\diamond DT)}(\tilde{A})$ *is a fuzzy interval and its* α*-level sets are bounded closed intervals given by*

$$\left(\widetilde{f'}^{(\diamond DT)}(\tilde{A})\right)_\alpha = f'\left(\tilde{A}_\alpha\right) = \left[\min_{x\in\tilde{A}_\alpha} f'(x), \max_{x\in\tilde{A}_\alpha} f'(x)\right]$$

for $\alpha \in [0,1]$.

Proof. Using the continuity of f', we can obtain the desired results by following Theorem 6.2.6 immediately. ■

Definition 6.4.7 Let $f : \mathbb{R} \to \mathbb{R}$ be a real-valued function, and let $\tilde{A}$ be a fuzzy interval. We say that the **($\star$DT)-fuzzy derivative** $\widetilde{f'}^{(\star DT)}(\tilde{A})$ at $\tilde{A}$ exists when the real-valued function f is differentiable at any point in the 0-level set $\tilde{A}_0$ of $\tilde{A}$. The membership function of $\widetilde{f'}^{(\star DT)}(\tilde{A})$ is defined as follows. Let

$$\zeta^L(\alpha) = \min\left\{f'\left(\tilde{A}_\alpha^L\right), f'\left(\tilde{A}_\alpha^U\right)\right\} \text{ and } \zeta^U(\alpha) = \max\left\{f'\left(\tilde{A}_\alpha^L\right), f'\left(\tilde{A}_\alpha^U\right)\right\},$$

and let

$$M_\alpha = \bigcup_{\alpha\le\beta\le 1}\left[\zeta^L(\beta), \zeta^U(\beta)\right].$$

The membership function of $\widetilde{f'}^{(\star DT)}(\tilde{A})$ is defined by

$$\xi_{\widetilde{f'}^{(\star DT)}(\tilde{A})}(r) = \sup_{0<\alpha\le 1} \alpha \cdot \chi_{M_\alpha}(r).$$

It is clear to see that $\{M_\alpha : 0 < \alpha \le 1\}$ is a nested family. The ($\star$DT)-fuzzy derivative $\widetilde{f'}^{(\star DT)}(\tilde{A})$ is just a fuzzy set in $\mathbb{R}$. In order to make $\widetilde{f'}^{(\star DT)}(\tilde{A})$ to be a fuzzy interval, we need more assumptions given below.

Proposition 6.4.8 *Let* $f : \mathbb{R} \to \mathbb{R}$ *be a real-valued function, and let* $\tilde{A}$ *be a fuzzy interval. Suppose that the real-valued function* f *is differentiable at any point in the* 0*-level set* $\tilde{A}_0$ *of* $\tilde{A}$*, and that the family* $\{M_\alpha : 0 < \alpha \le 1\}$ *is given by*

$$M_\alpha = \bigcup_{\alpha\le\beta\le 1}\left[\zeta^L(\beta), \zeta^U(\beta)\right],$$

where

$$\zeta^L(\alpha) = \min\left\{f'\left(\tilde{A}_\alpha^L\right), f'\left(\tilde{A}_\alpha^U\right)\right\} \text{ and } \zeta^U(\alpha) = \max\left\{f'\left(\tilde{A}_\alpha^L\right), f'\left(\tilde{A}_\alpha^U\right)\right\}.$$

We also assume that ζ^L *is lower semi-continuous on* $[0,1]$ *and* ζ_j^U *is upper semi-continuous on* $[0,1]$*. Then, the ($\star$DT)-fuzzy derivative* $\widetilde{f'}^{(\star DT)}(\tilde{A})$ *is a fuzzy interval and its* α*-level sets are bounded closed intervals given by*

$$\begin{aligned}\left(\widetilde{f'}^{(\star DT)}(\tilde{A})\right)_\alpha &= M_\alpha = \bigcup_{\alpha\le\beta\le 1}[\zeta^L(\beta), \zeta^U(\beta)] = \left[\min_{\alpha\le\beta\le 1}\zeta^L(\beta), \max_{\alpha\le\beta\le 1}\zeta^U(\beta)\right]\\ &= \left[\min_{\alpha\le\beta\le 1}\min\left\{f'\left(\tilde{A}_\beta^L\right), f'\left(\tilde{A}_\beta^L\right)\right\}, \max_{\alpha\le\beta\le 1}\max\left\{f'\left(\tilde{A}_\beta^L\right), f'\left(\tilde{A}_\beta^L\right)\right\}\right]\\ &= \left[\min\left\{\min_{\alpha\le\beta\le 1} f'\left(\tilde{A}_\beta^L\right), \min_{\alpha\le\beta\le 1} f'\left(\tilde{A}_\beta^U\right)\right\},\right.\end{aligned}$$

$$\max\left\{\max_{\{\beta\in[0,1]:\beta\geq\alpha\}} f'\left(\tilde{A}_\beta^L\right), \max_{\{\beta\in[0,1]:\beta\geq\alpha\}} f'\left(\tilde{A}_\beta^U\right)\right\}\Bigg]$$

for $\alpha \in [0,1]$.

Proof. The desired results follow immediately from Theorem 6.2.7. ∎

Definition 6.4.9 Let $f : \mathbb{R} \to \mathbb{R}$ be a real-valued function, and let $\tilde{A}$ be a fuzzy interval. We say that the **(†DT)-fuzzy derivative** $\tilde{f'}^{(\dagger DT)}(\tilde{A})$ at $\tilde{A}$ exists when the real-valued function f is differentiable at any point in the 0-level set $\tilde{A}_0$ of $\tilde{A}$. The membership function of $\tilde{f'}^{(\dagger DT)}(\tilde{A})$ is defined as follows. Let

$$M_\alpha = \left[\min\left\{f'\left(\tilde{A}_\alpha^L\right), f'\left(\tilde{A}_\alpha^U\right)\right\}, \max\left\{f'\left(\tilde{A}_\alpha^L\right), f'\left(\tilde{A}_\alpha^U\right)\right\}\right].$$

The membership function of $\tilde{f'}^{(\dagger DT)}(\tilde{A})$ is defined by

$$\xi_{\tilde{f'}^{(\dagger DT)}(\tilde{A})}(r) = \sup_{0<\alpha\leq 1} \alpha \cdot \chi_{M_\alpha}(r).$$

We see that $\{M_\alpha : 0 < \alpha \leq 1\}$ is not a nested family. The (†DT)-fuzzy derivative $\tilde{f'}^{(\dagger DT)}(\tilde{A})$ is just a fuzzy set in $\mathbb{R}$. In order to make $\tilde{f'}^{(\star DT)}(\tilde{A})$ a fuzzy interval, we need more assumptions, given below.

Proposition 6.4.10 *Let $f : \mathbb{R} \to \mathbb{R}$ be a real-valued function, and let $\tilde{A}$ be a canonical fuzzy interval. Suppose that the real-valued function f is differentiable at any point in the 0-level set $\tilde{A}_0$ of $\tilde{A}$, and that f' is continuous on $\tilde{A}_0$. Then, given a family $\{M_\alpha : 0 < \alpha \leq 1\}$ defined by*

$$M_\alpha = \left[\min\left\{f'\left(\tilde{A}_\alpha^L\right), f'\left(\tilde{A}_\alpha^U\right)\right\}, \max\left\{f'\left(\tilde{A}_\alpha^L\right), f'\left(\tilde{A}_\alpha^U\right)\right\}\right],$$

the (†DT)-fuzzy derivative $\tilde{f'}^{(\dagger DT)}(\tilde{A})$ is a fuzzy interval and its α-level sets are bounded closed intervals given by

$$\left(\tilde{f'}^{(\dagger DT)}(\tilde{A})\right)_\alpha = \bigcup_{\alpha\leq\beta\leq 1} M_\beta^{(j)}$$

$$= \left[\min_{\alpha\leq\beta\leq 1}\min\left\{f'\left(\tilde{A}_\beta^L\right), f'\left(\tilde{A}_\beta^L\right)\right\}, \max_{\alpha\leq\beta\leq 1}\max\left\{f'\left(\tilde{A}_\beta^L\right), f'\left(\tilde{A}_\beta^L\right)\right\}\right]$$

$$= \left[\min\left\{\min_{\alpha\leq\beta\leq 1} f'\left(\tilde{A}_\beta^L\right), \min_{\alpha\leq\beta\leq 1} f'\left(\tilde{A}_\beta^U\right)\right\}, \max\left\{\max_{\alpha\leq\beta\leq 1} f'\left(\tilde{A}_\beta^L\right), \max_{\alpha\leq\beta\leq 1} f'\left(\tilde{A}_\beta^U\right)\right\}\right]$$

for $\alpha \in [0,1]$.

Proof. The desired results follow immediately from Theorem 6.2.10. ∎

Let $f : \mathbb{R} \to \mathbb{R}$ be a real-valued function, and let $\tilde{A}$ be a fuzzy interval. Suppose that the real-valued function f is differentiable at any point in the 0-level set $\tilde{A}_0$ of $\tilde{A}$. The equivalence and fuzziness are presented below.

- Suppose that that f' is continuous on the 0-level set $\tilde{A}_0$. Propositions 6.4.4 and 6.4.6 says that

$$\widetilde{f'}^{(EP)}(\tilde{A}) = \widetilde{f'}^{(\diamond DT)}(\tilde{A}).$$

Regarding the fuzziness, part (iii) of Theorem 6.3.7 says that

$$\left(\widetilde{f'}^{(\star DT)}(\tilde{A})\right)_\alpha \subseteq \left(\widetilde{f'}^{(\diamond DT)}(\tilde{A})\right)_\alpha = \left(\widetilde{f'}^{(EP)}(\tilde{A})\right)_\alpha$$

for all $\alpha \in (0,1]$, which also means that $\widetilde{f'}^{(\diamond DT)}(\tilde{A})$ is fuzzier than $\widetilde{f'}^{(\star DT)}(\tilde{A})$.

- Under the assumptions of Propositions 6.4.8 and 6.4.10, we see that

$$\widetilde{f'}^{(\star DT)}(\tilde{A}) = \widetilde{f'}^{(\dagger DT)}(\tilde{A}).$$

Regarding the fuzziness, suppose that $\tilde{A}$ is taken to be a canonical fuzzy interval. Part (iv) of Theorem 6.3.7 says that

$$\left(\widetilde{f'}^{(\dagger DT)}(\tilde{A})\right)_\alpha \subseteq \left(\widetilde{f'}^{(\star DT)}(\tilde{A})\right)_\alpha,$$

which also means that $\widetilde{f'}^{(\star DT)}(\tilde{A})$ is fuzzier than $\widetilde{f'}^{(\dagger DT)}(\tilde{A})$.

- Suppose that that f' is continuous on the 0-level set $\tilde{A}_0$, and that $\tilde{A}$ is taken to be a canonical fuzzy interval. Then, we have

$$\left(\widetilde{f'}^{(\dagger DT)}(\tilde{A})\right)_\alpha \subseteq \left(\widetilde{f'}^{(\star DT)}(\tilde{A})\right)_\alpha \subseteq \left(\widetilde{f'}^{(\diamond DT)}(\tilde{A})\right)_\alpha = \left(\widetilde{f'}^{(EP)}(\tilde{A})\right)_\alpha$$

for all $\alpha \in (0,1]$. Therefore, in real applications, we prefer to take $\widetilde{f'}^{(\dagger DT)}(\tilde{A})$, which has less fuzziness than any others.

6.5 Integrals of Fuzzy Functions

Two types of integrals of fuzzy functions will be studied. One considers the fuzzy functions defined on a measurable set E, and the other one considers the fuzzified real-valued function defined on $\mathfrak{F}(\mathbb{R})$. We can also apply this methodology to study fuzzy integral equations.

6.5.1 Lebesgue Integrals on a Measurable Set

Let $\tilde{f} : E \to \mathfrak{F}(\mathbb{R})$ be a fuzzy function defined on a measure space (E, ν). It is clear to see that $\tilde{f}_\alpha^L \leq \tilde{f}_\alpha^U$ for all $\alpha \in [0,1]$. Suppose that $\tilde{f}_\alpha^L$ and $\tilde{f}_\alpha^U$ are Lebesgue-integrable on E. Then, we can consider the following closed interval

$$M_\alpha = \left[\int_E \tilde{f}_\alpha^L(x)d\nu, \int_E \tilde{f}_\alpha^U(x)d\nu\right], \tag{6.45}$$

which consists of two Lebesgue integrals. According to the expression in the decomposition theorem, the Lebesgue integral is defined below.

Definition 6.5.1 Let $\tilde{f} : E \to \mathfrak{F}(\mathbb{R})$ be a fuzzy function defined on a measure space (E, ν). We say that $\tilde{f}$ is **fuzzy Lebesgue-integrable** on E when the real-valued functions $\tilde{f}_\alpha^L$ and

$\tilde{f}_{\alpha}^{U}$ are Lebesgue-integrable on E for all $\alpha \in [0,1]$. The membership function of the **fuzzy Lebesgue integral** $\int_E \tilde{f}(x)dv$ is defined by

$$\xi_{\int_E \tilde{f}(x)dv}(r) = \sup_{0<\alpha\leq 1} \alpha \cdot \chi_{M_\alpha}(r),$$

where the closed interval M_α is given in (6.45).

We are going to claim that the fuzzy Lebesgue integral $\int_E \tilde{f}(x)dv$ is a fuzzy interval and to present its α-level sets.

Proposition 6.5.2 *Let $\tilde{f} : E \to \mathfrak{F}(\mathbb{R})$ be a fuzzy function defined on a measure space (E, v). Suppose that $\tilde{f}$ is fuzzy Lebesgue-integrable on E. Then, the fuzzy Lebesgue integral $\int_E \tilde{f}(x)dv$ is a fuzzy interval with the α-level set given by*

$$\left(\int_E \tilde{f}(x)dv\right)_\alpha = \left[\int_E \tilde{f}_{\alpha}^{L}(x)dv, \int_E \tilde{f}_{\alpha}^{U}(x)dv\right]$$

for $\alpha \in [0,1]$.

Proof. We define two real-valued functions as follows

$$\eta^L(\alpha) = \int_E \tilde{f}_{\alpha}^{L}(x)dv \text{ and } \eta^U(\alpha) = \int_E \tilde{f}_{\alpha}^{U}(x)dv.$$

Given any fixed $x \in E$, since $\tilde{f}(x)$ is a fuzzy interval, it follows that $\tilde{f}_{\alpha}^{L}(x)$ and $\tilde{f}_{\alpha}^{U}(x)$ are left-continuous with respect to α; that is to say, for $\alpha \in (0,1]$ and $\alpha_n \uparrow \alpha$, we have

$$\tilde{f}_{\alpha_n}^{L}(x) \to \tilde{f}_{\alpha}^{L}(x) \text{ and } \tilde{f}_{\alpha_n}^{U}(x) \to \tilde{f}_{\alpha}^{U}(x)$$

pointwise on E. When $x \in E$ is fixed, we also see that $\tilde{f}_{\alpha}^{L}(x)$ is increasing with respect to α and $\tilde{f}_{\alpha}^{U}(x)$ is decreasing with respect to α. Using the *Monotone Convergence Theorem* for Legesgue integrals, we have

$$\eta^L(\alpha) = \int_E \tilde{f}_{\alpha}^{L}(x)dv = \lim_{n\to\infty} \int_E \tilde{f}_{\alpha_n}^{L}(x)dv = \lim_{n\to\infty} \eta^L(\alpha_n)$$

and

$$\eta^U(\alpha) = \int_E \tilde{f}_{\alpha}^{U}(x)dv = \lim_{n\to\infty} \int_E \tilde{f}_{\alpha_n}^{U}(x)dv = \lim_{n\to\infty} \eta^U(\alpha_n),$$

which say that the functions η^L and η^U are left-continuous on $(0,1]$. Theorem 5.5.1 says that $\int_E \tilde{f}(x)dv$ is a fuzzy interval with the α-level sets given by

$$\left(\int_E \tilde{f}(x)dv\right)_\alpha = M_\alpha = \left[\int_E \tilde{f}_{\alpha}^{L}(x)dv, \int_E \tilde{f}_{\alpha}^{U}(x)dv\right]$$

for $\alpha \in (0,1]$ and the 0-level set given by

$$\left(\int_E \tilde{f}(x)dv\right)_0 = \text{cl}\left(\bigcup_{0<\alpha\leq 1} M_\alpha\right) = \text{cl}\left(\bigcup_{0<\alpha\leq 1} \left[\int_E \tilde{f}_{\alpha}^{L}(x)dv, \int_E \tilde{f}_{\alpha}^{U}(x)dv\right]\right).$$

Finally, we are going to claim

$$\text{cl}\left(\bigcup_{0<\alpha\leq 1} \left[\int_E \tilde{f}_{\alpha}^{L}(x)dv, \int_E \tilde{f}_{\alpha}^{U}(x)dv\right]\right) = \left[\int_E \tilde{f}_{0}^{L}(x)dv, \int_E \tilde{f}_{0}^{U}(x)dv\right]. \tag{6.46}$$

Given any fixed $x \in E$, we see that $\tilde{f}_\alpha^L(x)$ and $\tilde{f}_\alpha^U(x)$ are right-continuous at $\alpha = 0$ by referring to Proposition 2.3.4; that is to say, for $\alpha_n \downarrow 0$, we have

$$\tilde{f}_{\alpha_n}^L(x) \to \tilde{f}_0^L(x) \text{ and } \tilde{f}_{\alpha_n}^U(x) \to \tilde{f}_0^U(x)$$

pointwise on E. Using the *Monotone Convergence Theorem* for Legesgue integrals again, we have

$$\int_E \tilde{f}_0^L(x)dv = \lim_{n\to\infty} \int_E \tilde{f}_{\alpha_n}^L(x)dv \text{ and } \int_E \tilde{f}_0^U(x)dv = \lim_{n\to\infty} \int_E \tilde{f}_{\alpha_n}^U(x)dv. \tag{6.47}$$

It is clear that, for $\alpha < \beta$, we have

$$\left[\int_E \tilde{f}_\beta^L(x)dv, \int_E \tilde{f}_\beta^U(x)dv\right] \subseteq \left[\int_E \tilde{f}_\alpha^L(x)dv, \int_E \tilde{f}_\alpha^U(x)dv\right]. \tag{6.48}$$

Using (6.47) and (6.48), we obtain (6.46). This completes the proof. ■

6.5.2 Fuzzy Riemann Integrals Using the Expression in the Decomposition Theorem

We are going to study the fuzzy Riemann integral using the expression in the decomposition theorem. Let $\tilde{A}$ and $\tilde{B}$ be two fuzzy intervals. We write

$$\tilde{A} \preccurlyeq \tilde{B} \text{ when } \tilde{A}_\alpha^U \leq \tilde{B}_\alpha^L \text{ for all } \alpha \in [0,1].$$

In order to propose the fuzzy Riemann integrals from $\tilde{A}$ to $\tilde{B}$, we need to assume $\tilde{A} \preccurlyeq \tilde{B}$.

Definition 6.5.3 Let $f : \mathbb{R} \to \mathbb{R}$ be a real-valued function. Given two fuzzy intervals $\tilde{A}$ and $\tilde{B}$ satisfying $\tilde{A} \preccurlyeq \tilde{B}$, we say that the **($\diamond$DT)-fuzzy Riemann integral** ($\diamond$DT) $\int_{\tilde{A}}^{\tilde{B}} f(x)dx$ exists when the real-valued function f is Riemann-integrable on the closed interval $[\tilde{A}_0^L, \tilde{B}_0^U]$, where $\tilde{A}_0^L$ is the left endpoint of the 0-level set $\tilde{A}_0$ and $\tilde{B}_0^U$ is the right endpoint of the 0-level set $\tilde{B}_0$. The membership function of ($\diamond$DT)-fuzzy Riemann integral ($\diamond$DT) $\int_{\tilde{A}}^{\tilde{B}} f(x)dx$ is defined as follows. We consider the function

$$F(a, b) = \int_a^b f(x)dx$$

defined on $\mathbb{R}^2$. The membership function of ($\diamond$DT) $\int_{\tilde{A}}^{\tilde{B}} f(x)dx$ is defined by

$$\xi_{(\diamond\mathrm{DT}) \int_{\tilde{A}}^{\tilde{B}} f(x)dx}(r) = \sup_{0<\alpha\leq 1} \alpha \cdot \chi_{M_\alpha}(r),$$

where

$$M_\alpha = F\left(\tilde{A}_\alpha, \tilde{B}_\alpha\right) = \left\{\int_a^b f(x)dx : a \in \tilde{A}_\alpha \text{ and } b \in \tilde{B}_\alpha\right\} \tag{6.49}$$

for $\alpha \in (0,1]$.

For $\tilde{A} \preccurlyeq \tilde{B}$ and $\alpha \in (0,1]$, it is clear to see that

$$a \in \tilde{A}_\alpha \text{ and } b \in \tilde{B}_\alpha \text{ imply } a \leq b,$$

which says that (6.49) is well defined. The ($\diamond$DT)-fuzzy Riemann integral ($\diamond$DT) $\int_{\tilde{A}}^{\tilde{B}} f(x)dx$ is just a fuzzy set in $\mathbb{R}$. In order to make ($\diamond$DT) $\int_{\tilde{A}}^{\tilde{B}} f(x)dx$ a fuzzy interval, we need the assumption of continuity.

Proposition 6.5.4 *Let $\tilde{A}$ and $\tilde{B}$ be two canonical fuzzy intervals satisfying $\tilde{A} \preccurlyeq \tilde{B}$, and let $f : \mathbb{R} \to \mathbb{R}$ be a real-valued function such that it is Riemann-integrable on the closed interval $[\tilde{A}_0^L, \tilde{B}_0^U]$. Suppose that the function*

$$F(a,b) = \int_a^b f(x)dx$$

defined on $\mathbb{R}^2$ is continuous. Then, given a family $\{M_\alpha : 0 < \alpha \leq 1\}$ defined by

$$M_\alpha = F\left(\tilde{A}_\alpha, \tilde{B}_\alpha\right) = \left\{ \int_a^b f(x)dx : a \in \tilde{A}_\alpha \text{ and } b \in \tilde{B}_\alpha \right\},$$

the ($\diamond$DT)-fuzzy Riemann integral ($\diamond$DT) $\int_{\tilde{A}}^{\tilde{B}} f(x)dx$ is a fuzzy interval and its α-level sets are bounded closed intervals given by

$$\left((\diamond DT) \int_{\tilde{A}}^{\tilde{B}} f(x)dx \right)_\alpha = \left[\min_{(a,b)\in \tilde{A}_\alpha \times \tilde{B}_\alpha} F(a,b),\ \min_{(a,b)\in \tilde{A}_\alpha \times \tilde{B}_\alpha} F(a,b) \right]$$
$$= \left[\min_{(a,b)\in \tilde{A}_\alpha \times \tilde{B}_\alpha} \int_a^b f(x)dx,\ \min_{(a,b)\in \tilde{A}_\alpha \times \tilde{B}_\alpha} \int_a^b f(x)dx \right]$$

for $\alpha \in [0,1]$.

Proof. The desired results follow immediately from Theorem 6.2.6. ∎

Let $\tilde{A}$ and $\tilde{B}$ be two fuzzy intervals. We write

$\tilde{A} \preccurlyeq \tilde{B}$ when $\tilde{A}_\alpha^L \leq \tilde{B}_\alpha^L$ and $\tilde{A}_\alpha^U \leq \tilde{B}_\alpha^U$ for all $\alpha \in [0,1]$.

It is clear to see that

$\tilde{A} \preccurlyeq \tilde{B}$ implies $\tilde{A} \preccurlyeq \tilde{B}$.

In order to propose the other two types of fuzzy Riemann integrals from $\tilde{A}$ to $\tilde{B}$, we can just assume $\tilde{A} \preccurlyeq \tilde{B}$.

Definition 6.5.5 Let $f : \mathbb{R} \to \mathbb{R}$ be a real-valued function. Given two fuzzy intervals $\tilde{A}$ and $\tilde{B}$ satisfying $\tilde{A} \preccurlyeq \tilde{B}$, we say that the **($\star$DT)-fuzzy Riemann integral** ($\star$DT) $\int_{\tilde{A}}^{\tilde{B}} f(x)dx$ exists when the real-valued function f is Riemann-integrable on the closed interval $[\tilde{A}_0^L, \tilde{B}_0^U]$. The membership function of ($\star$DT)-fuzzy Riemann integral ($\star$DT) $\int_{\tilde{A}}^{\tilde{B}} f(x)dx$ is defined as follows. Let

$$\zeta^L(\alpha) = \min\left\{ \int_{\tilde{A}_\alpha^L}^{\tilde{B}_\alpha^L} f(x)dx,\ \int_{\tilde{A}_\alpha^U}^{\tilde{B}_\alpha^U} f(x)dx \right\}$$

and

$$\zeta^U(\alpha) = \max\left\{ \int_{\tilde{A}_\alpha^L}^{\tilde{B}_\alpha^L} f(x)dx,\ \int_{\tilde{A}_\alpha^U}^{\tilde{B}_\alpha^U} f(x)dx \right\},$$

and let

$$M_\alpha = \bigcup_{\alpha \leq \beta \leq 1} \left[\zeta^L(\beta), \zeta^U(\beta)\right].$$

The membership function of $(\star\text{DT})\int_{\tilde{A}}^{\tilde{B}} f(x)dx$ is defined by

$$\xi_{(\star\text{DT})\int_{\tilde{A}}^{\tilde{B}} f(x)dx}(r) = \sup_{0<\alpha\leq 1} \alpha \cdot \chi_{M_\alpha}(r).$$

We consider the function

$$F(a,b) = \int_a^b f(x)dx$$

defined on $\mathbb{R}^2$. Then, we see that

$$M_\alpha = \bigcup_{\alpha\leq\beta\leq 1}\left[\min\left\{F\left(\tilde{A}_\beta^L,\tilde{B}_\beta^L\right),F\left(\tilde{A}_\beta^U,\tilde{B}_\beta^U\right)\right\}\max\left\{F\left(\tilde{A}_\beta^L,\tilde{B}_\beta^L\right),F\left(\tilde{A}_\beta^U,\tilde{B}_\beta^U\right)\right\}\right],$$

and that $\{M_\alpha : 0 < \alpha \leq 1\}$ is a nested family. The $(\star\text{DT})$-fuzzy Riemann integral $(\star\text{DT})\int_{\tilde{A}}^{\tilde{B}} f(x)dx$ is just a fuzzy set in $\mathbb{R}$. In order to make $(\star\text{DT})\int_{\tilde{A}}^{\tilde{B}} f(x)dx$ a fuzzy interval, we need more assumptions, given below.

Proposition 6.5.6 *Let $\tilde{A}$ and $\tilde{B}$ be two fuzzy intervals satisfying $\tilde{A} \preccurlyeq \tilde{B}$, and let $f : \mathbb{R} \to \mathbb{R}$ be a real-valued function such that it is Riemann-integrable on the closed interval $[\tilde{A}_0^L, \tilde{B}_0^U]$. Then, given a family $\{M_\alpha : 0 < \alpha \leq 1\}$ defined by*

$$M_\alpha = \bigcup_{\alpha\leq\beta\leq 1}\left[\zeta^L(\beta),\zeta^U(\beta)\right],$$

where

$$\zeta^L(\alpha) = \min\left\{F\left(\tilde{A}_\alpha^L,\tilde{B}_\alpha^L\right),F\left(\tilde{A}_\alpha^U,\tilde{B}_\alpha^U\right)\right\} \text{ and } \zeta^U(\alpha) = \max\left\{F\left(\tilde{A}_\alpha^L,\tilde{B}_\alpha^L\right),F\left(\tilde{A}_\alpha^U,\tilde{B}_\alpha^U\right)\right\}.$$

We also assume that ζ^L is lower semi-continuous on [0,1] and ζ_j^U is upper semi-continuous on [0,1]. Then, the $(\star DT)$-fuzzy Riemann integral $(\star DT)\int_{\tilde{A}}^{\tilde{B}} f(x)dx$ is a fuzzy interval and its α-level sets are bounded closed intervals given by

$$\begin{aligned}
\left((\star DT)\int_{\tilde{A}}^{\tilde{B}} f(x)dx\right)_\alpha &= \left[\min_{\alpha\leq\beta\leq 1}\zeta^L(\beta), \max_{\alpha\leq\beta\leq 1}\zeta^U(\beta)\right]\\
&= \left[\min_{\alpha\leq\beta\leq 1}\min\left\{\int_{\tilde{A}_\beta^L}^{\tilde{B}_\beta^L} f(x)dx, \int_{\tilde{A}_\beta^U}^{\tilde{B}_\beta^U} f(x)dx\right\}, \max_{\alpha\leq\beta\leq 1}\max\left\{\int_{\tilde{A}_\beta^L}^{\tilde{B}_\beta^L} f(x)dx, \int_{\tilde{A}_\beta^U}^{\tilde{B}_\beta^U} f(x)dx\right\}\right]\\
&= \left[\min\left\{\min_{\alpha\leq\beta\leq 1}\int_{\tilde{A}_\beta^L}^{\tilde{B}_\beta^L} f(x)dx, \min_{\alpha\leq\beta\leq 1}\int_{\tilde{A}_\beta^U}^{\tilde{B}_\beta^U} f(x)dx\right\},\right.\\
&\qquad \left.\max\left\{\max_{\alpha\leq\beta\leq 1}\int_{\tilde{A}_\beta^L}^{\tilde{B}_\beta^L} f(x)dx, \max_{\alpha\leq\beta\leq 1}\int_{\tilde{A}_\beta^U}^{\tilde{B}_\beta^U} f(x)dx\right\}\right]
\end{aligned}$$

for $\alpha \in [0,1]$.

Proof. The desired results follow immediately from Theorem 6.2.7. ∎

Definition 6.5.7 Let $f : \mathbb{R} \to \mathbb{R}$ be a real-valued function. Given two fuzzy intervals $\tilde{A}$ and $\tilde{B}$ satisfying $\tilde{A} \preccurlyeq \tilde{B}$, we say that the **(†DT)-fuzzy Riemann integral** $(\dagger\text{DT})\int_{\tilde{A}}^{\tilde{B}} f(x)dx$

exists when the real-valued function f is Riemann-integrable on the closed interval $[\tilde{A}_0^L, \tilde{B}_0^U]$. The membership function of (†DT)-fuzzy Riemann integral (†DT) $\int_{\tilde{A}}^{\tilde{B}} f(x)dx$ is defined as follows. Let

$$\zeta^L(\alpha) = \min\left\{ \int_{\tilde{A}_\alpha^L}^{\tilde{B}_\alpha^L} f(x)dx, \int_{\tilde{A}_\alpha^U}^{\tilde{B}_\alpha^U} f(x)dx \right\}$$

and

$$\zeta^U(\alpha) = \max\left\{ \int_{\tilde{A}_\alpha^L}^{\tilde{B}_\alpha^L} f(x)dx, \int_{\tilde{A}_\alpha^U}^{\tilde{B}_\alpha^U} f(x)dx \right\},$$

and let

$$M_\alpha = [\zeta^L(\alpha), \zeta^U(\alpha)].$$

The membership function of (†DT) $\int_{\tilde{A}}^{\tilde{B}} f(x)dx$ is defined by

$$\xi_{(\dagger\mathrm{DT})\int_{\tilde{A}}^{\tilde{B}} f(x)dx}(r) = \sup_{0<\alpha\leq 1} \alpha \cdot \chi_{M_\alpha}(r).$$

We consider the function

$$F(a, b) = \int_a^b f(x)dx$$

defined on $\mathbb{R}^2$. Then, we see that

$$M_\alpha = \left[\min\left\{F\left(\tilde{A}_\alpha^L, \tilde{B}_\alpha^L\right), F\left(\tilde{A}_\alpha^U, \tilde{B}_\alpha^U\right)\right\} \max\left\{F\left(\tilde{A}_\alpha^L, \tilde{B}_\alpha^L\right), F\left(\tilde{A}_\alpha^U, \tilde{B}_\alpha^U\right)\right\}\right]$$

and that $\{M_\alpha : 0 < \alpha \leq 1\}$ is not a nested family. The (†DT)-fuzzy Riemann integral (†DT) $\int_{\tilde{A}}^{\tilde{B}} f(x)dx$ is just a fuzzy set in $\mathbb{R}$. In order to make (†DT) $\int_{\tilde{A}}^{\tilde{B}} f(x)dx$ a fuzzy interval, we need more assumptions given below.

Proposition 6.5.8 *Let $\tilde{A}$ and $\tilde{B}$ be two canonical fuzzy intervals satisfying $\tilde{A} \preccurlyeq \tilde{B}$. Let $f : \mathbb{R} \to \mathbb{R}$ be a real-valued function such that it is Riemann-integrable on the closed interval $[\tilde{A}_0^L, \tilde{B}_0^U]$, and that the function*

$$F(a, b) = \int_a^b f(x)dx$$

defined on $\mathbb{R}^2$ is continuous. Then, given a family $\{M_\alpha : 0 < \alpha \leq 1\}$ defined by

$$M_\alpha = \left[\min\left\{F\left(\tilde{A}_\alpha^L, \tilde{B}_\alpha^L\right), F\left(\tilde{A}_\alpha^U, \tilde{B}_\alpha^U\right)\right\} \max\left\{F\left(\tilde{A}_\alpha^L, \tilde{B}_\alpha^L\right), F\left(\tilde{A}_\alpha^U, \tilde{B}_\alpha^U\right)\right\}\right],$$

the (†DT)-fuzzy Riemann integral (†DT) $\int_{\tilde{A}}^{\tilde{B}} f(x)dx$ is a fuzzy interval and its α-level sets are bounded closed intervals given by

$$\left((\dagger DT)\int_{\tilde{A}}^{\tilde{B}} f(x)dx\right)_\alpha = \left[\min_{\alpha\leq\beta\leq 1} \zeta^L(\beta), \max_{\alpha\leq\beta\leq 1} \zeta^U(\beta)\right]$$

$$= \left[\min_{\alpha\leq\beta\leq 1} \min\left\{ \int_{\tilde{A}_\beta^L}^{\tilde{B}_\beta^L} f(x)dx, \int_{\tilde{A}_\beta^U}^{\tilde{B}_\beta^U} f(x)dx \right\},\right.$$

$$\left.\max_{\alpha\leq\beta\leq 1} \max\left\{ \int_{\tilde{A}_\beta^L}^{\tilde{B}_\beta^L} f(x)dx, \int_{\tilde{A}_\beta^U}^{\tilde{B}_\beta^U} f(x)dx \right\}\right]$$

$$= \left[\min \left\{ \min_{\alpha \le \beta \le 1} \int_{\tilde{A}^L_\beta}^{\tilde{B}^L_\beta} f(x)dx, \min_{\alpha \le \beta \le 1} \int_{\tilde{A}^U_\beta}^{\tilde{B}^U_\beta} f(x)dx \right\}, \right.$$
$$\left. \max \left\{ \max_{\alpha \le \beta \le 1} \int_{\tilde{A}^L_\beta}^{\tilde{B}^L_\beta} f(x)dx, \max_{\alpha \le \beta \le 1} \int_{\tilde{A}^U_\beta}^{\tilde{B}^U_\beta} f(x)dx \right\} \right]$$

for $\alpha \in [0,1]$.

Proof. The desired results follow immediately from Theorem 6.2.10. ∎

6.5.3 Fuzzy Riemann Integrals Using the Extension Principle

Now, we are going to study the fuzzy Riemann integrals using the extension principle. Suppose that $\tilde{A} \preccurlyeq \tilde{B}$. For $\alpha \in [0,1]$, given any $r \in \mathbb{R}$, we define the set-valued functions

$$E_\alpha(r) = \left\{ (a,b) \in \mathbb{R}^2 : r = \int_a^b f(x)dx \text{ for } (a,b) \in \tilde{A}_\alpha \times \tilde{B}_\alpha \right\} \tag{6.50}$$

and

$$E(r) = \bigcup_{0 \le \alpha \le 1} E_\alpha(r). \tag{6.51}$$

For $\tilde{A} \preccurlyeq \tilde{B}$, recall that

$$a \in \tilde{A}_\alpha \text{ and } b \in \tilde{B}_\alpha \text{ imply } a \le b,$$

which says that (6.50) is well defined. The formal definition is given below

Definition 6.5.9 Let $f : \mathbb{R} \to \mathbb{R}$ be a real-valued function. Given two fuzzy intervals $\tilde{A}$ and $\tilde{B}$ satisfying $\tilde{A} \preccurlyeq \tilde{B}$, we say that the **(EP)-fuzzy Riemann integral** (EP) $\int_{\tilde{A}}^{\tilde{B}} f(x)dx$ exists when the real-valued function f is Riemann-integrable on the closed interval $[\tilde{A}^L_0, \tilde{B}^U_0]$. We define a set-valued function E given in (6.51). The membership function of (EP)-fuzzy Riemann integral (EP) $\int_{\tilde{A}}^{\tilde{B}} f(x)dx$ is defined by

$$\xi_{(\mathrm{EP})\int_{\tilde{A}}^{\tilde{B}} f(x)dx}(r) = \sup_{(a,b) \in E(r)} \min \left\{ \xi_{\tilde{A}}(a), \xi_{\tilde{B}}(b) \right\}.$$

We notice that the α-level sets of (EP) $\int_{\tilde{A}}^{\tilde{B}} f(x)dx$ cannot be obtained from Theorem 4.1.2, unless there exists a function F defined on $\mathbb{R}^2$ satisfying

$$E(r) = \bigcup_{0 \le \alpha \le 1} E_\alpha(r) = \left\{ (a,b) \in \mathbb{R}^2 : r = F(a,b) \right\}.$$

It seems that we are not able to find such function F at this stage. In this case, the α-level sets of (EP) $\int_{\tilde{A}}^{\tilde{B}} f(x)dx$ remains open, and its proof is left as an exercise for the reader.

Suppose that we define

$$F(a,b) = \int_a^b f(x)dx$$

on $\mathbb{R}^2$ for $a \le b$. If $a > b$, then $F(a,b)$ is interpreted as

$$F(a,b) = \int_a^b f(x)dx = -\int_b^a f(x)dx$$

In this case, we can also define a set-valued function

$$E^*(r) = \left\{(a,b) \in \mathbb{R}^2 : r = F(a,b)\right\} = \left\{(a,b) \in \mathbb{R}^2 : r = \int_a^b f(x)dx\right\}.$$

Under these settings, we can define a so-called pseudo-fuzzy Riemann integral as follows.

Definition 6.5.10 Let $f : \mathbb{R} \to \mathbb{R}$ be a real-valued function. Given two fuzzy intervals $\tilde{A}$ and $\tilde{B}$, we say that the **(EP)-pseudo-fuzzy Riemann integral** $(\star\text{EP}) \int_{\tilde{A}}^{\tilde{B}} f(x)dx$ exists when the real-valued function f is Riemann-integrable on the closed interval $[\tilde{A}_0^L, \tilde{B}_0^U]$. The membership function of $(\star$EP)-fuzzy Riemann integral $(\star\text{EP}) \int_{\tilde{A}}^{\tilde{B}} f(x)dx$ is defined by

$$\begin{aligned}\xi_{(\star\text{EP}) \int_{\tilde{A}}^{\tilde{B}} f(x)dx}(r) &= \sup_{(a,b)\in E^*(r)} \min\left\{\xi_{\tilde{A}}(a), \xi_{\tilde{B}}(b)\right\} \\ &= \sup_{\{(a,b)\in\mathbb{R}^2 : r=\int_a^b f(x)dx\}} \min\left\{\xi_{\tilde{A}}(a), \xi_{\tilde{B}}(b)\right\}.\end{aligned}$$

We notice that the (EP)-pseudo-fuzzy Riemann integral does not assume $\tilde{A} \preccurlyeq \tilde{B}$ or $\tilde{A} \preccurlyeq \tilde{B}$. We also see that the (EP)-pseudo-fuzzy Riemann integral $(\star\text{EP}) \int_{\tilde{A}}^{\tilde{B}} f(x)dx$ is a fuzzification of function F. Suppose that the function F is continuous. Then, part (iii) of Theorem 4.1.2 is applicable to obtain the α-level sets of $(\star\text{EP}) \int_{\tilde{A}}^{\tilde{B}} f(x)dx$ as follows

$$\begin{aligned}\left((\star\text{EP}) \int_{\tilde{A}}^{\tilde{B}} f(x)dx\right)_\alpha &= F\left(\tilde{A}_\alpha, \tilde{B}_\alpha\right) = \left\{F(a,b) : a \in \tilde{A}_\alpha \text{ and } b \in \tilde{B}_\alpha\right\} \\ &= \left[\min_{(a,b)\in\tilde{A}_\alpha\times\tilde{B}_\alpha} F(a,b), \max_{(a,b)\in\tilde{A}_\alpha\times\tilde{B}_\alpha} F(a,b)\right] \\ &= \left[\min_{(a,b)\in\tilde{A}_\alpha\times\tilde{B}_\alpha} \int_a^b f(x)dx, \max_{(a,b)\in\tilde{A}_\alpha\times\tilde{B}_\alpha} \int_a^b f(x)dx\right] \qquad (6.52)\end{aligned}$$

for $\alpha \in [0,1]$.

By referring to Proposition 6.5.4, we notice that

$$\left((\star\text{EP}) \int_{\tilde{A}}^{\tilde{B}} f(x)dx\right)_\alpha \neq \left((\diamond\text{DT}) \int_{\tilde{A}}^{\tilde{B}} f(x)dx\right)_\alpha$$

in general. The reason is that we may have $a > b$ in (6.52). However, when we further assume that $\tilde{A} \preccurlyeq \tilde{B}$, we have

$$\left((\star\text{EP}) \int_{\tilde{A}}^{\tilde{B}} f(x)dx\right)_\alpha = \left((\diamond\text{DT}) \int_{\tilde{A}}^{\tilde{B}} f(x)dx\right)_\alpha$$

for all $\alpha \in [0,1]$ under the assumptions of Proposition 6.5.4. The above results are summarized in the following proposition.

Proposition 6.5.11 *Let $\tilde{A}$ and $\tilde{B}$ be two fuzzy intervals, and let $f : \mathbb{R} \to \mathbb{R}$ be a real-valued function such that it is Riemann-integrable on the closed interval $[\tilde{A}_0^L, \tilde{B}_0^U]$. Suppose that the function*

$$F(a,b) = \int_a^b f(x)dx$$

defined on $\mathbb{R}^2$ is continuous. Then, the (EP)-pseudo-fuzzy Riemann integral $(\star EP)\int_{\tilde{A}}^{\tilde{B}} f(x)dx$ is a fuzzy interval and its α-level sets are bounded closed intervals given by

$$\left((\star EP)\int_{\tilde{A}}^{\tilde{B}} f(x)dx\right)_{\alpha} = \left[\min_{(a,b)\in\tilde{A}_{\alpha}\times\tilde{B}_{\alpha}} F(a,b),\ \min_{(a,b)\in\tilde{A}_{\alpha}\times\tilde{B}_{\alpha}} F(a,b)\right]$$

$$= \left[\min_{(a,b)\in\tilde{A}_{\alpha}\times\tilde{B}_{\alpha}} \int_a^b f(x)dx,\ \min_{(a,b)\in\tilde{A}_{\alpha}\times\tilde{B}_{\alpha}} \int_a^b f(x)dx\right]$$

for $\alpha \in [0,1]$. We further assume that $\tilde{A}$ and $\tilde{B}$ are taken to be canonical fuzzy intervals satisfying $\tilde{A} \preccurlyeq \tilde{B}$. Then, we have

$$(\star EP)\int_{\tilde{A}}^{\tilde{B}} f(x)dx = (\diamond DT)\int_{\tilde{A}}^{\tilde{B}} f(x)dx.$$

7

Arithmetics of Fuzzy Sets

Now, we want to introduce the arithmetic operations between fuzzy sets. Given any $a, b \in \mathbb{R}$, we see that

$$\chi_{\{a+b\}}(z) = \begin{cases} 1 & \text{if } z = a + b \\ 0 & \text{if } z \neq a + b. \end{cases} = \sup_{\{(x,y):z=x+y\}} \min \left\{ \chi_{\{a\}}(x), \chi_{\{b\}}(y) \right\}.$$

In general, let $\circ$ be any binary operations $+, -, \times, /$ between two real numbers a and b. Then, we have

$$\chi_{\{a \circ b\}}(z) = \begin{cases} 1 & \text{if } z = a \circ b \\ 0 & \text{if } z \neq a \circ b. \end{cases} = \sup_{\{(x,y):z=x\circ y\}} \min \left\{ \chi_{\{a\}}(x), \chi_{\{b\}}(y) \right\}, \tag{7.1}$$

where $b \neq 0$ when a/b is considered.

7.1 Arithmetics of Fuzzy Sets in $\mathbb{R}$

Let $\odot$ denote any one of the four basic arithmetic operations $\oplus, \ominus, \otimes, \oslash$ between fuzzy sets $\tilde{A}$ and $\tilde{B}$ in $\mathbb{R}$ with membership functions $\xi_{\tilde{A}}$ and $\xi_{\tilde{B}}$, respectively. Then, according to (7.1), the membership function of $\tilde{A} \odot \tilde{B}$ is defined by

$$\xi_{\tilde{A}\odot\tilde{B}}(z) = \sup_{\{(x,y):z=x\circ y\}} \min\{\xi_{\tilde{A}}(x), \xi_{\tilde{B}}(y)\} \text{ for all } z \in \mathbb{R} \tag{7.2}$$

by replacing the characteristic functions with the membership functions. More precisely, for all $z \in \mathbb{R}$, we define

$$\xi_{\tilde{A}\oplus\tilde{B}}(z) = \sup_{\{(x,y):z=x+y\}} \min\left\{\xi_{\tilde{A}}(x), \xi_{\tilde{B}}(y)\right\};$$

$$\xi_{\tilde{A}\ominus\tilde{B}}(z) = \sup_{\{(x,y):z=x-y\}} \min\left\{\xi_{\tilde{A}}(x), \xi_{\tilde{B}}(y)\right\};$$

$$\xi_{\tilde{A}\otimes\tilde{B}}(z) = \sup_{\{(x,y):z=x*y\}} \min\left\{\xi_{\tilde{A}}(x), \xi_{\tilde{B}}(y)\right\};$$

$$\xi_{\tilde{A}\oslash\tilde{B}}(z) = \sup_{\{(x,y):z=x/y,y\neq 0\}} \min\left\{\xi_{\tilde{A}}(x), \xi_{\tilde{B}}(y)\right\}.$$

Mathematical Foundations of Fuzzy Sets, First Edition. Hsien-Chung Wu.

For any $\lambda \in \mathbb{R}$, we define $\lambda\tilde{A}$ as $\tilde{1}_{\{\lambda\}} \otimes \tilde{A}$. Therefore, we have

$$\xi_{\lambda\tilde{A}}(z) = \xi_{\tilde{1}_{\{\lambda\}}\otimes\tilde{A}}(z) = \sup_{\{(x,y):z=x*y\}} \min\left\{\xi_{\tilde{A}}(x), \tilde{1}_{\{\lambda\}}(y)\right\}$$

$$= \sup_{\{x:\lambda x=z\}} \xi_{\tilde{A}}(x) = \begin{cases} \xi_{\tilde{A}}(z/\lambda) & \text{if } \lambda \neq 0 \\ \sup\limits_{x\in\mathbb{R}} \xi_{\tilde{A}}(x) & \text{if } \lambda = 0 = z. \end{cases}$$

For $\lambda = 0$ and $z \neq 0$, we define $\xi_{\lambda\tilde{A}}(z) = 0$. In other words, we have

$$\xi_{\lambda\tilde{A}}(z) = \begin{cases} \xi_{\tilde{A}}(z/\lambda) & \text{if } \lambda \neq 0 \\ \sup\limits_{x\in\mathbb{R}} \xi_{\tilde{A}}(x) & \text{if } \lambda = 0 = z, \\ 0 & \text{if } \lambda = 0 \text{ and } z \neq 0. \end{cases} \tag{7.3}$$

On the other hand, we also have

$$\xi_{\tilde{A}\oplus(-\tilde{B})}(z) = \sup_{\{(x,y):x+y=z\}} \min\left\{\xi_{\tilde{A}}(x), \xi_{-\tilde{B}}(y)\right\}$$

$$= \sup_{\{(x,y):x+y=z\}} \min\left\{\xi_{\tilde{A}}(x), \xi_{\tilde{B}}(-y)\right\} \text{ (using (7.3))}$$

$$= \sup_{\{(x,y):x-y=z\}} \min\left\{\xi_{\tilde{A}}(x), \xi_{\tilde{B}}(y)\right\} = \xi_{\tilde{A}\ominus\tilde{B}}(z).$$

This shows that $\tilde{A} \ominus \tilde{B} = \tilde{A} \oplus (-\tilde{B})$.

Theorem 7.1.1 *Let $\tilde{A}$ and $\tilde{B}$ be two fuzzy sets in $\mathbb{R}$ with membership functions $\xi_{\tilde{A}}$ and $\xi_{\tilde{B}}$, respectively. Suppose that the arithmetic operations $\odot \in \{\oplus, \ominus, \otimes\}$ correspond to the operations $\circ \in \{+, -, *\}$, respectively. Then, the following statements hold true.*

(i) We have the following inclusions

$$(\tilde{A} \odot \tilde{B})_\alpha \supseteq \tilde{A}_\alpha \circ \tilde{B}_\alpha \text{ for all } \alpha \in [0, 1].$$

(ii) Suppose that the membership functions $\xi_{\tilde{A}}$ and $\xi_{\tilde{B}}$ are upper semi-continuous. Then, we have

$$(\tilde{A} \odot \tilde{B})_\alpha = \tilde{A}_\alpha \circ \tilde{B}_\alpha \text{ for all } \alpha \in (0, 1].$$

(iii) Suppose that the membership functions $\xi_{\tilde{A}}$ and $\xi_{\tilde{B}}$ are upper semi-continuous, and that the supports $\tilde{A}_{0+}$ and $\tilde{B}_{0+}$ are bounded. Then, we have

$$(\tilde{A} \odot \tilde{B})_\alpha = \tilde{A}_\alpha \circ \tilde{B}_\alpha \text{ for all } \alpha \in [0, 1].$$

Proof. To prove part (i), for $\alpha \in (0, 1]$ and $z_\alpha \in \tilde{A}_\alpha \circ \tilde{B}_\alpha$, there exist $x_\alpha \in \tilde{A}_\alpha$ and $y_\alpha \in \tilde{B}_\alpha$ satisfying $z_\alpha = x_\alpha \circ y_\alpha$ for $\circ \in \{+, -, *\}$, where $\xi_{\tilde{A}}(x_\alpha) \geq \alpha$ and $\xi_{\tilde{B}}(y_\alpha) \geq \alpha$. Therefore, we have

$$\xi_{\tilde{A}\odot\tilde{B}}(z_\alpha) = \sup_{\{(x,y):z_\alpha=x\circ y\}} \min\{\xi_{\tilde{A}}(x), \xi_{\tilde{B}}(y)\} \geq \min\{\xi_{\tilde{A}}(x_\alpha), \xi_{\tilde{B}}(y_\alpha)\} \geq \alpha,$$

which says that $z_\alpha \in (\tilde{A} \odot \tilde{B})_\alpha$. This shows the inclusion $\tilde{A}_\alpha \circ \tilde{B}_\alpha \subseteq (\tilde{A} \odot \tilde{B})_\alpha$ for $\alpha \in (0, 1]$.

For $\alpha = 0$ and $z_0 \in \tilde{A}_0 \circ \tilde{B}_0$, there also exist $x_0 \in \tilde{A}_0$ and $y_0 \in \tilde{B}_0$ satisfying $z_0 = x_0 \circ y_0$ for $\circ \in \{+, -, *\}$. Since

$$\tilde{A}_0 = \text{cl}\left(\{x \in \mathbb{R} : \xi_{\tilde{A}}(x) > 0\}\right) \text{ and } \tilde{B}_0 = \text{cl}\left(\{y \in \mathbb{R} : \xi_{\tilde{B}}(y) > 0\}\right),$$

there exist sequence $\{x_n\}_{n=1}^{\infty}$ in $\{x \in \mathbb{R} : \xi_{\tilde{A}}(x) > 0\}$ and sequence $\{y_n\}_{n=1}^{\infty}$ in $\{y \in \mathbb{R} : \xi_{\tilde{B}}(y) > 0\}$ satisfying $x_n \to x_0$ and $y_n \to y_0$ as $n \to \infty$. Let $z_n = x_n \circ y_n$. Then, we see that $z_n \to x_0 \circ y_0 = z_0$, since the binary operation $\circ \in \{+, -, *\}$ can be regarded as a continuous function $h(x, y) = x \circ y$. We also have

$$\xi_{\tilde{A}\odot\tilde{B}}(z_n) = \sup_{\{(x,y):z_n=x\circ y\}} \min\{\xi_{\tilde{A}}(x), \xi_{\tilde{B}}(y)\} \geq \min\{\xi_{\tilde{A}}(x_n), \xi_{\tilde{B}}(y_n)\} > 0,$$

which says that $z_n \in \{z \in \mathbb{R} : \xi_{\tilde{A}\odot\tilde{B}}(z) > 0\}$. Since $z_n \to z_0$, it means that

$$z_0 \in \text{cl}\left(\{z \in \mathbb{R} : \xi_{\tilde{A}\odot\tilde{B}}(z) > 0\}\right) = (\tilde{A} \odot \tilde{B})_0.$$

This shows the inclusion $\tilde{A}_0 \circ \tilde{B}_0 \subseteq (\tilde{A} \odot \tilde{B})_0$. Therefore, we conclude that

$$\tilde{A}_\alpha \circ \tilde{B}_\alpha \subseteq (\tilde{A} \odot \tilde{B})_\alpha \text{ for } \alpha \in [0, 1].$$

To prove part (ii), in order to prove the other direction of inclusion, we further assume that the membership functions $\xi_{\tilde{A}}$ and $\xi_{\tilde{B}}$ are upper semi-continuous; that is, the nonempty α-level sets $\tilde{A}_\alpha$ and $\tilde{B}_\alpha$ are closed subsets of $\mathbb{R}$ for all $\alpha \in [0, 1]$. Given any $\alpha \in (0, 1]$ and $z_\alpha \in (\tilde{A} \odot \tilde{B})_\alpha$, we have

$$\sup_{\{(x,y):z_\alpha=x\circ y\}} \min\{\xi_{\tilde{A}}(x), \xi_{\tilde{B}}(y)\} = \xi_{\tilde{A}\odot\tilde{B}}(z_\alpha) \geq \alpha. \tag{7.4}$$

Since z_α is finite, it is clear to see that

$$F \equiv \{(x, y) : z_\alpha = x \circ y\}$$

is a bounded subset of $\mathbb{R}^2$. We also see that the function $h(x, y) = x \circ y$ is continuous on $\mathbb{R}^2$. Since the singleton $\{z_\alpha\}$ is a closed subset of $\mathbb{R}$, it follows that the inverse image $F = h^{-1}(\{z_\alpha\})$ of $\{z_\alpha\}$ is also a closed subset of $\mathbb{R}^2$. This says that F is a closed and bounded subset of $\mathbb{R}^2$. Now, we want to show that the function

$$f(x, y) = \min\{\xi_{\tilde{A}}(x), \xi_{\tilde{B}}(y)\}$$

is upper semi-continuous, i.e. we want to show that

$$\{(x, y) : f(x, y) \geq \alpha\}$$

is a closed subset of $\mathbb{R}^2$ for any $\alpha \in \mathbb{R}$.

- For $\alpha \in (0, 1]$, i.e. $\tilde{A}_\alpha \neq \emptyset$ and $\tilde{B}_\alpha \neq \emptyset$, we have

$$\begin{aligned}\{(x, y) : f(x, y) \geq \alpha\} &= \left\{(x, y) : \xi_{\tilde{A}}(x) \geq \alpha \text{ and } \xi_{\tilde{B}}(y) \geq \alpha\right\} \\ &= \left\{(x, y) : x \in \tilde{A}_\alpha \text{ and } y \in \tilde{B}_\alpha\right\} = \tilde{A}_\alpha \times \tilde{B}_\alpha,\end{aligned}$$

  which is a closed subset of $\mathbb{R}^2$, since $\tilde{A}_\alpha$ and $\tilde{B}_\alpha$ are closed subsets of $\mathbb{R}$.
- If $\alpha \leq 0$, then $\{(x, y) : f(x, y) \geq \alpha\} = \mathbb{R}^2$ is a closed subset of $\mathbb{R}^2$.
- If $\alpha > 1$, then $\tilde{A}_\alpha = \emptyset$ and $\tilde{B}_\alpha = \emptyset$, which implies $\{(x, y) : f(x, y) \geq \alpha\} = \emptyset$, which is also a closed subset of $\mathbb{R}^2$.

Therefore, the function $f(x,y)$ is indeed upper semi-continuous. By Proposition 1.4.4, the function f assumes its maximum on F; that is, from (7.4), we have

$$\max_{(x,y)\in F} f(x,y) = \max_{\{(x,y):z_\alpha = x\circ y\}} f(x,y) = \sup_{\{(x,y):z_\alpha = x\circ y\}} f(x,y) \geq \alpha. \tag{7.5}$$

In other words, there exists $(x_\alpha, y_\alpha) \in F$ satisfying $z_\alpha = x_\alpha \circ y_\alpha$ and

$$\min\{\xi_{\tilde{A}}(x_\alpha), \xi_{\tilde{B}}(y_\alpha)\} = f(x_\alpha, y_\alpha) = \max_{(x,y)\in F} f(x,y) \geq \alpha,$$

i.e. $\xi_{\tilde{A}}(x_\alpha) \geq \alpha$ and $\xi_{\tilde{B}}(y_\alpha) \geq \alpha$. Therefore, we obtain $x_\alpha \in \tilde{A}_\alpha$ and $y_\alpha \in \tilde{B}_\alpha$, which says that $z_\alpha \in \tilde{A}_\alpha \circ \tilde{B}_\alpha$, i.e. $(\tilde{A} \odot \tilde{B})_\alpha \subseteq \tilde{A}_\alpha \circ \tilde{B}_\alpha$ for all $\alpha \in (0,1]$. Using part (i), we obtain the desired equalities.

To prove part (iii), for $\alpha = 0$ and

$$z_0 \in (\tilde{A} \odot \tilde{B})_0 = \text{cl}\left((\tilde{A} \odot \tilde{B})_{0+}\right) = \text{cl}\left(\{z \in \mathbb{R} : \xi_{\tilde{A}\odot\tilde{B}}(z) > 0\}\right),$$

there exists a sequence $\{z_n\}_{n=1}^\infty$ in the set $\{z \in \mathbb{R} : \xi_{\tilde{A}\odot\tilde{B}}(z) > 0\}$ satisfying $z_n \to z_0$ as $n \to \infty$. Using the above same arguments by referring to (7.5), we also have

$$0 < \xi_{\tilde{A}\odot\tilde{B}}(z_n) = \sup_{\{(x,y):z_n = x\circ y\}} \min\{\xi_{\tilde{A}}(x), \xi_{\tilde{B}}(y)\} = \max_{\{(x,y):z_n = x\circ y\}} \min\{\xi_{\tilde{A}}(x), \xi_{\tilde{B}}(y)\}.$$

Therefore, there exist x_n and y_n satisfying $z_n = x_n \circ y_n$ and

$$\min\{\xi_{\tilde{A}}(x_n), \xi_{\tilde{B}}(y_n)\} = \max_{\{(x,y):z_n = x\circ y\}} \min\{\xi_{\tilde{A}}(x), \xi_{\tilde{B}}(y)\} > 0,$$

i.e. $\xi_{\tilde{A}}(x_n) > 0$ and $\xi_{\tilde{B}}(y_n) > 0$. This shows that the sequences $\{x_n\}_{n=1}^\infty$ and $\{y_n\}_{n=1}^\infty$ are in the supports $\tilde{A}_{0+}$ and $\tilde{B}_{0+}$, respectively. Since $\tilde{A}_{0+}$ and $\tilde{B}_{0+}$ are bounded, i.e. $\{x_n\}_{n=1}^\infty$ and $\{y_n\}_{n=1}^\infty$ are bounded sequences, there exist convergent subsequences $\{x_{n_k}\}_{k=1}^\infty$ and $\{y_{n_k}\}_{k=1}^\infty$ of $\{x_n\}_{n=1}^\infty$ and $\{y_n\}_{n=1}^\infty$, respectively. In other words, we have $x_{n_k} \to x_0$ and $y_{n_k} \to y_0$ as $k \to \infty$, where $x_0 \in \text{cl}(\tilde{A}_{0+}) = \tilde{A}_0$ and $y_0 \in \text{cl}(\tilde{B}_{0+}) = \tilde{B}_0$. Let $z_{n_k} = x_{n_k} \circ y_{n_k}$. Then $\{z_{n_k}\}_{k=1}^\infty$ is a subsequence of $\{z_n\}_{n=1}^\infty$, i.e. $z_{n_k} \to z_0$ as $k \to \infty$. Since

$$z_0 = \lim_{k\to\infty} z_{n_k} = \lim_{k\to\infty} (x_{n_k} \circ y_{n_k}) = \left(\lim_{k\to\infty} x_{n_k}\right) \circ \left(\lim_{k\to\infty} y_{n_k}\right) = x_0 \circ y_0,$$

which shows that $z_0 \in \tilde{A}_0 \circ \tilde{B}_0$. Therefore, we obtain the inclusion $(\tilde{A} \odot \tilde{B})_0 \subseteq \tilde{A}_0 \circ \tilde{B}_0$. Using parts (i) and (ii), we obtain the desired equalities. This completes the proof. ■

7.1.1 Arithmetics of Fuzzy Intervals

Let $\tilde{A}$ and $\tilde{B}$ be fuzzy intervals. Since the α-level sets of fuzzy intervals are closed intervals, in order to discuss the α-level sets of arithmetic $\tilde{A} \odot \tilde{B}$ for $\odot \in \{\oplus, \ominus, \otimes, \oslash\}$, we need to introduce the arithmetic operations between closed intervals.

Let $\circ$ denote any of the four arithmetic operations between closed intervals: addition $+$, subtraction $-$, multiplication $\times$, and division $/$. Then, the arithmetic operation between two closed intervals $[a,b]$ and $[d,e]$ is defined by

$$[a,b] \circ [d,e] = \{x \circ y : a \leq x \leq b \text{ and } d \leq y \leq e\}.$$

The division $[a,b]/[d,e]$ of closed intervals should be read as

$$[a,b]/[d,e] = \{x/y : a \leq x \leq b \text{ and } d \leq y \leq e \text{ with } y \neq 0\}. \tag{7.6}$$

We can see that

$$[a,b]+[d,e]=[a+d,b+e],$$
$$[a,b]-[d,e]=[a-e,b-d],$$
$$[a,b]\times[d,e]=[\min\{ad,ae,bd,be\},\max\{ad,ae,bd,be\}].$$

Suppose that $0\notin[d,e]$. Then, we also have

$$\begin{aligned}[a,b]/[d,e]&=[a,b]\cdot[1/e.1/d]\\&=[\min\{a/d,a/e,b/d,b/e\},\max\{a/d,a/e,b/d,b/e\}].\end{aligned}$$

Suppose that $0\in[d,e]$. Then $[a,b]/[d,e]$ cannot be a closed interval, which just has the form shown in (7.6).

Let $\mathbf{0}=[0,0]$ and $\mathbf{1}=[1,1]$. Assume that $A=[a_1,a_2]$, $B=[b_1,b_2]$, $C=[c_1,c_2]$ are closed intervals. Arithmetic operations on closed intervals satisfy some useful properties as follows:

- (Commutativity) $A+B=B+A$ and $A\times B=B\times A$;
- (Associativity) $(A+B)+C=A+(B+C)$ and $(A\times B)\times C=A\times(B\times C)$;
- (Identity) $A=\mathbf{0}+A=A+\mathbf{0}$ and $A=\mathbf{1}\times A=A\times\mathbf{1}$;
- (Subdistributivity) $A\times(B+C)\subseteq A\times B+A\times C$;
- (Distributivity) if $bc\geq 0$ for every $b\in B$ and $c\in C$, then

  $$A\times(B+C)=A\times B+A\times C;$$

  furthermore, if $A=[a,a]$, then

  $$A\times(B+C)=a\times(B+C)=a\times B+a\times C=A\times B+A\times C;$$

- $0\in A-A$ and $1\in A/A$;
- if $A\subseteq E$ and $B\subseteq F$, then $A\circ B\subseteq E\circ F$, where "$\circ$" is any arithmetic operation.

Theorem 7.1.2 *Let $\tilde{A}$ and $\tilde{B}$ be two fuzzy intervals. Then, we have the following properties.*

(i) *Suppose that the binary operation $\odot$ is any one of $\{\oplus,\ominus,\otimes\}$. Then $\tilde{A}\odot\tilde{B}$ is a fuzzy interval, and we have*

$$(\tilde{A}\odot\tilde{B})_\alpha=\tilde{A}_\alpha\circ\tilde{B}_\alpha \text{ for any } \alpha\in[0,1]. \tag{7.7}$$

More precisely, we have

$$(\tilde{A}\oplus\tilde{B})_\alpha=\left[\tilde{A}^L_\alpha+\tilde{B}^L_\alpha,\tilde{A}^U_\alpha+\tilde{B}^U_\alpha\right],$$
$$(\tilde{A}\ominus\tilde{B})_\alpha=\left[\tilde{A}^L_\alpha-\tilde{B}^U_\alpha,\tilde{A}^U_\alpha-\tilde{B}^L_\alpha\right],$$
$$(\tilde{A}\otimes\tilde{B})_\alpha=\Big[\min\left\{\tilde{A}^L_\alpha\tilde{B}^L_\alpha,\tilde{A}^L_\alpha\tilde{B}^U_\alpha,\tilde{A}^U_\alpha\tilde{B}^L_\alpha,\tilde{A}^U_\alpha\tilde{B}^U_\alpha\right\},\max\left\{\tilde{A}^L_\alpha\tilde{B}^L_\alpha,\tilde{A}^L_\alpha\tilde{B}^U_\alpha,\tilde{A}^U_\alpha\tilde{B}^L_\alpha,\tilde{A}^U_\alpha\tilde{B}^U_\alpha\right\}\Big]$$

for any $\alpha\in[0,1]$.

(ii) *Suppose that the binary operation $\odot$ is taken to be $\oslash$. Then, we have*

$$(\tilde{A}\oslash\tilde{B})_\alpha=\left\{x/y: x\in\tilde{A}_\alpha \text{ and } y\in\tilde{B}_\alpha \text{ with } y\neq 0\right\} \tag{7.8}$$

for any $\alpha \in [0,1]$. *Moreover, we have*

$$(\tilde{A} \oslash \tilde{B})_\alpha = \begin{cases} \tilde{A}_\alpha / \tilde{B}_\alpha & \text{if } 0 \notin \tilde{B}_\alpha \\ \mathbb{R} & \text{if } 0 \in \tilde{B}_\alpha, \end{cases} \tag{7.9}$$

where

$$\tilde{A}_\alpha / \tilde{B}_\alpha = \left[\min\left\{\tilde{A}^L_\alpha/\tilde{B}^L_\alpha, \tilde{A}^L_\alpha/\tilde{B}^U_\alpha, \tilde{A}^U_\alpha/\tilde{B}^L_\alpha, \tilde{A}^U_\alpha/\tilde{B}^U_\alpha\right\},\right.$$
$$\left.\max\left\{\tilde{A}^L_\alpha/\tilde{B}^L_\alpha, \tilde{A}^L_\alpha/\tilde{B}^U_\alpha, \tilde{A}^U_\alpha/\tilde{B}^L_\alpha, \tilde{A}^U_\alpha/\tilde{B}^U_\alpha\right\}\right].$$

Suppose that $\tilde{B}$ *is a positive or negative fuzzy interval (Definition 2.3.6) satisfying* $0 \notin \tilde{B}_0$. *Then* $\tilde{A} \oslash \tilde{B}$ *is a fuzzy interval and* $(\tilde{A} \oslash \tilde{B})_\alpha = \tilde{A}_\alpha/\tilde{B}_\alpha$ *for any* $\alpha \in [0,1]$, *where the* 0*-level sets are taken into account. Note that if there exists* $\hat{\alpha} \in [0,1]$ *satisfying* $0 \in \tilde{B}_{\hat{\alpha}}$, *then* $\tilde{A} \oslash \tilde{B}$ *is just a fuzzy set in* $\mathbb{R}$ *with the* α*-level sets given in* (7.8).

Proof. To prove part (i), the equality (7.7) follows immediately from part (iii) of Theorem 7.1.1. Using Theorem 4.4.18, we see that $\tilde{A} \odot \tilde{B}$ is a fuzzy interval by considering the fuzzification of crisp function $f(x,y) = x \circ y$.

To prove part (ii), for any $\alpha \in [0,1]$, let

$$A_\alpha = \left\{x/y : x \in \tilde{A}_\alpha \text{ and } y \in \tilde{B}_\alpha \text{ with } y \neq 0\right\}.$$

Here $\tilde{B}_\alpha$ is allowed to contain 0. For $\alpha \in (0,1]$ and $z_0 \in A_\alpha$, there exist $x_0 \in \tilde{A}_\alpha$ and $y_0 \in \tilde{B}_\alpha$ satisfying $z_0 = x_0/y_0$, where $\xi_{\tilde{A}}(x_0) \geq \alpha$ and $\xi_{\tilde{B}}(y_0) \geq \alpha$. Therefore, we have

$$\xi_{\tilde{A}\oslash\tilde{B}}(z_0) = \sup_{\{(x,y):z_0=x/y\}} \min\left\{\xi_{\tilde{A}}(x), \xi_{\tilde{B}}(y)\right\} \geq \min\left\{\xi_{\tilde{A}}(x_0), \xi_{\tilde{B}}(y_0)\right\} \geq \alpha,$$

which says that $z_0 \in (\tilde{A} \oslash \tilde{B})_\alpha$. This shows that $A_\alpha \subseteq (\tilde{A} \oslash \tilde{B})_\alpha$. Now, for $\alpha = 0$ and $z_0 \in A_0$, there exist $x_0 \in \tilde{A}_0$ and $y_0 \in \tilde{B}_0$ satisfying $z_0 = x_0/y_0$. Since

$$\tilde{A}_0 = \mathrm{cl}\left(\{x \in \mathbb{R} : \xi_{\tilde{A}}(x) > 0\}\right) \text{ and } \tilde{B}_0 = \mathrm{cl}\left(\{y \in \mathbb{R} : \xi_{\tilde{B}}(y) > 0\}\right),$$

there exist sequence $\{x_n\}_{n=1}^\infty$ in $\{x \in \mathbb{R} : \xi_{\tilde{A}}(x) > 0\}$ and sequence $\{y_n\}_{n=1}^\infty$ in $\{y \in \mathbb{R} : \xi_{\tilde{B}}(y) > 0\}$ satisfying $x_n \to x_0$ and $y_n \to y_0$ as $n \to \infty$. Let $z_n = x_n/y_n$, where $\xi_{\tilde{A}}(x_n) > 0$ and $\xi_{\tilde{B}}(y_n) > 0$. Then, we see that $z_n \to x_0/y_0 = z_0$, since the binary operation "/" is continuous. We also have

$$\xi_{\tilde{A}\oslash\tilde{B}}(z_n) = \sup_{\{(x,y):z_n=x/y\}} \min\left\{\xi_{\tilde{A}}(x), \xi_{\tilde{B}}(y)\right\} \geq \min\left\{\xi_{\tilde{A}}(x_n), \xi_{\tilde{B}}(y_n)\right\} > 0,$$

which says that $z_n \in \{z \in \mathbb{R} : \xi_{\tilde{A}\oslash\tilde{B}}(z) > 0\}$. Therefore, we conclude that

$$z_0 \in \mathrm{cl}\left(\{z \in \mathbb{R} : \xi_{\tilde{A}\oslash\tilde{B}}(z) > 0\}\right) = (\tilde{A} \oslash \tilde{B})_0.$$

This shows that $A_0 \subseteq (\tilde{A} \oslash \tilde{B})_0$. Therefore, we conclude that $A_\alpha \subseteq (\tilde{A} \oslash \tilde{B})_\alpha$ for $\alpha \in [0,1]$.

On the other hand, for any $\alpha \in (0,1]$, assume $z_0 \in (\tilde{A} \oslash \tilde{B})_\alpha$, i.e.

$$\sup_{\{(x,y):z_0=x/y\}} \min\left\{\xi_{\tilde{A}}(x), \xi_{\tilde{B}}(y)\right\} \geq \alpha. \tag{7.10}$$

We consider the function $f : \mathbb{R} \times (\mathbb{R}\backslash\{0\}) \to \mathbb{R}$ by $f(x,y) = x/y$ and the set $F = \{(x,y) : z_0 = x/y\}$. Since $|z_0| < \infty$, i.e. $|x|, |y| < \infty$, we see that F is a bounded subset of $\mathbb{R} \times (\mathbb{R}\backslash\{0\})$. Here $\mathbb{R}\backslash\{0\}$ is endowed with a subspace topology of $\mathbb{R}$; that is, any open set in $\mathbb{R}$ not containing 0 is also an open subset of $\mathbb{R}\backslash\{0\}$. We can see that f is continuous on $\mathbb{R} \times (\mathbb{R}\backslash\{0\})$.

This also says that $F = f^{-1}(\{z_0\})$ is a closed subset of $\mathbb{R} \times (\mathbb{R}\backslash\{0\})$; that is, F is a closed and bounded subset of $\mathbb{R} \times (\mathbb{R}\backslash\{0\})$. By the definition of $\tilde{A} \oslash \tilde{B}$, we see that the function $g(x,y) = \min\{\xi_{\tilde{A}}(x), \xi_{\tilde{B}}(y)\}$ is defined on $\mathbb{R} \times (\mathbb{R}\backslash\{0\})$. We want to show that g is upper semi-continuous, i.e. we want to show that $\{(x,y) : g(x,y) \geq \alpha\}$ is a closed subset of $\mathbb{R} \times (\mathbb{R}\backslash\{0\})$ for any $\alpha \in (0,1]$. Now, we have

$$\begin{aligned}\{(x,y) : g(x,y) \geq \alpha\} &= \{(x,y) : \min\{\xi_{\tilde{A}}(x), \xi_{\tilde{B}}(y)\} \geq \alpha\} \\ &= \left\{(x,y) : \xi_{\tilde{A}}(x) \geq \alpha \text{ and } \xi_{\tilde{B}}(y) \geq \alpha\right\} \\ &= \left\{(x,y) : x \in \tilde{A}_\alpha \text{ and } y \in \tilde{B}_\alpha\right\} \\ &= \tilde{A}_\alpha \times \tilde{B}_\alpha,\end{aligned}$$

where $0 \notin \tilde{B}_\alpha = [\tilde{B}^L_\alpha, \tilde{B}^U_\alpha]$. The complement of $\tilde{B}_\alpha$ is given by

$$\begin{cases} (-\infty,0) \cup (0,\tilde{B}^L_\alpha) \cup (\tilde{B}^U_\alpha,\infty) & \text{if } \tilde{B}^L_\alpha > 0 \\ (-\infty,\tilde{B}^L_\alpha) \cup (\tilde{B}^U_\alpha,0) \cup (0,\infty) & \text{if } \tilde{B}^U_\alpha < 0, \end{cases}$$

which is an open subset of $\mathbb{R}\backslash\{0\}$. This shows that $\tilde{B}_\alpha$ is also a closed subset of $\mathbb{R}\backslash\{0\}$; that is, $\tilde{A}_\alpha \times \tilde{B}_\alpha$ is also a closed subset of $\mathbb{R} \times (\mathbb{R}\backslash\{0\})$. Therefore, the function g is indeed upper semi-continuous. By Proposition 1.4.4, the function g assumes its maximum on F, i.e.

$$\max_{\{(x,y):z_0=x/y\}} g(x,y) = \sup_{\{(x,y):z_0=x/y\}} g(x,y) \geq \alpha \tag{7.11}$$

by (7.10). In other words, there exists $(x_0,y_0) \in F$ satisfying

$$g(x_0,y_0) = \min\{\xi_{\tilde{A}}(x_0), \xi_{\tilde{B}}(y_0)\} \geq \alpha,$$

i.e. $\xi_{\tilde{A}}(x_0) \geq \alpha$ and $\xi_{\tilde{B}}(y_0) \geq \alpha$, where $z_0 = x_0/y_0$. Therefore, we obtain $x_0 \in \tilde{A}_\alpha$ and $y_0 \in \tilde{B}_\alpha$. We conclude that $z_0 \in A_\alpha$, i.e. $(\tilde{A} \oslash \tilde{B})_\alpha \subseteq A_\alpha$. Now, for $\alpha = 0$, we assume that $z_0 \in (\tilde{A} \oslash \tilde{B})_0$. Then, there exists sequence $\{z_n\}_{n=1}^\infty$ in $\{z \in \mathbb{R} : \xi_{\tilde{A}\oslash\tilde{B}}(z) > 0\}$ satisfying $z_n \to z_0$ as $n \to \infty$. Using the same arguments above by referring to (7.11), we also have

$$0 < \xi_{\tilde{A}\oslash\tilde{B}}(z_n) = \sup_{\{(x,y):z_n=x/y\}} \min\left\{\xi_{\tilde{A}}(x), \xi_{\tilde{B}}(y)\right\} = \max_{\{(x,y):z_n=x/y\}} \min\left\{\xi_{\tilde{A}}(x), \xi_{\tilde{B}}(y)\right\}.$$

Therefore, the same arguments above also says that there exist x_n and y_n satisfying $z_n = x_n/y_n$, $\xi_{\tilde{A}}(x_n) > 0$ and $\xi_{\tilde{B}}(y_n) > 0$. This shows that $\{x_n\}_{n=1}^\infty$ and $\{y_n\}_{n=1}^\infty$ are sequences in the closed and bounded sets $\tilde{A}_0$ and $\tilde{B}_0$, respectively. Therefore, there exist convergent subsequences $\{x_{n_k}\}_{k=1}^\infty$ and $\{y_{n_k}\}_{k=1}^\infty$ of $\{x_n\}_{n=1}^\infty$ and $\{y_n\}_{n=1}^\infty$, respectively. We also write $x_{n_k} \to x_0$ and $y_{n_k} \to y_0$ as $k \to \infty$, where $x_0 \in \tilde{A}_0$ and $y_0 \in \tilde{B}_0$. Let $z_{n_k} = x_{n_k}/y_{n_k}$. Then, we see that $\{z_{n_k}\}_{k=1}^\infty$ is a subsequence of $\{z_n\}_{n=1}^\infty$, i.e. $z_{n_k} \to z_0$ as $k \to \infty$. Since

$$z_0 = \lim_{k\to\infty} z_{n_k} = \lim_{k\to\infty} (x_{n_k}/y_{n_k}) = x_0/y_0,$$

i.e. $z_0 \in A_0$, which shows that $(\tilde{A} \oslash \tilde{B})_0 \subseteq A_0$. Therefore, we conclude that $(\tilde{A} \oslash \tilde{B})_\alpha = A_\alpha$ for any $\alpha \in [0,1]$. The formula (7.9) is also obvious.

Finally, we want to show that $\tilde{A} \oslash \tilde{B}$ is indeed a fuzzy interval. We need to check the conditions in Definition 2.3.1. Since $\tilde{A}$ and $\tilde{B}$ are fuzzy intervals, there exists x^* and y^* satisfying $\xi_{\tilde{A}}(x^*) = 1$ and $\xi_{\tilde{B}}(y^*) = 1$. Let $z^* = x^*/y^*$. Then, we see that

$$\xi_{\tilde{A}\oslash\tilde{B}}(z^*) = \sup_{\{(x,y):z^*=x/y\}} \min\left\{\xi_{\tilde{A}}(x), \xi_{\tilde{B}}(y)\right\} = 1.$$

This proves the normality. Since $\tilde{A}_\alpha$ and $\tilde{B}_\alpha$ are closed intervals for all $\alpha \in [0,1]$, if $0 \notin \tilde{B}_\alpha$, then we see that

$$(\tilde{A} \oslash \tilde{B})_\alpha = A_\alpha = \tilde{A}_\alpha / \tilde{B}_\alpha$$

is also a closed interval, i.e. a closed, bounded and convex set in $\mathbb{R}$. Suppose that $\tilde{B}$ is a positive or negative fuzzy interval satisfying $0 \notin \tilde{B}_0$. Then $0 \notin \tilde{B}_\alpha$ for all $\alpha \in [0,1]$. In this case, $(\tilde{A} \oslash \tilde{B})_\alpha$ is a closed interval for each $\alpha \in [0,1]$; that is, the membership function $\xi_{\tilde{A} \oslash \tilde{B}}$ is upper semi-continuous and quasi-concave. This shows that $\tilde{A} \oslash \tilde{B}$ is indeed a fuzzy interval, and the proof is complete. ■

Let A and B be two closed, bounded and convex sets in $\mathbb{R}^m$. Suppose that there exists a closed, bounded and convex set C in $\mathbb{R}^m$ satisfying $A = B + C$. Then C is called the **Hukuhara difference** of A and B. We also write $C = A \ominus B$ (see Banks and Jacobs [7]). Inspired by this concept, we can also define the Hukuhara difference between two fuzzy intervals. Let $\tilde{A}$ and $\tilde{B}$ be two fuzzy intervals. Suppose that there exists a fuzzy interval $\tilde{C}$ satisfying $\tilde{C} \oplus \tilde{B} = \tilde{A}$ (note that the fuzzy addition is commutative). Then $\tilde{C}$ is unique. In this case, $\tilde{C}$ is called the **Hukuhara difference** of $\tilde{A}$ and $\tilde{B}$ and is denoted by $\tilde{A} \ominus_H \tilde{B}$ (see Puri and Ralescu [89]). The following proposition is very useful for considering the differentiation of a fuzzy function.

Proposition 7.1.3 *Let $\tilde{A}$ and $\tilde{B}$ be two fuzzy intervals. Then, we have the following properties.*

(i) *Suppose that the Hukuhara difference $\tilde{C} = \tilde{A} \ominus_H \tilde{B}$ exists. Then*

$$\tilde{C}^L_\alpha = \tilde{A}^L_\alpha - \tilde{B}^L_\alpha \text{ and } \tilde{C}^U_\alpha = \tilde{A}^U_\alpha - \tilde{B}^U_\alpha$$

for all $\alpha \in [0,1]$.

(ii) *Suppose that $\tilde{A}^L_\alpha - \tilde{B}^L_\alpha \leq \tilde{A}^U_\alpha - \tilde{B}^U_\alpha$, $\tilde{A}^L_\alpha - \tilde{B}^L_\alpha \leq \tilde{A}^L_\beta - \tilde{B}^L_\beta$ and $\tilde{A}^U_\beta - \tilde{B}^U_\beta \leq \tilde{A}^U_\alpha - \tilde{B}^U_\alpha$ for $\alpha < \beta$. Then, the Hukuhara difference $\tilde{C} = \tilde{A} \ominus_H \tilde{B}$ exists.*

Proof. Part (i) follows immediately from part (i) of Theorem 7.1.2. To prove part (ii), let

$$A_\alpha = [\tilde{A}^L_\alpha - \tilde{B}^L_\alpha, \tilde{A}^U_\alpha - \tilde{B}^U_\alpha].$$

Then $A_\beta \subseteq A_\alpha$ for $\alpha < \beta$. By Proposition 2.3.4, we see that the functions

$$\eta^L(\alpha) \equiv \tilde{A}^L_\alpha - \tilde{B}^L_\alpha \text{ and } \eta^U(\alpha) \equiv \tilde{A}^U_\alpha - \tilde{B}^U_\alpha$$

are left-continuous on $(0,1]$. Therefore, we have $\bigcap_{n=1}^\infty A_{\alpha_n} = A_\alpha$ for $0 \leq \alpha_n \uparrow \alpha$. Theorem 5.4.7 says that we can generate a fuzzy interval $\tilde{C}$ satisfying

$$\tilde{C}^L_\alpha = \tilde{A}^L_\alpha - \tilde{B}^L_\alpha \text{ and } \tilde{C}^U_\alpha = \tilde{A}^U_\alpha - \tilde{B}^U_\alpha.$$

This completes the proof. ■

Recall Example 2.2.4, the membership function of a trapezoidal fuzzy interval is given by

$$\xi_{\tilde{A}}(r) = \begin{cases} (r - a^L)/(a_1 - a^L) & \text{if } a^L \leq r \leq a_1 \\ 1 & \text{if } a_1 < r \leq a_2 \\ (a^U - r)/(a^U - a_2) & \text{if } a_2 < r \leq a^U \\ 0 & \text{otherwise,} \end{cases}$$

which is denoted by $\tilde{A} = (a^L, a_1, a_2, a^U)$. We also have

$$\tilde{A}_\alpha^L = (1-\alpha)a^L + \alpha a_1 \text{ and } \tilde{A}_\alpha^U = (1-\alpha)a^U + \alpha a_2, \tag{7.12}$$

where $\tilde{A}_\alpha = \left[\tilde{A}_\alpha^L, \tilde{A}_\alpha^U\right]$. Let $\tilde{A} = (a^L, a_1, a_2, a^U)$ and $\tilde{B} = (b^L, b_1, b_2, b^U)$ be two trapezoidal fuzzy intervals. We can show that $\tilde{A} \oplus \tilde{B}$ is also a trapezoidal fuzzy interval. Moreover, we have

$$\tilde{A} \oplus \tilde{B} = (a^L + b^L, a_1 + b_1, a_2 + b_2, a^U + b^U). \tag{7.13}$$

Suppose that $\tilde{A} = (a^L, a_1, a_2, a^U)$ and $\tilde{B} = (b^L, b_1, b_2, b^U)$ are two trapezoidal fuzzy intervals. We write $\tilde{A} \preceq \tilde{B}$ when $a^L \le b^L$, $a_1 \le b_1$, $a_2 \le b_2$ and $a^U \le b^U$. Then, we have the following easy consequence.

Proposition 7.1.4 *Suppose that* $\tilde{A} = (a^L, a_1, a_2, a^U)$, $\tilde{B} = (b^L, b_1, b_2, b^U)$, $\tilde{C} = (c^L, c_1, c_2, c^U)$, *and* $\tilde{D} = (d^L, d_1, d_2, d^U)$ *are trapezoidal fuzzy intervals. If* $\tilde{A} \preceq \tilde{C}$ *and* $\tilde{B} \preceq \tilde{D}$, *then* $\tilde{A} \oplus \tilde{B} \preceq \tilde{C} \oplus \tilde{D}$. *This shows that the ordering "$\preceq$" is compatible with the fuzzy addition.*

Proof. From (7.13), we see that

$$\tilde{A} \oplus \tilde{B} = (a^L + b^L, a_1 + b_1, a_2 + b_2, a^U + b^U)$$

and

$$\tilde{C} \oplus \tilde{D} = (c^L + d^L, c_1 + d_1, c_2 + d_2, c^U + d^U).$$

The result follows from definition immediately. ∎

Proposition 7.1.5 *Let* $\tilde{A} = (a^L, a_1, a_2, a^U)$ *and* $\tilde{B} = (b^L, b_1, b_2, b^U)$ *be two trapezoidal fuzzy intervals satisfying* $\tilde{B} \preceq \tilde{A}$. *Then* $\widetilde{\max}\{\tilde{A}, \tilde{B}\} = \tilde{A}$.

Proof. From Example 4.1.4, we have

$$\left(\widetilde{\max}\{\tilde{A}, \tilde{B}\}\right)_\alpha^L = \max\left\{\tilde{A}_\alpha^L, \tilde{B}_\alpha^L\right\} \text{ and } \left(\widetilde{\max}\{\tilde{A}, \tilde{B}\}\right)_\alpha^U = \max\left\{\tilde{A}_\alpha^U, \tilde{B}_\alpha^U\right\}.$$

Therefore, from (7.12), we see that

$$\left(\widetilde{\max}\{\tilde{A}, \tilde{B}\}\right)_\alpha^L = \max\left\{(1-\alpha)a^L + \alpha a_1, (1-\alpha)b^L + \alpha b_1\right\} = (1-\alpha)a^L + \alpha a_1 = \tilde{A}_\alpha^L$$

and

$$\left(\widetilde{\max}\{\tilde{A}, \tilde{B}\}\right)_\alpha^U = \max\left\{(1-\alpha)a^U + \alpha a_2, (1-\alpha)b^U + \alpha b_2\right\} = (1-\alpha)a^U + \alpha a_2 = \tilde{A}_\alpha^U$$

since $\tilde{B} \preceq \tilde{A}$. This completes the proof. ∎

Now, we also have the following easy and interesting consequence.

Corollary 7.1.6 *Suppose that* $\tilde{A}$ *and* $\tilde{B}$ *are two trapezoidal fuzzy intervals with the same right and left spread values, respectively, i.e.* $\tilde{A} = (a_1 - h, a_1, a_2, a_2 + k)$ *and* $\tilde{B} = (b_1 - h, b_1, b_2, b_2 + k)$ *for some positive values* h *and* k. *If* $a_1 \ge b_1$ *and* $a_2 \ge b_2$, *then* $\widetilde{\max}\{\tilde{A}, \tilde{B}\} = \tilde{A}$.

7.1.2 Arithmetics Using EP and DT

Let $\circ$ denote the arithmetics of real numbers with $\circ = +, -, \times$ or $/$. We define a two-variable real-valued function $f : \mathbb{R}^2 \to \mathbb{R}$ as

$$f(x, y) = x \circ y,$$

where $y \neq 0$ when $\circ$ is taken as the division $/$. Let $\tilde{A}$ and $\tilde{B}$ be two fuzzy intervals. We can use the fuzzified functions $\tilde{f}^{(EP)}$ and $\tilde{f}^{(DT)}$ defined in (6.1) and (6.8), respectively, to study the different types of arithmetics between $\tilde{A}$ and $\tilde{B}$.

- Based on the extension principle, the arithmetics of $\tilde{A}$ and $\tilde{B}$ is given by

$$\tilde{f}^{(EP)}(\tilde{A}, \tilde{B}) = \tilde{A} \odot_{EP} \tilde{B}. \tag{7.14}$$

- Based on the expression in the decomposition theorem, the arithmetics of $\tilde{A}$ and $\tilde{B}$ is given by

$$\tilde{f}^{(DT)}(\tilde{A}, \tilde{B}) = \tilde{A} \odot_{DT} \tilde{B}. \tag{7.15}$$

More precisely, we can consider

$$\tilde{f}^{(\diamond DT)}(\tilde{A}, \tilde{B}) = \tilde{A} \odot^{\diamond}_{DT} \tilde{B} \tag{7.16}$$

by referring to Subsection 6.2.1, consider

$$\tilde{f}^{(\star DT)}(\tilde{A}, \tilde{B}) = \tilde{A} \odot^{\star}_{DT} \tilde{B} \tag{7.17}$$

by referring to Subsection 6.2.2, and consider

$$\tilde{f}^{(\dagger DT)}(\tilde{A}, \tilde{B}) = \tilde{A} \odot^{\dagger}_{DT} \tilde{B} \tag{7.18}$$

by referring to Subsection 6.2.3.

We are gong to study the relationship among $\tilde{A} \odot_{EP} \tilde{B}, \tilde{A} \odot^{\diamond}_{DT} \tilde{B}, \tilde{A} \odot^{\star}_{DT} \tilde{B}$, and $\tilde{A} \odot^{\dagger}_{DT} \tilde{B}$.

7.1.2.1 Addition of Fuzzy Intervals

Given any two fuzzy intervals $\tilde{A}$ and $\tilde{B}$, according to (7.14) and (7.15) by considering the function $f(x, y) = x + y$, we can obtain four additions of $\tilde{A}$ and $\tilde{B}$ as follows

$$\tilde{A} \oplus_{EP} \tilde{B}, \quad \tilde{A} \oplus^{\diamond}_{DT} \tilde{B}, \quad \tilde{A} \oplus^{\star}_{DT} \tilde{B}, \text{ and } \tilde{A} \oplus^{\dagger}_{DT} \tilde{B}.$$

The membership function of $\tilde{A} \oplus_{EP} \tilde{B}$ is given by

$$\xi_{\tilde{A} \oplus_{EP} \tilde{B}}(z) = \sup_{\{(x,y): z=x+y\}} \mathfrak{A}\left(\xi_{\tilde{A}}(x), \xi_{\tilde{B}}(y)\right).$$

Suppose that we take

$$\mathfrak{A}\left(\alpha_1, \alpha_2\right) = \min\left\{\alpha_1, \alpha_2\right\}.$$

Then, by referring to Theorem 6.1.2, the α-level sets are given by

$$\begin{aligned}
\left(\tilde{A} \oplus_{EP} \tilde{B}\right)_\alpha &= \left[\min_{(x,y)\in\tilde{A}_\alpha\times\tilde{B}_\alpha} f(x, y), \max_{(x,y)\in\tilde{A}_\alpha\times\tilde{B}_\alpha} f(x, y)\right] \\
&= \left[\min_{(x,y)\in[\tilde{A}^L_\alpha,\tilde{A}^U_\alpha]\times[\tilde{B}^L_\alpha,\tilde{B}^U_\alpha]} (x + y), \max_{(x,y)\in[\tilde{A}^L_\alpha,\tilde{A}^U_\alpha]\times[\tilde{B}^L_\alpha,\tilde{B}^U_\alpha]} (x + y)\right] \\
&= \left[\tilde{A}^L_\beta + \tilde{B}^L_\beta, \tilde{A}^U_\beta + \tilde{B}^U_\beta\right]
\end{aligned} \tag{7.19}$$

for all $\alpha \in [0, 1]$.

By referring to Subsection 6.2.1, we take

$$M_\alpha^+ = \left\{x+y : x \in \tilde{A}_\alpha \text{ and } y \in \tilde{B}_\alpha\right\} = \left[\tilde{A}_\alpha^L + \tilde{B}_\alpha^L, \tilde{A}_\alpha^U + \tilde{B}_\alpha^U\right]$$

to define $\tilde{A} \oplus_{DT}^{\diamond} \tilde{B}$. From Theorem 6.2.6, the α-level sets have the same form as (7.19) and are given by

$$\left(\tilde{A} \oplus_{DT}^{\diamond} \tilde{B}\right)_\alpha = \left[\tilde{A}_\beta^L + \tilde{B}_\beta^L, \tilde{A}_\beta^U + \tilde{B}_\beta^U\right]$$

for all $\alpha \in [0,1]$. It means that, when we take

$$\mathfrak{A}\left(\alpha_1, \alpha_2\right) = \min\left\{\alpha_1, \alpha_2\right\},$$

we have the following equivalence

$$\tilde{A} \oplus_{EP} \tilde{B} = \tilde{A} \oplus_{DT}^{\diamond} \tilde{B}. \tag{7.20}$$

By referring to (6.18) and (6.19) in Subsection 6.2.2, we take

$$M_\alpha^+ = \bigcup_{\alpha \le \beta \le 1} M_\beta,$$

where M_β is a bounded closed interval given by

$$\begin{aligned} M_\beta &= \left[\min\left\{\tilde{A}_\beta^L + \tilde{B}_\beta^L, \tilde{A}_\beta^U + \tilde{B}_\beta^U\right\}, \max\left\{\tilde{A}_\beta^L + \tilde{B}_\beta^L, \tilde{A}_\beta^U + \tilde{B}_\beta^U\right\}\right] \\ &= \left[\tilde{A}_\beta^L + \tilde{B}_\beta^L, \tilde{A}_\beta^U + \tilde{B}_\beta^U\right] \end{aligned}$$

to define $\tilde{A} \oplus_{DT}^{\star} \tilde{B}$. This says that

$$M_\alpha^+ = \bigcup_{\alpha \le \beta \le 1} \left[\tilde{A}_\beta^L + \tilde{B}_\beta^L, \tilde{A}_\beta^U + \tilde{B}_\beta^U\right].$$

Let $\zeta^L(\beta) = A_\beta^L + B_\beta^L$ and $\zeta^U(\beta) = A_\beta^U + B_\beta^U$. Proposition 2.3.4 say that ζ^L is lower semi-continuous on $[0,1]$ and ζ^U is upper semi-continuous on $[0,1]$. Then, for $\alpha \in [0,1]$, we can obtain

$$\begin{aligned} M_\alpha^+ &= \bigcup_{\alpha \le \beta \le 1} \left[\zeta^L(\beta), \zeta^U(\beta)\right] = \left[\min_{\alpha \le \beta \le 1} \zeta^L(\beta), \max_{\alpha \le \beta \le 1} \zeta^U(\beta)\right] \\ &= \left[\zeta^L(\alpha), \zeta^U(\alpha)\right] = \left[A_\alpha^L + B_\alpha^L, A_\alpha^U + B_\alpha^U\right]. \end{aligned} \tag{7.21}$$

From Theorem 6.2.7, the α-level sets are given by

$$\left(\tilde{A} \oplus_{DT}^{\star} \tilde{B}\right)_\alpha = M_\alpha^+ = \left[A_\alpha^L + B_\alpha^L, A_\alpha^U + B_\alpha^U\right] = \left(\tilde{A} \oplus_{DT}^{\diamond} \tilde{B}\right)_\alpha$$

for all $\alpha \in [0,1]$.

By referring to Subsection 6.2.3, we take

$$\begin{aligned} M_\alpha^+ &= \left[\min\left\{\tilde{A}_\alpha^L + \tilde{B}_\alpha^L, \tilde{A}_\alpha^U + \tilde{B}_\alpha^U\right\}, \max\left\{\tilde{A}_\alpha^L + \tilde{B}_\alpha^L, \tilde{A}_\alpha^U + \tilde{B}_\alpha^U\right\}\right] \\ &= \left[\tilde{A}_\alpha^L + \tilde{B}_\alpha^L, \tilde{A}_\alpha^U + \tilde{B}_\alpha^U\right] \end{aligned}$$

to define $\tilde{A} \oplus_{DT}^{\dagger} \tilde{B}$. By referring to (7.21), we see that

$$\tilde{A} \oplus_{DT}^{\diamond} \tilde{B} = \tilde{A} \oplus_{DT}^{\star} \tilde{B} = \tilde{A} \oplus_{DT}^{\dagger} \tilde{B}.$$

In this case, we simply write $\tilde{A} \oplus_{DT} \tilde{B}$ and, based on the expression in the decomposition theorem, its membership function is given by

$$\xi_{\tilde{A} \oplus_{DT} \tilde{B}}(z) = \sup_{0<\alpha\leq 1} \alpha \cdot \chi_{M_\alpha^+}(z).$$

Suppose that we take

$$\mathfrak{A}\left(\alpha_1, \alpha_2\right) = \min\left\{\alpha_1, \alpha_2\right\}.$$

Theorem 6.3.1 says that

$$\tilde{A} \oplus_{EP} \tilde{B} = \tilde{A} \oplus_{DT} \tilde{B}.$$

Without considering their equivalence, Theorem 6.3.7 says that $\tilde{A} \oplus_{EP} \tilde{B}$ is fuzzier than $\tilde{A} \oplus_{DT} \tilde{B}$. In this case, we prefer to use $\tilde{A} \oplus_{DT} \tilde{B}$ in real applications by referring to Remark 6.3.8.

7.1.2.2 Difference of Fuzzy Intervals

Given any two fuzzy intervals $\tilde{A}$ and $\tilde{B}$, according to (7.14) and (7.15) by considering the function $f(x, y) = x - y$, we can obtain four differences of $\tilde{A}$ and $\tilde{B}$ as follows

$$\tilde{A} \ominus_{EP} \tilde{B}, \quad \tilde{A} \ominus_{DT}^{\diamond} \tilde{B}, \quad \tilde{A} \ominus_{DT}^{\star} \tilde{B}, \text{ and } \tilde{A} \ominus_{DT}^{\dagger} \tilde{B}.$$

The membership function of $\tilde{A} \ominus_{EP} \tilde{B}$ is given by

$$\xi_{\tilde{A} \ominus_{EP} \tilde{B}}(z) = \sup_{\{(x,y):z=x-y\}} \mathfrak{A}\left(\xi_{\tilde{A}}(x), \xi_{\tilde{B}}(y)\right).$$

Suppose that we take

$$\mathfrak{A}\left(\alpha_1, \alpha_2\right) = \min\left\{\alpha_1, \alpha_2\right\}.$$

Then, by referring to Theorem 6.1.2, the α-level sets are given by

$$\begin{aligned}
\left(\tilde{A} \ominus_{EP} \tilde{B}\right)_\alpha &= \left[\min_{(x,y)\in\tilde{A}_\alpha\times\tilde{B}_\alpha} f(x,y), \max_{(x,y)\in\tilde{A}_\alpha\times\tilde{B}_\alpha} f(x,y)\right] \\
&= \left[\min_{(x,y)\in[\tilde{A}_\alpha^L,\tilde{A}_\alpha^U]\times[\tilde{B}_\alpha^L,\tilde{B}_\alpha^U]} (x-y), \max_{(x,y)\in[\tilde{A}_\alpha^L,\tilde{A}_\alpha^U]\times[\tilde{B}_\alpha^L,\tilde{B}_\alpha^U]} (x-y)\right] \\
&= \left[\tilde{A}_\beta^L - \tilde{B}_\beta^U, \tilde{A}_\beta^U - \tilde{B}_\beta^L\right] \qquad (7.22)
\end{aligned}$$

for all $\alpha \in [0, 1]$.

By referring to Subsection 6.2.1, we take

$$M_\alpha^- = \left\{x - y : x \in \tilde{A}_\alpha \text{ and } y \in \tilde{B}_\alpha\right\} = \left[\tilde{A}_\alpha^L - \tilde{B}_\alpha^U, \tilde{A}_\alpha^U - \tilde{B}_\alpha^L\right]$$

to define $\tilde{A} \ominus_{DT}^{\diamond} \tilde{B}$. From Theorem 6.2.6, the α-level sets have the same form as (7.22) and are given by

$$\left(\tilde{A} \ominus_{DT}^{\diamond} \tilde{B}\right)_\alpha = \left[\tilde{A}_\beta^L - \tilde{B}_\beta^U, \tilde{A}_\beta^U - \tilde{B}_\beta^L\right]$$

for all $\alpha \in [0, 1]$. It means that, when we take

$$\mathfrak{A}\left(\alpha_1, \alpha_2\right) = \min\left\{\alpha_1, \alpha_2\right\},$$

we have the following equivalence

$$\tilde{A} \ominus_{EP} \tilde{B} = \tilde{A} \ominus_{DT}^{\diamond} \tilde{B}.$$

By referring to (6.18) and (6.19) in Subsection 6.2.2, we take

$$M_\alpha^- = \bigcup_{\alpha \le \beta \le 1} M_\beta,$$

where M_β is a bounded closed interval given by

$$M_\beta = \left[\min\left\{\tilde{A}_\beta^L - \tilde{B}_\beta^L, \tilde{A}_\beta^U - \tilde{B}_\beta^U\right\}, \max\left\{\tilde{A}_\beta^L - \tilde{B}_\beta^L, \tilde{A}_\beta^U - \tilde{B}_\beta^U\right\}\right]$$

to define $\tilde{A} \ominus_{DT}^{\star} \tilde{B}$. Let

$$\eta^L(\beta) = \tilde{A}_\beta^L - \tilde{B}_\beta^L \text{ and } \eta^U(\beta) = \tilde{A}_\beta^U - \tilde{B}_\beta^U.$$

Since the semi-continuities of $\tilde{A}_\alpha^L, \tilde{B}_\alpha^L, \tilde{A}_\alpha^U$ and $\tilde{B}_\alpha^U$ with respect to α are not sufficient to guarantee the semi-continuities of η^L and η^U, we assume that $\tilde{A}$ and $\tilde{B}$ are taken to be canonical fuzzy intervals. In this case, the functions η^L and η^U are continuous on $[0,1]$. Let

$$\zeta^L(\beta) = \min\left\{\eta^L(\beta), \eta^U(\beta)\right\} \text{ and } \zeta^U(\beta) = \max\left\{\eta^L(\beta), \eta^U(\beta)\right\}.$$

Then, the functions ζ^L and ζ^U are also continuous on $[0,1]$. From Theorem 6.2.7, the α-level sets are given by

$$\begin{aligned}
\left(\tilde{A} \ominus_{DT}^{\star} \tilde{B}\right)_\alpha &= M_\alpha^- = \bigcup_{\alpha \le \beta \le 1} \left[\zeta^L(\beta), \zeta^U(\beta)\right] = \left[\min_{\alpha \le \beta \le 1} \zeta^L(\beta), \max_{\alpha \le \beta \le 1} \zeta^U(\beta)\right] \\
&= \left[\min_{\alpha \le \beta \le 1} \min\left\{\tilde{A}_\beta^L - \tilde{B}_\beta^L, \tilde{A}_\beta^U - \tilde{B}_\beta^U\right\}, \max_{\alpha \le \beta \le 1} \max\left\{\tilde{A}_\beta^L - \tilde{B}_\beta^L, \tilde{A}_\beta^U - \tilde{B}_\beta^U\right\}\right] \\
&= \left[\min\left\{\min_{\alpha \le \beta \le 1}\left(\tilde{A}_\beta^L - \tilde{B}_\beta^L\right), \min_{\alpha \le \beta \le 1}\left(\tilde{A}_\beta^U - \tilde{B}_\beta^U\right)\right\},\right. \\
&\qquad \left.\max\left\{\max_{\alpha \le \beta \le 1}\left(\tilde{A}_\beta^L - \tilde{B}_\beta^L\right), \max_{\alpha \le \beta \le 1}\left(\tilde{A}_\beta^U - \tilde{B}_\beta^U\right)\right\}\right]
\end{aligned}$$

for all $\alpha \in [0,1]$.

By referring to Subsection 6.2.3, we take

$$M_\alpha^- = M_\alpha = \left[\min\left\{\tilde{A}_\alpha^L - \tilde{B}_\alpha^L, \tilde{A}_\alpha^U - \tilde{B}_\alpha^U\right\}, \max\left\{\tilde{A}_\alpha^L - \tilde{B}_\alpha^L, \tilde{A}_\alpha^U - \tilde{B}_\alpha^U\right\}\right]$$

to define $\tilde{A} \ominus_{DT}^{\dagger} \tilde{B}$. We assume that $\tilde{A}$ and $\tilde{B}$ are taken to be canonical fuzzy intervals. From part (ii) of Theorem 6.2.10, the α-level sets are given by

$$\begin{aligned}
\left(\tilde{A} \ominus_{DT}^{\dagger} \tilde{B}\right)_\alpha &= \bigcup_{\alpha \le \beta \le 1} M_\alpha^- = \left[\min_{\alpha \le \beta \le 1}\left\{f\left(\tilde{A}_\beta^L, \tilde{B}_\beta^L\right), f\left(\tilde{A}_\beta^U, \tilde{B}_\beta^U\right)\right\},\right. \\
&\qquad \left.\max_{\alpha \le \beta \le 1}\left\{f\left(\tilde{A}_\beta^L, \tilde{B}_\beta^L\right), f\left(\tilde{A}_\beta^U, \tilde{B}_\beta^U\right)\right\}\right] \\
&= \left[\min_{\alpha \le \beta \le 1} \min\left\{\tilde{A}_\beta^L - \tilde{B}_\beta^L, \tilde{A}_\beta^U - \tilde{B}_\beta^U\right\}, \max_{\alpha \le \beta \le 1} \max\left\{\tilde{A}_\beta^L - \tilde{B}_\beta^L, \tilde{A}_\beta^U - \tilde{B}_\beta^U\right\}\right] \\
&= \left[\min\left\{\min_{\alpha \le \beta \le 1}\left(\tilde{A}_\beta^L - \tilde{B}_\beta^L\right), \min_{\alpha \le \beta \le 1}\left(\tilde{A}_\beta^U - \tilde{B}_\beta^U\right)\right\},\right. \\
&\qquad \left.\max\left\{\max_{\alpha \le \beta \le 1}\left(\tilde{A}_\beta^L - \tilde{B}_\beta^L\right), \max_{\alpha \le \beta \le 1}\left(\tilde{A}_\beta^U - \tilde{B}_\beta^U\right)\right\}\right]
\end{aligned}$$

for all $\alpha \in [0,1]$. Therefore, when $\tilde{A}$ and $\tilde{B}$ are taken to be canonical fuzzy intervals, we have the following equivalence

$$\tilde{A} \ominus^{\star}_{DT} \tilde{B} = \tilde{A} \ominus^{\dagger}_{DT} \tilde{B}.$$

Without considering their equivalence, Theorem 6.3.7 says that $\tilde{A} \ominus_{EP} \tilde{B}$ is fuzzier than $\tilde{A} \ominus^{\dagger}_{DT} \tilde{B}$. In this case, we prefer to use $\tilde{A} \ominus^{\dagger}_{DT} \tilde{B}$ in real applications by referring to Remark 6.3.8.

7.1.2.3 Multiplication of Fuzzy Intervals

Given any two fuzzy intervals $\tilde{A}$ and $\tilde{B}$, according to (7.14) and (7.15) by considering the function $f(x,y) = x \cdot y$, we can obtain four multiplications of $\tilde{A}$ and $\tilde{B}$ as follows

$$\tilde{A} \otimes_{EP} \tilde{B}, \quad \tilde{A} \otimes^{\diamond}_{DT} \tilde{B}, \quad \tilde{A} \otimes^{\star}_{DT} \tilde{B}, \text{ and } \tilde{A} \otimes^{\dagger}_{DT} \tilde{B}.$$

The membership function of $\tilde{A} \otimes_{EP} \tilde{B}$ is given by

$$\xi_{\tilde{A} \otimes_{EP} \tilde{B}}(z) = \sup_{\{(x,y):z=x\cdot y\}} \mathfrak{A}\left(\xi_{\tilde{A}}(x), \xi_{\tilde{B}}(y)\right).$$

Suppose that we take

$$\mathfrak{A}\left(\alpha_1, \alpha_2\right) = \min\left\{\alpha_1, \alpha_2\right\}.$$

Then, by referring to Theorem 6.1.2, the α-level sets are given by

$$\begin{aligned}
\left(\tilde{A} \otimes_{EP} \tilde{B}\right)_\alpha &= \left[\min_{(x,y)\in\tilde{A}_\alpha\times\tilde{B}_\alpha} f(x,y), \max_{(x,y)\in\tilde{A}_\alpha\times\tilde{B}_\alpha} f(x,y)\right] \\
&= \left[\min_{(x,y)\in[\tilde{A}^L_\alpha,\tilde{A}^U_\alpha]\times[\tilde{B}^L_\alpha,\tilde{B}^U_\alpha]} x\cdot y, \max_{(x,y)\in[\tilde{A}^L_\alpha,\tilde{A}^U_\alpha]\times[\tilde{B}^L_\alpha,\tilde{B}^U_\alpha]} x\cdot y\right] \\
&= \left[\min\left\{\tilde{A}^L_\alpha\tilde{B}^L_\alpha, \tilde{A}^L_\alpha\tilde{B}^U_\alpha, \tilde{A}^U_\alpha\tilde{B}^L_\alpha, \tilde{A}^U_\alpha\tilde{B}^U_\alpha\right\}, \max\left\{\tilde{A}^L_\alpha\tilde{B}^L_\alpha, \tilde{A}^L_\alpha\tilde{B}^U_\alpha, \tilde{A}^U_\alpha\tilde{B}^L_\alpha, \tilde{A}^U_\alpha\tilde{B}^U_\alpha\right\}\right] \quad (7.23)
\end{aligned}$$

for all $\alpha \in [0,1]$.

By referring to Subsection 6.2.1, we take

$$\begin{aligned}
M^{\times}_\alpha &= \left\{x \cdot y : x \in \tilde{A}_\alpha \text{ and } y \in \tilde{B}_\alpha\right\} \\
&= \left[\min\left\{\tilde{A}^L_\alpha\tilde{B}^L_\alpha, \tilde{A}^L_\alpha\tilde{B}^U_\alpha, \tilde{A}^U_\alpha\tilde{B}^L_\alpha, \tilde{A}^U_\alpha\tilde{B}^U_\alpha\right\}, \max\left\{\tilde{A}^L_\alpha\tilde{B}^L_\alpha, \tilde{A}^L_\alpha\tilde{B}^U_\alpha, \tilde{A}^U_\alpha\tilde{B}^L_\alpha, \tilde{A}^U_\alpha\tilde{B}^U_\alpha\right\}\right]
\end{aligned}$$

to define $\tilde{A} \otimes^{\diamond}_{DT} \tilde{B}$. From Theorem 6.2.6, the α-level sets have the same form as (7.23) and are given by

$$\begin{aligned}
\left(\tilde{A} \otimes^{\diamond}_{DT} \tilde{B}\right)_\alpha = \Big[&\min\left\{\tilde{A}^L_\alpha\tilde{B}^L_\alpha, \tilde{A}^L_\alpha\tilde{B}^U_\alpha, \tilde{A}^U_\alpha\tilde{B}^L_\alpha, \tilde{A}^U_\alpha\tilde{B}^U_\alpha\right\}, \\
&\max\left\{\tilde{A}^L_\alpha\tilde{B}^L_\alpha, \tilde{A}^L_\alpha\tilde{B}^U_\alpha, \tilde{A}^U_\alpha\tilde{B}^L_\alpha, \tilde{A}^U_\alpha\tilde{B}^U_\alpha\right\}\Big]
\end{aligned}$$

for all $\alpha \in [0,1]$. It means that, when we take

$$\mathfrak{A}\left(\alpha_1, \alpha_2\right) = \min\left\{\alpha_1, \alpha_2\right\},$$

we have the following equivalence

$$\tilde{A} \otimes_{EP} \tilde{B} = \tilde{A} \otimes^{\diamond}_{DT} \tilde{B}. \quad (7.24)$$

By referring to (6.18) and (6.19) in Subsection 6.2.2, we take

$$M_\alpha^\times = \bigcup_{\alpha\le\beta\le1} M_\beta,$$

where M_β is a bounded closed interval given by

$$M_\beta = \left[\min\left\{\tilde{A}_\beta^L\tilde{B}_\beta^L, \tilde{A}_\beta^U\tilde{B}_\beta^U\right\}, \max\left\{\tilde{A}_\beta^L\tilde{B}_\beta^L, \tilde{A}_\beta^U\tilde{B}_\beta^U\right\}\right]$$

to define $\tilde{A} \otimes_{DT}^\star \tilde{B}$. Let

$$\eta^L(\beta) = \tilde{A}_\beta^L\tilde{B}_\beta^L \text{ and } \eta^U(\beta) = \tilde{A}_\beta^U\tilde{B}_\beta^U.$$

We assume that $\tilde{A}$ and $\tilde{B}$ are taken to be canonical fuzzy intervals. In this case, the functions η^L and η^U are continuous on $[0, 1]$. Let

$$\zeta^L(\beta) = \min\left\{\eta^L(\beta), \eta^U(\beta)\right\} \text{ and } \zeta^U(\beta) = \max\left\{\eta^L(\beta), \eta^U(\beta)\right\}.$$

Then, the functions ζ^L and ζ^U are also continuous on $[0, 1]$. From Theorem 6.2.7, the α-level sets are given by

$$\begin{aligned}
\left(\tilde{A} \otimes_{DT}^\star \tilde{B}\right)_\alpha &= M_\alpha^\times = \bigcup_{\alpha\le\beta\le1} \left[\zeta^L(\beta), \zeta^U(\beta)\right] = \left[\min_{\alpha\le\beta\le1} \zeta^L(\beta), \max_{\alpha\le\beta\le1} \zeta^U(\beta)\right] \\
&= \left[\min_{\alpha\le\beta\le1} \min\left\{\tilde{A}_\beta^L\tilde{B}_\beta^L, \tilde{A}_\beta^U\tilde{B}_\beta^U\right\}, \max_{\alpha\le\beta\le1} \max\left\{\tilde{A}_\beta^L\tilde{B}_\beta^L, \tilde{A}_\beta^U\tilde{B}_\beta^U\right\}\right] \\
&= \left[\min\left\{\min_{\alpha\le\beta\le1}\left(\tilde{A}_\beta^L\tilde{B}_\beta^L\right), \min_{\alpha\le\beta\le1}\left(\tilde{A}_\beta^U\tilde{B}_\beta^U\right)\right\},\right. \\
&\qquad \left.\max\left\{\max_{\alpha\le\beta\le1}\left(\tilde{A}_\beta^L\tilde{B}_\beta^L\right), \max_{\alpha\le\beta\le1}\left(\tilde{A}_\beta^U\tilde{B}_\beta^U\right)\right\}\right]
\end{aligned}$$

for all $\alpha \in [0, 1]$.

By referring to Subsection 6.2.3, we take

$$M_\alpha^\times = M_\alpha = \left[\min\left\{\tilde{A}_\alpha^L\tilde{B}_\alpha^L, \tilde{A}_\alpha^U\tilde{B}_\alpha^U\right\}, \max\left\{\tilde{A}_\alpha^L\tilde{B}_\alpha^L, \tilde{A}_\alpha^U\tilde{B}_\alpha^U\right\}\right]$$

to define $\tilde{A} \otimes_{DT}^\dagger \tilde{B}$. We assume that $\tilde{A}$ and $\tilde{B}$ are taken to be canonical fuzzy intervals. From part (ii) of Theorem 6.2.10, the α-level sets are given by

$$\begin{aligned}
\left(\tilde{A} \otimes_{DT}^\dagger \tilde{B}\right)_\alpha &= \bigcup_{\alpha\le\beta\le1} M_\alpha^\times \\
&= \left[\min_{\alpha\le\beta\le1}\left\{f\left(\tilde{A}_\beta^L, \tilde{B}_\beta^L\right), f\left(\tilde{A}_\beta^U, \tilde{B}_\beta^U\right)\right\}, \max_{\alpha\le\beta\le1}\left\{f\left(\tilde{A}_\beta^L, \tilde{B}_\beta^L\right), f\left(\tilde{A}_\beta^U, \tilde{B}_\beta^U\right)\right\}\right] \\
&= \left[\min_{\alpha\le\beta\le1} \min\left\{\tilde{A}_\beta^L\tilde{B}_\beta^L, \tilde{A}_\beta^U\tilde{B}_\beta^U\right\}, \max_{\alpha\le\beta\le1} \max\left\{\tilde{A}_\beta^L\tilde{B}_\beta^L, \tilde{A}_\beta^U\tilde{B}_\beta^U\right\}\right] \qquad (7.25) \\
&= \left[\min\left\{\min_{\alpha\le\beta\le1}\left(\tilde{A}_\beta^L\tilde{B}_\beta^L\right), \min_{\alpha\le\beta\le1}\left(\tilde{A}_\beta^U\tilde{B}_\beta^U\right)\right\},\right. \\
&\qquad \left.\max\left\{\max_{\alpha\le\beta\le1}\left(\tilde{A}_\beta^L\tilde{B}_\beta^L\right), \max_{\alpha\le\beta\le1}\left(\tilde{A}_\beta^U\tilde{B}_\beta^U\right)\right\}\right]
\end{aligned}$$

for all $\alpha \in [0,1]$. Therefore, when $\tilde{A}$ and $\tilde{B}$ are taken to be canonical fuzzy intervals, we have the equivalence

$$\tilde{A} \otimes_{DT}^{\star} \tilde{B} = \tilde{A} \otimes_{DT}^{\dagger} \tilde{B}. \tag{7.26}$$

Without considering their equivalence, Theorem 6.3.7 says that $\tilde{A} \otimes_{EP} \tilde{B}$ is fuzzier than $\tilde{A} \otimes_{DT}^{\dagger} \tilde{B}$. In this case, we prefer to use $\tilde{A} \otimes_{DT}^{\dagger} \tilde{B}$ in real applications by referring to Remark 6.3.8.

Let $x \in \mathbb{R}$ be a real number. Recall that the crisp number with value x is defined by

$$1_{\{x\}}(r) = \begin{cases} 1 & r = x \\ 0 & r \neq x. \end{cases}$$

Then, it is clear to see that $1_{\{x\}} \in \mathfrak{F}(\mathbb{R})$.

Example 7.1.7 Let $\tilde{A}^{(1)}, \ldots, \tilde{A}^{(n)}$ be two fuzzy intervals. We can define a fuzzy function as follows:

$$\tilde{f}(\mathbf{x}) = \left(\tilde{A}^{(1)} \otimes 1_{\{x_1\}}\right) \oplus \cdots \oplus \left(\tilde{A}^{(n)} \otimes 1_{\{x_n\}}\right), \tag{7.27}$$

where $\otimes$ can be taken as $\otimes_{EP}$ or $\otimes_{DT}$, and $\oplus$ can be taken as $\oplus_{EP}$ or $\oplus_{DT}$. From (7.23) and (7.25), for $i = 1, \ldots, n$ and $\alpha \in [0,1]$, we have

$$\begin{aligned}\left(\tilde{A}^{(i)} \otimes_{EP} 1_{\{x_i\}}\right)_\alpha &= \left[\min\left\{\tilde{A}_{i\alpha}^{L} x_i, \tilde{A}_{i\alpha}^{U} x_i\right\}, \max\left\{\tilde{A}_{i\alpha}^{L} x_i, \tilde{A}_{i\alpha}^{U} x_i\right\}\right] \\ &= \begin{cases} \left[\tilde{A}_{i\alpha}^{L} x_i, \tilde{A}_{i\alpha}^{U} x_i\right] & \text{if } x_i \geq 0 \\ \left[\tilde{A}_{i\alpha}^{U} x_i, \tilde{A}_{i\alpha}^{L} x_i\right] & \text{if } x_i < 0. \end{cases}\end{aligned}$$

and

$$\begin{aligned}\left(\tilde{A}^{(i)} \otimes_{DT} 1_{\{x_i\}}\right)_\alpha &= \left[\min_{\alpha \leq t \leq 1} \min\left\{\tilde{A}_{it}^{L} x_i, \tilde{A}_{it}^{U} x_i\right\}, \max_{\alpha \leq t \leq 1} \max\left\{\tilde{A}_{it}^{L} x_i, \tilde{A}_{it}^{U} x_i\right\}\right] \\ &= \begin{cases} \left[\tilde{A}_{i\alpha}^{L} x_i, \tilde{A}_{i\alpha}^{U} x_i\right] & \text{if } x_i \geq 0 \\ \left[\tilde{A}_{i\alpha}^{U} x_i, \tilde{A}_{i\alpha}^{L} x_i\right] & \text{if } x_i < 0. \end{cases}\end{aligned}$$

Therefore, we obtain

$$\tilde{A}^{(i)} \otimes_{EP} 1_{\{x_i\}} = \tilde{A}^{(i)} \otimes_{DT} 1_{\{x_i\}}$$

for all $i = 1, \ldots, n$. Since $\oplus_{EP} = \oplus_{DT}$, from (7.27), we have

$$\tilde{f}_\alpha^L(\mathbf{x}) = \sum_{i=1}^{n} \min\left\{\tilde{A}_{i\alpha}^{L} x_i, \tilde{A}_{i\alpha}^{U} x_i\right\} \text{ and } \tilde{f}_\alpha^U(\mathbf{x}) = \sum_{i=1}^{n} \max\left\{\tilde{A}_{i\alpha}^{L} x_i, \tilde{A}_{i\alpha}^{U} x_i\right\}.$$

Given any two fuzzy intervals $\tilde{A}$ and $\tilde{B}$, according to (7.14) and (7.15) by considering the function $f(x,y) = x/y$ for $y \neq 0$, we can obtain four divisions of $\tilde{A}$ and $\tilde{B}$ as follows

$$\tilde{A} \oslash_{EP} \tilde{B}, \quad \tilde{A} \oslash_{DT}^{\diamond} \tilde{B}, \quad \tilde{A} \oslash_{DT}^{\star} \tilde{B}, \text{ and } \tilde{A} \oslash_{DT}^{\dagger} \tilde{B}.$$

Their α-level sets can be similarly obtained by referring to the multiplication of fuzzy intervals when we assume $0 \notin \tilde{B}_\alpha$ for all $\alpha \in [0,1]$. The details are left for the readers.

7.2 Arithmetics of Fuzzy Vectors

Let $\tilde{A}$ and $\tilde{B}$ be two fuzzy sets in $\mathbb{R}^m$. In this case, we can only consider $\tilde{A} \oplus \tilde{B}$ and $\lambda\tilde{A}$, where $\lambda \in \mathbb{R}$. Therefore, we obtain

$$\xi_{\tilde{A}\oplus\tilde{B}}(z) = \sup_{\{(x,y)\in\mathbb{R}^m\times\mathbb{R}^m : x+y=z\}} \min\left\{\xi_{\tilde{A}}(x), \xi_{\tilde{B}}(y)\right\}$$

and

$$\xi_{\lambda\tilde{A}}(z) = \begin{cases} \xi_{\tilde{A}}(x/\lambda) & \text{if } \lambda \neq 0 \\ \sup\limits_{x\in\mathbb{R}^m} \xi_{\tilde{A}}(x) & \text{if } \lambda = 0 \text{ and } z = \theta, \\ 0 & \text{if } \lambda = 0 \text{ and } z \neq \theta \end{cases} \tag{7.28}$$

by referring to (7.3), where θ is the zero vector of $\mathbb{R}^m$. We can also show that

$$\tilde{A} \ominus \tilde{B} = \tilde{A} \oplus (-\tilde{B}).$$

Proposition 7.2.1 *Let $\tilde{A}$ and $\tilde{B}$ be two fuzzy sets in $\mathbb{R}^m$, and let $\lambda \in \mathbb{R}$. Suppose that $\lambda \neq 0$. Then, we have*

$$\lambda(\tilde{A} \oplus \tilde{B}) = \lambda\tilde{A} \oplus \lambda\tilde{B}.$$

Suppose that $\lambda = 0$ and $z \neq \theta$. Then, we have

$$\xi_{\lambda(\tilde{A}\oplus\tilde{B})}(z) = \xi_{\lambda\tilde{A}\oplus\lambda\tilde{B}}(z) = 0.$$

Proof. For $\lambda \neq 0$ and for all $z \in \mathbb{R}^m$, we have

$$\xi_{\lambda(\tilde{A}\oplus\tilde{B})}(z) = \xi_{\tilde{A}\oplus\tilde{B}}(z/\lambda) = \sup_{\{(x,y):x+y=z/\lambda\}} \min\left\{\xi_{\tilde{A}}(x), \xi_{\tilde{B}}(y)\right\}$$

and

$$\begin{aligned} \xi_{\lambda\tilde{A}\oplus\lambda\tilde{B}}(z) &= \sup_{\{(x,y):x+y=z\}} \min\left\{\xi_{\widetilde{\lambda A}}(x), \xi_{\widetilde{\lambda B}}(y)\right\} = \sup_{\{(x,y):x+y=z\}} \min\left\{\xi_{\tilde{A}}(x/\lambda), \xi_{\tilde{B}}(y/\lambda)\right\} \\ &= \sup_{\{(u,v):\lambda u+\lambda v=z\}} \min\left\{\xi_{\tilde{A}}(u), \xi_{\tilde{B}}(v)\right\}. \end{aligned}$$

Therefore, we obtain $\xi_{\lambda(\tilde{A}\oplus\tilde{B})}(z) = \xi_{\lambda\tilde{A}\oplus\lambda\tilde{B}}(z)$.

For $\lambda = 0$ and $z \neq \theta$, we have $\xi_{\lambda(\tilde{A}\oplus\tilde{B})}(z) = 0$ and

$$\xi_{\lambda\tilde{A}\oplus\lambda\tilde{B}}(z) = \sup_{\{(x,y):x+y=z\}} \min\left\{\xi_{\widetilde{\lambda A}}(x), \xi_{\widetilde{\lambda B}}(y)\right\} = 0,$$

since x and y cannot be the zero vector θ simultaneously. This completes the proof. ■

Proposition 7.2.2 *Let $\tilde{A}$ be a fuzzy set in $\mathbb{R}^m$. For any $\lambda_1, \lambda_2 \in \mathbb{R}$, we have*

$$\lambda_1(\lambda_2\tilde{A}) = (\lambda_1\lambda_2)\tilde{A}.$$

Proof. If $\lambda_1 \neq 0$ and $\lambda_2 \neq 0$, then

$$\xi_{\lambda_1(\lambda_2\tilde{A})}(z) = \xi_{\lambda_2\tilde{A}}(z/\lambda_1) = \xi_{\tilde{A}}(z/\lambda_1\lambda_2) = \xi_{(\lambda_1\lambda_2)\tilde{A}}(z)$$

for all $z \in \mathbb{R}^m$. Now, using (7.28), we consider the following cases.

- For $\lambda_1 = 0 = \lambda_2$, we have

$$\xi_{\lambda_1(\lambda_2\tilde{A})}(\theta) = \sup_{z\in\mathbb{R}^m} \xi_{\lambda_2\tilde{A}}(z) = \sup_{z\in\mathbb{R}^m} \begin{cases} 0 & \text{if } z \neq \theta \\ \sup_{x\in\mathbb{R}^m} \xi_{\tilde{A}}(x) & \text{if } z = \theta \end{cases}$$

$$= \sup_{x\in\mathbb{R}^m} \xi_{\tilde{A}}(x) = \xi_{(\lambda_1\lambda_2)\tilde{A}}(\theta).$$

If $z \neq \theta$, we immediately have $\xi_{\lambda_1(\lambda_2\tilde{A})}(z) = 0 = \xi_{(\lambda_1\lambda_2)\tilde{A}}(z)$.

- For $\lambda_1 = 0$ and $\lambda_2 \neq 0$, we have

$$\xi_{\lambda_1(\lambda_2\tilde{A})}(\theta) = \sup_{z\in\mathbb{R}^m} \xi_{\lambda_2\tilde{A}}(z) = \sup_{z\in\mathbb{R}^m} \xi_{\tilde{A}}(z/\lambda_2) = \sup_{x\in\mathbb{R}^m} \xi_{\tilde{A}}(x) = \xi_{(\lambda_1\lambda_2)\tilde{A}}(\theta).$$

If $z \neq \theta$, we immediately have $\xi_{\lambda_1(\lambda_2\tilde{A})}(z) = 0 = \xi_{(\lambda_1\lambda_2)\tilde{A}}(z)$.

- For $\lambda_1 \neq 0$ and $\lambda_2 = 0$, we have

$$\xi_{\lambda_1(\lambda_2\tilde{A})}(\theta) = \xi_{\lambda_2\tilde{A}}(\theta/\lambda_1) = \xi_{\lambda_2\tilde{A}}(\theta) = \sup_{x\in\mathbb{R}^m} \xi_{\tilde{A}}(x) = \xi_{(\lambda_1\lambda_2)\tilde{A}}(\theta).$$

If $z \neq \theta$, since $z/\lambda_1 \neq \theta$, we immediately have

$$\xi_{\lambda_1(\lambda_2\tilde{A})}(z) = \xi_{\lambda_2\tilde{A}}(z/\lambda_1) = 0 = \xi_{(\lambda_1\lambda_2)\tilde{A}}(z).$$

This completes the proof. ∎

Next, we are going to consider the fuzzy vectors in Definition 2.3.1.

Theorem 7.2.3 *We have the following results.*

(i) *Suppose that $\tilde{A} \in \mathfrak{F}(\mathbb{R}^m)$ and $\tilde{B} \in \mathfrak{F}(\mathbb{R}^m)$ are two fuzzy vectors in $\mathbb{R}^m$. Then $\tilde{A} \oplus \tilde{B}$ is also a fuzzy vector in $\mathbb{R}^m$ satisfying*

$$(\tilde{A} \oplus \tilde{B})_\alpha = \tilde{A}_\alpha + \tilde{B}_\alpha$$

for any $\alpha \in [0, 1]$.

(ii) *Suppose that $\tilde{\lambda}$ is a fuzzy interval, and that $\tilde{A}$ is a fuzzy vector in $\mathbb{R}^m$. Then, we have*

$$\left(\tilde{\lambda} \otimes \tilde{A}\right)_\alpha = \tilde{\lambda}_\alpha \tilde{A}_\alpha$$

for any $\alpha \in [0, 1]$. We further assume that $\tilde{\lambda}$ is taken to be a nonnegative or nonpositive fuzzy interval. Then $\tilde{\lambda} \otimes \tilde{A}$ is also a fuzzy vector in $\mathbb{R}^m$.

(iii) *Suppose that $\lambda \in \mathbb{R}$ and $\tilde{A}$ is a fuzzy vector in $\mathbb{R}^m$. Then $\lambda\tilde{A} = \tilde{1}_{\{\lambda\}} \otimes \tilde{A}$ is also a fuzzy vector in $\mathbb{R}^m$, where $\tilde{1}_{\{\lambda\}}$ is a crisp number with value λ.*

Proof. We take the function $f : \mathbb{R}^m \times \mathbb{R}^m \to \mathbb{R}^m$ by $(x_1, x_2) \mapsto x_1 + x_2$. Then, we see that f is a continuous and onto crisp function under the assumption that $\mathbb{R}^m$ is taken to be a topological vector space. Applying Corollary 4.4.13, we see that

$$(\tilde{A} \oplus \tilde{B})_\alpha = \left(\tilde{f}(\tilde{A}, \tilde{B})\right)_\alpha = f(\tilde{A}_\alpha, \tilde{B}_\alpha) = \tilde{A}_\alpha + \tilde{B}_\alpha$$

for all $\alpha \in [0, 1]$. For any convex subsets A_1, A_2 of $\mathbb{R}^m$, it is not hard to show that $f(A_1, A_2) = A_1 + A_2$ is also a convex set in $\mathbb{R}^m$. Therefore, from Theorem 4.2.7, we conclude that $\tilde{f}(\tilde{A}, \tilde{B}) = \tilde{A} \oplus \tilde{B} \in \mathfrak{F}(\mathbb{R}^m)$.

We take the function $f : \mathbb{R} \times \mathbb{R}^m \to \mathbb{R}^m$ by $(\lambda, x) \mapsto \lambda x$. Then, we see that f is continuous and onto crisp function under the assumption that $\mathbb{R}^m$ is taken to be a topological vector space. Applying Corollary 4.4.13, we obtain

$$(\tilde{\lambda} \otimes \tilde{A})_\alpha = \tilde{\lambda}_\alpha \tilde{A}_\alpha$$

for any $\alpha \in [0,1]$. For the case of nonnegative fuzzy interval $\tilde{\lambda}$, in order to claim $\tilde{\lambda} \otimes \tilde{A} \in \mathfrak{F}(\mathbb{R}^m)$ by applying Theorem 4.2.7 and Remark 2.3.7, we need to show that $f(A_1, A_2)$ is a convex set in $\mathbb{R}^m$ for any convex subsets A_1 and A_2 of $\mathbb{R}_+$ and $\mathbb{R}^m$, respectively. We also see that

$$f(A_1, A_2) = \{\lambda x : \lambda \in A_1 \text{ and } x \in A_2\} = \bigcup_{\lambda \in A_1} \lambda A_2.$$

For $y_1, y_2 \in f(A_1, A_2)$, there exist $\lambda_1, \lambda_2 \in A_1$ and $a_1, a_2 \in A_2$ satisfying $y_1 = \lambda_1 a_1$ and $y_2 = \lambda_2 a_2$, where $\lambda_1, \lambda_2 \in \mathbb{R}_+$. For any $\alpha \in (0,1)$, let $\lambda_3 = \alpha\lambda_1 + (1-\alpha)\lambda_2$. Then, we see that $\lambda_3 \in A_1$, since A_1 is convex. Since A_2 is also convex, a convex combination says that

$$\frac{\alpha\lambda_1}{\lambda_3} a_1 + \frac{(1-\alpha)\lambda_2}{\lambda_3} a_2 \in A_2.$$

Therefore, we obtain

$$\alpha y_1 + (1-\alpha) y_2 = \alpha\lambda_1 a_1 + (1-\alpha)\lambda_2 a_2 \in \lambda_3 A_2 \subseteq f(A_1, A_2),$$

which says that $f(A_1, A_2)$ is indeed a convex set in $\mathbb{R}^m$. For the case of nonpositive fuzzy interval $\tilde{\lambda}$, since $-A_2$ is also a convex set in $\mathbb{R}^m$, we still can show that $f(A_1, A_2)$ is a convex set in $\mathbb{R}^m$ by using Remark 2.3.7 and the above arguments. Also, for the case of crisp number $\tilde{\lambda} = \tilde{1}_{\{\lambda\}}$, since the α-level sets $(\tilde{1}_{\{\lambda\}})_\alpha = \{\lambda\}$ are convex sets in $\mathbb{R}$ for all $\alpha \in [0,1]$, the above arguments can also show that $\tilde{1}_{\{\lambda\}} \otimes \tilde{A} \in \mathfrak{F}(\mathbb{R}^m)$. This completes the proof. ∎

In what follows, we shall consider the vectors of fuzzy sets in $\mathbb{R}$. The purpose is to study the addition and difference of vectors of fuzzy sets. Suppose that the vectors $\tilde{\mathbf{A}}$ and $\tilde{\mathbf{B}}$ consist of fuzzy sets in $\mathbb{R}$ given by

$$\tilde{\mathbf{A}} = \left(\tilde{A}^{(1)}, \ldots, \tilde{A}^{(n)}\right) \text{ and } \tilde{\mathbf{B}} = \left(\tilde{B}^{(1)}, \ldots, \tilde{B}^{(n)}\right),$$

where $\tilde{A}^{(i)}$ and $\tilde{B}^{(i)}$ are fuzzy sets in $\mathbb{R}$ for $i = 1, \ldots, n$. Then, we are going to study the addition $\tilde{\mathbf{A}} \oplus \tilde{\mathbf{B}}$ and the difference $\tilde{\mathbf{A}} \ominus \tilde{\mathbf{B}}$.

The addition $\tilde{A}^{(i)} \oplus \tilde{B}^{(i)}$ and the difference $\tilde{A}^{(i)} \ominus \tilde{B}^{(i)}$ regarding the components can be realized as shown above. Let $\xi_{\tilde{A}^{(i)}}$ and $\xi_{\tilde{B}^{(i)}}$ be the membership functions of $\tilde{A}^{(i)}$ and $\tilde{B}^{(i)}$, respectively, and let $\odot$ denote any one of the arithmetic operations $\oplus, \ominus, \otimes$ between $\tilde{A}^{(i)}$ and $\tilde{B}^{(i)}$. According to the extension principle, the membership function of $\tilde{A}^{(i)} \odot \tilde{B}^{(i)}$ is defined by

$$\xi_{\tilde{A}^{(i)} \odot \tilde{B}^{(i)}}(z) = \sup_{\{(x,y):z=x\circ y\}} \min\{\xi_{\tilde{A}^{(i)}}(x), \xi_{\tilde{B}^{(i)}}(y)\}$$

for all $z \in \mathbb{R}$, where the arithmetic operations $\odot \in \{\oplus, \ominus, \otimes\}$ correspond to the arithmetic operations $\circ \in \{+, -, *\}$. More generally, the membership function of $\tilde{A}^{(i)} \odot_{\mathfrak{T}} \tilde{B}^{(i)}$ is given by

$$\xi_{\tilde{A}^{(i)} \odot_{\mathfrak{T}} \tilde{B}^{(i)}}(z) = \sup_{\{(x,y):z=x\circ y\}} t(\xi_{\tilde{A}^{(i)}}(x), \xi_{\tilde{B}^{(i)}}(y))$$

for all $z \in \mathbb{R}$, where t is a t-norm. It is well known that the minimum function min is a t-norm.

In this section, we are going to consider the general function rather than using t-norms. In this case, the membership function of $\tilde{A}^{(i)} \odot_{\mathfrak{A}} \tilde{B}^{(i)}$ will be given by

$$\xi_{\tilde{A}^{(i)} \odot_{\mathfrak{A}} \tilde{B}^{(i)}}(z) = \sup_{\{(x,y):z=x\circ y\}} \mathfrak{A}\left(\xi_{\tilde{A}^{(i)}}(x), \xi_{\tilde{B}^{(i)}}(y)\right) \tag{7.29}$$

for all $z \in \mathbb{R}$, where $\mathfrak{A}$ is a function from $[0,1] \times [0,1]$ into $[0,1]$ without satisfying some required conditions.

According to the arithmetic operations (7.29), the addition $\tilde{\mathbf{A}} \oplus_{\mathfrak{A}} \tilde{\mathbf{B}}$ and the difference $\tilde{\mathbf{A}} \ominus_{\mathfrak{A}} \tilde{\mathbf{B}}$ can be naturally defined as

$$\tilde{\mathbf{A}} \oplus_{\mathfrak{A}} \tilde{\mathbf{B}} = \left(\tilde{A}^{(1)} \oplus_{\mathfrak{A}} \tilde{B}^{(1)}, \ldots, \tilde{A}^{(n)} \oplus_{\mathfrak{A}} \tilde{B}^{(n)}\right)$$

and

$$\tilde{\mathbf{A}} \ominus_{\mathfrak{A}} \tilde{\mathbf{B}} = \left(\tilde{A}^{(1)} \ominus_{\mathfrak{A}} \tilde{B}^{(1)}, \ldots, \tilde{A}^{(n)} \ominus_{\mathfrak{A}} \tilde{B}^{(n)}\right).$$

We can see that $\tilde{\mathbf{A}} \oplus_{\mathfrak{A}} \tilde{\mathbf{B}}$ and $\tilde{\mathbf{A}} \ominus_{\mathfrak{A}} \tilde{\mathbf{B}}$ are vectors of fuzzy sets. However, their membership functions cannot be obtained directly from (7.29) regarding the components. The main purpose of this section is to propose two methodologies to define the membership functions of $\tilde{\mathbf{A}} \oplus_{\mathfrak{A}} \tilde{\mathbf{B}}$ and $\tilde{\mathbf{A}} \ominus_{\mathfrak{A}} \tilde{\mathbf{B}}$.

7.2.1 Arithmetics Using the Extension Principle

Let $\mathbf{x} = (x_1, \ldots, x_n)$ and $\mathbf{y} = (y_1, \ldots, y_n)$ be two vectors in $\mathbb{R}^m$. Then, the arithmetics of vectors $\mathbf{x}$ and $\mathbf{y}$ are given by

$$\mathbf{x} + \mathbf{y} = (x_1 + y_1, \ldots, x_n + y_n)$$

and

$$\mathbf{x} - \mathbf{y} = (x_1 - y_1, \ldots, x_n - y_n).$$

Let $\tilde{\mathbf{A}}$ and $\tilde{\mathbf{B}}$ be two vectors of fuzzy sets in $\mathbb{R}$ given by

$$\tilde{\mathbf{A}} = \left(\tilde{A}^{(1)}, \tilde{A}^{(2)}, \ldots, \tilde{A}^{(n)}\right) \text{ and } \tilde{\mathbf{B}} = \left(\tilde{B}^{(1)}, \tilde{B}^{(2)}, \ldots, \tilde{B}^{(n)}\right).$$

Based on the extension principle, we are going to study the arithmetics of $\tilde{\mathbf{A}}$ and $\tilde{\mathbf{B}}$ by considering the addition $\tilde{\mathbf{A}} \oplus_{EP} \tilde{\mathbf{B}}$ and the difference $\tilde{\mathbf{A}} \ominus_{EP} \tilde{\mathbf{B}}$. Given a function $\mathfrak{A} : [0,1]^{2n} \to [0,1]$, for each $\mathbf{z} \in \mathbb{R}^m$ and for the operation $\odot \in \{\oplus, \ominus\}$ corresponding to the operation $\circ \in \{+,-\}$, the membership function of $\tilde{\mathbf{A}} \odot_{EP} \tilde{\mathbf{B}}$ is given by

$$\xi_{\tilde{\mathbf{A}} \odot_{EP} \tilde{\mathbf{B}}}(\mathbf{z}) = \sup_{\{(\mathbf{x},\mathbf{y}):\mathbf{z}=\mathbf{x}\circ\mathbf{y}\}} \mathfrak{A}\left(\xi_{\tilde{A}^{(1)}}(x_1), \ldots, \xi_{\tilde{A}^{(n)}}(x_n), \xi_{\tilde{B}^{(1)}}(y_1), \ldots, \xi_{\tilde{B}^{(n)}}(y_n)\right). \tag{7.30}$$

7.2.2 Arithmetics Using the Expression in the Decomposition Theorem

In what follows, we shall use the expression in the decomposition theorem presented in Theorem 2.2.13 to study the arithmetics of vector of fuzzy intervals. Let $\tilde{\mathbf{A}}$ and $\tilde{\mathbf{B}}$ be two vectors of fuzzy intervals with components $\tilde{A}^{(i)}$ and $\tilde{B}^{(i)}$, respectively, for $i = 1, \ldots, n$. Given $\alpha \in [0,1]$, the α-level sets of $\tilde{A}^{(i)}$ and $\tilde{B}^{(i)}$ are nonempty and denoted by

$$\tilde{A}_\alpha^{(i)} \equiv \left[\tilde{A}_{i\alpha}^L, \tilde{A}_{i\alpha}^U\right] \text{ and } \tilde{B}_\alpha^{(i)} \equiv \left[\tilde{B}_{i\alpha}^L, \tilde{B}_{i\alpha}^U\right].$$

For convenience, we write

$$\left(\tilde{A}_{1\alpha}^L, \tilde{A}_{2\alpha}^L, \ldots, \tilde{A}_{n\alpha}^L\right) = \tilde{\mathbf{A}}_\alpha^L \in \mathbb{R}^m \text{ and } \left(\tilde{A}_{1\alpha}^U, \tilde{A}_{2\alpha}^U, \ldots, \tilde{A}_{n\alpha}^U\right) = \tilde{\mathbf{A}}_\alpha^U \in \mathbb{R}^m. \tag{7.31}$$

We also write

$$\tilde{\mathbf{A}}_\alpha = \tilde{A}_\alpha^{(1)} \times \cdots \times \tilde{A}_\alpha^{(n)} = \left[\tilde{A}_{1\alpha}^L, \tilde{A}_{1\alpha}^U\right] \times \cdots \times \left[\tilde{A}_{n\alpha}^L, \tilde{A}_{n\alpha}^U\right] \tag{7.32}$$

and

$$\tilde{\mathbf{B}}_\alpha = \tilde{B}_\alpha^{(1)} \times \cdots \times \tilde{B}_\alpha^{(n)} = \left[\tilde{B}_{1\alpha}^L, \tilde{B}_{1\alpha}^U\right] \times \cdots \times \left[\tilde{B}_{n\alpha}^L, \tilde{B}_{n\alpha}^U\right]. \tag{7.33}$$

In order to define the difference $\tilde{\mathbf{A}} \ominus_{DT} \tilde{\mathbf{B}}$, we consider the family $\{M_\alpha^- : 0 < \alpha \leq 1\}$ that is formed by applying the operation $\mathbf{x} - \mathbf{y}$ to the α-level sets $\tilde{A}_\alpha^{(i)}$ and $\tilde{B}_\alpha^{(i)}$ for $i = 1, \ldots, n$, where each M_α^- is a subset of $\mathbb{R}^m$. We shall study three different families described below.

- By referring to Subsection 6.2.1, we take

$$M_\alpha^- = \tilde{\mathbf{A}}_\alpha - \tilde{\mathbf{B}}_\alpha = \left\{\mathbf{x} - \mathbf{y} : \mathbf{x} \in \tilde{\mathbf{A}}_\alpha \text{ and } \mathbf{y} \in \tilde{\mathbf{B}}_\alpha\right\}$$

  to define $\tilde{\mathbf{A}} \ominus_{DT}^{\diamond} \tilde{\mathbf{B}}$.
- By referring to Subsection 6.2.2, we take

$$M_\alpha^- = \left(\bigcup_{\alpha \leq \beta \leq 1} M_\beta^{(1-)}\right) \times \cdots \times \left(\bigcup_{\alpha \leq \beta \leq 1} M_\beta^{(n-)}\right),$$

  where $M_\beta^{(i-)}$ are bounded closed intervals given by

$$M_\beta^{(i-)} = \left[\min\left\{\tilde{A}_{i\beta}^L - \tilde{B}_{i\beta}^L, \tilde{A}_{i\beta}^U - \tilde{B}_{i\beta}^U\right\}, \max\left\{\tilde{A}_{i\beta}^L - \tilde{B}_{i\beta}^L, \tilde{A}_{i\beta}^U - \tilde{B}_{i\beta}^U\right\}\right].$$

  for $i = 1, \ldots, n$ to define $\tilde{\mathbf{A}} \ominus_{DT}^{\star} \tilde{\mathbf{B}}$.
- By referring to Subsection 6.2.3, we take

$$M_\alpha^- = M_\alpha^{(1-)} \times \cdots \times M_\alpha^{(n-)},$$

  where $M_\alpha^{(i-)}$ are bounded closed intervals given by

$$M_\alpha^{(i-)} = \left[\min\left\{\tilde{A}_{i\alpha}^L - \tilde{B}_{i\alpha}^L, \tilde{A}_{i\alpha}^U - \tilde{B}_{i\alpha}^U\right\}, \max\left\{\tilde{A}_{i\alpha}^L - \tilde{B}_{i\alpha}^L, \tilde{A}_{i\alpha}^U - \tilde{B}_{i\alpha}^U\right\}\right]$$

  for $i = 1, \ldots, n$ to define $\tilde{\mathbf{A}} \ominus_{DT}^{\dagger} \tilde{\mathbf{B}}$.

For $\ominus_{DT} \in \{\ominus_{DT}^{\diamond}, \ominus_{DT}^{\star}, \ominus_{DT}^{\dagger}\}$, based on the expression in the decomposition theorem, the membership function of $\tilde{\mathbf{A}} \ominus_{DT} \tilde{\mathbf{B}}$ is defined by

$$\xi_{\tilde{\mathbf{A}} \ominus_{DT} \tilde{\mathbf{B}}}(\mathbf{z}) = \sup_{0 < \alpha \leq 1} \alpha \cdot \chi_{M_\alpha^-}(\mathbf{z}), \tag{7.34}$$

where M_α^- can be any one of the above three cases.

Example 7.2.4 Continued from Example 2.2.4, we consider the following trapezoidal fuzzy intervals

$$\tilde{A}^{(1)} = (1,2,3,4) \text{ and } \tilde{A}^{(2)} = (2,3,4,5)$$

and

$$\tilde{B}^{(1)} = (4,5,6,7) \text{ and } \tilde{B}^{(2)} = (3,4,5,6).$$

From (7.12), we have

$$\begin{aligned}
\tilde{A}^{(1)}_\alpha &= \left[\tilde{A}^L_{1\alpha}, \tilde{A}^U_{1\alpha}\right] = [(1-\alpha)+2\alpha, 4(1-\alpha)+3\alpha] = [1+\alpha, 4-\alpha]\\
\tilde{A}^{(2)}_\alpha &= \left[\tilde{A}^L_{2\alpha}, \tilde{A}^U_{2\alpha}\right] = [2(1-\alpha)+3\alpha, 5(1-\alpha)+4\alpha] = [2+\alpha, 5-\alpha]\\
\tilde{B}^{(1)}_\alpha &= \left[\tilde{B}^L_{1\alpha}, \tilde{B}^U_{1\alpha}\right] = [4(1-\alpha)+5\alpha, 7(1-\alpha)+6\alpha] = [4+\alpha, 7-\alpha]\\
\tilde{B}^{(2)}_\alpha &= \left[\tilde{B}^L_{2\alpha}, \tilde{B}^U_{2\alpha}\right] = [3(1-\alpha)+4\alpha, 6(1-\alpha)+5\alpha] = [3+\alpha, 6-\alpha].
\end{aligned}$$

Therefore, we obtain

$$\tilde{\mathbf{A}}_\alpha = \tilde{A}^{(1)}_\alpha \times \tilde{A}^{(2)}_\alpha = \left[\tilde{A}^L_{1\alpha}, \tilde{A}^U_{1\alpha}\right] \times \left[\tilde{A}^L_{n\alpha}, \tilde{A}^U_{n\alpha}\right] = [1+\alpha, 4-\alpha] \times [2+\alpha, 5-\alpha]$$

and

$$\tilde{\mathbf{B}}_\alpha = \tilde{B}^{(1)}_\alpha \times \tilde{B}^{(2)}_\alpha = \left[\tilde{B}^L_{1\alpha}, \tilde{B}^U_{1\alpha}\right] \times \left[\tilde{B}^L_{n\alpha}, \tilde{B}^U_{n\alpha}\right] = [4+\alpha, 7-\alpha] \times [3+\alpha, 6-\alpha].$$

We consider three families $\{M^-_\alpha : 0 < \alpha \le 1\}$ given below.

- We take

$$\begin{aligned}
M^-_\alpha = \tilde{\mathbf{A}}_\alpha - \tilde{\mathbf{B}}_\alpha &= \{(x_1,x_2)-(y_1,y_2) : (x_1,x_2) \in [1+\alpha, 4-\alpha] \times [2+\alpha, 5-\alpha]\\
&\qquad \text{and } (y_1,y_2) \in [4+\alpha, 7-\alpha] \times [3+\alpha, 6-\alpha]\}\\
&= [-6+2\alpha, -2\alpha] \times [-4+2\alpha, 2-2\alpha]
\end{aligned}$$

  for all $\alpha \in (0,1]$. The membership function of $\tilde{\mathbf{A}} \ominus^\circ_{DT} \tilde{\mathbf{B}}$ is given by

$$\xi_{\tilde{\mathbf{A}} \ominus^\circ_{DT} \tilde{\mathbf{B}}}(z_1,z_2) = \sup_{0<\alpha\le1} \alpha \cdot \chi_{M^-_\alpha}(z_1,z_2).$$

- We take

$$M^-_\alpha = \left(\bigcup_{\alpha\le\beta\le1} M^{(1-)}_\beta\right) \times \left(\bigcup_{\alpha\le\beta\le1} M^{(2-)}_\beta\right),$$

  where $M^{(1-)}_\beta$ and $M^{(2-)}_\beta$ are bounded closed intervals given by

$$\begin{aligned}
M^{(1-)}_\beta &= \left[\min\left\{\tilde{A}^L_{1\beta} - \tilde{B}^L_{1\beta}, \tilde{A}^U_{1\beta} - \tilde{B}^U_{1\beta}\right\}, \max\left\{\tilde{A}^L_{1\beta} - \tilde{B}^L_{1\beta}, \tilde{A}^U_{1\beta} - \tilde{B}^U_{1\beta}\right\}\right]\\
&= [\min\{(1+\beta)-(4+\beta), (4-\beta)-(7-\beta)\},\\
&\qquad \max\{(1+\beta)-(4+\beta), (4-\beta)-(7-\beta)\}]\\
&= [-3,-3] = \{-3\},
\end{aligned}$$

  which says that $M^{(1-)}_\beta$ is the singleton $\{-3\}$ for all $\beta \in [0,1]$. Similarly, we can obtain

$$M^{(2-)}_\beta = \left[\min\left\{\tilde{A}^L_{2\beta} - \tilde{B}^L_{2\beta}, \tilde{A}^U_{2\beta} - \tilde{B}^U_{2\beta}\right\}, \max\left\{\tilde{A}^L_{1\beta} - \tilde{B}^L_{1\beta}, \tilde{A}^U_{2\beta} - \tilde{B}^U_{2\beta}\right\}\right] = \{-1\}$$

  for all $\beta \in [0,1]$. Therefore, we obtain

$$M^-_\alpha = \{-3\} \times \{-1\} = \{(-3,-1)\} \tag{7.35}$$

for all $\alpha \in (0, 1]$, which is a singleton in $\mathbb{R}^2$. The membership function of $\tilde{\mathbf{A}} \ominus_{DT}^{\star} \tilde{\mathbf{B}}$ is given by

$$\xi_{\tilde{\mathbf{A}} \ominus_{DT}^{\star} \tilde{\mathbf{B}}}(z_1, z_2) = \sup_{0<\alpha\leq 1} \alpha \cdot \chi_{M_\alpha^-}(z_1, z_2) = \begin{cases} 1 & \text{if } (z_1, z_2) = (-3, -1) \\ 0 & \text{otherwise.} \end{cases}$$

- We take

$$M_\alpha^- = M_\alpha^{(1-)} \times M_\alpha^{(2-)},$$

where $M_\alpha^{(1-)}$ and $M_\alpha^{(2-)}$ are bounded closed intervals given by

$$M_\alpha^{(1-)} = \left[\min\left\{\tilde{A}_{1\alpha}^L - \tilde{B}_{1\alpha}^L, \tilde{A}_{1\alpha}^U - \tilde{B}_{1\alpha}^U\right\}, \max\left\{\tilde{A}_{1\alpha}^L - \tilde{B}_{1\alpha}^L, \tilde{A}_{1\alpha}^U - \tilde{B}_{1\alpha}^U\right\}\right] = \{-3\}$$

and

$$M_\alpha^{(2-)} = \left[\min\left\{\tilde{A}_{2\alpha}^L - \tilde{B}_{2\alpha}^L, \tilde{A}_{2\alpha}^U - \tilde{B}_{2\alpha}^U\right\}, \max\left\{\tilde{A}_{1\alpha}^L - \tilde{B}_{2\alpha}^L, \tilde{A}_{2\alpha}^U - \tilde{B}_{2\alpha}^U\right\}\right] = \{-1\}.$$

Therefore, we obtain

$$M_\alpha^- = \{-3\} \times \{-1\} = \{(-3, -1)\} \tag{7.36}$$

for all $\alpha \in (0, 1]$, which is a singleton in $\mathbb{R}^2$. The membership function of $\tilde{\mathbf{A}} \ominus_{DT}^{\dagger} \tilde{\mathbf{B}}$ is equal to membership function of $\tilde{\mathbf{A}} \ominus_{DT}^{\star} \tilde{\mathbf{B}}$.

In order to define the addition $\tilde{\mathbf{A}} \oplus_{DT} \tilde{\mathbf{B}}$, we consider the family $\{M_\alpha^+ : 0 < \alpha \leq 1\}$ that is formed by applying the operation $\mathbf{x} + \mathbf{y}$ to the α-level sets $\tilde{A}_\alpha^{(i)}$ and $\tilde{B}_\alpha^{(i)}$ for $i = 1, \ldots, n$, where each M_α^+ is a subset of $\mathbb{R}^m$. We shall study three different families described below.

- By referring to Subsection 6.2.1, we take

$$M_\alpha^+ = \tilde{\mathbf{A}}_\alpha + \tilde{\mathbf{B}}_\alpha = \left\{\mathbf{x} + \mathbf{y} : \mathbf{x} \in \tilde{\mathbf{A}}_\alpha \text{ and } \mathbf{y} \in \tilde{\mathbf{B}}_\alpha\right\}$$

to define $\tilde{\mathbf{A}} \oplus_{DT}^{\diamond} \tilde{\mathbf{B}}$.
- By referring to Subsection 6.2.2, we take

$$M_\alpha^+ = \left(\bigcup_{\alpha\leq\beta\leq 1} M_\beta^{(1+)}\right) \times \cdots \times \left(\bigcup_{\alpha\leq\beta\leq 1} M_\beta^{(n+)}\right),$$

where $M_\beta^{(i+)}$ are bounded closed intervals given by

$$M_\beta^{(i+)} = \left[a_{i\beta}^L + b_{i\beta}^L, a_{i\beta}^U + b_{i\beta}^U\right]$$

for $i = 1, \ldots, n$ to define $\tilde{\mathbf{A}} \oplus_{DT}^{\star} \tilde{\mathbf{B}}$.
- By referring to Subsection 6.2.3, we take

$$M_\alpha^+ = M_\alpha^{(1+)} \times \cdots \times M_\alpha^{(n+)},$$

where $M_\alpha^{(i+)}$ are bounded closed intervals given by

$$M_\alpha^{(i+)} = \left[a_{i\alpha}^L + b_{i\alpha}^L, a_{i\alpha}^U + b_{i\alpha}^U\right]$$

for $i = 1, \ldots, n$ to define $\tilde{\mathbf{A}} \oplus_{DT}^{\dagger} \tilde{\mathbf{B}}$.

For $\oplus_{DT} \in \{\oplus_{DT}^{\diamond}, \oplus_{DT}^{\star}, \oplus_{DT}^{\dagger}\}$, based on the expression in the decomposition theorem, the membership function of $\tilde{\mathbf{A}} \oplus_{DT} \tilde{\mathbf{B}}$ is defined by

$$\xi_{\tilde{\mathbf{A}} \oplus_{DT} \tilde{\mathbf{B}}}(\mathbf{z}) = \sup_{0<\alpha\leq 1} \alpha \cdot \chi_{M_\alpha^+}(\mathbf{z}), \tag{7.37}$$

where M_α^+ can be any one of the above three cases.

Example 7.2.5 Continued from Examples 2.2.4 and 7.2.4, we consider three families $\{M_\alpha^+ : 0 < \alpha \leq 1\}$ given below.

- We take

$$\begin{aligned} M_\alpha^+ = \tilde{\mathbf{A}}_\alpha + \tilde{\mathbf{B}}_\alpha &= \big\{(x_1,x_2)+(y_1,y_2) : (x_1,x_2) \in [1+\alpha, 4-\alpha] \times [2+\alpha, 5-\alpha] \\ &\qquad \text{and } (y_1,y_2) \in [4+\alpha, 7-\alpha] \times [3+\alpha, 6-\alpha]\big\} \\ &= [5+2\alpha, 11-2\alpha] \times [5+2\alpha, 11-2\alpha] \end{aligned}$$

  for all $\alpha \in (0,1]$. The membership function of $\tilde{\mathbf{A}} \oplus_{DT}^{\diamond} \tilde{\mathbf{B}}$ is given by

$$\xi_{\tilde{\mathbf{A}} \oplus_{DT}^{\diamond} \tilde{\mathbf{B}}}(z_1,z_2) = \sup_{0<\alpha\leq 1} \alpha \cdot \chi_{M_\alpha^+}(z_1,z_2).$$

- We take

$$M_\alpha^+ = \left(\bigcup_{\alpha\leq\beta\leq 1} M_\beta^{(1+)}\right) \times \left(\bigcup_{\alpha\leq\beta\leq 1} M_\beta^{(2+)}\right),$$

  where $M_\beta^{(1+)}$ and $M_\beta^{(2+)}$ are bounded closed intervals given by

$$\begin{aligned} M_\beta^{(1+)} &= \left[a_{1\beta}^L + b_{1\beta}^L, a_{1\beta}^U + b_{1\beta}^U\right] = [(1+\beta)+(4+\beta), (4-\beta)+(7-\beta)] \\ &= [5+2\beta, 11-2\beta] \end{aligned}$$

  for all $\beta \in [0,1]$. We also obtain

$$\begin{aligned} M_\beta^{(2+)} &= \left[a_{2\beta}^L + b_{2\beta}^L, a_{2\beta}^U + b_{2\beta}^U\right] = [(2+\beta)+(3+\beta), (5-\beta)+(6-\beta)] \\ &= [5+2\beta, 11-2\beta] \end{aligned}$$

  for all $\beta \in [0,1]$. Now, we have

$$\bigcup_{\alpha\leq\beta\leq 1} M_\beta^{(1+)} = \bigcup_{\alpha\leq\beta\leq 1} [5+2\beta, 11-2\beta] = [5+2\alpha, 11-2\alpha]$$

  and

$$\bigcup_{\alpha\leq\beta\leq 1} M_\beta^{(2+)} = \bigcup_{\alpha\leq\beta\leq 1} [5+2\beta, 11-2\beta] = [5+2\alpha, 11-2\alpha].$$

  Therefore, we obtain

$$M_\alpha^+ = [5+2\alpha, 11-2\alpha] \times [5+2\alpha, 11-2\alpha]$$

  for all $\alpha \in (0,1]$. The membership function of $\tilde{\mathbf{A}} \oplus_{DT}^{\star} \tilde{\mathbf{B}}$ is equal to the membership function of $\tilde{\mathbf{A}} \oplus_{DT}^{\diamond} \tilde{\mathbf{B}}$
- We take

$$M_\alpha^+ = M_\alpha^{(1+)} \times M_\alpha^{(2+)},$$

where $M_\alpha^{(1+)}$ and $M_\alpha^{(2+)}$ are bounded closed intervals given by

$$M_\alpha^{(1+)} = \left[a_{1\alpha}^L + b_{1\alpha}^L, a_{1\alpha}^U + b_{1\alpha}^U\right] = [(1+\alpha)+(4+\alpha),(4-\alpha)+(7-\alpha)]$$
$$= [5+2\alpha, 11-2\alpha]$$

and

$$M_\alpha^{(2+)} = \left[a_{2\alpha}^L + b_{2\alpha}^L, a_{2\alpha}^U + b_{2\alpha}^U\right] = [(2+\alpha)+(3+\alpha),(5-\alpha)+(6-\alpha)]$$
$$= [5+2\alpha, 11-2\alpha].$$

Therefore, we obtain

$$M_\alpha^+ = [5+2\alpha, 11-2\alpha] \times [5+2\alpha, 11-2\alpha]$$

for all $\alpha \in (0,1]$. The membership function of $\tilde{\mathbf{A}} \oplus_{DT}^{\dagger} \tilde{\mathbf{B}}$ is equal to membership function of $\tilde{\mathbf{A}} \oplus_{DT}^{\star} \tilde{\mathbf{B}}$.

In this example, we conclude that

$$\tilde{\mathbf{A}} \oplus_{DT}^{\diamond} \tilde{\mathbf{B}} = \tilde{\mathbf{A}} \oplus_{DT}^{\star} \tilde{\mathbf{B}} = \tilde{\mathbf{A}} \oplus_{DT}^{\dagger} \tilde{\mathbf{B}}.$$

7.3 Difference of Vectors of Fuzzy Intervals

Let $\tilde{\mathbf{A}}$ and $\tilde{\mathbf{B}}$ be two vectors of fuzzy intervals with components $\tilde{A}^{(i)}$ and $\tilde{B}^{(i)}$, respectively, for $i = 1, \ldots, n$. We are going to study the α-level set of $\tilde{\mathbf{A}} \ominus_{EP} \tilde{\mathbf{B}}$ that is obtained from the extension principle, and the α-level sets of $\tilde{\mathbf{A}} \ominus_{DT} \tilde{\mathbf{B}}$ for $\ominus_{DT} \in \{\ominus_{DT}^{\diamond}, \ominus_{DT}^{\star}, \ominus_{DT}^{\dagger}\}$ that are obtained from the expression in the decomposition theorem.

7.3.1 α-Level Sets of $\tilde{\mathbf{A}} \ominus_{EP} \tilde{\mathbf{B}}$

Given any function $\mathfrak{A} : [0,1]^{2n} \to [0,1]$, recall that the membership function of the difference $\tilde{\mathbf{A}} \ominus_{EP} \tilde{\mathbf{B}}$ is defined by

$$\xi_{\tilde{\mathbf{A}} \ominus_{EP} \tilde{\mathbf{B}}}(\mathbf{z}) = \sup_{\{(\mathbf{x},\mathbf{y}) : \mathbf{z}=\mathbf{x}-\mathbf{y}\}} \mathfrak{A}\left(\xi_{\tilde{A}^{(1)}}(x_1), \ldots, \xi_{\tilde{A}^{(n)}}(x_n), \xi_{\tilde{B}^{(1)}}(y_1), \ldots, \xi_{\tilde{B}^{(n)}}(y_n)\right)$$

for any $\mathbf{z} \in \mathbb{R}^m$. The α-level set $(\tilde{\mathbf{A}} \ominus_{EP} \tilde{\mathbf{B}})_\alpha$ of $\tilde{\mathbf{A}} \ominus_{EP} \tilde{\mathbf{B}}$ for $\alpha \in [0,1]$ can be obtained by applying Theorem 4.4.15 to the difference $\tilde{\mathbf{A}} \ominus_{EP} \tilde{\mathbf{B}}$, which is shown below. For each $\alpha \in (0,1]$, we have

$$\left(\tilde{\mathbf{A}} \ominus_{EP} \tilde{\mathbf{B}}\right)_\alpha = \left\{\mathbf{x}-\mathbf{y} : \mathfrak{A}\left(\xi_{\tilde{A}^{(1)}}(x_1), \cdots, \xi_{\tilde{A}^{(n)}}(x_n), \xi_{\tilde{B}^{(1)}}(y_1), \cdots, \xi_{\tilde{B}^{(n)}}(y_n)\right) \geq \alpha\right\} \tag{7.38}$$
$$= \left\{\left(x_1 - y_1, \ldots, x_n - y_n\right) : \right.$$
$$\left. \mathfrak{A}\left(\xi_{\tilde{A}^{(1)}}(x_1), \cdots, \xi_{\tilde{A}^{(n)}}(x_n), \xi_{\tilde{B}^{(1)}}(y_1), \cdots, \xi_{\tilde{B}^{(n)}}(y_n)\right) \geq \alpha\right\}.$$

The 0-level set is given by

$$\left(\tilde{\mathbf{A}} \ominus_{EP} \tilde{\mathbf{B}}\right)_0 = \tilde{\mathbf{A}}_0 - \tilde{\mathbf{B}}_0 = \left\{\mathbf{x}-\mathbf{y} : \mathbf{x} \in \tilde{\mathbf{A}}_0 \text{ and } \mathbf{y} \in \tilde{\mathbf{B}}_0\right\}.$$

Moreover, for each $\alpha \in [0,1]$, the α-level sets $(\tilde{\mathbf{A}} \ominus_{EP} \tilde{\mathbf{B}})_\alpha$ are convex, closed, and bounded subsets of $\mathbb{R}^m$.

Now, the function $\mathfrak{A} : [0,1]^{2n} \to [0,1]$ is given by

$$\mathfrak{A}\left(\alpha_1, \ldots, \alpha_{2n}\right) = \min\left\{\alpha_1, \ldots, \alpha_{2n}\right\}.$$

For each $\alpha \in (0,1]$, using (7.38), we have

$$\begin{aligned}
\left(\tilde{\mathbf{A}} \ominus_{EP} \tilde{\mathbf{B}}\right)_\alpha &= \left\{\mathbf{x} - \mathbf{y} : \min\left\{\xi_{\tilde{A}^{(1)}}(x_1), \cdots, \xi_{\tilde{A}^{(n)}}(x_n), \xi_{\tilde{B}^{(1)}}(y_1), \cdots, \xi_{\tilde{B}^{(n)}}(y_n)\right\} \geq \alpha\right\} \\
&= \left\{\mathbf{x} - \mathbf{y} : \xi_{\tilde{A}^{(i)}}(x_i) \geq \alpha \text{ and } \xi_{\tilde{B}^{(i)}}(y_i) \geq \alpha \text{ for each } i = 1, \ldots, n\right\} \\
&= \left\{\left(x_1 - y_1, \ldots, x_n - y_n\right) : x_i \in \tilde{A}^{(i)}_\alpha \equiv \left[\tilde{A}^L_{i\alpha}, \tilde{A}^U_{i\alpha}\right]\right. \\
&\qquad \left.\text{and } y_i \in \tilde{B}^{(i)}_\alpha \equiv \left[\tilde{B}^L_{i\alpha}, \tilde{B}^U_{i\alpha}\right] \text{ for each } i = 1, \ldots, n\right\} \\
&= \left[\tilde{A}^L_{1\alpha} - \tilde{B}^U_{1\alpha}, \tilde{A}^U_{1\alpha} - \tilde{B}^L_{1\alpha}\right] \times \cdots \times \left[\tilde{A}^L_{n\alpha} - \tilde{B}^U_{n\alpha}, \tilde{A}^U_{n\alpha} - \tilde{B}^L_{n\alpha}\right]. \qquad (7.39)
\end{aligned}$$

For the 0-level set, from (7.39) and (2.5), it is not difficult to show that

$$\begin{aligned}
\left(\tilde{\mathbf{A}} \ominus_{EP} \tilde{\mathbf{B}}\right)_0 &= \mathrm{cl}\left(\bigcup_{0<\alpha\leq 1} \left(\tilde{\mathbf{A}} \ominus_{EP} \tilde{\mathbf{B}}\right)_\alpha\right) \\
&= \left[\tilde{A}^L_{10} - \tilde{B}^U_{10}, \tilde{A}^U_{10} - \tilde{B}^L_{10}\right] \times \cdots \times \left[\tilde{A}^L_{n0} - \tilde{B}^U_{n0}, \tilde{A}^U_{n0} - \tilde{B}^L_{n0}\right].
\end{aligned}$$

Regarding the components $\tilde{A}^{(i)}$ and $\tilde{B}^{(i)}$, for $\alpha \in [0,1]$, we also have

$$\left(\tilde{A}^{(i)} \ominus_{EP} \tilde{B}^{(i)}\right)_\alpha = \left[\tilde{A}^L_{i\alpha} - \tilde{B}^U_{i\alpha}, \tilde{A}^U_{i\alpha} - \tilde{B}^L_{i\alpha}\right] \text{ for } i = 1, \ldots, n. \qquad (7.40)$$

Therefore, from (7.39) and (7.40), for $\alpha \in [0,1]$, we obtain

$$\left(\tilde{\mathbf{A}} \ominus_{EP} \tilde{\mathbf{B}}\right)_\alpha = \left(\tilde{A}^{(1)} \ominus_{EP} \tilde{B}^{(1)}\right)_\alpha \times \cdots \times \left(\tilde{A}^{(n)} \ominus_{EP} \tilde{B}^{(n)}\right)_\alpha.$$

The above results are summarized in the following theorem.

Theorem 7.3.1 *Let $\tilde{A}^{(1)}, \ldots, \tilde{A}^{(n)}$ and $\tilde{B}^{(1)}, \ldots, \tilde{B}^{(n)}$ be any fuzzy intervals. Suppose that the function $\mathfrak{A} : [0,1]^{2n} \to [0,1]$ is given by*

$$\mathfrak{A}\left(\alpha_1, \ldots, \alpha_{2n}\right) = \min\left\{\alpha_1, \ldots, \alpha_{2n}\right\}.$$

For each $\alpha \in [0,1]$, we have

$$\left(\tilde{A}^{(i)} \ominus_{EP} \tilde{B}^{(i)}\right)_\alpha = \left[\tilde{A}^L_{i\alpha} - \tilde{B}^U_{i\alpha}, \tilde{A}^U_{i\alpha} - \tilde{B}^L_{i\alpha}\right]$$

and

$$\left(\tilde{\mathbf{A}} \ominus_{EP} \tilde{\mathbf{B}}\right)_\alpha = \left(\tilde{A}^{(1)} \ominus_{EP} \tilde{B}^{(1)}\right)_\alpha \times \cdots \times \left(\tilde{A}^{(n)} \ominus_{EP} \tilde{B}^{(n)}\right)_\alpha.$$

Example 7.3.2 Continued from Examples 2.2.4 and 7.2.4, Theorem 7.3.1 says that

$$\begin{aligned}
\left(\tilde{A}^{(1)} \ominus_{EP} \tilde{B}^{(1)}\right)_\alpha &= \left[\tilde{A}^L_{1\alpha} - \tilde{B}^U_{1\alpha}, \tilde{A}^U_{1\alpha} - \tilde{B}^L_{1\alpha}\right] \\
&= [(1+\alpha) - (7-\alpha), (4-\alpha) - (4+\alpha)] = [-6+2\alpha, -2\alpha]
\end{aligned}$$

and

$$\begin{aligned}
\left(\tilde{A}^{(2)} \ominus_{EP} \tilde{B}^{(2)}\right)_\alpha &= \left[\tilde{A}^L_{2\alpha} - \tilde{B}^U_{2\alpha}, \tilde{A}^U_{2\alpha} - \tilde{B}^L_{2\alpha}\right] \\
&= [(2+\alpha) - (6-\alpha), (5-\alpha) - (3+\alpha)] = [-4+2\alpha, 2-2\alpha]
\end{aligned}$$

and

$$\left(\tilde{\mathbf{A}} \ominus_{EP} \tilde{\mathbf{B}}\right)_\alpha = \left(\tilde{A}^{(1)} \ominus_{EP} \tilde{B}^{(1)}\right)_\alpha \times \left(\tilde{A}^{(2)} \ominus_{EP} \tilde{B}^{(2)}\right)_\alpha$$
$$= [-6 + 2\alpha, -2\alpha] \times [-4 + 2\alpha, 2 - 2\alpha]$$

for $\alpha \in [0, 1]$.

7.3.2 α-Level Sets of $\tilde{\mathbf{A}} \ominus_{DT}^{\diamond} \tilde{\mathbf{B}}$

Let $\tilde{A}^{(1)}, \ldots, \tilde{A}^{(n)}$ and $\tilde{B}^{(1)}, \ldots, \tilde{B}^{(n)}$ be fuzzy intervals. By referring to Subsection 6.2.1, the family $\{M_\alpha^- : \alpha \in (0, 1]\}$ is given by $M_\alpha^- = \tilde{\mathbf{A}}_\alpha - \tilde{\mathbf{B}}_\alpha$. For $\alpha \in (0, 1]$, we have

$$\begin{aligned} M_\alpha^- &= \tilde{\mathbf{A}}_\alpha - \tilde{\mathbf{B}}_\alpha = \left\{\mathbf{x} - \mathbf{y} : \mathbf{x} \in \tilde{\mathbf{A}}_\alpha \text{ and } \mathbf{y} \in \tilde{\mathbf{B}}_\alpha\right\} \\ &= \left\{\left(x_1 - y_1, \ldots, x_1 - y_1\right) : x_i \in \tilde{A}_\alpha^{(i)} = \left[\tilde{A}_{i\alpha}^L, \tilde{A}_{i\alpha}^U\right] \text{ and } \right. \\ &\qquad \left. y_i \in \tilde{B}_\alpha^{(i)} = \left[\tilde{B}_{i\alpha}^L, \tilde{B}_{i\alpha}^U\right] \text{ for } i = 1, \ldots, n\right\} \\ &= \left[\tilde{A}_{1\alpha}^L - \tilde{B}_{1\alpha}^U, \tilde{A}_{1\alpha}^U - \tilde{B}_{1\alpha}^L\right] \times \cdots \times \left[\tilde{A}_{n\alpha}^L - \tilde{B}_{n\alpha}^U, \tilde{A}_{n\alpha}^U - \tilde{B}_{n\alpha}^L\right]. \end{aligned} \tag{7.41}$$

Based on the expression in the decomposition theorem, the membership function of $\tilde{\mathbf{A}} \ominus_{DT}^{\diamond} \tilde{\mathbf{B}}$ is given by

$$\xi_{\tilde{\mathbf{A}} \ominus_{DT}^{\diamond} \tilde{\mathbf{B}}}(\mathbf{z}) = \sup_{0<\alpha\leq 1} \alpha \cdot \chi_{M_\alpha^-}(\mathbf{z}). \tag{7.42}$$

The α-level sets $(\tilde{\mathbf{A}} \ominus_{DT}^{\diamond} \tilde{\mathbf{B}})_\alpha$ of $\tilde{\mathbf{A}} \ominus_{DT}^{\diamond} \tilde{\mathbf{B}}$ for $\alpha \in [0, 1]$ are presented below.

Proposition 7.3.3 *We have*

$$\left(\tilde{\mathbf{A}} \ominus_{DT}^{\diamond} \tilde{\mathbf{B}}\right)_\alpha = M_\alpha^- = \left[\tilde{A}_{1\alpha}^L - \tilde{B}_{1\alpha}^U, \tilde{A}_{1\alpha}^U - \tilde{B}_{1\alpha}^L\right] \times \cdots \times \left[\tilde{A}_{n\alpha}^L - \tilde{B}_{n\alpha}^U, \tilde{A}_{n\alpha}^U - \tilde{B}_{n\alpha}^L\right]$$

for each $\alpha \in [0, 1]$.

Proof. We first consider $\alpha \in (0, 1]$. Given any $\mathbf{z} \in M_\alpha^-$, we see that $\xi_{\tilde{\mathbf{A}} \ominus_{DT}^{\diamond} \tilde{\mathbf{B}}}(\mathbf{z}) \geq \alpha$ by (7.42). Therefore, we obtain $\mathbf{z} \in (\tilde{\mathbf{A}} \ominus_{DT}^{\diamond} \tilde{\mathbf{B}})_\alpha$, which proves the inclusion $M_\alpha^- \subseteq (\tilde{\mathbf{A}} \ominus_{DT}^{\diamond} \tilde{\mathbf{B}})_\alpha$.

For proving the other direction of inclusion, it is clear to see that $\{M_\alpha^- : 0 < \alpha \leq 1\}$ is a nested family. Given any $\mathbf{z} \in (\tilde{\mathbf{A}} \ominus_{DT}^{\diamond} \tilde{\mathbf{B}})_\alpha$, i.e. $\xi_{\tilde{\mathbf{A}} \ominus_{DT}^{\diamond} \tilde{\mathbf{B}}}(\mathbf{z}) \geq \alpha$, let $\hat{\alpha} = \xi_{\tilde{\mathbf{A}} \ominus_{DT}^{\diamond} \tilde{\mathbf{B}}}(\mathbf{z})$. Assume that $\hat{\alpha} > \alpha$. Let $\epsilon = \hat{\alpha} - \alpha > 0$. According to the concept of supremum, there exists α_0 satisfying $\mathbf{z} \in M_{\alpha_0}^-$ and $\hat{\alpha} - \epsilon < \alpha_0$, which implies $\alpha < \alpha_0$. This also says that $\mathbf{z} \in M_\alpha^-$, since $M_{\alpha_0}^- \subseteq M_\alpha^-$ by the nestedness.

Now, we assume that $\hat{\alpha} = \alpha$. We can consider a sequence $\{\alpha_s\}_{s=1}^\infty$ satisfying $0 < \alpha_s \uparrow \alpha$ with $\alpha > \alpha_s$ for all s. Since $\tilde{A}^{(i)}$ and $\tilde{B}^{(i)}$ are fuzzy intervals for $i = 1, \ldots, n$, we have

$$\tilde{A}_\alpha^{(i)} = \bigcap_{s=1}^{\infty} \tilde{A}_{\alpha_s}^{(i)} \text{ and } \tilde{B}_\alpha^{(i)} = \bigcap_{s=1}^{\infty} \tilde{B}_{\alpha_s}^{(i)} \text{ for } i = 1, \ldots, n$$

by referring to Proposition 2.2.8. Since

$$\tilde{\mathbf{A}}_\alpha = \tilde{A}_\alpha^{(1)} \times \cdots \times \tilde{A}_\alpha^{(n)} \text{ and } \tilde{\mathbf{B}}_\alpha = \tilde{B}_\alpha^{(1)} \times \cdots \times \tilde{B}_\alpha^{(n)},$$

we can obtain

$$\tilde{\mathbf{A}}_\alpha = \bigcap_{s=1}^{\infty} \tilde{\mathbf{A}}_{\alpha_s} \text{ and } \tilde{\mathbf{B}}_\alpha = \bigcap_{s=1}^{\infty} \tilde{\mathbf{B}}_{\alpha_s},$$

Since $M_\alpha^- = \tilde{\mathbf{A}}_\alpha - \tilde{\mathbf{B}}_\alpha$, we conclude that

$$M_\alpha^- = \bigcap_{s=1}^{\infty} M_{\alpha_s}^- \text{ for } 0 < \alpha_s \uparrow \alpha \text{ satisfying } \alpha > \alpha_s \text{ for all } s. \tag{7.43}$$

Let $\epsilon_s = \alpha - \alpha_s > 0$. According to the concept of supremum, there exists α_0 satisfying $\mathbf{z} \in M_{\alpha_0}^-$ and $\hat{\alpha} - \epsilon_s = \alpha - \epsilon_s < \alpha_0$, which implies $\alpha_0 > \alpha_s$. This also says that $\mathbf{z} \in M_{\alpha_s}^-$ by the nestedness for all s. Therefore, we conclude that $\mathbf{z} \in \bigcap_{s=1}^{\infty} M_{\alpha_s}^-$. From (7.43), it follows that $\mathbf{z} \in M_\alpha^-$. Therefore, for $\alpha \in (0, 1]$, we obtain

$$\left(\tilde{\mathbf{A}} \ominus_{DT}^{\diamond} \tilde{\mathbf{B}}\right)_\alpha = \left[\tilde{A}_{1\alpha}^L - \tilde{B}_{1\alpha}^U, \tilde{A}_{1\alpha}^U - \tilde{B}_{1\alpha}^L\right] \times \cdots \times \left[\tilde{A}_{n\alpha}^L - \tilde{B}_{n\alpha}^U, \tilde{A}_{n\alpha}^U - \tilde{B}_{n\alpha}^L\right].$$

For the 0-level set, by referring to (2.5) and using the nestedness, it is not difficult to show that

$$\begin{aligned}\left(\tilde{\mathbf{A}} \ominus_{DT}^{\diamond} \tilde{\mathbf{B}}\right)_0 &= \mathrm{cl}\left(\bigcup_{0<\alpha\leq 1} \left(\tilde{\mathbf{A}} \ominus_{DT}^{\diamond} \tilde{\mathbf{B}}\right)_\alpha\right) \\ &= \left[\tilde{A}_{10}^L - \tilde{B}_{10}^U, \tilde{A}_{10}^U - \tilde{B}_{10}^L\right] \times \cdots \times \left[\tilde{A}_{n0}^L - \tilde{B}_{n0}^U, \tilde{A}_{n0}^U - \tilde{B}_{n0}^L\right].\end{aligned}$$

This completes the proof. ■

Now, for $i = 1, \ldots, n$ and for $\alpha \in (0, 1]$, we take

$$M_\alpha^{(i-)} = \tilde{A}_\alpha^{(i)} - \tilde{B}_\alpha^{(i)} = \left[\tilde{A}_{i\alpha}^L, \tilde{A}_{i\alpha}^U\right] - \left[\tilde{B}_{i\alpha}^L, \tilde{B}_{i\alpha}^U\right] = \left[\tilde{A}_{i\alpha}^L - \tilde{B}_{i\alpha}^U, \tilde{A}_{i\alpha}^U - \tilde{B}_{i\alpha}^L\right]. \tag{7.44}$$

From (7.41), we see that

$$M_\alpha^- = M_\alpha^{(1-)} \times \cdots \times M_\alpha^{(n-)} \subset \mathbb{R}^m.$$

Let $\tilde{A}^{(i)} \ominus_{DT}^{\diamond} \tilde{B}^{(i)}$ be obtained using the expression in the decomposition theorem based on the family $\{M_\alpha^{(i-)} : 0 < \alpha \leq 1\}$ that is defined in (7.44). By referring to Proposition 7.3.3, we can similarly obtain

$$\left(\tilde{A}^{(i)} \ominus_{DT}^{\diamond} \tilde{B}^{(i)}\right)_\alpha = M_\alpha^{(i-)} = \left[\tilde{A}_{i\alpha}^L - \tilde{B}_{i\alpha}^U, \tilde{A}_{i\alpha}^U - \tilde{B}_{i\alpha}^L\right]$$

for $\alpha \in [0, 1]$ and $i = 1, \ldots, n$, which also implies

$$\left(\tilde{\mathbf{A}} \ominus_{DT}^{\diamond} \tilde{\mathbf{B}}\right)_\alpha = \left(\tilde{A}^{(1)} \ominus_{DT}^{\diamond} \tilde{B}^{(1)}\right)_\alpha \times \cdots \times \left(\tilde{A}^{(n)} \ominus_{DT}^{\diamond} \tilde{B}^{(n)}\right)_\alpha$$

for each $\alpha \in [0, 1]$. The above results are summarized below.

Theorem 7.3.4 *Let $\tilde{A}^{(1)}, \ldots, \tilde{A}^{(n)}$ and $\tilde{B}^{(1)}, \ldots, \tilde{B}^{(n)}$ be fuzzy intervals. Suppose that the family $\{M_\alpha^- : \alpha \in (0, 1]\}$ is given by $M_\alpha^- = \tilde{\mathbf{A}}_\alpha - \tilde{\mathbf{B}}_\alpha$. For each $\alpha \in [0, 1]$, we have*

$$\left(\tilde{A}^{(i)} \ominus_{DT}^{\diamond} \tilde{B}^{(i)}\right)_\alpha = \left[\tilde{A}_{i\alpha}^L - \tilde{B}_{i\alpha}^U, \tilde{A}_{i\alpha}^U - \tilde{B}_{i\alpha}^L\right]$$

and

$$\left(\tilde{\mathbf{A}} \ominus_{DT}^{\diamond} \tilde{\mathbf{B}}\right)_\alpha = \left(\tilde{A}^{(1)} \ominus_{DT}^{\diamond} \tilde{B}^{(1)}\right)_\alpha \times \cdots \times \left(\tilde{A}^{(n)} \ominus_{DT}^{\diamond} \tilde{B}^{(n)}\right)_\alpha.$$

Example 7.3.5 Continued from Example 7.2.4, Theorem 7.3.4 says that

$$\left(\tilde{\mathbf{A}} \ominus_{DT}^{\diamond} \tilde{\mathbf{B}}\right)_\alpha = \left[\tilde{A}_{1\alpha}^L - \tilde{B}_{1\alpha}^U, \tilde{A}_{1\alpha}^U - \tilde{B}_{1\alpha}^L\right] \times \left[\tilde{A}_{2\alpha}^L - \tilde{B}_{2\alpha}^U, \tilde{A}_{2\alpha}^U - \tilde{B}_{2\alpha}^L\right]$$
$$= [-6 + 2\alpha, -2\alpha] \times [-4 + 2\alpha, 2 - 2\alpha]$$

for $\alpha \in [0, 1]$.

7.3.3 α-Level Sets of $\tilde{\mathbf{A}} \ominus_{DT}^{\star} \tilde{\mathbf{B}}$

Let $\tilde{A}^{(1)}, \dots, \tilde{A}^{(n)}$ and $\tilde{B}^{(1)}, \dots, \tilde{B}^{(n)}$ be any fuzzy intervals. By referring to Subsection 6.2.2, the family $\{M_\alpha^- : 0 < \alpha \le 1\}$ is taken by

$$M_\alpha^- = \left(\bigcup_{\alpha \le \beta \le 1} M_\beta^{(1-)}\right) \times \cdots \times \left(\bigcup_{\alpha \le \beta \le 1} M_\beta^{(n-)}\right), \tag{7.45}$$

where $M_\beta^{(i-)}$ are bounded closed intervals given by

$$M_\beta^{(i-)} = \left[\min\left\{\tilde{A}_{i\beta}^L - \tilde{B}_{i\beta}^L, \tilde{A}_{i\beta}^U - \tilde{B}_{i\beta}^U\right\}, \max\left\{\tilde{A}_{i\beta}^L - \tilde{B}_{i\beta}^L, \tilde{A}_{i\beta}^U - \tilde{B}_{i\beta}^U\right\}\right]$$

for $i = 1, \dots, n$. Based on the expression in the decomposition theorem, the membership function of $\tilde{\mathbf{A}} \ominus_{DT}^{\star} \tilde{\mathbf{B}}$ is given by

$$\xi_{\tilde{\mathbf{A}} \ominus_{DT}^{\star} \tilde{\mathbf{B}}}(\mathbf{z}) = \sup_{0 < \alpha \le 1} \alpha \cdot \chi_{M_\alpha^-}(\mathbf{z}). \tag{7.46}$$

We are going to study the α-level sets $(\tilde{\mathbf{A}} \ominus_{DT}^{\star} \tilde{\mathbf{B}})_\alpha$ of $\tilde{\mathbf{A}} \ominus_{DT}^{\star} \tilde{\mathbf{B}}$ for $\alpha \in [0, 1]$.

For $i = 1, \dots, n$, we write

$$N_\alpha^{(i-)} \equiv \bigcup_{\alpha \le \beta \le 1} M_\beta^{(i-)}.$$

It is clear to see that $\{N_\alpha^{(i-)} : 0 < \alpha \le 1\}$ is a nested family. Using the nestedness, we can show that

$$N_\alpha^{(i-)} = \bigcap_{s=1}^{\infty} N_{\alpha_s}^{(i-)} \text{ for } 0 < \alpha_s \uparrow \alpha \text{ with } \alpha_s < \alpha \text{ for all } s. \tag{7.47}$$

From (7.45), we also see that

$$M_\alpha^- = N_\alpha^{(1-)} \times \cdots \times N_\alpha^{(n-)}.$$

Using (7.47), we can also obtain

$$M_\alpha^- = \bigcap_{s=1}^{\infty} M_{\alpha_s}^- \text{ for } 0 < \alpha_s \uparrow \alpha \text{ with } \alpha_s < \alpha \text{ for all } s. \tag{7.48}$$

By applying (7.48) to the argument in the proof of Proposition 7.3.3, we can obtain

$$\left(\tilde{\mathbf{A}} \ominus_{DT}^{\star} \tilde{\mathbf{B}}\right)_\alpha = M_\alpha^- = N_\alpha^{(1-)} \times \cdots \times N_\alpha^{(n-)}$$

for any $\alpha \in [0, 1]$.

Now, we consider the difference $\tilde{A}^{(i)} \ominus_{DT}^{\star} \tilde{B}^{(i)}$ of components $\tilde{A}^{(i)}$ and $\tilde{B}^{(i)}$ for $i = 1, \dots, n$. Using the expression in the decomposition theorem, the membership function of $\tilde{A}^{(i)} \ominus_{DT}^{\star} \tilde{B}^{(i)}$ is defined by

$$\xi_{\tilde{A}^{(i)} \ominus_{DT}^{\star} \tilde{B}^{(i)}}(\mathbf{z}) = \sup_{0 < \alpha \le 1} \alpha \cdot \chi_{N_\alpha^{(i-)}}(\mathbf{z}).$$

We shall also study the α-level sets $(\tilde{A}^{(i)} \ominus^{\star}_{DT} \tilde{B}^{(i)})_\alpha$ of $\tilde{A}^{(i)} \ominus^{\star}_{DT} \tilde{B}^{(i)}$ for $\alpha \in [0,1]$. Using the argument in the proof of Proposition 7.3.3 again, we can obtain

$$\left(\tilde{A}^{(i)} \ominus^{\star}_{DT} \tilde{B}^{(i)}\right)_\alpha = N_\alpha^{(i-)}$$

for any $\alpha \in [0,1]$.

In order to obtain the compact form of the α-level sets, we assume that $\tilde{A}^{(1)},\dots,\tilde{A}^{(n)}$ and $\tilde{B}^{(1)},\dots,\tilde{B}^{(n)}$ are taken to be canonical fuzzy intervals. Let

$$\zeta_i^L(\beta) = \min\left\{\tilde{A}^L_{i\beta} - \tilde{B}^L_{i\beta}, \tilde{A}^U_{i\beta} - \tilde{B}^U_{i\beta}\right\} \text{ and } \zeta_i^U(\beta) = \max\left\{\tilde{A}^L_{i\beta} - \tilde{B}^L_{i\beta}, \tilde{A}^U_{i\beta} - \tilde{B}^U_{i\beta}\right\}.$$

Then $M_\beta^{(i-)} = [\zeta_i^L(\beta), \zeta_i^U(\beta)]$. We also see that ζ_i^L and ζ_i^U are continuous functions on $[0,1]$. Then, for $\alpha \in (0,1]$, we can obtain

$$\begin{aligned} N_\alpha^{(i-)} &= \bigcup_{\alpha\le\beta\le1} M_\beta^{(i-)} = \bigcup_{\alpha\le\beta\le1} [\zeta_i^L(\beta), \zeta_i^U(\beta)] \qquad (7.49)\\ &= \left[\min_{\alpha\le\beta\le1} \zeta_i^L(\beta), \max_{\alpha\le\beta\le1} \zeta_i^U(\beta)\right]\\ &= \left[\min_{\alpha\le\beta\le1} \min\left\{\tilde{A}^L_{i\beta} - \tilde{B}^L_{i\beta}, \tilde{A}^U_{i\beta} - \tilde{B}^U_{i\beta}\right\}, \max_{\alpha\le\beta\le1} \max\left\{\tilde{A}^L_{i\beta} - \tilde{B}^L_{i\beta}, \tilde{A}^U_{i\beta} - \tilde{B}^U_{i\beta}\right\}\right]. \end{aligned}$$

The above results are summarized below.

Theorem 7.3.6 *Let $\tilde{A}^{(1)},\dots,\tilde{A}^{(n)}$ and $\tilde{B}^{(1)},\dots,\tilde{B}^{(n)}$ be any fuzzy intervals. Suppose that the family $\{M_\alpha^- : \alpha \in (0,1]\}$ is taken by*

$$M_\alpha^- = \left(\bigcup_{\alpha\le\beta\le1} M_\beta^{(1-)}\right) \times\cdots\times \left(\bigcup_{\alpha\le\beta\le1} M_\beta^{(n-)}\right),$$

where $M_\beta^{(i-)}$ are bounded closed intervals given by

$$M_\beta^{(i-)} = \left[\min\left\{\tilde{A}^L_{i\beta} - \tilde{B}^L_{i\beta}, \tilde{A}^U_{i\beta} - \tilde{B}^U_{i\beta}\right\}, \max\left\{\tilde{A}^L_{i\beta} - \tilde{B}^L_{i\beta}, \tilde{A}^U_{i\beta} - \tilde{B}^U_{i\beta}\right\}\right]$$

for $i = 1,\dots,n$. For each $\alpha \in [0,1]$, we have

$$\left(\tilde{A}^{(i)} \ominus^{\star}_{DT} \tilde{B}^{(i)}\right)_\alpha = \bigcup_{\alpha\le\beta\le1} M_\beta^{(i-)}$$

and

$$\left(\tilde{\mathbf{A}} \ominus^{\star}_{DT} \tilde{\mathbf{B}}\right)_\alpha = \left(\tilde{A}^{(1)} \ominus^{\star}_{DT} \tilde{B}^{(1)}\right)_\alpha \times\cdots\times \left(\tilde{A}^{(n)} \ominus^{\star}_{DT} \tilde{B}^{(n)}\right)_\alpha.$$

We further ssume that $\tilde{A}^{(1)},\dots,\tilde{A}^{(n)}$ and $\tilde{B}^{(1)},\dots,\tilde{B}^{(n)}$ are taken to be canonical fuzzy intervals. Then, for $i = 1,\dots,n$, we have

$$\begin{aligned} \bigcup_{\alpha\le\beta\le1} M_\beta^{(i-)} &= \left[\min_{\alpha\le\beta\le1} \min\left\{\tilde{A}^L_{i\beta} - \tilde{B}^L_{i\beta}, \tilde{A}^U_{i\beta} - \tilde{B}^U_{i\beta}\right\}, \max_{\alpha\le\beta\le1} \max\left\{\tilde{A}^L_{i\beta} - \tilde{B}^L_{i\beta}, \tilde{A}^U_{i\beta} - \tilde{B}^U_{i\beta}\right\}\right]\\ &= \left[\min\left\{\min_{\alpha\le\beta\le1}\left(\tilde{A}^L_{i\beta} - \tilde{B}^L_{i\beta}\right), \min_{\alpha\le\beta\le1}\left(\tilde{A}^U_{i\beta} - \tilde{B}^U_{i\beta}\right)\right\},\right.\\ &\qquad \left.\max\left\{\max_{\alpha\le\beta\le1}\left(\tilde{A}^L_{i\beta} - \tilde{B}^L_{i\beta}\right), \max_{\alpha\le\beta\le1}\left(\tilde{A}^U_{i\beta} - \tilde{B}^U_{i\beta}\right)\right\}\right], \end{aligned}$$

which are bounded closed intervals.

Example 7.3.7 Continued from Example 7.2.4 by referring to (7.35), Theorem 7.3.6 says that

$$\left(\tilde{\mathbf{A}} \ominus_{DT}^{\star} \tilde{\mathbf{B}}\right)_\alpha = M_\alpha^- = \{(-3,-1)\}$$

for $\alpha \in [0,1]$.

7.3.4 α-Level Sets of $\tilde{\mathbf{A}} \ominus_{DT}^{\dagger} \tilde{\mathbf{B}}$

Let $\tilde{A}^{(1)}, \ldots, \tilde{A}^{(n)}$ and $\tilde{B}^{(1)}, \ldots, \tilde{B}^{(n)}$ be any fuzzy intervals. By referring to Subsection 6.2.3, the family $\{M_\alpha^- : 0 < \alpha \leq 1\}$ is given by

$$M_\alpha^- = M_\alpha^{(1-)} \times \cdots \times M_\alpha^{(n-)},$$

where $M_\alpha^{(i-)}$ are bounded closed intervals given by

$$M_\alpha^{(i-)} = \left[\min\left\{\tilde{A}_{i\alpha}^L - \tilde{B}_{i\alpha}^L, \tilde{A}_{i\alpha}^U - \tilde{B}_{i\alpha}^U\right\}, \max\left\{\tilde{A}_{i\alpha}^L - \tilde{B}_{i\alpha}^L, \tilde{A}_{i\alpha}^U - \tilde{B}_{i\alpha}^U\right\}\right]$$

for $i = 1, \ldots, n$. Based on the expression in the decomposition theorem, the membership functions of $\tilde{\mathbf{A}} \ominus_{DT}^{\dagger} \tilde{\mathbf{B}}$ and $\tilde{A}^{(i)} \ominus_{DT}^{\dagger} \tilde{B}^{(i)}$ for $i = 1, \ldots, n$ are given by

$$\xi_{\tilde{\mathbf{A}} \ominus_{DT}^{\dagger} \tilde{\mathbf{B}}}(\mathbf{z}) = \sup_{0<\alpha\leq 1} \alpha \cdot \chi_{M_\alpha^-}(\mathbf{z}) \tag{7.50}$$

for any $\mathbf{z} \in \mathbb{R}^m$ and

$$\xi_{\tilde{A}^{(i)} \ominus_{DT}^{\dagger} \tilde{B}^{(i)}}(z) = \sup_{0<\alpha\leq 1} \alpha \cdot \chi_{M_\alpha^{(i-)}}(z)$$

for any $z \in \mathbb{R}$, respectively. We are going to study the α-level sets $(\tilde{\mathbf{A}} \ominus_{DT}^{\dagger} \tilde{\mathbf{B}})_\alpha$ of $\tilde{\mathbf{A}} \ominus_{DT}^{\dagger} \tilde{\mathbf{B}}$ for $\alpha \in [0,1]$, and the α-level sets $(\tilde{A}^{(i)} \ominus_{DT}^{\dagger} \tilde{B}^{(i)})_\alpha$ of $\tilde{A}^{(i)} \ominus_{DT}^{\dagger} \tilde{B}^{(i)}$ for $\alpha \in [0,1]$.

Now, we assume that $\tilde{A}^{(1)}, \ldots, \tilde{A}^{(n)}$ and $\tilde{B}^{(1)}, \ldots, \tilde{B}^{(n)}$ are taken to be canonical fuzzy intervals. We are going to claim

$$\left(\tilde{\mathbf{A}} \ominus_{DT}^{\dagger} \tilde{\mathbf{B}}\right)_\alpha = \bigcup_{\alpha\leq\beta\leq 1} M_\beta^- \text{ for } \alpha \in (0,1]. \tag{7.51}$$

Let

$$\zeta_i^L(\alpha) = \min\left\{\tilde{A}_{i\alpha}^L - \tilde{B}_{i\alpha}^L, \tilde{A}_{i\alpha}^U - \tilde{B}_{i\alpha}^U\right\} \text{ and } \zeta_i^U(\alpha) = \max\left\{\tilde{A}_{i\alpha}^L - \tilde{B}_{i\alpha}^L, \tilde{A}_{i\alpha}^U - \tilde{B}_{i\alpha}^U\right\}.$$

Then $M_\alpha^{(i-)} = [\zeta_i^L(\alpha), \zeta_i^U(\alpha)]$. We also see that ζ_i^L and ζ_i^U are continuous functions on $[0,1]$. Using Propositions 1.3.9 and 1.3.10, given any fixed $\mathbf{x} = (x_1, \ldots, x_n) \in \mathbb{R}^m$, the following functions

$$\zeta(\alpha) = \alpha \cdot \chi_{M_\alpha}(\mathbf{x}) \text{ and } \zeta_i(\alpha) = \alpha \cdot \chi_{M_\alpha^{(i-)}}(x_i).$$

are upper semi-continuous on $[0,1]$ for $i = 1, \ldots, n$.

Suppose that $\mathbf{z} \in (\tilde{\mathbf{A}} \ominus_{DT}^{\dagger} \tilde{\mathbf{B}})_\alpha$ and $\mathbf{z} \notin M_\beta^-$ for all β satisfying $\beta \geq \alpha$. Then $\beta \cdot \chi_{M_\beta^-}(\mathbf{z}) < \alpha$ for all $\beta \in [0,1]$. Since $\zeta(\beta) = \beta \cdot \chi_{M_\beta^-}(\mathbf{z})$ is upper semi-continuous on $[0,1]$ as described above, the supremum of the function ζ is attained by Proposition 1.4.4. This says that

$$0 < \xi_{\tilde{\mathbf{A}} \ominus_{DT}^{\dagger} \tilde{\mathbf{B}}}(\mathbf{z}) = \sup_{\beta\in[0,1]} \zeta(\beta) = \sup_{\beta\in[0,1]} \beta \cdot \chi_{M_\beta^-}(\mathbf{z}) = \max_{\beta\in[0,1]} \beta \cdot \chi_{M_\beta^-}(\mathbf{z}) = \beta^* \cdot \chi_{M_{\beta^*}}(\mathbf{z}) < \alpha$$

for some $\beta^* \in [0,1]$ with $\beta^* \neq 0$, which violates $\mathbf{z} \in (\tilde{\mathbf{A}} \ominus_{DT}^{\dagger} \tilde{\mathbf{B}})_\alpha$. Therefore, there exists β_0 satisfying $\beta_0 \geq \alpha$ and $\mathbf{z} \in M_{\beta_0}^-$, which shows the following inclusion

$$\left(\tilde{\mathbf{A}} \ominus_{DT}^{\dagger} \tilde{\mathbf{B}}\right)_\alpha \subseteq \bigcup_{\alpha \leq \beta \leq 1} M_\beta^-.$$

On the other hand, the following inclusion

$$\bigcup_{\alpha \leq \beta \leq 1} M_\beta^- \subseteq \left\{ \mathbf{z} \in \mathbb{R}^m : \sup_{\beta \in [0,1]} \beta \cdot \chi_{M_\beta^-}(\mathbf{z}) \geq \alpha \right\}$$
$$= \{\mathbf{z} \in \mathbb{R}^m : \xi_{\tilde{\mathbf{A}} \ominus_{DT}^{\dagger} \tilde{\mathbf{B}}}(\mathbf{z}) \geq \alpha\} = (\tilde{\mathbf{A}} \ominus_{DT}^{\dagger} \tilde{\mathbf{B}})_\alpha$$

is obvious. This shows (7.51). We can similarly obtain

$$\left(\tilde{A}^{(i)} \ominus_{DT}^{\dagger} \tilde{B}^{(i)}\right)_\alpha = \bigcup_{\alpha \leq \beta \leq 1\}} M_\beta^{(i-)} \text{ for } \alpha \in (0,1].$$

The above results are summarized below.

Theorem 7.3.8 *Let $\tilde{A}^{(1)}, \ldots, \tilde{A}^{(n)}$ and $\tilde{B}^{(1)}, \ldots, \tilde{B}^{(n)}$ be canonical fuzzy intervals. Suppose that the family $\{M_\alpha : \alpha \in (0,1]\}$ is given by*

$$M_\alpha^- = M_\alpha^{(1-)} \times \cdots \times M_\alpha^{(n-)},$$

where $M_\alpha^{(i-)}$ are bounded closed intervals given by

$$M_\alpha^{(i-)} = \left[\min\left\{\tilde{A}_{i\alpha}^L - \tilde{B}_{i\alpha}^L, \tilde{A}_{i\alpha}^U - \tilde{B}_{i\alpha}^U\right\}, \max\left\{\tilde{A}_{i\alpha}^L - \tilde{B}_{i\alpha}^L, \tilde{A}_{i\alpha}^U - \tilde{B}_{i\alpha}^U\right\}\right]$$

for $i = 1, \ldots, n$.

(i) *For each $\alpha \in (0,1]$, we have*

$$\left(\tilde{A}^{(i)} \ominus_{DT}^{\dagger} \tilde{B}^{(i)}\right)_\alpha = \bigcup_{\alpha \leq \beta \leq 1} M_\beta^{(i-)}$$

and the 0-level set

$$\left(\tilde{A}^{(i)} \ominus_{DT}^{\dagger} \tilde{B}^{(i)}\right)_0 = cl\left(\bigcup_{0<\alpha \leq 1} \left(\tilde{A}^{(i)} \ominus_{DT}^{\dagger} \tilde{B}^{(i)}\right)_\alpha\right).$$

Moreover, for $i = 1, \ldots, n$, we have

$$\bigcup_{\alpha \leq \beta \leq 1} M_\beta^{(i-)} = \left[\min_{\alpha \leq \beta \leq 1} \min\left\{\tilde{A}_{i\beta}^L - \tilde{B}_{i\beta}^L, \tilde{A}_{i\beta}^U - \tilde{B}_{i\beta}^U\right\}, \max_{\alpha \leq \beta \leq 1} \max\left\{\tilde{A}_{i\beta}^L - \tilde{B}_{i\beta}^L, \tilde{A}_{i\beta}^U - \tilde{B}_{i\beta}^U\right\}\right]$$
$$= \left[\min\left\{\min_{\alpha \leq \beta \leq 1}\left(\tilde{A}_{i\beta}^L - \tilde{B}_{i\beta}^L\right), \min_{\alpha \leq \beta \leq 1}\left(\tilde{A}_{i\beta}^U - \tilde{B}_{i\beta}^U\right)\right\},\right.$$
$$\left.\max\left\{\max_{\alpha \leq \beta \leq 1}\left(\tilde{A}_{i\beta}^L - \tilde{B}_{i\beta}^L\right), \max_{\alpha \leq \beta \leq 1}\left(\tilde{A}_{i\beta}^U - \tilde{B}_{i\beta}^U\right)\right\}\right],$$

which are bounded closed intervals.

(ii) *For each $\alpha \in (0,1]$, we have*

$$\left(\tilde{\mathbf{A}} \ominus_{DT}^{\dagger} \tilde{\mathbf{B}}\right)_\alpha = \bigcup_{\alpha \leq \beta \leq 1} M_\beta^- = \bigcup_{\alpha \leq \beta \leq 1} \left(M_\beta^{(1-)} \times \cdots \times M_\beta^{(n-)}\right)$$

and the 0*-level set*

$$\left(\tilde{\mathbf{A}} \ominus_{DT}^{\dagger} \tilde{\mathbf{B}}\right)_0 = cl\left(\bigcup_{0<\alpha\leq 1}\left(\tilde{\mathbf{A}} \ominus_{DT}^{\dagger} \tilde{\mathbf{B}}\right)_\alpha\right).$$

Remark 7.3.9 We notice that

$$\left(\tilde{\mathbf{A}} \ominus_{DT}^{\dagger} \tilde{\mathbf{B}}\right)_\alpha \neq \left(\tilde{A}^{(1)} \ominus_{DT}^{\dagger} \tilde{B}^{(1)}\right)_\alpha \times \cdots \times \left(\tilde{A}^{(n)} \ominus_{DT}^{\dagger} \tilde{B}^{(n)}\right)_\alpha \text{ for each } \alpha \in [0,1]$$

in general.

Example 7.3.10 Continued from Example 7.2.4 by referring to (7.36), we have $M_\alpha^- = \{(-3,-1)\}$. Part (i) of Theorem 7.3.8 says that

$$\left(\tilde{\mathbf{A}} \ominus_{DT}^{\dagger} \tilde{\mathbf{B}}\right)_\alpha = \bigcup_{\alpha\leq\beta\leq 1} M_\beta^- = \bigcup_{\alpha\leq\beta\leq 1} \{(-3,-1)\} = \{(-3,-1)\}$$

for $\alpha \in [0,1]$.

7.3.5 The Equivalences and Fuzziness

Next, we are going to present the equivalence between $\tilde{\mathbf{A}} \ominus_{EP} \tilde{\mathbf{B}}$ and $\tilde{\mathbf{A}} \ominus_{DT}^{\diamond} \tilde{\mathbf{B}}$ in Theorems 7.3.1 and 7.3.4, respectively.

Theorem 7.3.11 *Let* $\tilde{A}^{(1)}, \ldots, \tilde{A}^{(n)}$ *and* $\tilde{B}^{(1)}, \ldots, \tilde{B}^{(n)}$ *be fuzzy intervals. Suppose that* $\tilde{\mathbf{A}} \ominus_{EP} \tilde{\mathbf{B}}$ *and* $\tilde{\mathbf{A}} \ominus_{DT}^{\diamond} \tilde{\mathbf{B}}$ *are obtained from Theorems 7.3.1 and 7.3.4, respectively. Then, we have*

$$\tilde{\mathbf{A}} \ominus_{EP} \tilde{\mathbf{B}} = \tilde{\mathbf{A}} \ominus_{DT}^{\diamond} \tilde{\mathbf{B}}.$$

Moreover, for $\alpha \in [0,1]$*, we have*

$$\begin{aligned}(\tilde{\mathbf{A}} \ominus_{EP} \tilde{\mathbf{B}})_\alpha &= (\tilde{\mathbf{A}} \ominus_{DT}^{\diamond} \tilde{\mathbf{B}})_\alpha \\ &= \left[\tilde{A}_{1\alpha}^L - \tilde{B}_{1\alpha}^U, \tilde{A}_{1\alpha}^U - \tilde{B}_{1\alpha}^L\right] \times \cdots \times \left[\tilde{A}_{n\alpha}^L - \tilde{B}_{n\alpha}^U, \tilde{A}_{n\alpha}^U - \tilde{B}_{n\alpha}^L\right].\end{aligned} \tag{7.52}$$

Proof. The equality (7.52) follows immediately from Theorems 7.3.1 and 7.3.4, which also says that $\tilde{\mathbf{A}} \ominus_{EP} \tilde{\mathbf{B}} = \tilde{\mathbf{A}} \ominus_{DT}^{\diamond} \tilde{\mathbf{B}}$. This completes the proof. ■

We are not able to study the equivalences among $\tilde{\mathbf{A}} \ominus_{EP} \tilde{\mathbf{B}}$, $\tilde{\mathbf{A}} \ominus_{DT}^{\star} \tilde{\mathbf{B}}$, and $\tilde{\mathbf{A}} \ominus_{DT}^{\dagger} \tilde{\mathbf{B}}$. However, we can study their fuzziness by considering their α-level sets. Recall Definition 6.3.4, we say that $\tilde{A}$ is fuzzier than $\tilde{B}$ when $\tilde{B}_\alpha \subseteq \tilde{A}_\alpha$ for all $\alpha \in (0,1]$.

Let $\tilde{A}^{(1)}, \ldots, \tilde{A}^{(n)}$ and $\tilde{B}^{(1)}, \ldots, \tilde{B}^{(n)}$ be canonical fuzzy intervals, and let $\tilde{\mathbf{A}} \ominus_{DT}^{\star} \tilde{\mathbf{B}}$ and $\tilde{\mathbf{A}} \ominus_{DT}^{\dagger} \tilde{\mathbf{B}}$ be obtained from Theorems 7.3.6 and 7.3.8, respectively. For each $\alpha \in (0,1]$, we have

$$\left(\tilde{\mathbf{A}} \ominus_{DT}^{\dagger} \tilde{\mathbf{B}}\right)_\alpha = \bigcup_{\alpha\leq\beta\leq 1}\left(M_\beta^{(1-)} \times \cdots \times M_\beta^{(n-)}\right)$$

and

$$(\tilde{\mathbf{A}} \ominus_{DT}^{\star} \tilde{\mathbf{B}})_\alpha = \left(\bigcup_{\alpha\leq\beta\leq 1} M_\beta^{(1-)}\right) \times \cdots \times \left(\bigcup_{\alpha\leq\beta\leq 1} M_\beta^{(n-)}\right).$$

Since the following inclusion

$$\bigcup_{\alpha\leq\beta\leq 1}\left(M_\beta^{(1-)}\times\cdots\times M_\beta^{(n-)}\right)\subseteq\left(\bigcup_{\alpha\leq\beta\leq 1}M_\beta^{(1-)}\right)\times\cdots\times\left(\bigcup_{\alpha\leq\beta\leq 1}M_\beta^{(n-)}\right)$$

is obvious, it follows that

$$\left(\tilde{\mathbf{A}}\ominus_{DT}^{\dagger}\tilde{\mathbf{B}}\right)_\alpha\subseteq\left(\tilde{\mathbf{A}}\ominus_{DT}^{\star}\tilde{\mathbf{B}}\right)_\alpha \text{ for each } \alpha\in(0,1],$$

which says that $\tilde{\mathbf{A}}\ominus_{DT}^{\star}\tilde{\mathbf{B}}$ is fuzzier than $\tilde{\mathbf{A}}\ominus_{DT}^{\dagger}\tilde{\mathbf{B}}$.

On the other hand, from Theorems 7.3.11 and 7.3.6, for each $\alpha\in(0,1]$, we also have

$$\begin{aligned}\bigcup_{\alpha\leq\beta\leq 1}M_\beta^{(i-)} &= \left[\min_{\alpha\leq\beta\leq 1}\min\left\{\tilde{A}_{i\beta}^L-\tilde{B}_{i\beta}^L,\tilde{A}_{i\beta}^U-\tilde{B}_{i\beta}^U\right\},\max_{\alpha\leq\beta\leq 1}\max\left\{\tilde{A}_{i\beta}^L-\tilde{B}_{i\beta}^L,\tilde{A}_{i\beta}^U-\tilde{B}_{i\beta}^U\right\}\right]\\ &\subseteq\left[\min_{\alpha\leq\beta\leq 1}\left(\tilde{A}_{i\beta}^L-\tilde{B}_{i\beta}^U\right),\max_{\alpha\leq\beta\leq 1}\left(\tilde{A}_{i\beta}^U-\tilde{B}_{i\beta}^L\right)\right]\\ &\subseteq\left[\tilde{A}_{i\alpha}^L-\tilde{B}_{i\alpha}^U,\tilde{A}_{i\alpha}^U-\tilde{B}_{i\alpha}^L\right].\end{aligned}$$

It follows that

$$\left(\tilde{\mathbf{A}}\ominus_{DT}^{\star}\tilde{\mathbf{B}}\right)_\alpha\subseteq\left(\tilde{\mathbf{A}}\ominus_{DT}^{\diamond}\tilde{\mathbf{B}}\right)_\alpha \text{ for each } \alpha\in(0,1],$$

which says that $\tilde{\mathbf{A}}\ominus_{DT}^{\diamond}\tilde{\mathbf{B}}$ is fuzzier than $\tilde{\mathbf{A}}\ominus_{DT}^{\star}\tilde{\mathbf{B}}$. The above results are summarized below.

Theorem 7.3.12 *Let $\tilde{A}^{(1)},\ldots,\tilde{A}^{(n)}$ and $\tilde{B}^{(1)},\ldots,\tilde{B}^{(n)}$ be canonical fuzzy intervals. Suppose that $\tilde{\mathbf{A}}\ominus_{EP}\tilde{\mathbf{B}}$, $\tilde{\mathbf{A}}\ominus_{DT}^{\diamond}\tilde{\mathbf{B}}$, $\tilde{\mathbf{A}}\ominus_{DT}^{\star}\tilde{\mathbf{B}}$ and $\tilde{\mathbf{A}}\ominus_{DT}^{\dagger}\tilde{\mathbf{B}}$ are obtained from Theorems 7.3.1, 7.3.4, 7.3.6 and 7.3.8, respectively. Then*

$$\left(\tilde{\mathbf{A}}\ominus_{DT}^{\dagger}\tilde{\mathbf{B}}\right)_\alpha\subseteq\left(\tilde{\mathbf{A}}\ominus_{DT}^{\star}\tilde{\mathbf{B}}\right)_\alpha\subseteq\left(\tilde{\mathbf{A}}\ominus_{DT}^{\diamond}\tilde{\mathbf{B}}\right)_\alpha=\left(\tilde{\mathbf{A}}\ominus_{EP}\tilde{\mathbf{B}}\right)_\alpha$$

for each $\alpha\in[0,1]$. In other words, $\tilde{\mathbf{A}}\ominus_{DT}^{\star}\tilde{\mathbf{B}}$ is fuzzier than $\tilde{\mathbf{A}}\ominus_{DT}^{\dagger}\tilde{\mathbf{B}}$, and $\tilde{\mathbf{A}}\ominus_{DT}^{\diamond}\tilde{\mathbf{B}}$ is fuzzier than $\tilde{\mathbf{A}}\ominus_{DT}^{\star}\tilde{\mathbf{B}}$.

Theorem 7.3.12 says that, when $\tilde{A}^{(1)},\ldots,\tilde{A}^{(n)}$ and $\tilde{B}^{(1)},\ldots,\tilde{B}^{(n)}$ are taken to be canonical fuzzy intervals, we may prefer to pick $\tilde{\mathbf{A}}\ominus_{DT}^{\dagger}\tilde{\mathbf{B}}$, which has less fuzziness in applications.

7.4 Addition of Vectors of Fuzzy Intervals

Let $\tilde{\mathbf{A}}$ and $\tilde{\mathbf{B}}$ be two vectors of fuzzy intervals with components $\tilde{A}^{(i)}$ and $\tilde{B}^{(i)}$, respectively, for $i=1,\ldots,n$. We are going to study the α-level set of $\tilde{\mathbf{A}}\oplus_{EP}\tilde{\mathbf{B}}$, which is obtained from the extension principle, and the α-level sets of $\tilde{\mathbf{A}}\oplus_{DT}\tilde{\mathbf{B}}$ for $\oplus_{DT}\in\{\oplus_{DT}^{\diamond},\oplus_{DT}^{\star},\oplus_{DT}^{\dagger}\}$ that are obtained from the expression in the decomposition theorem.

7.4.1 α-Level Sets of $\tilde{\mathbf{A}}\oplus_{EP}\tilde{\mathbf{B}}$

Given any function $\mathfrak{A}:[0,1]^{2n}\to[0,1]$, the membership function of addition $\tilde{\mathbf{A}}\oplus_{EP}\tilde{\mathbf{B}}$ is defined by

$$\xi_{\tilde{\mathbf{A}}\oplus_{EP}\tilde{\mathbf{B}}}(\mathbf{z})=\sup_{\{(\mathbf{x},\mathbf{y}):\mathbf{z}=\mathbf{x}+\mathbf{y}\}}\mathfrak{A}\left(\xi_{\tilde{A}^{(1)}}(x_1),\ldots,\xi_{\tilde{A}^{(n)}}(x_n),\xi_{\tilde{B}^{(1)}}(y_1),\ldots,\xi_{\tilde{B}^{(n)}}(y_n)\right)$$

for any $\mathbf{z} \in \mathbb{R}^m$. The α-level set $(\tilde{\mathbf{A}} \oplus_{EP} \tilde{\mathbf{B}})_\alpha$ of $\tilde{\mathbf{A}} \oplus_{EP} \tilde{\mathbf{B}}$ for $\alpha \in [0,1]$ can be obtained by applying Theorem 4.4.15 to the addition $\tilde{\mathbf{A}} \oplus_{EP} \tilde{\mathbf{B}}$, which is shown below. For each $\alpha \in (0,1]$, we have

$$\begin{aligned}\left(\tilde{\mathbf{A}} \oplus_{EP} \tilde{\mathbf{B}}\right)_\alpha &= \left\{\mathbf{x}+\mathbf{y} : \mathfrak{A}\left(\xi_{\tilde{A}^{(1)}}(x_1), \cdots, \xi_{\tilde{A}^{(n)}}(x_n), \xi_{\tilde{B}^{(1)}}(y_1), \cdots, \xi_{\tilde{B}^{(n)}}(y_n)\right) \geq \alpha\right\} \\ &= \left\{\left(x_1+y_1, \ldots, x_n+y_n\right) : \right. \\ &\qquad \left. \mathfrak{A}\left(\xi_{\tilde{A}^{(1)}}(x_1), \cdots, \xi_{\tilde{A}^{(n)}}(x_n), \xi_{\tilde{B}^{(1)}}(y_1), \cdots, \xi_{\tilde{B}^{(n)}}(y_n)\right) \geq \alpha\right\}.\end{aligned} \tag{7.53}$$

The 0-level set is given by

$$\left(\tilde{\mathbf{A}} \oplus_{EP} \tilde{\mathbf{B}}\right)_0 = \tilde{\mathbf{A}}_0 + \tilde{\mathbf{B}}_0 = \left\{\mathbf{x}+\mathbf{y} : \mathbf{x} \in \tilde{\mathbf{A}}_0 \text{ and } \mathbf{y} \in \tilde{\mathbf{B}}_0\right\}.$$

Moreover, for each $\alpha \in [0,1]$, the α-level sets $(\tilde{\mathbf{A}} \oplus_{EP} \tilde{\mathbf{B}})_\alpha$ are convex, closed, and bounded subsets of $\mathbb{R}^m$.

Suppose the function $\mathfrak{A} : [0,1]^{2n} \to [0,1]$ is given by

$$\mathfrak{A}\left(\alpha_1, \ldots, \alpha_{2n}\right) = \min\left\{\alpha_1, \ldots, \alpha_{2n}\right\}.$$

For each $\alpha \in (0,1]$, using (7.53), we have

$$\begin{aligned}\left(\tilde{\mathbf{A}} \oplus_{EP} \tilde{\mathbf{B}}\right)_\alpha &= \left\{\mathbf{x}+\mathbf{y} : \min\left\{\xi_{\tilde{A}^{(1)}}(x_1), \cdots, \xi_{\tilde{A}^{(n)}}(x_n), \xi_{\tilde{B}^{(1)}}(y_1), \cdots, \xi_{\tilde{B}^{(n)}}(y_n)\right\} \geq \alpha\right\} \\ &= \left\{\mathbf{x}+\mathbf{y} : \xi_{\tilde{A}^{(i)}}(x_i) \geq \alpha \text{ and } \xi_{\tilde{B}^{(i)}}(y_i) \geq \alpha \text{ for each } i = 1, \ldots, n\right\} \\ &= \left\{\left(x_1+y_1, \ldots, x_n+y_n\right) : x_i \in \tilde{A}^{(i)}_\alpha \equiv \left[\tilde{A}^L_{i\alpha}, \tilde{A}^U_{i\alpha}\right]\right. \\ &\qquad \left. \text{and } y_i \in \tilde{B}^{(i)}_\alpha \equiv \left[\tilde{B}^L_{i\alpha}, \tilde{B}^U_{i\alpha}\right] \text{ for each } i = 1, \ldots, n\right\} \\ &= \left[\tilde{A}^L_{1\alpha} + \tilde{B}^L_{1\alpha}, \tilde{A}^U_{1\alpha} + \tilde{B}^U_{1\alpha}\right] \times \cdots \times \left[\tilde{A}^L_{n\alpha} + \tilde{B}^L_{n\alpha}, \tilde{A}^U_{n\alpha} + \tilde{B}^U_{n\alpha}\right].\end{aligned} \tag{7.54}$$

For the 0-level set, from (7.54) and (2.5), it is not difficult to show that

$$\begin{aligned}\left(\tilde{\mathbf{A}} \oplus_{EP} \tilde{\mathbf{B}}\right)_0 &= \mathrm{cl}\left(\bigcup_{0<\alpha\leq 1} \left(\tilde{\mathbf{A}} \oplus_{EP} \tilde{\mathbf{B}}\right)_\alpha\right) \\ &= \left[\tilde{A}^L_{10} + \tilde{B}^L_{10}, \tilde{A}^U_{10} + \tilde{B}^U_{10}\right] \times \cdots \times \left[\tilde{A}^L_{n0} + \tilde{B}^L_{n0}, \tilde{A}^U_{n0} + \tilde{B}^U_{n0}\right].\end{aligned}$$

Regarding the components $\tilde{A}^{(i)}$ and $\tilde{B}^{(i)}$, for each $\alpha \in [0,1]$, we also have

$$\left(\tilde{A}^{(i)} \oplus_{EP} \tilde{B}^{(i)}\right)_\alpha = \left[\tilde{A}^L_{i\alpha} + \tilde{B}^L_{i\alpha}, \tilde{A}^U_{i\alpha} + \tilde{B}^U_{i\alpha}\right] \text{ for } i = 1, \ldots, n. \tag{7.55}$$

Therefore, from (7.54) and (7.55), for $\alpha \in [0,1]$, we obtain

$$\left(\tilde{\mathbf{A}} \oplus_{EP} \tilde{\mathbf{B}}\right)_\alpha = \left(\tilde{A}^{(1)} \oplus_{EP} \tilde{B}^{(1)}\right)_\alpha \times \cdots \times \left(\tilde{A}^{(n)} \oplus_{EP} \tilde{B}^{(n)}\right)_\alpha.$$

The above results are summarized in the following theorem.

Theorem 7.4.1 *Let $\tilde{A}^{(1)}, \ldots, \tilde{A}^{(n)}$ and $\tilde{B}^{(1)}, \ldots, \tilde{B}^{(n)}$ be fuzzy intervals. Suppose that the function $\mathfrak{A} : [0,1]^{2n} \to [0,1]$ is given by*

$$\mathfrak{A}\left(\alpha_1, \ldots, \alpha_{2n}\right) = \min\left\{\alpha_1, \ldots, \alpha_{2n}\right\}.$$

For each $\alpha \in [0,1]$, we have

$$\left(\tilde{A}^{(i)} \oplus_{EP} \tilde{B}^{(i)}\right)_\alpha = \left[\tilde{A}^L_{i\alpha} + \tilde{B}^L_{i\alpha}, \tilde{A}^U_{i\alpha} + \tilde{B}^U_{i\alpha}\right]$$

and

$$(\tilde{\mathbf{A}} \oplus_{EP} \tilde{\mathbf{B}})_\alpha = \left(\tilde{A}^{(1)} \oplus_{EP} \tilde{B}^{(1)}\right)_\alpha \times \cdots \times \left(\tilde{A}^{(n)} \oplus_{EP} \tilde{B}^{(n)}\right)_\alpha.$$

Example 7.4.2 Continued from Examples 2.2.4 and 7.2.4, Theorem 7.4.1 says that

$$\begin{aligned}\left(\tilde{A}^{(1)} \oplus_{EP} \tilde{B}^{(1)}\right)_\alpha &= \left[\tilde{A}^L_{1\alpha} + \tilde{B}^L_{1\alpha}, \tilde{A}^U_{1\alpha} + \tilde{B}^U_{1\alpha}\right] \\ &= [(1+\alpha) + (4+\alpha), (4-\alpha) + (7-\alpha)] = [5+2\alpha, 11-2\alpha]\end{aligned}$$

and

$$\begin{aligned}\left(\tilde{A}^{(2)} \oplus_{EP} \tilde{B}^{(2)}\right)_\alpha &= \left[\tilde{A}^L_{2\alpha} + \tilde{B}^L_{2\alpha}, \tilde{A}^U_{2\alpha} + \tilde{B}^U_{2\alpha}\right] \\ &= [(2+\alpha) + (3+\alpha), (5-\alpha) + (6-\alpha)] = [5+2\alpha, 11-2\alpha]\end{aligned}$$

and

$$\begin{aligned}(\tilde{\mathbf{A}} \oplus_{EP} \tilde{\mathbf{B}})_\alpha &= \left(\tilde{A}^{(1)} \oplus_{EP} \tilde{B}^{(1)}\right)_\alpha \times \left(\tilde{A}^{(2)} \oplus_{EP} \tilde{B}^{(2)}\right)_\alpha \\ &= [5+2\alpha, 11-2\alpha] \times [5+2\alpha, 11-2\alpha]\end{aligned}$$

for $\alpha \in [0,1] = [0,1]$.

7.4.2 α-Level Sets of $\tilde{\mathbf{A}} \oplus_{DT} \tilde{\mathbf{B}}$

Let $\tilde{A}^{(1)}, \ldots, \tilde{A}^{(n)}$ and $\tilde{B}^{(1)}, \ldots, \tilde{B}^{(n)}$ be fuzzy intervals. By referring to Subsection 6.2.1, the family $\{M^+_\alpha : \alpha \in (0,1]\}$ is given by $M^+_\alpha = \tilde{\mathbf{A}}_\alpha + \tilde{\mathbf{B}}_\alpha$. For any $\alpha \in (0,1]$, we have

$$\begin{aligned}M^+_\alpha = \tilde{\mathbf{A}}_\alpha + \tilde{\mathbf{B}}_\alpha &= \left\{\mathbf{x} + \mathbf{y} : \mathbf{x} \in \tilde{\mathbf{A}}_\alpha \text{ and } \mathbf{y} \in \tilde{\mathbf{B}}_\alpha\right\} \\ &= \Big\{ \left(x_1 + y_1, \ldots, x_1 + y_1\right) : x_i \in \tilde{A}^{(i)}_\alpha = \left[\tilde{A}^L_{i\alpha}, \tilde{A}^U_{i\alpha}\right] \\ &\qquad \text{and } y_i \in \tilde{B}^{(i)}_\alpha = \left[\tilde{B}^L_{i\alpha}, \tilde{B}^U_{i\alpha}\right] \text{ for } i = 1, \ldots, n\Big\} \\ &= \left[\tilde{A}^L_{1\alpha} + \tilde{B}^L_{1\alpha}, \tilde{A}^U_{1\alpha} + \tilde{B}^U_{1\alpha}\right] \times \cdots \times \left[\tilde{A}^L_{n\alpha} + \tilde{B}^L_{n\alpha}, \tilde{A}^U_{n\alpha} + \tilde{B}^U_{n\alpha}\right]. \end{aligned} \tag{7.56}$$

Based on the expression in the decomposition theorem, the membership function of $\tilde{\mathbf{A}} \oplus^\diamond_{DT} \tilde{\mathbf{B}}$ is given by

$$\xi_{\tilde{\mathbf{A}} \oplus^\diamond_{DT} \tilde{\mathbf{B}}}(\mathbf{z}) = \sup_{0<\alpha\leq 1} \alpha \cdot \chi_{M^+_\alpha}(\mathbf{z}). \tag{7.57}$$

Using the similar argument in the proof of Proposition 7.3.3, we can obtain the α-level sets $(\tilde{\mathbf{A}} \oplus^\diamond_{DT} \tilde{\mathbf{B}})_\alpha$ of $\tilde{\mathbf{A}} \oplus^\diamond_{DT} \tilde{\mathbf{B}}$ given by

$$\left(\tilde{\mathbf{A}} \oplus^\diamond_{DT} \tilde{\mathbf{B}}\right)_\alpha = M^+_\alpha = \left[\tilde{A}^L_{1\alpha} + \tilde{B}^L_{1\alpha}, \tilde{A}^U_{1\alpha} + \tilde{B}^U_{1\alpha}\right] \times \cdots \times \left[\tilde{A}^L_{n\alpha} + \tilde{B}^L_{n\alpha}, \tilde{A}^U_{n\alpha} + \tilde{B}^U_{n\alpha}\right]$$

for $\alpha \in [0,1]$.

Now, for $i = 1, \ldots, n$ and for $\alpha \in (0,1]$, we take

$$M^{(i+)}_\alpha = \tilde{A}^{(i)}_\alpha + \tilde{B}^{(i)}_\alpha = \left[\tilde{A}^L_{i\alpha}, \tilde{A}^U_{i\alpha}\right] + \left[\tilde{B}^L_{i\alpha}, \tilde{B}^U_{i\alpha}\right] = \left[\tilde{A}^L_{i\alpha} + \tilde{B}^L_{i\alpha}, \tilde{A}^U_{i\alpha} + \tilde{B}^U_{i\alpha}\right]. \tag{7.58}$$

Then, for $\alpha \in [0,1]$, from (7.56), we see that

$$M^+_\alpha = M^{(1+)}_\alpha \times \cdots \times M^{(n+)}_\alpha \subset \mathbb{R}^m.$$

Let $\tilde{A}^{(i)} \oplus_{DT}^{\diamond} \tilde{B}^{(i)}$ be obtained using the expression in the decomposition theorem based on the family $\{M_\alpha^{(i+)} : 0 < \alpha \leq 1\}$ that is defined in (7.58). For $\alpha \in [0,1]$, we can similarly obtain

$$\left(\tilde{A}^{(i)} \oplus_{DT}^{\diamond} \tilde{B}^{(i)}\right)_\alpha = M_\alpha^{(i+)} = \left[\tilde{A}_{i\alpha}^L + \tilde{B}_{i\alpha}^L, \tilde{A}_{i\alpha}^U + \tilde{B}_{i\alpha}^U\right] \text{ for } i = 1, \ldots, n,$$

which also implies

$$\left(\tilde{\mathbf{A}} \oplus_{DT}^{\diamond} \tilde{\mathbf{B}}\right)_\alpha = \left(\tilde{A}^{(1)} \oplus_{DT}^{\diamond} \tilde{B}^{(1)}\right)_\alpha \times \cdots \times \left(\tilde{A}^{(n)} \oplus_{DT}^{\diamond} \tilde{B}^{(n)}\right)_\alpha.$$

The above results are summarized below.

Theorem 7.4.3 *Let $\tilde{A}^{(1)}, \ldots, \tilde{A}^{(n)}$ and $\tilde{B}^{(1)}, \ldots, \tilde{B}^{(n)}$ be fuzzy intervals. Suppose that the family $\{M_\alpha^+ : \alpha \in (0,1]\}$ is given by $M_\alpha^+ = \tilde{\mathbf{A}}_\alpha + \tilde{\mathbf{B}}_\alpha$. Then, the following statements hold true.*

(i) *For each $\alpha \in [0,1]$, we have*

$$\left(\tilde{A}^{(i)} \oplus_{DT}^{\diamond} \tilde{B}^{(i)}\right)_\alpha = \left[\tilde{A}_{i\alpha}^L + \tilde{B}_{i\alpha}^L, \tilde{A}_{i\alpha}^U + \tilde{B}_{i\alpha}^U\right]$$

and

$$\left(\tilde{\mathbf{A}} \oplus_{DT}^{\diamond} \tilde{\mathbf{B}}\right)_\alpha = \left(\tilde{A}^{(1)} \oplus_{DT}^{\diamond} \tilde{B}^{(1)}\right)_\alpha \times \cdots \times \left(\tilde{A}^{(n)} \oplus_{DT}^{\diamond} \tilde{B}^{(n)}\right)_\alpha.$$

(ii) *Suppose that the function $\mathfrak{A} : [0,1]^{2n} \to [0,1]$ is given by*

$$\mathfrak{A}\left(\alpha_1, \ldots, \alpha_{2n}\right) = \min\left\{\alpha_1, \ldots, \alpha_{2n}\right\}.$$

Then, we have

$$\tilde{\mathbf{A}} \oplus_{DT}^{\diamond} \tilde{\mathbf{B}} = \tilde{\mathbf{A}} \oplus_{EP} \tilde{\mathbf{B}},$$

where $\tilde{\mathbf{A}} \oplus_{EP} \tilde{\mathbf{B}}$ is obtained from Theorem 7.4.1.

By referring to Subsection 6.2.2, the family $\{M_\alpha^+ : 0 < \alpha \leq 1\}$ is given by

$$M_\alpha^+ = \left(\bigcup_{\alpha \leq \beta \leq 1} M_\beta^{(1+)}\right) \times \cdots \times \left(\bigcup_{\alpha \leq \beta \leq 1} M_\beta^{(n+)}\right),$$

where $M_\beta^{(i+)}$ are bounded closed intervals given by

$$\begin{aligned} M_\beta^{(i+)} &= \left[\min\left\{\tilde{A}_{i\beta}^L + \tilde{B}_{i\beta}^L, \tilde{A}_{i\beta}^U + \tilde{B}_{i\beta}^U\right\}, \max\left\{\tilde{A}_{i\beta}^L + \tilde{B}_{i\beta}^L, \tilde{A}_{i\beta}^U + \tilde{B}_{i\beta}^U\right\}\right] \\ &= \left[A_{i\alpha}^L + B_{i\alpha}^L, A_{i\alpha}^U + B_{i\alpha}^U\right] \end{aligned}$$

for $i = 1, \ldots, n$. Based on the expression in the decomposition theorem, the membership function of $\tilde{\mathbf{A}} \oplus_{DT}^{\star} \tilde{\mathbf{B}}$ is given by

$$\xi_{\tilde{\mathbf{A}} \oplus_{DT}^{\star} \tilde{\mathbf{B}}}(\mathbf{z}) = \sup_{0 < \alpha \leq 1} \alpha \cdot \chi_{M_\alpha^+}(\mathbf{z}).$$

Then, we have the following results.

Theorem 7.4.4 *Let $\tilde{A}^{(1)}, \ldots, \tilde{A}^{(n)}$ and $\tilde{B}^{(1)}, \ldots, \tilde{B}^{(n)}$ be fuzzy intervals. Suppose that the family $\{M_\alpha : \alpha \in (0,1]\}$ is given by*

$$M_\alpha^+ = \left(\bigcup_{\alpha \leq \beta \leq 1} M_\beta^{(1+)}\right) \times \cdots \times \left(\bigcup_{\alpha \leq \beta \leq 1} M_\beta^{(n+)}\right), \tag{7.59}$$

where $M_\beta^{(i+)}$ are bounded closed intervals given by

$$M_\beta^{(i+)} = \left[a_{i\beta}^L + b_{i\beta}^L, a_{i\beta}^U + b_{i\beta}^U\right]$$

for $i = 1, \ldots, n$. Then

$$\tilde{\mathbf{A}} \oplus_{DT}^{\star} \tilde{\mathbf{B}} = \tilde{\mathbf{A}} \oplus_{DT}^{\diamond} \tilde{\mathbf{B}}.$$

Proof. Let $\zeta_i^L(\beta) = a_{i\beta}^L + b_{i\beta}^L$ and $\zeta_i^U(\beta) = a_{i\beta}^U + b_{i\beta}^U$. Then $M_\beta^{(i)} = [\zeta_i^L(\beta), \zeta_i^U(\beta)]$. Proposition 2.3.4 say that ζ_i^L is lower semi-continuous on $[0,1]$ and ζ_i^U is upper semi-continuous on $[0,1]$. Then, for $\alpha \in (0,1]$, we can obtain

$$\begin{aligned}
\bigcup_{\alpha \le \beta \le 1} M_\beta^{(i+)} &= \left[\min_{\alpha \le \beta \le 1} \zeta_i^L(\beta), \max_{\alpha \le \beta \le 1} \zeta_i^U(\beta)\right] \\
&= \left[\min_{\alpha \le \beta \le 1} \left(a_{i\beta}^L + b_{i\beta}^L\right), \max_{\alpha \le \beta \le 1} \left(a_{i\beta}^U + b_{i\beta}^U\right)\right] \\
&= \left[a_{i\alpha}^L + b_{i\alpha}^L, a_{i\alpha}^U + b_{i\alpha}^U\right] = M_\alpha^{(i)}.
\end{aligned}$$

Therefore, by referring to (7.59), we have

$$M_\alpha^+ = \left[\tilde{A}_{1\alpha}^L + \tilde{B}_{1\alpha}^L, \tilde{A}_{1\alpha}^U + \tilde{B}_{1\alpha}^U\right] \times \cdots \times \left[\tilde{A}_{n\alpha}^L + \tilde{B}_{n\alpha}^L, \tilde{A}_{n\alpha}^U + \tilde{B}_{n\alpha}^U\right],$$

which is the same as (7.56). Therefore, we obtain

$$\tilde{\mathbf{A}} \oplus_{DT}^{\star} \tilde{\mathbf{B}} = \tilde{\mathbf{A}} \oplus_{DT}^{\diamond} \tilde{\mathbf{B}}.$$

This completes the proof. ■

By referring to Subsection 6.2.3, the family $\{M_\alpha^+ : 0 < \alpha \le 1\}$ is given by

$$M_\alpha^+ = M_\alpha^{(1+)} \times \cdots \times M_\alpha^{(n+)},$$

where $M_\alpha^{(i+)}$ are bounded closed intervals given by

$$\begin{aligned}
M_\alpha^{(i+)} &= \left[\min\left\{\tilde{A}_{i\alpha}^L + \tilde{B}_{i\alpha}^L, \tilde{A}_{i\alpha}^U + \tilde{B}_{i\alpha}^U\right\}, \max\left\{\tilde{A}_{i\alpha}^L + \tilde{B}_{i\alpha}^L, \tilde{A}_{i\alpha}^U + \tilde{B}_{i\alpha}^U\right\}\right] \\
&= \left[A_{i\alpha}^L + B_{i\alpha}^L, A_{i\alpha}^U + B_{i\alpha}^U\right]
\end{aligned}$$

for $i = 1, \ldots, n$. Based on the expression in the decomposition theorem, the membership functions of $\tilde{\mathbf{A}} \oplus_{DT}^{\dagger} \tilde{\mathbf{B}}$ is given by

$$\xi_{\tilde{\mathbf{A}} \oplus_{DT}^{\dagger} \tilde{\mathbf{B}}}(\mathbf{z}) = \sup_{0 < \alpha \le 1} \alpha \cdot \chi_{M_\alpha^+}(\mathbf{z})$$

for any $\mathbf{z} \in \mathbb{R}^m$. However, in this case, we need to consider canonical fuzzy intervals rather than fuzzy intervals.

Theorem 7.4.5 *Let $\tilde{A}^{(1)}, \ldots, \tilde{A}^{(n)}$ and $\tilde{B}^{(1)}, \ldots, \tilde{B}^{(n)}$ be canonical fuzzy intervals. Suppose that the family $\{M_\alpha : \alpha \in (0,1]\}$ is given by*

$$M_\alpha^+ = M_\alpha^{(1+)} \times \cdots \times M_\alpha^{(n+)},$$

where $M_\alpha^{(i+)}$ are bounded closed intervals given by

$$M_\alpha^{(i+)} = \left[A_{i\alpha}^L + B_{i\alpha}^L, A_{i\alpha}^U + B_{i\alpha}^U\right]$$

for $i = 1, \ldots, n$. *Then, we have*

$$\tilde{\mathbf{A}} \oplus^{\dagger}_{DT} \tilde{\mathbf{B}} = \tilde{\mathbf{A}} \oplus^{\star}_{DT} \tilde{\mathbf{B}} = \tilde{\mathbf{A}} \oplus^{\diamond}_{DT} \tilde{\mathbf{B}}.$$

Proof. For each $i = 1, \ldots, n$, it is clear to see that

$$\bigcup_{\alpha \leq \beta \leq 1} M^{(i+)}_{\beta} = \left[A^L_{i\alpha} + B^L_{i\alpha}, A^U_{i\alpha} + B^U_{i\alpha}\right] = M^{(i+)}_{\alpha}.$$

For each $\alpha \in (0, 1]$, using the similar argument of Theorem 7.3.8, we can obtain

$$\begin{aligned}
\left(\tilde{\mathbf{A}} \oplus^{\dagger}_{DT} \tilde{\mathbf{B}}\right)_{\alpha} &= \bigcup_{\alpha \leq \beta \leq 1} M^{+}_{\beta} = \bigcup_{\alpha \leq \beta \leq 1} \left(M^{(1+)}_{\beta} \times \cdots \times M^{(n+)}_{\beta}\right) \\
&\bigcup_{\alpha \leq \beta \leq 1} \left(\left[A^L_{1\beta} + B^L_{1\beta}, A^U_{1\beta} + B^U_{1\beta}\right] \times \cdots \times \left[A^L_{n\beta} + B^L_{n\beta}, A^U_{n\beta} + B^U_{n\beta}\right]\right) \\
&= \left[A^L_{1\alpha} + B^L_{1\alpha}, A^U_{1\alpha} + B^U_{1\alpha}\right] \times \cdots \times \left[A^L_{n\alpha} + B^L_{n\alpha}, A^U_{n\alpha} + B^U_{n\alpha}\right] \\
&= \left(\tilde{\mathbf{A}} \oplus^{\star}_{DT} \tilde{\mathbf{B}}\right)_{\alpha} = \left(\tilde{\mathbf{A}} \oplus^{\diamond}_{DT} \tilde{\mathbf{B}}\right)_{\alpha} = \left(\tilde{\mathbf{A}} \oplus_{EP} \tilde{\mathbf{B}}\right)_{\alpha}.
\end{aligned}$$

This completes the proof. ■

Example 7.4.6 Using Theorem 7.4.5 and Example 7.4.2, we see that

$$\left(\tilde{\mathbf{A}} \oplus^{\diamond}_{DT} \tilde{\mathbf{B}}\right)_{\alpha} = \left(\tilde{\mathbf{A}} \oplus^{\star}_{DT} \tilde{\mathbf{B}}\right)_{\alpha} = \left(\tilde{\mathbf{A}} \oplus^{\dagger}_{DT} \tilde{\mathbf{B}}\right)_{\alpha} = [5 + 2\alpha, 11 - 2\alpha] \times [5 + 2\alpha, 11 - 2\alpha]$$

for $\alpha \in [0, 1]$.

7.5 Arithmetic Operations Using Compatibility and Associativity

Let $\tilde{A}^{(1)}, \ldots, \tilde{A}^{(n)}$ be fuzzy sets in $\mathbb{R}$. We consider the operations $\odot_i \in \{\oplus, \ominus, \otimes\}$ for $i = 1, \ldots, n-1$. Given a function $\mathfrak{A}_n : [0,1]^n \to [0,1]$ defined on the product set $[0,1]^n$, the membership function of $\tilde{A} = \tilde{A}^{(1)} \odot_1 \cdots \odot_{n-1} \tilde{A}^{(n)}$ is defined by

$$\xi_{\tilde{A}}(z) = \xi_{\tilde{A}^{(1)} \odot_1 \cdots \odot_{n-1} \tilde{A}^{(n)}}(z) = \sup_{\{(x_1,\ldots,x_n): z = x_1 \circ_1 \cdots \circ_{n-1} x_n\}} \mathfrak{A}_n\left(\xi_{\tilde{A}^{(1)}}(x_1), \ldots, \xi_{\tilde{A}^{(n)}}(x_n)\right), \tag{7.60}$$

where the way of calculation $\tilde{A}^{(1)} \odot_1 \cdots \odot_{n-1} \tilde{A}^{(n)}$ for $\odot_i \in \{\oplus, \ominus, \otimes\}$ and $i = 1, \ldots, n-1$ is based on the way of calculation of $x_1 \circ_1 \cdots \circ_{n-1} x_n$ for $\circ_i \in \{+, -, *\}$ and $i = 1, \ldots, n-1$. In order to emphasize that the expressions $\tilde{A}^{(1)} \odot_1 \cdots \odot_{n-1} \tilde{A}^{(n)}$ depends on the function $\mathfrak{A}_n$, we sometimes write it as

$$\tilde{A}^{(1)} \odot_1 \cdots \odot_{n-1} \tilde{A}^{(n)}(\mathfrak{A}_n).$$

The case of $\circ = \div$ will not be considered, since the division of x/y should be avoided when $y = 0$, which will make the proofs more complicated. However, the case of $\circ = \div$ can be similarly discussed according to the arguments of the case of $\circ = *$ by carefully avoiding division by zero. Therefore, the case of $\circ = \div$ is left for the readers.

When the function is given by

$$\mathfrak{A}_n(\alpha_1, \ldots, \alpha_n) = \min\left\{\alpha_1, \ldots, \alpha_n\right\},$$

by referring to (7.60), we use the notation

$$\tilde{A}^{(1)}\boxdot_1 \cdots \boxdot_{n-1}\tilde{A}^{(n)} \tag{7.61}$$

and its membership function is given by

$$\xi_{\tilde{A}^{(1)}\boxdot_1\cdots\boxdot_{n-1}\tilde{A}^{(n)}}(z) = \sup_{\{(x_1,\ldots,x_n):z=x_1\circ_1\cdots\circ_{n-1}x_n\}} \min\left\{\xi_{\tilde{A}^{(1)}}(x_1),\ldots,\xi_{\tilde{A}^{(n)}}(x_n)\right\}, \tag{7.62}$$

where $\boxdot_i \in \{\boxplus, \boxminus, \boxtimes\}$ for $i = 1, \ldots, n-1$.

For the purpose of considering the priority of calculations, we can insert parentheses in the expression $\tilde{A}^{(1)}\odot_1 \cdots \odot_{n-1}\tilde{A}^{(n)}$. The membership functions regarding the expression including parentheses are defined to be the same form of (7.60) in which the parentheses in the expression $x_1 \circ_1 \cdots \circ_{n-1} x_n$ correspond to those of $\tilde{A}^{(1)}\odot_1 \cdots \odot_{n-1}\tilde{A}^{(n)}$. The following example presents the form of inserting parentheses.

Example 7.5.1 Let $\tilde{A}^{(1)}, \ldots, \tilde{A}^{(7)}$ be fuzzy sets in $\mathbb{R}$. The membership functions of

$$\tilde{B} \equiv \tilde{A}^{(1)} \otimes \left(\tilde{A}^{(2)} \oplus \tilde{A}^{(3)}\right) \ominus \left(\tilde{A}^{(4)} \otimes \left(\tilde{A}^{(5)} \oplus \tilde{A}^{(6)} \ominus \tilde{A}^{(7)}\right)\right)$$

and

$$\tilde{C} \equiv \tilde{A}^{(1)} \otimes \tilde{A}^{(2)} \oplus \tilde{A}^{(3)} \ominus \tilde{A}^{(4)} \otimes \tilde{A}^{(5)} \oplus \tilde{A}^{(6)} \ominus \tilde{A}^{(7)}$$

are given by

$$\xi_{\tilde{B}}(z) = \sup_{\{(x_1,\ldots,x_7):z=x_1*(x_2+x_3)-(x_4*(x_5+x_6-x_7))\}} \mathfrak{A}_7\left(\xi_{\tilde{A}^{(1)}}(x_1),\ldots,\xi_{\tilde{A}^{(7)}}(x_7)\right)$$

and

$$\xi_{\tilde{C}}(z) = \sup_{\{(x_1,\ldots,x_7):z=x_1*x_2+x_3-x_4*x_5+x_6-x_7\}} \mathfrak{A}_7\left(\xi_{\tilde{A}^{(1)}}(x_1),\ldots,\xi_{\tilde{A}^{(7)}}(x_7)\right),$$

respectively. We see that $\tilde{B} \neq \tilde{C}$. Since

$$x_1 * x_2 + x_3 - x_4 * x_5 + x_6 - x_7 = \left(x_1 * x_2\right) + x_3 - \left(x_4 * x_5\right) + x_6 - x_7,$$

it follows that

$$\tilde{C} = \left(\tilde{A}^{(1)} \otimes \tilde{A}^{(2)}\right) \oplus \tilde{A}^{(3)} \ominus \left(\tilde{A}^{(4)} \otimes \tilde{A}^{(5)}\right) \oplus \tilde{A}^{(6)} \ominus \tilde{A}^{(7)}.$$

7.5.1 Compatibility

Now, we are going to study the compatibility with the arithmetic operations of α-level sets. Let $A_1, \ldots, A_n$ be the subsets of $\mathbb{R}$. We write

$$A_1 \circ_1 \cdots \circ_{n-1} A_n = \left\{x_1 \circ_1 \cdots \circ_{n-1} x_n : x_i \in A_i \text{ for } i = 1, \ldots, n\right\},$$

where the arithmetic operations $\circ_i \in \{+, -, *\}$ for $i = 1, \ldots, n-1$. Let $\tilde{A} = \tilde{A}^{(1)}\odot_1 \cdots \odot_{n-1}\tilde{A}^{(n)}$ be defined in (7.60). We propose the concept of compatibility below.

Definition 7.5.2 Let $\tilde{A}^{(1)}, \ldots, \tilde{A}^{(n)}$ be fuzzy sets in $\mathbb{R}$. We consider the arithmetic operations $\odot_i \in \{\oplus, \ominus, \otimes\}$ that correspond to the arithmetic operations $\circ_i \in \{+, -, *\}$ for $i = 1, \ldots, n-1$.

- We say that the function $\mathfrak{A}_n : [0,1]^n \to [0,1]$ is **compatible with arithmetic operations of α-level sets** when
$$\left(\tilde{A}^{(1)} \odot_1 \cdots \odot_{n-1} \tilde{A}^{(n)}\right)_\alpha = \tilde{A}^{(1)}_\alpha \circ_1 \cdots \circ_{n-1} \tilde{A}^{(n)}_\alpha \text{ for all } \alpha \in (0,1].$$
- We say that the function $\mathfrak{A}_n : [0,1]^n \to [0,1]$ is **strongly compatible with arithmetic operations of α-level sets** when
$$\left(\tilde{A}^{(1)} \odot_1 \cdots \odot_{n-1} \tilde{A}^{(n)}\right)_\alpha = \tilde{A}^{(1)}_\alpha \circ_1 \cdots \circ_{n-1} \tilde{A}^{(n)}_\alpha \text{ for all } \alpha \in [0,1].$$

Note that the strong compatibility includes the case of $\alpha = 0$. We are going to present the sufficient conditions to guarantee the compatibility with the arithmetic operations of α-level sets.

Theorem 7.5.3 *Let $\tilde{A}^{(1)}, \ldots, \tilde{A}^{(n)}$ be fuzzy sets in $\mathbb{R}$. Suppose that the arithmetic operations $\odot_i \in \{\oplus, \ominus, \otimes\}$ correspond to the arithmetic operations $\circ_i \in \{+, -, *\}$ for $i = 1, \ldots, n-1$. Then, we have the following properties.*

(i) Suppose that, for any $\alpha \in (0,1]$, the function $\mathfrak{A}_n$ satisfies the following condition:
$$\alpha_i \geq \alpha \text{ for all } i = 1, \ldots, n \text{ imply } \mathfrak{A}_n(\alpha_1, \ldots, \alpha_n) \geq \alpha. \tag{7.63}$$
Then, we have the following inclusion
$$\tilde{A}^{(1)}_\alpha \circ_1 \cdots \circ_{n-1} \tilde{A}^{(n)}_\alpha \subseteq \left(\tilde{A}^{(1)} \odot_1 \cdots \odot_{n-1} \tilde{A}^{(n)}\right)_\alpha$$
for each $\alpha \in [0,1]$.

(ii) Suppose that the membership functions of $\tilde{A}^{(i)}$ are upper semi-continuous for all $i = 1, \ldots, n$, and that the function $\mathfrak{A}_n$ satisfies the following conditions: for any $\alpha \in (0,1]$,
$$\mathfrak{A}_n(\alpha_1, \ldots, \alpha_n) \geq \alpha \text{ if and only if } \alpha_i \geq \alpha \text{ for all } i = 1, \ldots, n. \tag{7.64}$$
Then
$$\mathfrak{A}_n\left(\alpha_1, \ldots, \alpha_n\right) = \min\left\{\alpha_1, \ldots, \alpha_n\right\} \tag{7.65}$$
and
$$\left(\tilde{A}^{(1)} \odot_1 \cdots \odot_{n-1} \tilde{A}^{(n)}\right)_\alpha = \tilde{A}^{(1)}_\alpha \circ_1 \cdots \circ_{n-1} \tilde{A}^{(n)}_\alpha \tag{7.66}$$
for each $\alpha \in (0,1]$. We further assume that the supports $\tilde{A}^{(i)}_{0+}$ are bounded for all $i = 1, \ldots, n$. Then, regarding the 0-level sets, we have
$$\left(\tilde{A}^{(1)} \odot_1 \cdots \odot_{n-1} \tilde{A}^{(n)}\right)_0 = \tilde{A}^{(1)}_0 \circ_1 \cdots \circ_{n-1} \tilde{A}^{(n)}_0. \tag{7.67}$$

Proof. To prove part (i), for $\alpha \in (0,1]$ and $z_\alpha \in \tilde{A}^{(1)}_\alpha \circ_1 \cdots \circ_{n-1} \tilde{A}^{(n)}_\alpha$, there exist $x^{(i)}_\alpha \in \tilde{A}^{(i)}_\alpha$ for $i = 1, \ldots, n$ satisfying $z_\alpha = x^{(1)}_\alpha \circ_1 \cdots \circ_{n-1} x^{(n)}_\alpha$, where $\xi_{\tilde{A}^{(i)}}(x^{(i)}_\alpha) \geq \alpha$ for $i = 1, \ldots, n$. Then, we have
$$\begin{aligned}
\xi_{\tilde{A}^{(1)} \odot_1 \cdots \odot_{n-1} \tilde{A}^{(n)}}(z_\alpha) &= \sup_{\{(a_1,\ldots,a_n): z_\alpha = a_1 \circ_1 \cdots \circ_{n-1} a_n\}} \mathfrak{A}_n\left(\xi_{\tilde{A}^{(1)}}(a_1), \ldots, \xi_{\tilde{A}^{(n)}}(a_n)\right) \\
&\geq \mathfrak{A}_n\left(\xi_{\tilde{A}^{(1)}}(x^{(1)}_\alpha), \ldots, \xi_{\tilde{A}^{(n)}}(x^{(n)}_\alpha)\right) \\
&\geq \alpha \text{ (using the assumption (7.63) of } \mathfrak{A}_n\text{),}
\end{aligned}$$

which says that $z_\alpha \in (\tilde{A}^{(1)} \odot_1 \cdots \odot_{n-1} \tilde{A}^{(n)})_\alpha$. This shows the inclusion

$$\tilde{A}^{(1)}_\alpha \circ_1 \cdots \circ_{n-1} \tilde{A}^{(n)}_\alpha \subseteq \left(\tilde{A}^{(1)} \odot_1 \cdots \odot_{n-1} \tilde{A}^{(n)}\right)_\alpha$$

for each $\alpha \in (0, 1]$.

Now, for $\alpha = 0$ and $z_0 \in \tilde{A}^{(1)}_0 \circ_1 \cdots \circ_{n-1} \tilde{A}^{(n)}_0$, there also exist $x^{(i)}_0 \in \tilde{A}^{(i)}_0$ for $i = 1, \ldots, n$ satisfying $z_0 = x^{(1)}_0 \circ_1 \cdots \circ_{n-1} x^{(n)}_0$. Since

$$\tilde{A}^{(i)}_0 = \mathrm{cl}\left(\{x \in \mathbb{R} : \xi_{\tilde{A}^{(i)}}(x) > 0\}\right),$$

according to the concept of closure, there exists a sequence $\{x^{(i)}_m\}_{m=1}^\infty$ in the set $\{x \in \mathbb{R} : \xi_{\tilde{A}^{(i)}}(x) > 0\}$ satisfying $x^{(i)}_m \to x^{(i)}_0$ as $m \to \infty$ for $i = 1, \ldots, n$. Let $z_m = x^{(1)}_m \circ_1 \cdots \circ_{n-1} x^{(n)}_m$. Then, we see that

$$z_m \to x^{(1)}_0 \circ_1 \cdots \circ_{n-1} x^{(n)}_0 = z_0 \text{ as } m \to \infty, \tag{7.68}$$

since the binary operations $\circ_i \in \{+, -, *\}$ for $i = 1, \ldots, n$ can be regarded as continuous functions. For any $\alpha_i \in (0, 1]$, let $\alpha = \min\{\alpha_1, \ldots, \alpha_n\}$. The assumption (7.63) of $\mathfrak{A}_n$ says that

$$\mathfrak{A}_n(\alpha_1, \ldots, \alpha_n) \geq \alpha > 0.$$

Therefore, the following statement holds true:

$$0 < \alpha_i \in [0, 1] \text{ for all } i = 1, \ldots, n \text{ imply } \mathfrak{A}_n(\alpha_1, \ldots, \alpha_n) > 0. \tag{7.69}$$

Now, we have

$$\begin{aligned} \xi_{\tilde{A}^{(1)} \odot_1 \cdots \odot_{n-1} \tilde{A}^{(n)}}(z_m) &= \sup_{\{(a_1, \ldots, a_n) : z_m = a_1 \circ_1 \cdots \circ_{n-1} a_n\}} \mathfrak{A}_n\left(\xi_{\tilde{A}^{(1)}}(a_1), \ldots, \xi_{\tilde{A}^{(n)}}(a_n)\right) \\ &\geq \mathfrak{A}_n\left(\xi_{\tilde{A}^{(1)}}(x^{(1)}_m), \ldots, \xi_{\tilde{A}^{(n)}}(x^{(n)}_m)\right) > 0 \text{ (using (7.69))}, \end{aligned}$$

which says that

$$z_m \in \{z \in \mathbb{R} : \xi_{\tilde{A}^{(1)} \odot_1 \cdots \odot_{n-1} \tilde{A}^{(n)}}(z) > 0\}.$$

Since $z_m \to z_0$ as $m \to \infty$ from (7.68), it means that

$$z_0 \in \mathrm{cl}\left(\{z \in \mathbb{R} : \xi_{\tilde{A}^{(1)} \odot_1 \cdots \odot_{n-1} \tilde{A}^{(n)}}(z) > 0\}\right) = \left(\tilde{A}^{(1)} \odot_1 \cdots \odot_{n-1} \tilde{A}^{(n)}\right)_0.$$

Therefore, we obtain the inclusion

$$\tilde{A}^{(1)}_0 \circ_1 \cdots \circ_{n-1} \tilde{A}^{(n)}_0 \subseteq \left(\tilde{A}^{(1)} \odot_1 \cdots \odot_{n-1} \tilde{A}^{(n)}\right)_0.$$

This proves the desired inclusion.

To prove part (ii), the equality (7.65) can be realized from Proposition 1.4.6. In order to prove the other direction of inclusion, we further assume that the membership functions of $\tilde{A}^{(i)}$ are upper semi-continuous for all $i = 1, \ldots, n$; that is, the nonempty α-level sets $\tilde{A}^{(i)}_\alpha$ are closed subsets of $\mathbb{R}$ for all $\alpha \in [0, 1]$ and $i = 1, \ldots, n$. Given any $\alpha \in (0, 1]$ and $z_\alpha \in (\tilde{A}^{(1)} \odot_1 \cdots \odot_{n-1} \tilde{A}^{(n)})_\alpha$, we have

$$\sup_{\{(a_1, \ldots, a_n) : z_\alpha = a_1 \circ_1 \cdots \circ_{n-1} a_n\}} \mathfrak{A}_n\left(\xi_{\tilde{A}^{(1)}}(a_1), \ldots, \xi_{\tilde{A}^{(n)}}(a_n)\right) = \xi_{\tilde{A}^{(1)} \odot_1 \cdots \odot_{n-1} \tilde{A}^{(n)}}(z_\alpha) \geq \alpha. \tag{7.70}$$

Since z_α is finite, it is clear to see that

$$F \equiv \left\{(a_1, \ldots, a_n) : z_\alpha = a_1 \circ_1 \cdots \circ_{n-1} a_n\right\}$$

is a bounded subset of $\mathbb{R}^m$. We also see that the function

$$\eta(a_1, \dots, a_n) = a_1 \circ_1 \cdots \circ_{n-1} a_n$$

is continuous on $\mathbb{R}^m$. Since the singleton $\{z_\alpha\}$ is a closed subset of $\mathbb{R}$, it follows that the inverse image $F = \eta^{-1}(\{z_\alpha\})$ of $\{z_\alpha\}$ is also a closed subset of $\mathbb{R}^m$ by the continuity of η. This says that F is a bounded and closed subset of $\mathbb{R}^m$. Now, we want to show that the function

$$f\left(a_1, \dots, a_n\right) = \mathfrak{A}_n\left(\xi_{\tilde{A}^{(1)}}(a_1), \dots, \xi_{\tilde{A}^{(n)}}(a_n)\right)$$

is upper semi-continuous, i.e. we want to show that

$$\left\{\left(a_1, \dots, a_n\right) : f\left(a_1, \dots, a_n\right) \geq \alpha\right\}$$

is a closed subset of $\mathbb{R}^m$ for any $\alpha \in \mathbb{R}$.

- If $\alpha \leq 0$, then
$$\left\{\left(a_1, \dots, a_n\right) : f\left(a_1, \dots, a_n\right) \geq \alpha\right\} = \mathbb{R}^m$$
is a closed subset of $\mathbb{R}^m$.
- If $\alpha > 1$, then
$$\left\{\left(a_1, \dots, a_n\right) : f\left(a_1, \dots, a_n\right) \geq \alpha\right\} = \emptyset$$
is also a closed subset of $\mathbb{R}^m$.
- If $\alpha \in (0, 1]$, then
$$\begin{aligned}
&\left\{\left(a_1, \dots, a_n\right) : f\left(a_1, \dots, a_n\right) \geq \alpha\right\}\\
&= \left\{\left(a_1, \dots, a_n\right) : \mathfrak{A}_n\left(\xi_{\tilde{A}^{(1)}}(a_1), \dots, \xi_{\tilde{A}^{(n)}}(a_n)\right) \geq \alpha\right\}\\
&= \left\{\left(a_1, \dots, a_n\right) : \xi_{\tilde{A}^{(i)}}(a_i) \geq \alpha \text{ for all } i = 1, \dots, n\right\} \text{ (using (7.64))}\\
&= \left\{\left(a_1, \dots, a_n\right) : a_i \in \tilde{A}_\alpha^{(i)} \text{ for all } i = 1, \dots, n\right\} = \tilde{A}_\alpha^{(1)} \times \cdots \times \tilde{A}_\alpha^{(n)},
\end{aligned}$$
which is a closed subset of $\mathbb{R}^m$, since $\tilde{A}_\alpha^{(i)}$ are closed subsets of $\mathbb{R}$ for all $i = 1, \dots, n$.

Therefore, the function $f(a_1, \dots, a_n)$ is indeed upper semi-continuous. Proposition 1.4.4 says that the function f assumes its maximum on F; that is, from (7.70), we have

$$\begin{aligned}
\max_{(a_1,\dots,a_n)\in F} f(a_1, \dots, a_n) &= \max_{\{(a_1,\dots,a_n): z_\alpha = a_1 \circ_1 \cdots \circ_{n-1} a_n\}} f(a_1, \dots, a_n)\\
&= \sup_{\{(a_1,\dots,a_n): z_\alpha = a_1 \circ_1 \cdots \circ_{n-1} a_n\}} f(a_1, \dots, a_n) \geq \alpha. \qquad (7.71)
\end{aligned}$$

In other words, there exists $(a_1^*, \dots, a_n^*) \in F$ satisfying $z_\alpha = a_1^* \circ_1 \cdots \circ_{n-1} a_n^*$ and

$$\mathfrak{A}_n\left(\xi_{\tilde{A}^{(1)}}(a_1^*), \dots, \xi_{\tilde{A}^{(n)}}(a_n^*)\right) = f\left(a_1^*, \dots, a_n^*\right) = \max_{(a_1,\dots,a_n)\in F} f(a_1, \dots, a_n) \geq \alpha.$$

Using assumption (7.64), we obtain $\xi_{\tilde{A}^{(i)}}(a_i^*) \geq \alpha$, i.e. $a_i^* \in \tilde{A}_\alpha^{(i)}$ for all $i = 1, \dots, n$, which says that $z_\alpha \in \tilde{A}_\alpha^{(1)} \circ_1 \cdots \circ_{n-1} \tilde{A}_\alpha^{(n)}$, i.e.

$$\left(\tilde{A}^{(1)} \odot_1 \cdots \odot_{n-1} \tilde{A}^{(n)}\right)_\alpha \subseteq \tilde{A}_\alpha^{(1)} \circ_1 \cdots \circ_{n-1} \tilde{A}_\alpha^{(n)}.$$

for all $\alpha \in (0, 1]$. Using part (i), we obtain the desired equality (7.66).

For $\alpha = 0$, we further assume that the supports $\tilde{A}^{(i)}_{0+}$ are bounded for all $i = 1, \ldots, n$. Suppose that $\mathfrak{A}_n(\alpha_1 \ \ldots, \alpha_n) > 0$ for $\alpha_i \in [0, 1]$ and $i = 1, \ldots, n$. Since $[0, 1]$ is an interval beginning from 0, using the denseness of $\mathbb{R}$, there exists $\alpha \in (0, 1]$ satisfying

$$\mathfrak{A}_n(\alpha_1 \ \ldots, \alpha_n) \geq \alpha > 0.$$

From the assumption (7.64), the following statement holds true:

$$\mathfrak{A}_n(\alpha_1 \ \ldots, \alpha_n) > 0 \text{ for } \alpha_i \in [0, 1] \text{ and } i = 1, \ldots, n \text{ imply } \alpha_i > 0 \text{ for all } i = 1, \ldots, n. \tag{7.72}$$

Now, regarding the 0-level set, we have

$$\begin{aligned} z_0 \in \left(\tilde{A}^{(1)} \odot_1 \cdots \odot_{n-1} \tilde{A}^{(n)}\right)_0 &= \mathrm{cl}\left(\left(\tilde{A}^{(1)} \odot_1 \cdots \odot_{n-1} \tilde{A}^{(n)}\right)_{0+}\right) \\ &= \mathrm{cl}\left(\left\{z \in \mathbb{R} : \xi_{\tilde{A}^{(1)} \odot_1 \cdots \odot_{n-1} \tilde{A}^{(n)}}(z) > 0\right\}\right). \end{aligned}$$

Therefore, there exists a sequence $\{z_m\}_{m=1}^{\infty}$ in the set $\{z \in \mathbb{R} : \xi_{\tilde{A}^{(1)} \odot_1 \cdots \odot_{n-1} \tilde{A}^{(n)}}(z) > 0\}$ satisfying $z_m \to z_0$ as $m \to \infty$. Using the same arguments above by referring to (7.71), we can obtain

$$\begin{aligned} 0 < \xi_{\tilde{A}^{(1)} \odot_1 \cdots \odot_{n-1} \tilde{A}^{(n)}}(z_m) &= \sup_{\{(a_1, \ldots, a_n) : z_m = a_1 \circ_1 \cdots \circ_{n-1} a_n\}} \mathfrak{A}_n\left(\xi_{\tilde{A}^{(1)}}(a_1), \ldots, \xi_{\tilde{A}^{(n)}}(a_n)\right) \\ &= \max_{\{(a_1, \ldots, a_n) : z_m = a_1 \circ_1 \cdots \circ_{n-1} a_n\}} \mathfrak{A}_n\left(\xi_{\tilde{A}^{(1)}}(a_1), \ldots, \xi_{\tilde{A}^{(n)}}(a_n)\right). \end{aligned}$$

Therefore, there exist $a_{1m}, \ldots, a_{nm}$ satisfying $z_m = a_{1m} \circ_1 \cdots \circ_{n-1} a_{nm}$ and

$$\begin{aligned} &\mathfrak{A}_n\left(\xi_{\tilde{A}^{(1)}}(a_{1m}), \ldots, \xi_{\tilde{A}^{(n)}}(a_{nm})\right) \\ &= \max_{\{(a_1, \ldots, a_n) : z_m = a_1 \circ_1 \cdots \circ_{n-1} a_n\}} \mathfrak{A}_n\left(\xi_{\tilde{A}^{(1)}}(a_1), \ldots, \xi_{\tilde{A}^{(n)}}(a_n)\right) > 0. \end{aligned}$$

Using assumption (7.72), we have $\xi_{\tilde{A}^{(i)}}(a_{im}) > 0$ for all $i = 1, \ldots, n$. This shows that the sequence $\{a_{im}\}_{m=1}^{\infty}$ is in the support $\tilde{A}^{(i)}_{0+}$ for all $i = 1, \ldots, n$. Since each $\tilde{A}^{(i)}_{0+}$ is bounded for $i = 1, \ldots, n$, i.e. $\{a_{im}\}_{m=1}^{\infty}$ is a bounded sequence, there exists a convergent subsequence $\{a_{im_k}\}_{k=1}^{\infty}$ of $\{a_{im}\}_{m=1}^{\infty}$. In other words, we have $a_{im_k} \to a_{i0}$ as $k \to \infty$ for all $i = 1, \ldots, n$, where $a_{i0} \in \mathrm{cl}(\tilde{A}^{(i)}_{0+}) = \tilde{A}^{(i)}_0$ for all $i = 1, \ldots, n$. Let $z_{m_k} = a_{1m_k} \circ_1 \cdots \circ_{n-1} a_{nm_k}$. Then $\{z_{m_k}\}_{k=1}^{\infty}$ is a subsequence of $\{z_m\}_{n=1}^{\infty}$, i.e. $z_{m_k} \to z_0$ as $k \to \infty$. Since

$$\begin{aligned} z_0 &= \lim_{k \to \infty} z_{m_k} = \lim_{k \to \infty} \left(a_{1m_k} \circ_1 \cdots \circ_{n-1} a_{nm_k}\right) \\ &= \left(\lim_{k \to \infty} a_{1m_k}\right) \circ_1 \cdots \circ_{n-1} \left(\lim_{k \to \infty} a_{nm_k}\right) = a_{10} \circ_1 \cdots \circ_{n-1} a_{n0}, \end{aligned}$$

which shows that

$$z_0 \in \tilde{A}^{(1)}_0 \circ_1 \cdots \circ_{n-1} \tilde{A}^{(n)}_0.$$

Therefore, we obtain the inclusion

$$\left(\tilde{A}^{(1)} \odot_1 \cdots \odot_{n-1} \tilde{A}^{(n)}\right)_0 \subseteq \tilde{A}^{(1)}_0 \circ_1 \cdots \circ_{n-1} \tilde{A}^{(n)}_0.$$

Using part (i), we obtain the desired equality (7.67). This completes the proof. ■

By referring to (7.61), we have the following interesting results.

Theorem 7.5.4 *Let $\tilde{A}^{(1)}, \ldots, \tilde{A}^{(n)}$ be fuzzy sets in $\mathbb{R}$. Consider the arithmetic operations $\odot_i \in \{\oplus, \ominus, \otimes\}$ correspond to the arithmetic operations $\circ_i \in \{+, -, *\}$ for*

$i = 1, \ldots, n-1$. *Suppose that the function* $\mathfrak{A}_n$ *is taken to be the minimum function, given by*

$$\mathfrak{A}_n\left(\alpha_1, \ldots, \alpha_n\right) = \min\left\{\alpha_1, \ldots, \alpha_n\right\}.$$

Then, we have the following properties.

(i) *Suppose that the membership functions of* $\tilde{A}^{(i)}$ *are upper semi-continuous for all* $i = 1, \ldots, n$. *Then, the function* $\mathfrak{A}_n$ *is compatible with arithmetic operations of* α*-level sets; that is,*

$$\left(\tilde{A}^{(1)} \boxdot_1 \cdots \boxdot_{n-1} \tilde{A}^{(n)}\right)_\alpha = \tilde{A}^{(1)}_\alpha \circ_1 \cdots \circ_{n-1} \tilde{A}^{(n)}_\alpha \tag{7.73}$$

for all $\alpha \in (0, 1]$.

(ii) *Suppose that the membership functions of* $\tilde{A}^{(i)}$ *are upper semi-continuous, and that the supports* $\tilde{A}^{(i)}_{0+}$ *are bounded for all* $i = 1, \ldots, n$. *Then, the function* $\mathfrak{A}_n$ *is strongly compatible with arithmetic operations of* α*-level sets; that is, the equality holds true for all* $\alpha \in [0, 1]$.

Proof. To prove part (i), the equality (7.73) follows from part (ii) of Theorem 7.5.3 immediately. Part (ii) follows from part (ii) of Theorem 7.5.3 and part (i) of this theorem. This completes the proof. ■

7.5.2 Associativity

In what follows, we are going to study the associativity regarding the arithmetic operations of fuzzy sets in $\mathbb{R}$. Let $\tilde{A}$ and $\tilde{B}$ be two fuzzy sets in $\mathbb{R}$. It is clear to see that if the function $\mathfrak{A}_2 : [0, 1]^2 \to [0, 1]$ is commutative in the sense of

$$\mathfrak{A}_2(\alpha_1, \alpha_2) = \mathfrak{A}_2(\alpha_2, \alpha_1),$$

then

$$\tilde{A} \odot \tilde{B} = \tilde{B} \odot \tilde{A} \text{ for } \odot \in \{\oplus, \otimes\}.$$

Let $\tilde{C}$ be another fuzzy set in $\mathbb{R}$. Since the associativity does not necessarily hold true, we see that

$$(\tilde{A} \odot_1 \tilde{B}) \odot_2 \tilde{C} \neq \tilde{A} \odot_1 (\tilde{B} \odot_2 \tilde{C})$$

in general for $\odot_1, \odot_2 \in \{\oplus, \ominus\}$. Therefore, we are going to present the sufficient conditions to guarantee the associativity.

Proposition 7.5.5 *Let* $\tilde{A}, \tilde{B}, \tilde{C}$ *be three fuzzy sets in* $\mathbb{R}$. *Suppose that the function* $\mathfrak{A}_2 : [0, 1]^2 \to [0, 1]$ *satisfies the following condition*

$$\mathfrak{A}_2(\alpha_1, \alpha_2) \leq \alpha \text{ if and only if } \alpha_1 \leq \alpha \text{ or } \alpha_2 \leq \alpha, \tag{7.74}$$

where $\alpha \in [0, 1]$ *and* $\mathfrak{A}_2$ *is not necessarily commutative, i.e.* $\mathfrak{A}_2(\alpha_1, \alpha_2) \neq \mathfrak{A}_2(\alpha_2, \alpha_1)$ *in general. Then, for the arithmetic operations* $\odot_1, \odot_2 \in \{\oplus, \ominus, \otimes\}$, *the membership function of* $(\tilde{A} \odot_1 \tilde{B}) \odot_2 \tilde{C}$ *can be rewritten as the following simple form:*

$$\xi_{(\tilde{A} \odot_1 \tilde{B}) \odot_2 \tilde{C}}(z) = \sup_{\{(d,c):z=d\circ_2 c\}} \mathfrak{A}_2\left(\sup_{\{(a,b):d=a\circ_1 b\}} \mathfrak{A}_2\left(\xi_{\tilde{A}}(a), \xi_{\tilde{B}}(b)\right), \xi_{\tilde{C}}(c)\right)$$

$$= \sup_{\{(a,b,c):z=(a\circ_1 b)\circ_2 c\}} \mathfrak{A}_2\left(\mathfrak{A}_2\left(\xi_{\tilde{A}}(a), \xi_{\tilde{B}}(b)\right), \xi_{\tilde{C}}(c)\right) \tag{7.75}$$

$$= \sup_{\{(a,b,c):z=(a\circ_1 b)\circ_2 c\}} \mathfrak{A}_2\left(\xi_{\tilde{C}}(c), \mathfrak{A}_2\left(\xi_{\tilde{A}}(a), \xi_{\tilde{B}}(b)\right)\right),$$

where

$$\mathfrak{A}_2\left(\mathfrak{A}_2\left(\xi_{\tilde{A}}(a), \xi_{\tilde{B}}(b)\right), \xi_{\tilde{C}}(c)\right) \neq \mathfrak{A}_2\left(\xi_{\tilde{C}}(c), \mathfrak{A}_2\left(\xi_{\tilde{A}}(a), \xi_{\tilde{B}}(b)\right)\right)$$

in general.

Proof. We first have the membership function of $\tilde{A} \odot_1 \tilde{B}$ given by

$$\xi_{\tilde{A}\odot_1\tilde{B}}(d) = \sup_{\{(a,b):d=a\circ_1 b\}} \mathfrak{A}_2\left(\xi_{\tilde{A}}(a), \xi_{\tilde{B}}(b)\right).$$

Therefore, the membership function of $(\tilde{A} \odot_1 \tilde{B}) \odot_2 \tilde{C}$ is given by

$$\xi_{(\tilde{A}\odot_1\tilde{B})\odot_2\tilde{C}}(z) = \sup_{\{(d,c):z=d\circ_2 c\}} \mathfrak{A}_2\left(\xi_{\tilde{A}\odot_1\tilde{B}}(d), \xi_{\tilde{C}}(c)\right)$$

$$= \sup_{\{(d,c):z=d\circ_2 c\}} \mathfrak{A}_2\left(\sup_{\{(a,b):d=a\circ_1 b\}} \mathfrak{A}_2\left(\xi_{\tilde{A}}(a), \xi_{\tilde{B}}(b)\right), \xi_{\tilde{C}}(c)\right).$$

Let $r = \xi_{(\tilde{A}\odot_1\tilde{B})\odot_2\tilde{C}}(z)$. Then

$$r \geq \mathfrak{A}_2\left(\sup_{\{(a,b):d=a\circ_1 b\}} \mathfrak{A}_2\left(\xi_{\tilde{A}}(a), \xi_{\tilde{B}}(b)\right), \xi_{\tilde{C}}(c)\right) \text{ for all } (d,c) \text{ satisfying } z = d \circ_2 c,$$

which says that, by using (7.74),

$$r \geq \xi_{\tilde{C}}(c) \text{ or } r \geq \sup_{\{(a,b):d=a\circ_1 b\}} \mathfrak{A}_2\left(\xi_{\tilde{A}}(a), \xi_{\tilde{B}}(b)\right) \text{ for all } (d,c) \text{ satisfying } z = d \circ_2 c.$$

Therefore, we obtain

$$r \geq \xi_{\tilde{C}}(c) \text{ or } r \geq \mathfrak{A}_2\left(\xi_{\tilde{A}}(a), \xi_{\tilde{B}}(b)\right) \text{ for all } (a,b,c) \text{ satisfying } z = (a \circ_1 b) \circ_2 c,$$

i.e. by using (7.74) again,

$$r \geq \mathfrak{A}_2\left(\mathfrak{A}_2\left(\xi_{\tilde{A}}(a), \xi_{\tilde{B}}(b)\right), \xi_{\tilde{C}}(c)\right) \text{ for all } (a,b,c) \text{ satisfying } z = (a \circ_1 b) \circ_2 c$$

and

$$r \geq \mathfrak{A}_2\left(\xi_{\tilde{C}}(c), \mathfrak{A}_2\left(\xi_{\tilde{A}}(a), \xi_{\tilde{B}}(b)\right)\right) \text{ for all } (a,b,c) \text{ satisfying } z = (a \circ_1 b) \circ_2 c,$$

which also say that

$$r \geq \sup_{\{(a,b,c):z=(a\circ_1 b)\circ_2 c\}} \mathfrak{A}_2\left(\mathfrak{A}_2\left(\xi_{\tilde{A}}(a), \xi_{\tilde{B}}(b)\right), \xi_{\tilde{C}}(c)\right)$$

and

$$r \geq \sup_{\{(a,b,c):z=(a\circ_1 b)\circ_2 c\}} \mathfrak{A}_2\left(\xi_{\tilde{C}}(c), \mathfrak{A}_2\left(\xi_{\tilde{A}}(a), \xi_{\tilde{B}}(b)\right)\right).$$

On the other hand, let

$$s = \sup_{\{(a,b,c):z=(a\circ_1 b)\circ_2 c\}} \mathfrak{A}_2\left(\mathfrak{A}_2\left(\xi_{\tilde{A}}(a), \xi_{\tilde{B}}(b)\right), \xi_{\tilde{C}}(c)\right).$$

Then, we have

$$s \geq \mathfrak{A}_2\left(\mathfrak{A}_2\left(\xi_{\tilde{A}}(a), \xi_{\tilde{B}}(b)\right), \xi_{\tilde{C}}(c)\right) \text{ for all } (a,b,c) \text{ satisfying } z = (a \circ_1 b) \circ_2 c,$$

which implies, by using (7.74),

$$s \geq \xi_{\tilde{C}}(c) \text{ or } s \geq \mathfrak{A}_2\left(\xi_{\tilde{A}}(a), \xi_{\tilde{B}}(b)\right) \text{ for all } (d,c) \text{ satisfying } z = d \circ_2 c \text{ and } d = a \circ_1 b.$$

Therefore, we obtain

$$s \geq \xi_{\tilde{C}}(c) \text{ or } s \geq \sup_{\{(a,b):d=a\circ_1 b\}} \mathfrak{A}_2\left(\xi_{\tilde{A}}(a), \xi_{\tilde{B}}(b)\right) \text{ for all } (d,c) \text{ satisfying } z = d \circ_2 c,$$

which also says that, by using (7.74) again,

$$s \geq \mathfrak{A}_2\left(\sup_{\{(a,b):d=a\circ_1 b\}} \mathfrak{A}_2\left(\xi_{\tilde{A}}(a), \xi_{\tilde{B}}(b)\right), \xi_{\tilde{C}}(c)\right) \text{ for all } (d,c) \text{ satisfying } z = d \circ_2 c.$$

This shows that

$$s \geq \sup_{\{(d,c):z=d\circ_2 c\}} \mathfrak{A}_2\left(\sup_{\{(a,b):d=a\circ_1 b\}} \mathfrak{A}_2\left(\xi_{\tilde{A}}(a), \xi_{\tilde{B}}(b)\right), \xi_{\tilde{C}}(c)\right) = \xi_{(\tilde{A}\odot_1\tilde{B})\odot_2\tilde{C}}(z).$$

We can similarly show that

$$\sup_{\{(a,b,c):z=(a\circ_1 b)\circ_2 c\}} \mathfrak{A}_2\left(\xi_{\tilde{C}}(c), \mathfrak{A}_2\left(\xi_{\tilde{A}}(a), \xi_{\tilde{B}}(b)\right)\right) \geq \xi_{(\tilde{A}\odot_1\tilde{B})\odot_2\tilde{C}}(z).$$

Therefore, we obtain the desired equalities. This completes the proof. ■

Remark 7.5.6 In Proposition 7.5.5, although

$$\mathfrak{A}_2\left(\mathfrak{A}_2\left(\tilde{A}(a), \tilde{B}(b)\right), \tilde{C}(c)\right) \neq \mathfrak{A}_2\left(\tilde{C}(c), \mathfrak{A}_2\left(\tilde{A}(a), \tilde{B}(b)\right)\right)$$

in general, we always have the following equality

$$\begin{aligned} &\sup_{\{(a,b,c):z=(a\circ_1 b)\circ_2 c\}} \mathfrak{A}_2\left(\mathfrak{A}_2\left(\tilde{A}(a), \tilde{B}(b)\right), \tilde{C}(c)\right) \\ &= \sup_{\{(a,b,c):z=(a\circ_1 b)\circ_2 c\}} \mathfrak{A}_2\left(\tilde{C}(c), \mathfrak{A}_2\left(\tilde{A}(a), \tilde{B}(b)\right)\right). \end{aligned}$$

Proposition 7.5.7 *Let $\tilde{A}, \tilde{B}, \tilde{C}$ be three fuzzy sets in $\mathbb{R}$. Suppose that the function $\mathfrak{A}_2 : [0,1]^2 \to [0,1]$ satisfies the following condition*

$$\mathfrak{A}_2(\alpha_1, \alpha_2) \leq \alpha \text{ if and only if } \alpha_1 \leq \alpha \text{ or } \alpha_2 \leq \alpha, \tag{7.76}$$

where $\alpha \in [0,1]$ and $\mathfrak{A}_2$ is not necessarily commutative. Then, for the arithmetic operations $\odot_1, \odot_2 \in \{\oplus, \ominus, \otimes\}$, the membership function of $\tilde{A} \odot_1 (\tilde{B} \odot_2 \tilde{C})$ can be rewritten as the following simple form:

$$\begin{aligned} \xi_{\tilde{A}\odot_1(\tilde{B}\odot_2\tilde{C})}(z) &= \sup_{\{(a,d):z=a\circ_1 d\}} \mathfrak{A}_2\left(\xi_{\tilde{A}}(a), \sup_{\{(b,c):d=b\circ_2 c\}} \mathfrak{A}_2\left(\xi_{\tilde{B}}(b), \xi_{\tilde{C}}(c)\right)\right) \\ &= \sup_{\{(a,b,c):z=a\circ_1(b\circ_2 c)\}} \mathfrak{A}_2\left(\xi_{\tilde{A}}(a), \mathfrak{A}_2\left(\xi_{\tilde{B}}(b), \xi_{\tilde{C}}(c)\right)\right) \qquad (7.77) \\ &= \sup_{\{(a,b,c):z=a\circ_1(b\circ_2 c)\}} \mathfrak{A}_2\left(\mathfrak{A}_2\left(\xi_{\tilde{B}}(b), \xi_{\tilde{C}}(c)\right), \xi_{\tilde{A}}(a)\right), \end{aligned}$$

where

$$\mathfrak{A}_2\left(\xi_{\tilde{A}}(a), \mathfrak{A}_2\left(\xi_{\tilde{B}}(b), \xi_{\tilde{C}}(c)\right)\right) \neq \mathfrak{A}_2\left(\mathfrak{A}_2\left(\xi_{\tilde{B}}(b), \xi_{\tilde{C}}(c)\right), \xi_{\tilde{A}}(a)\right)$$

in general.

Proof. Let $r = \xi_{\tilde{A}\odot_1(\tilde{B}\odot_2\tilde{C})}(z)$. Then

$$r \geq \mathfrak{A}_2\left(\xi_{\tilde{A}}(a), \sup_{\{(b,c):d=b\circ_2 c\}} \mathfrak{A}_2\left(\xi_{\tilde{B}}(b), \xi_{\tilde{C}}(c)\right)\right) \text{ for all } (a,d) \text{ satisfying } z = a \circ_1 d,$$

which says that, by using (7.76),

$$r \geq \xi_{\tilde{A}}(a) \text{ or } r \geq \sup_{\{(b,c):d=c\circ_2 c\}} \mathfrak{A}_2\left(\xi_{\tilde{B}}(b), \xi_{\tilde{C}}(c)\right) \text{ for all } (a,d) \text{ satisfying } z = a \circ_1 d.$$

Therefore, we obtain

$$r \geq \xi_{\tilde{A}}(a) \text{ or } r \geq \mathfrak{A}_2\left(\xi_{\tilde{B}}(b), \xi_{\tilde{C}}(c)\right) \text{ for all } (a,b,c) \text{ satisfying } z = a \circ_1 (b \circ_2 c),$$

i.e. by using (7.76) again,

$$r \geq \mathfrak{A}_2\left(\xi_{\tilde{A}}(a), \mathfrak{A}_2\left(\xi_{\tilde{B}}(b), \xi_{\tilde{C}}(c)\right)\right) \text{ for all } (a,b,c) \text{ satisfying } z = a \circ_1 (b \circ_2 c)$$

and

$$r \geq \mathfrak{A}_2\left(\mathfrak{A}_2\left(\xi_{\tilde{B}}(b), \xi_{\tilde{C}}(c)\right), \xi_{\tilde{A}}(a)\right) \text{ for all } (a,b,c) \text{ satisfying } z = a \circ_1 (b \circ_2 c),$$

which also say that

$$r \geq \sup_{\{(a,b,c):z=a\circ_1(b\circ_2 c)\}} \mathfrak{A}_2\left(\xi_{\tilde{A}}(a), \mathfrak{A}_2\left(\xi_{\tilde{B}}(b), \xi_{\tilde{C}}(c)\right)\right)$$

and

$$r \geq \sup_{\{(a,b,c):z=a\circ_1(b\circ_2 c)\}} \mathfrak{A}_2\left(\mathfrak{A}_2\left(\xi_{\tilde{B}}(b), \xi_{\tilde{C}}(c)\right), \xi_{\tilde{A}}(a)\right).$$

On the other hand, let

$$s = \sup_{\{(a,b,c):z=a\circ_1(b\circ_2 c)\}} \mathfrak{A}_2\left(\xi_{\tilde{A}}(a), \mathfrak{A}_2\left(\xi_{\tilde{B}}(b), \xi_{\tilde{C}}(c)\right)\right).$$

Then, we have

$$s \geq \mathfrak{A}_2\left(\xi_{\tilde{A}}(a), \mathfrak{A}_2\left(\xi_{\tilde{B}}(b), \xi_{\tilde{C}}(c)\right)\right) \text{ for all } (a,b,c) \text{ satisfying } z = a \circ_1 (b \circ_2 c),$$

which implies, by using (7.76),

$$s \geq \xi_{\tilde{A}}(a) \text{ or } s \geq \mathfrak{A}_2\left(\xi_{\tilde{B}}(b), \xi_{\tilde{C}}(c)\right) \text{ for all } (a,d) \text{ satisfying } z = a \circ_1 d \text{ and } d = b \circ_1 c.$$

Therefore, we obtain

$$s \geq \xi_{\tilde{A}}(a) \text{ or } s \geq \sup_{\{(b,c):d=b\circ_2 c\}} \mathfrak{A}_2\left(\xi_{\tilde{B}}(b), \xi_{\tilde{C}}(c)\right) \text{ for all } (a,d) \text{ satisfying } z = a \circ_1 d,$$

which also says that, by using (7.76) again,

$$s \geq \mathfrak{A}_2\left(\xi_{\tilde{A}}(a), \sup_{\{(b,c):d=b\circ_2 c\}} \mathfrak{A}_2\left(\xi_{\tilde{B}}(b), \xi_{\tilde{C}}(c)\right)\right) \text{ for all } (a,d) \text{ satisfying } z = a \circ_1 d.$$

This shows that

$$s \geq \sup_{\{(a,d):z=a\circ_1 d\}} \mathfrak{A}_2\left(\xi_{\tilde{A}}(a), \sup_{\{(b,c):d=b\circ_2 c\}} \mathfrak{A}_2\left(\xi_{\tilde{B}}(b), \xi_{\tilde{C}}(c)\right)\right) = \xi_{\tilde{A}\odot_1(\tilde{B}\odot_2\tilde{C})}(z).$$

We can similarly show that

$$\sup_{\{(a,b,c):z=a\circ_1(b\circ_2 c)\}} \mathfrak{A}_2\left(\mathfrak{A}_2\left(\xi_{\tilde{B}}(b), \xi_{\tilde{C}}(c)\right), \xi_{\tilde{A}}(a)\right) \geq \xi_{\tilde{A}\odot_1(\tilde{B}\odot_2\tilde{C})}(z).$$

Therefore, we obtain the desired equalities. This completes the proof. ■

Proposition 7.5.8 *Let $\tilde{A}, \tilde{B}, \tilde{C}$ be three fuzzy sets in $\mathbb{R}$. Suppose that the function $\mathfrak{A}_2 : [0,1]^2 \to [0,1]$ satisfies the following conditions*

$$\mathfrak{A}_2\left(\mathfrak{A}_2\left(\alpha_1, \alpha_2\right), \alpha_3\right) = \mathfrak{A}_2\left(\alpha_1, \mathfrak{A}_2\left(\alpha_2, \alpha_3\right)\right) \tag{7.78}$$

and

$$\mathfrak{A}_2(\alpha_1, \alpha_2) \leq \alpha \text{ if and only if } \alpha_1 \leq \alpha \text{ or } \alpha_2 \leq \alpha, \tag{7.79}$$

where $\alpha \in [0,1]$ and $\mathfrak{A}_2$ is not necessarily commutative. Then, for $\odot_1, \odot_2 \in \{\oplus, \ominus\}$, we have

$$(\tilde{A} \odot_1 \tilde{B}) \odot_2 \tilde{C} = \tilde{A} \odot_1 (\tilde{B} \odot_2 \tilde{C}).$$

Proof. From Propositions 7.5.5 and 7.5.7, we have

$$\xi_{(\tilde{A}\odot_1\tilde{B})\odot_2\tilde{C}}(z) = \sup_{\{(a,b,c):z=(a\circ_1 b)\circ_2 c\}} \mathfrak{A}_2\left(\mathfrak{A}_2\left(\xi_{\tilde{A}}(a), \xi_{\tilde{B}}(b)\right), \xi_{\tilde{C}}(c)\right)$$

and

$$\xi_{\tilde{A}\odot_1(\tilde{B}\odot_2\tilde{C})}(z) = \sup_{\{(a,b,c):z=a\circ_1(b\circ_2 c)\}} \mathfrak{A}_2\left(\xi_{\tilde{A}}(a), \mathfrak{A}_2\left(\xi_{\tilde{B}}(b), \xi_{\tilde{C}}(c)\right)\right).$$

Since we consider the arithmetic operations $\circ_1, \circ_2 \in \{+, -\}$, it follows that

$$(a \circ_1 b) \circ_2 c = a \circ_1 (b \circ_2 c).$$

Therefore, using (7.78), we complete the proof. ■

We can also say that the function $\mathfrak{A}_2$ is **associative** when the condition (7.78) is satisfied. Given any three fuzzy sets $\tilde{A}, \tilde{B}, \tilde{C}$ in $\mathbb{R}$, there are two ways to define the arithmetic operations $\tilde{A} \odot_1 \tilde{B} \odot_2 \tilde{C}$ for $\odot_1, \odot_2 \in \{\oplus, \ominus\}$, which are given below.

- Using the function $\mathfrak{A}_2$ twice, the arithmetic operations can be defined as

  $$\tilde{A}^{\circ} \equiv \tilde{A} \odot_1 \tilde{B} \odot_2 \tilde{C} \equiv \left(\tilde{A} \odot_1 \tilde{B}\right) \odot_2 \tilde{C},$$

  whose membership function $\xi_{\tilde{A}^{\circ}}$ has been studied in Propositions 7.5.5, 7.5.7 and 7.5.8.
- We directly consider a function $\mathfrak{A}_3 : [0,1]^3 \to [0,1]$ rather than applying the function $\mathfrak{A}_2 : [0,1]^2 \to [0,1]$ twice as presented above. In this case, the membership function of arithmetic operations $\tilde{A}^{\dagger} \equiv \tilde{A} \odot_1 \tilde{B} \odot_2 \tilde{C}$ is given by

  $$\xi_{\tilde{A}^{\dagger}}(z) = \xi_{\tilde{A}\odot_1\tilde{B}\odot_2\tilde{C}}(z) = \sup_{\{(x_1,x_2,x_3):z=x_1\circ_1 x_2\circ_2 x_3\}} \mathfrak{A}_3\left(\xi_{\tilde{A}}(x_1), \xi_{\tilde{B}}(x_2), \xi_{\tilde{C}}(x_3)\right).$$

It is clear to see that $\xi_{\tilde{A}^\circ} \neq \xi_{\tilde{A}^\dagger}$ in general, since $\tilde{A}^\circ$ is obtained by using the function $\mathfrak{A}_2$ twice and $\tilde{A}^\dagger$ is obtained by directly using the function $\mathfrak{A}_3$. Next, we shall present the equality $\tilde{A}^\circ = \tilde{A}^\dagger$ under some suitable settings. We recall that the notation $\tilde{A}^{(1)} \odot_1 \cdots \odot_{n-1} \tilde{A}^{(n)}(\mathfrak{A}_n)$ means that the expression $\tilde{A}^{(1)} \odot_1 \cdots \odot_{n-1} \tilde{A}^{(n)}$ depends on the function $\mathfrak{A}_n$, which can refer to (7.60).

Proposition 7.5.9 *Let $\tilde{A}, \tilde{B}, \tilde{C}$ be three fuzzy sets in $\mathbb{R}$. Suppose that the function $\mathfrak{A}_2 : [0,1]^2 \to [0,1]$ satisfies the following condition*

$$\mathfrak{A}_2(\alpha_1, \alpha_2) \leq \alpha \text{ if and only if } \alpha_1 \leq \alpha \text{ or } \alpha_2 \leq \alpha,$$

where $\alpha \in [0,1]$ and $\mathfrak{A}_2$ is not necessarily commutative. Based on the function $\mathfrak{A}_2$, we define the function $\mathfrak{A}_3^L : [0,1]^3 \to [0,1]$ as follows

$$\mathfrak{A}_3^L\left(\alpha_1, \alpha_2, \alpha_3\right) = \mathfrak{A}_2\left(\mathfrak{A}_2\left(\alpha_1, \alpha_2\right), \alpha_3\right). \tag{7.80}$$

Then, we have the following properties.

(i) For $\odot_1, \odot_2 \in \{\oplus, \ominus\}$, based on the function $\mathfrak{A}_2$, we define the arithmetic operations $(\tilde{A} \odot_1 \tilde{B}) \odot_2 \tilde{C}$ as described in Proposition 7.5.5. Based on the function $\mathfrak{A}_3^L$, we define the membership functions of $\tilde{D}^L \equiv \tilde{A} \odot_1 \tilde{B} \odot_2 \tilde{C}(\mathfrak{A}_3^L)$ as

$$\xi_{\tilde{D}^L}(z) = \sup_{\{(a,b,c): z = a \circ_1 b \circ_2 c\}} \mathfrak{A}_3^L\left(\xi_{\tilde{A}}(a), \xi_{\tilde{B}}(b), \xi_{\tilde{C}}(c)\right) \text{ for } \circ_1, \circ_2 \in \{+, -\}.$$

Then, we have

$$(\tilde{A} \odot_1 \tilde{B}) \odot_2 \tilde{C} = \tilde{A} \odot_1 \tilde{B} \odot_2 \tilde{C}(\mathfrak{A}_3^L) = \tilde{D}^L.$$

(ii) For $\odot \in \{\oplus, \ominus\}$, based on the function $\mathfrak{A}_2$, we define the arithmetic operations $(\tilde{A} \otimes \tilde{B}) \odot \tilde{C}$ as described in Proposition 7.5.5. Based on the function $\mathfrak{A}_3^L$, we define the membership function of $\tilde{E}^L \equiv \tilde{A} \otimes \tilde{B} \odot \tilde{C}(\mathfrak{A}_3^L)$ as

$$\xi_{\tilde{E}^L}(z) = \sup_{\{(a,b,c): z = a * b \circ c\}} \mathfrak{A}_3^L\left(\xi_{\tilde{A}}(a), \xi_{\tilde{B}}(b), \xi_{\tilde{C}}(c)\right) \text{ for } \circ \in \{+, -\}.$$

Then, we have

$$(\tilde{A} \otimes \tilde{B}) \odot \tilde{C} = \tilde{A} \otimes \tilde{B} \odot \tilde{C}(\mathfrak{A}_3^L) = \tilde{E}^L.$$

(iii) Based on the function $\mathfrak{A}_2$, we define the arithmetic operations $(\tilde{A} \otimes \tilde{B}) \otimes \tilde{C}$ as described in Proposition 7.5.5. Based on the function $\mathfrak{A}_3^L$, we define the membership function of $\tilde{F}^L \equiv \tilde{A} \otimes \tilde{B} \otimes \tilde{C}(\mathfrak{A}_3^L)$ as

$$\xi_{\tilde{F}^L}(z) = \sup_{\{(a,b,c): z = a * b * c\}} \mathfrak{A}_3^L\left(\xi_{\tilde{A}}(a), \xi_{\tilde{B}}(b), \xi_{\tilde{C}}(c)\right).$$

Then, we have

$$(\tilde{A} \otimes \tilde{B}) \otimes \tilde{C} = \tilde{A} \otimes \tilde{B} \otimes \tilde{C}(\mathfrak{A}_3^L) = \tilde{F}^L.$$

Proof. To prove part (i), we have

$$\xi_{(\tilde{A} \odot_1 \tilde{B}) \odot_2 \tilde{C}}(z) = \sup_{\{(a,b,c): z = (a \circ_1 b) \circ_2 c\}} \mathfrak{A}_2\left(\mathfrak{A}_2\left(\xi_{\tilde{A}}(a), \xi_{\tilde{B}}(b)\right), \xi_{\tilde{C}}(c)\right) \text{ (uisng (7.75))}$$

$$= \sup_{\{(a,b,c):z=a\circ_1 b\circ_2 c\}} \mathfrak{A}_3^L\left(\xi_{\tilde{A}}(a), \xi_{\tilde{B}}(b), \xi_{\tilde{C}}(c)\right) \text{ (uisng (7.80))}$$
$$= \xi_{\tilde{D}^L}(z),$$

where

$$(a \circ_1 b) \circ_2 c = a \circ_1 b \circ_2 c = a \circ_1 (b \circ_2 c) = b \circ_1 (a \circ_2 c) = \cdots \text{ etc.},$$

since the arithmetic operations $\circ_1$, $\circ_2 \in \{+,-\}$ for real numbers are commutative and associative.

To prove part (ii), we also have

$$\xi_{(\tilde{A}\otimes\tilde{B})\odot\tilde{C}}(z) = \sup_{\{(a,b,c):z=(a*b)\circ c\}} \mathfrak{A}_3^L\left(\xi_{\tilde{A}}(a), \xi_{\tilde{B}}(b), \xi_{\tilde{C}}(c)\right) \text{ (using the arguments of part (i))}$$
$$= \sup_{\{(a,b,c):z=a*b\circ c\}} \mathfrak{A}_3^L\left(\xi_{\tilde{A}}(a), \xi_{\tilde{B}}(b), \xi_{\tilde{C}}(c)\right) = \xi_{\tilde{E}^L}(z),$$

where the fact of $(a * b) \circ c = a * b \circ c$ is used for $\circ \in \{+,-\}$. We can similarly obtain part (iii) by proving $\xi_{(\tilde{A}\otimes\tilde{B})\otimes\tilde{C}}(z) = \xi_{\tilde{F}^L}(z)$. This completes the proof. ■

Proposition 7.5.10 *Let $\tilde{A}, \tilde{B}, \tilde{C}$ be three fuzzy sets in $\mathbb{R}$. Suppose that the function $\mathfrak{A}_2 : [0,1]^2 \to [0,1]$ satisfies the following condition*

$$\mathfrak{A}_2(\alpha_1, \alpha_2) \leq \alpha \textit{ if and only if } \alpha_1 \leq \alpha \textit{ or } \alpha_2 \leq \alpha,$$

where $\alpha \in [0,1]$ and $\mathfrak{A}_2$ is not necessarily commutative. Based on the function $\mathfrak{A}_2$, we define the function $\mathfrak{A}_3^R : [0,1]^3 \to [0,1]$ as follows

$$\mathfrak{A}_3^R\left(\alpha_1, \alpha_2, \alpha_3\right) = \mathfrak{A}_2\left(\alpha_1, \mathfrak{A}_2\left(\alpha_2, \alpha_3\right)\right). \tag{7.81}$$

Then, we have the following properties.

(i) For $\odot_1, \odot_2 \in \{\oplus, \ominus\}$, based on the function $\mathfrak{A}_2$, we define the arithmetic operations $\tilde{A} \odot_1 (\tilde{B} \odot_2 \tilde{C})$ as described in Proposition 7.5.7. Based on the function $\mathfrak{A}_3^R$, we define the membership functions of $\tilde{D}^R \equiv \tilde{A} \odot_1 \tilde{B} \odot_2 \tilde{C}(\mathfrak{A}_3^R)$ as

$$\xi_{\tilde{D}^R}(z) = \sup_{\{(a,b,c):z=a\circ_1 b\circ_2 c\}} \mathfrak{A}_3^R\left(\xi_{\tilde{A}}(a), \xi_{\tilde{B}}(b), \xi_{\tilde{C}}(c)\right) \textit{ for } \circ_1, \circ_2 \in \{+,-\}.$$

Then, we have

$$\tilde{A} \odot_1 (\tilde{B} \odot_2 \tilde{C}) = \tilde{A} \odot_1 \tilde{B} \odot_2 \tilde{C}(\mathfrak{A}_3^R) = \tilde{D}^R.$$

(ii) For $\odot \in \{\oplus, \ominus\}$, based on the function $\mathfrak{A}_2$, we define the arithmetic operations $\tilde{A} \odot (\tilde{B} \otimes \tilde{C})$ as described in Proposition 7.5.7. Based on the function $\mathfrak{A}_3^R$, we define the membership function of $\tilde{E}^R \equiv \tilde{A} \otimes \tilde{B} \odot \tilde{C}(\mathfrak{A}_3^R)$ as

$$\xi_{\tilde{E}^R}(z) = \sup_{\{(a,b,c):z=a\circ b*c\}} \mathfrak{A}_3^R\left(\xi_{\tilde{A}}(a), \xi_{\tilde{B}}(b), \xi_{\tilde{C}}(c)\right) \textit{ for } \circ \in \{+,-\}.$$

Then, we have

$$\tilde{A} \odot (\tilde{B} \otimes \tilde{C}) = \tilde{A} \odot \tilde{B} \otimes \tilde{C}(\mathfrak{A}_3^R) = \tilde{E}^R.$$

(iii) Based on the function $\mathfrak{A}_2$, we define the arithmetic operations $\tilde{A} \otimes (\tilde{B} \otimes \tilde{C})$ as described in Proposition 7.5.7. Based on the function $\mathfrak{A}_3^R$, we define the membership function of $\tilde{F}^R \equiv \tilde{A} \otimes \tilde{B} \otimes \tilde{C}(\mathfrak{A}_3^R)$ as

$$\xi_{\tilde{F}^R}(z) = \sup_{\{(a,b,c):z=a*b*c\}} \mathfrak{A}_3^R\left(\xi_{\tilde{A}}(a), \xi_{\tilde{B}}(b), \xi_{\tilde{C}}(c)\right).$$

Then,we have

$$\tilde{A} \otimes (\tilde{B} \otimes \tilde{C}) = \tilde{A} \otimes \tilde{B} \otimes \tilde{C}(\mathfrak{A}_3^R) = \tilde{F}^R.$$

Proof. To prove part (i), we have

$$\begin{aligned}\xi_{\tilde{A} \odot_1 (\tilde{B} \odot_2 \tilde{C})}(z) &= \sup_{\{(a,b,c):z=a\circ_1(b\circ_2 c)\}} \mathfrak{A}_2\left(\xi_{\tilde{A}}(a), \mathfrak{A}_2\left(\xi_{\tilde{B}}(b), \xi_{\tilde{C}}(c)\right)\right) \text{ (uisng (7.77))}\\ &= \sup_{\{(a,b,c):z=a\circ_1 b\circ_2 c\}} \mathfrak{A}_3^R\left(\xi_{\tilde{A}}(a), \xi_{\tilde{B}}(b), \xi_{\tilde{C}}(c)\right) \text{ (uisng (7.81))}\\ &= \xi_{\tilde{D}^R}(z).\end{aligned}$$

We can similarly obtain

$$\xi_{\tilde{A}\otimes(\tilde{B}\otimes\tilde{C})}(z) = \xi_{\tilde{E}^R}(z) \text{ and } \xi_{\tilde{A}\odot(\tilde{B}\otimes\tilde{C})}(z) = \xi_{\tilde{F}^R}(z)$$

by using the fact of $a \circ (b * c) = a \circ b * c$ for $\circ \in \{+, -\}$. This completes the proof. ∎

Recall that the function $\mathfrak{A}_2$ is associative if and only if the conditions (7.80) and (7.81) are identical. We also say that the functions $\mathfrak{A}_3^L$ and $\mathfrak{A}_3^R$ in Propositions 7.5.9 and 7.5.10 are left-generated and right-generated by $\mathfrak{A}_2$, respectively.

Proposition 7.5.11 *Let $\tilde{A}, \tilde{B}, \tilde{C}$ be three fuzzy sets in $\mathbb{R}$. Suppose that the function $\mathfrak{A}_2 : [0,1]^2 \to [0,1]$ satisfies the following condition*

$$\mathfrak{A}_2(\alpha_1, \alpha_2) \leq \alpha \textit{ if and only if } \alpha_1 \leq \alpha \textit{ or } \alpha_2 \leq \alpha,$$

where $\alpha \in [0,1]$ and $\mathfrak{A}_2$ is not necessarily commutative. Let the functions $\mathfrak{A}_3^L$ and $\mathfrak{A}_3^R$ be left-generated and right-generated by $\mathfrak{A}_2$, respectively. Suppose that any one of the following conditions is satisfied:

- *the function $\mathfrak{A}_2$ is associative;*
- *the function $\mathfrak{A}_2$ is compatible with arithmetic operations of α-level sets.*

Then, for $\odot_1, \odot_2 \in \{\oplus, \ominus\}$, we have

$$(\tilde{A} \odot_1 \tilde{B})\odot_2\tilde{C} = \tilde{A} \odot_1 (\tilde{B} \odot_2 \tilde{C}) = \tilde{A} \odot_1 \tilde{B} \odot_2 \tilde{C}(\mathfrak{A}_3^L) = \tilde{A} \odot_1 \tilde{B} \odot_2 \tilde{C}(\mathfrak{A}_3^R) \tag{7.82}$$

and

$$(\tilde{A} \otimes \tilde{B}) \otimes \tilde{C} = \tilde{A} \otimes (\tilde{B} \otimes \tilde{C}) = \tilde{A} \otimes \tilde{B} \otimes \tilde{C}(\mathfrak{A}_3^L) = \tilde{A} \otimes \tilde{B} \otimes \tilde{C}(\mathfrak{A}_3^R). \tag{7.83}$$

In this case, we simply write

$$(\tilde{A} \odot_1 \tilde{B})\odot_2\tilde{C} = \tilde{A} \odot_1 (\tilde{B} \odot_2 \tilde{C}) = \tilde{A} \odot_1 \tilde{B} \odot_2 \tilde{C}$$

and

$$(\tilde{A} \otimes \tilde{B}) \otimes \tilde{C} = \tilde{A} \otimes (\tilde{B} \otimes \tilde{C}) = \tilde{A} \otimes \tilde{B} \otimes \tilde{C}.$$

Proof. Suppose that the function $\mathfrak{A}_2$ is associative; that is, the conditions (7.80) and (7.81) are identical. This says that $\mathfrak{A}_3^L = \mathfrak{A}_3^R$. Therefore, from Propositions 7.5.9 and 7.5.10, we obtain the desired equalities (7.82) and (7.83).

Now, we assume that $\mathfrak{A}_2$ is compatible with arithmetic operations of α-level sets. Then, for any $\alpha \in (0,1]$, we have

$$\begin{aligned}\left(\tilde{A} \odot_1 \tilde{B} \odot_2 \tilde{C}(\mathfrak{A}_3^L)\right)_\alpha &= \left(\left(\tilde{A} \odot_1 \tilde{B}\right) \odot_2 \tilde{C}\right)_\alpha \text{ (using Proposition 7.5.9)}\\ &= \left(\tilde{A} \odot_1 \tilde{B}\right)_\alpha \circ_2 \tilde{C}_\alpha = \left(\tilde{A}_\alpha \circ_1 \tilde{B}_\alpha\right) \circ_2 \tilde{C}_\alpha \text{ (using compatibility for } \mathfrak{A}_2)\\ &= \tilde{A}_\alpha \circ_1 \left(\tilde{B}_\alpha \circ_2 \tilde{C}_\alpha\right) = \tilde{A}_\alpha \circ_1 \left(\tilde{B} \odot_2 \tilde{C}\right)_\alpha = \left(\tilde{A} \odot_1 \left(\tilde{B} \odot_2 \tilde{C}\right)\right)_\alpha\\ &= \left(\tilde{A} \odot_1 \tilde{B} \odot_2 \tilde{C}(\mathfrak{A}_3^R)\right)_\alpha \text{ (using Proposition 7.5.10).}\end{aligned}$$

Therefore, we obtain the desired equalities (7.82). Finally, we can similarly prove the case of operation $\otimes$. This completes the proof. ■

We notice that Propositions 7.5.9, 7.5.10 and 7.5.11 can be inductively extended by considering more than three fuzzy sets in $\mathbb{R}$.

Remark 7.5.12 From Propositions 7.5.9 and 7.5.10, we have

$$\tilde{A} \otimes \tilde{B} \odot \tilde{C}(\mathfrak{A}_3^L) = (\tilde{A} \otimes \tilde{B}) \odot \tilde{C} \text{ and } \tilde{A} \odot \tilde{B} \otimes \tilde{C}(\mathfrak{A}_3^R) = \tilde{A} \odot (\tilde{B} \otimes \tilde{C}). \tag{7.84}$$

Using the proofs of Propositions 7.5.9 and 7.5.10, we can also obtain the membership functions of $\tilde{A} \otimes (\tilde{B} \odot \tilde{C})$ and $(\tilde{A} \odot \tilde{B}) \otimes \tilde{C}$, which are given by

$$\begin{aligned}\xi_{\tilde{A}\otimes(\tilde{B}\odot\tilde{C})}(z) &= \sup_{\{(a,b,c):z=a*(b\circ c)\}} \mathfrak{A}_3^R\left(\xi_{\tilde{A}}(a), \xi_{\tilde{B}}(b), \xi_{\tilde{C}}(c)\right)\\ &\neq \sup_{\{(a,b,c):z=a*b\circ c\}} \mathfrak{A}_3^R\left(\xi_{\tilde{A}}(a), \xi_{\tilde{B}}(b), \xi_{\tilde{C}}(c)\right)\end{aligned}$$

and

$$\begin{aligned}\xi_{(\tilde{A}\odot\tilde{B})\otimes\tilde{C}}(z) &= \sup_{\{(a,b,c):z=(a\circ b)*c\}} \mathfrak{A}_3^L\left(\xi_{\tilde{A}}(a), \xi_{\tilde{B}}(b), \xi_{\tilde{C}}(c)\right)\\ &\neq \sup_{\{(a,b,c):z=a\circ b*c\}} \mathfrak{A}_3^L\left(\xi_{\tilde{A}}(a), \xi_{\tilde{B}}(b), \xi_{\tilde{C}}(c)\right),\end{aligned}$$

respectively, for $\circ \in \{+,-\}$. Therefore, in general, we have

$$\tilde{A} \otimes \tilde{B} \odot \tilde{C}(\mathfrak{A}_3^L) \neq \tilde{A} \otimes (\tilde{B} \odot \tilde{C}) \text{ and } \tilde{A} \odot \tilde{B} \otimes \tilde{C}(\mathfrak{A}_3^R) \neq (\tilde{A} \odot \tilde{B}) \otimes \tilde{C}. \tag{7.85}$$

The expressions (7.84) and (7.85) say that the operation $\otimes$ has the highest priority for firstly performing calculation.

Recall the left-generated function $\mathfrak{A}_n^L$ and right-generated function $\mathfrak{A}_n^R$ in Definition 3.4.1. We present an example.

Example 7.5.13 Let $\tilde{A}^{(1)}, \ldots, \tilde{A}^{(7)}$ be fuzzy sets in $\mathbb{R}$. Suppose that we take the function $\mathfrak{A}_2$ as

$$\mathfrak{A}_2(\alpha_1, \alpha_2) = \min\left\{\alpha_1, \alpha_2\right\}.$$

Then, it is clear to see $\mathfrak{A}_7^L = \mathfrak{A}_7^R \equiv \mathfrak{A}_7$ that is given by

$$\mathfrak{A}_7(\alpha_1, \ldots, \alpha_7) = \min\left\{\alpha_1, \ldots, \alpha_7\right\}.$$

According to Propositions 7.5.8–7.5.12, the membership functions of

$$\tilde{B} \equiv \tilde{A}^{(1)} \boxtimes \left(\tilde{A}^{(2)} \boxplus \tilde{A}^{(3)}\right) \boxminus \left(\tilde{A}^{(4)} \boxtimes \left(\tilde{A}^{(5)} \boxplus \tilde{A}^{(6)} \boxminus \tilde{A}^{(7)}\right)\right)$$

and

$$\tilde{C} \equiv \tilde{A}^{(1)} \boxtimes \tilde{A}^{(2)} \boxplus \tilde{A}^{(3)} \boxminus \tilde{A}^{(4)} \boxtimes \tilde{A}^{(5)} \boxplus \tilde{A}^{(6)} \boxminus \tilde{A}^{(7)}$$

are given by

$$\xi_{\tilde{B}}(z) = \sup_{\{(x_1,\ldots,x_7):z=x_1*(x_2+x_3)-(x_4*(x_5+x_6-x_7))\}} \min\left\{\xi_{\tilde{A}^{(1)}}(x_1),\ldots,\xi_{\tilde{A}^{(7)}}(x_7)\right\}$$

and

$$\xi_{\tilde{C}}(z) = \sup_{\{(x_1,\ldots,x_7):z=x_1*x_2+x_3-x_4*x_5+x_6-x_7\}} \min\left\{\xi_{\tilde{A}^{(1)}}(x_1),\ldots,\xi_{\tilde{A}^{(7)}}(x_7)\right\},$$

respectively. We see that $\tilde{B} \neq \tilde{C}$ and

$$\tilde{C} = \left(\tilde{A}^{(1)} \boxtimes \tilde{A}^{(2)}\right) \boxplus \tilde{A}^{(3)} \boxminus \left(\tilde{A}^{(4)} \boxtimes \tilde{A}^{(5)}\right) \boxplus \tilde{A}^{(6)} \boxminus \tilde{A}^{(7)}.$$

Suppose that their membership functions are upper semi-continuous, and that their supports are bounded. Using Theorem 7.5.4, the α-level sets are given by

$$\tilde{B}_\alpha = \tilde{A}^{(1)}_\alpha * \left(\tilde{A}^{(2)}_\alpha + \tilde{A}^{(3)}_\alpha\right) - \left(\tilde{A}^{(4)}_\alpha * \left(\tilde{A}^{(5)}_\alpha + \tilde{A}^{(6)}_\alpha - \tilde{A}^{(7)}_\alpha\right)\right)$$

and

$$\begin{aligned}\tilde{C}_\alpha &= \tilde{A}^{(1)}_\alpha * \tilde{A}^{(2)}_\alpha + \tilde{A}^{(3)}_\alpha - \tilde{A}^{(4)}_\alpha * \tilde{A}^{(5)}_\alpha + \tilde{A}^{(6)}_\alpha - \tilde{A}^{(7)}_\alpha \\ &= \left(\tilde{A}^{(1)}_\alpha * \tilde{A}^{(2)}_\alpha\right) + \tilde{A}^{(3)}_\alpha - \left(\tilde{A}^{(4)}_\alpha * \tilde{A}^{(5)}_\alpha\right) + \tilde{A}^{(6)}_\alpha - \tilde{A}^{(7)}_\alpha\end{aligned}$$

for all $\alpha \in [0, 1]$.

7.5.3 Computational Procedure

The following example shows the way of general computational procedure for any arithmetic expressions.

Example 7.5.14 Let $\tilde{A}^{(1)},\ldots,\tilde{A}^{(10)}$ be fuzzy sets in $\mathbb{R}$. Given a function $\mathfrak{A}_2$ satisfying the following condition

$$\mathfrak{A}_2(\alpha_1, \alpha_2) \leq \alpha \text{ if and only if } \alpha_1 \leq \alpha \text{ or } \alpha_2 \leq \alpha, \tag{7.86}$$

we assume that the functions $\mathfrak{A}^L_3$ and $\mathfrak{A}^L_4$ are left-generated by $\mathfrak{A}_2$. In order to calculate the following expression

$$\left(\tilde{A}^{(1)} \ominus \tilde{A}^{(2)}\right) \otimes \left(\tilde{A}^{(3)} \ominus \tilde{A}^{(4)} \oplus \tilde{A}^{(5)}\right) \ominus \left(\tilde{A}^{(6)} \otimes \left(\tilde{A}^{(7)} \oplus \tilde{A}^{(8)} \ominus \tilde{A}^{(9)} \ominus \tilde{A}^{(10)}\right)\right), \tag{7.87}$$

the computational procedure are presented below.

- Calculate $\tilde{B}^{(1)} \equiv \tilde{A}^{(1)} \ominus \tilde{A}^{(2)}$ using the function $\mathfrak{A}_2$. The membership function of $\tilde{B}^{(1)}$ is given by

$$\xi_{\tilde{B}^{(1)}}(z) = \sup_{\{(x_1,x_2):z=x_1-x_2\}} \mathfrak{A}_2\left(\xi_{\tilde{A}^{(1)}}(x_1), \xi_{\tilde{A}^{(2)}}(x_2)\right).$$

- Calculate $\tilde{B}^{(2)} \equiv \tilde{A}^{(3)} \ominus \tilde{A}^{(4)} \oplus \tilde{A}^{(5)}$ using the function $\mathfrak{A}_3^L$, which is left-generated by $\mathfrak{A}_2$, i.e.

$$\mathfrak{A}_3^L(\alpha_1, \alpha_2, \alpha_3) = \mathfrak{A}_2(\mathfrak{A}_2(\alpha_1, \alpha_2), \alpha_3).$$

Using Proposition 7.5.9, the membership function of $\tilde{B}^{(2)}$ is given by

$$\begin{aligned}\xi_{\tilde{B}^{(2)}}(z) &= \sup_{\{(x_3,x_4,x_5):z=x_3-x_4+x_5\}} \mathfrak{A}_3^L\left(\xi_{\tilde{A}^{(3)}}(x_3), \xi_{\tilde{A}^{(4)}}(x_4), \xi_{\tilde{A}^{(5)}}(x_5)\right) \\ &= \sup_{\{(x_3,x_4,x_5):z=x_3-x_4+x_5\}} \mathfrak{A}_2\left(\mathfrak{A}_2\left(\xi_{\tilde{A}^{(3)}}(x_3), \xi_{\tilde{A}^{(4)}}(x_4)\right), \xi_{\tilde{A}^{(5)}}(x_5)\right).\end{aligned}$$

- Calculate $\tilde{B}^{(3)} \equiv \tilde{A}^{(7)} \oplus \tilde{A}^{(8)} \ominus \tilde{A}^{(9)} \ominus \tilde{A}^{(10)}$ using the function $\mathfrak{A}_4^L$, which is left-generated by $\mathfrak{A}_2$, i.e.

$$\mathfrak{A}_4^L\left(\alpha_1, \alpha_2, \alpha_3, \alpha_4\right) = \mathfrak{A}_2\left(\mathfrak{A}_2\left(\mathfrak{A}_2\left(\alpha_1, \alpha_2\right), \alpha_3\right), \alpha_4\right).$$

Using Proposition 7.5.9 repeatedly, the membership function of $\tilde{B}^{(3)}$ is given by

$$\begin{aligned}\xi_{\tilde{B}^{(3)}}(z) &= \sup_{\{(x_7,x_8,x_9,x_{10}):z=x_7+x_8-x_9-x_{10}\}} \mathfrak{A}_4^L\left(\xi_{\tilde{A}^{(7)}}(x_7), \xi_{\tilde{A}^{(8)}}(x_8), \xi_{\tilde{A}^{(9)}}(x_9), \xi_{\tilde{A}^{(10)}}(x_{10})\right) \\ &= \sup_{\{(x_7,x_8,x_9,x_{10}):z=x_7+x_8-x_9-x_{10}\}} \mathfrak{A}_2\left(\mathfrak{A}_2\left(\mathfrak{A}_2\left(\xi_{\tilde{A}^{(7)}}(x_7), \xi_{\tilde{A}^{(8)}}(x_8)\right), \xi_{\tilde{A}^{(9)}}(x_9)\right), \xi_{\tilde{A}^{(10)}}(x_{10})\right).\end{aligned}$$

- Calculate $\tilde{B}^{(4)} \equiv \tilde{A}^{(6)} \otimes \tilde{B}^{(3)}$ using the function $\mathfrak{A}_2$. The membership function of $\tilde{B}^{(4)}$ is given by

$$\xi_{\tilde{B}^{(4)}}(z) = \sup_{\{(x_6,b_3):z=x_6\cdot b_3\}} \mathfrak{A}_2\left(\xi_{\tilde{A}^{(6)}}(x_6), \xi_{\tilde{B}^{(3)}}(b_3)\right).$$

- Finally, calculate $\tilde{B}^L = \tilde{B}^{(1)} \otimes \tilde{B}^{(2)} \ominus \tilde{B}^{(4)}$ using the function $\mathfrak{A}_3^L$, which is left-generated by $\mathfrak{A}_2$. Using Proposition 7.5.9, the membership function of $\tilde{B}^L$ is given by

$$\begin{aligned}\xi_{\tilde{B}^L}(z) &= \sup_{\{(b_1,b_2,b_4):z=b_1\cdot b_2-b_4\}} \mathfrak{A}_3^L\left(\xi_{\tilde{B}^{(1)}}(b_1), \xi_{\tilde{B}^{(2)}}(b_2), \xi_{\tilde{B}^{(4)}}(b_4)\right) \\ &= \sup_{\{(b_1,b_2,b_4):z=b_1\cdot b_2-b_4\}} \mathfrak{A}_2\left(\mathfrak{A}_2\left(\xi_{\tilde{B}^{(1)}}(b_1), \xi_{\tilde{B}^{(2)}}(b_2)\right), \xi_{\tilde{B}^{(4)}}(b_4)\right).\end{aligned} \tag{7.88}$$

We can see that the membership function of $\xi_{\tilde{B}^L}$ is very complicated when the membership functions of $\tilde{B}^{(1)}, \tilde{B}^{(2)}, \tilde{B}^{(4)}$ are substituted into it. By inductively using Propositions 7.5.5 and 7.5.9, we see that the membership function of $\tilde{B}^L$ can be rewritten as the following simple form

$$\xi_{\tilde{B}^L}(z) = \sup_{\{(x_1,\ldots,x_{10}):z=f(x_1,\ldots,x_{10})\}} \mathfrak{A}_{10}^L\left(\xi_{\tilde{A}^{(1)}}(x_1), \ldots, \xi_{\tilde{A}^{(10)}}(x_{10})\right), \tag{7.89}$$

where

$$f(x_1, \ldots, x_{10}) = \left(x_1 - x_2\right) \cdot \left(x_3 - x_4 + x_5\right) - \left(x_6 \cdot \left(x_7 + x_8 - x_9 - x_{10}\right)\right)$$

and $\mathfrak{A}_{10}^L$ is left-generated by $\mathfrak{A}_2$.

We can also assume that the functions $\mathfrak{A}_3^R$ and $\mathfrak{A}_4^R$ are right-generated by $\mathfrak{A}_2$. Therefore, using the above computational procedure, we can obtain the membership function of $\tilde{B}^R$ that is also very complicated. Using Propositions 7.5.7 and 7.5.10, we can similarly calculate the arithmetic expression (7.87) to obtain $\tilde{B}^R$, whose simple form of membership function is given by

$$\xi_{\tilde{B}^R}(z) = \sup_{\{(x_1,\ldots,x_{10}):z=f(x_1,\ldots,x_{10})\}} \mathfrak{A}^R_{10}\left(\xi_{\tilde{A}^{(1)}}(x_1),\ldots,\xi_{\tilde{A}^{(10)}}(x_{10})\right), \tag{7.90}$$

where the function f is the same as above and $\mathfrak{A}^R_{10}$ is right-generated by $\mathfrak{A}_2$. In general, we see that $\tilde{B}^L \neq \tilde{B}^R$. However, if the function $\mathfrak{A}_2$ is associative, then $\tilde{B}^L = \tilde{B}^R$.

On the other hand, without assuming condition (7.86), we can calculate the arithmetic expression (7.87) using the form (7.60) by directly defining a function $\mathfrak{A}_{10}$ that is different from $\mathfrak{A}^L_{10}$ and $\mathfrak{A}^U_{10}$.

Next, we want to calculate the following expression

$$\tilde{A}^{(1)} \ominus \tilde{A}^{(2)} \otimes \tilde{A}^{(3)} \otimes \tilde{A}^{(4)} \oplus \tilde{A}^{(5)} \ominus \tilde{A}^{(6)} \otimes \tilde{A}^{(7)} \oplus \tilde{A}^{(8)} \otimes \tilde{A}^{(9)} \otimes \tilde{A}^{(10)}.$$

Since the operation $\otimes$ has the highest priority for performing calculation, it means that we equivalently calculate the following expression

$$\tilde{A}^{(1)} \ominus \left(\tilde{A}^{(2)} \otimes \tilde{A}^{(3)} \otimes \tilde{A}^{(4)}\right) \oplus \tilde{A}^{(5)} \ominus \left(\tilde{A}^{(6)} \otimes \tilde{A}^{(7)}\right) \oplus \left(\tilde{A}^{(8)} \otimes \tilde{A}^{(9)} \otimes \tilde{A}^{(10)}\right). \tag{7.91}$$

Based on the function $\mathfrak{A}_2$, using Propositions 7.5.9 and 7.5.10, we can similarly use the above computational procedure to calculate the expression (7.91) by separately considering the appropriate aggregation functions $\mathfrak{A}^L_n$ or $\mathfrak{A}^R_n$ for $n \geq 3$.

Next, we want to study compatibility for aggregation functions with arithmetic operations. Continued from Example 7.5.14, we provide a comprehensive example regarding the compatibility with the arithmetic operations.

Example 7.5.15 Let $\tilde{A}^{(1)},\ldots,\tilde{A}^{(10)}$ be fuzzy sets in $\mathbb{R}$. Suppose that the function $\mathfrak{A}_2$ satisfying condition (7.86) is compatible with arithmetic operations of α-level sets. This means that, for $\odot \in \{\oplus,\ominus,\otimes\}$, the membership function of $\tilde{A} \odot \tilde{B}$ is defined by using $\mathfrak{A}_2$ and satisfies

$$\left(\tilde{A} \odot \tilde{B}\right)_\alpha = \tilde{A}_\alpha \circ \tilde{B}_\alpha \text{ for } \alpha \in (0,1],$$

where the operation $\circ \in \{+,-,\times\}$. Let us consider the arithmetic expression (7.87). We have shown that the membership functions (7.88) and (7.89) are identical. We are going to claim that the function $\mathfrak{A}^L_{10}$ that is left-generated by $\mathfrak{A}_2$ is also compatible with arithmetic operations of α-level sets. Now, we follow the steps of the computational procedure presented in Example 7.5.14 to obtain the α-level sets of expression (7.87) for $\alpha \in (0,1]$.

- Since $\tilde{B}^{(1)} \equiv \tilde{A}^{(1)} \ominus \tilde{A}^{(2)}$ uses the function $\mathfrak{A}_2$, the compatibility of $\mathfrak{A}_2$ says that

$$\tilde{B}^{(1)}_\alpha = \left(\tilde{A}^{(1)} \ominus \tilde{A}^{(2)}\right)_\alpha = \tilde{A}^{(1)}_\alpha - \tilde{A}^{(2)}_\alpha \text{ for } \alpha \in (0,1].$$

- Since $\tilde{B}^{(2)} \equiv \tilde{A}^{(3)} \ominus \tilde{A}^{(4)} \oplus \tilde{A}^{(5)}$ uses the function $\mathfrak{A}^L_3$, Proposition 7.5.9 shows $\tilde{B}^{(2)} = (\tilde{A}^{(3)} \ominus \tilde{A}^{(4)}) \oplus \tilde{A}^{(5)}$, which is obtained by using $\mathfrak{A}_2$ twice. Then, the compatibility of $\mathfrak{A}_2$ says that

$$\begin{aligned}\tilde{B}^{(2)}_\alpha &= \left(\tilde{A}^{(3)} \ominus \tilde{A}^{(4)} \oplus \tilde{A}^{(5)}\right)_\alpha = \left(\left(\tilde{A}^{(3)} \ominus \tilde{A}^{(4)}\right) \oplus \tilde{A}^{(5)}\right)_\alpha \\ &= \left(\tilde{A}^{(3)} \ominus \tilde{A}^{(4)}\right)_\alpha + \tilde{A}^{(5)}_\alpha = \tilde{A}^{(3)}_\alpha - \tilde{A}^{(4)}_\alpha + \tilde{A}^{(5)}_\alpha \text{ for } \alpha \in (0,1].\end{aligned}$$

- Since $\tilde{B}^{(3)} \equiv \tilde{A}^{(7)} \oplus \tilde{A}^{(8)} \ominus \tilde{A}^{(9)} \ominus \tilde{A}^{(10)}$ uses the function $\mathfrak{A}_4^L$, Repeatedly using Proposition 7.5.9, we have

$$\tilde{B}^{(3)} = \left(\left(\tilde{A}^{(7)} \oplus \tilde{A}^{(8)}\right) \ominus \tilde{A}^{(9)}\right) \ominus \tilde{A}^{(10)},$$

which is obtained by using $\mathfrak{A}_2$ three times. Then, the compatibility of $\mathfrak{A}_2$ says that

$$\begin{aligned}
\tilde{B}_\alpha^{(3)} &= \left(\tilde{A}^{(7)} \oplus \tilde{A}^{(8)} \ominus \tilde{A}^{(9)} \ominus \tilde{A}^{(10)}\right)_\alpha = \left(\left(\left(\tilde{A}^{(7)} \oplus \tilde{A}^{(8)}\right) \ominus \tilde{A}^{(9)}\right) \ominus \tilde{A}^{(10)}\right)_\alpha \\
&= \left(\left(\tilde{A}^{(7)} \oplus \tilde{A}^{(8)}\right) \ominus \tilde{A}^{(9)}\right)_\alpha - \tilde{A}_\alpha^{(10)} = \left(\tilde{A}^{(7)} \oplus \tilde{A}^{(8)}\right)_\alpha - \tilde{A}_\alpha^{(9)} - \tilde{A}_\alpha^{(10)} \\
&= \tilde{A}_\alpha^{(7)} + \tilde{A}_\alpha^{(8)} - \tilde{A}_\alpha^{(9)} - \tilde{A}_\alpha^{(10)} \text{ for } \alpha \in (0,1].
\end{aligned}$$

- Since $\tilde{B}^{(4)} \equiv \tilde{A}^{(6)} \otimes \tilde{B}^{(3)}$ uses the function $\mathfrak{A}_2$, the compatibility of $\mathfrak{A}_2$ says that

$$\begin{aligned}
\tilde{B}_\alpha^{(4)} &= \left(\tilde{A}^{(6)} \otimes \tilde{B}^{(3)}\right)_\alpha = \tilde{A}_\alpha^{(6)} \times \tilde{B}_\alpha^{(3)} \\
&= \tilde{A}_\alpha^{(6)} \times \left(\tilde{A}_\alpha^{(7)} + \tilde{A}_\alpha^{(8)} - \tilde{A}_\alpha^{(9)} - \tilde{A}_\alpha^{(10)}\right) \text{ for } \alpha \in (0,1].
\end{aligned}$$

- Finally, since $\tilde{B}^L \equiv \tilde{B}^{(1)} \otimes \tilde{B}^{(2)} \ominus \tilde{B}^{(4)}$ uses the function $\mathfrak{A}_3^L$, Proposition 7.5.9 shows $\tilde{B}^L = (\tilde{B}^{(1)} \otimes \tilde{B}^{(2)}) \ominus \tilde{B}^{(4)}$ that is obtained by using $\mathfrak{A}_2$ twice. Then, the compatibility of $\mathfrak{A}_2$ says that

$$\begin{aligned}
\tilde{B}_\alpha^L &= \left(\tilde{B}^{(1)} \otimes \tilde{B}^{(2)} \ominus \tilde{B}^{(4)}\right)_\alpha = \left(\left(\tilde{B}^{(1)} \otimes \tilde{B}^{(2)}\right) \ominus \tilde{B}^{(4)}\right)_\alpha \\
&= \left(\tilde{B}^{(1)} \otimes \tilde{B}^{(2)}\right)_\alpha - \tilde{B}_\alpha^{(4)} = \tilde{B}_\alpha^{(1)} \times \tilde{B}_\alpha^{(2)} - \tilde{B}_\alpha^{(4)} \\
&= \left(\tilde{A}_\alpha^{(1)} - \tilde{A}_\alpha^{(2)}\right) \times \left(\tilde{A}_\alpha^{(3)} - \tilde{A}_\alpha^{(4)} + \tilde{A}_\alpha^{(5)}\right) - \left(\tilde{A}_\alpha^{(6)} \times \left(\tilde{A}_\alpha^{(7)} + \tilde{A}_\alpha^{(8)} - \tilde{A}_\alpha^{(9)} - \tilde{A}_\alpha^{(10)}\right)\right)
\end{aligned} \tag{7.92}$$

for $\alpha \in (0,1]$.

We see that the expression (7.92) is the α-level set of membership function presented in (7.88). Since the membership functions (7.88) and (7.89) are identical, we see that the α-level set presented in (7.92) is also the α-level set of membership function presented in (7.89). This shows that the function $\mathfrak{A}_{10}^L$, which is left-generated by $\mathfrak{A}_2$ is compatible with arithmetic operations of α-level sets.

Using the functions $\mathfrak{A}_3^R$ and $\mathfrak{A}_4^R$ that are right-generated by $\mathfrak{A}_2$, we can similarly obtain the α-level set $\tilde{B}_\alpha^R$ of $\tilde{B}^R$ whose membership function is given in (7.90). Using the above computational procedure, it is clear to see that $\tilde{B}_\alpha^R = \tilde{B}_\alpha^L$ for all $\alpha \in (0,1]$, which says that $\tilde{B}^R = \tilde{B}^L$. This equality can also be realized by inductively (repeatedly) using Proposition 7.5.11. On the other hand, the function $\mathfrak{A}_{10}^R$, which is right-generated by $\mathfrak{A}_2$ is compatible with arithmetic operations of α-level sets. Although we have $\tilde{B}^R = \tilde{B}^L$, by referring to (7.89) and (7.90), $\mathfrak{A}_{10}^L$ and $\mathfrak{A}_{10}^R$ are not necessarily identical.

In general, let $\tilde{A}^{(1)}, \ldots, \tilde{A}^{(n)}$ be fuzzy sets in $\mathbb{R}$. Given a function $\mathfrak{A}_2$ and arithmetic operations $\odot_i \in \{\oplus, \ominus, \otimes\}$ for $i = 1, \ldots, n$, we want to calculate the arithmetic expression

$$\tilde{A}^{(1)} \odot_1 \cdots \odot_{n-1} \tilde{A}^{(n)}$$

based on the functions $\mathfrak{A}_n^L$ and $\mathfrak{A}_n^R$, which are left-generated and right-generated by $\mathfrak{A}_2$, respectively, where the arithmetic expression can be implicitly assumed to include parentheses. We also write

$$\tilde{B}^L \equiv \tilde{A}^{(1)} \odot_1 \cdots \odot_{n-1} \tilde{A}^{(n)}(\mathfrak{A}_n^L) \text{ and } \tilde{B}^R \equiv \tilde{A}^{(1)} \odot_1 \cdots \odot_{n-1} \tilde{A}^{(n)}(\mathfrak{A}_n^R);$$

that is, $\tilde{B}^L$ and $\tilde{B}^R$ are obtained using $\mathfrak{A}_n^L$ and $\mathfrak{A}_n^R$, respectively. Then, we have the following interesting result

Theorem 7.5.16 *Let $\tilde{A}^{(1)}, \ldots, \tilde{A}^{(n)}$ be fuzzy sets in $\mathbb{R}$. Suppose that the function $\mathfrak{A}_2 : [0,1]^2 \to [0,1]$ satisfies the following condition*

$$\mathfrak{A}_2(\alpha_1, \alpha_2) \leq \alpha \textit{ if and only if } \alpha_1 \leq \alpha \textit{ or } \alpha_2 \leq \alpha,$$

where $\alpha \in [0,1]$ and $\mathfrak{A}_2$ is not necessarily commutative. Let $\mathfrak{A}_n^L$ and $\mathfrak{A}_n^R$ be functions that are left-generated and right-generated by $\mathfrak{A}_2$, respectively. For the arithmetic operations $\odot_i \in \{\oplus, \ominus, \otimes\}$ for $i = 1, \ldots, n$, we write

$$\tilde{B}^L \equiv \tilde{A}^{(1)} \odot_1 \cdots \odot_{n-1} \tilde{A}^{(n)}(\mathfrak{A}_n^L) \textit{ and } \tilde{B}^R \equiv \tilde{A}^{(1)} \odot_1 \cdots \odot_{n-1} \tilde{A}^{(n)}(\mathfrak{A}_n^R),$$

where the parentheses are implicitly assumed to be included. Assume that $\mathfrak{A}_2$ is compatible with arithmetic operations of α-level sets. Then, the functions $\mathfrak{A}_n^L$ and $\mathfrak{A}_n^R$ are also compatible with arithmetic operations of α-level sets. Moreover, we have $\tilde{B}^R = \tilde{B}^L$.

Proof. By referring to Example 7.5.15 and inductively using Proposition 7.5.11, we can obtain the desired results. ■

7.6 Binary Operations

Let $\tilde{A}$ and $\tilde{B}$ be two fuzzy intervals. Suppose that the binary operation $\odot$ is any one of $\{\oplus, \ominus, \otimes\}$. Theorem 7.1.2 says that $\tilde{A} \odot \tilde{B}$ is also a fuzzy interval satisfying

$$(\tilde{A} \odot \tilde{B})_\alpha = \tilde{A}_\alpha \circ \tilde{B}_\alpha \text{ for any } \alpha \in [0,1],$$

where the operation $\tilde{A} \odot \tilde{B}$ is based on the extension principle using a minimum function.

In general, given any two fuzzy sets $\tilde{A}$ and $\tilde{B}$ in $\mathbb{R}$, without using the minimum function, we want to study the operation $\tilde{A} \odot_{\mathfrak{L}} \tilde{B}$ based on their α-level sets. In other words, we want to study whether there exists a fuzzy set $\tilde{C}$ in $\mathbb{R}$ satisfying

$$\tilde{C}_\alpha = \tilde{A}_\alpha \circ \tilde{B}_\alpha$$

for all $\alpha \in [0,1]$, where $\circ$ is a binary operation. When the existence of $\tilde{C}$ is guaranteed, we write $\tilde{C} = \tilde{A} \odot_{\mathfrak{L}} \tilde{B}$. Some sufficient conditions are needed to guarantee the existence of such $\tilde{C}$. When the existence of $\tilde{C}$ is guaranteed, according to the decomposition theorem presented in Theorem 2.2.13, the membership function of $\tilde{C} = \tilde{A} \odot_{\mathfrak{L}} \tilde{B}$ is given by

$$\xi_{\tilde{C}}(z) = \xi_{\tilde{A} \odot_{\mathfrak{L}} \tilde{B}}(z) = \sup_{\alpha \in [0,1]} \alpha \cdot \chi_{\tilde{A}_\alpha \circ \tilde{B}_\alpha}(z) = \sup_{\alpha \in [0,1]} \alpha \cdot \chi_{\tilde{C}_\alpha}(z).$$

In what follows, we shall consider three types of binary operations between $\tilde{A}$ and $\tilde{B}$.

7.6.1 First Type of Binary Operation

In the first type of binary operation, we consider the binary operation $\odot_{\mathfrak{g}}$ between two fuzzy sets $\tilde{A}$ and $\tilde{B}$ in $\mathbb{R}^m$ and write $\tilde{A} \odot_{\mathfrak{g}} \tilde{B}$. We also consider the binary operation $\circ$ between α-level sets $\tilde{A}_\alpha$ and $\tilde{B}_\alpha$ of $\tilde{A}$ and $\tilde{B}$, respectively, and write $\tilde{A}_\alpha \circ \tilde{B}_\alpha$. Suppose that there exists a fuzzy set $\tilde{C}$ in $\mathbb{R}^m$ satisfying $\tilde{C}_\alpha = \tilde{A}_\alpha \circ \tilde{B}_\alpha$ for all $\alpha \in [0,1]$, where $\circ$ is a pre-defined binary operation. Then, we define $\tilde{A} \odot_{\mathfrak{g}} \tilde{B} = \tilde{C}$. The first type of binary operation is defined as follows:

$$\tilde{A} \odot_{\mathfrak{g}} \tilde{B} = \tilde{C} \text{ when } \tilde{C}_\alpha = \tilde{A}_\alpha \circ \tilde{B}_\alpha \text{ for each } \alpha \in [0,1]. \tag{7.93}$$

In other words, the binary operation $\tilde{A} \odot_{\mathfrak{g}} \tilde{B}$ is based on the pre-defined binary operation $\tilde{A}_\alpha \circ \tilde{B}_\alpha$ for all $\alpha \in [0,1]$. Some special cases of the binary operation $\tilde{A} \odot_{\mathfrak{g}} \tilde{B}$ are presented below.

- According to (7.93) by taking the binary operations $\odot_{\mathfrak{g}} = \oplus_{\mathfrak{g}}$ and $\circ = +$, Puri and Ralescu [92] and Diamond and Kloeden [21] considered the addition

$$\tilde{A} \oplus_{\mathfrak{g}} \tilde{B} = \tilde{C} \text{ when } \tilde{C}_\alpha = \tilde{A}_\alpha + \tilde{B}_\alpha = \left\{a+b : a \in \tilde{A}_\alpha \text{ and } b \in \tilde{B}_\alpha\right\} \tag{7.94}$$

  for each $\alpha \in [0,1]$. This means that if there exists a fuzzy set $\tilde{C}$ satisfying $\tilde{C}_\alpha = \tilde{A}_\alpha + \tilde{B}_\alpha$ for all $\alpha \in [0,1]$, then we define $\tilde{A} \oplus_{\mathfrak{g}} \tilde{B} = \tilde{C}$. It is well known that the existence of such fuzzy set $\tilde{C}$ is always guaranteed.
- According to (7.93) by taking the binary operations $\odot_{\mathfrak{g}} = \ominus_{\mathfrak{g}}$ and $\circ = -$, we consider the difference

$$\tilde{A} \ominus_{\mathfrak{g}} \tilde{B} = \tilde{C} \text{ when } \tilde{C}_\alpha = \tilde{A}_\alpha - \tilde{B}_\alpha = \left\{a-b : a \in \tilde{A}_\alpha \text{ and } b \in \tilde{B}_\alpha\right\} \tag{7.95}$$

  for each $\alpha \in [0,1]$. This means that if there exists a fuzzy set $\tilde{C}$ satisfying $\tilde{C}_\alpha = \tilde{A}_\alpha - \tilde{B}_\alpha$ for all $\alpha \in [0,1]$, then we define $\tilde{A} \ominus_{\mathfrak{g}} \tilde{B} = \tilde{C}$. It is also well known that the existence of such fuzzy set $\tilde{C}$ is always guaranteed.
- Let $\tilde{A}$ and $\tilde{B}$ be two fuzzy intervals. Since the α-level sets $\tilde{A}_\alpha$ and $\tilde{B}_\alpha$ are bounded closed intervals given by $\tilde{A}_\alpha = [\tilde{A}^L_\alpha, \tilde{A}^U_\alpha]$ and $\tilde{B}_\alpha = [\tilde{B}^L_\alpha, \tilde{B}^U_\alpha]$, respectively, for $\alpha \in [0,1]$, we define the binary operation $\circ = \boxminus_H$ as

$$\tilde{A}_\alpha \boxminus_H \tilde{B}_\alpha = \left[\min\left\{\tilde{A}^L_\alpha - \tilde{B}^L_\alpha, \tilde{A}^U_\alpha - \tilde{B}^U_\alpha\right\}, \max\left\{\tilde{A}^L_\alpha - \tilde{B}^L_\alpha, \tilde{A}^U_\alpha - \tilde{B}^U_\alpha\right\}\right]. \tag{7.96}$$

  According to (7.93), by taking $\odot_{\mathfrak{g}} = \ominus_H$, we define the so-called **Hausdorff difference** $\tilde{A} \ominus_H \tilde{B}$ between fuzzy intervals $\tilde{A}$ and $\tilde{B}$ as

$$\tilde{A} \ominus_H \tilde{B} = \tilde{C} \text{ when } \tilde{C}_\alpha = \tilde{A}_\alpha \boxminus_H \tilde{B}_\alpha \text{ for each } \alpha \in [0,1]. \tag{7.97}$$

  This means that if there exists a fuzzy interval $\tilde{C}$ satisfying $\tilde{C}_\alpha = \tilde{A}_\alpha \boxminus_H \tilde{B}_\alpha$ for all $\alpha \in [0,1]$, then we define $\tilde{A} \ominus_H \tilde{B} = \tilde{C}$. The existence of such fuzzy interval $\tilde{C}$ will be studied in this section.
- Bede and Stefanini [8] considered the so-called **generalized difference** $\tilde{C}$ (which is called the **type-I-generalized difference** in this section) of fuzzy intervals $\tilde{A}$ and $\tilde{B}$. By referring to (7.96), we define the binary operation $\circ = \boxminus_{G_1}$ as

$$\tilde{A}_\alpha \boxminus_{G_1} \tilde{B}_\alpha = \text{cl}\left(\bigcup_{\alpha \le \beta \le 1} \tilde{A}_\beta \boxminus_H \tilde{B}_\beta\right) \tag{7.98}$$

$$= \text{cl}\left(\bigcup_{\alpha \le \beta \le 1} \left[\min\left\{\tilde{A}^L_\beta - \tilde{B}^L_\beta, \tilde{A}^U_\beta - \tilde{B}^U_\beta\right\}, \max\left\{\tilde{A}^L_\beta - \tilde{B}^L_\beta, \tilde{A}^U_\beta - \tilde{B}^U_\beta\right\}\right]\right). \tag{7.99}$$

According to (7.93), the type-I-generalized difference is defined by

$$\tilde{A} \ominus_{G_1} \tilde{B} = \tilde{C} \text{ when } \tilde{C}_\alpha = \tilde{A}_\alpha \boxminus_{G_1} \tilde{B}_\alpha \text{ for each } \alpha \in [0,1]. \tag{7.100}$$

This means that if there exists a fuzzy interval $\tilde{C}$ satisfying $\tilde{C}_\alpha = \tilde{A}_\alpha \boxminus_{G_1} \tilde{B}_\alpha$ for all $\alpha \in [0,1]$, then we define $\tilde{A} \ominus_{G_1} \tilde{B} = \tilde{C}$. The existence of such fuzzy interval $\tilde{C}$ is not discussed in Bede and Stefanini [8]. In this section, we shall study the existence of such fuzzy interval $\tilde{C}$ in a more general sense.

- Gomes and Barros [42] modified (7.98) to define another so-called **generalized difference** $\tilde{C}$ (which is called the **type-II-generalized difference** in this section) of fuzzy intervals $\tilde{A}$ and $\tilde{B}$. The convex hull of a set A is denoted by conv(A). We define the binary operation $\circ = \boxminus_{G_2}$ as follows

$$\tilde{A}_\alpha \boxminus_{G_2} \tilde{B}_\alpha = \text{cl}\left(\text{conv}\left(\bigcup_{\alpha \le \beta \le 1} \tilde{A}_\beta \boxminus_H \tilde{B}_\beta\right)\right) \tag{7.101}$$

$$= \left[\inf_{\alpha \le \beta \le 1} \min\left\{\tilde{A}^L_\beta - \tilde{B}^L_\beta, \tilde{A}^U_\beta - \tilde{B}^U_\beta\right\}, \sup_{\alpha \le \beta \le 1} \max\left\{\tilde{A}^L_\beta - \tilde{B}^L_\beta, \tilde{A}^U_\beta - \tilde{B}^U_\beta\right\}\right]. \tag{7.102}$$

According to (7.93), the type-II-generalized difference is defined by

$$\tilde{A} \ominus_{G_2} \tilde{B} = \tilde{C} \text{ when } \tilde{C}_\alpha = \tilde{A}_\alpha \boxminus_{G_2} \tilde{B}_\alpha \text{ for each } \alpha \in [0,1]. \tag{7.103}$$

This means that if there exists a fuzzy interval $\tilde{C}$ satisfying $\tilde{C}_\alpha = \tilde{A}_\alpha \boxminus_{G_2} \tilde{B}_\alpha$ for all $\alpha \in [0,1]$, then we define $\tilde{A} \ominus_{G_2} \tilde{B} = \tilde{C}$. The existence of such fuzzy interval $\tilde{C}$ was not presented in Gomes and Barros [42]. In this section, we shall study the existence of such fuzzy interval $\tilde{C}$ in a more general sense.

Example 7.6.1 We consider the triangular fuzzy number $\tilde{A} = (a^L, a, a^U)$ with membership function given by

$$\xi_{\tilde{A}}(r) = \begin{cases} \dfrac{r - a^L}{a - a^L} & \text{if } a^L \le r \le a \\ \dfrac{a^U - r}{a^U - a} & \text{if } a < r \le a^U \\ 0 & \text{otherwise.} \end{cases}$$

Then, the α-level set is given by

$$\tilde{A}_\alpha = \left[(1-\alpha)\cdot a^L + \alpha \cdot a, (1-\alpha)\cdot a^U + \alpha \cdot a\right];$$

that is,

$$\tilde{A}^L_\alpha = (1-\alpha)\cdot a^L + \alpha \cdot a \text{ and } \tilde{A}^U_\alpha = (1-\alpha)\cdot a^U + \alpha \cdot a.$$

Let $\tilde{B} = (b^L, b, b^U)$ be another triangular fuzzy number. Then

$$\tilde{A}^L_\alpha - \tilde{B}^L_\alpha = (1-\alpha)\cdot a^L + \alpha \cdot a - (1-\alpha)\cdot b^L - \alpha \cdot b = (1-\alpha)\cdot\left(a^L - b^L\right) + \alpha \cdot (a - b)$$

and

$$\begin{aligned} \tilde{A}^U_\alpha - \tilde{B}^U_\alpha &= (1-\alpha)\cdot a^U + \alpha \cdot a - (1-\alpha)\cdot b^U - \alpha \cdot b \\ &= (1-\alpha)\cdot\left(a^U - b^U\right) + \alpha \cdot (a - b). \end{aligned}$$

According to (7.99), we have the following cases.

- Suppose that $a^L - b^L \leq a^U - b^U$. Then, for each $\alpha \in [0,1]$,

$$\tilde{A}_\alpha \boxminus_{G_1} \tilde{B}_\alpha$$
$$= \text{cl}\left(\bigcup_{\beta \leq \beta \leq 1} \left[(1-\beta)\cdot\left(a^L - b^L\right) + \beta\cdot(a-b), (1-\beta)\cdot\left(a^U - b^U\right) + \beta\cdot(a-b)\right] \right).$$

- Suppose that $a^L - b^L > a^U - b^U$. Then, for each $\alpha \in [0,1]$,

$$\tilde{A}_\alpha \boxminus_{G_1} \tilde{B}_\alpha$$
$$= \text{cl}\left(\bigcup_{\alpha \leq \beta \leq 1} \left[(1-\beta)\cdot\left(a^U - b^U\right) + \beta\cdot(a-b), (1-\beta)\cdot\left(a^L - b^L\right) + \beta\cdot(a-b)\right] \right).$$

According to (7.102), we have the following cases.

- Suppose that $a^L - b^L \leq a^U - b^U$. Then, for each $\alpha \in [0,1]$,

$$\tilde{A}_\alpha \boxminus_{G_2} \tilde{B}_\alpha = \left[\inf_{\alpha \leq \beta \leq 1} \left((1-\beta)\cdot\left(a^L - b^L\right) + \beta\cdot(a-b)\right), \right.$$
$$\left. \sup_{\alpha \leq \beta \leq 1} \left((1-\beta)\cdot\left(a^U - b^U\right)\right) + \beta\cdot(a-b) \right].$$

- Suppose that $a^L - b^L > a^U - b^U$. Then, for each $\alpha \in [0,1]$,

$$\tilde{A}_\alpha \boxminus_{G_2} \tilde{B}_\alpha = \left[\inf_{\alpha \leq \beta \leq 1} \left((1-\beta)\cdot\left(a^U - b^U\right) + \beta\cdot(a-b)\right), \right.$$
$$\left. \sup_{\alpha \leq \beta \leq 1} \left((1-\beta)\cdot\left(a^L - b^L\right)\right) + \beta\cdot(a-b) \right].$$

This example will be continued in the subsequent examples.

The existence of fuzzy set $\tilde{C}$ in $\mathbb{R}^m$ presented in (7.93) should be verified as follows. Suppose that such fuzzy set $\tilde{C}$ in $\mathbb{R}^m$ exists. Then, according to the decomposition theorem presented in Theorem 2.2.13, the membership function $\xi_{\tilde{C}}$ can be expressed as

$$\xi_{\tilde{C}}(x) = \sup_{\alpha \in [0,1]} \alpha \cdot \chi_{\tilde{C}_\alpha}(x) = \sup_{0 < \alpha \leq 1} \alpha \cdot \chi_{\tilde{C}_\alpha}(x). \tag{7.104}$$

According to (7.93), we must have

$$\tilde{C}_\alpha = \tilde{A}_\alpha \circ \tilde{B}_\alpha \text{ for all } \alpha \in [0,1]. \tag{7.105}$$

Therefore, the fuzzy set $\tilde{C}$ in $\mathbb{R}^m$ exists if and only if the membership function of $\tilde{C}$ shown in (7.104) and its α-level set $\tilde{C}_\alpha$ shown in (7.105) are consistent. More precisely, the fuzzy set $\tilde{C}$ in $\mathbb{R}^m$ exists if and only if the following equalities are satisfied:

$$\tilde{C}_0 = \text{cl}\left(\left\{x : \xi_{\tilde{C}}(x) > 0\right\}\right) = \text{cl}\left(\left\{x : \sup_{\alpha \in [0,1]} \alpha \cdot \chi_{\tilde{C}_\alpha}(x) > 0\right\}\right) = \tilde{A}_0 \circ \tilde{B}_0 \tag{7.106}$$

and

$$\tilde{C}_\alpha = \left\{x : \xi_{\tilde{C}}(x) \geq \alpha\right\} = \left\{x : \sup_{\alpha \in [0,1]} \alpha \cdot \chi_{\tilde{C}_\alpha}(x) \geq \alpha\right\} = \tilde{A}_\alpha \circ \tilde{B}_\alpha \text{ for all } \alpha \in (0,1]. \tag{7.107}$$

In this section, we shall investigate the sufficient conditions to guarantee the equalities (7.106) and (7.107), which presents the existence of the binary operation $\tilde{A} \odot_{\mathfrak{L}} \tilde{B}$.

Let us go back to check the generalized differences $\tilde{A} \ominus_{G_1} \tilde{B}$ and $\tilde{A} \ominus_{G_2} \tilde{B}$ presented in (7.100) and (7.103), which were proposed by Bede and Stefanini [8] and Gomes and Barros [42], respectively. The important issue is to know the membership functions of $\tilde{A} \ominus_{G_1} \tilde{B}$ and $\tilde{A} \ominus_{G_2} \tilde{B}$. However, these researchers did not present the precise expressions of the membership functions of $\tilde{A} \ominus_{G_1} \tilde{B}$ and $\tilde{A} \ominus_{G_2} \tilde{B}$. Therefore, the existence of such fuzzy intervals was not studied by them.

Suppose that the generalized differences $\tilde{A} \ominus_{G_1} \tilde{B} = \tilde{C}^{(1)}$ and $\tilde{A} \ominus_{G_2} \tilde{B} = \tilde{C}^{(2)}$ exist. Then, from (7.100) and (7.103), the α-level sets $\tilde{C}^{(1)}_\alpha$ and $\tilde{C}^{(2)}_\alpha$ of $\tilde{C}^{(1)}$ and $\tilde{C}^{(2)}$ are, respectively, given by

$$\tilde{C}^{(1)}_\alpha = \tilde{A}_\alpha \boxminus_{G_1} \tilde{B}_\alpha \text{ and } \tilde{C}^{(2)}_\alpha = \tilde{A}_\alpha \boxminus_{G_2} \tilde{B}_\alpha. \tag{7.108}$$

According to the decomposition theorem presented in Theorem 2.2.13, the membership functions $\xi_{\tilde{C}^{(1)}}$ and $\xi_{\tilde{C}^{(2)}}$ of the type-I-generalized difference and type-II-generalized difference, respectively, can be expressed as

$$\xi_{\tilde{C}^{(1)}}(x) = \sup_{\alpha\in[0,1]} \alpha \cdot \chi_{\tilde{C}^{(1)}_\alpha}(x) \text{ and } \xi_{\tilde{C}^{(2)}}(x) = \sup_{\alpha\in[0,1]} \alpha \cdot \chi_{\tilde{C}^{(2)}_\alpha}(x).$$

According to the equalities (7.106) and (7.107), we consider the following two cases.

- The type-I-generalized difference $\tilde{C}^{(1)}$ exists if and only if the following equalities are satisfied: by referring to (7.108)

$$\begin{aligned} \tilde{C}^{(1)}_0 &= \mathrm{cl}\left(\left\{x : \xi_{\tilde{C}^{(1)}}(x) > 0\right\}\right) \\ &= \mathrm{cl}\left(\left\{x : \sup_{\alpha\in[0,1]} \alpha \cdot \chi_{\tilde{C}^{(1)}}(x) > 0\right\}\right) = \tilde{A}_0 \boxminus_{G_1} \tilde{B}_0 \end{aligned} \tag{7.109}$$

  and

$$\begin{aligned} \tilde{C}^{(1)}_\alpha &= \left\{x : \xi_{\tilde{C}^{(1)}}(x) \geq \alpha\right\} \\ &= \left\{x : \sup_{\alpha\in[0,1]} \alpha \cdot \chi_{\tilde{C}^{(1)}}(x) \geq \alpha\right\} = \tilde{A}_\alpha \boxminus_{G_1} \tilde{B}_\alpha \text{ for all } \alpha \in (0,1]. \end{aligned} \tag{7.110}$$

- The type-II-generalized difference $\tilde{C}^{(2)}$ exists if and only if the following equalities are satisfied: by referring to (7.108) again

$$\begin{aligned} \tilde{C}^{(2)}_0 &= \mathrm{cl}\left(\left\{x : \xi_{\tilde{C}^{(2)}}(x) > 0\right\}\right) \\ &= \mathrm{cl}\left(\left\{x : \sup_{\alpha\in[0,1]} \alpha \cdot \chi_{\tilde{C}^{(2)}}(x) > 0\right\}\right) = \tilde{A}_0 \boxminus_{G_2} \tilde{B}_0 \end{aligned} \tag{7.111}$$

  and

$$\begin{aligned} \tilde{C}^{(2)}_\alpha &= \left\{x : \xi_{\tilde{C}^{(2)}}(x) \geq \alpha\right\} \\ &= \left\{x : \sup_{\alpha\in[0,1]} \alpha \cdot \chi_{\tilde{C}^{(2)}}(x) \geq \alpha\right\} = \tilde{A}_\alpha \boxminus_{G_2} \tilde{B}_\alpha \text{ for all } \alpha \in (0,1]. \end{aligned} \tag{7.112}$$

The equalities (7.109)–(7.112) were not mentioned and studied by Bede and Stefanini [8] and Gomes and Barros [42]. In this section, we shall investigate the sufficient conditions to guarantee the equalities (7.109)-(7.112), which means to guarantee the existence of generalized differences.

On the other hand, according to the decomposition theorem, the membership function $\xi_{\tilde{C}}$ of the Hausdorff difference $\tilde{C} = \tilde{A} \ominus_H \tilde{B}$ presented in (7.97) can also be expressed as

$$\xi_{\tilde{C}}(x) = \sup_{\alpha \in [0,1]} \alpha \cdot \chi_{\tilde{C}_\alpha}(x).$$

Therefore, the Hausdorff difference $\tilde{C}$ exists if and only if

$$\tilde{C}_\alpha = \mathrm{cl}\left(\left\{x : \xi_{\tilde{C}}(x) > 0\right\}\right) = \mathrm{cl}\left(\left\{x : \sup_{\alpha \in [0,1]} \alpha \cdot \chi_{\tilde{C}_\alpha}(x) > 0\right\}\right) = \tilde{A}_0 \boxminus_H \tilde{B}_0 \tag{7.113}$$

and

$$\tilde{C}_\alpha = \left\{x : \xi_{\tilde{C}}(x) \geq \alpha\right\} = \left\{x : \sup_{\alpha \in [0,1]} \alpha \cdot \chi_{\tilde{C}_\alpha}(x) \geq \alpha\right\} = \tilde{A}_\alpha \boxminus_H \tilde{B}_\alpha \text{ for all } \alpha \in (0,1]. \tag{7.114}$$

In this section, we shall also investigate the sufficient conditions to guarantee the equalities (7.113) and (7.114), which also means to guarantee the existence of the Hausdorff difference $\tilde{A} \ominus_H \tilde{B}$.

7.6.2 Second Type of Binary Operation

As we mentioned before, the 0-level set is defined by (2.4). Otherwise, the 0-level set is defined to be the whole set $\mathbb{R}^m$. That is to say, the definitions of 0-level sets are different. In order to avoid these conflicts, we are going to propose a second type of binary operation without considering the 0-level sets $\tilde{A}_0$ and $\tilde{B}_0$.

Suppose that there exists a fuzzy set $\tilde{C}^*$ in $\mathbb{R}^m$ satisfying $\tilde{C}^*_\alpha = \tilde{A}_\alpha \circ \tilde{B}_\alpha$ for all $\alpha \in (0,1]$, where $\alpha = 0$ is not included. The second type of binary operation is defined as follows:

$$\tilde{A} \odot^*_{\mathfrak{L}} \tilde{B} = \tilde{C}^* \text{ when } \tilde{C}^*_\alpha = \tilde{A}_\alpha \circ \tilde{B}_\alpha \text{ for each } \alpha \in (0,1]. \tag{7.115}$$

According to the decomposition theorem, the membership function $\xi_{\tilde{C}^*}$ can be expressed as

$$\xi_{\tilde{C}^*}(x) = \sup_{\alpha \in [0,1]} \alpha \cdot \chi_{\tilde{C}^*_\alpha}(x) = \sup_{0<\alpha\leq 1} \alpha \cdot \chi_{\tilde{C}^*_\alpha}(x). \tag{7.116}$$

Therefore, the fuzzy set $\tilde{C}^*$ in $\mathbb{R}^m$ exists if and only if

$$\tilde{C}^*_\alpha = \left\{x : \xi_{\tilde{C}^*}(x) \geq \alpha\right\} = \left\{x : \sup_{\alpha \in [0,1]} \alpha \cdot \chi_{\tilde{C}^*_\alpha}(x) \geq \alpha\right\} = \tilde{A}_\alpha \circ \tilde{B}_\alpha \text{ for all } \alpha \in (0,1], \tag{7.117}$$

where $\alpha = 0$ is not included.

Example 7.6.2 Continued from Example 7.6.1, we can consider $\tilde{A}_\alpha \boxminus_{G_1} \tilde{B}_\alpha$ and $\tilde{A}_\alpha \boxminus_{G_2} \tilde{B}_\alpha$ for each $\alpha \in (0,1]$, where $\alpha = 0$ is not included. The purpose is to find a fuzzy set $\tilde{C}^*$ satisfying $\tilde{C}^*_\alpha = \tilde{A}_\alpha \boxminus_{G_1} \tilde{B}_\alpha$ for all $\alpha \in (0,1]$. If this fuzzy set $\tilde{C}^*$ exists, then we define $\tilde{A} \ominus^*_{G_1} \tilde{B} = \tilde{C}^*$ and, according to the decomposition theorem, its membership function is given by

$$\xi_{\tilde{A} \ominus^*_{G_1} \tilde{B}}(x) = \xi_{\tilde{C}}(x) = \sup_{0<\alpha\leq 1} \alpha \cdot \chi_{\tilde{C}^*_\alpha}(x) = \sup_{0<\alpha\leq 1} \alpha \cdot \chi_{\tilde{A}_\alpha \boxminus_{G_1} \tilde{B}_\alpha}(x).$$

Similarly, we can also try to find a fuzzy set $\tilde{C}^*$ satisfying $\tilde{C}^*_\alpha = \tilde{A}_\alpha \boxminus_{G_2} \tilde{B}_\alpha$ for all $\alpha \in (0,1]$. If this fuzzy number $\tilde{C}^*$ exists, then we define $\tilde{A} \ominus^*_{G_2} \tilde{B} = \tilde{C}^*$ and its membership function can also be obtained using the decomposition theorem.

It is clear that the existence of $\tilde{A} \odot_{\mathfrak{L}} \tilde{B}$ implies the existence of $\tilde{A} \odot_{\mathfrak{L}}^{*} \tilde{B}$. The converse is not necessarily true. However, we have the following interesting result.

Proposition 7.6.3 *Suppose that the existence of* $\tilde{A} \odot_{\mathfrak{L}} \tilde{B}$ *is guaranteed. Then*

$$\tilde{A} \odot_{\mathfrak{L}} \tilde{B} = \tilde{A} \odot_{\mathfrak{L}}^{*} \tilde{B}$$

and

$$\tilde{C}_0 = \tilde{C}_0^{*} = cl\left(\bigcup_{0<\alpha\leq 1} \tilde{A}_\alpha \circ \tilde{B}_\alpha\right) = \tilde{A}_0 \circ \tilde{B}_0.$$

Proof. We first note that the existence of $\tilde{C} = \tilde{A} \odot_{\mathfrak{L}} \tilde{B}$ and $\tilde{C}^{*} = \tilde{A} \odot_{\mathfrak{L}}^{*} \tilde{B}$ are simultaneously guaranteed. We also have

$$\left\{x : \xi_{\tilde{C}^{*}}(x) \geq \alpha\right\} = \tilde{C}_\alpha^{*} = \tilde{A}_\alpha \circ \tilde{B}_\alpha = \tilde{C}_\alpha = \left\{x : \xi_{\tilde{C}}(x) \geq \alpha\right\}$$

for all $\alpha \in (0, 1]$. The result follows from Proposition 1.4.8 immediately. ■

Proposition 7.6.3 says that if $\tilde{A} \odot_{\mathfrak{L}} \tilde{B} = \tilde{C}$ exists then $\tilde{C} = \tilde{C}^{*}$; that is, the first and second type of binary operations are equivalent. In this case, it is convenient to consider the second type of binary operation without considering the 0-level sets. However the sufficient conditions regarding the existence of $\tilde{A} \odot_{\mathfrak{L}} \tilde{B} = \tilde{C}$ and $\tilde{A} \odot_{\mathfrak{L}}^{*} \tilde{B} = \tilde{C}^{*}$ can be different. Therefore, we need to study them separately.

Example 7.6.4 Continued from Examples 7.6.1 and 7.6.2, if $\tilde{C}$ in Example 7.6.1 exists, then Proposition 7.6.3 says that $\tilde{C} = \tilde{C}^{*}$, where $\tilde{C}^{*}$ is given in Example 7.6.2. As a matter of fact, the fuzzy set $\tilde{C}$ exists, which will be realized in the subsequent discussion.

The sufficient conditions for guaranteeing the existence of $\tilde{A} \odot_{\mathfrak{L}}^{*} \tilde{B}$ and $\tilde{A} \odot_{\mathfrak{L}} \tilde{B}$ will be provided in the subsequent discussion. We shall also claim that the sufficient conditions for guaranteeing the existence of $\tilde{A} \odot_{\mathfrak{L}}^{*} \tilde{B}$ is weaker than that of $\tilde{A} \odot_{\mathfrak{L}} \tilde{B}$, which will be presented in Theorems 7.6.14 and 7.6.19 below, since the binary operation $\tilde{A} \odot_{\mathfrak{L}}^{*} \tilde{B}$ does not include the 0-level sets.

7.6.3 Third Type of Binary Operation

For the third type of binary operation, let

$$M_\alpha = \tilde{A}_\alpha \circ \tilde{B}_\alpha \text{ for all } \alpha \in (0, 1],$$

where $\alpha = 0$ is not included. Using the expression in the decomposition theorem, we consider the more general binary operation $\tilde{A} \odot_{\mathfrak{G}} \tilde{B}$ in which the membership function is defined by

$$\xi_{\tilde{A} \odot_{\mathfrak{G}} \tilde{B}}(x) = \sup_{0<\alpha\leq 1} \alpha \cdot \chi_{M_\alpha}(x), \tag{7.118}$$

where M_0 is not considered. It is clear that this general binary operation $\tilde{A} \odot_{\mathfrak{G}} \tilde{B}$ always exists.

Given two fuzzy intervals $\tilde{A}$ and $\tilde{B}$, by referring to (7.96), let

$$M_\alpha = \tilde{A}_\alpha \boxminus_H \tilde{B}_\alpha = \left[\min\left\{\tilde{A}^L_\alpha - \tilde{B}^L_\alpha, \tilde{A}^U_\alpha - \tilde{B}^U_\alpha\right\}, \max\left\{\tilde{A}^L_\alpha - \tilde{B}^L_\alpha, \tilde{A}^U_\alpha - \tilde{B}^U_\alpha\right\}\right]$$

for all $\alpha \in (0,1]$. According to (7.118), we can define the so-called **natural Hausdorff difference** $\tilde{A} \ominus_{NH} \tilde{B}$ with membership function given by

$$\xi_{\tilde{A} \ominus_{NH} \tilde{B}}(x) = \sup_{0<\alpha\le 1} \alpha \cdot \chi_{M_\alpha}(x). \tag{7.119}$$

The natural Hausdorff difference $\tilde{A} \ominus_{NH} \tilde{B}$ always exists. However, the existence of the Hausdorff difference $\tilde{A} \ominus_H \tilde{B}$ should be verified using (7.113) and (7.114).

Recall that the definitions of generalized differences $\tilde{A} \ominus_{G_1} \tilde{B}$ and $\tilde{A} \ominus_{G_2} \tilde{B}$ are also based on (7.96). Therefore, their existence should be verified using (7.109) and (7.110), and (7.111) and (7.112), respectively. In this chapter, we shall investigate the relationships among the different types of differences $\tilde{A} \ominus_{G_1} \tilde{B}$, $\tilde{A} \ominus_{G_2} \tilde{B}$, $\tilde{A} \ominus_H \tilde{B}$, and $\tilde{A} \ominus_{NH} \tilde{B}$.

Example 7.6.5 Continued from Example 7.6.1, we can define any binary operation $M_\alpha = \tilde{A}_\alpha \boxminus \tilde{B}_\alpha$ for all $\alpha \in (0,1]$. Then, we can consider the general difference $\tilde{A} \ominus_G \tilde{B}$ whose membership function is given by

$$\xi_{\tilde{A} \ominus_G \tilde{B}}(x) = \sup_{0<\alpha\le 1} \alpha \cdot \chi_{M_\alpha}(x) = \sup_{0<\alpha\le 1} \alpha \cdot \chi_{\tilde{A}_\alpha \boxminus \tilde{B}_\alpha}(x).$$

This general difference $\tilde{A} \ominus_G \tilde{B}$ always exists. In particular, we define

$$\begin{aligned} M_\alpha = \tilde{A}_\alpha \boxminus_H \tilde{B}_\alpha &= \left[\min\left\{\tilde{A}^L_\alpha - \tilde{B}^L_\alpha, \tilde{A}^U_\alpha - \tilde{B}^U_\alpha\right\}, \max\left\{\tilde{A}^L_\alpha - \tilde{B}^L_\alpha, \tilde{A}^U_\alpha - \tilde{B}^U_\alpha\right\}\right] \\ &= \left[(1-\alpha) \cdot \min\left\{a^L - b^L, a^U - b^U\right\} + \alpha \cdot (a-b), (1-\alpha)\right. \\ &\quad \left. \cdot \max\left\{a^L - b^L, a^U - b^U\right\} + \alpha \cdot (a-b)\right] \end{aligned}$$

for all $\alpha \in (0,1]$. Then, we can define the natural Hausdorff difference $\tilde{A} \ominus_{NH} \tilde{B}$ with membership function given by (7.119), which always exists. We also notice that this family $\{M_\alpha : 0 < \alpha \le 1\}$ is not nested.

Theorem 7.6.6 *Let $\tilde{A}$ and $\tilde{B}$ be any two fuzzy sets in $\mathbb{R}^m$, and let $M_\alpha = \tilde{A}_\alpha \circ \tilde{B}_\alpha$ for all $\alpha \in (0,1]$, where $\alpha = 0$ is not included and $\circ$ denotes an arbitrary binary operation between the α-level sets $\tilde{A}_\alpha$ and $\tilde{B}_\alpha$. Suppose that, for any fixed $x \in \mathbb{R}^m$, the function $\zeta_x(\alpha) = \alpha \cdot \chi_{M_\alpha}(x)$ is upper semi-continuous on $[0,1]$. Then, the α-level set of $\tilde{A} \odot_{\mathfrak{G}} \tilde{B}$ is given by*

$$\left(\tilde{A} \odot_{\mathfrak{G}} \tilde{B}\right)_\alpha = \bigcup_{\alpha\le\beta\le 1} M_\beta$$

for every $\alpha \in (0,1]$ and

$$\left(\tilde{A} \odot_{\mathfrak{G}} \tilde{B}\right)_{0+} = \bigcup_{0<\alpha\le 1} \left(\tilde{A} \odot_{\mathfrak{G}} \tilde{B}\right)_\alpha = \bigcup_{0<\alpha\le 1} M_\alpha.$$

Proof. By referring to (7.118), the result follows immediately from Theorem 5.4.9. ∎

Remark 7.6.7 Let $\tilde{A}$ and $\tilde{B}$ be two fuzzy sets in $\mathbb{R}^m$, and let $M_\alpha = \tilde{A}_\alpha \circ \tilde{B}_\alpha$ for all $\alpha \in (0,1]$, where $\alpha = 0$ is not included and $\circ$ denotes an arbitrary binary operation between the α-level sets $\tilde{A}_\alpha$ and $\tilde{B}_\alpha$. For any $\alpha \in (0,1]$ and any increasing sequence $\{\alpha_n\}_{n=1}^{\infty}$ in $(0,1]$ with $\alpha_n \uparrow \alpha$, Proposition 2.2.8 says that

$$\bigcap_{n=1}^{\infty} \tilde{A}_{\alpha_n} = A_\alpha \text{ and } \bigcap_{n=1}^{\infty} \tilde{B}_{\alpha_n} = B_\alpha.$$

Let $M_\alpha = \tilde{A}_\alpha \circ \tilde{B}_\alpha$ for all $\alpha \in (0,1]$. Then, the following inclusion

$$\bigcap_{n=1}^{\infty} M_{\alpha_n} \subseteq M_\alpha$$

is satisfied if and only if the following inclusion

$$\bigcap_{n=1}^{\infty} \left(\tilde{A}_{\alpha_n} \circ \tilde{B}_{\alpha_n}\right) \subseteq \tilde{A}_\alpha \circ \tilde{B}_\alpha = \left(\bigcap_{n=1}^{\infty} \tilde{A}_{\alpha_n}\right) \circ \left(\bigcap_{n=1}^{\infty} \tilde{B}_{\alpha_n}\right)$$

is satisfied.

Theorem 7.6.8 *Let $\tilde{A}$ and $\tilde{B}$ be any two fuzzy sets in $\mathbb{R}^m$, let $M_\alpha = \tilde{A}_\alpha \circ \tilde{B}_\alpha$ for all $\alpha \in (0,1]$, where $\alpha = 0$ is not included, and let $\circ$ denote an arbitrary binary operation between the α-level sets $\tilde{A}_\alpha$ and $\tilde{B}_\alpha$. Suppose that, for any $\alpha \in (0,1]$ and any increasing sequence $\{\alpha_n\}_{n=1}^{\infty}$ in $(0,1]$ with $\alpha_n \uparrow \alpha$, the following inclusion is satisfied:*

$$\bigcap_{n=1}^{\infty} \left(\tilde{A}_{\alpha_n} \circ \tilde{B}_{\alpha_n}\right) \subseteq \tilde{A}_\alpha \circ \tilde{B}_\alpha = \left(\bigcap_{n=1}^{\infty} \tilde{A}_{\alpha_n}\right) \circ \left(\bigcap_{n=1}^{\infty} \tilde{B}_{\alpha_n}\right). \tag{7.120}$$

Then, the α-level set of $\tilde{A} \odot_{\mathfrak{G}} \tilde{B}$ is given by

$$\left(\tilde{A} \odot_{\mathfrak{G}} \tilde{B}\right)_\alpha = \bigcup_{\alpha \le \beta \le 1} M_\beta$$

for every $\alpha \in (0,1]$ and

$$\left(\tilde{A} \odot_{\mathfrak{G}} \tilde{B}\right)_{0+} = \bigcup_{0<\alpha\le 1} \left(\tilde{A} \odot_{\mathfrak{G}} \tilde{B}\right)_\alpha = \bigcup_{0<\alpha\le 1} M_\alpha.$$

Proof. Since $\tilde{A}_\beta \subseteq \tilde{A}_\alpha$ and $\tilde{B}_\beta \subseteq \tilde{B}_\alpha$ for $\alpha < \beta$, it follows that $M_\beta \subseteq M_\alpha$ for $\alpha < \beta$. Therefore, the result follows immediately from Theorem 7.6.6, Remark 7.6.7 and Proposition 5.2.5. ■

Example 7.6.9 Continued from Example 7.6.5, we see that the natural Hausdorff difference $\tilde{A} \ominus_{NH} \tilde{B}$ always exists with membership function given in (7.119). Let

$$l_\alpha = (1-\alpha) \cdot \min\left\{a^L - b^L, a^U - b^U\right\} + \alpha \cdot (a - b)$$

and

$$u_\alpha = (1-\alpha) \cdot \max\left\{a^L - b^L, a^U - b^U\right\} + \alpha \cdot (a - b).$$

The family $\{M_\alpha : 0 < \alpha \le 1\}$ given by

$$M_\alpha = \tilde{A}_\alpha \boxminus_H \tilde{B}_\alpha = \left[\min\left\{\tilde{A}_\alpha^L - \tilde{B}_\alpha^L, \tilde{A}_\alpha^U - \tilde{B}_\alpha^U\right\}, \max\left\{\tilde{A}_\alpha^L - \tilde{B}_\alpha^L, \tilde{A}_\alpha^U - \tilde{B}_\alpha^U\right\}\right] = \left[l_\alpha, u_\alpha\right] \tag{7.121}$$

is not nested. Given any $\alpha \in (0,1]$ and any increasing sequence $\{\alpha_n\}_{n=1}^{\infty}$ in $(0,1]$ with $\alpha_n \uparrow \alpha$, it is clear to see

$$l_{\alpha_n} \to l_\alpha \text{ and } u_{\alpha_n} \to u_\alpha \text{ as } n \to \infty.$$

Given $x \in \tilde{A}_{\alpha_n} \boxminus_H \tilde{B}_{\alpha_n}$ for all n, it means that $l_{\alpha_n} \leq x \leq u_{\alpha_n}$ for all n. By taking the limit as $n \to \infty$, we see that $l_\alpha \leq x \leq u_\alpha$, i.e. $x \in \tilde{A}_\alpha \boxminus_H \tilde{B}_\alpha$. This says that the inclusion (7.120) is satisfied. Theorem 7.6.8 says that the α-level set of the natural Hausdorff difference $\tilde{A} \ominus_{NH} \tilde{B}$ is given by

$$\left(\tilde{A} \ominus_{NH} \tilde{B}\right)_\alpha = \bigcup_{\alpha \leq \beta \leq 1} M_\beta = \bigcup_{\alpha \leq \beta \leq 1} \tilde{A}_\alpha \boxminus_H \tilde{B}_\alpha = \bigcup_{\alpha \leq \beta \leq 1} \left[l_\alpha, u_\alpha\right]$$

for $\alpha \in (0,1]$.

Recall that the binary operation $\tilde{A} \odot_{\mathfrak{G}} \tilde{B}$ always exists. The binary operation $\tilde{A} \odot \tilde{B}$ can only exist when (7.106) and (7.107) are satisfied, and the binary operation $\tilde{A} \odot^* \tilde{B}$ can only exist when (7.117) is satisfied.

7.6.4 Existence and Equivalence

Recall that the binary operation $\tilde{A} \odot_{\mathfrak{G}} \tilde{B}$ always exists. The binary operation $\tilde{A} \odot \tilde{B}$ can only exist when (7.106) and (7.107) are satisfied, and the binary operation $\tilde{A} \odot^* \tilde{B}$ can only exist when (7.117) is satisfied. The existence will be presented below.

Theorem 7.6.10 ***(Equivalence)*** *Given any two fuzzy sets $\tilde{A}$ and $\tilde{B}$ in $\mathbb{R}^m$, we have the following equivalences.*

(i) *Suppose that the binary operation $\tilde{A} \odot_{\mathfrak{L}} \tilde{B}$ exists. Then, we have*

$$\tilde{A} \odot_{\mathfrak{L}} \tilde{B} = \tilde{A} \odot_{\mathfrak{G}} \tilde{B}.$$

(ii) *Suppose that the binary operation $\tilde{A} \odot_{\mathfrak{L}}^* \tilde{B}$ exists. Then, we have*

$$\tilde{A} \odot_{\mathfrak{L}}^* \tilde{B} = \tilde{A} \odot_{\mathfrak{G}} \tilde{B}.$$

(iii) *Suppose that the binary operation $\tilde{A} \odot_{\mathfrak{L}} \tilde{B}$ and $\tilde{A} \odot_{\mathfrak{L}}^* \tilde{B}$ both exist. Then*

$$\tilde{A} \odot_{\mathfrak{L}} \tilde{B} = \tilde{A} \odot_{\mathfrak{L}}^* \tilde{B} = \tilde{A} \odot_{\mathfrak{G}} \tilde{B}.$$

Proof. To prove part (i), the existence of $\tilde{A} \odot_{\mathfrak{L}} \tilde{B}$ says that (7.106) and (7.107) are satisfied. Since

$$\tilde{C}_\beta = M_\beta = \tilde{A}_\beta \circ \tilde{B}_\beta \text{ for all } \beta \in (0,1],$$

from (7.104) and (7.118), it follows that

$$\left\{x \in \mathbb{R}^m : \xi_{\tilde{A} \odot_{\mathfrak{L}} \tilde{B}}(x) = \sup_{\beta \in [0,1]} \beta \cdot \chi_{\tilde{C}_\beta}(x) \geq \alpha\right\}$$

$$= \left\{x \in \mathbb{R}^m : \xi_{\tilde{A} \odot_{\mathfrak{G}} \tilde{B}}(x) = \sup_{\beta \in [0,1]} \beta \cdot \chi_{M_\beta}(x) \geq \alpha\right\}.$$

Using Proposition 1.4.8, we obtain the desired equality. To prove part (ii), since the existence of $\tilde{A} \odot^*_{\mathfrak{g}} \tilde{B}$ says that

$$\tilde{C}^*_\beta = M_\beta = \tilde{A}_\beta \circ \tilde{B}_\beta \text{ for all } \beta \in (0,1],$$

from (7.116) and (7.118), we obtain the desired equality. Finally, part (iii) follows from Proposition 7.6.3 immediately, and the proof is complete. ■

Proposition 7.6.11 *Let $\tilde{A}$ and $\tilde{B}$ be two fuzzy sets in $\mathbb{R}^m$, and let $M_\alpha = \tilde{A}_\alpha \circ \tilde{B}_\alpha$ for all $\alpha \in (0,1]$, where $\alpha = 0$ is not included and $\circ$ denotes an arbitrary binary operation between the α-level sets $\tilde{A}_\alpha$ and $\tilde{B}_\alpha$. Suppose that the existence of $\tilde{A} \odot^*_{\mathfrak{g}} \tilde{B}$ is guaranteed; that is, there exists a fuzzy set $\tilde{C}^*$ in $\mathbb{R}^m$ such that (7.117) is satisfied. Then $\{M_\alpha : 0 < \alpha \leq 1\}$ is a nested family.*

Proof. By definition, we have $\tilde{A} \odot^*_{\mathfrak{g}} \tilde{B} = \tilde{C}^*$. Since (7.117) is satisfied, we also have

$$M_\alpha = \tilde{A}_\alpha \circ \tilde{B}_\alpha = \tilde{C}^*_\alpha$$

for all $\alpha \in (0,1]$. Using (2.2), the proof is complete. ■

Theorem 7.6.12 *(**Nonexistence for** $\tilde{A} \odot^*_{\mathfrak{g}} \tilde{B}$) Let $\tilde{A}$ and $\tilde{B}$ be two fuzzy sets in $\mathbb{R}^m$, and let $M_\alpha = \tilde{A}_\alpha \circ \tilde{B}_\alpha$ for all $\alpha \in (0,1]$, where $\alpha = 0$ is not included and $\circ$ denotes an arbitrary binary operation between the α-level sets $\tilde{A}_\alpha$ and $\tilde{B}_\alpha$. Suppose that the family $\{M_\alpha : 0 < \alpha \leq 1\}$ is not nested. Then, the binary operation $\tilde{A} \odot^*_{\mathfrak{g}} \tilde{B}$ does not exist.*

Proof. The result follows immediately from the contrapositive of Proposition 7.6.11. ■

Example 7.6.13 Let $\tilde{A}$ and $\tilde{B}$ be two fuzzy intervals in $\mathbb{R}$. By referring to (7.121), we define

$$M_\alpha = \tilde{A}_\alpha \boxminus \tilde{B}_\alpha = \left[\min\left\{\tilde{A}^L_\alpha - \tilde{B}^L_\alpha, \tilde{A}^U_\alpha - \tilde{B}^U_\alpha\right\}, \max\left\{\tilde{A}^L_\alpha - \tilde{B}^L_\alpha, \tilde{A}^U_\alpha - \tilde{B}^U_\alpha\right\}\right]$$

for $\alpha \in (0,1]$. Then this family $\{M_\alpha : 0 < \alpha \leq 1\}$ is not nested in general. Therefore, using Theorem 7.6.12, the difference $\tilde{A} \ominus^* \tilde{B}$ does not exist. In other words, we cannot find a fuzzy interval $\tilde{C}^*$ satisfying $\tilde{C}^*_\alpha = \tilde{A}_\alpha \boxminus \tilde{B}_\alpha$ for each $\alpha \in (0,1]$.

Theorem 7.6.14 *(**Existence for** $\tilde{A} \odot^*_{\mathfrak{g}} \tilde{B}$) Let $\tilde{A}$ and $\tilde{B}$ be two fuzzy sets in $\mathbb{R}^m$, and let $M_\alpha = \tilde{A}_\alpha \circ \tilde{B}_\alpha$ for all $\alpha \in (0,1]$, where $\alpha = 0$ is not included and $\circ$ denotes an arbitrary binary operation between the α-level sets $\tilde{A}_\alpha$ and $\tilde{B}_\alpha$. Suppose that the following conditions are satisfied:*

- *$\{M_\alpha : 0 < \alpha \leq 1\}$ is a nested family;*
- *for any fixed $x \in \mathbb{R}^m$, the function $\zeta_x(\alpha) = \alpha \cdot \chi_{M_\alpha}(x)$ is upper semi-continuous on $[0,1]$.*

Then, there exists a fuzzy set $\tilde{C}^*$ in $\mathbb{R}^m$ with membership function defined by

$$\xi_{\tilde{C}^*}(x) = \sup_{0<\alpha\leq 1} \alpha \cdot \chi_{M_\alpha}(x) \tag{7.122}$$

satisfying $\tilde{C}^*_\alpha = M_\alpha = \tilde{A}_\alpha \circ \tilde{B}_\alpha$ for each $\alpha \in (0,1]$ and

$$\tilde{C}^*_0 = cl\left(\bigcup_{0<\alpha\leq 1} M_\alpha\right) = cl\left(\bigcup_{0<\alpha\leq 1} \tilde{A}_\alpha \circ \tilde{B}_\alpha\right).$$

This guarantees the existence of $\tilde{C}^* = \tilde{A} \odot^*_{\mathfrak{L}} \tilde{B}$. *Moreover, we have*

$$\tilde{C}^* = \tilde{A} \odot^*_{\mathfrak{L}} \tilde{B} = \tilde{A} \odot_{\mathfrak{G}} \tilde{B}. \tag{7.123}$$

Proof. Since $\{M_\alpha : 0 < \alpha \leq 1\}$ is a nested family, we have the equality

$$M_\alpha = \bigcup_{\alpha \leq \beta \leq 1} M_\beta \text{ for } \alpha > 0.$$

From (7.118) and (7.122), we have

$$\begin{aligned}\tilde{A}_\alpha \circ \tilde{B}_\alpha = M_\alpha = \bigcup_{\alpha \leq \beta \leq 1} M_\beta &= (\tilde{A} \odot_{\mathfrak{G}} \tilde{B})_\alpha \\ &\text{(using Theorem 5.4.9)} \\ &= \tilde{C}^*_\alpha\end{aligned}$$

for all $\alpha \in (0, 1]$, which shows the existence of $\tilde{A} \odot^*_{\mathfrak{L}} \tilde{B} = \tilde{C}^*$ by referring to (7.117). In this case, we also have $\tilde{C}^*_\alpha = (\tilde{A} \odot^*_{\mathfrak{L}} \tilde{B})_\alpha$ for all $\alpha \in (0, 1]$. The 0-level set $\tilde{C}^*_0$ is given by

$$\tilde{C}^*_0 = \text{cl}\left(\{x \in \mathbb{R}^m : \xi_{\tilde{C}^*}(x) > 0\}\right) = \text{cl}\left(\bigcup_{0<\alpha\leq 1} M_\alpha\right) = \text{cl}\left(\bigcup_{0<\alpha\leq 1} \tilde{A}_\alpha \circ \tilde{B}_\alpha\right).$$

Finally, using part (ii) of Theorem 7.6.10, we obtain

$$\tilde{A} \odot^*_{\mathfrak{L}} \tilde{B} = \tilde{A} \odot_{\mathfrak{G}} \tilde{B},$$

and the proof is complete. ■

Theorem 7.6.14 says that, given $\tilde{A}$ and $\tilde{B}$, using the family $\{\tilde{A}_\alpha \circ \tilde{B}_\alpha : 0 < \alpha \leq 1\}$, we can generate a fuzzy set $\tilde{C}^*$ in $\mathbb{R}^m$ with membership function given by

$$\xi_{\tilde{C}^*}(x) = \sup_{0<\alpha\leq 1} \alpha \cdot \chi_{\tilde{A}_\alpha \circ \tilde{B}_\alpha}(x)$$

satisfying $\tilde{C}^*_\alpha = \tilde{A}_\alpha \circ \tilde{B}_\alpha$ for all $\alpha \in (0, 1]$. In this case, we obtain the binary operation $\tilde{A} \odot^*_{\mathfrak{L}} \tilde{B} = \tilde{C}^*$. Moreover, we also have the equalities (7.123).

Remark 7.6.15 The second condition of Theorem 7.6.14 assumes that the function ζ_x is upper semi-continuous on [0, 1]. This condition can be replaced by the following inclusion

$$\bigcap_{n=1}^{\infty} \left(\tilde{A}_{\alpha_n} \circ \tilde{B}_{\alpha_n}\right) \subseteq \tilde{A}_\alpha \circ \tilde{B}_\alpha = \left(\bigcap_{n=1}^{\infty} \tilde{A}_{\alpha_n}\right) \circ \left(\bigcap_{n=1}^{\infty} \tilde{B}_{\alpha_n}\right) \tag{7.124}$$

for $\alpha_n \uparrow \alpha$ in $(0, 1]$. This replacement can be realized by Remark 7.6.7 and Proposition 5.2.5.

Example 7.6.16 From Example 7.6.1, we defined $M_\alpha = \tilde{A}_\alpha \boxminus_{G_2} \tilde{B}_\alpha$ for $\alpha \in (0, 1]$. Let

$$l_\alpha = \inf_{\alpha \leq \beta \leq 1} \left((1-\beta) \cdot \min\left\{a^L - b^L, a^U - b^U\right\} + \beta \cdot (a - b)\right)$$

and

$$u_\alpha = \sup_{\alpha \leq \beta \leq 1} \left((1-\beta) \cdot \max\left\{a^L - b^L, a^U - b^U\right\} + \beta \cdot (a - b)\right).$$

Then

$$M_\alpha = \tilde{A}_\alpha \boxminus_{G_2} \tilde{B}_\alpha = [l_\alpha, u_\alpha] \text{ for all } \alpha \in (0,1].$$

It is obvious that l_α is increasing with respect to α on $(0,1]$ and u_α is decreasing with respect to α on $(0,1]$, which says that this family $\{M_\alpha : 0 < \alpha \leq 1\}$ is nested. Since the functions involved in this example are all continuous, given any $\alpha \in (0,1]$ and any increasing sequence $\{\alpha_n\}_{n=1}^{\infty}$ in $(0,1]$ with $\alpha_n \uparrow \alpha$, it is not hard to prove $l_{\alpha_n} \to l_\alpha$ and $u_{\alpha_n} \to u_\alpha$. For $x \in \tilde{A}_{\alpha_n} \boxminus_{G_2} \tilde{B}_{\alpha_n}$ for all n, it means that $l_{\alpha_n} \leq x \leq u_{\alpha_n}$ for all n. By taking the limit as $n \to \infty$, we see that $l_\alpha \leq x \leq u_\alpha$, i.e. $x \in \tilde{A}_\alpha \boxminus_{G_2} \tilde{B}_\alpha$. This says that the inclusion (7.124) is satisfied. Using Theorem 7.6.14 and Remark 7.6.15, it follows that the difference $\tilde{A}_\alpha \ominus_{G_2} \tilde{B} = \tilde{C}^*$ exists satisfying $\tilde{C}^*_\alpha = \tilde{A}_\alpha \boxminus_{G_2} \tilde{B}_\alpha$ for all $\alpha \in (0,1]$.

Proposition 7.6.17 *Let $\tilde{A}$ and $\tilde{B}$ be two fuzzy sets in $\mathbb{R}^m$, and let $M_\alpha = \tilde{A}_\alpha \circ \tilde{B}_\alpha$ for all $\alpha \in [0,1]$, where $\circ$ denotes an arbitrary binary operation between the α-level sets $\tilde{A}_\alpha$ and $\tilde{B}_\alpha$. Suppose that the existence of $\tilde{A} \odot_{\mathfrak{g}} \tilde{B}$ is guaranteed; that is, there exists a fuzzy set $\tilde{C}$ in $\mathbb{R}^m$ such that (7.106) and (7.107) are satisfied. Then $\{M_\alpha : 0 \leq \alpha \leq 1\}$ is a nested family.*

Proof. By definition, we have $\tilde{A} \odot_{\mathfrak{g}} \tilde{B} = \tilde{C}$. Since (7.106) and (7.107) are satisfied, we also have

$$M_\alpha = \tilde{A}_\alpha \circ \tilde{B}_\alpha = \tilde{C}_\alpha$$

for all $\alpha \in [0,1]$. Using (2.2), the proof is complete. ■

Theorem 7.6.18 *(**Nonexistence for** $\tilde{A} \odot_{\mathfrak{g}} \tilde{B}$) Let $\tilde{A}$ and $\tilde{B}$ be two fuzzy sets in $\mathbb{R}^m$, and let $M_\alpha = \tilde{A}_\alpha \circ \tilde{B}_\alpha$ for all $\alpha \in [0,1]$, where $\circ$ denotes an arbitrary binary operation between the α-level sets $\tilde{A}_\alpha$ and $\tilde{B}_\alpha$. Suppose that the family $\{M_\alpha : 0 \leq \alpha \leq 1\}$ is not nested. Then, the binary operation $\tilde{A} \odot_{\mathfrak{g}} \tilde{B}$ does not exist.*

Proof. The result follows immediately from the contrapositive of Proposition 7.6.17. ■

Theorem 7.6.19 *(**Existence for** $\tilde{A} \odot_{\mathfrak{g}} \tilde{B}$) Let $\tilde{A}$ and $\tilde{B}$ be two fuzzy sets in $\mathbb{R}^m$, and let $M_\alpha = \tilde{A}_\alpha \circ \tilde{B}_\alpha$ for all $\alpha \in [0,1]$, where $\circ$ denotes an arbitrary binary operation between the α-level sets $\tilde{A}_\alpha$ and $\tilde{B}_\alpha$. Suppose that the following conditions are satisfied:*

- *$\{M_\alpha : 0 \leq \alpha \leq 1\}$ is a nested family;*
- *for any fixed $x \in \mathbb{R}^m$, the function $\zeta_x(\alpha) = \alpha \cdot \chi_{M_\alpha}(x)$ is upper semi-continuous on $[0,1]$;*
- *the following equality is satisfied:*

$$cl\left(\bigcup_{0<\alpha\leq 1} \tilde{A}_\alpha \circ \tilde{B}_\alpha\right) = \tilde{A}_0 \circ \tilde{B}_0. \tag{7.125}$$

Then, there exists a fuzzy set $\tilde{C}$ in $\mathbb{R}^m$ with membership function defined by

$$\xi_{\tilde{C}}(x) = \sup_{\alpha\in[0,1]} \alpha \cdot \chi_{M_\alpha}(x) = \sup_{0<\alpha\leq 1} \alpha \cdot \chi_{M_\alpha}(x) \tag{7.126}$$

satisfying

$$\tilde{C}_\alpha = M_\alpha = \tilde{A}_\alpha \circ \tilde{B}_\alpha$$

for each $\alpha \in [0,1]$. *This guarantees the existence of* $\tilde{C} = \tilde{A} \odot_{\mathfrak{L}} \tilde{B}$. *Moreover, the existence of* $\tilde{A} \odot^{*}_{\mathfrak{L}} \tilde{B}$ *is also guaranteed and we have*

$$\tilde{C} = \tilde{A} \odot_{\mathfrak{L}} \tilde{B} = \tilde{A} \odot^{*}_{\mathfrak{L}} \tilde{B} = \tilde{A} \odot_{\mathfrak{G}} \tilde{B}. \tag{7.127}$$

Proof. The existence of $\tilde{A} \odot^{*}_{\mathfrak{L}} \tilde{B}$ is guaranteed immediately from Theorem 7.6.14. We also have the equality

$$M_\alpha = \bigcup_{\alpha \le \beta \le 1} M_\beta \text{ for } \alpha > 0$$

by the nestedness. From (7.118) and (7.126), using Theorem 5.4.9, we have

$$\tilde{A}_\alpha \circ \tilde{B}_\alpha = M_\alpha = \bigcup_{\alpha \le \beta \le 1} M_\beta = (\tilde{A} \odot_{\mathfrak{G}} \tilde{B})_\alpha = \tilde{C}_\alpha$$

for all $\alpha \in (0,1]$. For $\alpha = 0$, using the equality (7.125), we have

$$\tilde{C}_0 = \mathrm{cl}\left(\{x \in \mathbb{R}^m : \xi_{\tilde{C}}(x) > 0\}\right) = \mathrm{cl}\left(\bigcup_{0<\alpha\le 1} M_\alpha\right) = \mathrm{cl}\left(\bigcup_{0<\alpha\le 1} \tilde{A}_\alpha \circ \tilde{B}_\alpha\right)$$
$$= \tilde{A}_0 \circ \tilde{B}_0 = M_0.$$

This shows the existence of $\tilde{A} \odot_{\mathfrak{L}} \tilde{B}$ by referring to (7.106) and (7.107). Finally, using part (iii) of Theorem 7.6.10, the proof is complete. ■

Remark 7.6.20 We see that Theorem 7.6.19 considers the 0-level sets $\tilde{A}_0$ and $\tilde{B}_0$ by defining $M_0 = \tilde{A}_0 \circ \tilde{B}_0$. However, Theorem 7.6.14 does not consider the set M_0. In other words, if M_0 is not taken into account, then we can just discuss the existence of $\tilde{A} \odot^{*}_{\mathfrak{L}} \tilde{B}$ as shown in Theorem 7.6.14, and we cannot discuss the existence of $\tilde{A} \odot_{\mathfrak{L}} \tilde{B}$ as shown in Theorem 7.6.19.

Example 7.6.21 Continued from Example 7.6.16, we defined $M_\alpha = \tilde{A}_\alpha \boxminus_{G_2} \tilde{B}_\alpha$ for $\alpha \in [0,1]$, where $\alpha = 0$ is included by defining

$$l_0 = \inf_{0 \le \beta \le 1} \left((1-\beta) \cdot \min\left\{a^L - b^L, a^U - b^U\right\} + \beta \cdot (a-b)\right)$$

and

$$u_0 = \sup_{0 \le \beta \le 1} \left((1-\beta) \cdot \max\left\{a^L - b^L, a^U - b^U\right\} + \beta \cdot (a-b)\right).$$

Then

$$M_\alpha = \tilde{A}_\alpha \boxminus_{G_2} \tilde{B}_\alpha = \left[l_\alpha, u_\alpha\right] \text{ for all } \alpha \in [0,1].$$

It is obvious that this family $\{M_\alpha : 0 \le \alpha \le 1\}$ is nested. In order to use Theorem 7.6.19, from the argument of Example 7.6.16, it remains to check the equality (7.125), which means to check

$$\mathrm{cl}\left(\bigcup_{0<\alpha\le 1} M_\alpha\right) = M_0, \text{ i.e. } \mathrm{cl}\left(\bigcup_{0<\alpha\le 1} \left[l_\alpha, u_\alpha\right]\right) = \left[l_0, u_0\right]. \tag{7.128}$$

Since l_α is increasing with respect to α on $[0,1]$ and u_α is decreasing with respect to α on $[0,1]$, it follows that the limits

$$\lim_{\alpha \to 0} l_\alpha = \sup_{0<\alpha\le 1} l_\alpha \text{ and } \lim_{\alpha \to 0} u_\alpha = \inf_{0<\alpha\le 1} u_\alpha$$

exist. Therefore, we obtain

$$\text{cl}\left(\bigcup_{0<\alpha\leq 1}[l_\alpha, u_\alpha]\right) = \left[\lim_{\alpha\to 0} l_\alpha, \lim_{\alpha\to 0} u_\alpha\right] = \left[\sup_{0<\alpha\leq 1} l_\alpha, \inf_{0<\alpha\leq 1} u_\alpha\right].$$

Given a sequence $\alpha_n \downarrow 0$ for each $\alpha_n \in (0,1]$, it is not hard to prove $l_{\alpha_n} \to l_0$ and $u_{\alpha_n} \to u_0$ as $n \to \infty$. This shows that the equality (7.128) is satisfied. Therefore Theorem 7.6.19 says that the difference $\tilde{A}\ominus_{G_2}\tilde{B} = \tilde{C}$ exists, satisfying

$$\tilde{C}_\alpha = \tilde{A}_\alpha \boxminus_{G_2} \tilde{B}_\alpha \text{ for all } \alpha \in [0,1],$$

which includes 0.

7.6.5 Equivalent Arithmetic Operations on Fuzzy Sets in $\mathbb{R}$

Let $\tilde{A}$ and $\tilde{B}$ be two fuzzy sets in $\mathbb{R}$. Up to now, we have presented four ways to define the arithmetics of fuzzy sets in $\mathbb{R}$, which are summarized below.

- The conventional way of defining the arithmetic $\tilde{A} \odot \tilde{B}$ is according to the extension principle given by

$$\xi_{\tilde{A}\odot\tilde{B}}(z) = \sup_{\{(x,y):z=x\circ y\}} \min\left\{\xi_{\tilde{A}}(x), \xi_{\tilde{B}}(y)\right\}.$$

- The general arithmetic $\tilde{A} \odot_{\mathfrak{A}} \tilde{B}$ using the function $\mathfrak{A}$ is given by

$$\xi_{\tilde{A}\odot_{\mathfrak{A}}\tilde{B}}(z) = \sup_{\{(x,y):z=x\circ y\}} \mathfrak{A}\left(\xi_{\tilde{A}}(x), \xi_{\tilde{B}}(y)\right).$$

- The arithmetic $\tilde{A} \odot_{\mathfrak{L}} \tilde{B}$ is according to (7.93) based on the α-level sets.
- The general arithmetic $\tilde{A} \odot_{\mathfrak{G}} \tilde{B}$ according to (7.118) is based on the α-level sets by considering $M_\alpha = \tilde{A}_\alpha \circ \tilde{B}_\alpha$ for all $\alpha \in (0,1]$.

We are going to provide some suitable conditions to demonstrate their equivalences. We first present an example to demonstrate the above four arithmetics by considering addition.

Example 7.6.22 We consider the triangular fuzzy number $\tilde{A} = (a_L, a, a_U)$ with membership function given by

$$\xi_{\tilde{A}}(r) = \begin{cases} (r-a_L)/(a-a_L) & \text{if } a_L \leq r \leq a \\ (a_U - r)/(a_U - a) & \text{if } a < r \leq a_U \\ 0 & \text{otherwise.} \end{cases}$$

The α-level set of $\tilde{A}$ is a bounded closed interval given by

$$\tilde{A}_\alpha = \left[(1-\alpha)a_L + \alpha a, (1-\alpha)a_U + \alpha a\right].$$

Let $\tilde{A} = (a_L, a, a_U)$ and $\tilde{B} = (b_L, b, b_U)$ be two triangular fuzzy numbers. The four different additions of $\tilde{A}$ and $\tilde{B}$ are given below.

- It is well known that

$$\tilde{A} \oplus \tilde{B} = (a_L, a, a_U) \oplus (b_L, b, b_U) = (a_L + b_L, a + b, a_U + b_U).$$

- Given a function $\mathfrak{A} : [0,1] \times [0,1] \to [0,1]$, the membership function of $\tilde{A} \oplus_{\mathfrak{A}} \tilde{B}$ is given by

$$\xi_{\tilde{A} \oplus_{\mathfrak{A}} \tilde{B}}(z) = \sup_{\{(x,y):z=x+y\}} \mathfrak{A}\left(\xi_{\tilde{A}}(x), \xi_{\tilde{B}}(y)\right).$$

- Regarding the α-level sets, we have

$$\begin{aligned} M_\alpha &= \tilde{A}_\alpha + \tilde{B}_\alpha \\ &= \left[(1-\alpha)a_L + \alpha a, (1-\alpha)a_U + \alpha a\right] + \left[(1-\alpha)b_L + \alpha b, (1-\alpha)b_U + \alpha b\right] \\ &= \left[(1-\alpha)(a_L + b_L) + \alpha(a+b), (1-\alpha)(a_U + b_U) + \alpha(a+b)\right]. \end{aligned}$$

The membership function of $\tilde{A} \oplus_{\mathfrak{G}} \tilde{B}$ is given by

$$\xi_{\tilde{A} \oplus_{\mathfrak{G}} \tilde{B}}(x) = \sup_{0<\alpha\leq 1} \alpha \cdot \chi_{M_\alpha}(x).$$

- We have

$$\tilde{A}_\alpha + \tilde{B}_\alpha = \left[(1-\alpha)(a_L + b_L) + \alpha(a+b), (1-\alpha)(a_U + b_U) + \alpha(a+b)\right].$$

There exists a triangular fuzzy number

$$\tilde{C} = (a_L + b_L, a + b, a_U + b_U)$$

satisfying (7.94), i.e.

$$\tilde{C}_\alpha = \tilde{A}_\alpha + \tilde{B}_\alpha$$

for all $\alpha \in [0,1]$, which says that

$$\tilde{A} \oplus_{\mathfrak{L}} \tilde{B} = \tilde{C} = (a_L + b_L, a + b, a_U + b_U).$$

In other words, we also have

$$\tilde{A} \oplus_{\mathfrak{L}} \tilde{B} = \tilde{A} \oplus \tilde{B}.$$

Theorem 7.6.23 *Let $\tilde{A}$ and $\tilde{B}$ be two fuzzy sets in $\mathbb{R}$. For $\circ \in \{+,-,\times\}$, suppose that the following conditions are satisfied.*

- *$\tilde{A}_0 \circ \tilde{B}_0$ is a closed subset of $\mathbb{R}$ for the arithmetics of 0-level sets $\tilde{A}_0$ and $\tilde{B}_0$.*
- *For any increasing and convergent sequence $\{\alpha_n\}_{n=1}^\infty$ in $[0,1]$, the following inclusion is satisfied:*

$$\bigcap_{n=1}^{\infty}\left(\tilde{A}_{\alpha_n} \circ \tilde{B}_{\alpha_n}\right) \subseteq \left(\bigcap_{n=1}^{\infty} \tilde{A}_{\alpha_n}\right) \circ \left(\bigcap_{n=1}^{\infty} \tilde{B}_{\alpha_n}\right). \tag{7.129}$$

Then, there exists a fuzzy set $\tilde{C}$ in $\mathbb{R}$ with membership function defined by

$$\xi_{\tilde{C}}(x) = \sup_{\alpha \in (0,1]} \alpha \cdot \chi_{M_\alpha}(x) \tag{7.130}$$

satisfying $\tilde{C}_\alpha = \tilde{A}_\alpha \circ \tilde{B}_\alpha$ for each $\alpha \in [0,1]$. This guarantees the existence of $\tilde{C} = \tilde{A} \odot_{\mathfrak{L}} \tilde{B}$ for $\odot \in \{\oplus, \ominus, \otimes\}$. Moreover, we have

$$\tilde{C} = \tilde{A} \odot_{\mathfrak{L}} \tilde{B} = \tilde{A} \odot_{\mathfrak{G}} \tilde{B}.$$

Proof. We define $M_\alpha = \tilde{A}_\alpha \circ \tilde{B}_\alpha$ for all $\alpha \in [0,1]$. Since $\tilde{A}_\beta \subseteq \tilde{A}_\alpha$ and $\tilde{B}_\beta \subseteq \tilde{B}_\alpha$ for $\alpha, \beta \in [0,1]$ with $\alpha < \beta$, we have $M_\beta \subseteq M_\alpha$ for $\alpha, \beta \in [0,1]$ with $\alpha < \beta$. In particular, we have $M_\alpha \subseteq M_0$ for all $\alpha \in (0,1]$, i.e.

$$\text{cl}(M_{0+}) = \text{cl}\left(\bigcup_{\alpha \in (0,1]} M_\alpha\right) \subseteq \text{cl}(M_0) = M_0, \tag{7.131}$$

since $M_0 = \tilde{A}_0 \circ \tilde{B}_0$ is a closed subset of $\mathbb{R}$ by the assumption. For $\alpha_n \uparrow \alpha$ with $\alpha \in (0,1]$ and $\{\alpha_n\}_{n=1}^\infty \subset [0,1]$, from Proposition 2.2.8, we have the following inclusion

$$M_\alpha = \tilde{A}_\alpha \circ \tilde{B}_\alpha = \left(\bigcap_{n=1}^\infty \tilde{A}_{\alpha_n}\right) \circ \left(\bigcap_{n=1}^\infty \tilde{B}_{\alpha_n}\right) \subseteq \bigcap_{n=1}^\infty \left(\tilde{A}_{\alpha_n} \circ \tilde{B}_{\alpha_n}\right) = \bigcap_{n=1}^\infty M_{\alpha_n}.$$

Using the assumption (7.129), we have $M_\alpha = \bigcap_{n=1}^\infty M_{\alpha_n}$ for $\alpha_n \uparrow \alpha$, where $\alpha \in (0,1]$ and $\{\alpha_n\}_{n=1}^\infty \subset [0,1]$. Next, we are going to claim $M_0 \subseteq \text{cl}(M_{0+})$. Given any $x \in M_0$, we have $x = a \circ b$ for some

$$a \in \tilde{A}_0 = \text{cl}\left(\bigcup_{\alpha \in (0,1]} \tilde{A}_\alpha\right) \text{ and } b \in \tilde{B}_0 = \text{cl}\left(\bigcup_{\alpha \in (0,1]} \tilde{B}_\alpha\right).$$

Therefore, using the concept of closure, there exist two sequences $\{a_n\}_{n=1}^\infty$ and $\{b_n\}_{n=1}^\infty$ that converge to a and b, respectively, and

$$\{a_n\}_{n=1}^\infty \subset \bigcup_{\alpha \in (0,1]} \tilde{A}_\alpha \text{ and } \{b_n\}_{n=1}^\infty \subset \bigcup_{\alpha \in (0,1]} \tilde{B}_\alpha.$$

Let $x_n = a_n \circ b_n$. Then, we see that the sequence $\{x_n\}_{n=1}^\infty$ is convergent given by

$$x = \lim_{n\to\infty} x_n = \lim_{n\to\infty} a_n \circ b_n = a \circ b.$$

Now, we have $a_n \in \tilde{A}_{\alpha_n}$ and $b_n \in \tilde{B}_{\beta_n}$ for some α_n and β_n in $(0,1]$. Let $\gamma_n = \min\{\alpha_n, \beta_n\}$, since the family of α-level sets of a fuzzy set is nested, we have

$$x_n = a_n \circ b_n \in \tilde{A}_{\alpha_n} \circ \tilde{B}_{\beta_n} \subseteq \tilde{A}_{\gamma_n} \circ \tilde{B}_{\gamma_n} \subset \bigcup_{\alpha \in (0,1]} \left(\tilde{A}_\alpha \circ \tilde{B}_\alpha\right),$$

which says that

$$x \in \text{cl}\left(\bigcup_{\alpha \in (0,1]} \left(\tilde{A}_\alpha \circ \tilde{B}_\alpha\right)\right) = \text{cl}\left(\bigcup_{\alpha \in (0,1]} M_\alpha\right) = \text{cl}(M_{0+})$$

by referring to (7.131). Therefore, we indeed have the inclusion $M_0 \subseteq \text{cl}(M_{0+})$. Using (7.131), we obtain $M_0 = \text{cl}(M_{0+})$. We consider the membership function in (7.130). Theorem 5.4.7 says that $\tilde{C}_\alpha = M_\alpha$ for all $\alpha \in (0,1]$ and

$$\tilde{C}_0 = \text{cl}\left(\bigcup_{\alpha \in (0,1]} M_\alpha\right) = \text{cl}(M_{0+}) = M_0.$$

Therefore, we obtain $\tilde{C}_\alpha = M_\alpha = \tilde{A}_\alpha \circ \tilde{B}_\alpha$ for all $\alpha \in [0,1]$, which shows the existence of $\tilde{C} = \tilde{A} \odot_{\mathfrak{g}} \tilde{B}$. From (7.118), we also see that

$$\xi_{\tilde{C}}(x) = \sup_{\alpha \in (0,1]} \alpha \cdot \chi_{M_\alpha}(x) = \xi_{\tilde{A} \odot_{\mathfrak{g}} \tilde{B}}(x).$$

This completes the proof. ∎

Proposition 7.6.24 *Let $\tilde{A}$ and $\tilde{B}$ be two fuzzy sets in $\mathbb{R}$. For $\odot \in \{\oplus, \ominus, \otimes\}$, the following statements hold true.*

(i) *We have*

$$\left(\tilde{A} \odot \tilde{B}\right)_\alpha \subseteq \left(\tilde{A} \odot_{\mathfrak{G}} \tilde{B}\right)_\alpha \text{ for all } \alpha \in [0,1].$$

(ii) *Suppose that the function $\mathfrak{A} : [0,1] \times [0,1] \to [0,1]$ satisfies the following condition. For each $\alpha \in (0,1]$,*

$$\mathfrak{A}(\alpha_1, \alpha_2) \geq \alpha \text{ implies } \alpha_1 \geq \alpha \text{ and } \alpha_2 \geq \alpha. \tag{7.132}$$

Then, we have

$$\left(\tilde{A} \odot_{\mathfrak{A}} \tilde{B}\right)_\alpha \subseteq \left(\tilde{A} \odot_{\mathfrak{G}} \tilde{B}\right)_\alpha \text{ for all } \alpha \in [0,1].$$

Proof. We first prove part (ii). For $\alpha \in (0,1]$, given any $z \in (\tilde{A} \odot_{\mathfrak{A}} \tilde{B})_\alpha$, we have

$$r = \xi_{\tilde{A} \odot_{\mathfrak{A}} \tilde{B}}(z) = \sup_{\{(x,y): z = x \circ y\}} \mathfrak{A}\left(\xi_{\tilde{A}}(x), \xi_{\tilde{B}}(y)\right) \geq \alpha.$$

By the definition of supremum, given any $\epsilon > 0$, there exists (x^*, y^*) satisfying $z = x^* \circ y^*$ and

$$\mathfrak{A}\left(\xi_{\tilde{A}}(x^*), \xi_{\tilde{B}}(y^*)\right) > r - \epsilon,$$

which implies

$$\xi_{\tilde{A}}(x^*) \geq r - \epsilon \text{ and } \xi_{\tilde{B}}(y^*) \geq r - \epsilon$$

by the assumption (7.132). This says that

$$x^* \in \tilde{A}_{r-\epsilon} \text{ and } y^* \in \tilde{B}_{r-\epsilon}.$$

Let $M_\alpha = \tilde{A}_\alpha \circ \tilde{B}_\alpha$ for all $\alpha \in [0,1]$. Then, we obtain

$$z = x^* \circ y^* \in \tilde{A}_{r-\epsilon} \circ \tilde{B}_{r-\epsilon} = M_{r-\epsilon},$$

which implies

$$\xi_{\tilde{A} \odot_{\mathfrak{G}} \tilde{B}}(z) = \sup_{0<\alpha\leq 1} \alpha \cdot \chi_{M_\alpha}(z) = \sup_{\alpha \in [0,1]} \alpha \cdot \chi_{M_\alpha}(z) \geq r - \epsilon.$$

Since ϵ can be any positive number, we must have

$$\xi_{\tilde{A} \odot_{\mathfrak{G}} \tilde{B}}(z) = \sup_{0<\alpha\leq 1} \alpha \cdot \chi_{M_\alpha}(z) = \sup_{\alpha \in [0,1]} \alpha \cdot \chi_{M_\alpha}(z) \geq r \geq \alpha,$$

which shows that $z \in (\tilde{A} \odot_{\mathfrak{G}} \tilde{B})_\alpha$. Therefore, we obtain

$$\left(\tilde{A} \odot_{\mathfrak{A}} \tilde{B}\right)_\alpha \subseteq (\tilde{A} \odot_{\mathfrak{G}} \tilde{B})_\alpha \text{ for all } \alpha \in (0,1].$$

Now, we also have

$$\left(\tilde{A} \odot_{\mathfrak{A}} \tilde{B}\right)_0 = \text{cl}\left(\bigcup_{0<\alpha\leq 1} \left(\tilde{A} \odot_{\mathfrak{A}} \tilde{B}\right)_\alpha\right) \subseteq \text{cl}\left(\bigcup_{0<\alpha\leq 1} \left(\tilde{A} \odot_{\mathfrak{G}} \tilde{B}\right)_\alpha\right) = \left(\tilde{A} \odot_{\mathfrak{G}} \tilde{B}\right)_0.$$

Part (i) can be similarly obtained from part (ii), since $\tilde{A} \odot \tilde{B} = \tilde{A} \odot_{\mathfrak{A}} \tilde{B}$ when $\mathfrak{A}(x,y) = \min\{x,y\}$. This completes the proof. ■

Proposition 7.6.25 *Let $\tilde{A}$ and $\tilde{B}$ be two fuzzy sets in $\mathbb{R}$. For $\circ \in \{+,-,\times\}$, let $M_\alpha = \tilde{A}_\alpha \circ \tilde{B}_\alpha$ for all $\alpha \in [0,1]$. Suppose that, given any fixed $x \in \mathbb{R}^m$, the function $\eta_x(\alpha) = \alpha \cdot \chi_{M_\alpha}(x)$ is upper semi-continuous on $[0,1]$. Then, for $\odot \in \{\oplus, \ominus, \otimes\}$, the following statements hold true.*

(i) We have

$$\left(\tilde{A} \odot_{\mathfrak{G}} \tilde{B}\right)_\alpha \subseteq \left(\tilde{A} \odot \tilde{B}\right)_\alpha \text{ for all } \alpha \in [0,1].$$

(ii) Suppose that, for each $\alpha \in (0,1]$,

$$\alpha_1 \geq \alpha \text{ and } \alpha_2 \geq \alpha \text{ imply } \mathfrak{A}(\alpha_1,\alpha_2) \geq \alpha.$$

Then, we have

$$\left(\tilde{A} \odot_{\mathfrak{G}} \tilde{B}\right)_\alpha \subseteq \left(\tilde{A} \odot_{\mathfrak{A}} \tilde{B}\right)_\alpha \text{ for all } \alpha \in [0,1].$$

Proof. It suffices to prove part (ii). For any fixed $\beta \in (0,1]$, if $z \in (\tilde{A} \odot_{\mathfrak{G}} \tilde{B})_\beta$, then we have

$$\sup_{\alpha\in[0,1]} \eta_z(\alpha) = \sup_{\alpha\in[0,1]} \alpha \cdot \chi_{M_\alpha}(z) \geq \beta. \tag{7.133}$$

Suppose that $z \notin M_\alpha$ for all $\alpha \geq \beta$. Then, we have

$$\sup_{\alpha\in[0,1]} \eta_z(\alpha) = \sup_{\alpha\in[0,1]} \alpha \cdot \chi_{M_\alpha}(z) \leq \beta. \tag{7.134}$$

Suppose that (7.134) holds true for the case of equality. According to the upper semi-continuity of $\eta_z(\alpha)$, the supremum β in (7.134) is attained by some α in $[0,1]$, which contradicts $z \notin M_\alpha$ for all $\alpha \geq \beta$. Therefore, we must have

$$\sup_{\alpha\in[0,1]} \alpha \cdot \chi_{M_\alpha}(z) < \beta,$$

which also contradicts (7.133). Therefore, there exists $\alpha_0 \geq \beta$ satisfying $z \in M_{\alpha_0}$. This also says that $z = x \circ y$ with $x \in \tilde{A}_{\alpha_0}$ and $y \in \tilde{B}_{\alpha_0}$, i.e.

$$\xi_{\tilde{A}}(x) \geq \alpha_0 \text{ and } \xi_{\tilde{B}}(y) \geq \alpha_0,$$

which imply

$$\mathfrak{A}(\xi_{\tilde{A}}(x), \xi_{\tilde{B}}(y)) \geq \alpha_0$$

by the assumption of $\mathfrak{A}$. Therefore, we obtain

$$\xi_{\tilde{A} \odot_{\mathfrak{A}} \tilde{B}}(z) = \sup_{\{(x,y):z=x\circ y\}} \mathfrak{A}\left(\xi_{\tilde{A}}(x), \xi_{\tilde{B}}(y)\right) \geq \alpha_0 \geq \beta,$$

which shows that $z \in (\tilde{A} \odot_{\mathfrak{A}} \tilde{B})_\beta$, i.e.

$$\left(\tilde{A} \odot_{\mathfrak{G}} \tilde{B}\right)_\alpha \subseteq \left(\tilde{A} \odot_{\mathfrak{A}} \tilde{B}\right)_\alpha \text{ for all } \alpha \in (0,1].$$

Now, we also have

$$\left(\tilde{A} \odot_{\mathfrak{G}} \tilde{B}\right)_0 = \text{cl}\left(\bigcup_{0<\alpha\leq 1} \left(\tilde{A} \odot_{\mathfrak{G}} \tilde{B}\right)_\alpha\right) \subseteq \text{cl}\left(\bigcup_{0<\alpha\leq 1} \left(\tilde{A} \odot_{\mathfrak{A}} \tilde{B}\right)_\alpha\right) = \left(\tilde{A} \odot_{\mathfrak{A}} \tilde{B}\right)_0.$$

Part (i) can be similarly obtained from part (ii), since $\tilde{A} \odot \tilde{B} = \tilde{A} \odot_{\mathfrak{A}} \tilde{B}$ when $\mathfrak{A}(x,y) = \min\{x,y\}$. This completes the proof. ∎

Theorem 7.6.26 *Let $\tilde{A}$ and $\tilde{B}$ be two fuzzy sets in $\mathbb{R}$. For $\circ \in \{+, -, \times\}$, let $M_\alpha = \tilde{A}_\alpha \circ \tilde{B}_\alpha$ for all $\alpha \in [0, 1]$. Suppose that, given any fixed $x \in \mathbb{R}^m$, the function $\eta_x(\alpha) = \alpha \cdot \chi_{M_\alpha}(x)$ is upper semi-continuous on $[0, 1]$. Then, for $\odot \in \{\oplus, \ominus, \otimes\}$, we have*

$$\tilde{A} \odot_{\mathfrak{G}} \tilde{B} = \tilde{A} \odot \tilde{B}.$$

Proof. The results follow immediately from Proposition 2.2.15, part (i) of Proposition 7.6.24 and part (i) of Proposition 7.6.25. ∎

We remark that Theorem 7.6.26 can also be obtained using part (ii) of Proposition 7.6.24 and part (ii) of Proposition 7.6.25 by taking

$$\mathfrak{A}(\alpha_1, \alpha_2) = \min\{\alpha_1, \alpha_2\}.$$

The reason is as follows. From Proposition 1.4.6, when

$$\mathfrak{A}(\alpha_1, \alpha_2) \geq \alpha \text{ if and only if } \alpha_1 \geq \alpha \text{ and } \alpha_2 \geq \alpha,$$

we must have

$$\mathfrak{A}(\alpha_1, \alpha_2) = \min\{\alpha_1, \alpha_2\}.$$

Lemma 7.6.27 *Let $\tilde{A}$ and $\tilde{B}$ be two fuzzy sets in $\mathbb{R}$. For $\circ \in \{+, -, \times\}$, let $M_\alpha = \tilde{A}_\alpha \circ \tilde{B}_\alpha$ for all $\alpha \in [0, 1]$. Suppose that, for any increasing and convergent sequence $\{\alpha_n\}_{n=1}^\infty$ in $[0, 1]$, the following inclusion is satisfied:*

$$\bigcap_{n=1}^{\infty} \left(\tilde{A}_{\alpha_n} \circ \tilde{B}_{\alpha_n}\right) \subseteq \left(\bigcap_{n=1}^{\infty} \tilde{A}_{\alpha_n}\right) \circ \left(\bigcap_{n=1}^{\infty} \tilde{B}_{\alpha_n}\right).$$

Then, given any $x \in \mathbb{R}^m$, the following set

$$F_r = \left\{\alpha \in [0, 1] : \alpha \cdot \chi_{M_\alpha}(x) \geq r\right\}$$

is closed for each $r \in [0, 1]$; that is, the function $\eta_x(\alpha) = \alpha \cdot \chi_{M_\alpha}(x)$ is upper semi-continuous on $[0, 1]$.

Proof. If $r = 0$, then $F_0 = [0, 1]$ is closed. Therefore, we assume that $r \in (0, 1]$. For each $\alpha \in \text{cl}(F_r)$, since $r > 0$, we have $\alpha > 0$. According to the concept of closure, there exists a sequence $\{\alpha_n\}_{n=1}^\infty$ satisfying $\alpha_n \to \alpha$ and $\alpha_n \in F_r$ for all n, i.e. $\alpha_n \geq r$ and $x \in M_{\alpha_n}$ for all n. We consider the following two cases.

- Suppose that $\alpha_n \downarrow \alpha$, i.e. $\alpha \leq \alpha_n$ for all n. Then, we have $x \in M_\alpha$, since $M_{\alpha_n} \subseteq M_\alpha$ for all n by part (i) of Proposition 2.2.8. This says that $\alpha \in F_r$, since $\alpha = \lim_n \alpha_n \geq r$.
- Suppose that $\alpha_n \uparrow \alpha$. Then, using Proposition 2.2.8 and the assumption, we have

$$\bigcap_{n=1}^{\infty} M_{\alpha_n} = \bigcap_{n=1}^{\infty} \left(\tilde{A}_{\alpha_n} \circ \tilde{B}_{\alpha_n}\right) \subseteq \bigcap_{n=1}^{\infty} \tilde{A}_{\alpha_n} \circ \bigcap_{n=1}^{\infty} \tilde{B}_{\alpha_n} = \tilde{A}_\alpha \circ \tilde{B}_\alpha = M_\alpha.$$

  Since $x \in M_{\alpha_n}$ for all n, it follows that $x \in M_\alpha$. This says that $\alpha \in F_r$, since

$$\alpha = \lim_n \alpha_n \geq r.$$

Therefore, we conclude that $\text{cl}(F_r) \subseteq F_r$, which means that F_r is closed, and the proof is complete. ∎

Lemma 7.6.28 *Let $\tilde{A}$ and $\tilde{B}$ be two fuzzy sets in $\mathbb{R}$ such that the α-level sets $\tilde{A}_\alpha$ and $\tilde{B}_\alpha$ are closed and bounded subsets of $\mathbb{R}$ for all $\alpha \in [0,1]$. Then, for $\circ \in \{+,-,\times\}$, given any increasing and convergent sequence $\{\alpha_n\}_{n=1}^\infty$ in $[0,1]$, we have the following inclusion*

$$\bigcap_{n=1}^{\infty}\left(\tilde{A}_{\alpha_n} \circ \tilde{B}_{\alpha_n}\right) \subseteq \left(\bigcap_{n=1}^{\infty}\tilde{A}_{\alpha_n}\right) \circ \left(\bigcap_{n=1}^{\infty}\tilde{B}_{\alpha_n}\right).$$

Proof. Given any $x \in \bigcap_{n=1}^{\infty}(\tilde{A}_{\alpha_n} \circ \tilde{B}_{\alpha_n})$, for each $n \in \mathbb{N}$, we have $x = y_{\alpha_n} \circ z_{\alpha_n}$ for some $y_{\alpha_n} \in \tilde{A}_{\alpha_n} \subseteq \tilde{A}_0$ and $z_{\alpha_n} \in \tilde{B}_{\alpha_n} \subseteq \tilde{B}_0$. For convenience, we simply write $y_{\alpha_n} \equiv y_n$ and $z_{\alpha_n} \equiv z_n$. Since we the sequences $\{y_n\}_{n=1}^\infty$ and $\{z_n\}_{n=1}^\infty$ are bounded, there exists a convergent subsequence $\{y_{n_i}\}_{i=1}^\infty$ of $\{y_n\}_{n=1}^\infty$ which converges to some y. Now, we consider the subsequence $\{z_{n_i}\}_{i=1}^\infty$ of $\{z_n\}_{n=1}^\infty$. There also exists a convergent subsequence $\{z_{n_{i_j}}\}_{j=1}^\infty$ of $\{z_{n_i}\}_{i=1}^\infty$ which converges to some z. This says that the subsequences $\{y_{n_{i_j}}\}_{j=1}^\infty$ and $\{z_{n_{i_j}}\}_{j=1}^\infty$ converge to y and z, respectively. Then, the sequence $\{x = y_{n_{i_j}} \circ z_{n_{i_j}}\}_{j=1}^\infty$ converges to $y \circ z$. The uniqueness of the limit says that $x = y \circ z$.

Now, we are going to claim that $y \in \bigcap_{n=1}^\infty \tilde{A}_{\alpha_n}$ and $z \in \bigcap_{n=1}^\infty \tilde{B}_{\alpha_n}$. Suppose that there exists α_m satisfying $y \notin \tilde{A}_{\alpha_m}$. Since $y_n \in \tilde{A}_{\alpha_n}$ for all n and the family $\{\tilde{A}_{\alpha_n}\}$ is nested, it follows that

$$\{y_m, y_{m+1}, y_{m+2}, \ldots\} \subseteq \tilde{A}_{\alpha_m}.$$

Since $\{y_m, y_{m+1}, y_{m+2}, \ldots\}$ is a subset of the sequence $\{y_n\}_{n=1}^\infty$, and the subsequence $\{y_{n_{i_j}}\}_{j=1}^\infty$ converges to y, there is also a subsequence $\{y_{m_{i_j}}\}_{j=1}^\infty$ of $\{y_m, y_{m+1}, y_{m+2}, \ldots,\}$, where $m_{i_j} \geq m$, which converges to y. Since $\tilde{A}_{\alpha_m}$ is closed, we have $y \in \tilde{A}_{\alpha_m}$, which contradicts $y \notin \tilde{A}_{\alpha_m}$. Therefore, we must have $y \in \tilde{A}_{\alpha_n}$ for all α_n, i.e. $y \in \bigcap_{n=1}^\infty \tilde{A}_{\alpha_n}$. We can similarly show that $z \in \bigcap_{n=1}^\infty \tilde{B}_{\alpha_n}$. This completes the proof. ∎

Theorem 7.6.29 *Let $\tilde{A}$ and $\tilde{B}$ be two fuzzy sets in $\mathbb{R}$ such that the α-level sets $\tilde{A}_\alpha$ and $\tilde{B}_\alpha$ are closed and bounded subsets of $\mathbb{R}$ for all $\alpha \in [0,1]$. Then, for $\odot \in \{\oplus, \ominus, \otimes\}$, the following statements hold true.*

(i) For $\circ \in \{+,-,\times\}$, there exists a fuzzy set $\tilde{C}$ in $\mathbb{R}$ with membership function defined by

$$\xi_{\tilde{C}}(x) = \sup_{\alpha \in (0,1]} \alpha \cdot \chi_{M_\alpha}(x)$$

satisfying $\tilde{C}_\alpha = \tilde{A}_\alpha \circ \tilde{B}_\alpha$ for each $\alpha \in [0,1]$. This guarantees the existence of $\tilde{C} = \tilde{A} \odot_{\mathfrak{L}} \tilde{B}$. Moreover, we have

$$\tilde{C} = \tilde{A} \odot_{\mathfrak{L}} \tilde{B} = \tilde{A} \odot_{\mathfrak{G}} \tilde{B}.$$

(ii) We have

$$\tilde{A} \odot_{\mathfrak{L}} \tilde{B} = \tilde{A} \odot_{\mathfrak{G}} \tilde{B} = \tilde{A} \odot \tilde{B}$$

satisfying

$$\left(\tilde{A} \odot_{\mathfrak{L}} \tilde{B}\right)_\alpha = \left(\tilde{A} \odot_{\mathfrak{G}} \tilde{B}\right)_\alpha = \left(\tilde{A} \odot \tilde{B}\right)_\alpha = \tilde{A}_\alpha \circ \tilde{B}_\alpha$$

for all $\alpha \in [0,1]$.

Proof. To prove part (i), we first claim that $\tilde{A}_0 \circ \tilde{B}_0$ is a closed subset of $\mathbb{R}$. Given any convergent sequence $\{x_n\}_{n=1}^{\infty}$ in $\tilde{A}_0 \circ \tilde{B}_0$, we can write $x_n = a_n \circ b_n$ for some $a_n \in \tilde{A}_0$ and $b_n \in \tilde{B}_0$. Since $\tilde{A}_0$ and $\tilde{B}_0$ are closed and bounded subsets of $\mathbb{R}$, there exist two subsequences $\{a_{n_i}\}_{i=1}^{\infty}$ and $\{b_{n_i}\}_{i=1}^{\infty}$ of $\{a_n\}_{n=1}^{\infty}$ and $\{b_n\}_{n=1}^{\infty}$, respectively, that converge to some $a \in \tilde{A}_0$ and $b \in \tilde{B}_0$, respectively. Therefore, it follows that the sequence $\{x_n\}_{n=1}^{\infty}$ converges to $a \circ b \in \tilde{A}_0 \circ \tilde{B}_0$, which shows that $\tilde{A}_0 \circ \tilde{B}_0$ is a closed subset of $\mathbb{R}$. Using Lemma 7.6.28 and Theorem 7.6.23, we obtain part (i). Using Lemmas 7.6.27 and 7.6.28, part (ii) can be obtained immediately by combining part (i) and Theorem 7.6.26. This completes the proof. ■

7.6.6 Equivalent Additions of Fuzzy Sets in $\mathbb{R}^m$

Regarding the m-dimensional Euclidean space $\mathbb{R}^m$, we can only consider the vector addition and scalar multiplication in $\mathbb{R}^m$. Let $\tilde{A}$ and $\tilde{B}$ be two fuzzy sets in $\mathbb{R}^m$. We consider four ways to define the addition of fuzzy sets in $\mathbb{R}^m$ as follows.

- The conventional way of defining the addition $\tilde{A} \oplus \tilde{B}$ is according to the extension principle given by

$$\xi_{\tilde{A}\oplus\tilde{B}}(z) = \sup_{\{(x,y):z=x+y\}} \min\left\{\xi_{\tilde{A}}(x), \xi_{\tilde{B}}(y)\right\}.$$

- The general addition $\tilde{A} \oplus_{\mathfrak{A}} \tilde{B}$ using the function $\mathfrak{A}$ is defined by

$$\xi_{\tilde{A}\oplus_{\mathfrak{A}}\tilde{B}}(z) = \sup_{\{(x,y):z=x\circ y\}} \mathfrak{A}\left(\xi_{\tilde{A}}(x), \xi_{\tilde{B}}(y)\right).$$

- The addition $\tilde{A} \oplus_{\mathfrak{L}} \tilde{B}$ according to (7.94) is based on the α-level sets.
- The general addition $\tilde{A} \oplus_{\mathfrak{G}} \tilde{B}$ is according to (7.118) based on the α-level sets by considering $M_\alpha = \tilde{A}_\alpha + \tilde{B}_\alpha$ for all $\alpha \in (0, 1]$.

We are going to provide some suitable conditions to demonstrate their equivalences. Let X be a subset of m-dimensional Euclidean space $\mathbb{R}^m$. Recall that

$$x \in \text{cl}(X) \text{ if and only if } N_x \cap X \neq \emptyset$$

for any neighborhood N_x of x.

Theorem 7.6.30 *Let $\tilde{A}$ and $\tilde{B}$ be two fuzzy sets in $\mathbb{R}^m$. Suppose that the following conditions are satisfied.*

- *The addition $\tilde{A}_0 + \tilde{B}_0$ of the 0-level sets $\tilde{A}_0$ and $\tilde{B}_0$ is a closed subset of $\mathbb{R}^m$.*
- *For any increasing and convergent sequence $\{\alpha_n\}_{n=1}^{\infty}$ in $[0, 1]$, the following inclusion is satisfied:*

$$\bigcap_{n=1}^{\infty}\left(\tilde{A}_{\alpha_n} + \tilde{B}_{\alpha_n}\right) \subseteq \bigcap_{n=1}^{\infty}\tilde{A}_{\alpha_n} + \bigcap_{n=1}^{\infty}\tilde{B}_{\alpha_n}. \tag{7.135}$$

Then, there exists a fuzzy set $\tilde{C}$ in U with membership function defined by

$$\xi_{\tilde{C}}(x) = \sup_{\alpha\in(0,1]} \alpha \cdot \chi_{M_\alpha}(x)$$

satisfying $\tilde{C}_\alpha = \tilde{A}_\alpha + \tilde{B}_\alpha$ *for each* $\alpha \in [0,1]$. *This guarantees the existence of* $\tilde{C} = \tilde{A} \oplus_{\mathfrak{L}} \tilde{B}$. *Moreover, we have*

$$\tilde{C} = \tilde{A} \oplus_{\mathfrak{L}} \tilde{B} = \tilde{A} \oplus_{\mathfrak{G}} \tilde{B}.$$

Proof. We define $M_\alpha = \tilde{A}_\alpha + \tilde{B}_\alpha$ for all $\alpha \in [0,1]$. Since $\tilde{A}_\beta \subseteq \tilde{A}_\alpha$ and $\tilde{B}_\beta \subseteq \tilde{B}_\alpha$ for $\alpha, \beta \in [0,1]$ with $\alpha < \beta$, we have $M_\beta \subseteq M_\alpha$ for $\alpha, \beta \in [0,1]$ with $\alpha < \beta$. In particular, we have $M_\alpha \subseteq M_0$ for all $\alpha \in (0,1]$, i.e.

$$\text{cl}(M_{0+}) = \text{cl}\left(\bigcup_{\alpha\in(0,1]} M_\alpha\right) \subseteq \text{cl}(M_0) = M_0, \tag{7.136}$$

since $M_0 = \tilde{A}_0 \circ \tilde{B}_0$ is a closed subset of $\mathbb{R}^m$ by the assumption. For $\alpha_n \uparrow \alpha$ with $\alpha \in (0,1]$ and $\{\alpha_n\}_{n=1}^\infty \subset [0,1]$, from Proposition 2.2.8, we have the following inclusion

$$M_\alpha = \tilde{A}_\alpha + \tilde{B}_\alpha = \bigcap_{n=1}^\infty \tilde{A}_{\alpha_n} + \bigcap_{n=1}^\infty \tilde{B}_{\alpha_n} \subseteq \bigcap_{n=1}^\infty \left(\tilde{A}_{\alpha_n} + \tilde{B}_{\alpha_n}\right) = \bigcap_{n=1}^\infty M_{\alpha_n}.$$

Using the condition (7.135), we have $M_\alpha = \bigcap_{n=1}^\infty M_{\alpha_n}$ for $\alpha_n \uparrow \alpha$, where $\alpha \in (0,1]$ and $\{\alpha_n\}_{n=1}^\infty \subset [0,1]$. Next, we are going to claim $M_0 \subseteq \text{cl}(M_{0+})$. For $x = y + z \in M_0$, we are going to show that, given any neighborhood N_x of x,

$$N_x \bigcap \left(\bigcup_{\alpha\in(0,1]} (\tilde{A}_\alpha + \tilde{B}_\alpha)\right) \neq \emptyset, \tag{7.137}$$

where

$$y \in \text{cl}\left(\bigcup_{\alpha\in(0,1]} \tilde{A}_\alpha\right) \text{ and } z \in \text{cl}\left(\bigcup_{\alpha\in(0,1]} \tilde{B}_\alpha\right).$$

Since $N_x - x$ is a neighborhood of zero vector θ, there exists another neighborhood V of θ satisfying

$$V + V \subseteq N_x - x, \text{ i.e. } (V + y) + (V + z) \subseteq N_x.$$

Since $V + y$ is a neighborhood of y, and

$$y \in \text{cl}\left(\bigcup_{\alpha\in(0,1]} \tilde{A}_\alpha\right) \text{ i.e. } (V + y) \cap \left(\bigcup_{\alpha\in(0,1]} \tilde{A}_\alpha\right) \neq \emptyset,$$

there exists $y_0 \in V + y$ and $y_0 \in \tilde{A}_{\alpha_0}$ for some $\alpha_0 > 0$. Similarly, since $V + z$ is a neighborhood of z, and

$$z \in \text{cl}\left(\bigcup_{\alpha\in(0,1]} \tilde{B}_\alpha\right),$$

there exists $z_0 \in V + z$ and $z_0 \in \tilde{B}_{\beta_0}$ for some $\beta_0 > 0$. We immediately have

$$y_0 + z_0 \in (V + y) + (V + z) \subseteq N_x.$$

Let $\alpha^* = \min\{\alpha_0, \beta_0\} > 0$. Since $0 < \alpha^* \le \alpha_0$ and $0 < \alpha^* \le \beta_0$, we have $y_0 \in \tilde{A}_{\alpha_0} \subseteq \tilde{A}_{\alpha^*}$ and $z_0 \in \tilde{B}_{\beta_0} \subseteq \tilde{B}_{\alpha^*}$, which says that

$$y_0 + z_0 \in \bigcup_{\alpha\in(0,1]} (\tilde{A}_\alpha + \tilde{B}_\alpha),$$

and proves (7.137), i.e. $x \in \text{cl}(M_{0+})$. Therefore, we indeed have the inclusion $M_0 \subseteq \text{cl}(M_{0+})$. Using (7.136), we obtain $M_0 = \text{cl}(M_{0+})$. We consider the membership function

$$\xi_{\tilde{C}}(x) = \sup_{\alpha \in (0,1]} \alpha \cdot \chi_{M_\alpha}(x).$$

Theorem 5.4.7 says that $\tilde{C}_\alpha = M_\alpha$ for all $\alpha \in (0,1]$ and

$$\tilde{C}_0 = \text{cl}\left(\bigcup_{\alpha \in (0,1]} M_\alpha\right) = \text{cl}(M_{0+}) = M_0.$$

Therefore, we obtain $\tilde{C}_\alpha = M_\alpha = \tilde{A}_\alpha + \tilde{B}_\alpha$ for all $\alpha \in [0,1]$, which shows the existence of $\tilde{C} = \tilde{A} \oplus_{\mathfrak{L}} \tilde{B}$. From (7.118), we also see that

$$\xi_{\tilde{C}}(x) = \sup_{\alpha \in (0,1]} \alpha \cdot \chi_{M_\alpha}(x) = \xi_{\tilde{A} \oplus_{\mathfrak{G}} \tilde{B}}(x).$$

This completes the proof. ∎

Remark 7.6.31 We have the following observations.

- Let $\{\alpha_n\}_{n=1}^{\infty}$ be any increasing and convergent sequence in $[0,1]$. Suppose that the inclusion (7.135) holds true. Then, we have

  $$\bigcap_{n=1}^{\infty}\left(\tilde{A}_{\alpha_n} + \tilde{B}_{\alpha_n}\right) = \bigcap_{n=1}^{\infty}\tilde{A}_{\alpha_n} + \bigcap_{n=1}^{\infty}\tilde{B}_{\alpha_n},$$

  since the inclusion

  $$\bigcap_{n=1}^{\infty}\tilde{A}_{\alpha_n} + \bigcap_{n=1}^{\infty}\tilde{B}_{\alpha_n} \subseteq \bigcap_{n=1}^{\infty}\left(\tilde{A}_{\alpha_n} + \tilde{B}_{\alpha_n}\right)$$

  is always true. Indeed, we have

  $$\bigcap_{n=1}^{\infty}\tilde{A}_{\alpha_n} \subseteq \tilde{A}_{\alpha_n} \text{ and } \bigcap_{n=1}^{\infty}\tilde{B}_{\alpha_n} \subseteq \tilde{B}_{\alpha_n},$$

  which imply the desired equality.
- It is well known that if E is a closed subset of $\mathbb{R}^m$ and F is a closed and bounded subset of $\mathbb{R}^m$, then $E + F$ is a closed subset of $\mathbb{R}^m$ by Kelley and Namioka [55]. Therefore, if either the 0-level set $\tilde{A}_0$ or the 0-level set $\tilde{B}_0$ is closed and bounded, then we have that $\tilde{A}_0 + \tilde{B}_0$ is closed, i.e. the first condition in Theorem 7.6.30 is satisfied.

Proposition 7.6.32 *Let $\tilde{A}$ and $\tilde{B}$ be two fuzzy sets in $\mathbb{R}^m$. The following statements hold true.*

(i) *We have*

$$\left(\tilde{A} \oplus \tilde{B}\right)_\alpha \subseteq \left(\tilde{A} \oplus_{\mathfrak{G}} \tilde{B}\right)_\alpha \text{ for all } \alpha \in [0,1].$$

(ii) *Suppose that the function $\mathfrak{A} : [0,1] \times [0,1] \to [0,1]$ satisfies the following condition. For each $\alpha \in (0,1]$,*

$$\mathfrak{A}(\alpha_1, \alpha_2) > \alpha \text{ implies } \alpha_1 \geq \alpha \text{ and } \alpha_2 \geq \alpha.$$

Then, we have

$$\left(\tilde{A} \oplus_{\mathfrak{A}} \tilde{B}\right)_\alpha \subseteq \left(\tilde{A} \oplus_{\mathfrak{G}} \tilde{B}\right)_\alpha \text{ for all } \alpha \in [0,1].$$

Proof. The arguments in the proof of Proposition 7.6.24 are still valid when the operation $\odot$ is replaced by the addition $\oplus$. ■

Proposition 7.6.33 *Let $\tilde{A}$ and $\tilde{B}$ be two fuzzy sets in $\mathbb{R}^m$, and let $M_\alpha = \tilde{A}_\alpha + \tilde{B}_\alpha$ for all $\alpha \in [0,1]$. Suppose that, given any fixed $x \in \mathbb{R}^m$, the function $\eta_x(\alpha) = \alpha \cdot \chi_{M_\alpha}(x)$ is upper semi-continuous on $[0,1]$. Then, the following statements hold true.*

(i) We have

$$\left(\tilde{A} \oplus_{\mathfrak{G}} \tilde{B}\right)_\alpha \subseteq \left(\tilde{A} \oplus \tilde{B}\right)_\alpha \text{ for all } \alpha \in [0,1].$$

(ii) Suppose that, for each $\alpha \in (0,1]$,

$$\alpha_1 \geq \alpha \text{ and } \alpha_2 \geq \alpha \text{ imply } \mathfrak{A}(\alpha_1, \alpha_2) \geq \alpha.$$

Then, we have

$$\left(\tilde{A} \oplus_{\mathfrak{G}} \tilde{B}\right)_\alpha \subseteq \left(\tilde{A} \oplus_{\mathfrak{A}} \tilde{B}\right)_\alpha \text{ for all } \alpha \in [0,1].$$

Proof. The arguments in the proof of Proposition 7.6.25 are still valid when the operation $\odot$ is replaced by the addition $\oplus$. ■

Theorem 7.6.34 *Let $\tilde{A}$ and $\tilde{B}$ be two fuzzy sets in $\mathbb{R}^m$, and let $M_\alpha = \tilde{A}_\alpha + \tilde{B}_\alpha$ for all $\alpha \in [0,1]$. Suppose that, given any fixed $x \in \mathbb{R}^m$, the function $\eta_x(\alpha) = \alpha \cdot \chi_{M_\alpha}(x)$ is upper semi-continuous on $[0,1]$. Then, we have*

$$\tilde{A} \oplus_{\mathfrak{G}} \tilde{B} = \tilde{A} \oplus \tilde{B}.$$

Proof. The result follows immediately from Proposition 2.2.15, part (i) of Proposition 7.6.32, and part (i) of Proposition 7.6.33. ■

Theorem 7.6.34 is too general to be used in a real-world situation, since the assumption for the upper semi-continuity of function $\eta_x(\alpha) = \alpha \cdot \chi_{M_\alpha}(x)$ on $[0,1]$ may not be easy to check. Now, we are going to provide some useful sufficient conditions to guarantee the upper semi-continuity of function $\eta_x(\alpha) = \alpha \cdot \chi_{M_\alpha}(x)$ on $[0,1]$.

Lemma 7.6.35 *Let $\tilde{A}$ and $\tilde{B}$ be two fuzzy sets in $\mathbb{R}^m$, and let $M_\alpha = \tilde{A}_\alpha + \tilde{B}_\alpha$ for all $\alpha \in [0,1]$. Suppose that, for any increasing and convergent sequence $\{\alpha_n\}_{n=1}^\infty$ in $[0,1]$, the following inclusion is satisfied:*

$$\bigcap_{n=1}^\infty \left(\tilde{A}_{\alpha_n} + \tilde{B}_{\alpha_n}\right) \subseteq \bigcap_{n=1}^\infty \tilde{A}_{\alpha_n} + \bigcap_{n=1}^\infty \tilde{B}_{\alpha_n}. \tag{7.138}$$

Then, given any $x \in \mathbb{R}^m$, the following set

$$F_r = \left\{\alpha \in [0,1] : \alpha \cdot \chi_{M_\alpha}(x) \geq r\right\}$$

is closed for each $r \in [0,1]$; that is, the function $\eta_x(\alpha) = \alpha \cdot \chi_{M_\alpha}(x)$ is upper semi-continuous on $[0,1]$.

Proof. The arguments in the proof of Lemma 7.6.27 are still valid when the operation $\circ$ is replaced by the addition $+$. ■

Let $\{x_n\}_{n=1}^{\infty}$ be a sequence in $\mathbb{R}^m$. Recall that the sequence $\{x_n\}_{n=1}^{\infty}$ converges to x if and only if, given any neighborhood N_x of x, there is an integer m satisfying $x_n \in N$ for all $n \geq m$.

Lemma 7.6.36 *Let $\tilde{A}$ and $\tilde{B}$ be two fuzzy sets in $\mathbb{R}^m$. Suppose that $\tilde{A}_\alpha$ and $\tilde{B}_\alpha$ are closed and bounded in $\mathbb{R}^m$ for all $\alpha \in [0,1]$. Then, for any increasing and convergent sequence $\{\alpha_n\}_{n=1}^{\infty}$ in $[0,1]$, we have the following inclusion*

$$\bigcap_{n=1}^{\infty}\left(\tilde{A}_{\alpha_n} + \tilde{B}_{\alpha_n}\right) \subseteq \bigcap_{n=1}^{\infty}\tilde{A}_{\alpha_n} + \bigcap_{n=1}^{\infty}\tilde{B}_{\alpha_n}.$$

Proof. Given any $x \in \bigcap_{n=1}^{\infty}(\tilde{A}_{\alpha_n} + \tilde{B}_{\alpha_n})$, for each $n \in \mathbb{N}$, we have $x = y_{\alpha_n} + z_{\alpha_n}$ for some $y_{\alpha_n} \in \tilde{A}_{\alpha_n} \subseteq \tilde{A}_0$ and $z_{\alpha_n} \in \tilde{B}_{\alpha_n} \subseteq \tilde{B}_0$. For convenience, we simply write $y_{\alpha_n} \equiv y_n$ and $z_{\alpha_n} \equiv z_n$. Since the 0-level sets $\tilde{A}_0$ and $\tilde{B}_0$ are bounded, it says that the bounded sequences $\{y_n\}_{n=1}^{\infty}$ and $\{z_n\}_{n=1}^{\infty}$ have convergent subsequences. Let $\{y_{n_i}\}_{i=1}^{\infty}$ be a subsequence of $\{y_n\}_{n=1}^{\infty}$ that converges to y. Using the same indices, we consider the subsequence $\{z_{n_i}\}_{i=1}^{\infty}$ of $\{z_n\}_{n=1}^{\infty}$. Therefore, there also exists a subsequence $\{z_{n_{i_j}}\}_{j=1}^{\infty}$ of $\{z_{n_i}\}_{i=1}^{\infty}$ that converges to z. This also says that the subsequences $\{y_{n_{i_j}}\}_{j=1}^{\infty}$ and $\{z_{n_{i_j}}\}_{j=1}^{\infty}$ converge to y and z, respectively.

Now, we are going to claim that the subsequence $\{y_{n_{i_j}} + z_{n_{i_j}}\}_{j=1}^{\infty}$ converges to $y+z$. Let N be any neighborhood of $y+z$. Then, we see that $N - y - z$ is a neighborhood of the zero vector θ. Therefore, there exists another neighborhood V of θ satisfying

$$V + V \subseteq N - y - z, \text{ i.e. } V + V + y + z \subseteq N.$$

Since $V + y$ and $V + z$ are the neighborhoods of y and z, respectively, by the convergence, there exist integers m_0 and m_1 satisfying $y_{n_{i_j}} \in V + y$ for $n_{i_j} \geq m_0$ and $z_{n_{i_j}} \in V + z$ for $n_{i_j} \geq m_1$. Let $m = \max\{m_0, m_1\}$. We obtain

$$y_{n_{i_j}} + z_{n_{i_j}} \in V + V + y + z \subseteq N \text{ for } n_{i_j} \geq m,$$

which implies $y_{n_{i_j}} + z_{n_{i_j}} \in N$ for $n_{i_j} \geq m$. Therefore, we conclude that the subsequence $\{y_{n_{i_j}} + z_{n_{i_j}}\}_{j=1}^{\infty}$ converges to $y+z$. The uniqueness of the limit says that $x = y + z$. Finally, using the same argument in the proof of Lemma 7.6.28, we can show that $y \in \bigcap_{n=1}^{\infty}\tilde{A}_{\alpha_n}$ and $z \in \bigcap_{n=1}^{\infty}\tilde{B}_{\alpha_n}$. This completes the proof. ■

Theorem 7.6.37 *Let $\tilde{A}$ and $\tilde{B}$ be two fuzzy sets in $\mathbb{R}^m$. Suppose that $\tilde{A}_\alpha$ and $\tilde{B}_\alpha$ are closed and bounded in $\mathbb{R}^m$ for all $\alpha \in [0,1]$. Then, we have*

$$\tilde{A} \oplus_{\mathfrak{L}} \tilde{B} = \tilde{A} \oplus_{\mathfrak{G}} \tilde{B} = \tilde{A} \oplus \tilde{B}$$

satisfying

$$\left(\tilde{A} \oplus_{\mathfrak{L}} \tilde{B}\right)_\alpha = \left(\tilde{A} \oplus_{\mathfrak{G}} \tilde{B}\right)_\alpha = \left(\tilde{A} \oplus \tilde{B}\right)_\alpha = \tilde{A}_\alpha + \tilde{B}_\alpha$$

for all $\alpha \in [0,1]$.

Proof. Since the 0-level sets $\tilde{A}_0$ and $\tilde{B}_0$ are closed, from Remark 7.6.31, it follows that $\tilde{A}_0 + \tilde{B}_0$ is also closed. Using Lemmas 7.6.35 and 7.6.36, the results can be obtained immediately by combining Theorems 7.6.30 and 7.6.34. ■

7.7 Hausdorff Differences

The difference of fuzzy intervals can be used to define the differentiation of fuzzy functions. Let A and B be any two nonempty sets in $\mathbb{R}^m$. The **Hausdorff metric** of A and B is defined by

$$d_H(A,B) = \max\left\{ \sup_{a\in A}\inf_{b\in B} \| a-b \|, \sup_{b\in B}\inf_{a\in A} \| a-b \| \right\},$$

where $a = (a_1, \ldots, a_n)$ and $b = (b_1, \ldots, b_n)$ with the norm

$$\| a-b \| = \sqrt{(a_1-b_1)^2 + \cdots + (a_n-b_n)^2}.$$

Given two fuzzy sets $\tilde{A}$ and $\tilde{B}$ in $\mathbb{R}$, inspired by the form of Hausdorff metric, we shall propose many types of Hausdorff differences between $\tilde{A}$ and $\tilde{B}$ without considering a norm.

7.7.1 Fair Hausdorff Difference

Let $\tilde{A}$ and $\tilde{B}$ be two fuzzy sets in $\mathbb{R}$. The α-level sets $\tilde{A}_\alpha$ and $\tilde{B}_\alpha$ are not necessarily bounded closed intervals. For each $\alpha \in [0,1]$, we define

$$\underline{d}_H\left(\tilde{A}_\alpha, \tilde{B}_\alpha\right) = \min\left\{ \sup_{a\in\tilde{A}_\alpha}\inf_{b\in\tilde{B}_\alpha} (a-b), \sup_{b\in\tilde{B}_\alpha}\inf_{a\in\tilde{A}_\alpha} (a-b) \right\} \tag{7.139}$$

and

$$\bar{d}_H\left(\tilde{A}_\alpha, \tilde{B}_\alpha\right) = \max\left\{ \sup_{a\in\tilde{A}_\alpha}\inf_{b\in\tilde{B}_\alpha} (a-b), \sup_{b\in\tilde{B}_\alpha}\inf_{a\in\tilde{A}_\alpha} (a-b) \right\}. \tag{7.140}$$

We also see that

$$\sup_{a\in\tilde{A}_\alpha}\inf_{b\in\tilde{B}_\alpha} (a-b) = \sup_{a\in\tilde{A}_\alpha}\left[a + \inf_{b\in\tilde{B}_\alpha}(-b) \right] = \sup_{a\in\tilde{A}_\alpha} a + \inf_{b\in\tilde{B}_\alpha}(-b) = \sup_{a\in\tilde{A}_\alpha} a - \sup_{b\in\tilde{B}_\alpha} b \tag{7.141}$$

and

$$\sup_{b\in\tilde{B}_\alpha}\inf_{a\in\tilde{A}_\alpha} (a-b) = \sup_{b\in\tilde{B}_\alpha}\left[\left(\inf_{a\in\tilde{A}_\alpha} a \right) - b \right] = \sup_{b\in\tilde{B}_\alpha}(-b) + \inf_{a\in\tilde{A}_\alpha} a = \inf_{a\in\tilde{A}_\alpha} a - \inf_{b\in\tilde{B}_\alpha} b. \tag{7.142}$$

Based on $\underline{d}_H(\tilde{A}_\alpha, \tilde{B}_\alpha)$ and $\bar{d}_H(\tilde{A}_\alpha, \tilde{B}_\alpha)$ shown in (7.139) and (7.140), respectively, we define

$$\tilde{A}_\alpha \boxminus_{FH} \tilde{B}_\alpha = \left[\underline{d}_H\left(\tilde{A}_\alpha, \tilde{B}_\alpha\right), \bar{d}_H\left(\tilde{A}_\alpha, \tilde{B}_\alpha\right)\right] \tag{7.143}$$

to be a bounded closed interval for each $\alpha \in [0,1]$. The family

$$\{M_\alpha \equiv \tilde{A}_\alpha \boxminus_{FH} \tilde{B}_\alpha : 0 \leq \alpha \leq 1\}$$

is not necessarily nested. According to the binary operation presented in (7.93), the **fair Hausdorff difference** between $\tilde{A}$ and $\tilde{B}$ is defined by

$$\tilde{A} \ominus_{FH} \tilde{B} = \tilde{C} \text{ when } \tilde{C}_\alpha = \tilde{A}_\alpha \boxminus_{FH} \tilde{B}_\alpha \text{ for each } \alpha \in [0,1]. \tag{7.144}$$

However, the existence of such $\tilde{C}$ is not guaranteed.

Without considering the 0-level sets $\tilde{A}_0$ and $\tilde{B}_0$, according to the binary operation presented in (7.115), the **fair *-Hausdorff difference** between $\tilde{A}$ and $\tilde{B}$ is defined by

$$\tilde{A} \ominus_{FH^*} \tilde{B} = \tilde{C}^* \text{ when } \tilde{C}^*_\alpha = \tilde{A}_\alpha \boxminus_{FH} \tilde{B}_\alpha \text{ for all } \alpha \in (0,1]. \tag{7.145}$$

Also, the existence of such $\tilde{C}^*$ is not guaranteed. The sufficient conditions for guaranteeing the existence of $\tilde{A} \ominus_{FH^*} \tilde{B}$ are weaker than those of $\tilde{A} \ominus_{FH} \tilde{B}$ by referring to Theorems 7.6.14 and 7.6.19. Proposition 7.6.3 also says that if the existence of $\tilde{A} \ominus_{FH^*} \tilde{B}$ and $\tilde{A} \ominus_{FH} \tilde{B}$ are guaranteed, then $\tilde{A} \ominus_{FH^*} \tilde{B} = \tilde{A} \ominus_{FH} \tilde{B}$.

According to the binary operation presented in (7.118), we can define the corresponding **generalized fair Hausdorff difference** between $\tilde{A}$ and $\tilde{B}$, which will always exist, as follows. Let $M_\alpha = \tilde{A}_\alpha \boxminus_{FH} \tilde{B}_\alpha$ be given in (7.143) for all $\alpha \in (0, 1]$. Using (7.118), we consider the generalized fair Hausdorff difference $\tilde{A} \ominus_{GFH} \tilde{B}$ with membership function defined by

$$\xi_{\tilde{A} \ominus_{GFH} \tilde{B}}(x) = \sup_{0<\alpha\leq 1} \alpha \cdot \chi_{M_\alpha}(x). \tag{7.146}$$

This says that $\tilde{A} \ominus_{GFH} \tilde{B}$ always exists.

Example 7.7.1 Let $\tilde{A}$ and $\tilde{B}$ be two fuzzy intervals. Then

$$\tilde{A}_\alpha = \left[\tilde{A}^L_\alpha, \tilde{A}^U_\alpha\right] \text{ and } \tilde{B}_\alpha = \left[\tilde{B}^L_\alpha, \tilde{B}^U_\alpha\right].$$

According to (7.141) and (7.142), we have

$$\sup_{a\in\tilde{A}_\alpha} \inf_{b\in\tilde{B}_\alpha} (a-b) = \tilde{A}^U_\alpha - \tilde{B}^U_\alpha \text{ and } \sup_{b\in\tilde{B}_\alpha} \inf_{a\in\tilde{A}_\alpha} (a-b) = \tilde{A}^L_\alpha - \tilde{B}^L_\alpha. \tag{7.147}$$

Let us recall the Hausdorff difference defined in (7.97). According to (7.143) and (7.96), it follows that

$$\tilde{A}_\alpha \boxminus_{FH} \tilde{B}_\alpha = \left[\min\left\{\tilde{A}^L_\alpha - \tilde{B}^L_\alpha, \tilde{A}^U_\alpha - \tilde{B}^U_\alpha\right\}, \max\left\{\tilde{A}^L_\alpha - \tilde{B}^L_\alpha, \tilde{A}^U_\alpha - \tilde{B}^U_\alpha\right\}\right] = \tilde{A}_\alpha \boxminus_H \tilde{B}_\alpha. \tag{7.148}$$

Therefore, we obtain

$$\tilde{A} \ominus_{FH} \tilde{B} = \tilde{A} \ominus_H \tilde{B},$$

which says that the Hausdorff difference defined in (7.97) is a special case of the fair Hasudorff difference defined in (7.144).

Now, we study the existence of fair Hausdorff difference by considering

$$\begin{aligned} M_\alpha = \tilde{A}_\alpha \boxminus_{FH} \tilde{B}_\alpha &= \left[\underline{d}_H\left(\tilde{A}_\alpha, \tilde{B}_\alpha\right), \bar{d}_H\left(\tilde{A}_\alpha, \tilde{B}_\alpha\right)\right] \\ &= \left[\min\left\{\sup_{a\in\tilde{A}_\alpha} \inf_{b\in\tilde{B}_\alpha} (a-b), \sup_{b\in\tilde{B}_\alpha} \inf_{a\in\tilde{A}_\alpha} (a-b)\right\},\right. \\ &\qquad \left.\max\left\{\sup_{a\in\tilde{A}_\alpha} \inf_{b\in\tilde{B}_\alpha} (a-b), \sup_{b\in\tilde{B}_\alpha} \inf_{a\in\tilde{A}_\alpha} (a-b)\right\}\right] \end{aligned}$$

for all $\alpha \in (0, 1]$ by referring to (7.139), (7.140) and (7.143). We see that the family $\{M_\alpha : 0 < \alpha \leq 1\}$ is not necessarily nested.

Let $\tilde{A}$ and $\tilde{B}$ be two fuzzy sets in $\mathbb{R}$. For convenience, we also define

$$l(\alpha) \equiv \min\left\{\sup_{a\in\tilde{A}_\alpha} \inf_{b\in\tilde{B}_\alpha} (a-b), \sup_{b\in\tilde{B}_\alpha} \inf_{a\in\tilde{A}_\alpha} (a-b)\right\} \tag{7.149}$$

and

$$u(\alpha) \equiv \max \left\{ \sup_{a\in\tilde{A}_\alpha} \inf_{b\in\tilde{B}_\alpha} (a-b), \sup_{b\in\tilde{B}_\alpha} \inf_{a\in\tilde{A}_\alpha} (a-b) \right\} \tag{7.150}$$

for $\alpha \in [0,1]$.

Lemma 7.7.2 *Let $\tilde{A}$ and $\tilde{B}$ be two fuzzy sets in $\mathbb{R}$. Then, the functions l and u defined in (7.149) and (7.150), respectively, are right-continuous at 0.*

Proof. Using the denseness of $\mathbb{R}$, there exists a sequence $\{\alpha_n\}_{n=1}^{\infty}$ in $(0,1]$ satisfying $\alpha_n \downarrow 0$ and $\alpha_n > 0$ for all n. Using Proposition 2.2.8, we have

$$\tilde{A}_{0+} = \bigcup_{n=1}^{\infty} \tilde{A}_{\alpha_n} \text{ and } \tilde{B}_{0+} = \bigcup_{n=1}^{\infty} \tilde{B}_{\alpha_n}.$$

Since $\tilde{A}_0 = \text{cl}(\tilde{A}_{0+})$ and $\tilde{B}_0 = \text{cl}(\tilde{B}_{0+})$, from Propositions 1.2.10 and 1.2.7, we have

$$\lim_{n\to\infty} \sup_{a\in\tilde{A}_{\alpha_n}} \inf_{b\in\tilde{B}_{\alpha_n}} (a-b) = \sup_{a\in\tilde{A}_{0+}} \inf_{b\in\tilde{B}_{0+}} (a-b) = \sup_{a\in\tilde{A}_0} \inf_{b\in\tilde{B}_0} (a-b)$$

and

$$\lim_{n\to\infty} \sup_{b\in\tilde{B}_{\alpha_n}} \inf_{a\in\tilde{A}_{\alpha_n}} (a-b) = \sup_{b\in\tilde{B}_{0+}} \inf_{a\in\tilde{A}_{0+}} (a-b) = \sup_{b\in\tilde{B}_0} \inf_{a\in\tilde{A}_0} (a-b),$$

which shows that

$$l(\alpha_n) \to l(0) \text{ and } u(\alpha_n) \to u(0) \text{ as } n \to \infty.$$

This completes the proof. ∎

Theorem 7.7.3 *(**Existence for** $\tilde{A} \ominus_{FH^*} \tilde{B}$) Let $\tilde{A}$ and $\tilde{B}$ be two fuzzy sets in $\mathbb{R}$, and let $M_\alpha = \tilde{A}_\alpha \boxminus_{FH} \tilde{B}_\alpha$ for all $\alpha \in (0,1]$. Suppose that $\{M_\alpha : 0 < \alpha \leq 1\}$ is a nested family. Then, there exists a fuzzy set $\tilde{C}^*$ in $\mathbb{R}$ with membership function defined by*

$$\xi_{\tilde{C}^*}(x) = \sup_{0<\alpha\leq 1} \alpha \cdot \chi_{M_\alpha}(x)$$

satisfying

$$\tilde{C}^*_\alpha = M_\alpha = \tilde{A}_\alpha \boxminus_{FH} \tilde{B}_\alpha \textit{ for each } \alpha \in (0,1]$$

and

$$\tilde{C}^*_0 = M_0 = cl\left(\bigcup_{0<\alpha\leq 1} M_\alpha \right) = cl\left(\bigcup_{0<\alpha\leq 1} \tilde{A}_\alpha \boxminus_{FH} \tilde{B}_\alpha \right).$$

This guarantees the existence of $\tilde{C}^* = \tilde{A} \ominus_{FH^*} \tilde{B}$. Moreover, we have

$$\tilde{C}^* = \tilde{A} \ominus_{FH^*} \tilde{B} = \tilde{A} \ominus_{GFH} \tilde{B}.$$

Proof. From Theorem 7.6.14 and Lemma 7.6.35, it suffices to prove the inclusion (7.138). Given an increasing and convergent sequence $\{\alpha_n\}_{n=1}^{\infty}$ in $[0,1]$ with $\alpha_n \uparrow \alpha$, we write

$$\tilde{A}_{\alpha_n} = A_n \text{ and } \tilde{B}_{\alpha_n} = B_n$$

and

$$\tilde{A}_\alpha = A \text{ and } \tilde{B}_\alpha = B.$$

Part (ii) of Proposition 2.2.8 says that

$$\bigcap_{n=1}^{\infty} A_n = A \text{ and } \bigcap_{n=1}^{\infty} B_n = B.$$

For $x \in \bigcap_{n=1}^{\infty} M_{\alpha_n}$, it follows that

$$\min\left\{ \sup_{a\in A_n} \inf_{b\in B_n} (a-b), \sup_{b\in B_n} \inf_{a\in A_n} (a-b) \right\} \le x$$
$$\le \max\left\{ \sup_{a\in A_n} \inf_{b\in B_n} (a-b), \sup_{b\in B_n} \inf_{a\in A_n} (a-b) \right\}$$

for all n. By taking limit on each side and using Proposition 1.2.9, we obtain

$$\min\left\{ \sup_{a\in A} \inf_{b\in B} (a-b), \sup_{b\in B} \inf_{a\in A} (a-b) \right\} \le x$$
$$\le \max\left\{ \sup_{a\in A} \inf_{b\in B} (a-b), \sup_{b\in B} \inf_{a\in A} (a-b) \right\},$$

i.e. $x \in M_\alpha$, which proves the inclusion (7.138). Using Theorem 7.6.14 and Lemma 7.6.35, we complete the proof. ∎

Theorem 7.7.4 ***(Existence for*** $\tilde{A} \ominus_{FH} \tilde{B}$***)*** *Let $\tilde{A}$ and $\tilde{B}$ be two fuzzy sets in $\mathbb{R}$, and let $M_\alpha = \tilde{A}_\alpha \boxminus_{FH} \tilde{B}_\alpha$ for all $\alpha \in [0,1]$. Suppose that the following conditions are satisfied:*

- *$\{M_\alpha : 0 \le \alpha \le 1\}$ is a nested family;*
- *the following equality is satisfied:*

$$cl\left(\bigcup_{0<\alpha\le 1} \tilde{A}_\alpha \boxminus_{FH} \tilde{B}_\alpha \right) = \tilde{A}_0 \boxminus_{FH} \tilde{B}_0. \tag{7.151}$$

Then, there exists a fuzzy set $\tilde{C}$ in $\mathbb{R}$ with membership function defined by

$$\xi_{\tilde{C}}(x) = \sup_{\alpha\in[0,1]} \alpha \cdot \chi_{M_\alpha}(x) = \sup_{0<\alpha\le 1} \alpha \cdot \chi_{M_\alpha}(x)$$

satisfying

$$\tilde{C}_\alpha = M_\alpha = \tilde{A}_\alpha \boxminus_{FH} \tilde{B}_\alpha \text{ for each } \alpha \in [0,1].$$

This guarantees the existence of $\tilde{C} = \tilde{A} \ominus_{FH} \tilde{B}$. Moreover, the existence of $\tilde{C} = \tilde{A} \ominus_{FH^*} \tilde{B}$ is also guaranteed and we have

$$\tilde{C} = \tilde{A} \ominus_{FH} \tilde{B} = \tilde{A} \ominus_{FH^*} \tilde{B} = \tilde{A} \ominus_{GFH} \tilde{B}.$$

Proof. By applying Theorem 7.6.19 to the argument of the proof of Theorem 7.7.3, we can obtain the desired results. ∎

Example 7.7.5 Continued from Example 7.6.1 by considering the triangular fuzzy numbers, from (7.148) in Example 7.7.1, we have

$$\tilde{A}_\alpha \boxminus_{FH} \tilde{B}_\alpha = \left[\min\left\{\tilde{A}_\alpha^L - \tilde{B}_\alpha^L, \tilde{A}_\alpha^U - \tilde{B}_\alpha^U\right\}, \max\left\{\tilde{A}_\alpha^L - \tilde{B}_\alpha^L, \tilde{A}_\alpha^U - \tilde{B}_\alpha^U\right\}\right].$$

We define

$$l(\alpha) = (1-\alpha)\cdot\min\left\{a^L - b^L, a^U - b^U\right\} + \alpha\cdot(a-b)$$

and

$$u(\alpha) = (1-\alpha)\cdot\max\left\{a^L - b^L, a^U - b^U\right\} + \alpha\cdot(a-b)$$

on $[0,1]$. Then

$$M_\alpha = \tilde{A}_\alpha \boxminus_{FH} \tilde{B}_\alpha = \tilde{A}_\alpha \boxminus_{H} \tilde{B}_\alpha = \left[l(\alpha), u(\alpha)\right].$$

Considering the derivatives of functions l and u, if

$$l'(\alpha) = a - b - \min\left\{a^L - b^L, a^U - b^U\right\} \geq 0$$

and

$$u'(\alpha) = a - b - \max\left\{a^L - b^L, a^U - b^U\right\} \leq 0,$$

then l is an increasing function and u is a decreasing function. In other words, if

$$\min\left\{a^L - b^L, a^U - b^U\right\} \leq a - b \leq \max\left\{a^L - b^L, a^U - b^U\right\}, \tag{7.152}$$

then $\{M_\alpha : 0 \leq \alpha \leq 1\}$ is a nested family. Now, we assume that the inequalities (7.152) are satisfied. In order to use Theorem 7.7.4, it remains to check the equality (7.151), which means to check

$$\mathrm{cl}\left(\bigcup_{0<\alpha\leq 1} M_\alpha\right) = M_0, \text{ i.e. } \mathrm{cl}\left(\bigcup_{0<\alpha\leq 1}\left[l(\alpha), u(\alpha)\right]\right) = \left[l_0, u_0\right]. \tag{7.153}$$

Since l is increasing and u is decreasing on $[0,1]$, it follows that the limits

$$\lim_{\alpha\to 0} l(\alpha) = \sup_{0<\alpha\leq 1} l(\alpha) \text{ and } \lim_{\alpha\to 0} u(\alpha) = \inf_{0<\alpha\leq 1} u(\alpha)$$

exist. Therefore, we obtain

$$\mathrm{cl}\left(\bigcup_{0<\alpha\leq 1}\left[l(\alpha), u(\alpha)\right]\right) = \left[\lim_{\alpha\to 0} l(\alpha), \lim_{\alpha\to 0} u(\alpha)\right] = \left[\sup_{0<\alpha\leq 1} l(\alpha), \inf_{0<\alpha\leq 1} u(\alpha)\right].$$

Given a sequence $\alpha_n \downarrow 0$ for each $\alpha_n \in (0,1]$, it is not hard to prove

$$l(\alpha_n) \to l(0) \text{ and } u(\alpha_n) \to u(0) \text{ as } n \to \infty.$$

This shows that the equality (7.153) is satisfied. Therefore, Theorem 7.7.4 says that the fair Hausdorff difference $\tilde{A} \ominus_{FH} \tilde{B} = \tilde{C}$ exists satisfying

$$\tilde{C}_\alpha = \tilde{A}_\alpha \boxminus_{HF} \tilde{B}_\alpha \text{ for all } \alpha \in [0,1].$$

In other words, given any two triangular fuzzy numbers

$$\tilde{A} = (a^L, a, a^U) \text{ and } \tilde{B} = (b^L, b, b^U),$$

if the inequalities (7.152) are satisfied, then the fair Hausdorff difference $\tilde{A} \ominus_{FH} \tilde{B}$ exists.

We say that the triangular fuzzy number $\tilde{A} = (a^L, a, a^U)$ is symmetric when

$$a - a^L = a^U - a.$$

We are going to claim that if $\tilde{A}$ and $\tilde{B}$ are symmetric triangular fuzzy numbers, then the inequalities (7.152) hold true. Now, we have

$$a - a^L = a^U - a \equiv k_a \text{ and } b - b^L = b^U - b \equiv k_b.$$

Let $k = k_a - k_b$. Then

$$a - b = a^L - b^L + k = a^U - b^U - k.$$

We consider the following cases.

- Suppose that $k \geq 0$. Then $a^L - b^L \leq a^U - b^U$. Therefore, we obtain

$$\begin{aligned}\min\left\{a^L - b^L, a^U - b^U\right\} &= a^L - b^L = a - b - k \leq a - b \\ &\leq a - b + k = a^U - b^U = \max\left\{a^L - b^L, a^U - b^U\right\}.\end{aligned}$$

- Suppose that $k < 0$. Then $a^L - b^L \geq a^U - b^U$. Therefore, we obtain

$$\begin{aligned}\min\left\{a^L - b^L, a^U - b^U\right\} &= a^U - b^U = a - b + k \leq a - b \\ &\leq a - b - k = a^L - b^L = \max\left\{a^L - b^L, a^U - b^U\right\}.\end{aligned}$$

The above cases conclude that the inequalities (7.152) are satisfied. In other words, if $\tilde{A}$ and $\tilde{B}$ are symmetric triangular fuzzy numbers, then the fair Hausdorff difference $\tilde{A} \ominus_{FH} \tilde{B}$ always exists.

7.7.2 Composite Hausdorff Difference

Let $\tilde{A}$ and $\tilde{B}$ be two fuzzy sets in $\mathbb{R}$. We shall propose a new difference that extends the so-called type-I-generalized difference proposed by Bede and Stefanini [8] in (7.98). According to (7.143), we define

$$\tilde{A}_\alpha \boxminus_{CH} \tilde{B}_\alpha = \text{cl}\left(\bigcup_{\alpha \leq \beta \leq 1} \tilde{A}_\beta \boxminus_{FH} \tilde{B}_\beta\right) \tag{7.154}$$

$$\begin{aligned}&= \text{cl}\left(\bigcup_{\alpha \leq \beta \leq 1}\left[\min\left\{\sup_{a \in \tilde{A}_\beta} \inf_{b \in \tilde{B}_\beta} (a - b), \sup_{b \in \tilde{B}_\beta} \inf_{a \in \tilde{A}_\beta} (a - b)\right\},\right.\right. \\ &\qquad \left.\left.\max\left\{\sup_{a \in \tilde{A}_\beta} \inf_{b \in \tilde{B}_\beta} (a - b), \sup_{b \in \tilde{B}_\beta} \inf_{a \in \tilde{A}_\beta} (a - b)\right\}\right]\right).\end{aligned} \tag{7.155}$$

It is clear to see that the family

$$\{M_\alpha \equiv \tilde{A}_\alpha \boxminus_{CH} \tilde{B}_\alpha : 0 \leq \alpha \leq 1\}$$

is automatically nested. According to the binary operation in (7.93), the **composite Hausdorff difference** between $\tilde{A}$ and $\tilde{B}$ is defined by

$$\tilde{A} \ominus_{CH} \tilde{B} = \tilde{C} \text{ when } \tilde{C}_\alpha = \tilde{A}_\alpha \boxminus_{CH} \tilde{B}_\alpha \text{ for each } \alpha \in [0, 1]. \tag{7.156}$$

However, the existence of such $\tilde{C}$ is not guaranteed.

Without considering the 0-level sets $\tilde{A}_0$ and $\tilde{B}_0$, according to the binary operation in (7.115), the **composite *-Hausdorff difference** between $\tilde{A}$ and $\tilde{B}$ is defined by

$$\tilde{A} \ominus^*_{CH} \tilde{B} = \tilde{C}^* \text{ when } \tilde{C}^*_\alpha = \tilde{A}_\alpha \boxminus_{CH} \tilde{B}_\alpha \text{ for all } \alpha \in (0,1]. \tag{7.157}$$

The existence of such $\tilde{C}^*$ is also not guaranteed. The sufficient conditions for guaranteeing the existence of $\tilde{A} \ominus^*_{CH} \tilde{B}$ are weaker than those of $\tilde{A} \ominus_{CH} \tilde{B}$ by referring to Theorems 7.6.14 and 7.6.19. Proposition 7.6.3 also says that if the existence of $\tilde{A} \ominus^*_{CH} \tilde{B}$ and $\tilde{A} \ominus_{CH} \tilde{B}$ are guaranteed, then we have

$$\tilde{A} \ominus^*_{CH} \tilde{B} = \tilde{A} \ominus_{CH} \tilde{B}.$$

According to the binary operation in (7.118), we can define the corresponding **generalized composite Hausdorff difference** between $\tilde{A}$ and $\tilde{B}$ that will always exist as follows. Let $M_\alpha = \tilde{A}_\alpha \boxminus_{CH} \tilde{B}_\alpha$ be given in (7.154) for all $\alpha \in (0,1]$. Using (7.118), we consider the generalized composite Hausdorff difference $\tilde{A} \ominus_{GCH} \tilde{B}$ with membership function defined by

$$\xi_{\tilde{A} \ominus_{GCH} \tilde{B}}(x) = \sup_{0<\alpha\leq 1} \alpha \cdot \chi_{M_\alpha}(x). \tag{7.158}$$

This says that $\tilde{A} \ominus_{GCH} \tilde{B}$ always exists.

Example 7.7.6 Let $\tilde{A}$ and $\tilde{B}$ be two fuzzy intervals. Then, according to (7.155) and (7.147), we have

$$\tilde{A}_\alpha \boxminus_{CH} \tilde{B}_\alpha = \text{cl}\left(\bigcup_{\alpha\leq\beta\leq 1} \left[\min\left\{\tilde{A}^L_\beta - \tilde{B}^L_\beta, \tilde{A}^U_\beta - \tilde{B}^U_\beta\right\}, \max\left\{\tilde{A}^L_\beta - \tilde{B}^L_\beta, \tilde{A}^U_\beta - \tilde{B}^U_\beta\right\}\right]\right),$$

which is also equal to $\tilde{A}_\alpha \boxminus_{G_1} \tilde{B}_\alpha$ by referring to (7.99). Therefore, we obtain

$$\tilde{A} \ominus_{CH} \tilde{B} = \tilde{A} \ominus_{G_1} \tilde{B},$$

which means that the so-called type-I-generalized difference proposed by Bede and Stefanini [8] in (7.100) is a special case of the composite Hausdorff difference.

Theorem 7.7.7 ***(Existence for*** $\tilde{A} \ominus^*_{CH} \tilde{B}$***)*** *Let $\tilde{A}$ and $\tilde{B}$ be two fuzzy sets in $\mathbb{R}$, and let $M_\alpha = \tilde{A}_\alpha \boxminus_{CH} \tilde{B}_\alpha$ for all $\alpha \in (0,1]$. Then, there exists a fuzzy set $\tilde{C}^*$ in $\mathbb{R}$ with membership function defined by*

$$\xi_{\tilde{C}^*}(x) = \sup_{0<\alpha\leq 1} \alpha \cdot \chi_{M_\alpha}(x),$$

satisfying

$$\tilde{C}^*_\alpha = M_\alpha = \tilde{A}_\alpha \boxminus_{CH} \tilde{B}_\alpha \textit{ for each } \alpha \in (0,1]$$

and

$$\tilde{C}^*_0 = cl\left(\bigcup_{0<\alpha\leq 1} M_\alpha\right) = cl\left(\bigcup_{0<\alpha\leq 1} \tilde{A}_\alpha \boxminus_{CH} \tilde{B}_\alpha\right). \tag{7.159}$$

This guarantees the existence of $\tilde{C}^* = \tilde{A} \ominus^*_{CH} \tilde{B}$. Moreover, we have

$$\tilde{C}^* = \tilde{A} \ominus^*_{CH} \tilde{B} = \tilde{A} \ominus_{GCH} \tilde{B}.$$

Proof. From (7.154), we have

$$M_\alpha = \tilde{A}_\alpha \boxminus_{CH} \tilde{B}_\alpha = \mathrm{cl}\left(\bigcup_{\alpha \le \beta \le 1} \tilde{A}_\beta \boxminus_{FH} \tilde{B}_\beta \right) \text{ for } 0 < \alpha \le 1.$$

It is clear that $\{M_\alpha : 0 < \alpha \le 1\}$ is a nested family. We are going to check that the conditions in Theorem 5.4.7 are satisfied. For any increasing and convergent sequence $\alpha_n \uparrow \alpha$, we want to claim $\bigcap_{n=1}^\infty M_{\alpha_n} \subseteq M_\alpha$. From (7.155), we have

$$M_\alpha = \tilde{A}_\alpha \boxminus_{CH} \tilde{B}_\alpha = \mathrm{cl}\left(\bigcup_{\alpha \le \beta \le 1} \tilde{A}_\beta \boxminus_{FH} \tilde{B}_\beta \right) = \mathrm{cl}\left(\bigcup_{\alpha \le \beta \le 1} [l(\beta), u(\beta)] \right),$$

where the functions l and u are defined in (7.149) and (7.150), respectively. Let

$$C_\alpha = \bigcup_{\alpha \le \beta \le 1} [l(\beta), u(\beta)] .$$

Then $M_\alpha = \mathrm{cl}(C_\alpha)$. Using Proposition 1.2.9 and parts (i) and (ii) of Proposition 2.2.8, if $\beta_n \uparrow \beta$ as $n \to \infty$, then

$$l(\beta_n) \to l(\beta) \text{ and } u(\beta_n) \to u(\beta) \text{ as } n \to \infty. \tag{7.160}$$

For $x \in M_{\alpha_n}$, there exists a sequence $\{x_k^{(n)}\}_{k=1}^\infty$ in C_{α_n} satisfying $x_k^{(n)} \to x$ as $k \to \infty$. This also says that, for each fixed k,

$$x_k^{(n)} \in \left[l(\beta_n), u(\beta_n)\right] \text{ for some } \beta_n \text{ satisfying } 0 \le \alpha_n \le \beta_n \le 1.$$

Therefore, we obtain a bounded sequence $\{\beta_n\}_{n=1}^\infty$. Since each bounded sequence has a convergent subsequence, and each convergent sequence has an increasing and convergent subsequence, it follows that there exists a subsequence $\{\beta_{n_r}\}_{r=1}^\infty$ of $\{\beta_n\}_{n=1}^\infty$ satisfying $\beta_{n_r} \uparrow \beta$ as $r \to \infty$. We also see that $\{x_k^{(n_r)}\}_{r=1}^\infty$ is a bounded subsequence of $\{x_k^{(n)}\}_{n=1}^\infty$ for each fixed k. Therefore, there exists a convergent subsequence $\{x_k^{(n_{r_s})}\}_{s=1}^\infty$ of $\{x_k^{(n_r)}\}_{r=1}^\infty$ satisfying $x_k^{(n_{r_s})} \to x_k$ as $s \to \infty$ for each fixed k. Then, we have

$$l(\beta_{n_{r_s}}) \le x_k^{(n_{r_s})} \le u(\beta_{n_{r_s}}) \text{ for } \alpha_{n_{r_s}} \le \beta_{n_{r_s}} \le 1.$$

By taking $s \to \infty$, from (7.160), we obtain

$$l(\beta) \le x_k \le u(\beta) \text{ for } \alpha \le \beta \le 1,$$

which says that the sequence $\{x_k\}_{k=1}^\infty$ is in C_α. Since $x_k^{(n)} \to x$ as $k \to \infty$ for each fixed n and $x_k^{(n_{r_s})} \to x_k$ as $s \to \infty$ for each fixed k, it follows that

$$|x_k - x| \le \left|x_k - x_k^{(n_{r_s})}\right| + \left|x_k^{(n_{r_s})} - x\right| \to 0 \text{ as } s \to \infty \text{ and } k \to \infty,$$

which also says that $x_k \to x$ as $k \to \infty$. This shows that $x \in \mathrm{cl}(C_\alpha) = M_\alpha$. Therefore, we obtain the desired inclusion $\bigcap_{n=1}^\infty M_{\alpha_n} \subseteq M_\alpha$. Using Theorem 5.4.7, we conclude that (7.157) is satisfied, which also shows the existence of $\tilde{A} \ominus_{CH}^* \tilde{B} = \tilde{C}^*$. Since $\tilde{C}_\beta^* = M_\beta$ for all $\beta \in (0, 1]$, from (7.158), it follows that

$$\left\{ x \in \mathbb{R} : \xi_{\tilde{A} \ominus_{CH}^* \tilde{B}}(x) = \sup_{0 < \beta \le 1} \beta \cdot \chi_{\tilde{C}_\beta^*}(x) \ge \alpha \right\}$$

$$= \left\{ x \in \mathbb{R} : \xi_{\tilde{A} \ominus_{GCH} \tilde{B}}(x) = \sup_{0 < \beta \le 1} \beta \cdot \chi_{M_\beta}(x) \ge \alpha \right\}.$$

Using Proposition 1.4.8, we obtain

$$\xi_{\tilde{A}\ominus^{*}_{CH}\tilde{B}}(x) = \xi_{\tilde{A}\ominus_{GCH}\tilde{B}}(x)$$

for all $x \in \mathbb{R}$, i.e. $\tilde{A} \ominus^{*}_{CH} \tilde{B} = \tilde{A} \ominus_{GCH} \tilde{B}$. This completes the proof. ■

The existence for $\tilde{A} \ominus^{*}_{CH} \tilde{B}$ as shown in Theorem 7.7.7 above does not consider the 0-level set $\tilde{A}_0$ and $\tilde{B}_0$ as $M_0 = \tilde{A}_0 \circ \tilde{B}_0$. However, the existence for $\tilde{A} \ominus_{CH} \tilde{B}$ needs to consider $M_0 = \tilde{A}_0 \circ \tilde{B}_0$, which will be shown below.

Lemma 7.7.8 *The following inclusion holds true:*

$$\mathrm{cl}\left(\bigcup_{0<\alpha\leq 1} A_\alpha\right) \subseteq \mathrm{cl}\left(\bigcup_{0<\alpha\leq 1}\left(\mathrm{cl}\left(\bigcup_{\alpha\leq\beta\leq 1} A_\beta\right)\right)\right).$$

Proof. Suppose that

$$x \in \mathrm{cl}\left(\bigcup_{0<\alpha\leq 1} A_\alpha\right).$$

Then, there exists a sequence $\{x_n\}_{n=1}^{\infty}$ in $\bigcup_{0<\alpha\leq 1} A_\alpha$ satisfying $x_n \to x$ as $n \to \infty$. Now, we have

$$x_n \in \bigcup_{0<\alpha\leq 1} A_\alpha \subseteq \bigcup_{0<\alpha\leq 1}\bigcup_{\alpha\leq\beta\leq 1} A_\beta \subseteq \bigcup_{0<\alpha\leq 1}\left(\mathrm{cl}\left(\bigcup_{\alpha\leq\beta\leq 1} A_\beta\right)\right) \text{ for all } n,$$

which shows the desired inclusion, since $x_n \to x$ as $n \to \infty$. This completes the proof. ■

Theorem 7.7.9 ***(Existence for*** $\tilde{A} \ominus_{CH} \tilde{B}$***)*** *Let* $\tilde{A}$ *and* $\tilde{B}$ *be two fuzzy sets in* $\mathbb{R}$*, and let* $M_\alpha = \tilde{A}_\alpha \boxminus_{CH} \tilde{B}_\alpha$ *for all* $\alpha \in [0,1]$*. Then, there exists a fuzzy set* $\tilde{C}$ *in* $\mathbb{R}$ *with membership function defined by*

$$\xi_{\tilde{C}}(x) = \sup_{\alpha\in[0,1]} \alpha \cdot \chi_{M_\alpha}(x) = \sup_{0<\alpha\leq 1} \alpha \cdot \chi_{M_\alpha}(x)$$

satisfying

$$\tilde{C}_\alpha = M_\alpha = \tilde{A}_\alpha \boxminus_{CH} \tilde{B}_\alpha \text{ for each } \alpha \in [0,1].$$

This guarantees the existence of $\tilde{C} = \tilde{A} \ominus_{CH} \tilde{B}$*. Moreover, the existence of* $\tilde{C} = \tilde{A} \ominus^{*}_{CH} \tilde{B}$ *is also guaranteed, and we have*

$$\tilde{C} = \tilde{A} \ominus_{CH} \tilde{B} = \tilde{A} \ominus^{*}_{CH} \tilde{B} = \tilde{A} \ominus_{GCH} \tilde{B}. \tag{7.161}$$

Proof. From Theorem 7.7.7 by considering $\tilde{C}^{*}$ as $\tilde{C}$, we have

$$\tilde{C}_0 = \mathrm{cl}\left(\bigcup_{0<\alpha\leq 1} M_\alpha\right) \text{ and } \tilde{C}_\alpha = M_\alpha \text{ for } 0 < \alpha \leq 1. \tag{7.162}$$

In order to claim the existence of $\tilde{C} = \tilde{A} \ominus_{CH} \tilde{B}$, we are going to consider the 0-level sets. Therefore, it remains to prove

$$\tilde{C}_0 = \mathrm{cl}\left(\bigcup_{0<\alpha\leq 1} M_\alpha\right) = M_0,$$

where

$$M_0 = \tilde{A}_0 \boxminus_{CH} \tilde{B}_0 = \mathrm{cl}\left(\bigcup_{0\le\beta\le1} \tilde{A}_\beta \boxminus_{FH} \tilde{B}_\beta\right)$$

is a closed set. Since $M_\alpha \subseteq M_0$ for all $\alpha > 0$, we have $\bigcup_{0<\alpha\le1} M_\alpha \subseteq M_0$, which implies

$$\mathrm{cl}\left(\bigcup_{0<\alpha\le1} M_\alpha\right) \subseteq \mathrm{cl}\left(M_0\right) = M_0.$$

Next, we want to prove the other direction of inclusion. Let

$$l^* = \sup_{0<\alpha\le1} l(\alpha) \text{ and } u^* = \inf_{0<\alpha\le1} u(\alpha).$$

For $y \in (l(0), u(0))$, we consider the following cases.

- Suppose that $y < l^*$. Then, there exists $\epsilon > 0$ satisfying $y + \epsilon < l^*$ by taking $\epsilon < l^* - y$. According to the concept of supremum, there exists $\alpha_l^* \in (0, 1]$ satisfying $l^* - \epsilon \le l(\alpha_l^*)$. Therefore, we obtain

$$l(0) < y < l^* - \epsilon \le l(\alpha_l^*),$$

  which says that $l(0) < y < l(\alpha_l^*)$ and $\alpha_l^* \ne 0$. Using Proposition 1.4.3, there exists $\bar{\alpha} \in (0, \alpha_l^*)$ satisfying $y = l(\bar{\alpha})$. This says that $y \in [l(\bar{\alpha}), u(\bar{\alpha})]$ for $\bar{\alpha} > 0$.
- Suppose that $y > u^*$. Then, there exists $\epsilon > 0$ satisfying $y - \epsilon > u^*$ by taking $\epsilon < y - u^*$. According to the concept of infimum, there exists $\alpha_u^* \in (0, 1]$ satisfying $u^* + \epsilon \ge u(\alpha_u^*)$. Therefore, we obtain

$$u(0) > y > u^* + \epsilon \ge u(\alpha_u^*),$$

  which says that $u(0) > y > u(\alpha_u^*)$ and $\alpha_u^* \ne 0$. Using Proposition 1.4.3, there exists $\bar{\alpha} \in (0, \alpha_u^*)$ satisfying $y = u(\bar{\alpha})$. This says that $y \in [l(\bar{\alpha}), u(\bar{\alpha})]$ for $\bar{\alpha} > 0$.
- Suppose that $l^* \le y \le u^*$. Since $l^* \ge l(\alpha)$ and $u^* \le u(\alpha)$ for all $\alpha \in (0, 1]$. We can take some $\bar{\alpha} > 0$ satisfying

$$l(\bar{\alpha}) \le l^* \le y \le u^* \le u(\bar{\alpha}).$$

  This says that $y \in [l(\bar{\alpha}), u(\bar{\alpha})]$ for $\bar{\alpha} > 0$.

The above cases imply that

$$(l(0), u(0)) \subseteq \bigcup_{0<\alpha\le1} \left[l(\alpha), u(\alpha)\right]. \tag{7.163}$$

Suppose that $x \in [l(0), u(0)]$. Since $[l(0), u(0)] = \mathrm{cl}(l(0), u(0))$, there exists a sequence $\{x_n\}_{n=1}^\infty$ in the open interval $(l(0), u(0))$ satisfying $x_n \to x$. From (7.163), we see that

$$\left\{x_n\right\}_{n=1}^\infty \subseteq \bigcup_{0<\alpha\le1} \left[l(\alpha), u(\alpha)\right],$$

which shows that

$$x \in \mathrm{cl}\left(\bigcup_{0<\alpha\le1} \left[l(\alpha), u(\alpha)\right]\right) \text{ and } [l(0), u(0)] \subseteq \mathrm{cl}\left(\bigcup_{0<\alpha\le1} \left[l(\alpha), u(\alpha)\right]\right). \tag{7.164}$$

Therefore, we obtain

$$M_0 = \mathrm{cl}\left(\bigcup_{0\le\alpha\le 1} \tilde{A}_\alpha \boxminus_{FH} \tilde{B}_\alpha\right) = \mathrm{cl}\left(\bigcup_{0\le\alpha\le 1} [l(\alpha), u(\alpha)]\right)$$

$$= [l(0), u(0)] \bigcup \mathrm{cl}\left(\bigcup_{0<\alpha\le 1} [l(\alpha), u(\alpha)]\right)$$

$$\text{(using the fact of } \mathrm{cl}(A \cup B) = \mathrm{cl}(A) \cup \mathrm{cl}(B))$$

$$= \mathrm{cl}\left(\bigcup_{0<\alpha\le 1} [l(\alpha), u(\alpha)]\right) \text{ (using (7.164))}$$

$$= \mathrm{cl}\left(\bigcup_{0<\alpha\le 1} \tilde{A}_\alpha \boxminus_{FH} \tilde{B}_\alpha\right) \subseteq \mathrm{cl}\left(\bigcup_{0<\alpha\le 1} \left(\mathrm{cl}\left(\bigcup_{\alpha\le\beta\le 1} \tilde{A}_\beta \boxminus_{FH} \tilde{B}_\beta\right)\right)\right)$$

$$\text{(using Lemma 7.7.8)}$$

$$= \mathrm{cl}\left(\bigcup_{0<\alpha\le 1} M_\alpha\right) \text{ (using (7.162)).}$$

This shows $\tilde{C}_\alpha = M_\alpha$ for all $0 \le \alpha \le 1$ and the existence of $\tilde{C} = \tilde{A} \ominus_{CH} \tilde{B}$. Using Proposition 1.4.8, we can also obtain the equalities (7.161) by referring to the proof of Theorem 7.7.7. This completes the proof. ■

Theorems 7.7.7 and 7.7.9 say that the composite Hausdorff differences $\tilde{A} \ominus_{CH}^{*} \tilde{B}$ and $\tilde{A} \ominus_{CH} \tilde{B}$ always exist without needing any sufficient conditions. Moreover, the equalities (7.161) says that the three differences $\tilde{A} \ominus_{CH}^{*} \tilde{B}$, $\tilde{A} \ominus_{CH} \tilde{B}$, and $\tilde{A} \ominus_{GCH} \tilde{B}$ are identical.

7.7.3 Complete Composite Hausdorff Difference

Let $\tilde{A}$ and $\tilde{B}$ be two fuzzy sets in $\mathbb{R}$. We shall propose two new differences that extend the so-called type-II-generalized difference proposed by Gomes and Barros [42] in (7.101) and (7.102). According to (7.101), we define

$$\tilde{A}_\alpha \boxminus_{CCH_1} \tilde{B}_\alpha = \mathrm{cl}\left(\mathrm{conv}\left(\bigcup_{\alpha\le\beta\le 1} \tilde{A}_\beta \boxminus_{FH} \tilde{B}_\beta\right)\right) \tag{7.165}$$

$$= \mathrm{cl}\left(\mathrm{conv}\left(\bigcup_{\alpha\le\beta\le 1}\left[\min\left\{\sup_{a\in\tilde{A}_\beta}\inf_{b\in\tilde{B}_\beta}(a-b), \sup_{b\in\tilde{B}_\beta}\inf_{a\in\tilde{A}_\beta}(a-b)\right\},\right.\right.\right.$$
$$\left.\left.\left.\max\left\{\sup_{a\in\tilde{A}_\beta}\inf_{b\in\tilde{B}_\beta}(a-b), \sup_{b\in\tilde{B}_\beta}\inf_{a\in\tilde{A}_\beta}(a-b)\right\}\right]\right)\right). \tag{7.166}$$

It is clear to see that the family

$$\{M_\alpha \equiv \tilde{A}_\alpha \boxminus_{CCH_1} \tilde{B}_\alpha : 0 \le \alpha \le 1\}$$

is automatically nested. According to (7.102), we define

$$\tilde{A}_\alpha \boxminus_{CCH_2} \tilde{B}_\alpha = \left[\inf_{\alpha\le\beta\le 1} \underline{d}_H\left(\tilde{A}_\beta, \tilde{B}_\beta\right), \sup_{\alpha\le\beta\le 1} \bar{d}_H\left(\tilde{A}_\beta, \tilde{B}_\beta\right)\right] \tag{7.167}$$

$$= \left[\inf_{\alpha \leq \beta \leq 1} \min \left\{ \sup_{a \in \tilde{A}_\beta} \inf_{b \in \tilde{B}_\beta} (a - b), \sup_{b \in \tilde{B}_\beta} \inf_{a \in \tilde{A}_\beta} (a - b) \right\}, \right.$$

$$\left. \sup_{\alpha \leq \beta \leq 1} \max \left\{ \sup_{a \in \tilde{A}_\beta} \inf_{b \in \tilde{B}_\beta} (a - b), \sup_{b \in \tilde{B}_\beta} \inf_{a \in \tilde{A}_\beta} (a - b) \right\} \right]. \quad (7.168)$$

It is also clear to see that the family

$$\{M_\alpha \equiv \tilde{A}_\alpha \boxminus_{CCH_2} \tilde{B}_\alpha : 0 \leq \alpha \leq 1\}$$

is automatically nested. In general, for any fuzzy sets $\tilde{A}$ and $\tilde{B}$ in $\mathbb{R}$, the differences $\tilde{A}_\alpha \boxminus_{CCH_1} \tilde{B}_\alpha$ and $\tilde{A}_\alpha \boxminus_{CCH_2} \tilde{B}_\alpha$ are not necessarily identical.

According to the binary operation in (7.93), the **type-I-complete composite Hausdorff difference** between $\tilde{A}$ and $\tilde{B}$ is defined by

$$\tilde{A} \ominus_{CCH_1} \tilde{B} = \tilde{C} \text{ when } \tilde{C}_\alpha = \tilde{A}_\alpha \boxminus_{CCH_1} \tilde{B}_\alpha \text{ for each } \alpha \in [0, 1] \quad (7.169)$$

and the **type-II-complete composite Hausdorff difference** between $\tilde{A}$ and $\tilde{B}$ is defined by

$$\tilde{A} \ominus_{CCH_2} \tilde{B} = \tilde{C} \text{ when } \tilde{C}_\alpha = \tilde{A}_\alpha \boxminus_{CCH_2} \tilde{B}_\alpha \text{ for each } \alpha \in [0, 1]. \quad (7.170)$$

However, the existence of such $\tilde{C}$ is not guaranteed.

Without considering the 0-level sets $\tilde{A}_0$ and $\tilde{B}_0$, according to the binary operation in (7.115), the **type-I-complete composite *-Hausdorff difference** between $\tilde{A}$ and $\tilde{B}$ is defined by

$$\tilde{A} \ominus^*_{CCH_1} \tilde{B} = \tilde{C}^* \text{ when } \tilde{C}^*_\alpha = \tilde{A}_\alpha \boxminus_{CCH_1} \tilde{B}_\alpha \text{ for all } \alpha \in (0, 1], \quad (7.171)$$

and the **type-II-complete composite *-Hausdorff difference** between $\tilde{A}$ and $\tilde{B}$ is defined by

$$\tilde{A} \ominus^*_{CCH_2} \tilde{B} = \tilde{C}^* \text{ when } \tilde{C}^*_\alpha = \tilde{A}_\alpha \boxminus_{CCH_2} \tilde{B}_\alpha \text{ for all } \alpha \in (0, 1]. \quad (7.172)$$

The existence of such $\tilde{C}$ is also not guaranteed. The sufficient conditions for guaranteeing the existence of $\tilde{A} \ominus^*_{CCH_1} \tilde{B}$ are weaker than those of $\tilde{A} \ominus_{CCH_1} \tilde{B}$, and sufficient conditions for guaranteeing the existence of $\tilde{A} \ominus^*_{CCH_2} \tilde{B}$ are also weaker than those of $\tilde{A} \ominus_{CCH_2} \tilde{B}$ by referring to Theorems 7.6.14 and 7.6.19. From Proposition 7.6.3, we also have the following observations.

- Suppose that the existence of $\tilde{A} \ominus^*_{CCH_1} \tilde{B}$ and $\tilde{A} \ominus_{CCH_1} \tilde{B}$ are guaranteed. Then

  $$\tilde{A} \ominus^*_{CCH_1} \tilde{B} = \tilde{A} \ominus_{CCH_1} \tilde{B}.$$

- Suppose that the existence of $\tilde{A} \ominus^*_{CCH_2} \tilde{B}$ and $\tilde{A} \ominus_{CCH_2} \tilde{B}$ are guaranteed. Then

  $$\tilde{A} \ominus^*_{CCH_2} \tilde{B} = \tilde{A} \ominus_{CCH_2} \tilde{B}.$$

According to the binary operation in (7.118), we can define the corresponding generalized Hausdorff difference between $\tilde{A}$ and $\tilde{B}$, which will always exist, as follows.

- Let $M_\alpha = \tilde{A}_\alpha \boxminus_{CCH_1} \tilde{B}_\alpha$ be given in (7.165) for all $\alpha \in (0, 1]$. Using (7.118), we consider the **generalized type-I-complete composite Hausdorff difference** $\tilde{A} \ominus_{GCCH_1} \tilde{B}$ with membership function defined by

  $$\xi_{\tilde{A} \ominus_{GCCH_1} \tilde{B}}(x) = \sup_{0 < \alpha \leq 1} \alpha \cdot \chi_{M_\alpha}(x). \quad (7.173)$$

- Let $M_\alpha = \tilde{A}_\alpha \boxminus_{CCH_2} \tilde{B}_\alpha$ be given in (7.167) for all $\alpha \in (0, 1]$. Using (7.118), we consider the **generalized type-II-complete composite Hausdorff difference** $\tilde{A} \ominus_{GCCH_2} \tilde{B}$ with membership function defined by

$$\xi_{\tilde{A}\ominus_{GCCH_2}\tilde{B}}(x) = \sup_{0<\alpha\leq 1} \alpha \cdot \chi_{M_\alpha}(x). \tag{7.174}$$

We see that $\tilde{A} \ominus_{GCCH_1} \tilde{B}$ and $\tilde{A} \ominus_{GCCH_2} \tilde{B}$ always exist.

Example 7.7.10 Let $\tilde{A}$ and $\tilde{B}$ be two fuzzy intervals. Then, according to (7.165) and (7.148), we have

$$\tilde{A}_\alpha \boxminus_{CCH_1} \tilde{B}_\alpha = \text{cl}\left(\text{conv}\left(\bigcup_{\alpha\leq\beta\leq 1} \tilde{A}_\beta \boxminus_{FH} \tilde{B}_\beta\right)\right) = \text{cl}\left(\text{conv}\left(\bigcup_{\alpha\leq\beta\leq 1} \tilde{A}_\beta \boxminus_{H} \tilde{B}_\beta\right)\right).$$

According to (7.168) and (7.147), we also have

$$\tilde{A}_\alpha \boxminus_{CCH_2} \tilde{B}_\alpha = \left[\inf_{\alpha\leq\beta\leq 1} \min\left\{\tilde{A}^L_\beta - \tilde{B}^L_\beta, \tilde{A}^U_\beta - \tilde{B}^U_\beta\right\}, \sup_{\alpha\leq\beta\leq 1} \max\left\{\tilde{A}^L_\beta - \tilde{B}^L_\beta, \tilde{A}^U_\beta - \tilde{B}^U_\beta\right\}\right].$$

Using (7.101) and (7.102), we obtain

$$\tilde{A}_\alpha \boxminus_{CCH_1} \tilde{B}_\alpha = \tilde{A}_\alpha \boxminus_{CCH_2} \tilde{B}_\alpha = \tilde{A}_\alpha \boxminus_{G_2} \tilde{B}_\alpha.$$

Therefore, we obtain

$$\tilde{A} \ominus_{CCH_1} \tilde{B} = \tilde{A} \ominus_{CCH_2} \tilde{B} = \tilde{A} \ominus_{G_2} \tilde{B}, \tag{7.175}$$

which means that the so-called type-II-generalized difference proposed by Gomes and Barros [42] in (7.103) is a special case of the complete composite Hausdorff difference.

Now, we present the existence of type-I-complete composite Hausdorff difference.

Theorem 7.7.11 *(**Existence for** $\tilde{A} \ominus^*_{CCH_1} \tilde{B}$) Let $\tilde{A}$ and $\tilde{B}$ be two fuzzy sets in $\mathbb{R}$, and let $M_\alpha = \tilde{A}_\alpha \boxminus_{CCH_1} \tilde{B}_\alpha$ for all $\alpha \in (0, 1]$. Suppose that the function l defined in (7.149) is concave on $[0, 1]$ and the function u defined in (7.150) is convex on $[0, 1]$. Then, there exists a fuzzy set $\tilde{C}^*$ in $\mathbb{R}$ with membership function defined by*

$$\xi_{\tilde{C}^*}(x) = \sup_{0<\alpha\leq 1} \alpha \cdot \chi_{M_\alpha}(x),$$

satisfying

$$\tilde{C}^*_\alpha = M_\alpha = \tilde{A}_\alpha \boxminus_{CCH_1} \tilde{B}_\alpha \text{ for each } \alpha \in (0, 1]$$

and

$$\tilde{C}^*_0 = M_0 = \text{cl}\left(\bigcup_{0<\alpha\leq 1} M_\alpha\right) = \text{cl}\left(\bigcup_{0<\alpha\leq 1} \tilde{A}_\alpha \boxminus_{CCH_1} \tilde{B}_\alpha\right). \tag{7.176}$$

This guarantees the existence of $\tilde{C}^* = \tilde{A} \ominus^*_{CCH_1} \tilde{B}$. Moreover, we have

$$\tilde{C}^* = \tilde{A} \ominus^*_{CCH_1} \tilde{B} = \tilde{A} \ominus_{GCCH_1} \tilde{B}. \tag{7.177}$$

Proof. From (7.165), we have

$$M_\alpha = \tilde{A}_\alpha \boxminus_{CCH_1} \tilde{B}_\alpha = \text{cl}\left(\text{conv}\left(\bigcup_{\alpha\leq\beta\leq 1} \tilde{A}_\beta \boxminus_{FH} \tilde{B}_\beta\right)\right) \text{ for } 0 < \alpha \leq 1. \tag{7.178}$$

It is clear that $\{M_\alpha : 0 < \alpha \leq 1\}$ is a nested family. We are going to check that the conditions in Theorem 5.4.7 are satisfied. For any increasing and convergent sequence $\alpha_n \uparrow \alpha$, we want to claim $\bigcap_{n=1}^{\infty} M_{\alpha_n} \subseteq M_\alpha$. From (7.166), we have

$$M_\alpha = \tilde{A}_\alpha \boxminus_{CCH_1} \tilde{B}_\alpha = \text{cl}\left(\text{conv}\left(\bigcup_{\alpha\leq\beta\leq 1} \tilde{A}_\beta \boxminus_{FH} \tilde{B}_\beta\right)\right)$$
$$= \text{cl}\left(\text{conv}\left(\bigcup_{\alpha\leq\beta\leq 1} [l(\beta), u(\beta)]\right)\right),$$

where the functions l and u are defined in (7.149) and (7.150), respectively. Let

$$C_\alpha = \bigcup_{\alpha\leq\beta\leq 1} [l(\beta), u(\beta)].$$

Then $M_\alpha = \text{cl}(\text{conv}(C_\alpha))$. Using Proposition 1.2.9 and parts (i) and (ii) of Proposition 2.2.8, if $\beta_n \uparrow \beta$ as $n \to \infty$, then

$$l(\beta_n) \to l(\beta) \text{ and } u(\beta_n) \to u(\beta) \tag{7.179}$$

as $n \to \infty$. For $x \in M_{\alpha_n}$, there exists a sequence $\{x_k^{(n)}\}_{k=1}^{\infty}$ in $\text{conv}(C_{\alpha_n})$ satisfying $x_k^{(n)} \to x$ as $k \to \infty$, where each $x_k^{(n)}$ is a convex combination of elements in C_{α_n}, i.e.

$$x_k^{(n)} = \sum_{m=1}^{s_k^{(n)}} \lambda_{mk}^{(n)} y_{mk}^{(n)} \text{ with } \sum_{m=1}^{s_k^{(n)}} \lambda_{mk}^{(n)} = 1 \text{ and } y_{mk}^{(n)} \in \left[l\left(\beta_{mk}^{(n)}\right), u\left(\beta_{mk}^{(n)}\right)\right],$$

where $0 \leq \lambda_{mk}^{(n)} \leq 1$ and $\alpha_n \leq \beta_{mk}^{(n)} \leq 1$ for all m, n and k. We also see that

$$x_k^{(n)} \in \left[\sum_{m=1}^{s_k^{(n)}} \lambda_{mk}^{(n)} l\left(\beta_{mk}^{(n)}\right), \sum_{m=1}^{s_k^{(n)}} \lambda_{mk}^{(n)} u\left(\beta_{mk}^{(n)}\right)\right]. \tag{7.180}$$

For each fixed k, we define

$$\beta_n = \sum_{m=1}^{s_k^{(n)}} \lambda_{mk}^{(n)} \beta_{mk}^{(n)}.$$

It is clear that $\alpha_n \leq \beta_n \leq 1$ by the fact of $\sum_{m=1}^{s_k^{(n)}} \lambda_{mk}^{(n)} = 1$. Since l is concave and u is convex, we have

$$\sum_{m=1}^{s_k^{(n)}} \lambda_{mk}^{(n)} l\left(\beta_{mk}^{(n)}\right) \geq l\left(\sum_{m=1}^{s_k^{(n)}} \lambda_{mk}^{(n)} \beta_{mk}^{(n)}\right) = l(\beta_n)$$

and

$$\sum_{m=1}^{s_k^{(n)}} \lambda_{mk}^{(n)} u\left(\beta_{mk}^{(n)}\right) \leq u\left(\sum_{m=1}^{s_k^{(n)}} \lambda_{mk}^{(n)} \beta_{mk}^{(n)}\right) = u(\beta_n).$$

For each fixed k, from (7.180), we see that $x_k^{(n)} \in [l(\beta_n), u(\beta_n)]$ with $0 \leq \alpha_n \leq \beta_n \leq 1$. Therefore, we obtain a bounded sequence $\{\beta_n\}_{n=1}^{\infty}$. It follows that there exists a convergent subsequence $\{\beta_{n_r}\}_{r=1}^{\infty}$ of $\{\beta_n\}_{n=1}^{\infty}$ satisfying $\beta_{n_r} \uparrow \beta$ as $r \to \infty$. We also see that $\{x_k^{(n_r)}\}_{r=1}^{\infty}$

is a bounded subsequence of $\{x_k^{(n)}\}_{k=1}^{\infty}$. Therefore, there exists a convergent subsequence $\{x_k^{(n_{r_s})}\}_{s=1}^{\infty}$ of $\{x_k^{(n_r)}\}_{r=1}^{\infty}$ satisfying $x_k^{(n_{r_s})} \to x_k$ as $s \to \infty$. Then, we have

$$l(\beta_{n_{r_s}}) \leq x_k^{(n_{r_s})} \leq u(\beta_{n_{r_s}}) \text{ for } \alpha_{n_{r_s}} \leq \beta_{n_{r_s}} \leq 1.$$

By taking $s \to \infty$, from (7.179), we obtain

$$l(\beta) \leq x_k \leq u(\beta) \text{ for } \alpha \leq \beta \leq 1,$$

which says that the sequence $\{x_k\}_{k=1}^{\infty}$ is in $C_\alpha \subseteq \mathrm{conv}(C_\alpha)$. Since

$$\left|x_k - x\right| \leq \left|x_k - x_k^{(n_{r_s})}\right| + \left|x_k^{(n_{r_s})} - x\right| \to 0 \text{ as } s \to \infty \text{ and } k \to \infty,$$

it says that $x_k \to x$ as $k \to \infty$, which implies $x \in \mathrm{cl}(\mathrm{conv}(C_\alpha)) = M_\alpha$. Therefore, we obtain the desired inclusion $\bigcap_{n=1}^{\infty} M_{\alpha_n} \subseteq M_\alpha$. Using Theorem 5.4.7, we conclude that (7.171) is satisfied, which also shows the existence of $\tilde{A} \ominus_{CCH_1}^{*} \tilde{B}$. Using Proposition 1.4.8, we can also obtain the equalities (7.177) by referring to the proof of Theorem 7.7.7. This completes the proof. ■

Theorem 7.7.12 *(Existence for $\tilde{A} \ominus_{CCH_1} \tilde{B}$) Let $\tilde{A}$ and $\tilde{B}$ be two fuzzy sets in $\mathbb{R}$, and let $M_\alpha = \tilde{A}_\alpha \boxminus_{CCH_1} \tilde{B}_\alpha$ for all $\alpha \in [0,1]$. Suppose that the following conditions are satisfied:*

- *the function l defined in (7.149) is concave on $[0,1]$ and the function u defined in (7.150) is convex on $[0,1]$;*
- *the following inclusions hold true:*

$$\mathrm{conv}\left(\bigcup_{0\leq\alpha\leq 1}[l(\alpha),u(\alpha)]\right) \subseteq [l(0),u(0)] \bigcup \mathrm{conv}\left(\bigcup_{0<\alpha\leq 1}[l(\alpha),u(\alpha)]\right) \tag{7.181}$$

and

$$\mathrm{conv}\left(\bigcup_{0<\alpha\leq 1}[l(\alpha),u(\alpha)]\right) \subseteq \bigcup_{0<\alpha\leq 1}\left(\mathrm{cl}\left(\mathrm{conv}\left(\bigcup_{\alpha\leq\beta\leq 1}[l(\beta),u(\beta)]\right)\right)\right). \tag{7.182}$$

Then, there exists a fuzzy set $\tilde{C}$ in $\mathbb{R}$ with membership function defined by

$$\xi_{\tilde{C}}(x) = \sup_{0<\alpha\leq 1} \alpha \cdot \chi_{M_\alpha}(x),$$

satisfying

$$\tilde{C}_\alpha = M_\alpha = \tilde{A}_\alpha \boxminus_{CCH_1} \tilde{B}_\alpha \text{ for each } \alpha \in [0,1].$$

This guarantees the existence of $\tilde{C} = \tilde{A} \ominus_{CCH_1} \tilde{B}$. Moreover, the existence of $\tilde{C} = \tilde{A} \ominus_{CCH_1}^{*} \tilde{B}$ is also guaranteed, and we have

$$\tilde{C} = \tilde{A} \ominus_{CCH_1} \tilde{B} = \tilde{A} \ominus_{CCH_1}^{*} \tilde{B} = \tilde{A} \ominus_{GCCH_1} \tilde{B}. \tag{7.183}$$

Proof. From Theorem 7.7.11 by considering $\tilde{C}^*$ as $\tilde{C}$, we have

$$\tilde{C}_0 = \mathrm{cl}\left(\bigcup_{0<\alpha\leq 1} M_\alpha\right) \text{ and } \tilde{C}_\alpha = M_\alpha \text{ for } 0 < \alpha \leq 1.$$

In order to claim the existence of $\tilde{C} = \tilde{A} \boxminus_{CCH_1} \tilde{B}$, we are going to consider the 0-level sets. Therefore, it remains to prove

$$\tilde{C}_0 = \mathrm{cl}\left(\bigcup_{0<\alpha\leq 1} M_\alpha\right) = M_0,$$

where

$$M_0 = \tilde{A}_0 \boxminus_{CCH_1} \tilde{B}_0 = \mathrm{cl}\left(\mathrm{conv}\left(\bigcup_{0\leq\gamma\leq 1} \tilde{A}_\gamma \boxminus_{FH} \tilde{B}_\gamma\right)\right)$$

is a closed set. Since $M_\alpha \subseteq M_0$ for all $\alpha > 0$, we have $\bigcup_{0<\alpha\leq 1} M_\alpha \subseteq M_0$, which implies

$$\mathrm{cl}\left(\bigcup_{0<\alpha\leq 1} M_\alpha\right) \subseteq \mathrm{cl}\left(M_0\right) = M_0.$$

Next, we want to prove the other direction of inclusion. From (7.164), we have

$$[l(0), u(0)] \subseteq \mathrm{cl}\left(\bigcup_{0<\alpha\leq 1} [l(\alpha), u(\alpha)]\right) \subseteq \mathrm{cl}\left(\mathrm{conv}\left(\bigcup_{0<\alpha\leq 1} [l(\alpha), u(\alpha)]\right)\right). \tag{7.184}$$

Therefore, we obtain

$$\begin{aligned}
M_0 &= \mathrm{cl}\left(\mathrm{conv}\left(\bigcup_{0\leq\alpha\leq 1} \tilde{A}_\alpha \boxminus_{FH} \tilde{B}_\alpha\right)\right) = \mathrm{cl}\left(\mathrm{conv}\left(\bigcup_{0\leq\alpha\leq 1} [l(\alpha), u(\alpha)]\right)\right) \\
&\subseteq \mathrm{cl}\left([l(0), u(0)] \bigcup \mathrm{conv}\left(\bigcup_{0<\alpha\leq 1} [l(\alpha), u(\alpha)]\right)\right) \text{ (using (7.181))} \\
&= [l(0), u(0)] \bigcup \mathrm{cl}\left(\mathrm{conv}\left(\bigcup_{0<\alpha\leq 1} [l(\alpha), u(\alpha)]\right)\right) \\
&\qquad \text{(using the fact that } \mathrm{cl}(A \cup B) = \mathrm{cl}(A) \cup \mathrm{cl}(B)) \\
&= \mathrm{cl}\left(\mathrm{conv}\left(\bigcup_{0<\alpha\leq 1} [l(\alpha), u(\alpha)]\right)\right) \text{ (using (7.184))} \\
&\subseteq \mathrm{cl}\left(\bigcup_{0<\alpha\leq 1}\left(\mathrm{cl}\left(\mathrm{conv}\left(\bigcup_{\alpha\leq\beta\leq 1} [l(\alpha), u(\alpha)]\right)\right)\right)\right) \text{ (using (7.182))} \\
&= \mathrm{cl}\left(\bigcup_{0<\alpha\leq 1}\left(\mathrm{cl}\left(\mathrm{conv}\left(\bigcup_{\alpha\leq\beta\leq 1} \tilde{A}_\beta \boxminus_{FH} \tilde{B}_\beta\right)\right)\right)\right) = \mathrm{cl}\left(\bigcup_{0<\alpha\leq 1} M_\alpha\right) \\
&\qquad \text{(using (7.178)).}
\end{aligned}$$

Therefore, we obtain $\tilde{C}_\alpha = M_\alpha$ for all $0 \leq \alpha \leq 1$. This shows the existence of $\tilde{C} = \tilde{A} \ominus_{CCH_1} \tilde{B}$. Using Proposition 1.4.8, we can also obtain the equalities (7.183) by referring to the proof of Theorem 7.7.7. This completes the proof. ■

Theorems 7.7.11 and 7.7.12 say that the existence of type-I complete composite Hausdorff difference $\tilde{A} \ominus^*_{CCH_1} \tilde{B}$ and $\tilde{A} \ominus_{CCH_1} \tilde{B}$ need some mild sufficient conditions. Next, we can show that the type-II complete composite Hausdorff difference $\tilde{A} \ominus^*_{CCH_2} \tilde{B}$ and $\tilde{A} \ominus_{CCH_2} \tilde{B}$ always exist without needing any sufficient conditions.

Although Theorems 7.7.11 and 7.7.12 need some sufficient conditions to guarantee the existence of $\tilde{A} \ominus^*_{CCH_1} \tilde{B}$ and $\tilde{A} \ominus_{CCH_1} \tilde{B}$ in general, it is still possible that, for some special kinds of fuzzy sets $\tilde{A}$ and $\tilde{B}$, the differences $\tilde{A} \ominus^*_{CCH_1} \tilde{B}$ and $\tilde{A} \ominus_{CCH_1} \tilde{B}$ always exist. For example, suppose that we take $\tilde{A}$ and $\tilde{B}$ as fuzzy intervals. Then

$$l(\alpha) \equiv \min \left\{ \sup_{a\in\tilde{A}_\alpha} \inf_{b\in\tilde{B}_\alpha} (a-b), \sup_{b\in\tilde{B}_\alpha} \inf_{a\in\tilde{A}_\alpha} (a-b) \right\} = \min \left\{ \tilde{A}^L_\alpha - \tilde{B}^L_\alpha, \tilde{A}^U_\alpha - \tilde{B}^U_\alpha \right\}$$

and

$$u(\alpha) \equiv \max \left\{ \sup_{a\in\tilde{A}_\alpha} \inf_{b\in\tilde{B}_\alpha} (a-b), \sup_{b\in\tilde{B}_\alpha} \inf_{a\in\tilde{A}_\alpha} (a-b) \right\} = \max \left\{ \tilde{A}^L_\alpha - \tilde{B}^L_\alpha, \tilde{A}^U_\alpha - \tilde{B}^U_\alpha \right\}.$$

Although the function l is not necessarily concave on $[0, 1]$ and the function u is not necessarily convex on $[0, 1]$, the difference $\tilde{A}\ominus_{CCH_1}\tilde{B}$ still always exists. The reason is as follows. From (7.175), we see that $\tilde{A}\ominus_{CCH_1}\tilde{B} = \tilde{A}\ominus_{CCH_2}\tilde{B}$. As we mentioned above, we can show that the type-II complete composite Hausdorff difference $\tilde{A}\ominus_{CCH_2}\tilde{B}$ always exists without needing any sufficient conditions. This also means that the difference $\tilde{A}\ominus_{CCH_1}\tilde{B}$ always exists. Next, we present the existence of type-II-complete composite Hausdorff differences.

Theorem 7.7.13 *(**Existence for** $\tilde{A} \ominus^*_{CCH_2} \tilde{B}$) Let $\tilde{A}$ and $\tilde{B}$ be two fuzzy sets in $\mathbb{R}$, and let $M_\alpha = \tilde{A}_\alpha \boxminus_{CCH_2} \tilde{B}_\alpha$ for all $\alpha \in (0, 1]$. Then, there exists a fuzzy set $\tilde{C}^*$ in $\mathbb{R}$ with membership function defined by*

$$\xi_{\tilde{C}^*}(x) = \sup_{0<\alpha\leq 1} \alpha \cdot \chi_{M_\alpha}(x),$$

satisfying

$$\tilde{C}^*_\alpha = M_\alpha = \tilde{A}_\alpha \boxminus_{CCH_2} \tilde{B}_\alpha \text{ for each } \alpha \in (0, 1]$$

and

$$\tilde{C}^*_0 = M_0 = cl\left(\bigcup_{0<\alpha\leq 1} M_\alpha \right) = cl\left(\bigcup_{0<\alpha\leq 1} \tilde{A}_\alpha \boxminus_{CCH_2} \tilde{B}_\alpha \right). \tag{7.185}$$

This guarantees the existence of $\tilde{C}^* = \tilde{A} \ominus^*_{CCH_2} \tilde{B}$. Moreover, we have

$$\tilde{C}^* = \tilde{A} \ominus^*_{CCH_2} \tilde{B} = \tilde{A}\ominus_{GCCH_2}\tilde{B}. \tag{7.186}$$

Proof. From (7.167), we have

$$M_\alpha = \tilde{A}_\alpha \boxminus_{CCH_2} \tilde{B}_\alpha = \left[\inf_{\alpha\leq\beta\leq 1} l(\beta), \sup_{\alpha\leq\beta\leq 1} u(\beta) \right] \text{ for } 0 < \alpha \leq 1, \tag{7.187}$$

where the functions l and u are defined in (7.149) and (7.150), respectively. It is clear that $\{M_\alpha : 0 < \alpha \leq 1\}$ is a nested family. We are going to check that the conditions in Theorem 5.4.7 are satisfied. For any increasing and convergent sequence $\alpha_n \uparrow \alpha$, we want to claim $\bigcap_{n=1}^\infty M_{\alpha_n} \subseteq M_\alpha$. Let $A_n = [\alpha_n, 1]$ and $A = [\alpha, 1]$. Then $A_{n+1} \subseteq A_n$. Suppose that $x \in A_n$ for all n. We have $\alpha_n \leq x \leq 1$ for all n, which implies $\alpha \leq x \leq 1$ by taking $n \to \infty$. Therefore, we obtain $\bigcap_{n=1}^\infty A_n = A$. Let

$$l_n = \inf_{\alpha_n\leq\beta\leq 1} l(\beta) = \inf_{\beta\in A_n} l(\beta) \text{ and } u_n = \sup_{\alpha_n\leq\beta\leq 1} u(\beta) = \sup_{\beta\in A_n} u(\beta).$$

Proposition 1.2.3 says that

$$\lim_{n\to\infty} l_n = \inf_{\beta\in A} l(\beta) = \inf_{\alpha\le\beta\le1} l(\beta) \text{ and } \lim_{n\to\infty} u_n = \sup_{\beta\in A} u(\beta) = \sup_{\alpha\le\beta\le1} u(\beta). \tag{7.188}$$

For $x \in \bigcap_{n=1}^{\infty} M_{\alpha_n}$, i.e. $l_n \le x \le u_n$ for all n, by taking $n \to \infty$, from (7.188), we obtain

$$\inf_{\alpha\le\beta\le1} l(\beta) = \lim_{n\to\infty} l_n \le x \le \lim_{n\to\infty} u_n = \sup_{\alpha\le\beta\le1} u(\beta),$$

which says that $x \in M_\alpha$. Therefore, we obtain the desired inclusion $\bigcap_{n=1}^{\infty} M_{\alpha_n} \subseteq M_\alpha$. Using Theorem 5.4.7, we conclude that (7.172) is satisfied, which also shows the existence of $\tilde{A} \ominus^*_{CCH_2} \tilde{B}$. Using Proposition 1.4.8, we can also obtain the equalities (7.186) by referring to the proof of Theorem 7.7.7. This completes the proof. ∎

Theorem 7.7.14 ***(Existence for** $\tilde{A}\ominus_{CCH_2}\tilde{B}$**)** *Let $\tilde{A}$ and $\tilde{B}$ be two fuzzy sets in $\mathbb{R}$, and let $M_\alpha = \tilde{A}_\alpha \boxminus_{CCH_2} \tilde{B}_\alpha$ for all $\alpha \in [0, 1]$. Then, there exists a fuzzy set $\tilde{C}$ in $\mathbb{R}$ with membership function defined by*

$$\xi_{\tilde{C}}(x) = \sup_{\alpha\in[0,1]} \alpha \cdot \chi_{M_\alpha}(x) = \sup_{0<\alpha\le1} \alpha \cdot \chi_{M_\alpha}(x),$$

satisfying

$$\tilde{C}_\alpha = M_\alpha = \tilde{A}_\alpha \boxminus_{CCH_2} \tilde{B}_\alpha \text{ for each } \alpha \in [0, 1].$$

This guarantees the existence of $\tilde{C} = \tilde{A}\ominus_{CCH_2}\tilde{B}$. Moreover, the existence of $\tilde{C} = \tilde{A} \ominus^*_{CCH_2} \tilde{B}$ is also guaranteed, and we have

$$\tilde{C} = \tilde{A}\ominus_{CCH_2}\tilde{B} = \tilde{A} \ominus^*_{CCH_2} \tilde{B} = \tilde{A}\ominus_{GCCH_2}\tilde{B}. \tag{7.189}$$

Proof. From Theorem 7.7.13 by considering $\tilde{C}^*$ as $\tilde{C}$, we have

$$\tilde{C}_0 = \text{cl}\left(\bigcup_{0<\alpha\le1} M_\alpha\right) \text{ and } \tilde{C}_\alpha = M_\alpha \text{ for } 0 < \alpha \le 1.$$

In order to claim the existence of $\tilde{C} = \tilde{A} \boxminus_{CCH_2} \tilde{B}$, we are going to consider the 0-level sets. Therefore, it remains to prove

$$\tilde{C}_0 = \text{cl}\left(\bigcup_{0<\alpha\le1} M_\alpha\right) = M_0,$$

where

$$M_0 = \tilde{A}_0 \boxminus_{CCH_2} \tilde{B}_0 = \left[\inf_{0\le\beta\le1} l(\beta), \sup_{0\le\beta\le1} u(\beta)\right]$$

is a closed set by referring to (7.167). Since $M_\alpha \subseteq M_0$ for all $\alpha > 0$, we have $\bigcup_{0<\alpha\le1} M_\alpha \subseteq M_0$, which implies

$$\text{cl}\left(\bigcup_{0<\alpha\le1} M_\alpha\right) \subseteq \text{cl}\,(M_0) = M_0.$$

Next, we want to prove the other direction of inclusion. For

$$x \in M_0 = \left[\inf_{0\le\beta\le1} l(\beta), \sup_{0\le\beta\le1} u(\beta)\right] = \text{cl}\left(\inf_{0\le\beta\le1} l(\beta), \sup_{0\le\beta\le1} u(\beta)\right),$$

there exists a sequence $\{x_n\}_{n=1}^{\infty}$ in the open interval

$$\left(\inf_{0\leq\beta\leq 1} l(\beta), \sup_{0\leq\beta\leq 1} u(\beta)\right)$$

satisfying $x_n \to x$ as $n \to \infty$, i.e.

$$\inf_{0\leq\beta\leq 1} l(\beta) < x_n < \sup_{0\leq\beta\leq 1} u(\beta) \text{ for all } n.$$

Let

$$l^* = \inf_{0<\beta\leq 1} l(\beta) \text{ and } u^* = \sup_{0<\beta\leq 1} u(\beta).$$

Then $l^* < x_n < u^*$ for all n. Now, for each fixed n, we consider the following cases.

- Let $\epsilon_n < x_n - l^*$. According to the concept of infimum, there exists $\beta_1 > 0$ satisfying $l^* + \epsilon_n > l(\beta_1)$. Therefore, we obtain $l(\beta_1) < x_n$.
- Let $\epsilon_n < u^* - x_n$. According to the concept of supremum, there exists $\beta_2 > 0$ satisfying $u^* - \epsilon_n < u(\beta_2)$. Therefore, we obtain $x_n < u^* - \epsilon < u(\beta_2)$.

The above cases imply that there exists $\beta_1 > 0$ and $\beta_2 > 0$ satisfying $l(\beta_1) < x_n < u(\beta_2)$. Let $\beta^* = \min\{\beta_1, \beta_2\}$. Then $\beta^* > 0$. Since $\beta^* \leq \beta_1 \leq 1$ and $\beta^* \leq \beta_2 \leq 1$, it follows that

$$\inf_{\beta^*\leq\beta\leq 1} l(\beta) \leq l(\beta_1) < x_n < u(\beta_2) \leq \sup_{\beta^*\leq\beta\leq 1} u(\beta),$$

which implies

$$x_n \in \bigcup_{0<\alpha\leq 1}\left[\inf_{\alpha\leq\beta\leq 1} l(\beta), \sup_{\alpha\leq\beta\leq 1} u(\beta)\right] \text{ for all } n.$$

This means that

$$x \in \text{cl}\left(\bigcup_{0<\alpha\leq 1}\left[\inf_{\alpha\leq\beta\leq 1} l(\beta), \sup_{\alpha\leq\beta\leq 1} u(\beta)\right]\right) = \text{cl}\left(\bigcup_{0<\alpha\leq 1} M_\alpha\right)$$

by referring to (7.187). Therefore, we obtain $\tilde{C}_\alpha = M_\alpha$ for all $0 \leq \alpha \leq 1$. This shows the existence of $\tilde{C} = \tilde{A}\ominus_{CCH_2}\tilde{B}$. Using Proposition 1.4.8, we can also obtain the equalities (7.189) by referring to the proof of Theorem 7.7.7. This completes the proof. ∎

Theorems 7.7.13 and 7.7.14 say that the type-II complete composite Hausdorff difference $\tilde{A} \ominus^*_{CCH_2} \tilde{B}$ and $\tilde{A}\ominus_{CCH_2}\tilde{B}$ always exist. Moreover, the equalities (7.189) says that the three differences $\tilde{A} \ominus^*_{CCH_2} \tilde{B}$, $\tilde{A}\ominus_{CCH_2}\tilde{B}$ and $\tilde{A}\ominus_{GCCH_2}\tilde{B}$ are identical.

7.8 Applications and Conclusions

In what follows, we present some simple applications to the gradual numbers proposed by Dubois and Prade, and Fortin and Fargier [29, 37] and fuzzy linear systems studied by Lodwick and Dubois [74].

7.8.1 Gradual Numbers

The concept of a gradual number was proposed by Dubois and Prade, and Fortin and Fargier [29, 37]. Each gradual number $\hat{r}$ corresponds to an assignment function $\mathcal{A}_{\hat{r}} : (0, 1] \to \mathbb{R}$. Given any two gradual numbers $\hat{r}_1$ and $\hat{r}_2$ with the corresponding assignment functions $\mathcal{A}_{\hat{r}_1}$

and $\mathcal{A}_{\hat{r}_2}$. Their difference $\hat{r}_1 \ominus \hat{r}_2$ is also a gradual number with the corresponding assignment function $\tilde{A}_{\hat{r}_1 \ominus \hat{r}_2}$ defined by

$$\mathcal{A}_{\hat{r}_1 \ominus \hat{r}_2}(\alpha) = \mathcal{A}_{\hat{r}_1}(\alpha) - \mathcal{A}_{\hat{r}_2}(\alpha) \text{ for } \alpha \in (0,1]. \tag{7.190}$$

Next, we are going to establish the relationship between the fair Hausdorff difference and the difference of gradual numbers shown in (7.190).

Given any two fuzzy intervals $\tilde{A}$ and $\tilde{B}$, we can generate two corresponding gradual numbers $\hat{r}_1$ and $\hat{r}_2$ with assignment functions defined by

$$\mathcal{A}_{\hat{r}_1}(\alpha) = \frac{1}{2}\left(\tilde{A}^L_\alpha + \tilde{A}^U_\alpha\right) \text{ and } \mathcal{A}_{\hat{r}_2}(\alpha) = \frac{1}{2}\left(\tilde{B}^L_\alpha + \tilde{B}^U_\alpha\right) \text{ for } \alpha \in (0,1], \tag{7.191}$$

where $\hat{r}_1$ and $\hat{r}_2$ are also called the gradual midpoint of $\tilde{A}$ and $\tilde{B}$, respectively. Suppose that the fair Hausdorff difference $\tilde{A} \ominus_{FH} \tilde{B} = \tilde{C}$ or fair $*$-Hausdorff difference $\tilde{A} \ominus_{FH^*} \tilde{B} = \tilde{C}^*$ exists. Then, the fuzzy intervals $\tilde{C}$ and $\tilde{C}^*$ can also generate the corresponding gradual midpoints $\hat{r}_3$ and $\hat{r}_3^*$ given by

$$\mathcal{A}_{\hat{r}_3}(\alpha) = \frac{1}{2}\left(\tilde{C}^L_\alpha + \tilde{C}^U_\alpha\right) \text{ and } \mathcal{A}_{\hat{r}_3^*}(\alpha) = \frac{1}{2}\left(\tilde{C}^{*L}_\alpha + \tilde{C}^{*U}_\alpha\right) \text{ for } \alpha \in (0,1].$$

We are going to claim that $\mathcal{A}_{\hat{r}_3} = \mathcal{A}_{\hat{r}_1 \ominus \hat{r}_2}$ or $\mathcal{A}_{\hat{r}_3^*} = \mathcal{A}_{\hat{r}_1 \ominus \hat{r}_2}$ by referring to (7.190). From (7.148) in Example 7.7.1, we see that

$$\tilde{C}^L_\alpha = \min\left\{\tilde{A}^L_\alpha - \tilde{B}^L_\alpha, \tilde{A}^U_\alpha - \tilde{B}^U_\alpha\right\} \text{ and } \tilde{C}^L_\alpha = \max\left\{\tilde{A}^L_\alpha - \tilde{B}^L_\alpha, \tilde{A}^U_\alpha - \tilde{B}^U_\alpha\right\},$$

which says that, for each $\alpha \in (0,1]$

$$\begin{aligned}\mathcal{A}_{\hat{r}_3}(\alpha) &= \frac{1}{2}\left(\min\left\{\tilde{A}^L_\alpha - \tilde{B}^L_\alpha, \tilde{A}^U_\alpha - \tilde{B}^U_\alpha\right\} + \max\left\{\tilde{A}^L_\alpha - \tilde{B}^L_\alpha, \tilde{A}^U_\alpha - \tilde{B}^U_\alpha\right\}\right)\\ &= \frac{1}{2}\left(\tilde{A}^L_\alpha - \tilde{B}^L_\alpha + \tilde{A}^U_\alpha - \tilde{B}^U_\alpha\right) = \mathcal{A}_{\hat{r}_1}(\alpha) - \mathcal{A}_{\hat{r}_2}(\alpha) \text{ (using (7.191)).}\end{aligned}$$

This shows that the corresponding gradual midpoint $\hat{r}_3$ of fair Hausdorff difference $\tilde{C} = \tilde{A} \ominus_{FH} \tilde{B}$ is equal to the difference $\hat{r}_1 \ominus \hat{r}_2$ of the corresponding gradual midpoints $\hat{r}_1$ and $\hat{r}_2$ of $\tilde{A}$ and $\tilde{B}$. We can similarly obtain $\mathcal{A}_{\hat{r}_3^*} = \mathcal{A}_{\hat{r}_1 \ominus \hat{r}_2}$. Finally, for the existence of the fair Hausdorff difference $\tilde{A} \ominus_{FH} \tilde{B} = \tilde{C}$, refer to Example 7.7.5. In particular, if $\tilde{A}$ and $\tilde{B}$ are symmetric triangular fuzzy numbers, then the fair Hausdorff difference $\tilde{A} \ominus_{FH} \tilde{B} = \tilde{C}$ always exists.

7.8.2 Fuzzy Linear Systems

Inspired by the form presented in (7.118), we can solve the fuzzy linear system as follows. By referring to Example 7 in Lodwick and Dubois [74], two fuzzy intervals $\tilde{A}$ and $\tilde{B}$ are considered with the corresponding α-level sets

$$\tilde{A}_\alpha = [2 + 5\alpha, 12 - 5\alpha] \text{ and } \tilde{B}_\alpha = [2 + 3\alpha, 12 - 3\alpha].$$

In order to solve the fuzzy equation $\tilde{A}\tilde{x} = \tilde{B}$, we consider solving the interval equations $\tilde{A}_\alpha x = \tilde{B}_\alpha$ for all $\alpha \in [0,1]$. Lodwick and Dubois [74] obtained the interval solution $x(\alpha)$

$$M_\alpha \equiv x(\alpha) = \left[\frac{2 + 3\alpha}{2 + 5\alpha}, \frac{12 - 3\alpha}{12 - 5\alpha}\right],$$

which depends on α. Unfortunately, this family $\{M_\alpha : \alpha \in [0,1]\}$ is not nested. It means that we cannot obtain a fuzzy solution $\tilde{x}$ genuinely from the family $\{M_\alpha : \alpha \in [0,1]\}$

satisfying $\tilde{x}_\alpha = M_\alpha$ for all $\alpha \in [0, 1]$. However, we can apply (7.118) to obtain a so-called *quasi-fuzzy solution* $\tilde{x}^*$ with membership function defined by

$$\xi_{\tilde{x}^*}(r) = \sup_{0<\alpha\leq 1} \alpha \cdot \chi_{M_\alpha}(r).$$

Using Theorem 7.6.8, the α-level set of $\tilde{x}^*$ is given by

$$\tilde{x}^*_\alpha = \bigcup_{\alpha\leq\beta\leq 1} M_\beta = \bigcup_{\alpha\leq\beta\leq 1} \left[\frac{2+3\beta}{2+5\beta}, \frac{12-3\beta}{12-5\beta}\right] \equiv \bigcup_{\alpha\leq\beta\leq 1} \left[l(\beta), u(\beta)\right],$$

for every $\alpha \in (0, 1]$ and the 0-level set of $\tilde{x}^*$ is given by

$$\tilde{x}^*_0 = \text{cl}\left(\tilde{x}^*_{0+}\right) = \text{cl}\left(\bigcup_{0<\alpha\leq 1} M_\alpha\right) = \text{cl}\left(\bigcup_{0<\alpha\leq 1} \left[\frac{2+3\alpha}{2+5\alpha}, \frac{12-3\alpha}{12-5\alpha}\right]\right)$$
$$= \text{cl}\left(\bigcup_{0<\alpha\leq 1} \left[l(\alpha), u(\alpha)\right]\right).$$

Since the function l is decreasing and the function u is increasing on $[0, 1]$, it follows that

$$\left[l(\alpha), u(\alpha)\right] \subseteq \left[l(\beta), u(\beta)\right]$$

for $\beta > \alpha$. The continuities of the functions l and u say that the α-level set of $\tilde{x}^*$ is given by

$$\tilde{x}^*_\alpha = \bigcup_{\alpha\leq\beta\leq 1} \left[l(\beta), u(\beta)\right] = \left[l(1), u(1)\right] = \left[\frac{5}{7}, \frac{9}{7}\right] \quad \text{for } \alpha \in (0, 1]$$

and the 0-level sets of $\tilde{x}^*$ is given by

$$\tilde{x}^*_0 = \text{cl}\left(\bigcup_{0<\alpha\leq 1} \left[l(\alpha), u(\alpha)\right]\right) = \text{cl}\left(\left[l(1), u(1)\right]\right) = \left[\frac{5}{7}, \frac{9}{7}\right].$$

Therefore, the membership function of $\tilde{x}^*$ is given by

$$\xi_{\tilde{x}^*}(r) = \begin{cases} 1 & \text{if } r \in [\frac{5}{7}, \frac{9}{7}] \\ 0 & \text{otherwise.} \end{cases}$$

In some sense, we can say that the quasi-fuzzy solution $\tilde{x}^*$ is in fact a closed interval $[\frac{5}{7}, \frac{9}{7}]$.

In general, we consider the fuzzy linear system $\tilde{A}\tilde{\mathbf{x}} = \tilde{\mathbf{B}}$, where $\tilde{A}$ is an $n \times n$ matrix whose entries are assumed to be fuzzy intervals and $\tilde{\mathbf{B}}$ is an m-dimensional vector whose components are also assumed to be fuzzy intervals. The purpose is to find an m-dimensional vector

$$\tilde{\mathbf{x}}^* = \left(\tilde{x}^{(1*)}, \tilde{x}^{(2*)}, \ldots, \tilde{x}^{(n*)}\right)$$

satisfying $\tilde{A}\tilde{\mathbf{x}}^* \approx \tilde{\mathbf{B}}$. Referring to Lodwick and Dubois [74], we can first obtain the interval solution

$$\mathbf{x}(\alpha) = \left(x_1(\alpha), x_2(\alpha), \ldots, x_n(\alpha)\right) \equiv \left(M^{(1)}_\alpha, M^{(2)}_\alpha, \ldots M^{(n)}_\alpha\right)$$

by solving the interval linear system $\tilde{A}_\alpha \mathbf{x} = \tilde{\mathbf{B}}_\alpha$, where each component $x_i(\alpha)$ is an interval for $i = 1, \ldots, n$ and $\alpha \in (0, 1]$. Now, using (7.118), we can obtain the quasi-fuzzy solution $\tilde{\mathbf{x}}^*$ in which the membership function of each component $\tilde{x}^{(i*)}$ is defined by

$$\xi_{\tilde{x}^{(i*)}}(r) = \sup_{0<\alpha\leq 1} \alpha \cdot \chi_{M^{(i)}_\alpha}(r) = \sup_{0<\alpha\leq 1} \alpha \cdot \chi_{x_i(\alpha)}(r)$$

for $i = 1, \ldots, n$. The α-level sets of $\tilde{x}^{(i*)}$ can be realized by referring to Theorem 7.6.8.

7.8.3 Summary and Conclusion

Three types of binary operations between any two fuzzy sets $\tilde{A}$ and $\tilde{B}$ in $\mathbb{R}^m$ are proposed in this chapter. The first type of binary operation is defined in (7.93), given by

$$\tilde{A} \odot_{\mathfrak{L}} \tilde{B} = \tilde{C} \text{ if and only if } \tilde{C}_\alpha = \tilde{A}_\alpha \circ \tilde{B}_\alpha \text{ for each } \alpha \in [0,1].$$

The addition proposed by Puri and Ralescu [92] and the difference proposed by Bede and Stefanini [8] and Gomes and Barros [42] are a special case of the first type binary operation. On the other hand, the second type of binary operation is defined in (7.115), given by

$$\tilde{A} \odot_{\mathfrak{L}}^{*} \tilde{B} = \tilde{C}^{*} \text{ if and only if } \tilde{C}_\alpha^{*} = \tilde{A}_\alpha \circ \tilde{B}_\alpha \text{ for each } \alpha \in (0,1].$$

The only difference between the first type and second type of binary operations is that the second type does not consider the 0-level set. The motivation for considering this kind of situation is based on the topological structure of universal set $\mathbb{R}^m$. As we mentioned before, when the universal set $\mathbb{R}^m$ is endowed with a topology, its 0-level set is defined by (2.4). Otherwise, its 0-level set is defined to be the whole set $\mathbb{R}^m$. In other words, the definitions of 0-level set are different. Therefore, we propose the second type of binary operation without considering the 0-level sets $\tilde{A}_0$ and $\tilde{B}_0$ for the purpose of avoiding these conflicts regarding the 0-level set. Theorem 7.6.10 presents the equivalence among these three types of binary operations under the different assumptions.

Since the existence of $\tilde{A} \odot_{\mathfrak{L}} \tilde{B}$ and $\tilde{A} \odot_{\mathfrak{L}}^{*} \tilde{B}$ cannot be guaranteed, the sufficient conditions are provided in this chapter to investigate their existence. Based on the expression in the decomposition theorem, the third type of binary operation is proposed in (7.118), given by

$$\xi_{\tilde{A} \odot_{\mathfrak{G}} \tilde{B}}(x) = \sup_{0<\alpha\leq 1} \alpha \cdot \chi_{M_\alpha}(x) = \sup_{\alpha\in[0,1]} \alpha \cdot \chi_{M_\alpha}(x),$$

where $M_\alpha = \tilde{A}_\alpha \circ \tilde{B}_\alpha$ for all $\alpha \in (0,1]$. We see that M_0 is not needed in this binary operation, which also resolves the conflicts mentioned above. The benefit of considering this general binary operation $\tilde{A} \odot_{\mathfrak{G}} \tilde{B}$ is that it always exists without needing any sufficient conditions to guarantee its existence.

- Theorem 7.6.14 says that, given $\tilde{A}$ and $\tilde{B}$, under some suitable conditions, we can construct a fuzzy set $\tilde{C}^{*}$ in $\mathbb{R}^m$ using the family $\{\tilde{A}_\alpha \circ \tilde{B}_\alpha : 0 < \alpha \leq 1\}$ in which M_0 is not included such that the membership function of $\tilde{C}$ is given by

$$\xi_{\tilde{C}^{*}}(x) = \sup_{0<\alpha\leq 1} \alpha \cdot \chi_{\tilde{A}_\alpha \circ \tilde{B}_\alpha}(x)$$

  and satisfies $\tilde{C}_\alpha^{*} = \tilde{A}_\alpha \circ \tilde{B}_\alpha$ for each $\alpha \in (0,1]$. Therefore, we obtain the second type of binary operation $\tilde{A} \odot_{\mathfrak{L}}^{*} \tilde{B} = \tilde{C}^{*}$. In this case, we also have that the second and third types of binary operation are equivalent by referring to the equalities (7.123).
- Theorem 7.6.19 says that, given $\tilde{A}$ and $\tilde{B}$, under some suitable conditions that are stronger than those of Theorem 7.6.14, we can construct a fuzzy set $\tilde{C}$ in $\mathbb{R}^m$ using the family $\{\tilde{A}_\alpha \circ \tilde{B}_\alpha : 0 \leq \alpha \leq 1\}$ in which M_0 is included such that the membership function of $\tilde{C}$ is given by

$$\xi_{\tilde{C}}(x) = \sup_{0<\alpha\leq 1} \alpha \cdot \chi_{\tilde{A}_\alpha \circ \tilde{B}_\alpha}(x) = \sup_{\alpha\in[0,1]} \alpha \cdot \chi_{\tilde{A}_\alpha \circ \tilde{B}_\alpha}(x)$$

  and satisfies $\tilde{C}_\alpha = \tilde{A}_\alpha \circ \tilde{B}_\alpha$ for each $\alpha \in [0,1]$. Therefore, we obtain the first type of binary operation $\tilde{A} \odot_{\mathfrak{L}} \tilde{B} = \tilde{C}$. In this case, we also have that all three types of binary operation are equivalent by referring to the equalities (7.127).

We also emphasize that, in Theorems 7.6.14 and 7.6.19, the existence of $\tilde{A} \odot_{\mathfrak{L}}^{*} \tilde{B}$ and $\tilde{A} \odot_{\mathfrak{L}} \tilde{B}$ need to assume the nested families $\{M_{\alpha} : 0 < \alpha \leq 1\}$ or $\{M_{\alpha} : 0 \leq \alpha \leq 1\}$.

The difference of fuzzy intervals can be used to define the differentiation of fuzzy-valued functions. In order to generalize the difference proposed by Bede and Stefanini [8] and Gomes and Barros [42], we propose many types of Hausdorff differences. We need to mention that the different types of Hausdorff differences are special cases of those three types of binary operations. Therefore, their existence can follow from Theorems 7.6.14 and 7.6.19.

- The composite Hausdorff difference proposed in (7.156) and (7.157) generalizes the difference proposed by Bede and Stefanini [8]. The difference in (7.156) and (7.157) considers the fuzzy sets in $\mathbb{R}$ and differ by including or not including the 0-level sets. However, the difference in Bede and Stefanini [8] considers the fuzzy intervals that are also fuzzy sets in $\mathbb{R}$ assuming some elegant conditions such that the α-level sets become the bounded closed intervals. Theorems 7.7.7 and 7.7.9 say that the composite Hausdorff differences $\tilde{A} \ominus_{CH}^{*} \tilde{B}$ and $\tilde{A} \ominus_{CH} \tilde{B}$ always exist. Moreover, the equalities (7.161) says that the three differences $\tilde{A} \ominus_{CH}^{*} \tilde{B}$, $\tilde{A} \ominus_{CH} \tilde{B}$, and $\tilde{A} \ominus_{GCH} \tilde{B}$ are identical.
- According to (7.166) and (7.167), two types of complete composite Hausdorff differences $\tilde{A} \ominus_{CCH_1} \tilde{B}$ and $\tilde{A} \ominus_{CCH_2} \tilde{B}$ are defined in (7.169) and (7.170), respectively. Without considering the 0-level sets, we also define another two types of differences $\tilde{A} \ominus_{CCH_1}^{*} \tilde{B}$ and $\tilde{A} \ominus_{CCH_2}^{*} \tilde{B}$ in (7.171) and (7.172), respectively. These differences generalize the difference proposed by Gomes and Barros [42]. In particular, suppose that the fuzzy sets $\tilde{A}$ and $\tilde{B}$ are taken to be the fuzzy intervals $\tilde{A}$ and $\tilde{B}$. Then

$$\tilde{A} \ominus_{CCH_1} \tilde{B} = \tilde{A} \ominus_{CCH_2} \tilde{B} \text{ and } \tilde{A} \ominus_{CCH_1}^{*} \tilde{B} = \tilde{A} \ominus_{CCH_2}^{*} \tilde{B},$$

  which are the differences proposed by Gomes and Barros [42].
  (a) Theorems 7.7.13 and 7.7.14 say that the type-II complete composite Hausdorff difference $\tilde{A} \ominus_{CCH_2}^{*} \tilde{B}$ and $\tilde{A} \ominus_{CCH_2} \tilde{B}$ always exist. Moreover, the equalities (7.189) says that the three differences $\tilde{A} \ominus_{CCH_2}^{*} \tilde{B}$, $\tilde{A} \ominus_{CCH_2} \tilde{B}$, and $\tilde{A} \ominus_{GCCH_2} \tilde{B}$ are identical.
  (b) Theorem 7.7.11 says that the existence of type-I complete composite Hausdorff difference $\tilde{A} \ominus_{CCH_1}^{*} \tilde{B}$ and $\tilde{A} \ominus_{CCH_1} \tilde{B}$ need to assume that the function l defined in (7.149) is concave on $[0, 1]$ and the function u defined in (7.150) is convex on $[0, 1]$. An extra condition is also needed to guarantee the existence of $\tilde{A} \ominus_{CCH_1} \tilde{B}$ by referring to Theorem 7.7.12.
- The families $\{M_{\alpha} : 0 < \alpha \leq 1\}$ or $\{M_{\alpha} : 0 \leq \alpha \leq 1\}$ taken in Theorems 7.7.7-7.7.14 are nested automatically. However, the families taken in Theorems 7.7.3 and 7.7.4 are not necessarily nested. Therefore, the existence of fair Hausdorff differences $\tilde{A} \ominus_{FH^{*}} \tilde{B}$ and $\tilde{A} \ominus_{FH} \tilde{B}$ need to assume that the corresponding family is nested. An extra condition is also needed to guarantee the existence of $\tilde{A} \ominus_{FH} \tilde{B}$ by referring to Theorem 7.7.4.

In this chapter, we introduce four types of Hausdorff differences between any two fuzzy sets $\tilde{A}$ and $\tilde{B}$ of $\mathbb{R}$. The composite Hausdorff difference $\tilde{A} \ominus_{CH} \tilde{B}$ and type-II complete composite Hausdorff difference $\tilde{A} \ominus_{CCH_2} \tilde{B}$ always exist without needing any sufficient condition. However, if $\tilde{A}$ and $\tilde{B}$ are taken to be the fuzzy intervals $\tilde{A}$ and $\tilde{B}$, then the type-I complete composite Hausdorff difference $\tilde{A} \ominus_{CCH_1} \tilde{B}$ always exists satisfying $\tilde{A} \ominus_{CCH_1} \tilde{B} = \tilde{A} \ominus_{CCH_2} \tilde{B}$. Finally, Example 7.7.5 says that if $\tilde{A}$ and $\tilde{B}$ are taken to be the symmetric triangular fuzzy numbers, then the fair Hausdorff difference $\tilde{A} \ominus_{FH} \tilde{B}$ also always exists. Therefore, the sufficient conditions provided in this chapter are reasonable for applications of differences between any two fuzzy intervals

8

Inner Product of Fuzzy Vectors

A fuzzy interval in $\mathbb{R}$ is a fuzzy set in $\mathbb{R}$ such that its α-level sets are bounded closed intervals. The purpose of this chapter is to study the inner product of vectors of fuzzy intervals using the extension principle and the expression in the decomposition theorem. Since the fuzzy linear optimization problems and fuzzy linear systems can be formulated in terms of inner product of fuzzy vectors, the results obtained in this chapter can be useful for studying the fuzzy linear optimization problems and fuzzy linear systems.

There are two types of inner product that will be studied in this chapter. The first type of inner product of fuzzy vectors is directly based on the inner product of vectors

$$\mathbf{x} = \left(x_1, \dots, x_n\right) \text{ and } \mathbf{y} = \left(y_1, \dots, y_n\right)$$

given by the following expression

$$\mathbf{x} \bullet \mathbf{y} = x_1 y_1 + \cdots + x_n y_n,$$

where $\mathbf{x}$ and $\mathbf{y}$ are two vectors in $\mathbb{R}^n$. The extension principle and the expression in the decomposition theorem will directly apply to the (conventional) inner product $\mathbf{x} \bullet \mathbf{y}$ given above without considering the addition and multiplication of fuzzy intervals.

The second type of inner product of fuzzy vectors will be based on the addition and multiplication of fuzzy intervals by considering the following expression

$$\left(\tilde{A}^{(1)} \otimes \tilde{B}^{(1)}\right) \oplus \cdots \oplus \left(\tilde{A}^{(n)} \otimes \tilde{B}^{(n)}\right),$$

where $\tilde{A}^{(i)}$ and $\tilde{B}^{(i)}$ are fuzzy intervals in $\mathbb{R}$ for $i = 1, \dots, n$. The main issue of second type is the addition and multiplication of fuzzy intervals. In this chapter, the addition and multiplication of fuzzy intervals will also be formulated based on the extension principle and the expression in the decomposition theorem. Therefore, different combinations of using different addition and multiplication will generate many different second type of inner product of fuzzy vectors. Their relationship will be established. Moreover, the relationship between the first type and second type of inner product will also be studied.

8.1 The First Type of Inner Product

Recall that the inner product of vectors $\mathbf{x}$ and $\mathbf{y}$ is given by

$$\mathbf{x} \bullet \mathbf{y} = x_1 y_1 + \cdots + x_n y_n,$$

where $\mathbf{x}$ and $\mathbf{y}$ are two vectors in $\mathbb{R}^n$.

Mathematical Foundations of Fuzzy Sets, First Edition. Hsien-Chung Wu.

Given any fuzzy intervals $\tilde{A}^{(1)},\ldots,\tilde{A}^{(n)}$ and $\tilde{B}^{(1)},\ldots,\tilde{B}^{(n)}$ in $\mathbb{R}$, for each $\alpha \in [0,1]$, the α-level sets of $\tilde{A}^{(i)}$ and $\tilde{B}^{(i)}$ are denoted by

$$\tilde{A}^{(i)}_\alpha \equiv \left[\tilde{A}^L_{i\alpha}, \tilde{A}^U_{i\alpha}\right] \text{ and } \tilde{B}^{(i)}_\alpha \equiv \left[\tilde{B}^L_{i\alpha}, \tilde{B}^U_{i\alpha}\right].$$

We also recall the notations in (7.31), (7.32), and (7.33).

Let $\tilde{\mathbf{A}}$ and $\tilde{\mathbf{B}}$ be two vectors of fuzzy intervals in $\mathbb{R}$ given by

$$\tilde{\mathbf{A}} = \left(\tilde{A}^{(1)}, \tilde{A}^{(2)}, \ldots, \tilde{A}^{(n)}\right) \text{ and } \tilde{\mathbf{B}} = \left(\tilde{B}^{(1)}, \tilde{B}^{(2)}, \ldots, \tilde{B}^{(n)}\right). \tag{8.1}$$

We shall study the inner product $\tilde{\mathbf{A}} \circledast_{EP} \tilde{\mathbf{B}}$ using the extension principle, and the inner product $\tilde{\mathbf{A}} \circledast_{DT} \tilde{\mathbf{B}}$ using the expression in the decomposition theorem.

8.1.1 Using the Extension Principle

Given two vectors $\tilde{\mathbf{A}}$ and $\tilde{\mathbf{B}}$ of fuzzy intervals in (8.1), the membership function of $\tilde{\mathbf{A}} \circledast_{EP} \tilde{\mathbf{B}}$ is defined by

$$\xi_{\tilde{\mathbf{A}} \circledast_{EP} \tilde{\mathbf{B}}}(z) = \sup_{\{(\mathbf{x},\mathbf{y}):z=\mathbf{x}\bullet\mathbf{y}\}} \min\left\{\xi_{\tilde{A}^{(1)}}(x_1), \ldots, \xi_{\tilde{A}^{(n)}}(x_n), \xi_{\tilde{B}^{(1)}}(y_1), \ldots, \xi_{\tilde{B}^{(n)}}(y_n)\right\} \tag{8.2}$$

for each $z \in \mathbb{R}$. The α-level set $(\tilde{\mathbf{A}} \circledast_{EP} \tilde{\mathbf{B}})_\alpha$ of $\tilde{\mathbf{A}} \circledast_{EP} \tilde{\mathbf{B}}$ for $\alpha \in [0,1]$ can be obtained by applying Proposition 4.4.15 to the inner product $\tilde{\mathbf{A}} \circledast_{EP} \tilde{\mathbf{B}}$, which is shown below. For each $\alpha \in (0,1]$, we have

$$\begin{aligned}
\left(\tilde{\mathbf{A}} \circledast_{EP} \tilde{\mathbf{B}}\right)_\alpha &= \left\{\mathbf{x}\bullet\mathbf{y} : \min\left\{\xi_{\tilde{A}^{(1)}}(x_1), \cdots, \xi_{\tilde{A}^{(n)}}(x_n), \xi_{\tilde{B}^{(1)}}(y_1), \cdots, \xi_{\tilde{B}^{(n)}}(y_n)\right\} \geq \alpha\right\} \\
&= \left\{\mathbf{x}\bullet\mathbf{y} : \xi_{\tilde{A}^{(i)}}(x_i) \geq \alpha \text{ and } \xi_{\tilde{B}^{(i)}}(y_i) \geq \alpha \text{ for each } i = 1, \ldots, n\right\} \\
&= \Big\{x_1y_1 + \cdots + x_ny_n : x_i \in \tilde{A}^{(i)}_\alpha \equiv \left[\tilde{A}^L_{i\alpha}, \tilde{A}^U_{i\alpha}\right] \\
&\qquad \text{and } y_i \in \tilde{B}^{(i)}_\alpha \equiv \left[\tilde{B}^L_{i\alpha}, \tilde{B}^U_{i\alpha}\right] \text{ for each } i = 1, \ldots, n\Big\} \\
&= \left[\min_{(\mathbf{x},\mathbf{y})\in(\tilde{\mathbf{A}}_\alpha,\tilde{\mathbf{B}}_\alpha)} (x_1y_1 + \cdots + x_ny_n), \max_{(\mathbf{x},\mathbf{y})\in(\tilde{\mathbf{A}}_\alpha,\tilde{\mathbf{B}}_\alpha)} (x_1y_1 + \cdots + x_ny_n)\right],
\end{aligned} \tag{8.3}$$

where $\tilde{\mathbf{A}}_\alpha$ and $\tilde{\mathbf{B}}_\alpha$ are given in (7.32) and (7.33). From (8.3), since the objective functions are separable, we see that

$$\begin{aligned}
\min_{(\mathbf{x},\mathbf{y})\in(\tilde{\mathbf{A}}_\alpha,\tilde{\mathbf{B}}_\alpha)} (x_1y_1 + \cdots + x_ny_n) &= \sum_{i=1}^{n} \left(\min_{(x_i,y_i)\in[\tilde{A}^L_{i\alpha},\tilde{A}^U_{i\alpha}]\times[\tilde{B}^L_{i\alpha},\tilde{B}^U_{i\alpha}]} x_iy_i\right) \\
&= \sum_{i=1}^{n} \left(\min\left\{\tilde{A}^L_{i\alpha}\tilde{B}^L_{i\alpha}, \tilde{A}^L_{i\alpha}\tilde{B}^U_{i\alpha}, \tilde{A}^U_{i\alpha}\tilde{B}^L_{i\alpha}, \tilde{A}^U_{i\alpha}\tilde{B}^U_{i\alpha}\right\}\right)
\end{aligned}$$

and

$$\begin{aligned}
\max_{(\mathbf{x},\mathbf{y})\in(\tilde{\mathbf{A}}_\alpha,\tilde{\mathbf{B}}_\alpha)} (x_1y_1 + \cdots + x_ny_n) &= \sum_{i=1}^{n} \left(\max_{(x_i,y_i)\in[\tilde{A}^L_{i\alpha},\tilde{A}^U_{i\alpha}]\times[\tilde{B}^L_{i\alpha},\tilde{B}^U_{i\alpha}]} x_iy_i\right) \\
&= \sum_{i=1}^{n} \left(\max\left\{\tilde{A}^L_{i\alpha}\tilde{B}^L_{i\alpha}, \tilde{A}^L_{i\alpha}\tilde{B}^U_{i\alpha}, \tilde{A}^U_{i\alpha}\tilde{B}^L_{i\alpha}, \tilde{A}^U_{i\alpha}\tilde{B}^U_{i\alpha}\right\}\right).
\end{aligned}$$

Now, we assume that $\tilde{A}^{(i)}$ and $\tilde{B}^{(i)}$ are taken to be the canonical fuzzy intervals in Definition 2.3.3 for $i = 1, \ldots, n$. For the 0-level set, from (8.3) and (2.5), using the nestedness and the continuities regarding the canonical fuzzy intervals, we have

$$\begin{aligned}
\left(\tilde{\mathbf{A}} \circledast_{EP} \tilde{\mathbf{B}}\right)_0 &= \mathrm{cl}\left(\bigcup_{0<\alpha\leq 1} \left(\tilde{\mathbf{A}} \circledast_{EP} \tilde{\mathbf{B}}\right)_\alpha\right)\\
&= \left[\min_{(\mathbf{x},\mathbf{y})\in(\tilde{\mathbf{A}}_0,\tilde{\mathbf{B}}_0)} \left(x_1y_1 + \cdots + x_ny_n\right), \max_{(\mathbf{x},\mathbf{y})\in(\tilde{\mathbf{A}}_0,\tilde{\mathbf{B}}_0)} \left(x_1y_1 + \cdots + x_ny_n\right)\right]\\
&= \left[\sum_{i=1}^{n}\left(\min_{(x_i,y_i)\in[\tilde{A}_{i0}^L,\tilde{A}_{i0}^U]\times[\tilde{B}_{i0}^L,\tilde{B}_{i0}^U]} x_iy_i\right), \sum_{i=1}^{n}\left(\max_{(x_i,y_i)\in[\tilde{A}_{i0}^L,\tilde{A}_{i0}^U]\times[\tilde{B}_{i0}^L,\tilde{B}_{i0}^U]} x_iy_i\right)\right]\\
&= \left[\sum_{i=1}^{n}\left(\min\left\{\tilde{A}_{i0}^L\tilde{B}_{i0}^L, \tilde{A}_{i0}^L\tilde{B}_{i0}^U, \tilde{A}_{i0}^U\tilde{B}_{i0}^L, \tilde{A}_{i0}^U\tilde{B}_{i0}^U\right\}\right),\right.\\
&\qquad \left.\sum_{i=1}^{n}\left(\max\left\{\tilde{A}_{i0}^L\tilde{B}_{i0}^L, \tilde{A}_{i0}^L\tilde{B}_{i0}^U, \tilde{A}_{i0}^U\tilde{B}_{i0}^L, \tilde{A}_{i0}^U\tilde{B}_{i0}^U\right\}\right)\right].
\end{aligned}$$

Recall that a fuzzy interval $\tilde{A}$ is nonnegative when $\tilde{A}_\alpha^L \geq 0$ for each $\alpha \in [0,1]$. Suppose that $\tilde{A}^{(1)}, \ldots, \tilde{A}^{(n)}$ and $\tilde{B}^{(1)}, \ldots, \tilde{B}^{(n)}$ are taken to be nonnegative fuzzy intervals. Then

$$\left(\tilde{\mathbf{A}} \circledast_{EP} \tilde{\mathbf{B}}\right)_\alpha = \left[\tilde{A}_{1\alpha}^L\tilde{B}_{1\alpha}^L + \cdots + \tilde{A}_{n\alpha}^L\tilde{B}_{n\alpha}^L, \tilde{A}_{1\alpha}^U\tilde{B}_{1\alpha}^U + \cdots + \tilde{A}_{n\alpha}^U\tilde{B}_{n\alpha}^U\right].$$

The above results are summarized in the following theorem.

Theorem 8.1.1 *Let $\tilde{A}^{(1)}, \ldots, \tilde{A}^{(n)}$ and $\tilde{B}^{(1)}, \ldots, \tilde{B}^{(n)}$ be any fuzzy intervals. Then, the α-level sets of $\tilde{\mathbf{A}} \circledast_{EP} \tilde{\mathbf{B}}$ are given by*

$$\begin{aligned}
\left(\tilde{\mathbf{A}} \circledast_{EP} \tilde{\mathbf{B}}\right)_\alpha &= \left[\min_{(\mathbf{x},\mathbf{y})\in(\tilde{\mathbf{A}}_\alpha,\tilde{\mathbf{B}}_\alpha)} \mathbf{x}\bullet\mathbf{y}, \max_{(\mathbf{x},\mathbf{y})\in(\tilde{\mathbf{A}}_\alpha,\tilde{\mathbf{B}}_\alpha)} \mathbf{x}\bullet\mathbf{y}\right]\\
&= \left[\sum_{i=1}^{n}\left(\min_{(x_i,y_i)\in[\tilde{A}_{i\alpha}^L,\tilde{A}_{i\alpha}^U]\times[\tilde{B}_{i\alpha}^L,\tilde{B}_{i\alpha}^U]} x_iy_i\right), \sum_{i=1}^{n}\left(\max_{(x_i,y_i)\in[\tilde{A}_{i\alpha}^L,\tilde{A}_{i\alpha}^U]\times[\tilde{B}_{i\alpha}^L,\tilde{B}_{i\alpha}^U]} x_iy_i\right)\right]\\
&= \left[\sum_{i=1}^{n}\left(\min\left\{\tilde{A}_{i\alpha}^L\tilde{B}_{i\alpha}^L, \tilde{A}_{i\alpha}^L\tilde{B}_{i\alpha}^U, \tilde{A}_{i\alpha}^U\tilde{B}_{i\alpha}^L, \tilde{A}_{i\alpha}^U\tilde{B}_{i\alpha}^U\right\}\right),\right.\\
&\qquad \left.\sum_{i=1}^{n}\left(\max\left\{\tilde{A}_{i\alpha}^L\tilde{B}_{i\alpha}^L, \tilde{A}_{i\alpha}^L\tilde{B}_{i\alpha}^U, \tilde{A}_{i\alpha}^U\tilde{B}_{i\alpha}^L, \tilde{A}_{i\alpha}^U\tilde{B}_{i\alpha}^U\right\}\right)\right]
\end{aligned}$$

for $\alpha \in (0,1]$, and the 0-level set is given by

$$\left(\tilde{\mathbf{A}} \circledast_{EP} \tilde{\mathbf{B}}\right)_0 = \mathrm{cl}\left(\bigcup_{0<\alpha\leq 1} \left(\tilde{\mathbf{A}} \circledast_{EP} \tilde{\mathbf{B}}\right)_\alpha\right).$$

where $\tilde{\mathbf{A}}_\alpha$ and $\tilde{\mathbf{B}}_\alpha$ are given in (7.32) and (7.33). Suppose that $\tilde{A}^{(1)}, \ldots, \tilde{A}^{(n)}$ and $\tilde{B}^{(1)}, \ldots, \tilde{B}^{(n)}$ are taken to be nonnegative fuzzy intervals. Then

$$\left(\tilde{\mathbf{A}} \circledast_{EP} \tilde{\mathbf{B}}\right)_\alpha = \left[\tilde{\mathbf{A}}_\alpha^L \bullet \tilde{\mathbf{B}}_\alpha^L, \tilde{\mathbf{A}}_\alpha^U \bullet \tilde{\mathbf{B}}_\alpha^U\right]$$

for $\alpha \in (0,1]$, *where* $\tilde{\mathbf{A}}_\alpha^L$, $\tilde{\mathbf{A}}_\alpha^U$, $\tilde{\mathbf{B}}_\alpha^L$ *and* $\tilde{\mathbf{B}}_\alpha^U$ *are given in* (7.31). *We further assume that* $\tilde{A}^{(1)},\ldots,\tilde{A}^{(n)}$ *and* $\tilde{B}^{(1)},\ldots,\tilde{B}^{(n)}$ *are taken to be canonical fuzzy intervals. Then, the* 0*-level set is given by*

$$\left(\tilde{\mathbf{A}} \circledast_{EP} \tilde{\mathbf{B}}\right)_0 = \left[\sum_{i=1}^{n}\left(\min\left\{\tilde{A}_{i0}^L\tilde{B}_{i0}^L, \tilde{A}_{i0}^L\tilde{B}_{i0}^U, \tilde{A}_{i0}^U\tilde{B}_{i0}^L, \tilde{A}_{i0}^U\tilde{B}_{i0}^U\right\}\right), \sum_{i=1}^{n}\left(\max\left\{\tilde{A}_{i0}^L\tilde{B}_{i0}^L, \tilde{A}_{i0}^L\tilde{B}_{i0}^U, \tilde{A}_{i0}^U\tilde{B}_{i0}^L, \tilde{A}_{i0}^U\tilde{B}_{i0}^U\right\}\right)\right].$$

Example 8.1.2 Continued from Examples 2.2.4 and 7.2.4, from (7.12), we have

$$\begin{aligned}
\tilde{A}_\alpha^{(1)} &= \left[\tilde{A}_{1\alpha}^L, \tilde{A}_{1\alpha}^U\right] = [(1-\alpha)+2\alpha, 4(1-\alpha)+3\alpha] = [1+\alpha, 4-\alpha]\\
\tilde{A}_\alpha^{(2)} &= \left[\tilde{A}_{2\alpha}^L, \tilde{A}_{2\alpha}^U\right] = [2(1-\alpha)+3\alpha, 5(1-\alpha)+4\alpha] = [2+\alpha, 5-\alpha]\\
\tilde{B}_\alpha^{(1)} &= \left[\tilde{B}_{1\alpha}^L, \tilde{B}_{1\alpha}^U\right] = [4(1-\alpha)+5\alpha, 7(1-\alpha)+6\alpha] = [4+\alpha, 7-\alpha]\\
\tilde{B}_\alpha^{(2)} &= \left[\tilde{B}_{2\alpha}^L, \tilde{B}_{2\alpha}^U\right] = [3(1-\alpha)+4\alpha, 6(1-\alpha)+5\alpha] = [3+\alpha, 6-\alpha].
\end{aligned}$$

Now, we obtain

$$\begin{aligned}
&\min\left\{\tilde{A}_{1\alpha}^L\tilde{B}_{1\alpha}^L, \tilde{A}_{1\alpha}^L\tilde{B}_{1\alpha}^U, \tilde{A}_{1\alpha}^U\tilde{B}_{1\alpha}^L, \tilde{A}_{1\alpha}^U\tilde{B}_{1\alpha}^U\right\} + \min\left\{\tilde{A}_{2\alpha}^L\tilde{B}_{1\alpha}^L, \tilde{A}_{2\alpha}^L\tilde{B}_{2\alpha}^U, \tilde{A}_{2\alpha}^U\tilde{B}_{2\alpha}^L, \tilde{A}_{2\alpha}^U\tilde{B}_{2\alpha}^U\right\}\\
&\quad = (1+\alpha)(4+\alpha) + (2+\alpha)(3+\alpha) = 10 + 10\alpha + \alpha^2
\end{aligned}$$

and

$$\begin{aligned}
&\max\left\{\tilde{A}_{1\alpha}^L\tilde{B}_{1\alpha}^L, \tilde{A}_{1\alpha}^L\tilde{B}_{1\alpha}^U, \tilde{A}_{1\alpha}^U\tilde{B}_{1\alpha}^L, \tilde{A}_{1\alpha}^U\tilde{B}_{1\alpha}^U\right\} + \max\left\{\tilde{A}_{2\alpha}^L\tilde{B}_{1\alpha}^L, \tilde{A}_{2\alpha}^L\tilde{B}_{2\alpha}^U, \tilde{A}_{2\alpha}^U\tilde{B}_{2\alpha}^L, \tilde{A}_{2\alpha}^U\tilde{B}_{2\alpha}^U\right\}\\
&\quad = (4-\alpha)(7-\alpha) + (5-\alpha)(6-\alpha) = 58 - 22\alpha + 2\alpha^2.
\end{aligned}$$

Therefore, Theorem 8.1.1 says that

$$\left(\tilde{\mathbf{A}} \circledast_{EP} \tilde{\mathbf{B}}\right)_\alpha = \left[10 + 10\alpha + \alpha^2, 58 - 22\alpha + 2\alpha^2\right]$$

for $\alpha \in [0,1]$.

Example 8.1.3 Continued from Example 8.1.2, we can consider the triangular fuzzy number that is a special case of trapezoidal fuzzy interval by taking $a_1 = a_2$. More precisely, the membership function is given by

$$\xi_{\tilde{A}}(r) = \begin{cases} (r-a^L)/(a-a^L) & \text{if } a^L \le r \le a\\ 1 & \text{if } r = a\\ (a^U - r)/(a^U - a) & \text{if } a < r \le a^U\\ 0 & \text{otherwise,}\end{cases}$$

which is denoted by $\tilde{A} = (a^L, a, a^U)$. For $\alpha \in [0,1]$, the α-level set $\tilde{A}_\alpha = \left[\tilde{A}_\alpha^L, \tilde{A}_\alpha^U\right]$ is given by

$$\tilde{A}_\alpha^L = (1-\alpha)a^L + \alpha a \text{ and } \tilde{A}_\alpha^U = (1-\alpha)a^U + \alpha a. \tag{8.4}$$

Now, we consider the following numerical triangular fuzzy numbers

$$\tilde{A}^{(1)} = (-1, 1, 3), \quad \tilde{A}^{(2)} = (-2, -1, 2) \text{ and } \tilde{A}^{(3)} = (01, 3)$$

and

$$\tilde{B}^{(1)} = (-10, 3), \quad \tilde{B}^{(2)} = (-2, -1, 1) \text{ and } \tilde{B}^{(3)} = (-1, 1, 3).$$

From (8.4), we also have

$$\tilde{A}_{\alpha}^{(1)} = \left[\tilde{A}_{1\alpha}^{L}, \tilde{A}_{1\alpha}^{U}\right] = [-(1-\alpha)+\alpha, 3(1-\alpha)+\alpha] = [2\alpha-1, 3-2\alpha]$$

$$\tilde{A}_{\alpha}^{(2)} = \left[\tilde{A}_{2\alpha}^{L}, \tilde{A}_{2\alpha}^{U}\right] = [-2(1-\alpha)-\alpha, 2(1-\alpha)-\alpha] = [-2+\alpha, 2-3\alpha]$$

$$\tilde{A}_{\alpha}^{(3)} = \left[\tilde{A}_{3\alpha}^{L}, \tilde{A}_{3\alpha}^{U}\right] = [\alpha, 3(1-\alpha)+\alpha] = [\alpha, 3-2\alpha]$$

$$\tilde{B}_{\alpha}^{(1)} = \left[\tilde{B}_{1\alpha}^{L}, \tilde{B}_{1\alpha}^{U}\right] = [-(1-\alpha), 3(1-\alpha)] = [\alpha-1, 3-3\alpha]$$

$$\tilde{B}_{\alpha}^{(2)} = \left[\tilde{B}_{2\alpha}^{L}, \tilde{B}_{2\alpha}^{U}\right] = [-2(1-\alpha)-\alpha, (1-\alpha)-\alpha] = [-2+\alpha, 1-2\alpha]$$

$$\tilde{B}_{\alpha}^{(3)} = \left[\tilde{B}_{3\alpha}^{L}, \tilde{B}_{3\alpha}^{U}\right] = [-(1-\alpha)+\alpha, 3(1-\alpha)+\alpha] = [2\alpha-1, 3-2\alpha].$$

Now, we obtain

$$\begin{aligned}
&\min\left\{\tilde{A}_{1\alpha}^{L}\tilde{B}_{1\alpha}^{L}, \tilde{A}_{1\alpha}^{L}\tilde{B}_{1\alpha}^{U}, \tilde{A}_{1\alpha}^{U}\tilde{B}_{1\alpha}^{L}, \tilde{A}_{1\alpha}^{U}\tilde{B}_{1\alpha}^{U}\right\} \\
&= \begin{cases} (\alpha-1)(3-2\alpha) & \text{if } \alpha \geq 0.5 \\ \min\{(\alpha-1)(3-2\alpha), (2\alpha-1)(3-3\alpha)\} & \text{if } \alpha < 0.5 \end{cases} \\
&= \left\{\begin{matrix} (\alpha-1)(3-2\alpha) \text{ if } \alpha \geq 0.5 \\ (\alpha-1)(3-2\alpha) \text{ if } \alpha < 0.5 \end{matrix}\right\} = (\alpha-1)(3-2\alpha)
\end{aligned}$$

$$\min\left\{\tilde{A}_{2\alpha}^{L}\tilde{B}_{2\alpha}^{L}, \tilde{A}_{2\alpha}^{L}\tilde{B}_{2\alpha}^{U}, \tilde{A}_{2\alpha}^{U}\tilde{B}_{2\alpha}^{L}, \tilde{A}_{2\alpha}^{U}\tilde{B}_{2\alpha}^{U}\right\} = \begin{cases} (2-3\alpha)(1-2\alpha) & \text{if } \alpha \geq 2/3 \\ (-2+\alpha)(2-3\alpha) & \text{if } \alpha < 2/3 \end{cases}$$

$$\min\left\{\tilde{A}_{3\alpha}^{L}\tilde{B}_{3\alpha}^{L}, \tilde{A}_{3\alpha}^{L}\tilde{B}_{3\alpha}^{U}, \tilde{A}_{3\alpha}^{U}\tilde{B}_{3\alpha}^{L}, \tilde{A}_{3\alpha}^{U}\tilde{B}_{3\alpha}^{U}\right\} = \begin{cases} \alpha(2\alpha-1) & \text{if } \alpha \geq 0.5 \\ (2\alpha-1)(3-2\alpha) & \text{if } \alpha < 0.5 \end{cases}$$

and

$$\begin{aligned}
&\max\left\{\tilde{A}_{1\alpha}^{L}\tilde{B}_{1\alpha}^{L}, \tilde{A}_{1\alpha}^{L}\tilde{B}_{1\alpha}^{U}, \tilde{A}_{1\alpha}^{U}\tilde{B}_{1\alpha}^{L}, \tilde{A}_{1\alpha}^{U}\tilde{B}_{1\alpha}^{U}\right\} \\
&= \begin{cases} (3-2\alpha)(3-3\alpha) & \text{if } \alpha \geq 0.5 \\ \max\{(2\alpha-1)(\alpha-1), (3-2\alpha)(3-3\alpha)\} & \text{if } \alpha < 0.5 \end{cases} \\
&= \left\{\begin{matrix} (3-2\alpha)(3-3\alpha) \text{ if } \alpha \geq 0.5 \\ (3-2\alpha)(3-3\alpha) \text{ if } \alpha < 0.5 \end{matrix}\right\} = (3-2\alpha)(3-3\alpha)
\end{aligned}$$

$$\max\left\{\tilde{A}_{2\alpha}^{L}\tilde{B}_{2\alpha}^{L}, \tilde{A}_{2\alpha}^{L}\tilde{B}_{2\alpha}^{U}, \tilde{A}_{2\alpha}^{U}\tilde{B}_{2\alpha}^{L}, \tilde{A}_{2\alpha}^{U}\tilde{B}_{2\alpha}^{U}\right\} = (-2+\alpha)^2$$

$$\max\left\{\tilde{A}_{3\alpha}^{L}\tilde{B}_{3\alpha}^{L}, \tilde{A}_{3\alpha}^{L}\tilde{B}_{3\alpha}^{U}, \tilde{A}_{3\alpha}^{U}\tilde{B}_{3\alpha}^{L}, \tilde{A}_{3\alpha}^{U}\tilde{B}_{3\alpha}^{U}\right\} = (3-2\alpha)^2.$$

In particular, for $\alpha \geq 2/3$, we have

$$\begin{aligned}
&\sum_{i=1}^{3}\left(\min\left\{\tilde{A}_{i\alpha}^{L}\tilde{B}_{i\alpha}^{L}, \tilde{A}_{i\alpha}^{L}\tilde{B}_{i\alpha}^{U}, \tilde{A}_{i\alpha}^{U}\tilde{B}_{i\alpha}^{L}, \tilde{A}_{i\alpha}^{U}\tilde{B}_{i\alpha}^{U}\right\}\right) \\
&= (\alpha-1)(3-2\alpha) + (2-3\alpha)(1-2\alpha) + \alpha(2\alpha-1) = 6\alpha^2 - 3\alpha - 1
\end{aligned}$$

and

$$\begin{aligned}
&\sum_{i=1}^{3}\left(\max\left\{\tilde{A}_{i\alpha}^{L}\tilde{B}_{i\alpha}^{L}, \tilde{A}_{i\alpha}^{L}\tilde{B}_{i\alpha}^{U}, \tilde{A}_{i\alpha}^{U}\tilde{B}_{i\alpha}^{L}, \tilde{A}_{i\alpha}^{U}\tilde{B}_{i\alpha}^{U}\right\}\right) \\
&= (3-2\alpha)(3-3\alpha) + (-2+\alpha)^2 + (3-2\alpha)^2 = 11\alpha^2 - 31\alpha + 22
\end{aligned}$$

Therefore, Theorem 8.1.1 says that

$$\left(\tilde{\mathbf{A}} \circledast_{EP} \tilde{\mathbf{B}}\right)_\alpha = \left[6\alpha^2 - 3\alpha - 1, 11\alpha^2 - 31\alpha + 22\right]$$

for $\alpha \geq 2/3$. We can similarly consider the cases of $0.5 \leq \alpha < 2/3$ and $0 \leq \alpha < 0.5$.

8.1.2 Using the Expression in the Decomposition Theorem

Let $\tilde{A}^{(i)}$ and $\tilde{B}^{(i)}$ be fuzzy intervals for $i = 1, \ldots, n$. Now, we are going to use the expression in the decomposition theorem to define three different inner products by considering three different families.

- We consider the family $\{M_\alpha^\bullet : 0 < \alpha \leq 1\}$ by taking
$$M_\alpha^\bullet = \tilde{\mathbf{A}}_\alpha \bullet \tilde{\mathbf{B}}_\alpha = \left\{\mathbf{x} \bullet \mathbf{y} : \mathbf{x} \in \tilde{\mathbf{A}}_\alpha \text{ and } \mathbf{y} \in \tilde{\mathbf{B}}_\alpha\right\} \tag{8.5}$$
to define the inner product $\tilde{\mathbf{A}} \circledast_{DT}^{\diamond} \tilde{\mathbf{B}}$, where $\tilde{\mathbf{A}}_\alpha$ and $\tilde{\mathbf{B}}_\alpha$ are given in (7.32) and (7.33).
- Let M_β be bounded closed intervals given by
$$M_\beta = \left[\min\left\{\tilde{\mathbf{A}}_\beta^L \bullet \tilde{\mathbf{B}}_\beta^L, \tilde{\mathbf{A}}_\beta^U \bullet \tilde{\mathbf{B}}_\beta^U\right\}, \max\left\{\tilde{\mathbf{A}}_\beta^L \bullet \tilde{\mathbf{B}}_\beta^L, \tilde{\mathbf{A}}_\beta^U \bullet \tilde{\mathbf{B}}_\beta^U\right\}\right],$$
where $\tilde{\mathbf{A}}_\alpha^L$, $\tilde{\mathbf{A}}_\alpha^U$, $\tilde{\mathbf{B}}_\alpha^L$ and $\tilde{\mathbf{B}}_\alpha^U$ are given in (7.31). We consider the family $\{M_\alpha^\bullet : 0 < \alpha \leq 1\}$ by taking
$$M_\alpha^\bullet = \bigcup_{\alpha \leq \beta \leq 1} M_\beta, \tag{8.6}$$
to define the inner product $\tilde{\mathbf{A}} \circledast_{DT}^{\star} \tilde{\mathbf{B}}$.
- We consider the family $\{M_\alpha^\bullet : 0 < \alpha \leq 1\}$ by directly taking
$$M_\alpha^\bullet = \left[\min\left\{\tilde{\mathbf{A}}_\alpha^L \bullet \tilde{\mathbf{B}}_\alpha^L, \tilde{\mathbf{A}}_\alpha^U \bullet \tilde{\mathbf{B}}_\alpha^U\right\}, \max\left\{\tilde{\mathbf{A}}_\alpha^L \bullet \tilde{\mathbf{B}}_\alpha^L, \tilde{\mathbf{A}}_\alpha^U \bullet \tilde{\mathbf{B}}_\alpha^U\right\}\right] \tag{8.7}$$
to define the inner product $\tilde{\mathbf{A}} \circledast_{DT}^{\dagger} \tilde{\mathbf{B}}$.

Using the expression in the decomposition theorem, for $\circledast_{DT} \in \{\circledast_{DT}^{\diamond}, \circledast_{DT}^{\star}, \circledast_{DT}^{\dagger}\}$, the membership function of $\tilde{\mathbf{A}} \circledast_{DT} \tilde{\mathbf{B}}$ is defined by

$$\xi_{\tilde{\mathbf{A}} \circledast_{DT} \tilde{\mathbf{B}}}(\mathbf{z}) = \sup_{0<\alpha\leq 1} \alpha \cdot \chi_{M_\alpha^\bullet}(\mathbf{z}), \tag{8.8}$$

where $M_\alpha^\bullet$ corresponds to the above three cases (8.5), (8.6) and (8.7). In what follows, we shall separately study those three different families $\{M_\alpha^\bullet : 0 < \alpha \leq 1\}$ given in (8.5), (8.6), and (8.7).

Example 8.1.4 Continued from Examples 2.2.4 and 7.2.4, we consider three families $\{M_\alpha^\bullet : 0 < \alpha \leq 1\}$ given below.

- We take
$$\begin{aligned}
M_\alpha^\bullet &= \tilde{\mathbf{A}}_\alpha \bullet \tilde{\mathbf{B}}_\alpha \\
&= \left\{(x_1, x_2) \bullet (y_1, y_2) : (x_1, x_2) \in [1+\alpha, 4-\alpha] \times [2+\alpha, 5-\alpha] \right.\\
&\qquad \left.\text{and } (y_1, y_2) \in [4+\alpha, 7-\alpha] \times [3+\alpha, 6-\alpha]\right\} \\
&= \left\{x_1y_1 + x_2y_2 : (x_1, x_2) \in [1+\alpha, 4-\alpha] \times [2+\alpha, 5-\alpha] \right.\\
&\qquad \left.\text{and } (y_1, y_2) \in [4+\alpha, 7-\alpha] \times [3+\alpha, 6-\alpha]\right\} \\
&= [(1+\alpha)(4+\alpha), (4-\alpha)(7-\alpha)] + [(2+\alpha)(3+\alpha), (5-\alpha)(6-\alpha)] \\
&= \left[10 + 10\alpha + \alpha^2, 58 - 22\alpha + \alpha^2\right]
\end{aligned}$$

for all $\alpha \in (0,1]$. The membership function of $\tilde{\mathbf{A}} \circledast^{\diamond}_{DT} \tilde{\mathbf{B}}$ is given by

$$\xi_{\tilde{\mathbf{A}} \circledast^{\diamond}_{DT} \tilde{\mathbf{B}}}(z_1, z_2) = \sup_{0<\alpha\leq 1} \alpha \cdot \chi_{M^{\bullet}_{\alpha}}(z_1, z_2).$$

- We take

$$M^{\bullet}_{\alpha} = \bigcup_{\alpha\leq\beta\leq 1} M_{\beta},$$

where M_{β} is a bounded closed interval given by

$$\begin{aligned}
M_{\beta} &= \left[\min\left\{\tilde{A}^L_{1\beta}\tilde{B}^L_{1\beta} + \tilde{A}^L_{2\beta}\tilde{B}^L_{2\beta}, \tilde{A}^U_{1\beta}\tilde{B}^U_{1\beta} + \tilde{A}^U_{2\beta}\tilde{B}^U_{2\beta}\right\},\right.\\
&\qquad \left.\max\left\{\tilde{A}^L_{1\beta}\tilde{B}^L_{1\beta} + \tilde{A}^L_{2\beta}\tilde{B}^L_{2\beta}, \tilde{A}^U_{1\beta}\tilde{B}^U_{1\beta} + \tilde{A}^U_{2\beta}\tilde{B}^U_{2\beta}\right\}\right]\\
&= [\min\{(1+\beta)(4+\beta) + (2+\beta)(3+\beta), (4-\beta)(7-\beta) + (5-\beta)(6-\beta)\},\\
&\qquad \max\{(1+\beta)(4+\beta) + (2+\beta)(3+\beta), (4-\beta)(7-\beta) + (5-\beta)(6-\beta)\}]\\
&= \left[\min\left\{10 + 10\beta + \beta^2, 58 - 22\beta + \beta^2\right\},\right.\\
&\qquad \left.\max\left\{10 + 10\beta + \beta^2, 58 - 22\beta + \beta^2\right\}\right]\\
&= \left[10 + 10\beta + \beta^2, 58 - 22\beta + \beta^2\right]
\end{aligned}$$

for all $\beta \in (0,1]$. Therefore, we obtain

$$\begin{aligned}
M^{\bullet}_{\alpha} &= \bigcup_{\alpha\leq\beta\leq 1} M_{\beta} = \bigcup_{\alpha\leq\beta\leq 1} \left[10 + 10\beta + \beta^2, 58 - 22\beta + \beta^2\right]\\
&= \left[10 + 10\alpha + \alpha^2, 58 - 22\alpha + \alpha^2\right].
\end{aligned}$$

The membership function of $\tilde{\mathbf{A}} \circledast^{\star}_{DT} \tilde{\mathbf{B}}$ is equal to the membership function of $\tilde{\mathbf{A}} \circledast^{\diamond}_{DT} \tilde{\mathbf{B}}$.

- We take

$$\begin{aligned}
M^{\bullet}_{\alpha} &= \left[\min\left\{\tilde{A}^L_{1\alpha}\tilde{B}^L_{1\alpha} + \tilde{A}^L_{2\alpha}\tilde{B}^L_{2\alpha}, \tilde{A}^U_{1\alpha}\tilde{B}^U_{1\alpha} + \tilde{A}^U_{2\alpha}\tilde{B}^U_{2\alpha}\right\},\right.\\
&\qquad \left.\max\left\{\tilde{A}^L_{1\alpha}\tilde{B}^L_{1\alpha} + \tilde{A}^L_{2\alpha}\tilde{B}^L_{2\alpha}, \tilde{A}^U_{1\alpha}\tilde{B}^U_{1\alpha} + \tilde{A}^U_{2\alpha}\tilde{B}^U_{2\alpha}\right\}\right].
\end{aligned}$$

Then, we obtain

$$M^{\bullet}_{\alpha} = \left[10 + 10\alpha + \alpha^2, 58 - 22\alpha + \alpha^2\right].$$

Therefore, the membership function of $\tilde{\mathbf{A}} \circledast^{\dagger}_{DT} \tilde{\mathbf{B}}$ is equal to the membership function of $\tilde{\mathbf{A}} \circledast^{\diamond}_{DT} \tilde{\mathbf{B}}$.

8.1.2.1 The Inner Product $\tilde{\mathbf{A}} \circledast^{\diamond}_{DT} \tilde{\mathbf{B}}$

We shall study the inner product $\tilde{\mathbf{A}} \circledast^{\diamond}_{DT} \tilde{\mathbf{B}}$ considering the family given in (8.5). Given any $\alpha \in (0,1]$, we have

$$\begin{aligned}
M^{\bullet}_{\alpha} &= \tilde{\mathbf{A}}_{\alpha} \bullet \tilde{\mathbf{B}}_{\alpha} = \left\{\mathbf{x} \bullet \mathbf{y} : \mathbf{x} \in \tilde{\mathbf{A}}_{\alpha} \text{ and } \mathbf{y} \in \tilde{\mathbf{B}}_{\alpha}\right\}\\
&= \left[\min_{(\mathbf{x},\mathbf{y})\in(\tilde{\mathbf{A}}_{\alpha},\tilde{\mathbf{B}}_{\alpha})} \mathbf{x} \bullet \mathbf{y}, \max_{(\mathbf{x},\mathbf{y})\in(\tilde{\mathbf{A}}_{\alpha},\tilde{\mathbf{B}}_{\alpha})} \mathbf{x} \bullet \mathbf{y}\right].
\end{aligned}$$

According to the expression in the decomposition theorem, the membership function of $\tilde{\mathbf{A}} \circledast^{\diamond}_{DT} \tilde{\mathbf{B}}$ is defined by

$$\xi_{\tilde{\mathbf{A}} \circledast^{\diamond}_{DT} \tilde{\mathbf{B}}}(z) = \sup_{0<\alpha\leq 1} \alpha \cdot \chi_{M^{\bullet}_{\alpha}}(z). \tag{8.9}$$

We have the following interesting results.

Theorem 8.1.5 *Let $\tilde{A}^{(i)}$ and $\tilde{B}^{(i)}$ be canonical fuzzy intervals for $i = 1, \ldots, n$, and let the family $\{M_\alpha^\bullet : 0 < \alpha \leq 1\}$be taken by $M_\alpha^\bullet = \tilde{\mathbf{A}}_\alpha \bullet \tilde{\mathbf{B}}_\alpha$. Then, for $\alpha \in [0,1]$, we have*

$$\left(\tilde{\mathbf{A}} \circledast_{DT}^{\diamond} \tilde{\mathbf{B}}\right)_\alpha = \left[\min_{(\mathbf{x},\mathbf{y})\in(\tilde{\mathbf{A}}_\alpha,\tilde{\mathbf{B}}_\alpha)} \mathbf{x}\bullet\mathbf{y},\ \max_{(\mathbf{x},\mathbf{y})\in(\tilde{\mathbf{A}}_\alpha,\tilde{\mathbf{B}}_\alpha)} \mathbf{x}\bullet\mathbf{y}\right].$$

When $\tilde{A}^{(i)}$ and $\tilde{B}^{(i)}$ are taken to be nonnegative canonical fuzzy intervals for $i = 1, \ldots, n$, we simply have

$$\left(\tilde{\mathbf{A}} \circledast_{DT}^{\diamond} \tilde{\mathbf{B}}\right)_\alpha = \left[\tilde{\mathbf{A}}_\alpha^L \bullet \tilde{\mathbf{B}}_\alpha^L, \tilde{\mathbf{A}}_\alpha^U \bullet \tilde{\mathbf{B}}_\alpha^U\right].$$

Proof. It is clear to see that $\{M_\alpha^\bullet : 0 < \alpha \leq 1\}$is a nested family in the sense of $M_\alpha^\bullet \subseteq M_\beta^\bullet$ for $\beta < \alpha$. Using the continuities regarding the canonical fuzzy intervals, we see that the family $\{M_\alpha^\bullet : 0 < \alpha \leq 1\}$ will continuously shrink when α increases on [0,1]. Therefore, for $\alpha \in (0,1]$, we have

$$M_\alpha^\bullet = \bigcap_{s=1}^{\infty} M_{\alpha_k}^\bullet \tag{8.10}$$

for $0 < \alpha_k \uparrow \alpha$ with $\alpha_k \in [0,1]$ for all k.

Next, we are going to show that

$$M_\alpha^\bullet = (\tilde{\mathbf{A}} \circledast_{DT}^{\diamond} \tilde{\mathbf{B}})_\alpha \text{ for } \alpha \in [0,1].$$

Given $\alpha \in (0,1]$ and any $\mathbf{z} \in M_\alpha^\bullet$, the expression (8.9) says that $\xi_{\tilde{\mathbf{A}}\circledast_{DT}^{\diamond}\tilde{\mathbf{B}}}(\mathbf{z}) \geq \alpha$, which implies $\mathbf{z} \in (\tilde{\mathbf{A}} \circledast_{DT}^{\diamond} \tilde{\mathbf{B}})_\alpha$ and proves the inclusion $M_\alpha^\bullet \subseteq (\tilde{\mathbf{A}} \circledast_{DT}^{\diamond} \tilde{\mathbf{B}})_\alpha$. To prove the other direction of inclusion, given any $\mathbf{z} \in (\tilde{\mathbf{A}} \circledast_{DT}^{\diamond} \tilde{\mathbf{B}})_\alpha$, it means that $\xi_{\tilde{\mathbf{A}}\circledast_{DT}^{\diamond}\tilde{\mathbf{B}}}(\mathbf{z}) \geq \alpha$. Let $\hat{\alpha} = \xi_{\tilde{\mathbf{A}}\circledast_{DT}^{\diamond}\tilde{\mathbf{B}}}(\mathbf{z})$. We consider the following cases.

- Assume that $\hat{\alpha} > \alpha$. Let $\epsilon = \hat{\alpha} - \alpha > 0$. By referring to (8.9), the concept of supremum says that there exists $\alpha_0 \in [0,1]$ satisfying $\mathbf{z} \in M_{\alpha_0}^\bullet$ and $\hat{\alpha} - \epsilon < \alpha_0$, which says that $\alpha < \alpha_0$. Therefore, we obtain $\mathbf{z} \in M_\alpha^\bullet$, since $M_{\alpha_0}^\bullet \subseteq M_\alpha^\bullet$ by the nestedness.
- Assume that $\hat{\alpha} = \alpha$. Since [0,1] is an interval with left endpoint 0, for any $\alpha \in (0,1]$, there exists a sequence $\{\alpha_k\}_{k=1}^\infty$ in [0,1] satisfying $0 < \alpha_k \uparrow \alpha$ with $\alpha_k \in [0,1]$ for all k. Let $\epsilon_k = \alpha - \alpha_k > 0$. By referring to (8.9), the concept of supremum says that there exists $\alpha_0 \in [0,1]$ satisfying $\mathbf{z} \in M_{\alpha_0}^\bullet$ and $\hat{\alpha} - \epsilon_k = \alpha - \epsilon_k < \alpha_0$, which implies $\alpha_0 > \alpha_k \in [0,1]$. The nestedness also says that $\mathbf{z} \in M_{\alpha_k}^\bullet$ for all k, i.e. $\mathbf{z} \in \bigcap_{k=1}^\infty M_{\alpha_k}^\bullet$. From (8.10), we obtain $\mathbf{z} \in M_\alpha^\bullet$.

The above two cases conclude that $(\tilde{\mathbf{A}} \circledast_{DT}^{\diamond} \tilde{\mathbf{B}})_\alpha \subseteq M_\alpha^\bullet$. Therefore, we obtain the equality

$$M_\alpha^\bullet = (\tilde{\mathbf{A}} \circledast_{DT}^{\diamond} \tilde{\mathbf{B}})_\alpha \text{ for } \alpha \in (0,1].$$

For the 0-level set, we also have

$$\begin{aligned}
\left(\tilde{\mathbf{A}} \circledast_{DT}^{\diamond} \tilde{\mathbf{B}}\right)_0 &= \text{cl}\left(\bigcup_{0<\alpha\leq 1\}} \left(\tilde{\mathbf{A}} \circledast_{DT}^{\diamond} \tilde{\mathbf{B}}\right)_\alpha\right) \text{ (referring to (2.5))}\\
&= \text{cl}\left(\bigcup_{0<\alpha\leq 1} M_\alpha^\bullet\right)\\
&= M_0^\bullet \text{ (using the nestedness and continuities in Definition 2.3.3).}
\end{aligned}$$

This completes the proof. ∎

Example 8.1.6 By referring to Example 8.1.2, Theorems 8.1.1 and 8.1.5 say that

$$\left(\tilde{\mathbf{A}} \circledast_{DT}^{\diamond} \tilde{\mathbf{B}}\right)_\alpha = \left(\tilde{\mathbf{A}} \circledast_{EP} \tilde{\mathbf{B}}\right)_\alpha = \left[10 + 10\alpha + \alpha^2, 58 - 22\alpha + 2\alpha^2\right]$$

for $\alpha \in [0,1]$. By referring to Example 8.1.3, we also have

$$\left(\tilde{\mathbf{A}} \circledast_{DT}^{\diamond} \tilde{\mathbf{B}}\right)_\alpha = \left(\tilde{\mathbf{A}} \circledast_{EP} \tilde{\mathbf{B}}\right)_\alpha = \left[6\alpha^2 - 3\alpha - 1, 11\alpha^2 - 31\alpha + 22\right]$$

for $\alpha \geq 2/3$.

8.1.2.2 The Inner Product $\tilde{\mathbf{A}} \circledast_{DT}^{\star} \tilde{\mathbf{B}}$

We shall study the inner product $\tilde{\mathbf{A}} \circledast_{DT}^{\star} \tilde{\mathbf{B}}$ considering the family given in (8.6). According to the expression in the decomposition theorem, the membership function of inner product $\tilde{\mathbf{A}} \circledast_{DT}^{\star} \tilde{\mathbf{B}}$ is defined by

$$\xi_{\tilde{\mathbf{A}} \circledast_{DT}^{\star} \tilde{\mathbf{B}}}(z) = \sup_{0<\alpha\leq 1} \alpha \cdot \chi_{M_\alpha^{\bullet}}(z). \tag{8.11}$$

We have the following interesting results.

Theorem 8.1.7 *Let $\tilde{A}^{(i)}$ and $\tilde{B}^{(i)}$ be canonical fuzzy intervals for $i = 1, \ldots, n$, and let the family $\{M_\alpha^{\bullet} : 0 < \alpha \leq 1\}$ be given by*

$$M_\alpha^{\bullet} = \left(\bigcup_{\alpha\leq\beta\leq 1} M_\beta\right),$$

where M_β is a bounded closed intervals given by

$$M_\beta = \left[\min\left\{\tilde{\mathbf{A}}_\beta^L \bullet \tilde{\mathbf{B}}_\beta^L, \tilde{\mathbf{A}}_\beta^U \bullet \tilde{\mathbf{B}}_\beta^U\right\}, \max\left\{\tilde{\mathbf{A}}_\beta^L \bullet \tilde{\mathbf{B}}_\beta^L, \tilde{\mathbf{A}}_\beta^U \bullet \tilde{\mathbf{B}}_\beta^U\right\}\right].$$

Then, for $\alpha \in [0,1]$, we have

$$\begin{aligned}
\left(\tilde{\mathbf{A}} \circledast_{DT}^{\star} \tilde{\mathbf{B}}\right)_\alpha &= M_\alpha^{\bullet} \\
&= \left[\min_{\alpha\leq\beta\leq 1} \min\left\{\tilde{\mathbf{A}}_\beta^L \bullet \tilde{\mathbf{B}}_\beta^L, \tilde{\mathbf{A}}_\beta^U \bullet \tilde{\mathbf{B}}_\beta^U\right\}, \max_{\alpha\leq\beta\leq 1} \max\left\{\tilde{\mathbf{A}}_\beta^L \bullet \tilde{\mathbf{B}}_\beta^L, \tilde{\mathbf{A}}_\beta^U \bullet \tilde{\mathbf{B}}_\beta^U\right\}\right] \\
&= \left[\min\left\{\min_{\alpha\leq\beta\leq 1} \tilde{\mathbf{A}}_\beta^L \bullet \tilde{\mathbf{B}}_\beta^L, \min_{\alpha\leq\beta\leq 1} \tilde{\mathbf{A}}_\beta^U \bullet \tilde{\mathbf{B}}_\beta^U\right\},\right. \\
&\qquad \left.\max\left\{\max_{\alpha\leq\beta\leq 1} \tilde{\mathbf{A}}_\beta^L \bullet \tilde{\mathbf{B}}_\beta^L, \max_{\alpha\leq\beta\leq 1} \tilde{\mathbf{A}}_\beta^U \bullet \tilde{\mathbf{B}}_\beta^U\right\}\right].
\end{aligned} \tag{8.12}$$

When $\tilde{A}^{(i)}$ and $\tilde{B}^{(i)}$ are taken to be nonnegative canonical fuzzy intervals for $i = 1, \ldots, n$, we simply have

$$\left(\tilde{\mathbf{A}} \circledast_{DT}^{\star} \tilde{\mathbf{B}}\right)_\alpha = \left[\tilde{\mathbf{A}}_\alpha^L \bullet \tilde{\mathbf{B}}_\alpha^L, \tilde{\mathbf{A}}_\alpha^U \bullet \tilde{\mathbf{B}}_\alpha^U\right].$$

Proof. We are going to show that $M_\alpha^{\bullet} = (\tilde{\mathbf{A}} \circledast_{DT}^{\star} \tilde{\mathbf{B}})_\alpha$ for $\alpha \in [0,1]$. By using (8.11) and the proof of Theorem 8.1.5, we can similarly obtain the inclusion $M_\alpha^{\bullet} \subseteq (\tilde{\mathbf{A}} \circledast_{DT}^{\star} \tilde{\mathbf{B}})_\alpha$.

On the other hand, we can see that $\{M_\alpha^{\bullet} : 0 < \alpha \leq 1\}$ is a nested family in the sense of $M_\alpha^{\bullet} \subseteq M_\beta^{\bullet}$ for $\beta < \alpha$. We define two functions ζ^L and ζ^U on [0,1] as follows:

$$\zeta^L(\beta) = \min\left\{\tilde{\mathbf{A}}_\beta^L \bullet \tilde{\mathbf{B}}_\beta^L, \tilde{\mathbf{A}}_\beta^U \bullet \tilde{\mathbf{B}}_\beta^U\right\} \text{ and } \zeta^U(\beta) = \max\left\{\tilde{\mathbf{A}}_\beta^L \bullet \tilde{\mathbf{B}}_\beta^L, \tilde{\mathbf{A}}_\beta^U \bullet \tilde{\mathbf{B}}_\beta^U\right\}.$$

It is clear to see that the functions ζ^L and ζ^U are continuous on [0,1] by the continuities regarding the canonical fuzzy intervals. We also see that $M_\beta = [\zeta^L(\beta), \zeta^U(\beta)]$. The continuities say that the family $\{M_\alpha^\bullet : \alpha \in [0,1] \text{ for } \alpha > 0\}$will continuously shrink when α increases on [0,1]. For $\alpha \in (0,1]$, it follows that

$$M_\alpha^\bullet = \bigcap_{k=1}^{\infty} M_{\alpha_k}^\bullet$$

for $0 < \alpha_k \uparrow \alpha$ with $\alpha_k \in [0,1]$ for all k. Using the proof of Theorem 8.1.5, we can similarly obtain the inclusion $(\tilde{\mathbf{A}} \circledast_{DT}^{\star} \tilde{\mathbf{B}})_\alpha \subseteq M_\alpha^\bullet$. Therefore, we have $M_\alpha^\bullet = (\tilde{\mathbf{A}} \circledast_{DT}^{\star} \tilde{\mathbf{B}})_\alpha$ for $\alpha \in (0,1]$. Moreover, for $\alpha \in (0,1]$, we have

$$\begin{aligned}
\left(\tilde{\mathbf{A}} \circledast_{DT}^{\star} \tilde{\mathbf{B}}\right)_\alpha = M_\alpha^\bullet = \bigcup_{\alpha\leq\beta\leq 1} M_\beta &= \left[\min_{\alpha\leq\beta\leq 1} \zeta^L(\beta), \max_{\alpha\leq\beta\leq 1} \zeta^U(\beta)\right] \\
&= \left[\min_{\alpha\leq\beta\leq 1} \min\left\{\tilde{\mathbf{A}}_\beta^L \bullet \tilde{\mathbf{B}}_\beta^L, \tilde{\mathbf{A}}_\beta^U \bullet \tilde{\mathbf{B}}_\beta^U\right\}, \max_{\alpha\leq\beta\leq 1} \min\left\{\tilde{\mathbf{A}}_\beta^L \bullet \tilde{\mathbf{B}}_\beta^L, \tilde{\mathbf{A}}_\beta^U \bullet \tilde{\mathbf{B}}_\beta^U\right\}\right].
\end{aligned}$$

For the 0-level set, we also have

$$\begin{aligned}
\left(\tilde{\mathbf{A}} \circledast_{DT}^{\star} \tilde{\mathbf{B}}\right)_0 &= \text{cl}\left(\bigcup_{0<\alpha\leq 1} \left(\tilde{\mathbf{A}} \circledast_{DT}^{\star} \tilde{\mathbf{B}}\right)_\alpha\right) \quad \text{(referring to (2.5))} \\
&= \text{cl}\left(\bigcup_{0<\alpha\leq 1} M_\alpha^\bullet\right) = \text{cl}\left(\bigcup_{0<\alpha\leq 1} \left[\min_{\alpha\leq\beta\leq 1} \zeta^L(\beta), \max_{\alpha\leq\beta\leq 1} \zeta^U(\beta)\right]\right) \\
&= \left[\min_{\{\beta\in[0,1]:\beta\geq 0\}} \zeta^L(\beta), \max_{\{\beta\in[0,1]:\beta\geq 0\}} \zeta^U(\beta)\right] \\
&\quad \text{(using the nestedness and the continuities of functions } \zeta^L \text{ and } \zeta^U) \\
&= \left[\min_{\{\beta\in[0,1]:\beta\geq 0\}} \min\left\{\tilde{\mathbf{A}}_\beta^L \bullet \tilde{\mathbf{B}}_\beta^L, \tilde{\mathbf{A}}_\beta^U \bullet \tilde{\mathbf{B}}_\beta^U\right\}, \max_{\{\beta\in[0,1]:\beta\geq 0\}} \min\left\{\tilde{\mathbf{A}}_\beta^L \bullet \tilde{\mathbf{B}}_\beta^L, \tilde{\mathbf{A}}_\beta^U \bullet \tilde{\mathbf{B}}_\beta^U\right\}\right].
\end{aligned}$$

Let $\eta^L(\beta) = \tilde{\mathbf{A}}_\beta^L \bullet \tilde{\mathbf{B}}_\beta^L$ and $\eta^L(\beta) = \tilde{\mathbf{A}}_\beta^U \bullet \tilde{\mathbf{B}}_\beta^U$. From the proof of Theorem 5.5.6, we have

$$\min_{\alpha\leq\beta\leq 1} \min\left\{\eta^L(\beta), \eta^U(\beta)\right\} = \min\left\{\min_{\alpha\leq\beta\leq 1} \eta^L(\beta), \min_{\alpha\leq\beta\leq 1} \eta^U(\beta),\right\}$$

and

$$\max_{\alpha\leq\beta\leq 1} \max\left\{\eta^L(\beta), \eta^U(\beta)\right\} = \max\left\{\max_{\alpha\leq\beta\leq 1} \eta^L(\beta), \max_{\alpha\leq\beta\leq 1} \eta^U(\beta),\right\}.$$

Therefore, we obtain (8.12). This completes the proof. ■

Example 8.1.8 Continued from Example 8.1.2, we can obtain

$$\min\left\{\tilde{\mathbf{A}}_\beta^L \bullet \tilde{\mathbf{B}}_\beta^L, \tilde{\mathbf{A}}_\beta^U \bullet \tilde{\mathbf{B}}_\beta^U\right\} = 10 + 10\beta + \beta^2$$

and

$$\max\left\{\tilde{\mathbf{A}}_\beta^L \bullet \tilde{\mathbf{B}}_\beta^L, \tilde{\mathbf{A}}_\beta^U \bullet \tilde{\mathbf{B}}_\beta^U\right\} = 58 - 22\beta + 2\beta^2.$$

Using (8.12), we have

$$\begin{aligned}
\left(\tilde{\mathbf{A}} \circledast_{DT}^{\star} \tilde{\mathbf{B}}\right)_\alpha &= \left[\min_{\{\beta\in[0,0.8]:\beta\geq\alpha\}} \left(10 + 10\beta + \beta^2\right), \max_{\{\beta\in[0,0.8]:\beta\geq\alpha\}} \left(58 - 22\beta + 2\beta^2\right)\right] \\
&= \left[10 + 10\alpha + \alpha^2, 58 - 22\alpha + 2\alpha^2\right]
\end{aligned}$$

for $\alpha \in [0,1]$.

Example 8.1.9 Continued from Example 8.1.3, we obtain

$$\min_{\alpha\le\beta\le 1}\left(\tilde{A}^L_{1\beta}\tilde{B}^L_{1\beta}+\tilde{A}^L_{2\beta}\tilde{B}^L_{2\beta}+\tilde{A}^L_{3\beta}\tilde{B}^L_{3\beta}\right)$$
$$=\min_{\alpha\le\beta\le 1}\left[(2\beta-1)(\beta-1)+(\beta-2)^2+\beta(2\beta-1)\right]=\min_{\alpha\le\beta\le 1}\left(5-8\beta+5\beta^2\right)$$
$$=\begin{cases}5-8\alpha+5\alpha^2 & \text{if } \alpha\ge 0.8\\ 5-8\cdot 0.8+5\cdot 0.8^2 & \text{if } \alpha<0.8\end{cases}=\begin{cases}5-8\alpha+5\alpha^2 & \text{if } \alpha\ge 0.8\\ 1.8 & \text{if } \alpha<0.8\end{cases}$$

and

$$\min_{\alpha\le\beta\le 1}\left(\tilde{A}^U_{1\beta}\tilde{B}^U_{1\beta}+\tilde{A}^U_{2\beta}\tilde{B}^U_{2\beta}+\tilde{A}^U_{3\beta}\tilde{B}^U_{3\beta}\right)$$
$$=\min_{\alpha\le\beta\le 1}\left[(3-2\beta)(3-3\beta)+(2-3\beta)(1-2\beta)+(3-2\beta)^2\right]$$
$$=\min_{\alpha\le\beta\le 1}\left(20-34\beta+16\beta^2\right)=20-34+16=2.$$

Therefore, we have

$$\min\left\{\min_{\alpha\le\beta\le 1}\left(\tilde{A}^L_{1\beta}\tilde{B}^L_{1\beta}+\tilde{A}^L_{2\beta}\tilde{B}^L_{2\beta}+\tilde{A}^L_{3\beta}\tilde{B}^L_{3\beta}\right),\min_{\alpha\le\beta\le 1}\left(\tilde{A}^U_{1\beta}\tilde{B}^U_{1\beta}+\tilde{A}^U_{2\beta}\tilde{B}^U_{2\beta}+\tilde{A}^U_{3\beta}\tilde{B}^U_{3\beta}\right)\right\}$$
$$=\begin{cases}\min\{2,5-8\alpha+5\alpha^2\} & \text{if } \alpha\ge 0.8\\ \min\{2,1.8\} & \text{if } \alpha<0.8\end{cases}=\begin{cases}5-8\alpha+5\alpha^2 & \text{if } \alpha\ge 0.8\\ 1.8 & \text{if } \alpha<0.8\end{cases}$$

Now, we can also obtain

$$\max_{\alpha\le\beta\le 1}\left(\tilde{A}^L_{1\beta}\tilde{B}^L_{1\beta}+\tilde{A}^L_{2\beta}\tilde{B}^L_{2\beta}+\tilde{A}^L_{3\beta}\tilde{B}^L_{3\beta}\right)=\max_{\alpha\le\beta\le 1}\left(5-8\beta+5\beta^2\right)$$
$$=\begin{cases}5-8+5 & \text{if } \alpha\ge 0.8\\ 5-8\alpha+5\alpha^2 & \text{if } \alpha<0.8\end{cases}=\begin{cases}2 & \text{if } \alpha\ge 0.8\\ 5-8\alpha+5\alpha^2 & \text{if } \alpha<0.8\end{cases}$$

and

$$\max_{\alpha\le\beta\le 1}\left(\tilde{A}^U_{1\beta}\tilde{B}^U_{1\beta}+\tilde{A}^U_{2\beta}\tilde{B}^U_{2\beta}+\tilde{A}^U_{3\beta}\tilde{B}^U_{3\beta}\right)=\max_{\alpha\le\beta\le 1}\left(20-34\beta+16\beta^2\right)=20-34\alpha+16\alpha^2.$$

Therefore, we have

$$\max\left\{\max_{\alpha\le\beta\le 1}\left(\tilde{A}^L_{1\beta}\tilde{B}^L_{1\beta}+\tilde{A}^L_{2\beta}\tilde{B}^L_{2\beta}+\tilde{A}^L_{3\beta}\tilde{B}^L_{3\beta}\right),\max_{\alpha\le\beta\le 1}\left(\tilde{A}^U_{1\beta}\tilde{B}^U_{1\beta}+\tilde{A}^U_{2\beta}\tilde{B}^U_{2\beta}+\tilde{A}^U_{3\beta}\tilde{B}^U_{3\beta}\right)\right\}$$
$$=\begin{cases}\max\{2,20-34\alpha+16\alpha^2\} & \text{if } \alpha\ge 0.8\\ \max\{5-8\alpha+5\alpha^2,20-34\alpha+16\alpha^2\} & \text{if } \alpha<0.8\end{cases}$$
$$=\left.\begin{cases}20-34\alpha+16\alpha^2 & \text{if } \alpha\ge 0.8\\ 20-34\alpha+16\alpha^2 & \text{if } \alpha<0.8\end{cases}\right\}=20-34\alpha+16\alpha^2.$$

Using (8.12), we obtain the α-level sets

$$\left(\tilde{\mathbf{A}}\circledast^{\star}_{DT}\tilde{\mathbf{B}}\right)_\alpha=\begin{cases}\left[5-8\alpha+5\alpha^2,20-34\alpha+16\alpha^2\right] & \text{if } \alpha\ge 0.8\\ \left[1.8,20-34\alpha+16\alpha^2\right] & \text{if } \alpha<0.8.\end{cases}$$

8.1.2.3 The Inner Product $\tilde{\mathbf{A}}\circledast^{\dagger}_{DT}\tilde{\mathbf{B}}$

We shall study the inner product $\tilde{\mathbf{A}}\circledast^{\dagger}_{DT}\tilde{\mathbf{B}}$ considering the family given in (8.7). According to the expression in the decomposition theorem, the membership function of inner product $\tilde{\mathbf{A}}\circledast^{\dagger}_{DT}\tilde{\mathbf{B}}$ is defined by

$$\xi_{\tilde{\mathbf{A}}\circledast^{\dagger}_{DT}\tilde{\mathbf{B}}}(z)=\sup_{0<\alpha\le 1}\alpha\cdot\chi_{M^{\bullet}_{\alpha}}(z). \tag{8.13}$$

Theorem 8.1.10 *Let $\tilde{A}^{(i)}$ and $\tilde{B}^{(i)}$ be canonical fuzzy intervals for $i = 1, \ldots, n$. The family $\{M_\alpha : 0 < \alpha \leq 1\}$ is given by*

$$M_\alpha^\bullet = \left[\min\left\{\tilde{\mathbf{A}}_\alpha^L \bullet \tilde{\mathbf{B}}_\alpha^L, \tilde{\mathbf{A}}_\alpha^U \bullet \tilde{\mathbf{B}}_\alpha^U\right\}, \max\left\{\tilde{\mathbf{A}}_\alpha^L \bullet \tilde{\mathbf{B}}_\alpha^L, \tilde{\mathbf{A}}_\alpha^U \bullet \tilde{\mathbf{B}}_\alpha^U\right\}\right].$$

Then, for $\alpha \in [0,1]$, we have

$$\begin{aligned}\left(\tilde{\mathbf{A}} \circledast_{DT}^{\dagger} \tilde{\mathbf{B}}\right)_\alpha &= \bigcup_{\alpha \leq \beta \leq 1} M_\beta^\bullet \\ &= \left[\min_{\alpha \leq \beta \leq 1} \min\left\{\tilde{\mathbf{A}}_\beta^L \bullet \tilde{\mathbf{B}}_\beta^L, \tilde{\mathbf{A}}_\beta^U \bullet \tilde{\mathbf{B}}_\beta^U\right\}, \max_{\alpha \leq \beta \leq 1} \max\left\{\tilde{\mathbf{A}}_\beta^L \bullet \tilde{\mathbf{B}}_\beta^L, \tilde{\mathbf{A}}_\beta^U \bullet \tilde{\mathbf{B}}_\beta^U\right\}\right].\end{aligned} \tag{8.14}$$

When $\tilde{A}^{(i)}$ and $\tilde{B}^{(i)}$ are taken to be nonnegative canonical fuzzy intervals for $i = 1, \ldots, n$, we have

$$\left(\tilde{\mathbf{A}} \circledast_{DT}^{\dagger} \tilde{\mathbf{B}}\right)_\alpha = \left[\tilde{\mathbf{A}}_\alpha^L \bullet \tilde{\mathbf{B}}_\alpha^L, \tilde{\mathbf{A}}_\alpha^U \bullet \tilde{\mathbf{B}}_\alpha^U\right].$$

Proof. We are going to show

$$\left(\tilde{\mathbf{A}} \circledast_{DT}^{\dagger} \tilde{\mathbf{B}}\right)_\alpha = \bigcup_{\alpha \leq \beta \leq 1} M_\beta^\bullet \text{ for } \alpha \in (0,1]. \tag{8.15}$$

Let

$$\zeta^L(\alpha) = \min_{\alpha \leq \beta \leq 1} \min\left\{\tilde{\mathbf{A}}_\beta^L \bullet \tilde{\mathbf{B}}_\beta^L, \tilde{\mathbf{A}}_\beta^U \bullet \tilde{\mathbf{B}}_\beta^U\right\} \text{ and } \zeta^U(\alpha) = \max_{\alpha \leq \beta \leq 1} \max\left\{\tilde{\mathbf{A}}_\beta^L \bullet \tilde{\mathbf{B}}_\beta^L, \tilde{\mathbf{A}}_\beta^U \bullet \tilde{\mathbf{B}}_\beta^U\right\}.$$

Then $M_\alpha^\bullet = [\zeta^L(\alpha), \zeta^U(\alpha)]$. The continuities regarding the canonical fuzzy intervals show that the functions ζ^L and ζ^U are continuous on [0,1]. Using Proposition 1.3.9, given any fixed $x \in \mathbb{R}$, the following function

$$\zeta(\alpha) = \begin{cases} 0, & \text{if } \alpha = 0 \\ \alpha \cdot \chi_{M_\alpha}(x), & \text{if } \alpha \in (0,1] \end{cases}$$

is upper semi-continuous on [0,1].

For $\alpha > 0$, given any $\mathbf{z} \in (\tilde{\mathbf{A}} \circledast_{DT}^{\dagger} \tilde{\mathbf{B}})_\alpha$ with $\mathbf{z} \notin M_\beta^\bullet$ for all $\beta \in [0,1]$ with $\beta \geq \alpha$, we see that $\beta \cdot \chi_{M_\beta^\bullet}(\mathbf{z}) < \alpha$ for all $\beta \in [0,1]$. Since [0,1] is a closed and bounded set and $\zeta(\beta) = \beta \cdot \chi_{M_\beta^\bullet}(\mathbf{z})$ is upper semi-continuous on [0,1] as described above, Proposition 1.4.4 says that the supremum of the function ζ is attained. Using (8.13), we have

$$\xi_{\tilde{\mathbf{A}} \circledast_{DT}^{\dagger} \tilde{\mathbf{B}}}(\mathbf{z}) = \sup_{\beta \in [0,1]} \zeta(\beta) = \sup_{\beta \in [0,1]} \beta \cdot \chi_{M_\beta^\bullet}(\mathbf{z}) = \max_{\beta \in [0,1]} \beta \cdot \chi_{M_\beta^\bullet}(\mathbf{z}) = \beta^* \cdot \chi_{M_{\beta^*}^\bullet}(\mathbf{z}) < \alpha$$

for some $\beta^* \in [0,1]$, which shows that $\mathbf{z} \notin (\tilde{\mathbf{A}} \circledast_{DT}^{\dagger} \tilde{\mathbf{B}})_\alpha$. This contradiction says that there exists $\beta_0 \in [0,1]$ with $\beta_0 \geq \alpha$ satisfying $\mathbf{z} \in M_{\beta_0}^\bullet$. Therefore, we have the following inclusion

$$\left(\tilde{\mathbf{A}} \circledast_{DT}^{\dagger} \tilde{\mathbf{B}}\right)_\alpha \subseteq \bigcup_{\alpha \leq \beta \leq 1} M_\beta^\bullet.$$

On the other hand, the following inclusion

$$\begin{aligned}\bigcup_{\alpha \leq \beta \leq 1} M_\beta^\bullet &\subseteq \left\{\mathbf{z} \in \mathbb{R}^m : \sup_{\beta \in [0,1]} \beta \cdot \chi_{M_\beta^\bullet}(\mathbf{z}) \geq \alpha\right\} \\ &= \left\{\mathbf{z} \in \mathbb{R}^m : \xi_{\tilde{\mathbf{A}} \circledast_{DT}^{\dagger} \tilde{\mathbf{B}}}(\mathbf{z}) \geq \alpha\right\} = \left(\tilde{\mathbf{A}} \circledast_{DT}^{\dagger} \tilde{\mathbf{B}}\right)_\alpha\end{aligned}$$

is obvious. This shows the equality (8.15). Using the continuities regarding the canonical fuzzy intervals, we can also obtain the equality (8.14).

For the 0-level set, we have

$$\left(\tilde{\mathbf{A}} \circledast^{\dagger}_{DT} \tilde{\mathbf{B}}\right)_0 = \mathrm{cl}\left(\bigcup_{\{\alpha\in[0,1]^{(\dagger DT)}_{\circledast}:\alpha>0\}} \left(\tilde{\mathbf{A}} \circledast^{\dagger}_{DT} \tilde{\mathbf{B}}\right)_\alpha\right) \text{ (referring to (2.5))}$$

$$= \left[\min_{\{\beta\in[0,1]:\beta\geq 0\}} \min\left\{\tilde{\mathbf{A}}^L_\beta \bullet \tilde{\mathbf{B}}^L_\beta, \tilde{\mathbf{A}}^U_\beta \bullet \tilde{\mathbf{B}}^U_\beta\right\}, \max_{\{\beta\in[0,1]:\beta\geq 0\}} \min\left\{\tilde{\mathbf{A}}^L_\beta \bullet \tilde{\mathbf{B}}^L_\beta, \tilde{\mathbf{A}}^U_\beta \bullet \tilde{\mathbf{B}}^U_\beta\right\}\right]$$

(using the nestedness, continuities and the equality (8.14)).

This completes the proof. ∎

Example 8.1.11 By referring to Example 8.1.8, Theorems 8.1.7 and 8.1.10 say that

$$\left(\tilde{\mathbf{A}} \circledast^{\dagger}_{DT} \tilde{\mathbf{B}}\right)_\alpha = \left(\tilde{\mathbf{A}} \circledast^{\star}_{DT} \tilde{\mathbf{B}}\right)_\alpha = \left[10 + 10\alpha + \alpha^2, 58 - 22\alpha + 2\alpha^2\right]$$

for $\alpha \in [0, 0.8]$. Moreover, we have $\left(\tilde{\mathbf{A}} \circledast^{\dagger}_{DT} \tilde{\mathbf{B}}\right)_\alpha = \emptyset$ for $\alpha \notin [0, 0.8]$. By referring to Example 8.1.9, we also have

$$\left(\tilde{\mathbf{A}} \circledast^{\dagger}_{DT} \tilde{\mathbf{B}}\right)_\alpha = \left(\tilde{\mathbf{A}} \circledast^{\star}_{DT} \tilde{\mathbf{B}}\right)_\alpha = \begin{cases} \left[5 - 8\alpha + 5\alpha^2, 20 - 34\alpha + 16\alpha^2\right] & \text{if } \alpha \geq 0.8 \\ \left[1.8, 20 - 34\alpha + 16\alpha^2\right] & \text{if } \alpha < 0.8. \end{cases}$$

8.1.3 The Equivalences and Fuzziness

The equivalences among $\tilde{\mathbf{A}} \circledast_{EP} \tilde{\mathbf{B}}$ and $\tilde{\mathbf{A}} \circledast_{DT} \tilde{\mathbf{B}}$ for $\circledast_{DT} \in \{\circledast^{\diamond}_{DT}, \circledast^{\star}_{DT}, \circledast^{\dagger}_{DT}\}$ will be presented below.

Theorem 8.1.12 *Let $\tilde{A}^{(i)}$ and $\tilde{B}^{(i)}$ be canonical fuzzy intervals for $i = 1, \ldots, n$. Suppose that the different inner products $\tilde{\mathbf{A}} \circledast_{EP} \tilde{\mathbf{B}}$ and $\tilde{\mathbf{A}} \circledast^{\diamond}_{DT} \tilde{\mathbf{B}}$ are obtained from Theorems 8.1.1 and 8.1.5, respectively. Then*

$$\tilde{\mathbf{A}} \circledast_{EP} \tilde{\mathbf{B}} = \tilde{\mathbf{A}} \circledast^{\diamond}_{DT} \tilde{\mathbf{B}}.$$

Moreover, for $\alpha \in [0,1]$, we have

$$\left(\tilde{\mathbf{A}} \circledast_{EP} \tilde{\mathbf{B}}\right)_\alpha = \left(\tilde{\mathbf{A}} \circledast^{\diamond}_{DT} \tilde{\mathbf{B}}\right)_\alpha = \left[\min_{(\mathbf{x},\mathbf{y})\in(\tilde{\mathbf{A}}_\alpha,\tilde{\mathbf{B}}_\alpha)} \mathbf{x} \bullet \mathbf{y}, \max_{(\mathbf{x},\mathbf{y})\in(\tilde{\mathbf{A}}_\alpha,\tilde{\mathbf{B}}_\alpha)} \mathbf{x} \bullet \mathbf{y}\right].$$

Theorem 8.1.13 *Let $\tilde{A}^{(i)}$ and $\tilde{B}^{(i)}$ be canonical fuzzy intervals for $i = 1, \ldots, n$. Suppose that the different inner products $\tilde{\mathbf{A}} \circledast^{\star}_{DT} \tilde{\mathbf{B}}$ and $\tilde{\mathbf{A}} \circledast^{\dagger}_{DT} \tilde{\mathbf{B}}$ are obtained from Theorems 8.1.7 and 8.1.10, respectively. Then*

$$\tilde{\mathbf{A}} \circledast^{\star}_{DT} \tilde{\mathbf{B}} = \tilde{\mathbf{A}} \circledast^{\dagger}_{DT} \tilde{\mathbf{B}}.$$

Moreover, for $\alpha \in [0,1]$, we have

$$\left(\tilde{\mathbf{A}} \circledast^{\star}_{DT} \tilde{\mathbf{B}}\right)_\alpha = \left(\tilde{\mathbf{A}} \circledast^{\dagger}_{DT} \tilde{\mathbf{B}}\right)_\alpha$$

$$= \left[\min_{\alpha\leq\beta\leq 1} \min\left\{\tilde{\mathbf{A}}^L_\beta \bullet \tilde{\mathbf{B}}^L_\beta, \tilde{\mathbf{A}}^U_\beta \bullet \tilde{\mathbf{B}}^U_\beta\right\}, \max_{\alpha\leq\beta\leq 1} \max\left\{\tilde{\mathbf{A}}^L_\beta \bullet \tilde{\mathbf{B}}^L_\beta, \tilde{\mathbf{A}}^U_\beta \bullet \tilde{\mathbf{B}}^U_\beta\right\}\right].$$

Theorem 8.1.14 *Let $\tilde{A}^{(i)}$ and $\tilde{B}^{(i)}$ be nonnegative canonical fuzzy intervals for $i = 1, \ldots, n$. Suppose that the different inner products $\tilde{\mathbf{A}} \circledast_{EP} \tilde{\mathbf{B}}$, $\tilde{\mathbf{A}} \circledast_{DT}^{\diamond} \tilde{\mathbf{B}}$, $\tilde{\mathbf{A}} \circledast_{DT}^{\star} \tilde{\mathbf{B}}$ and $\tilde{\mathbf{A}} \circledast_{DT}^{\dagger} \tilde{\mathbf{B}}$ are obtained from Theorems 8.1.1, 8.1.5, 8.1.7 and 8.1.10, respectively. Then*

$$\tilde{\mathbf{A}} \circledast_{EP} \tilde{\mathbf{B}} = \tilde{\mathbf{A}} \circledast_{DT}^{\diamond} \tilde{\mathbf{B}} = \tilde{\mathbf{A}} \circledast_{DT}^{\star} \tilde{\mathbf{B}} = \tilde{\mathbf{A}} \circledast_{DT}^{\dagger} \tilde{\mathbf{B}}.$$

Moreover, for $\alpha \in [0,1]$, we have

$$\left(\tilde{\mathbf{A}} \circledast_{EP} \tilde{\mathbf{B}}\right)_\alpha = \left(\tilde{\mathbf{A}} \circledast_{DT}^{\diamond} \tilde{\mathbf{B}}\right)_\alpha = \left(\tilde{\mathbf{A}} \circledast_{DT}^{\star} \tilde{\mathbf{B}}\right)_\alpha = \left(\tilde{\mathbf{A}} \circledast_{DT}^{\dagger} \tilde{\mathbf{B}}\right)_\alpha = \left[\tilde{\mathbf{A}}_\alpha^L \bullet \tilde{\mathbf{B}}_\alpha^L, \tilde{\mathbf{A}}_\alpha^U \bullet \tilde{\mathbf{B}}_\alpha^U\right].$$

The equivalence between $\tilde{\mathbf{A}} \circledast_{DT}^{\diamond} \tilde{\mathbf{B}}$ and $\tilde{\mathbf{A}} \circledast_{DT}^{\star} \tilde{\mathbf{B}}$ cannot be guaranteed. However, based on the α-level sets, we can compare their fuzziness by referring to Definition 6.3.4.

Theorem 8.1.15 *Let $\tilde{A}^{(i)}$ and $\tilde{B}^{(i)}$ be canonical fuzzy intervals for $i = 1, \ldots, n$. Suppose that $\tilde{\mathbf{A}} \circledast_{DT}^{\diamond} \tilde{\mathbf{B}}$ and $\tilde{\mathbf{A}} \circledast_{DT}^{\star} \tilde{\mathbf{B}}$ are obtained from Theorems 8.1.5 and 8.1.7, respectively. Then $\tilde{\mathbf{A}} \circledast_{DT}^{\diamond} \tilde{\mathbf{B}}$ is fuzzier than $\tilde{\mathbf{A}} \circledast_{DT}^{\star} \tilde{\mathbf{B}}$.*

Proof. Given any $\alpha \in (0,1]$, we see that

$$\min_{(\mathbf{x},\mathbf{y}) \in (\tilde{\mathbf{A}}_\alpha, \tilde{\mathbf{B}}_\alpha)} \mathbf{x} \bullet \mathbf{y} \le \min_{\alpha \le \beta \le 1} \min_{(\mathbf{x},\mathbf{y}) \in (\tilde{\mathbf{A}}_\beta, \tilde{\mathbf{B}}_\beta)} \mathbf{x} \bullet \mathbf{y} \le \min_{\alpha \le \beta \le 1} \min \left\{\tilde{\mathbf{A}}_\beta^L \bullet \tilde{\mathbf{B}}_\beta^L, \tilde{\mathbf{A}}_\beta^U \bullet \tilde{\mathbf{B}}_\beta^U\right\}$$

and

$$\max_{(\mathbf{x},\mathbf{y}) \in (\tilde{\mathbf{A}}_\alpha, \tilde{\mathbf{B}}_\alpha)} \mathbf{x} \bullet \mathbf{y} \ge \max_{\alpha \le \beta \le 1} \max_{(\mathbf{x},\mathbf{y}) \in (\tilde{\mathbf{A}}_\beta, \tilde{\mathbf{B}}_\beta)} \mathbf{x} \bullet \mathbf{y} \ge \max_{\alpha \le \beta \le 1} \max \left\{\tilde{\mathbf{A}}_\beta^L \bullet \tilde{\mathbf{B}}_\beta^L, \tilde{\mathbf{A}}_\beta^U \bullet \tilde{\mathbf{B}}_\beta^U\right\}.$$

From Theorems 8.1.12 and 8.1.13, we obtain the following inclusion

$$\left(\tilde{\mathbf{A}} \circledast_{DT}^{\star} \tilde{\mathbf{B}}\right)_\alpha \subseteq \left(\tilde{\mathbf{A}} \circledast_{DT}^{\diamond} \tilde{\mathbf{B}}\right)_\alpha \text{ for } \alpha \in (0,1].$$

This shows that $\tilde{\mathbf{A}} \circledast_{DT}^{\diamond} \tilde{\mathbf{B}}$ is fuzzier than $\tilde{\mathbf{A}} \circledast_{DT}^{\star} \tilde{\mathbf{B}}$, and the proof is complete. ■

8.2 The Second Type of Inner Product

The first type of inner product is directly based on the inner product of real vectors. Now, the second type of inner product will be based on the form of conventional inner product. First of all, we recall the addition and multiplication of fuzzy intervals.

Let $\tilde{A}$ and $\tilde{B}$ be two fuzzy intervals with membership functions $\xi_{\tilde{A}}$ and $\xi_{\tilde{B}}$, respectively. According to the extension principle, the membership function of multiplication $\tilde{A} \otimes_{EP} \tilde{B}$ is defined by

$$\xi_{\tilde{A} \otimes_{EP} \tilde{B}}(z) = \sup_{\{(x,y): z = x \cdot y\}} \min \left\{\xi_{\tilde{A}}(x), \xi_{\tilde{B}}(y)\right\}$$

for each $z \in \mathbb{R}$.

Let $\tilde{C}^{(1)}, \ldots, \tilde{C}^{(n)}$ be fuzzy intervals with membership functions $\xi_{\tilde{C}^{(1)}}, \ldots, \xi_{\tilde{C}^{(n)}}$, respectively. According to the extension principle, the membership function of addition

$$\tilde{C}^{(1)} \oplus_{EP} \cdots \oplus_{EP} \tilde{C}^{(n)}$$

is defined by

$$\xi_{\tilde{C}^{(1)} \oplus_{EP} \cdots \oplus_{EP} \tilde{C}^{(n)}}(z) = \sup_{\{(x_1, \ldots, x_n): z = x_1 + \cdots + x_n\}} \min \left\{\xi_{\tilde{C}^{(1)}}(x_1), \ldots, \xi_{\tilde{C}^{(n)}}(x_n)\right\} \tag{8.16}$$

for each $z \in \mathbb{R}$.

By referring to (8.5), (8.6), and (8.7), we can define the multiplication of $\tilde{A}$ and $\tilde{B}$ according to the expression in the decomposition theorem by considering three different families

$$\{M_\alpha^{\otimes} : \alpha \in [0,1] \cap [0,1] \text{ with } \alpha > 0\}$$

as follows. We can also refer to Subsubsection 7.1.2.3.

- In order to define the multiplication $\tilde{A} \otimes_{DT}^{\diamond} \tilde{B}$, we take

$$\begin{aligned} M_\alpha^{\otimes} &= \tilde{A}_\alpha \cdot \tilde{B}_\alpha = \left\{xy : x \in \tilde{A}_\alpha \text{ and } y \in \tilde{B}_\alpha\right\} \\ &= \left[\min\left\{\tilde{A}_\beta^L\tilde{B}_\beta^L, \tilde{A}_\beta^L\tilde{B}_\beta^U, \tilde{A}_\beta^U\tilde{B}_\beta^L, \tilde{A}_\beta^U\tilde{B}_\beta^U\right\}, \max\left\{\tilde{A}_\beta^L\tilde{B}_\beta^L, \tilde{A}_\beta^L\tilde{B}_\beta^U, \tilde{A}_\beta^U\tilde{B}_\beta^L, \tilde{A}_\beta^U\tilde{B}_\beta^U\right\}\right] \end{aligned} \tag{8.17}$$

  for $\alpha \in (0,1]$.
- In order to define the multiplication $\tilde{A} \otimes_{DT}^{\star} \tilde{B}$, let M_β be a closed interval given by

$$M_\beta = \left[\min\left\{\tilde{A}_\beta^L\tilde{B}_\beta^L, \tilde{A}_\beta^U\tilde{B}_\beta^U\right\}, \max\left\{\tilde{A}_\beta^L\tilde{B}_\beta^L, \tilde{A}_\beta^U\tilde{B}_\beta^U\right\}\right].$$

  We take

$$M_\alpha^{\otimes} = \bigcup_{\alpha \le \beta \le 1} M_\beta \tag{8.18}$$

  for $\alpha \in (0,1]$.
- In order to define the multiplication $\tilde{A} \otimes_{DT}^{\dagger} \tilde{B}$, we take

$$M_\alpha^{\otimes} = \left[\min\left\{\tilde{A}_\alpha^L\tilde{B}_\alpha^L, \tilde{A}_\alpha^U\tilde{B}_\alpha^U\right\}, \max\left\{\tilde{A}_\alpha^L\tilde{B}_\alpha^L, \tilde{A}_\alpha^U\tilde{B}_\alpha^U\right\}\right] \tag{8.19}$$

  for $\alpha \in (0,1]$.

Since the multiplications $\otimes_{EP}$ and $\otimes_{DT}^{\diamond}$ are equivalent by (7.24), it does not need to consider the multiplication $\otimes_{EP}$. Using the expression in the decomposition theorem, given $\otimes_{DT} \in \{\otimes_{DT}^{\diamond}, \otimes_{DT}^{\star}, \otimes_{DT}^{\dagger}\}$, the membership function of $\tilde{A}\otimes_{DT}\tilde{B}$ is defined by

$$\xi_{\tilde{A}\otimes_{DT}\tilde{B}}(z) = \sup_{0<\alpha\le 1} \alpha \cdot \chi_{M_\alpha^{\otimes}}(z), \tag{8.20}$$

where $M_\alpha^{\otimes}$ corresponds to the above three cases (8.17), (8.18), and (8.19).

We shall also define the addition of fuzzy intervals $\tilde{C}^{(1)}, \ldots, \tilde{C}^{(n)}$ based on the expression in the decomposition theorem by considering three different families $\{M_\alpha^{\oplus} : 0 < \alpha \le 1\}$ as follows. We can also refer to Subsubsection 7.1.2.1.

- In order to define the addition

$$\tilde{C}^{(1)} \oplus_{DT}^{\diamond} \cdots \oplus_{DT}^{\diamond} \tilde{C}^{(n)},$$

  we take

$$\begin{aligned} M_\alpha^{\oplus} &= \tilde{C}_\alpha^{(1)} + \cdots + \tilde{C}^{(n)} \\ &= \left\{x_1 + \cdots + x_n : x_i \in \tilde{C}_\alpha^{(i)} \text{ for } i = 1, \ldots, n\right\} \end{aligned} \tag{8.21}$$

$$= \left[\tilde{C}_{1\alpha}^L + \cdots + \tilde{C}_{n\alpha}^L, \tilde{C}_{1\alpha}^U + \cdots + \tilde{C}_{n\alpha}^U\right] \tag{8.22}$$

  for $\alpha \in (0,1]$, where $\tilde{C}_{i\alpha}^L = (\tilde{C}^{(i)})_\alpha^L$ and $\tilde{C}_{i\alpha}^U = (\tilde{C}^{(i)})_\alpha^U$ for $i = 1, \ldots, n$.
- In order to define the addition

$$\tilde{C}^{(1)} \oplus_{DT}^{\star} \cdots \oplus_{DT}^{\star} \tilde{C}^{(n)},$$

let M_β be a closed interval given by

$$M_\beta = \left[\min\left\{\tilde{C}^L_{1\alpha} + \cdots + \tilde{C}^L_{n\alpha}, \tilde{C}^U_{1\alpha} + \cdots + \tilde{C}^U_{n\alpha}\right\},\right.$$
$$\left.\max\left\{\tilde{C}^L_{1\alpha} + \cdots + \tilde{C}^L_{n\alpha}, \tilde{C}^U_{1\alpha} + \cdots + \tilde{C}^U_{n\alpha}\right\}\right]$$
$$= \left[\tilde{C}^L_{1\alpha} + \cdots + \tilde{C}^L_{n\alpha}, \tilde{C}^U_{1\alpha} + \cdots + \tilde{C}^U_{n\alpha}\right].$$

We take

$$M^\oplus_\alpha = \bigcup_{\alpha\leq\beta\leq 1} M_\beta \tag{8.23}$$

for $\alpha \in (0,1]$. It is clear to see that

$$M^\oplus_\alpha = \bigcup_{\alpha\leq\beta\leq 1} M_\beta$$
$$= \bigcup_{\alpha\leq\beta\leq 1} \left[\tilde{C}^L_{1\beta} + \cdots + \tilde{C}^L_{n\beta}, \tilde{C}^U_{1\beta} + \cdots + \tilde{C}^U_{n\beta}\right]$$
$$= \left[\tilde{C}^L_{1\alpha} + \cdots + \tilde{C}^L_{n\alpha}, \tilde{C}^U_{1\alpha} + \cdots + \tilde{C}^U_{n\alpha}\right].$$

- In order to define the addition

$$\tilde{C}^{(1)} \oplus^\dagger_{DT} \cdots \oplus^\dagger_{DT} \tilde{C}^{(n)},$$

we take

$$M^\oplus_\alpha = \left[\min\left\{\tilde{C}^L_{1\alpha} + \cdots + \tilde{C}^L_{n\alpha}, \tilde{C}^U_{1\alpha} + \cdots + \tilde{C}^U_{n\alpha}\right\},\right.$$
$$\left.\max\left\{\tilde{C}^L_{1\alpha} + \cdots + \tilde{C}^L_{n\alpha}, \tilde{C}^U_{1\alpha} + \cdots + \tilde{C}^U_{n\alpha}\right\}\right]$$
$$= \left[\tilde{C}^L_{1\alpha} + \cdots + \tilde{C}^L_{n\alpha}, \tilde{C}^U_{1\alpha} + \cdots + \tilde{C}^U_{n\alpha}\right] \tag{8.24}$$

for $\alpha \in (0,1]$.

Then, we see that

$$\tilde{C}^{(1)} \oplus^\circ_{DT} \cdots \oplus^\circ_{DT} \tilde{C}^{(n)} = \tilde{C}^{(1)} \oplus^\star_{DT} \cdots \oplus^\star_{DT} \tilde{C}^{(n)} = \tilde{C}^{(1)} \oplus^\dagger_{DT} \cdots \oplus^\dagger_{DT} \tilde{C}^{(n)}.$$

In this case, we simply write

$$\tilde{C}^{(1)} \oplus_{DT} \cdots \oplus_{DT} \tilde{C}^{(n)},$$

and its membership function is defined by

$$\xi_{\tilde{C}^{(1)} \oplus_{DT} \cdots \oplus_{DT} \tilde{C}^{(n)}}(z) = \sup_{0<\alpha\leq 1} \alpha \cdot \chi_{M^\oplus_\alpha}(z). \tag{8.25}$$

Since the additions $\oplus_{EP}$ and $\oplus_{DT}$ are equivalent by (7.20), it follows that

$$\tilde{C}^{(1)} \oplus_{ET} \cdots \oplus_{ET} \tilde{C}^{(n)} = \tilde{C}^{(1)} \oplus_{DT} \cdots \oplus_{DT} \tilde{C}^{(n)}. \tag{8.26}$$

In this case, we simply write

$$\tilde{C}^{(1)} \oplus \cdots \oplus \tilde{C}^{(n)}.$$

Now, we are in a position to define the second type of inner product of $\tilde{\mathbf{A}}$ and $\tilde{\mathbf{B}}$ as follows.

Definition 8.2.1 Let $\tilde{A}^{(k)}$ and $\tilde{B}^{(k)}$ be fuzzy intervals for $k = 1, \ldots, n$. The **inner product** between $\tilde{\mathbf{A}}$ and $\tilde{\mathbf{B}}$ is defined by

$$\tilde{\mathbf{A}} \circledcirc \tilde{\mathbf{B}} = \left(\tilde{A}^{(1)} \otimes_1 \tilde{B}^{(1)}\right) \oplus \cdots \oplus \left(\tilde{A}^{(n)} \otimes_n \tilde{B}^{(n)}\right), \tag{8.27}$$

where the multiplication

$$\otimes_j \in \{\otimes_{DT}^{\diamond}, \otimes_{DT}^{\star}, \otimes_{DT}^{\dagger}\} \text{ for } j = 1, \ldots, n. \tag{8.28}$$

The inner product $\tilde{\mathbf{A}} \circledcirc \tilde{\mathbf{B}}$ depends on the choice of multiplication according to (8.28). Therefore, it is completely different from the first type of inner product $\tilde{\mathbf{A}} \circledast \tilde{\mathbf{B}}$ for $\circledast \in \{\circledast_{EP}, \circledast_{DT}^{\diamond}, \circledast_{DT}^{\star}, \circledast_{DT}^{\dagger}\}$. We write

$$\tilde{C}^{(j)} = \tilde{A}^{(j)} \otimes_j \tilde{B}^{(j)} \text{ for } j = 1, \ldots, n.$$

Then, we have

$$\tilde{\mathbf{A}} \circledcirc \tilde{\mathbf{B}} = \tilde{C}^{(1)} \oplus \cdots \oplus \tilde{C}^{(n)} \tag{8.29}$$

and its membership function is given by

$$\begin{aligned} \xi_{\tilde{\mathbf{A}} \circledcirc \tilde{\mathbf{B}}}(z) &= \sup_{0<\alpha\leq 1} \alpha \cdot \chi_{M_\alpha^{\oplus}}(z) \\ &= \sup_{\{(x_1,\ldots,x_n):z=x_1+\cdots+x_n\}} \min\left\{\xi_{\tilde{C}^{(1)}}(x_1), \ldots, \xi_{\tilde{C}^{(n)}}(x_n)\right\} \end{aligned} \tag{8.30}$$

by referring to (8.25), (8.26) and (8.16).

The membership function $\xi_{\tilde{\mathbf{A}} \circledcirc \tilde{\mathbf{B}}}$ is in a very general situation, since the multiplication $\otimes_j$ for $j = 1, \ldots, n$ can be any operations in (8.28). However, we can use the Decomposition Theorem 2.2.13 to rewrite the membership function $\xi_{\tilde{\mathbf{A}} \circledcirc \tilde{\mathbf{B}}}$ using its α-level sets. Let $(\tilde{\mathbf{A}} \circledcirc \tilde{\mathbf{B}})_\alpha$ be the α-level set of $\tilde{\mathbf{A}} \circledcirc \tilde{\mathbf{B}}$. According to the Decomposition Theorem 2.2.13, the membership function is given by

$$\xi_{\tilde{\mathbf{A}} \circledcirc \tilde{\mathbf{B}}}(z) = \max_{0<\alpha\leq 1} \alpha \cdot \chi_{(\tilde{\mathbf{A}} \circledcirc \tilde{\mathbf{B}})_\alpha}(z).$$

Therefore, the purpose is to study the α-level set $(\tilde{\mathbf{A}} \circledcirc \tilde{\mathbf{B}})_\alpha$ in the subsequent discussion.

8.2.1 Using the Extension Principle

We shall present the α-level sets $(\tilde{\mathbf{A}} \circledcirc_{EP} \tilde{\mathbf{B}})_\alpha$. For any $\alpha \in [0,1]$, from Theorem 7.1.2, we have

$$\begin{aligned} \tilde{C}_\alpha^{(j)} &= \left(\tilde{A}^{(j)} \otimes_{EP} \tilde{B}^{(j)}\right)_\alpha = \tilde{A}_\alpha^{(j)} * \tilde{B}_\alpha^{(j)} = \left[\tilde{A}_{j\alpha}^L, \tilde{A}_{j\alpha}^U\right] * \left[\tilde{B}_{j\alpha}^L, \tilde{B}_{j\alpha}^U\right] \\ &= \left[\min\left\{\tilde{A}_{j\alpha}^L \tilde{B}_{j\alpha}^L, \tilde{A}_{j\alpha}^L \tilde{B}_{j\alpha}^U, \tilde{A}_{j\alpha}^U \tilde{B}_{j\alpha}^L, \tilde{A}_{j\alpha}^U \tilde{B}_{j\alpha}^U\right\}, \max\left\{\tilde{A}_{j\alpha}^L \tilde{B}_{j\alpha}^L, \tilde{A}_{j\alpha}^L \tilde{B}_{j\alpha}^U, \tilde{A}_{j\alpha}^U \tilde{B}_{j\alpha}^L, \tilde{A}_{j\alpha}^U \tilde{B}_{j\alpha}^U\right\}\right] \\ &\equiv \left[\tilde{C}_{j\alpha}^L, \tilde{C}_{j\alpha}^U\right]. \end{aligned} \tag{8.31}$$

and

$$\left(\tilde{\mathbf{A}} \circledcirc_{EP} \tilde{\mathbf{B}}\right)_\alpha = \tilde{C}_\alpha^{(1)} + \cdots + \tilde{C}_\alpha^{(n)} = \left[\tilde{C}_{1\alpha}^L + \cdots + \tilde{C}_{n\alpha}^L, \tilde{C}_{1\alpha}^U + \cdots + \tilde{C}_{n\alpha}^U\right] \tag{8.32}$$

$$\equiv \left[(\tilde{\mathbf{A}} \circledcirc_{EP} \tilde{\mathbf{B}})_\alpha^L, (\tilde{\mathbf{A}} \circledcirc_{EP} \tilde{\mathbf{B}})_\alpha^U\right], \tag{8.33}$$

where

$$\left(\tilde{\mathbf{A}} \circledcirc_{EP} \tilde{\mathbf{B}}\right)^L_\alpha = \sum_{j=1}^{n} \tilde{C}^L_{j\alpha} \text{ and } \left(\tilde{\mathbf{A}} \circledcirc_{EP} \tilde{\mathbf{B}}\right)^U_\alpha = \sum_{j=1}^{n} \tilde{C}^U_{j\alpha}. \tag{8.34}$$

Therefore, we can calculate the α-level sets $(\tilde{\mathbf{A}} \circledcirc_{EP} \tilde{\mathbf{B}})_\alpha$ according to the above formulas.

Example 8.2.2 Continued from Example 8.1.2, we first calculate $\tilde{C}^{(1)}_\alpha$ and $\tilde{C}^{(2)}_\alpha$ from (8.31). For $\alpha \in [0,1]$, we have

$$\tilde{C}^L_{1\alpha} = \min\left\{\tilde{A}^L_{1\alpha}\tilde{B}^L_{1\alpha}, \tilde{A}^L_{1\alpha}\tilde{B}^U_{1\alpha}, \tilde{A}^U_{1\alpha}\tilde{B}^L_{1\alpha}, \tilde{A}^U_{1\alpha}\tilde{B}^U_{1\alpha}\right\} = (1+\alpha)(4+\alpha) = 4 + 5\alpha + \alpha^2$$

and

$$\tilde{C}^L_{2\alpha} = \min\left\{\tilde{A}^L_{2\alpha}\tilde{B}^L_{2\alpha}, \tilde{A}^L_{2\alpha}\tilde{B}^U_{2\alpha}, \tilde{A}^U_{2\alpha}\tilde{B}^L_{2\alpha}, \tilde{A}^U_{2\alpha}\tilde{B}^U_{2\alpha}\right\} = (2+\alpha)(3+\alpha) = 6 + 5\alpha + \alpha^2.$$

We can similarly obtain

$$\tilde{C}^U_{1\alpha} = \max\left\{\tilde{A}^L_{1\alpha}\tilde{B}^L_{1\alpha}, \tilde{A}^L_{1\alpha}\tilde{B}^U_{1\alpha}, \tilde{A}^U_{1\alpha}\tilde{B}^L_{1\alpha}, \tilde{A}^U_{1\alpha}\tilde{B}^U_{1\alpha}\right\} = (4-\alpha)(7-\alpha) = 28 - 11\alpha + \alpha^2$$

and

$$\tilde{C}^U_{2\alpha} = \max\left\{\tilde{A}^L_{2\alpha}\tilde{B}^L_{2\alpha}, \tilde{A}^L_{2\alpha}\tilde{B}^U_{2\alpha}, \tilde{A}^U_{2\alpha}\tilde{B}^L_{2\alpha}, \tilde{A}^U_{2\alpha}\tilde{B}^U_{2\alpha}\right\} = (5-\alpha)(6-\alpha) = 30 - 11\alpha + \alpha^2.$$

From (8.33), we obtain

$$\left(\tilde{\mathbf{A}} \circledcirc_{EP} \tilde{\mathbf{B}}\right)_\alpha = \left[\left(\tilde{\mathbf{A}} \circledcirc_{EP} \tilde{\mathbf{B}}\right)^L_\alpha, \left(\tilde{\mathbf{A}} \circledcirc_{EP} \tilde{\mathbf{B}}\right)^U_\alpha\right] = \left[10 + 10\alpha + \alpha^2, 58 - 22\alpha + \alpha^2\right].$$

Example 8.2.3 Continued from Example 8.1.3, for $0.5 \le \alpha < 2/3$, we have

$$\begin{aligned}\left(\tilde{\mathbf{A}} \circledcirc_{EP} \tilde{\mathbf{B}}\right)^L_\alpha &= \sum_{i=1}^{3}\left(\min\left\{\tilde{A}^L_{i\alpha}\tilde{B}^L_{i\alpha}, \tilde{A}^L_{i\alpha}\tilde{B}^U_{i\alpha}, \tilde{A}^U_{i\alpha}\tilde{B}^L_{i\alpha}, \tilde{A}^U_{i\alpha}\tilde{B}^U_{i\alpha}\right\}\right) \\ &= (\alpha-1)(3-2\alpha) + (-2+\alpha)(2-3\alpha) + \alpha(2\alpha-1) = -3\alpha^2 + 12\alpha - 7\end{aligned}$$

and, for $0 \le \alpha < 0.5$, we have

$$\begin{aligned}&\sum_{i=1}^{3}\left(\min\left\{\tilde{A}^L_{i\alpha}\tilde{B}^L_{i\alpha}, \tilde{A}^L_{i\alpha}\tilde{B}^U_{i\alpha}, \tilde{A}^U_{i\alpha}\tilde{B}^L_{i\alpha}, \tilde{A}^U_{i\alpha}\tilde{B}^U_{i\alpha}\right\}\right) \\ &\quad = (\alpha-1)(3-2\alpha) + (-2+\alpha)(2-3\alpha) + (2\alpha-1)(3-2\alpha) = -9\alpha^2 + 21\alpha - 10.\end{aligned}$$

From (8.33), we obtain

$$\left(\tilde{\mathbf{A}} \circledcirc_{EP} \tilde{\mathbf{B}}\right)_\alpha = \begin{cases} \left[6\alpha^2 - 3\alpha - 1, 11\alpha^2 - 31\alpha + 22\right] & \text{if } \alpha \ge 2/3 \\ \left[-3\alpha^2 + 12\alpha - 7, 11\alpha^2 - 31\alpha + 22\right] & \text{if } 0.5 \le \alpha < 2/3 \\ \left[-9\alpha^2 + 21\alpha - 10, 11\alpha^2 - 31\alpha + 22\right] & \text{if } 0 \le \alpha < 0.5 \end{cases}$$

Let A, D, E be bounded closed intervals in $\mathbb{R}$. Then, we have the subdistributivity

$$A \times (D + E) \subseteq A \times D + A \times E. \tag{8.35}$$

We further assume that $d \cdot e \ge 0$ for every $d \in D$ and $e \in E$. Then, we have the distributivity

$$A \times (D + E) = A \times D + A \times E. \tag{8.36}$$

Let $\tilde{A}$ and $\tilde{B}$ be fuzzy intervals. We write $\tilde{A} \subseteq \tilde{B}$ when their α-level sets satisfy the inclusion $\tilde{A}_\alpha \subseteq \tilde{B}_\alpha$ for all $\alpha \in [0,1]$.

Proposition 8.2.4 *Let $\tilde{A}^{(j)}$, $\tilde{D}^{(j)}$ and $\tilde{E}^{(j)}$ be fuzzy intervals for $j = 1, \ldots, n$.*

(i) *We have the following subdistributivity*

$$\tilde{\mathbf{A}} \circledcirc_{EP} (\tilde{\mathbf{D}} \oplus_{EP} \tilde{\mathbf{E}}) \subseteq (\tilde{\mathbf{A}} \circledcirc_{EP} \tilde{\mathbf{D}}) \oplus_{EP} (\tilde{\mathbf{A}} \circledcirc_{EP} \tilde{\mathbf{E}}).$$

(ii) *For each fixed j, suppose that $\tilde{D}^{(j)}$ and $\tilde{E}^{(j)}$ are simultaneously nonnegative or nonpositive. Then, we have the following distributivity*

$$\tilde{\mathbf{A}} \circledcirc_{EP} (\tilde{\mathbf{D}} \oplus_{EP} \tilde{\mathbf{E}}) = (\tilde{\mathbf{A}} \circledcirc_{EP} \tilde{\mathbf{D}}) \oplus_{EP} (\tilde{\mathbf{A}} \circledcirc_{EP} \tilde{\mathbf{E}}).$$

Proof. Let $\tilde{\mathbf{B}} = \tilde{\mathbf{D}} \oplus_{EP} \tilde{\mathbf{E}}$. Then, regarding the components for $j = 1, \ldots, n$, we have

$$\tilde{B}^{(j)} = \tilde{D}^{(j)} \oplus_{EP} \tilde{E}^{(j)} \text{ and } \tilde{B}_\alpha^{(j)} = \tilde{D}_\alpha^{(j)} + \tilde{E}_\alpha^{(j)} \text{ for } \alpha \in [0,1], \tag{8.37}$$

where $\tilde{B}_\alpha^{(j)}$, $\tilde{D}_\alpha^{(j)}$ and $\tilde{E}_\alpha^{(j)}$ denote the α-level sets for $\alpha \in [0,1]$. From (8.31), we see that

$$\tilde{C}_\alpha^{(j)} = \tilde{A}_\alpha^{(j)} * \tilde{B}_\alpha^{(j)}. \tag{8.38}$$

Therefore, for $\alpha \in [0,1]$, we obtain

$$\begin{aligned}
\left(\tilde{\mathbf{A}} \circledcirc_{EP} (\tilde{\mathbf{D}} \oplus_{EP} \tilde{\mathbf{E}})\right)_\alpha &= \left(\tilde{\mathbf{A}} \circledcirc_{EP} \tilde{\mathbf{B}}\right)_\alpha = \tilde{C}_\alpha^{(1)} + \cdots + \tilde{C}_\alpha^{(n)} \text{ (using (8.32))} \\
&= \tilde{A}_\alpha^{(1)} * \left(\tilde{D}_\alpha^{(1)} + \tilde{E}_\alpha^{(1)}\right) + \cdots + \tilde{A}_\alpha^{(n)} * \left(\tilde{D}_\alpha^{(n)} + \tilde{E}_\alpha^{(n)}\right) \text{ (using (8.37) and (8.38))} \\
&\subseteq \left(\tilde{A}_\alpha^{(1)} * \tilde{D}_\alpha^{(1)} + \tilde{A}_\alpha^{(1)} * \tilde{E}_\alpha^{(1)}\right) + \cdots + \left(\tilde{A}_\alpha^{(n)} * \tilde{D}_\alpha^{(n)} + \tilde{A}_\alpha^{(n)} * \tilde{E}_\alpha^{(n)}\right) \text{ (using (8.35))} \\
&= \left(\tilde{A}_\alpha^{(1)} * \tilde{D}_\alpha^{(1)} + \cdots + \tilde{A}_\alpha^{(n)} * \tilde{D}_\alpha^{(n)}\right) + \left(\tilde{A}_\alpha^{(1)} * \tilde{E}_\alpha^{(1)} + \cdots + \tilde{A}_\alpha^{(n)} * \tilde{E}_\alpha^{(n)}\right) \\
&= (\tilde{\mathbf{A}} \circledcirc_{EP} \tilde{\mathbf{D}})_\alpha + (\tilde{\mathbf{A}} \circledcirc_{EP} \tilde{\mathbf{E}})_\alpha = \left((\tilde{\mathbf{A}} \circledcirc_{EP} \tilde{\mathbf{D}}) \oplus_{EP} (\tilde{\mathbf{A}} \circledcirc_{EP} \tilde{\mathbf{E}})\right)_\alpha.
\end{aligned}$$

This proves part (i). Applying (8.36) to the above argument, we can similarly obtain part (ii), and the proof is complete. ■

8.2.2 Using the Expression in the Decomposition Theorem

Now, we can take $\otimes_j = \otimes_{DT} \in \{\otimes_{DT}^{\diamond}, \otimes_{DT}^{\star}, \otimes_{DT}^{\dagger}\}$ in (8.28) for all $j = 1, \ldots, n$. In this case, by referring to (8.20), the membership functions of $\tilde{C}^{(j)} = \tilde{A}^{(j)} \otimes_{DT} \tilde{B}^{(j)}$ are given by

$$\xi_{\tilde{C}^{(j)}}(z) = \sup_{0<\alpha\leq 1} \alpha \cdot \chi_{M_\alpha^{\otimes_j}}(z), \tag{8.39}$$

for $j = 1, \ldots, n$. By referring to (8.30), the membership function of inner product (8.29) is given by

$$\xi_{\tilde{\mathbf{A}} \circledcirc_{DT} \tilde{\mathbf{B}}}(z) = \sup_{0<\alpha\leq 1} \alpha \cdot \chi_{M_\alpha^{\oplus}}(z), \tag{8.40}$$

where $M_\alpha^{\oplus}$ are given in (8.22), (8.23) or (8.24) with

$$\tilde{C}_{j\alpha}^L = \sum_{j=1}^n \tilde{A}_{j\alpha}^L + \sum_{j=1}^n \tilde{B}_{j\alpha}^L \text{ and } \tilde{C}_{j\alpha}^U = \sum_{j=1}^n \tilde{A}_{j\alpha}^U + \sum_{j=1}^n \tilde{B}_{j\alpha}^U.$$

Now, we assume that $\tilde{A}^{(k)}$ and $\tilde{B}^{(k)}$ are taken to be canonical fuzzy intervals for $k = 1, \ldots, n$. We shall present the α-level sets $(\tilde{\mathbf{A}} \odot_{DT} \tilde{\mathbf{B}})_\alpha$. From (7.26), we see that

$$\tilde{C}^{(j)} = \tilde{A}^{(j)} \otimes^{\star}_{DT} \tilde{B}^{(j)} = \tilde{A}^{(j)} \otimes^{\dagger}_{DT} \tilde{B}^{(j)}.$$

Then, we can obtain two different kinds of α-level sets $\tilde{C}^{(j)}_\alpha$ by referring to Subsubsection 7.1.2.3.

- Suppose that we take $\otimes_j = \circledast^{\diamond}_{DT}$ for $j = 1, \ldots, n$. For any $\alpha \in [0,1]$, we have

$$\begin{aligned}\tilde{C}^{(j)}_\alpha = \left(\tilde{A}^{(j)} \otimes^{\diamond}_{DT} \tilde{B}^{(j)}\right)_\alpha &= \left[\min_{(x,y)\in(\tilde{A}^{(j)}_\alpha,\tilde{B}^{(j)}_\alpha)} xy,\ \max_{(x,y)\in(\tilde{A}^{(j)}_\alpha,\tilde{B}^{(j)}_\alpha)} xy\right] \\ &= \left[\min\left\{\tilde{A}^L_{j\alpha}\tilde{B}^L_{j\alpha}, \tilde{A}^L_{j\alpha}\tilde{B}^U_{j\alpha}, \tilde{A}^U_{j\alpha}\tilde{B}^L_{j\alpha}, \tilde{A}^U_{j\alpha}\tilde{B}^U_{j\alpha}\right\},\right. \\ &\qquad \left.\max\left\{\tilde{A}^L_{j\alpha}\tilde{B}^L_{j\alpha}, \tilde{A}^L_{j\alpha}\tilde{B}^U_{j\alpha}, \tilde{A}^U_{j\alpha}\tilde{B}^L_{j\alpha}, \tilde{A}^U_{j\alpha}\tilde{B}^U_{j\alpha}\right\}\right] \\ &\equiv \left[\tilde{C}^L_{j\alpha}, \tilde{C}^U_{j\alpha}\right].\end{aligned} \tag{8.41}$$

- Suppose that we take $\otimes_j = \circledast^{\star}_{DT} = \circledast^{\dagger}_{DT}$ for $j = 1, \ldots, n$. For any $\alpha \in [0,1]$, we also have

$$\tilde{C}^{(j)}_\alpha = \left(\tilde{A}^{(j)} \otimes^{\star}_{DT} \tilde{B}^{(j)}\right)_\alpha = \left(\tilde{A}^{(j)} \otimes^{\dagger}_{DT} \tilde{B}^{(j)}\right)_\alpha \tag{8.42}$$

$$= \left[\min_{\alpha\le\beta\le 1} \min\left\{\tilde{A}^L_{j\beta}\tilde{B}^L_{j\beta}, \tilde{A}^U_{j\beta}\tilde{B}^U_{j\beta}\right\}, \max_{\alpha\le\beta\le 1} \max\left\{\tilde{A}^L_{j\beta}\tilde{B}^L_{j\beta}, \tilde{A}^U_{j\beta}\tilde{B}^U_{j\beta}\right\}\right] \tag{8.43}$$

$$= \left[\min\left\{\min_{\alpha\le\beta\le 1} \tilde{A}^L_{j\beta}\tilde{B}^L_{j\beta}, \min_{\alpha\le\beta\le 1} \tilde{A}^U_{j\beta}\tilde{B}^U_{j\beta}\right\}, \max\left\{\max_{\alpha\le\beta\le 1} \tilde{A}^L_{j\beta}\tilde{B}^L_{j\beta}, \max_{\alpha\le\beta\le 1} \tilde{A}^U_{j\beta}\tilde{B}^U_{j\beta}\right\}\right] \tag{8.44}$$

$$\equiv \left[\tilde{C}^L_{j\alpha}, \tilde{C}^U_{j\alpha}\right].$$

According to (8.29), for any $\alpha \in [0,1]$, we have

$$\begin{aligned}\left(\tilde{\mathbf{A}} \odot_{DT} \tilde{\mathbf{B}}\right)_\alpha &= \tilde{C}^{(1)}_\alpha + \cdots + \tilde{C}^{(n)}_\alpha = \left[\tilde{C}^L_{1\alpha} + \cdots + \tilde{C}^L_{n\alpha}, \tilde{C}^U_{1\alpha} + \cdots + \tilde{C}^U_{n\alpha}\right] \\ &\equiv \left[\left(\tilde{\mathbf{A}} \odot_{DT} \tilde{\mathbf{B}}\right)^L_\alpha, \left(\tilde{\mathbf{A}} \odot_{DT} \tilde{\mathbf{B}}\right)^U_\alpha\right],\end{aligned} \tag{8.45}$$

where the α-level sets $\tilde{C}^{(j)}_\alpha$ can be taken from (8.41) or (8.44) for $j = 1, \ldots, n$, and

$$\left(\tilde{\mathbf{A}} \odot_{DT} \tilde{\mathbf{B}}\right)^L_\alpha = \sum_{j=1}^{n} \tilde{C}^L_{j\alpha} \text{ and } \left(\tilde{\mathbf{A}} \odot_{DT} \tilde{\mathbf{B}}\right)^U_\alpha = \sum_{j=1}^{n} \tilde{C}^U_{j\alpha}. \tag{8.46}$$

Therefore, we can calculate two kinds of α-level sets $(\tilde{\mathbf{A}} \odot_{DT} \tilde{\mathbf{B}})_\alpha$ using (8.41) or (8.44) for $j = 1, \ldots, n$.

Example 8.2.5 Continued from Example 8.1.2, suppose that we take $\otimes_j = \circledast^{\star}_{DT} = \circledast^{\dagger}_{DT}$ for $j = 1, \ldots, n$. We first calculate $\tilde{C}^{(1)}_\alpha$ and $\tilde{C}^{(2)}_\alpha$ from (8.44). For $\alpha \in [0,1]$, we have

$$\begin{aligned}\tilde{C}^L_{1\alpha} &= \min\left\{\min_{\{\beta\in[0,0.8]:\beta\ge\alpha\}} \tilde{A}^L_{1\beta}\tilde{B}^L_{1\beta}, \min_{\{\beta\in[0,0.8]:\beta\ge\alpha\}} \tilde{A}^U_{1\beta}\tilde{B}^U_{1\beta}\right\} = \min\left\{\tilde{A}^L_{1\alpha}\tilde{B}^L_{1\alpha}, \tilde{A}^U_{1,0.8}\tilde{B}^U_{1,0.8}\right\} \\ &= \tilde{A}^L_{1\alpha}\tilde{B}^L_{1\alpha} = (1+\alpha)(4+\alpha) = 4 + 5\alpha + \alpha^2\end{aligned}$$

and

$$\tilde{C}_{2\alpha}^{L} = \min\left\{\min_{\{\beta\in[0,0.8]:\beta\geq\alpha\}} \tilde{A}_{2\beta}^{L}\tilde{B}_{2\beta}^{L}, \min_{\{\beta\in[0,0.8]:\beta\geq\alpha\}} \tilde{A}_{2\beta}^{U}\tilde{B}_{2\beta}^{U}\right\} = \min\left\{\tilde{A}_{2\alpha}^{L}\tilde{B}_{2\alpha}^{L}, \tilde{A}_{2,0.8}^{U}\tilde{B}_{2,0.8}^{U}\right\}$$
$$= \tilde{A}_{2\alpha}^{L}\tilde{B}_{2\alpha}^{L} = (2+\alpha)(3+\alpha) = 6+5\alpha+\alpha^2.$$

We can similarly obtain

$$\tilde{C}_{1\alpha}^{U} = (4-\alpha)(7-\alpha) = 28-11\alpha+\alpha^2$$

and

$$\tilde{C}_{2\alpha}^{U} = (5-\alpha)(6-\alpha) = 30-11\alpha+\alpha^2.$$

From (8.45) and (8.46), we obtain

$$\left(\tilde{\mathbf{A}} \circledcirc_{DT} \tilde{\mathbf{B}}\right)_{\alpha} = \left[\left(\tilde{\mathbf{A}} \circledcirc_{DT} \tilde{\mathbf{B}}\right)_{\alpha}^{L}, \left(\tilde{\mathbf{A}} \circledcirc_{DT} \tilde{\mathbf{B}}\right)_{\alpha}^{U}\right] = \left[10+10\alpha+\alpha^2, 58-22\alpha+\alpha^2\right].$$

Example 8.2.6 Continued from Example 8.1.3, suppose that we take $\otimes_j = \circledast_{DT}^{\star} = \circledast_{DT}^{\dagger}$ for $j = 1, \ldots, n$. We first calculate $\tilde{C}_{\alpha}^{(1)}$ and $\tilde{C}_{\alpha}^{(2)}$ from (8.43). Now, we have

$$\tilde{C}_{1\alpha}^{L} = \min_{\alpha\leq\beta\leq1} \min\left\{\tilde{A}_{1\beta}^{L}\tilde{B}_{1\beta}^{L}, \tilde{A}_{1\beta}^{U}\tilde{B}_{1\beta}^{U}\right\} = \min_{\alpha\leq\beta\leq1} \min\left\{(2\beta-1)(\beta-1), (3-2\beta)(3-3\beta)\right\}$$
$$= \min_{\alpha\leq\beta\leq1} (2\beta-1)(\beta-1) = \begin{cases} 2\alpha^2-3\alpha+1 & \text{if } \alpha\geq0.75 \\ 2(0.75)^2-3\cdot0.75+1 & \text{if } \alpha<0.75 \end{cases}$$

$$\tilde{C}_{1\alpha}^{U} = \max_{\alpha\leq\beta\leq1} \max\left\{\tilde{A}_{1\beta}^{L}\tilde{B}_{1\beta}^{L}, \tilde{A}_{1\beta}^{U}\tilde{B}_{1\beta}^{U}\right\} = \max_{\alpha\leq\beta\leq1} \max\left\{(2\beta-1)(\beta-1), (3-2\beta)(3-3\beta)\right\}$$
$$= \max_{\alpha\leq\beta\leq1} (3-2\beta)(3-3\beta) = 6\alpha^2-15\alpha+9$$

$$\tilde{C}_{2\alpha}^{L} = \min_{\alpha\leq\beta\leq1} \min\left\{\tilde{A}_{2\beta}^{L}\tilde{B}_{2\beta}^{L}, \tilde{A}_{2\beta}^{U}\tilde{B}_{2\beta}^{U}\right\} = \min_{\alpha\leq\beta\leq1} \min\left\{(\beta-2)^2, (2-3\beta)(1-2\beta)\right\}$$
$$= \min_{\alpha\leq\beta\leq1} (2-3\beta)(1-2\beta) = \begin{cases} 6\alpha^2-7\alpha+2 & \text{if } \alpha\geq7/12 \\ 6(7/12)^2-7(7/12)+2 & \text{if } \alpha<7/12 \end{cases}$$

$$\tilde{C}_{2\alpha}^{U} = \max_{\alpha\leq\beta\leq1} \max\left\{\tilde{A}_{2\beta}^{L}\tilde{B}_{2\beta}^{L}, \tilde{A}_{2\beta}^{U}\tilde{B}_{2\beta}^{U}\right\} = \max_{\alpha\leq\beta\leq1} \max\left\{(\beta-2)^2, (2-3\beta)(1-2\beta)\right\}$$
$$= \max_{\alpha\leq\beta\leq1} (\beta-2)^2 = \alpha^2-4\alpha+4$$

$$\tilde{C}_{3\alpha}^{L} = \min_{\alpha\leq\beta\leq1} \min\left\{\tilde{A}_{3\beta}^{L}\tilde{B}_{3\beta}^{L}, \tilde{A}_{3\beta}^{U}\tilde{B}_{3\beta}^{U}\right\} = \min_{\alpha\leq\beta\leq1} \min\left\{\beta(2\beta-1), (3-2\beta)^2\right\}$$
$$= \min_{\alpha\leq\beta\leq1} \beta(2\beta-1) = \begin{cases} 2\alpha^2-\alpha & \text{if } \alpha\geq0.25 \\ 2(0.25)^2-0.25 & \text{if } \alpha<0.25 \end{cases}$$

$$\tilde{C}_{3\alpha}^{U} = \max_{\alpha\leq\beta\leq1} \max\left\{\tilde{A}_{3\beta}^{L}\tilde{B}_{3\beta}^{L}, \tilde{A}_{3\beta}^{U}\tilde{B}_{3\beta}^{U}\right\} = \max_{\alpha\leq\beta\leq1} \max\left\{\beta(2\beta-1), (3-2\beta)^2\right\}$$
$$= \max_{\alpha\leq\beta\leq1} (3-2\beta)^2 = 4\alpha^2-12\alpha+9$$

Using (8.46), for $\alpha\in[0,1]$, we have

$$\left(\tilde{\mathbf{A}} \circledcirc_{DT} \tilde{\mathbf{B}}\right)_{\alpha}^{U} = \left(6\alpha^2-15\alpha+9\right) + \left(\alpha^2-4\alpha+4\right) + \left(4\alpha^2-12\alpha+9\right) = 10\alpha^2-31\alpha+22,$$

and, for $\alpha \in [0.75, 1]$, we have

$$\left(\tilde{\mathbf{A}} \circledcirc_{DT} \tilde{\mathbf{B}}\right)_\alpha^L = \left(2\alpha^2 - 3\alpha + 1\right) + \left(6\alpha^2 - 7\alpha + 2\right) + \left(2\alpha^2 - \alpha\right) = 10\alpha^2 - 11\alpha + 3.$$

From (8.45), for $\alpha \in [0.75, 1]$, we obtain

$$\left(\tilde{\mathbf{A}} \circledcirc_{DT} \tilde{\mathbf{B}}\right)_\alpha = \left[\left(\tilde{\mathbf{A}} \circledcirc_{DT} \tilde{\mathbf{B}}\right)_\alpha^L, \left(\tilde{\mathbf{A}} \circledcirc_{DT} \tilde{\mathbf{B}}\right)_\alpha^U\right] = \left[10\alpha^2 - 11\alpha + 3, 10\alpha^2 - 31\alpha + 22\right].$$

For the cases of $\alpha \in [7/12, 0.75)$, $\alpha \in [0.25, 7/12)$ and $\alpha \in [0, 0.25)$, we can similarly obtain the α-level sets $(\tilde{\mathbf{A}} \circledcirc_{DT} \tilde{\mathbf{B}})_\alpha$.

8.2.3 Comparison of Fuzziness

By referring to Definition 6.3.4, we are gong to compare the fuzziness between the first type of inner product $\tilde{\mathbf{A}} \circledast \tilde{\mathbf{B}}$ for $\circledast \in \{\circledast_{EP}, \circledast_{DT}^{\diamond}, \circledast_{DT}^{\star}, \circledast_{DT}^{\dagger}\}$ and the second type of inner product $\tilde{\mathbf{A}} \circledcirc \tilde{\mathbf{B}}$.

Theorem 8.2.7 *Let $\tilde{A}^{(k)}$ and $\tilde{B}^{(k)}$ be canonical fuzzy intervals for $k = 1, \ldots, n$. Suppose that the first type of inner products $\tilde{\mathbf{A}} \circledast_{EP} \tilde{\mathbf{B}}$ and $\tilde{\mathbf{A}} \circledast_{DT}^{\diamond} \tilde{\mathbf{B}}$ are obtained from Theorems 8.1.1 and 8.1.5, respectively. Then, we have the following properties.*

(i) *Assume that the second type of inner product is given by*

$$\tilde{\mathbf{A}} \circledcirc_{EP} \tilde{\mathbf{B}} = \left(\tilde{A}^{(1)} \otimes_{EP} \tilde{B}^{(1)}\right) \oplus \cdots \oplus \left(\tilde{A}^{(n)} \otimes_{EP} \tilde{B}^{(n)}\right). \tag{8.47}$$

Then, the first type of inner products $\tilde{\mathbf{A}} \circledast_{EP} \tilde{\mathbf{B}}$ and $\tilde{\mathbf{A}} \circledast_{DT}^{\diamond} \tilde{\mathbf{B}}$ are fuzzier than the second type of inner product $\tilde{\mathbf{A}} \circledcirc_{EP} \tilde{\mathbf{B}}$ in the sense of

$$\left(\tilde{\mathbf{A}} \circledcirc_{EP} \tilde{\mathbf{B}}\right)_\alpha \subseteq \left(\tilde{\mathbf{A}} \circledast_{EP} \tilde{\mathbf{B}}\right)_\alpha = \left(\tilde{\mathbf{A}} \circledast_{DT}^{\diamond} \tilde{\mathbf{B}}\right)_\alpha$$

for $\alpha \in [0,1]$.

(ii) *Assume that the second type of inner product is given by*

$$\tilde{\mathbf{A}} \circledcirc_{DT} \tilde{\mathbf{B}} = \left(\tilde{A}^{(1)} \otimes_{DT}^{\diamond} \tilde{B}^{(1)}\right) \oplus \cdots \oplus \left(\tilde{A}^{(n)} \otimes_{DT}^{\diamond} \tilde{B}^{(n)}\right). \tag{8.48}$$

Then, the first type of inner products $\tilde{\mathbf{A}} \circledast_{EP} \tilde{\mathbf{B}}$ and $\tilde{\mathbf{A}} \circledast_{DT}^{\diamond} \tilde{\mathbf{B}}$ are fuzzier than the second type of inner product $\tilde{\mathbf{A}} \circledcirc \tilde{\mathbf{B}}$ in the sense of

$$\left(\tilde{\mathbf{A}} \circledcirc_{DT} \tilde{\mathbf{B}}\right)_\alpha \subseteq \left(\tilde{\mathbf{A}} \circledast_{EP} \tilde{\mathbf{B}}\right)_\alpha = \left(\tilde{\mathbf{A}} \circledast_{DT}^{\diamond} \tilde{\mathbf{B}}\right)_\alpha$$

for $\alpha \in [0,1]$.

Proof. It is clear to see that

$$\min_{(\mathbf{x},\mathbf{y}) \in (\tilde{\mathbf{A}}_\alpha, \tilde{\mathbf{B}}_\alpha)} \left(x_1y_1 + \cdots + x_ny_n\right) = \sum_{j=1}^{n} \left(\min_{(\mathbf{x},\mathbf{y}) \in (\tilde{\mathbf{A}}_\alpha, \tilde{\mathbf{B}}_\alpha)} x_jy_j\right) \tag{8.49}$$

and

$$\max_{(\mathbf{x},\mathbf{y}) \in (\tilde{\mathbf{A}}_\alpha, \tilde{\mathbf{B}}_\alpha)} \left(x_1y_1 + \cdots + x_ny_n\right) = \sum_{j=1}^{n} \left(\max_{(\mathbf{x},\mathbf{y}) \in (\tilde{\mathbf{A}}_\alpha, \tilde{\mathbf{B}}_\alpha)} x_jy_j\right), \tag{8.50}$$

since their objective functions are separable. Now, we have

$$\min_{(\mathbf{x},\mathbf{y})\in(\tilde{\mathbf{A}}_\alpha,\tilde{\mathbf{B}}_\alpha)} x_jy_j \leq \min\left\{\tilde{A}^L_{j\alpha}\tilde{B}^L_{j\alpha}, \tilde{A}^L_{j\alpha}\tilde{B}^U_{j\alpha}, \tilde{A}^U_{j\alpha}\tilde{B}^L_{j\alpha}, \tilde{A}^U_{j\alpha}\tilde{B}^U_{j\alpha}\right\}$$

and

$$\max_{(\mathbf{x},\mathbf{y})\in(\tilde{\mathbf{A}}_\alpha,\tilde{\mathbf{B}}_\alpha)} x_jy_j \geq \max\left\{\tilde{A}^L_{j\alpha}\tilde{B}^L_{j\alpha}, \tilde{A}^L_{j\alpha}\tilde{B}^U_{j\alpha}, \tilde{A}^U_{j\alpha}\tilde{B}^L_{j\alpha}, \tilde{A}^U_{j\alpha}\tilde{B}^U_{j\alpha}\right\}$$

which imply, by referring (8.49) and (8.50),

$$\min_{(\mathbf{x},\mathbf{y})\in(\tilde{\mathbf{A}}_\alpha,\tilde{\mathbf{B}}_\alpha)} (x_1y_1 + \cdots + x_ny_n) \leq \sum_{j=1}^{n}\left(\min\left\{\tilde{A}^L_{j\alpha}\tilde{B}^L_{j\alpha}, \tilde{A}^L_{j\alpha}\tilde{B}^U_{j\alpha}, \tilde{A}^U_{j\alpha}\tilde{B}^L_{j\alpha}, \tilde{A}^U_{j\alpha}\tilde{B}^U_{j\alpha}\right\}\right) \tag{8.51}$$

and

$$\max_{(\mathbf{x},\mathbf{y})\in(\tilde{\mathbf{A}}_\alpha,\tilde{\mathbf{B}}_\alpha)} (x_1y_1 + \cdots + x_ny_n) \geq \sum_{j=1}^{n}\left(\max\left\{\tilde{A}^L_{j\alpha}\tilde{B}^L_{j\alpha}, \tilde{A}^L_{j\alpha}\tilde{B}^U_{j\alpha}, \tilde{A}^U_{j\alpha}\tilde{B}^L_{j\alpha}, \tilde{A}^U_{j\alpha}\tilde{B}^U_{j\alpha}\right\}\right). \tag{8.52}$$

To prove part (i), we obtain

$$\left(\tilde{\mathbf{A}} \circledast_{EP} \tilde{\mathbf{B}}\right)^L_\alpha = \left(\tilde{\mathbf{A}} \circledast^{\diamond}_{DT} \tilde{\mathbf{B}}\right)^L_\alpha = \min_{(\mathbf{x},\mathbf{y})\in(\tilde{\mathbf{A}}_\alpha,\tilde{\mathbf{B}}_\alpha)} \mathbf{x}\bullet\mathbf{y} \text{ (using Theorem 8.1.12)}$$

$$\leq \left(\tilde{A}^{(1)}\otimes_{EP}\tilde{B}^{(1)}\right)^L_\alpha + \cdots + \left(\tilde{A}^{(n)}\otimes_{EP}\tilde{B}^{(n)}\right)^L_\alpha = \left(\tilde{\mathbf{A}} \odot_{EP} \tilde{\mathbf{B}}\right)^L_\alpha$$

(using (8.31), (8.34), (8.47) and (8.51)).

Using (8.52), we can similarly obtain

$$\left(\tilde{\mathbf{A}} \circledast_{EP} \tilde{\mathbf{B}}\right)^U_\alpha = \left(\tilde{\mathbf{A}} \circledast^{\diamond}_{DT} \tilde{\mathbf{B}}\right)^U_\alpha = \max_{(\mathbf{x},\mathbf{y})\in(\tilde{\mathbf{A}}_\alpha,\tilde{\mathbf{B}}_\alpha)} \mathbf{x}\bullet\mathbf{y}$$

$$\geq \left(\tilde{A}^{(1)}\otimes_{EP}\tilde{B}^{(1)}\right)^U_\alpha + \cdots + \left(\tilde{A}^{(n)}\otimes_{EP}\tilde{B}^{(n)}\right)^U_\alpha = \left(\tilde{\mathbf{A}} \odot_{EP} \tilde{\mathbf{B}}\right)^U_\alpha.$$

For $\alpha \in [0,1]$, it follows that

$$\left(\tilde{\mathbf{A}} \odot_{EP} \tilde{\mathbf{B}}\right)_\alpha = \left[\left(\tilde{\mathbf{A}} \odot_{EP} \tilde{\mathbf{B}}\right)^L_\alpha, \left(\tilde{\mathbf{A}} \odot_{EP} \tilde{\mathbf{B}}\right)^U_\alpha\right] \text{ (using (8.33))}$$

$$\subseteq \left[\left(\tilde{\mathbf{A}} \circledast_{EP} \tilde{\mathbf{B}}\right)^L_\alpha, \left(\tilde{\mathbf{A}} \circledast_{EP} \tilde{\mathbf{B}}\right)^U_\alpha\right] = \left(\tilde{\mathbf{A}} \circledast_{EP} \tilde{\mathbf{B}}\right)_\alpha = \left(\tilde{\mathbf{A}} \circledast^{\diamond}_{DT} \tilde{\mathbf{B}}\right)_\alpha.$$

To prove part (ii), we obtain

$$\left(\tilde{\mathbf{A}} \circledast_{EP} \tilde{\mathbf{B}}\right)^L_\alpha = \left(\tilde{\mathbf{A}} \circledast^{\diamond}_{DT} \tilde{\mathbf{B}}\right)^L_\alpha = \min_{(\mathbf{x},\mathbf{y})\in(\tilde{\mathbf{A}}_\alpha,\tilde{\mathbf{B}}_\alpha)} \mathbf{x}\bullet\mathbf{y} \text{ (using Theorem 8.1.12)}$$

$$\leq \left(\tilde{A}^{(1)} \otimes^{\diamond}_{DT} \tilde{B}^{(1)}\right)^L_\alpha + \cdots + \left(\tilde{A}^{(n)} \otimes^{\diamond}_{DT} \tilde{B}^{(n)}\right)^L_\alpha = \left(\tilde{\mathbf{A}} \odot_{DT} \tilde{\mathbf{B}}\right)^L_\alpha$$

(using (8.41), (8.46), (8.48), and (8.51)).

Using (8.52), we can similarly obtain

$$\left(\tilde{\mathbf{A}} \circledast_{EP} \tilde{\mathbf{B}}\right)^U_\alpha = \left(\tilde{\mathbf{A}} \circledast^{\diamond}_{DT} \tilde{\mathbf{B}}\right)^U_\alpha = \max_{(\mathbf{x},\mathbf{y})\in(\tilde{\mathbf{A}}_\alpha,\tilde{\mathbf{B}}_\alpha)} \mathbf{x}\bullet\mathbf{y}$$

$$\geq \left(\tilde{A}^{(1)} \otimes^{\diamond}_{DT} \tilde{B}^{(1)}\right)^U_\alpha + \cdots + \left(\tilde{A}^{(n)} \otimes^{\diamond}_{DT} \tilde{B}^{(n)}\right)^U_\alpha = \left(\tilde{\mathbf{A}} \odot_{DT} \tilde{\mathbf{B}}\right)^U_\alpha.$$

For $\alpha \in [0,1]$, it follows that

$$\left(\tilde{\mathbf{A}} \odot_{DT} \tilde{\mathbf{B}}\right)_\alpha = \left[\left(\tilde{\mathbf{A}} \odot_{DT} \tilde{\mathbf{B}}\right)^L_\alpha, \left(\tilde{\mathbf{A}} \odot_{DT} \tilde{\mathbf{B}}\right)^U_\alpha\right] \text{ (using (8.45))}$$

$$\subseteq \left[\left(\tilde{\mathbf{A}} \circledast_{EP} \tilde{\mathbf{B}}\right)^L_\alpha, \left(\tilde{\mathbf{A}} \circledast_{EP} \tilde{\mathbf{B}}\right)^U_\alpha\right] = \left(\tilde{\mathbf{A}} \circledast_{EP} \tilde{\mathbf{B}}\right)_\alpha = \left(\tilde{\mathbf{A}} \circledast^{\diamond}_{DT} \tilde{\mathbf{B}}\right)_\alpha.$$

This completes the proof. ■

Theorem 8.2.8 *Let $\tilde{A}^{(k)}$ and $\tilde{B}^{(k)}$ be canonical fuzzy intervals for $k = 1, \ldots, n$. Suppose that the first type of inner products $\tilde{\mathbf{A}} \circledast^{\star}_{DT} \tilde{\mathbf{B}}$ and $\tilde{\mathbf{A}} \circledast^{\dagger}_{DT} \tilde{\mathbf{B}}$ are obtained from Theorems 8.1.7 and 8.1.10, respectively. We also assume that the second type of inner product is given by*

$$\tilde{\mathbf{A}} \circledcirc_{DT} \tilde{\mathbf{B}} = \left(\tilde{A}^{(1)} \otimes^{\star}_{DT} \tilde{B}^{(1)}\right) \oplus \cdots \oplus \left(\tilde{A}^{(n)} \otimes^{\star}_{DT} \tilde{B}^{(n)}\right) \tag{8.53}$$

or

$$\tilde{\mathbf{A}} \circledcirc_{DT} \tilde{\mathbf{B}} = \left(\tilde{A}^{(1)} \otimes^{\dagger}_{DT} \tilde{B}^{(1)}\right) \oplus \cdots \oplus \left(\tilde{A}^{(n)} \otimes^{\dagger}_{DT} \tilde{B}^{(n)}\right). \tag{8.54}$$

Then, the second type of inner product $\tilde{\mathbf{A}} \circledcirc_{DT} \tilde{\mathbf{B}}$ is fuzzier than the first type of inner products $\tilde{\mathbf{A}} \circledast^{\star}_{DT} \tilde{\mathbf{B}}$ and $\tilde{\mathbf{A}} \circledast^{\dagger}_{DT} \tilde{\mathbf{B}}$ in the sense of

$$\left(\tilde{\mathbf{A}} \circledast^{\star}_{DT} \tilde{\mathbf{B}}\right)_{\alpha} = \left(\tilde{\mathbf{A}} \circledast^{\dagger}_{DT} \tilde{\mathbf{B}}\right)_{\alpha} \subseteq \left(\tilde{\mathbf{A}} \circledcirc_{DT} \tilde{\mathbf{B}}\right)_{\alpha}$$

for $\alpha \in [0,1]$.

Proof. Now, we have

$$\tilde{A}^{L}_{1\beta}\tilde{B}^{L}_{1\beta} + \cdots + \tilde{A}^{L}_{n\beta}\tilde{B}^{L}_{n\beta} \geq \min\left\{\tilde{A}^{L}_{1\beta}\tilde{B}^{L}_{1\beta}, \tilde{A}^{U}_{1\beta}\tilde{B}^{U}_{1\beta}\right\} + \cdots + \min\left\{\tilde{A}^{L}_{n\beta}\tilde{B}^{L}_{1\beta}, \tilde{A}^{U}_{1\beta}\tilde{B}^{U}_{n\beta}\right\}$$

and

$$\tilde{A}^{U}_{1\beta}\tilde{B}^{U}_{1\beta} + \cdots + \tilde{A}^{U}_{n\beta}\tilde{B}^{U}_{n\beta} \geq \min\left\{\tilde{A}^{L}_{1\beta}\tilde{B}^{L}_{1\beta}, \tilde{A}^{U}_{1\beta}\tilde{B}^{U}_{1\beta}\right\} + \cdots + \min\left\{\tilde{A}^{L}_{n\beta}\tilde{B}^{L}_{1\beta}, \tilde{A}^{U}_{1\beta}\tilde{B}^{U}_{n\beta}\right\},$$

which imply

$$\begin{aligned}\min\left\{\tilde{\mathbf{A}}^{L}_{\alpha} \bullet \tilde{\mathbf{B}}^{L}_{\alpha}, \tilde{\mathbf{A}}^{U}_{\alpha} \bullet \tilde{\mathbf{B}}^{U}_{\alpha}\right\} &= \min\left\{\tilde{A}^{L}_{1\beta}\tilde{B}^{L}_{1\beta} + \cdots + \tilde{A}^{L}_{n\beta}\tilde{B}^{L}_{n\beta}, \tilde{A}^{U}_{1\beta}\tilde{B}^{U}_{1\beta} + \cdots + \tilde{A}^{U}_{n\beta}\tilde{B}^{U}_{n\beta}\right\} \\ &\geq \min\left\{\tilde{A}^{L}_{1\beta}\tilde{B}^{L}_{1\beta}, \tilde{A}^{U}_{1\beta}\tilde{B}^{U}_{1\beta}\right\} + \cdots + \min\left\{\tilde{A}^{L}_{n\beta}\tilde{B}^{L}_{1\beta}, \tilde{A}^{U}_{1\beta}\tilde{B}^{U}_{n\beta}\right\}.\end{aligned} \tag{8.55}$$

We can similarly obtain

$$\max\left\{\tilde{\mathbf{A}}^{L}_{\alpha} \bullet \tilde{\mathbf{B}}^{L}_{\alpha}, \tilde{\mathbf{A}}^{U}_{\alpha} \bullet \tilde{\mathbf{B}}^{U}_{\alpha}\right\} \leq \max\left\{\tilde{A}^{L}_{1\beta}\tilde{B}^{L}_{1\beta}, \tilde{A}^{U}_{1\beta}\tilde{B}^{U}_{1\beta}\right\} + \cdots + \max\left\{\tilde{A}^{L}_{n\beta}\tilde{B}^{L}_{1\beta}, \tilde{A}^{U}_{1\beta}\tilde{B}^{U}_{n\beta}\right\}. \tag{8.56}$$

Therefore, we have

$$\begin{aligned}\left(\tilde{\mathbf{A}} \circledast^{\star}_{DT} \tilde{\mathbf{B}}\right)^{L}_{\alpha} &= \left(\tilde{\mathbf{A}} \circledast^{\dagger}_{DT} \tilde{\mathbf{B}}\right)^{L}_{\alpha} \\ &= \min_{\alpha \leq \beta \leq 1} \min\left\{\tilde{\mathbf{A}}^{L}_{\alpha} \bullet \tilde{\mathbf{B}}^{L}_{\alpha}, \tilde{\mathbf{A}}^{U}_{\alpha} \bullet \tilde{\mathbf{B}}^{U}_{\alpha}\right\} \text{ (using Theorem 8.1.13)} \\ &\geq \min_{\alpha \leq \beta \leq 1}\left[\min\left\{\tilde{A}^{L}_{1\beta}\tilde{B}^{L}_{1\beta}, \tilde{A}^{U}_{1\beta}\tilde{B}^{U}_{1\beta}\right\} + \cdots + \min\left\{\tilde{A}^{L}_{n\beta}\tilde{B}^{L}_{1\beta}, \tilde{A}^{U}_{1\beta}\tilde{B}^{U}_{n\beta}\right\}\right] \text{ (using (8.55))} \\ &\geq \min_{\alpha \leq \beta \leq 1} \min\left\{\tilde{A}^{L}_{1\beta}\tilde{B}^{L}_{1\beta}, \tilde{A}^{U}_{1\beta}\tilde{B}^{U}_{1\beta}\right\} + \cdots + \min_{\alpha \leq \beta \leq 1} \min\left\{\tilde{A}^{L}_{n\beta}\tilde{B}^{L}_{1\beta}, \tilde{A}^{U}_{1\beta}\tilde{B}^{U}_{n\beta}\right\} \\ &= \left(\tilde{A}^{(1)} \otimes^{\star}_{DT} \tilde{B}^{(1)}\right)^{L}_{\alpha} + \cdots + \left(\tilde{A}^{(n)} \otimes^{\star}_{DT} \tilde{B}^{(n)}\right)^{L}_{\alpha} \text{ (using (8.42) and (8.44))} \\ &= \left(\tilde{A}^{(1)} \otimes^{\dagger}_{DT} \tilde{B}^{(1)}\right)^{L}_{\alpha} + \cdots + \left(\tilde{A}^{(n)} \otimes^{\dagger}_{DT} \tilde{B}^{(n)}\right)^{L}_{\alpha} \\ &= \left(\tilde{\mathbf{A}} \circledcirc_{DT} \tilde{\mathbf{B}}\right)^{L}_{\alpha} \text{ (using (8.46), (8.53), and (8.54)).}\end{aligned}$$

Using (8.56), we can similarly obtain

$$\begin{aligned}\left(\tilde{\mathbf{A}} \circledast_{DT}^{\star} \tilde{\mathbf{B}}\right)_{\alpha}^{U} &= \left(\tilde{\mathbf{A}} \circledast_{DT}^{\dagger} \tilde{\mathbf{B}}\right)_{\alpha}^{U} = \max_{\alpha\le\beta\le 1} \max\left\{\tilde{\mathbf{A}}_{\alpha}^{L} \bullet \tilde{\mathbf{B}}_{\alpha}^{L}, \tilde{\mathbf{A}}_{\alpha}^{U} \bullet \tilde{\mathbf{B}}_{\alpha}^{U}\right\}\\ &\le \left(\tilde{A}^{(1)} \otimes_{DT}^{\star} \tilde{B}^{(1)}\right)_{\alpha}^{U} + \cdots + \left(\tilde{A}^{(n)} \otimes_{DT}^{\star} \tilde{B}^{(n)}\right)_{\alpha}^{U}\\ &= \left(\tilde{A}^{(1)} \otimes_{DT}^{\dagger} \tilde{B}^{(1)}\right)_{\alpha}^{U} + \cdots + \left(\tilde{A}^{(n)} \otimes_{DT}^{\dagger} \tilde{B}^{(n)}\right)_{\alpha}^{U} = \left(\tilde{\mathbf{A}} \circledcirc_{DT} \tilde{\mathbf{B}}\right)_{\alpha}^{U}.\end{aligned}$$

For $\alpha \in [0,1]$, it follows that

$$\begin{aligned}\left(\tilde{\mathbf{A}} \circledast_{DT}^{\dagger} \tilde{\mathbf{B}}\right)_{\alpha} &= \left(\tilde{\mathbf{A}} \circledast_{DT}^{\star} \tilde{\mathbf{B}}\right)_{\alpha} = \left[\left(\tilde{\mathbf{A}} \circledast_{DT}^{\star} \tilde{\mathbf{B}}\right)_{\alpha}^{L}, \left(\tilde{\mathbf{A}} \circledast_{DT}^{\star} \tilde{\mathbf{B}}\right)_{\alpha}^{U}\right]\\ &\subseteq \left[\left(\tilde{\mathbf{A}} \circledcirc_{DT} \tilde{\mathbf{B}}\right)_{\alpha}^{L}, \left(\tilde{\mathbf{A}} \circledcirc_{DT} \tilde{\mathbf{B}}\right)_{\alpha}^{U}\right] = \left(\tilde{\mathbf{A}} \circledcirc_{DT} \tilde{\mathbf{B}}\right)_{\alpha}.\end{aligned}$$

This completes the proof. ∎

Two different types of inner product of fuzzy vectors have been completely studied in this chapter. Also, each type of inner product will be established using the extension principle and the expression in the decomposition theorem. In other words, we can establish four different kinds of inner product.

- Regarding the first type of inner product, the inner product $\tilde{\mathbf{A}} \circledast_{EP} \tilde{\mathbf{B}}$ using the extension principle has been established in (8.2). On the other hand, using the expression in the decomposition theorem, three inner products $\tilde{\mathbf{A}} \circledast_{DT}^{\diamond} \tilde{\mathbf{B}}$, $\tilde{\mathbf{A}} \circledast_{DT}^{\star} \tilde{\mathbf{B}}$, and $\tilde{\mathbf{A}} \circledast_{DT}^{\dagger} \tilde{\mathbf{B}}$ have been establish in (8.9), (8.11), and (8.13), respectively, based on three different families of sets in $\mathbb{R}$.
- Regarding the second type inner product, the inner product $\tilde{\mathbf{A}} \circledcirc_{EP} \tilde{\mathbf{B}}$ using the extension principle has been established in (8.30). On the other hand, the inner products $\tilde{\mathbf{A}} \circledcirc_{DT} \tilde{\mathbf{B}}$ using the expression in the decomposition theorem has been established in (8.40) in which three different families (8.22), (8.23), and (8.24) are implicitly considered.

The equivalences among these different inner products have also been investigated. For the cases of non-equivalences, the relations regarding their fuzziness have also been established. In a real-world situation, we prefer to take the inner product that has less fuzziness.

- Theorem 8.1.12 shows the equivalence

  $$\tilde{\mathbf{A}} \circledast_{EP} \tilde{\mathbf{B}} = \tilde{\mathbf{A}} \circledast_{DT}^{\diamond} \tilde{\mathbf{B}}.$$

  Theorem 8.1.13 shows the equivalence

  $$\tilde{\mathbf{A}} \circledast_{DT}^{\star} \tilde{\mathbf{B}} = \tilde{\mathbf{A}} \circledast_{DT}^{\dagger} \tilde{\mathbf{B}}.$$

  The equivalence between $\tilde{\mathbf{A}} \circledast_{DT}^{\diamond} \tilde{\mathbf{B}}$ and $\tilde{\mathbf{A}} \circledast_{DT}^{\star} \tilde{\mathbf{B}}$ cannot be established. Theorem 8.1.15 says that $\tilde{\mathbf{A}} \circledast_{DT}^{\diamond} \tilde{\mathbf{B}}$ is fuzzier than $\tilde{\mathbf{A}} \circledast_{DT}^{\star} \tilde{\mathbf{B}}$. In this case, we prefer to take $\tilde{\mathbf{A}} \circledast_{DT}^{\star} \tilde{\mathbf{B}}$ rather than $\tilde{\mathbf{A}} \circledast_{DT}^{\diamond} \tilde{\mathbf{B}}$ because of the issue of fuzziness.
- Theorem 8.2.7 says that when the second type of inner product $\tilde{\mathbf{A}} \circledcirc_{EP} \tilde{\mathbf{B}}$ is given by (8.47) or $\tilde{\mathbf{A}} \circledcirc_{DT} \tilde{\mathbf{B}}$ is given by (8.48), we prefer to take this second type of inner products rather than the first type of inner product $\tilde{\mathbf{A}} \circledast_{EP} \tilde{\mathbf{B}}$ and $\tilde{\mathbf{A}} \circledast_{DT}^{\diamond} \tilde{\mathbf{B}}$ because of the issue of fuzziness.

- Theorem 8.2.8 says that when the second type of inner product $\tilde{\mathbf{A}} \odot_{DT} \tilde{\mathbf{B}}$ is given by (8.53) or (8.54), we prefer to take the first type of inner products $\tilde{\mathbf{A}} \circledast_{DT}^{\star} \tilde{\mathbf{B}}$ and $\tilde{\mathbf{A}} \circledast_{DT}^{\dagger} \tilde{\mathbf{B}}$ rather than the second type of inner product $\tilde{\mathbf{A}} \circledcirc \tilde{\mathbf{B}}$ because of the issue of fuzziness.

Recall that a fuzzy interval $\tilde{A}$ is nonnegative when $\tilde{A}_\alpha^L$ and $\tilde{A}_\alpha^U$ are nonnegative real numbers for all $\alpha \in [0,1]$, and a fuzzy interval $\tilde{A}$ is nonpositive when $\tilde{A}_\alpha^L$ and $\tilde{A}_\alpha^U$ are nonpositive real numbers for all $\alpha \in [0,1]$. The fuzzy interval that is neither nonnegative nor nonpositive is called a mixed type of fuzzy interval, which means that its membership function will be across the y-axis. The fuzzy intervals that are taken to be nonnegative or nonpositive will simplify the calculation of inner product. For the case of mixed type of fuzzy intervals, Examples 8.1.3, 8.1.9, 8.2.3, and 8.2.6 have demonstrated the essence of calculation. In real-world applications, the fuzzy data can be strongly nonnegative or nonpositive, which says that the fuzzy intervals can be taken to be nonnegative or nonpositive. The mixed type of fuzzy interval means that its corresponding fuzzy data is around zero. When the concrete problem has less numbers of fuzzy data that are around zero, which seems to frequently occur in real-world applications, the less numbers of corresponding mixed type of fuzzy intervals can be taken. This situation will also simplify the calculation.

The potential application considering the inner product of fuzzy vectors is to solve the fuzzy linear programming problem as shown below

$$
\begin{aligned}
\text{(FLP)} \quad \max \quad & (\tilde{a}_1 \otimes \tilde{x}_1) \oplus (\tilde{a}_2 \otimes \tilde{x}_2) \oplus \cdots \oplus (\tilde{a}_n \otimes \tilde{x}_n) \\
\text{subject to} \quad & (\tilde{b}_{i1} \otimes \tilde{x}_1) \oplus (\tilde{b}_{i2} \otimes \tilde{x}_2) \oplus \cdots \oplus (\tilde{b}_{in} \otimes \tilde{x}_n) \preccurlyeq \tilde{c}_i \text{ for } i = 1, \ldots, m; \\
& \text{each } \tilde{x}_j \text{is a nonnegative fuzzy interval for } j = 1, \ldots, n,
\end{aligned}
$$

where the coefficients $\tilde{a}_j$, $\tilde{b}_{ij}$ and $\tilde{c}_i$ for $i = 1, \ldots, m$ and $j = 1, \ldots, n$ are taken to be fuzzy intervals. For most of the cases, those fuzzy intervals can be assumed to be nonnegative, since the coefficients of linear programming problem frequently represent the prices of items, budget of project, and benefit of corporation, which are always positive. Considering nonnegative fuzzy intervals in fuzzy linear programming problem will simplify the calculation of inner products of fuzzy vectors as explained above.

9

Gradual Elements and Gradual Sets

Let A be a (crisp) subset of $\mathbb{R}^m$. The concept of an element in A is well known by writing $x \in A$. Suppose that $\tilde{A}$ is a fuzzy set in $\mathbb{R}^m$. The main focus of this chapter is to consider the elements of fuzzy set $\tilde{A}$. The gradual elements will play the role to be regarded as the elements of $\tilde{A}$.

The concepts of gradual elements and gradual sets were introduced by Dubois and Prade [29, 30] and Fortin et al. [37], which were inspired by Goetschel [41] and Herencia and Lamata [47]. The gradual element is a function from (0,1] into $\mathbb{R}^m$, and the gradual set is a set-valued function from (0,1] into the hyperspace that consists of all sets in $\mathbb{R}^m$.

9.1 Gradual Elements and Gradual Sets

We denote by $\mathcal{P}(\mathbb{R}^m)$ the collection of all subsets of $\mathbb{R}^m$, which is also called the power set or hyperspace of $\mathbb{R}^m$.

Definition 9.1.1 The **gradual element** g in $\mathbb{R}^m$ is defined to be an assignment function $g : [0,1] \to \mathbb{R}^m$. If $m = 1$, then the gradual element is also called a **gradual number**. The **gradual set** $\mathfrak{G}$ in $\mathbb{R}^m$ is defined to be an assignment function $\mathfrak{G} : [0,1] \to \mathcal{P}(\mathbb{R}^m)$ such that each $\mathfrak{G}(\alpha)$ is a nonempty subset of $\mathbb{R}^m$ for $\alpha \in [0,1]$.

Definition 9.1.2 We say that the gradual set $\mathfrak{G}$ is **nested** when $\mathfrak{G}(\alpha) \subseteq \mathfrak{G}(\beta)$ for $\alpha, \beta \in [0,1]$ with $\alpha > \beta$. We say that the gradual set G is **strictly nested** when $\mathfrak{G}(\alpha) \subset \mathfrak{G}(\beta)$ and $\mathfrak{G}(\alpha) \neq \mathfrak{G}(\beta)$ for $\alpha, \beta \in [0,1]$ with $\alpha > \beta$.

Let $\tilde{A}$ be a fuzzy set in $\mathbb{R}^m$ with membership function $\xi_{\tilde{A}}$. We can generate a gradual set $\mathfrak{G}_{\tilde{A}}$ from $\tilde{A}$ by defining the assignment function $\mathfrak{G}_{\tilde{A}} : [0,1] \to \mathcal{P}(\mathbb{R}^m)$ as $\mathfrak{G}_{\tilde{A}}(\alpha) = \tilde{A}_\alpha$.

Example 9.1.3 Let $\tilde{A}$ be a fuzzy set in $\mathbb{R}$ with membership function given by

$$\xi_{\tilde{A}}(r) = \begin{cases} (r - a^L)/(a_1 - a^L) & \text{if } a^L \leq r \leq a_1 \\ 1 & \text{if } a_1 < r \leq a_2 \\ (a^U - r)/(a^U - a_2) & \text{if } a_2 < r \leq a^U \\ 0 & \text{otherwise.} \end{cases}$$

Mathematical Foundations of Fuzzy Sets, First Edition. Hsien-Chung Wu.

For $\alpha \in [0,1]$, the α-level set of $\tilde{A}$ is a bounded closed interval denoted by $\tilde{A}_\alpha = [\tilde{A}_\alpha^L, \tilde{A}_\alpha^U]$, where

$$\tilde{A}_\alpha^L = (1-\alpha)a^L + \alpha a_1 \text{ and } \tilde{A}_\alpha^U = (1-\alpha)a^U + \alpha a_2. \tag{9.1}$$

Therefore, the gradual set $\mathfrak{G}_{\tilde{A}} : [0,1] \to \mathcal{P}(\mathbb{R})$ generated from the fuzzy set $\tilde{A}$ is given by

$$\mathfrak{G}_{\tilde{A}}(\alpha) = \left[\tilde{A}_\alpha^L, \tilde{A}_\alpha^U\right] = \left[(1-\alpha)a^L + \alpha a_1, (1-\alpha)a^U + \alpha a_2\right]$$

for $\alpha \in [0,1]$.

On the other hand, given a gradual set $\mathfrak{G} : [0,1] \to \mathcal{P}(\mathbb{R}^m)$, we can generate a fuzzy set $\tilde{A}^{\mathfrak{G}}$ in $\mathbb{R}^m$ by using the expression in the decomposition theorem. The membership function of $\tilde{A}^{\mathfrak{G}}$ is then defined by

$$\xi_{\tilde{A}^{\mathfrak{G}}}(x) = \sup_{\alpha \in [0,1]} \alpha \cdot \chi_{\mathfrak{G}(\alpha)}(x). \tag{9.2}$$

Example 9.1.4 Given a gradual set $\mathfrak{G} : [0,1] \to \mathcal{P}(\mathbb{R})$ defined by

$$\mathfrak{G}(\alpha) = \left[(1-\alpha)x_1 + \alpha x_2, (1-\alpha)x_3 + \alpha x_4\right]$$

for $\alpha \in [0,1]$, where $x_1 < x_2 < x_3 < x_4$. It is clear to see that the gradual set $\mathfrak{G}$ is strictly nested. Then, we can generate a fuzzy set $\tilde{A}^{\mathfrak{G}}$ in $\mathbb{R}$ with membership function given by

$$\xi_{\tilde{A}^{\mathfrak{G}}}(x) = \sup_{\alpha \in [0,1]} \alpha \cdot \chi_{\mathfrak{G}(\alpha)}(x) = \sup_{\alpha \in [0,1]} \alpha \cdot \chi_{[(1-\alpha)x_1+\alpha x_2,(1-\alpha)x_3+\alpha x_4]}(x).$$

In the subsequent discussion, we are going to claim that the α-level set $\tilde{A}_\alpha^{\mathfrak{G}}$ of $\tilde{A}^{\mathfrak{G}}$ is equal to $\mathfrak{G}(\alpha)$, i.e. $\tilde{A}_\alpha^{\mathfrak{G}} = \mathfrak{G}(\alpha)$ for $\alpha \in [0,1]$.

Let Λ be an index set that can be an uncountable set. We consider a family $\{g_\lambda : \lambda \in \Lambda\}$ of gradual elements in $\mathbb{R}^m$. This family of gradual elements can generate a gradual set $\mathfrak{G}$ in $\mathbb{R}^m$ by

$$\mathfrak{G}(\alpha) = \left\{g_\lambda(\alpha) : \lambda \in \Lambda\right\} \text{ for } \alpha \in [0,1]. \tag{9.3}$$

Based on this gradual set $\mathfrak{G}$ in $\mathbb{R}^m$, we can also generate a fuzzy set $\tilde{A}^{\mathfrak{G}}$ in $\mathbb{R}^m$ with membership function defined in (9.2). In this case, we can also say that the family $\{g_\lambda : \lambda \in \Lambda\}$ of gradual elements generates the fuzzy set $\tilde{A}^{\mathfrak{G}}$.

Example 9.1.5 We take the index set $\Lambda = [0,1]$. Let

$$f_1(\alpha) = (1-\alpha)x_1 + \alpha x_2 \text{ and } f_2(\alpha) = (1-\alpha)x_3 + \alpha x_4$$

for $\alpha \in [0,1]$, where $x_1 < x_2 < x_3 < x_4$. Then $f_1(\alpha) \leq f_2(\alpha)$ for all $\alpha \in [0,1]$. For $\lambda \in \Lambda = [0,1]$, we define $g_\lambda : [0,1] \to \mathbb{R}$ by

$$g_\lambda(\alpha) = \lambda f_1(\alpha) + (1-\lambda)f_2(\alpha).$$

Then, we have a family $\{g_\lambda : \lambda \in [0,1]\}$ of gradual numbers. According to (9.3), we can generate a gradual set $\mathfrak{G} : [0,1] \to \mathcal{P}(\mathbb{R})$ given by

$$\mathfrak{G}(\alpha) = \left\{g_\lambda(\alpha) : \lambda \in [0,1]\right\} = \left[f_1(\alpha), f_2(\alpha)\right] = \left[(1-\alpha)x_1 + \alpha x_2, (1-\alpha)x_3 + \alpha x_4\right]$$

that is a bounded closed interval for $\alpha \in [0,1]$. This gradual set $\mathfrak{G}$ is the same as the gradual set in Example 9.1.4.

The gradual set $\mathfrak{G}$ is a set-valued function defined on [0,1] such that its function value $\mathfrak{G}(\alpha)$ is a subset of $\mathbb{R}^m$ for each $\alpha \in [0,1]$. According to the topic of set-valued analysis, the **selector** (or **selection function**) of set-valued function $\mathfrak{G}$ is a single-valued function $g : [0,1] \to \mathbb{R}^m$ defined on [0,1] by $g(\alpha) \in \mathfrak{G}(\alpha)$. In this case, we may write $g \in \mathfrak{G}$. It is clear to see that the selector of gradual set $\mathfrak{G}$ (i.e. set-valued function $\mathfrak{G}$) is a gradual element in $\mathbb{R}^m$. In some sense, we may say that the gradual set consists of gradual elements, which is similar to say that the (usual) set consists of (usual) elements.

Given a subset A of $\mathbb{R}^m$, the concept of elements of A can be realized in the usual sense by simply writing $a \in A$ when a is assumed to be an element of A. For the fuzzy set $\tilde{A}$ in $\mathbb{R}^m$, we plan to consider the concept of element of $\tilde{A}$ by also simply writing $\hat{a} \in \tilde{A}$, where the definition of $\hat{a}$ will be presented below.

Given a fuzzy set $\tilde{A}$ in $\mathbb{R}^m$, we can generate a gradual set $\mathfrak{G}_{\tilde{A}}$ given by $\mathfrak{G}_{\tilde{A}}(\alpha) = \tilde{A}_\alpha$. Therefore, we have the selector $\hat{a}$ of $\mathfrak{G}_{\tilde{A}}$ given by

$$\hat{a}(\alpha) \in \mathfrak{G}_{\tilde{A}}(\alpha) = \tilde{A}_\alpha \text{ for } \alpha \in [0,1].$$

We also see that the selector $\hat{a}$ is a gradual element in $\mathbb{R}^m$. This gradual element $\hat{a}$ can be regarded as an element of $\tilde{A}$ by simply writing $\hat{a} \in \tilde{A}$. The formal definition is given below.

Definition 9.1.6 Let $\tilde{A}$ be a fuzzy set in $\mathbb{R}^m$.

- We say that an **element** $\hat{a}$ is in $\tilde{A}$ when $\hat{a}$ is a gradual element in $\mathbb{R}^m$ satisfying $\hat{a}(\alpha) \in \tilde{A}_\alpha$ for each $\alpha \in [0,1]$. In this case, we also write $\hat{a} \in \tilde{A}$.
- We say that $\hat{A}$ is a subset of fuzzy set $\tilde{A}$ when $\hat{A}$ is a collection of gradual elements such that each $\hat{a}$ in $\hat{A}$ satisfies $\hat{a} \in \tilde{A}$. In this case, we also write $\hat{A} \subset \tilde{A}$.

Example 9.1.7 Consider the fuzzy set $\tilde{A}$ given in Example 9.1.3. The α-level set is given by

$$\tilde{A}_\alpha = \left[(1-\alpha)a^L + \alpha a_1, (1-\alpha)a^U + \alpha a_2\right] \equiv \left[\tilde{A}^L_\alpha, \tilde{A}^U_\alpha\right]$$

for $\alpha \in [0,1]$. We define the gradual number $\hat{a} : [0,1] \to \mathbb{R}$ by

$$\hat{a}(\alpha) = (1-\alpha)a^L + \alpha a_1 = \tilde{A}^L_\alpha$$

for $\alpha \in [0,1]$. Then $\hat{a} \in \tilde{A}$. In general, for $\lambda \in [0,1]$, we define

$$\hat{a}_\lambda(\alpha) = \lambda \tilde{A}^L_\alpha + (1-\lambda)\tilde{A}^U_\alpha.$$

Then $\hat{a}_\lambda \in \tilde{A}$ for $\lambda \in [0,1]$.

Since $\hat{a} \in \tilde{A}$ is a gradual element in $\mathbb{R}^m$, this gradual element $\hat{a}$ can also be regarded as a gradual set $\mathfrak{G}$ given by $\mathfrak{G}(\alpha) = \{\hat{a}(\alpha)\}$ that is a singleton for $\alpha \in [0,1]$. According to (9.2), we can generate a fuzzy set $\tilde{a}$ in $\mathbb{R}^m$ with membership function given by

$$\xi_{\tilde{a}}(x) = \sup_{\alpha\in[0,1]} \alpha \cdot \chi_{\mathfrak{G}(\alpha)}(x) = \sup_{\alpha\in[0,1]} \alpha \cdot \chi_{\{\hat{a}(\alpha)\}}(x) = \sup_{0<\alpha\leq 1} \alpha \cdot \chi_{\{\hat{a}(\alpha)\}}(x)$$

$$= \begin{cases} 0, & \text{if there is no } \alpha \in (0,1] \text{ satisfying } \hat{a}(\alpha) = x \\ \displaystyle\sup_{\{\alpha\in(0,1]:\hat{a}(\alpha)=x\}} \alpha, & \text{otherwise.} \end{cases}$$

Therefore, the formal definition of membership function of element $\hat{a} \in \tilde{A}$ is given below.

Definition 9.1.8 Let $\tilde{A}$ be a fuzzy set in $\mathbb{R}^m$. Given any gradual element $\hat{a} \in \tilde{A}$, the membership function of its corresponding fuzzy set $\tilde{a}$ is defined by

$$\xi_{\tilde{a}}(x) = \begin{cases} 0, & \text{if there is no } \alpha \in (0,1] \text{ satisfying } \hat{a}(\alpha) = x \\ \sup\limits_{\{\alpha \in (0,1] : \hat{a}(\alpha) = x\}} \alpha, & \text{otherwise.} \end{cases} \tag{9.4}$$

In what follows, we are going to claim that Definition 9.1.6 is well defined. A fuzzy set $\tilde{A}$ in $\mathbb{R}^m$ can be regarded as a family consisting of elements (i.e. gradual elements) in $\tilde{A}$. Therefore, according to (9.3), this family $\tilde{A}$ of gradual elements can generate a gradual set $\mathfrak{G}$ given by

$$\mathfrak{G}(\alpha) = \left\{\hat{a}(\alpha) : \hat{a} \in \tilde{A}\right\} \text{ for } \alpha \in [0,1], \tag{9.5}$$

Also, the fuzzy set $\tilde{A}$ can generate another gradual set $\mathfrak{G}_{\tilde{A}}$ given by

$$\mathfrak{G}_{\tilde{A}}(\alpha) = \tilde{A}_\alpha \text{ for } \alpha \in [0,1]. \tag{9.6}$$

On the other hand, according to (9.2), the gradual set $\mathfrak{G}$ in (9.5) can also generate another fuzzy set $\tilde{A}^{\mathfrak{G}}$ in $\mathbb{R}^m$. In order to claim the consistency of Definition 9.1.6, we need to show $\mathfrak{G}(\alpha) = \mathfrak{G}_{\tilde{A}}(\alpha)$ for all $\alpha \in [0,1]$ and $\tilde{A}^{\mathfrak{G}} = \tilde{A}$, which will be presented below.

Proposition 9.1.9 *Let $\tilde{A}$ be a fuzzy set in $\mathbb{R}^m$. Then, the following statements hold true.*

(i) *The gradual set $\mathfrak{G}$ generated by the family $\tilde{A}$ of gradual elements satisfies*

$$\mathfrak{G}(\alpha) = \tilde{A}_\alpha = \mathfrak{G}_{\tilde{A}}(\alpha) \text{ for each } \alpha \in [0,1].$$

(ii) *Let $\tilde{A}^{\mathfrak{G}}$ be a fuzzy set in $\mathbb{R}^m$ generated by the gradual set $\mathfrak{G}$ in part* (i). *Then $\tilde{A}^{\mathfrak{G}} = \tilde{A}$.*

Proof. To prove part (i), according to (9.5) and the definition of $\hat{a} \in \tilde{A}$, we see that $\mathfrak{G}(\alpha) \subseteq \tilde{A}_\alpha$ for $\alpha \in [0,1]$. On the other hand, given any fixed $\alpha \in [0,1]$ and any $x \in \tilde{A}_\alpha$, we define a function $\hat{a}$ on $[0,1]$ by

$$\hat{a}(\beta) = \begin{cases} x, & \text{if } \beta = \alpha \\ y \text{ for some } y \in \tilde{A}_\beta, & \text{if } \beta \neq \alpha. \end{cases}$$

Then it is clear to see that $\hat{a} \in \tilde{A}$. This shows that $\hat{a}(\alpha) = x \in \mathfrak{G}(\alpha)$, i.e. $\tilde{A}_\alpha \subseteq \mathfrak{G}(\alpha)$. Therefore, we obtain $\mathfrak{G}(\alpha) = \tilde{A}_\alpha$ for $\alpha \in [0,1]$.

To prove part (ii), from (9.2), the membership function of $\tilde{A}^{\mathfrak{G}}$ is given by

$$\xi_{\tilde{A}^{\mathfrak{G}}}(x) = \sup_{\alpha \in [0,1]} \alpha \cdot \chi_{\mathfrak{G}(\alpha)}(x).$$

Using part (i) of the Decomposition Theorem 2.2.13, the membership function of $\tilde{A}$ can be expressed as

$$\xi_{\tilde{A}}(x) = \sup_{\alpha \in [0,1]} \alpha \cdot \chi_{\tilde{A}_\alpha}(x).$$

Since $\mathfrak{G}(\alpha) = \tilde{A}_\alpha$ for each $\alpha \in [0,1]$ by part (i), we obtain $\tilde{A}^{\mathfrak{G}} = \tilde{A}$ by referring to their membership functions. This completes the proof. ■

Let $\tilde{A}$ be a fuzzy interval. Its α-level sets are bounded closed intervals given by $\tilde{A}_\alpha = [\tilde{A}_\alpha^L, \tilde{A}_\alpha^U]$ for $\alpha \in [0,1]$. We can generate two gradual numbers l and u from $[0,1]$ into $\mathbb{R}$ defined by $l(\alpha) = \tilde{A}_\alpha^L$ and $u(\alpha) = \tilde{A}_\alpha^U$. In other words, each fuzzy interval $\tilde{A}$ can be represented by a pair of gradual numbers (l, u). We can also generate a gradual interval $\mathfrak{I} : [0,1] \to \mathcal{P}(\mathbb{R})$ defined by

$$\mathfrak{I}(\alpha) = \tilde{A}_\alpha = \left[\tilde{A}_\alpha^L, \tilde{A}_\alpha^U\right] = \left[l(\alpha), u(\alpha)\right],$$

which also says that each fuzzy interval can be represented by a gradual interval $\mathfrak{I}$.

Let $\tilde{A}$ be a fuzzy interval represented by a pair of gradual numbers (l, u). Given a bounded closed interval $[a, b]$ in $\mathbb{R}$, an element $r \in [a, b]$ means $a \le r \le b$. Now, the fuzzy interval $\tilde{A}$ can be regarded as a kind of gradual closed interval with the gradual number l as the left endpoint and the gradual number u as the right endpoint of this gradual closed interval. By mimicking the belongness $r \in [a, b]$ in $\mathbb{R}$, Fortin et al. [37] consider the belongness $g \in \tilde{A}$ when g is a gradual number satisfying $l \le g \le u$, i.e. $l(\alpha) \le g(\alpha) \le u(\alpha)$ for all $\alpha \in [0,1]$. This belongness is consistent with Definition 9.1.6 in which $g \in \tilde{A}$ means

$$g(\alpha) \in \tilde{A}_\alpha = \left[\tilde{A}_\alpha^L, \tilde{A}_\alpha^U\right] = \left[l(\alpha), u(\alpha)\right] \tag{9.7}$$

for each $\alpha \in [0,1]$.

9.2 Fuzzification Using Gradual Numbers

Let $f : \mathbb{R}^m \to \mathbb{R}$ be a real-valued function defined on $\mathbb{R}^m$. Given gradual numbers g_i for $i = 1, \ldots, n$, by referring to Fortin et al. [37] and Sanchez et al. [101], we can generate a gradual number $f(g_1, \ldots, g_n)$ defined by

$$f\left(g_1, \ldots, g_n\right)(\alpha) = f\left(g_1(\alpha), \ldots, g_n(\alpha)\right)$$

for $\alpha \in [0,1]$.

Given any bounded closed intervals $A_i = [a_i^L, a_i^U]$ in $\mathbb{R}$ for $i = 1, \ldots, n$, we can apply the real-valued function f to the vector $(A_1, \ldots, A_n)$ of bounded closed intervals by defining

$$\begin{aligned} f\left(A_1, \ldots, A_n\right) &= \left\{f\left(x_1, \ldots, x_n\right) : x_i \in A_i \text{ for } i = 1, \ldots, n\right\} \\ &= \left\{f\left(x_1, \ldots, x_n\right) : a_i^L \le x_i \le a_i^U \text{ for } i = 1, \ldots, n\right\}. \end{aligned} \tag{9.8}$$

We see that $f(A_1, \ldots, A_n)$ is not necessarily a bounded closed interval.

Proposition 9.2.1 (Apostol [3]) *We consider the function $f : (M_1, d_1) \to (M_2, d_2)$ from the metric space (M_1, d_1) into the metric space (M_2, d_2). Let X be a connected subset of M_1. Suppose that f is continuous on X. Then $f(X)$ is a connected subset of M_2.*

Suppose that the real-valued function $f : \mathbb{R}^m \to \mathbb{R}$ is continuous. Given any bounded closed intervals $A_i = [a_i^L, a_i^U]$ in $\mathbb{R}$ for $i = 1, \ldots, n$, since $(A_1, \ldots, A_n)$ is a connected subset of $\mathbb{R}^m$, Proposition 9.2.1 says that $f(A_1, \ldots, A_n)$ is a connected subset of $\mathbb{R}$, which also means that $f(A_1, \ldots, A_n)$ is a bounded interval in $\mathbb{R}$. Therefore, it could be a bounded open interval, half-open interval, or closed interval in $\mathbb{R}$. The continuity of f also says that the supremum

$$a^U \equiv \sup f\left(A_1, \ldots, A_n\right) = \sup_{\{(x_1, \ldots, x_n) : a_i^L \le x_i \le a_i^U, i=1,\ldots,n\}} f\left(x_1, \ldots, x_n\right)$$

and the infimum

$$a^L \equiv \inf f\left(A_1, \dots, A_n\right) = \inf_{\{(x_1,\dots,x_n):\, a_i^L \le x_i \le a_i^U, i=1,\dots,n\}} f\left(x_1, \dots, x_n\right)$$

are attained. In other words, there exist $(x_1^*, \dots, x_n^*)$ and $(x_1^\circ, \dots, x_n^\circ)$ satisfying

$$a^U = f\left(x_1^*, \dots, x_n^*\right) \text{ for } a_i^L \le x_i^* \le a_i^U \text{ and } i = 1, \dots, n$$

and

$$a^L = f\left(x_1^\circ, \dots, x_n^\circ\right) \text{ for } a_i^L \le x_i^\circ \le a_i^U \text{ and } i = 1, \dots, n.$$

Therefore, we conclude that $f(A_1, \dots, A_n) = [a^L, a^U]$ is indeed a bounded closed interval when f is assumed to be continuous.

According to (9.3), we can generate a gradual set $\mathfrak{G}$ defined by

$$\begin{aligned}
\mathfrak{G}(\alpha) &= \left\{f\left(g_1, \dots, g_n\right)(\alpha) : g_i \in \tilde{A}^{(i)} \text{ for } i = 1, \dots, n\right\} \\
&= \left\{f\left(g_1(\alpha), \dots, g_n(\alpha)\right) : g_i(\alpha) \in \tilde{A}_\alpha^{(i)} \text{ for } i = 1, \dots, n\right\} \\
&= f\left(\tilde{A}_\alpha^{(1)}, \dots, \tilde{A}_\alpha^{(n)}\right).
\end{aligned}$$

According to (9.2), we can generate a fuzzy set $\tilde{f}(\tilde{A}^{(1)}, \dots, \tilde{A}^{(n)})$ with membership function given by

$$\xi_{\tilde{f}(\tilde{A}^{(1)},\dots,\tilde{A}^{(n)})}(x) = \sup_{\alpha\in[0,1]} \alpha \cdot \chi_{\mathfrak{G}(\alpha)}(x),$$

which is a fuzzification of f using gradual numbers.

9.3 Elements and Subsets of Fuzzy Intervals

The functions η_1 and η_2 in Theorem 5.5.6 are now assumed to be continuous and be defined on the whole unit interval [0,1]. Then, the functions

$$\zeta^L(\alpha) = \min\{\eta_1(\alpha), \eta_2(\alpha)\} \text{ and } \zeta^U(\alpha) = \max\{\eta_1(\alpha), \eta_2(\alpha)\}$$

are also continuous on [0,1]. Theorem 5.5.6 can be used to generate a fuzzy interval $\tilde{A}$ such that its α-level set is given by

$$\tilde{A}_\alpha = \mathfrak{G}(\alpha) = \left[\min_{\alpha\le\beta\le1} \zeta^L(\beta), \max_{\alpha\le\beta\le1} \zeta^U(\beta)\right]. \tag{9.9}$$

We define two functions l and u on [0,1] by

$$l(\alpha) = \min_{\alpha\le\beta\le1} \zeta^L(\beta) \tag{9.10}$$

and

$$u(\alpha) = \max_{\alpha\le\beta\le1} \zeta^U(\beta). \tag{9.11}$$

We can consider the elements of $\tilde{A}$ by saying $g \in \tilde{A}$, where g is a gradual number satisfying

$$g(\alpha) \in \tilde{A}_\alpha = \left[l(\alpha), u(\alpha)\right]$$

for $\alpha \in [0,1]$.

Given any fixed $\lambda \in [0,1]$, we consider the gradual number

$$\hat{a}_\lambda(\alpha) = \lambda l(\alpha) + (1-\lambda)u(\alpha) \text{ for } \alpha \in [0,1]. \tag{9.12}$$

Then, it is clear to see that

$$\hat{a}_\lambda(\alpha) \in \left[l(\alpha), u(\alpha)\right] = \tilde{A}_\alpha,$$

which says that each $\hat{a}_\lambda$ is an element of $\tilde{A}$ for $\lambda \in [0,1]$, i.e. $\hat{a}_\lambda \in \tilde{A}$ for all $\lambda \in [0,1]$. In other words, we have

$$\left\{\hat{a}_\lambda : \hat{a}_\lambda = \lambda l + (1-\lambda)u \text{ for } \lambda \in [0,1]\right\} \subset \tilde{A}$$

by referring to Definition 9.1.6 for the concept of subset of fuzzy interval $\tilde{A}$.

According to Definition 9.1.8, the gradual numbers $\hat{a}_\lambda \in \tilde{A}$ for $\lambda \in [0,1]$ can generate the corresponding fuzzy sets $\tilde{a}_\lambda$ with membership function given by

$$\xi_{\tilde{a}_\lambda}(x) = \begin{cases} 0, & \text{if there is no } \alpha \in (0,1] \text{ satisfying } \hat{a}_\lambda(\alpha) = x \\ \displaystyle\sup_{\{\alpha\in(0,1]:\hat{a}_\lambda(\alpha)=x\}} \alpha, & \text{otherwise.} \end{cases}$$

Since η_1 and η_2 are continuous on [0,1], the functions l and u in (9.10) and (9.11), respectively, are continuous on [0,1], which also says that each gradual number $\hat{a}_\lambda$ in (9.12) is continuous on [0,1]. According to Theorem 5.5.4 by considering $\zeta^L = \zeta^U = \hat{a}_\lambda$, the α-level set of $\tilde{a}_\lambda$ is given by

$$\begin{aligned}\left(\tilde{a}_\lambda\right)_\alpha &= \{x : \xi_{\hat{a}_\lambda}(x)(x) \geq \alpha\} = \left\{x : \sup_{0<\alpha\leq 1} \alpha \cdot \chi_{\{\hat{a}_\lambda(\alpha)\}}(x) \geq \alpha\right\} \\ &= \left[\min_{\alpha\leq\beta\leq 1} \hat{a}_\lambda(\beta), \max_{\alpha\leq\beta\leq 1} \hat{a}_\lambda(\beta)\right] \\ &= \left[\min_{\alpha\leq\beta\leq 1} (\lambda l(\beta) + (1-\lambda)u(\beta)), \max_{\alpha\leq\beta\leq 1} (\lambda l(\beta) + (1-\lambda)u(\beta))\right]\end{aligned}$$

for $\alpha \in [0,1]$.

From (9.10) and (9.11), we see that l is increasing and u is decreasing on [0,1]. Since $l \leq u$, for $\beta \geq \alpha$, we have

$$\lambda l(\beta) + (1-\lambda)u(\beta) \geq \lambda l(\beta) + (1-\lambda)l(\beta) = l(\beta) \geq l(\alpha)$$

and

$$\lambda l(\beta) + (1-\lambda)u(\beta) \leq \lambda u(\beta) + (1-\lambda)u(\beta) = u(\beta) \leq u(\alpha),$$

which imply

$$\min_{\alpha\leq\beta\leq 1} (\lambda l(\beta) + (1-\lambda)u(\beta)) \geq l(\alpha)$$

and

$$\max_{\alpha\leq\beta\leq 1} (\lambda l(\beta) + (1-\lambda)u(\beta)) \leq u(\alpha).$$

Therefore, by referring to (9.9), we obtain the inclusion

$$\left(\tilde{a}_\lambda\right)_\alpha \subseteq \tilde{A}_\alpha = \left[l(\alpha), u(\alpha)\right] \text{ for } \alpha \in [0,1],$$

which also says that $\tilde{a}_\lambda$ has less fuzziness than that of $\tilde{A}$, since the α-level set of $\tilde{a}_\lambda$ is narrower than the α-level set of $\tilde{A}$, which seems reasonable since $\hat{a}_\lambda$ is an element of $\tilde{A}$ saying that its corresponding fuzzy set cannot be fuzzier than their family set. In other words, the members should be more certain than the whole family. The above results are concluded in the following theorem.

Theorem 9.3.1 *Let $\eta_1 : [0,1] \to \mathbb{R}$ and $\eta_2 : [0,1] \to \mathbb{R}$ be two bounded and continuous functions defined on [0,1], and let*

$$\zeta^L(\alpha) = \min\{\eta_1(\alpha), \eta_2(\alpha)\} \text{ and } \zeta^U(\alpha) = \max\{\eta_1(\alpha), \eta_2(\alpha)\}.$$

We define a gradual set $\mathfrak{G}$ by the bounded closed interval

$$\mathfrak{G}(\alpha) = \left[\zeta^L(\alpha), \zeta^U(\alpha)\right] = \left[\min\{\eta_1(\alpha), \eta_2(\alpha)\}, \max\{\eta_1(\alpha), \eta_2(\alpha)\}\right]$$

for $\alpha \in [0,1]$. Then, we can generate a fuzzy interval $\tilde{A}$ in $\mathbb{R}$ with membership function defined by

$$\xi_{\tilde{A}}(x) = \sup_{\alpha\in(0,1]} \alpha \cdot \chi_{\mathfrak{G}(\alpha)}(x),$$

satisfying

$$\tilde{A}_\alpha = \mathfrak{G}(\alpha) = \left[\min_{\alpha\le\beta\le1} \zeta^L(\beta), \max_{\alpha\le\beta\le1} \zeta^U(\beta)\right].$$

Let l and u be two bounded functions defined on [0,1] as given in (9.10) and (9.11), respectively. Given any fixed $\lambda \in [0,1]$, the gradual number $\hat{a}_\lambda$ defined by

$$\hat{a}_\lambda(\alpha) = \lambda l(\alpha) + (1-\lambda)u(\alpha) \text{ for } \alpha \in [0,1]$$

is an element in $\tilde{A}$; that is, $\hat{a}_\lambda \in \tilde{A}$ for all $\lambda \in [0,1]$. Let $\tilde{a}_\lambda$ be a fuzzy set generated by the gradual number $\hat{a}_\lambda$. Then, the α-level sets of $\tilde{a}_\lambda$ are given by

$$\begin{aligned}\left(\tilde{a}_\lambda\right)_\alpha &= \left[\min_{\alpha\le\beta\le1} \hat{a}_\lambda(\beta), \max_{\alpha\le\beta\le1} \hat{a}_\lambda(\beta)\right]\\ &= \left[\min_{\alpha\le\beta\le1} \left(\lambda l(\beta) + (1-\lambda)u(\beta)\right), \max_{\alpha\le\beta\le1} \left(\lambda l(\beta) + (1-\lambda)u(\beta)\right)\right]\end{aligned}$$

for $\alpha \in [0,1]$, where $\tilde{a}_\lambda$ has less fuzziness than that of $\tilde{A}$ in the sense of

$$\left(\tilde{a}_\lambda\right)_\alpha \subseteq \tilde{A}_\alpha = \left[l(\alpha), u(\alpha)\right] \text{ for } \alpha \in [0,1].$$

Example 9.3.2 Given any real numbers a_1, a_2, b_1, b_2 satisfying $b_1 < a_1 < a_2 < b_2$, we define two bounded continuous functions η_1 and η_2 on [0,1] by

$$\eta_1(\alpha) = (1-\alpha)b_1 + \alpha a_1 \text{ and } \eta_2(\alpha) = (1-\alpha)b_2 + \alpha a_2.$$

Then, we have $\eta_1(\alpha) \le \eta_2(\alpha)$ for all $\alpha \in [0,1]$. We also see that η_1 is increasing and η_2 is decreasing on [0,1]. We also see that $\zeta^L = \eta_1$ and $\zeta^U = \eta_2$. Then, we define a gradual set $\mathfrak{G}$ by the bounded closed interval

$$\mathfrak{G}(\alpha) = \left[\zeta^L(\alpha), \zeta^U(\alpha)\right] = \left[(1-\alpha)b_1 + \alpha a_1, (1-\alpha)b_2 + \alpha a_2\right].$$

From (9.10) and (9.11), we have

$$l(\alpha) = \min_{\alpha\le\beta\le1} \eta_1(\beta) = \eta_1(\alpha)$$

and

$$u(\alpha) = \max_{\alpha \le \beta \le 1} \eta_2(\beta) = \eta_2(\alpha).$$

Then, we can generate a fuzzy interval $\tilde{A}$ in $\mathbb{R}$ with membership function defined by

$$\xi_{\tilde{A}}(x) = \sup_{0<\alpha\le 1} \alpha \cdot \chi_{\mathfrak{S}(\alpha)}(x),$$

satisfying $\tilde{A}_\alpha = [l(\alpha), u(\alpha)]$ for $\alpha \in [0,1]$. For any fixed $\lambda \in [0,1]$, we define the gradual number $\hat{a}_\lambda$ by

$$\hat{a}_\lambda(\beta) = \lambda(1-\beta)b_1 + \lambda\beta a_1 + (1-\lambda)(1-\beta)b_2 + (1-\lambda)\beta a_2 \equiv h\beta + k,$$

where

$$h = \lambda\left(a_1 - b_1\right) + (1-\lambda)\left(a_2 - b_2\right) \text{ and } k = \lambda b_1 + b_2 - \lambda b_2.$$

For $\alpha \in [0,1]$, the α-level sets $(\tilde{a}_\lambda)_\alpha$ of its corresponding fuzzy set $\tilde{a}_\lambda$ are given by

$$\begin{aligned}\left(\tilde{a}_\lambda\right)_\alpha &= \left[\min_{\alpha\le\beta\le 1} \hat{a}_\lambda(\beta), \max_{\alpha\le\beta\le 1} \hat{a}_\lambda(\beta)\right] = \begin{cases} \left[h\alpha + k, h + k\right] & \text{if } h \ge 0 \\ \left[h + k, h\alpha + k\right] & \text{if } h < 0 \end{cases} \\ &= \left[\min\left\{h\alpha + k, h + k\right\}, \max\left\{h\alpha + k, h + k\right\}\right].\end{aligned}$$

It is clear to see that $(\tilde{a}_\lambda)_\alpha \subseteq \tilde{A}_\alpha$ for all $\alpha \in [0,1]$.

9.4 Set Operations Using Gradual Elements

Let A and B be two (crisp) sets in $\mathbb{R}^m$. The intersection $A \cap B$ and union $A \cup B$ are given by

$$A \cap B = \{x : x \in A \text{ and } x \in B\} \text{ and } A \cup B = \{x : x \in A \text{ or } x \in B\}.$$

Suppose now that $\tilde{A}$ and $\tilde{B}$ are two fuzzy sets in $\mathbb{R}^m$. According to the belongness for fuzzy sets in Definition 9.1.6, we shall try to explain the intersection $\tilde{A} \cap \tilde{B}$ given by

$$\tilde{A} \cap \tilde{B} = \{\hat{x} : \hat{x} \in \tilde{A} \text{ and } \hat{x} \in \tilde{B}\}, \tag{9.13}$$

and the union $\tilde{A} \cup \tilde{B}$ given by

$$\tilde{A} \cup \tilde{B} = \{\hat{x} : \hat{x} \in \tilde{A} \text{ or } \hat{x} \in \tilde{B}\}. \tag{9.14}$$

The relationship with the conventional intersection and union using the minimum and maximum functions, respectively, will also be established.

9.4.1 Complement Set

Let $\tilde{A}$ be a fuzzy set in $\mathbb{R}^m$ with membership function $\xi_{\tilde{A}}$. Recall that the complement of $\tilde{A}$ is denoted by $\tilde{A}^c$ with membership function defined by

$$\xi_{\tilde{A}^c}(x) = 1 - \xi_{\tilde{A}}(x).$$

The strong α-level set of $\tilde{A}$ is given by

$$\tilde{A}_{\alpha+} = \left\{x \in \mathbb{R}^m : \xi_{\tilde{A}}(x) > \alpha\right\}$$

for $\alpha \in [0,1)$. We also recall that $\tilde{A}_{0+}$ is the support of $\tilde{A}$. It is clear that $\tilde{A}_{\alpha+} \subseteq \tilde{A}_\alpha$ for all $\alpha \in (0,1]$. Then, for $\alpha \in (0,1]$, the α-level set of $\tilde{A}^c$ is given by

$$\begin{aligned}\tilde{A}^c_\alpha &= \left\{x \in \mathbb{R}^m : \xi_{\tilde{A}^c}(x) \geq \alpha\right\} = \left\{x \in \mathbb{R}^m : 1 - \xi_{\tilde{A}}(x) \geq \alpha\right\} \\ &= \left\{x \in \mathbb{R}^m : \xi_{\tilde{A}}(x) \leq 1 - \alpha\right\} = \mathbb{R}^m \backslash \left\{x \in \mathbb{R}^m : \xi_{\tilde{A}}(x) > 1 - \alpha\right\} \\ &= \mathbb{R}^m \backslash \tilde{A}_{(1-\alpha)+} = \left[\tilde{A}_{(1-\alpha)+}\right]^c. \end{aligned} \tag{9.15}$$

We need to remark that $(\tilde{A}_\alpha)^c$ means the complement set of the α-level set $\tilde{A}_\alpha$ of $\tilde{A}$, which is different from the α-level set $\tilde{A}^c_\alpha$ of $\tilde{A}^c$. Moreover, for $\alpha < \beta$, since $\tilde{A}$ and $\tilde{A}^c$ are fuzzy sets in $\mathbb{R}^m$, the nestedness says that

$$\tilde{A}_{\beta+} \subseteq \tilde{A}_\beta \subseteq \tilde{A}_\alpha \text{ and } \tilde{A}^c_{\beta+} \subseteq \tilde{A}^c_\beta \subseteq \tilde{A}^c_\alpha,$$

which also says that

$$(\tilde{A}_{\beta+})^c \supseteq (\tilde{A}_\beta)^c \supseteq (\tilde{A}_\alpha)^c. \tag{9.16}$$

Next, we shall define the complement of $\tilde{A}$ based on the gradual element.

Let $\tilde{A}$ be a fuzzy set in $\mathbb{R}^m$ with membership function $\xi_{\tilde{A}}$. Inspired by (9.15), we consider the following family of gradual elements

$$\mathcal{A} = \left\{\hat{a} : \hat{a}(\alpha) \in \left[\tilde{A}_{(1-\alpha)+}\right]^c \text{ for all } \alpha \in [0,1]\right\}.$$

According to (9.3), this family can generate a gradual set $\mathfrak{G}^{\dagger c}$ defined by

$$\mathfrak{G}^{\dagger c}(\alpha) = \{\hat{a}(\alpha) : \hat{a} \in \mathcal{A}\} = \left\{\hat{a}(\alpha) : \hat{a}(\alpha) \in \left[\tilde{A}_{(1-\alpha)+}\right]^c \text{ for all } \alpha \in [0,1]\right\}. \tag{9.17}$$

According to (9.2), this gradual set $\mathfrak{G}^{\dagger c}$ can also generate a fuzzy set $\tilde{A}^{\dagger c}$ with membership function given by

$$\xi_{\tilde{A}^{\dagger c}}(x) = \sup_{\alpha \in (0,1]} \alpha \cdot \chi_{\mathfrak{G}^{\dagger c}(\alpha)}(x), \tag{9.18}$$

where the fuzzy set $\tilde{A}^{\dagger c}$ is defined to be a new type of complement of $\tilde{A}$. We are going to claim $\tilde{A}^{\dagger c} = \tilde{A}^c$, although $\tilde{A}^{\dagger c}$ is based on the gradual element and gradual set, and $\tilde{A}^c$ is based on the membership function.

Theorem 9.4.1 *Let $\tilde{A}$ be a fuzzy set in $\mathbb{R}^m$. Then, the gradual set $\mathfrak{G}^{\dagger c}$ generated by the family $\mathcal{A}$ of gradual elements satisfies*

$$\mathfrak{G}^{\dagger c}(\alpha) = \left[\tilde{A}_{(1-\alpha)+}\right]^c = \tilde{A}^c_\alpha \textit{ for each } \alpha \in [0,1].$$

Moreover, we have $\tilde{A}^{\dagger c} = \tilde{A}^c$.

Proof. According to (9.17), we see that $\mathfrak{G}^{\dagger c}(\alpha) \subseteq [\tilde{A}_{(1-\alpha)+}]^c$ for $\alpha \in [0,1]$. On the other hand, given any fixed $\alpha \in [0,1]$ and any $x \in [\tilde{A}_{(1-\alpha)+}]^c$, we define a function $\hat{a}$ on $[0,1]$ by

$$\hat{a}(\beta) = \begin{cases} x, & \text{if } \beta = \alpha \\ y \text{ for some } y \in [\tilde{A}_{(1-\beta)+}]^c, & \text{if } \beta \neq \alpha. \end{cases}$$

Then, it is clear to see that $\hat{a} \in \mathcal{A}$. This shows that

$$\hat{a}(\alpha) = x \in \mathfrak{G}^{\dagger c}(\alpha), \text{ i.e. } [\tilde{A}_{(1-\alpha)+}]^c \subseteq \mathfrak{G}^{\dagger c}(\alpha).$$

Therefore, we obtain

$$\mathfrak{G}^{\dagger c}(\alpha) = [\tilde{A}_{(1-\alpha)+}]^c \text{ for all } \alpha \in [0,1].$$

Using (9.15), we obtain the following equalities

$$\mathfrak{G}^{\dagger c}(\alpha) = \left[\tilde{A}_{(1-\alpha)+}\right]^c = \tilde{A}_\alpha^c \text{ for all } \alpha \in [0,1].$$

On the other hand, according to Theorem 2.2.13, the membership function of $\tilde{A}^c$ is given by

$$\xi_{\tilde{A}^c}(x) = \sup_{\alpha\in(0,1]} \alpha \cdot \chi_{\tilde{A}_\alpha^c}(x).$$

By referring to the membership function (9.18), since $\mathfrak{G}^{\dagger c}(\alpha) = \tilde{A}_\alpha^c$, it follows that $\tilde{A}^{\dagger c} = \tilde{A}^c$. This completes the proof. ■

9.4.2 Intersection and Union

Let $A_1, \ldots, A_n$ be (crisp) sets in $\mathbb{R}^m$. Their intersection is given by

$$A_1 \cap \ldots \cap A_n = \left\{x \in \mathbb{R}^m : x \in A_i \text{ for all } i = 1, \ldots, n\right\}.$$

Inspired by the above form, we are going to consider the intersection of fuzzy sets $\tilde{A}^{(1)}, \ldots, \tilde{A}^{(n)}$ in $\mathbb{R}^m$ using gradual elements. Now, we consider the following family

$$\left\{\hat{a} : \hat{a} \in \tilde{A}^{(i)} \text{ for all } i = 1, \ldots, n\right\},$$

which consists of all common gradual elements from $\tilde{A}^{(1)}, \ldots, \tilde{A}^{(n)}$. Then, this family can generate a gradual set $\mathfrak{G}^\cap$ defined by

$$\mathfrak{G}^\cap(\alpha) = \left\{\hat{a}(\alpha) : \hat{a} \in \tilde{A}^{(i)} \text{ for all } i = 1, \ldots, n\right\} \text{ for } \alpha \in [0,1]. \tag{9.19}$$

The union of (crisp) subsets $A_1, \ldots, A_n$ of $\mathbb{R}^m$ is given by

$$A_1 \cup \ldots \cup A_n = \left\{x \in \mathbb{R}^m : x \in A_i \text{ for some } i = 1, \ldots, n\right\}.$$

We can also define the union of $\tilde{A}^{(1)}, \ldots, \tilde{A}^{(n)}$ using gradual elements. Now, we consider the following family

$$\left\{\hat{a} : \hat{a} \in \tilde{A}^{(i)} \text{ for some } i = 1, \ldots, n\right\},$$

which consists of all gradual elements taken from some $\tilde{A}^{(1)}, \ldots, \tilde{A}^{(n)}$. Then, this family can also generate a gradual set $\mathfrak{G}^\cup$ defined by

$$\mathfrak{G}^\cup(\alpha) = \left\{\hat{a}(\alpha) : \hat{a} \in \tilde{A}^{(i)} \text{ for some } i = 1, \ldots, n\right\} \text{ for } \alpha \in [0,1]. \tag{9.20}$$

By definition, it is clear to see that

$$\hat{a}(\alpha) \in \mathfrak{G}^\cap(\alpha) \text{ implies } \hat{a}(\alpha) \in \tilde{A}_\alpha^{(1)} \cap \ldots \cap \tilde{A}_\alpha^{(n)} \text{ for } \alpha \in [0,1] \tag{9.21}$$

and

$$\hat{a}(\alpha) \in \mathfrak{G}^\cup(\alpha) \text{ implies } \hat{a}(\alpha) \in \tilde{A}_\alpha^{(1)} \cup \ldots \cup \tilde{A}_\alpha^{(n)} \text{ for } \alpha \in [0,1]. \tag{9.22}$$

Based on these two gradual sets $\mathfrak{G}^{\cap}$ and $\mathfrak{G}^{\cup}$, we can generate two fuzzy sets $\tilde{A}^{\cap}$ and $\tilde{A}^{\cup}$ in $\mathbb{R}^m$ with membership functions given by

$$\xi_{\tilde{A}^{\cap}}(x) = \sup_{\alpha\in(0,1]} \alpha \cdot \chi_{\mathfrak{G}^{\cap}(\alpha)}(x) \text{ and } \xi_{\tilde{A}^{\cup}}(x) = \sup_{\alpha\in(0,1]} \alpha \cdot \chi_{\mathfrak{G}^{\cup}(\alpha)}(x). \tag{9.23}$$

In this case, we define the intersection and union of $\tilde{A}^{(1)},\dots,\tilde{A}^{(n)}$ as

$$\tilde{A}^{(1)} \cap \dots \cap \tilde{A}^{(n)} = \tilde{A}^{\cap} \text{ and } \tilde{A}^{(1)} \cup \dots \cup \tilde{A}^{(n)} = \tilde{A}^{\cup},$$

where the membership functions of $\tilde{A}^{\cap}$ and $\tilde{A}^{\cup}$ are given in (9.23). This kind of intersection and union is based on the concept of gradual elements in fuzzy sets.

Example 9.4.2 Let $\tilde{A}^{(1)}$ and $\tilde{A}^{(2)}$ be two fuzzy sets with membership functions given by

$$\xi_{\tilde{A}^{(1)}}(x) = \begin{cases} x-1 & \text{if } 1 \le x \le 2 \\ 1 & \text{if } 2 < x < 3 \\ 4-x & \text{if } 3 \le x \le 4 \\ 0 & \text{otherwise,} \end{cases} \quad \text{and } \xi_{\tilde{A}^{(2)}}(x) = \begin{cases} x-2 & \text{if } 2 \le x \le 3 \\ 1 & \text{if } 3 < x < 4 \\ 5-x & \text{if } 4 \le x \le 5 \\ 0 & \text{otherwise.} \end{cases}$$

Then, the α-level sets of $\tilde{A}^{(1)}$ and $\tilde{A}^{(2)}$ are bounded closed intervals given by

$$\tilde{A}^{(1)}_{\alpha} = [1+\alpha, 4-\alpha] \text{ and } \tilde{A}^{(2)}_{\alpha} = [2+\alpha, 5-\alpha].$$

The gradual number $\hat{a} \in \tilde{A}^{(1)}$ must satisfy $\hat{a}(\alpha) \in \tilde{A}^{(1)}_{\alpha}$ for each $\alpha \in [0,1]$. For example, if we take

$$\hat{a}^{(1)}_1(\alpha) = 1+\alpha \text{ or } \hat{a}^{(1)}_2(\alpha) = 4-\alpha \text{ for } \alpha \in [0,1],$$

then $\hat{a}^{(1)}_1, \hat{a}^{(1)}_2 \in \tilde{A}^{(1)}$. Also, if we take

$$\hat{a}^{(2)}_1(\alpha) = 2+\alpha \text{ or } \hat{a}^{(2)}_2(\alpha) = 5-\alpha \text{ for } \alpha \in [0,1],$$

then $\hat{a}^{(2)}_1, \hat{a}^{(2)}_2 \in \tilde{A}^{(2)}$. Now, we consider two gradual sets given by

$$\mathfrak{G}^{\cap}(\alpha) = \left\{\hat{a}(\alpha) : \hat{a} \in \tilde{A}^{(1)} \text{ and } \hat{a} \in \tilde{A}^{(2)}\right\} \text{ for } \alpha \in [0,1]$$

and

$$\mathfrak{G}^{\cup}(\alpha) = \left\{\hat{a}(\alpha) : \hat{a} \in \tilde{A}^{(1)} \text{ or } \hat{a} \in \tilde{A}^{(2)}\right\} \text{ for } \alpha \in [0,1].$$

Then, it is clear to see that

$$\left\{\hat{a}^{(1)}_2(\alpha), \hat{a}^{(2)}_1(\alpha)\right\} \subset \mathfrak{G}^{\cap}(\alpha) \text{ for } \alpha \in [0,1]$$

and

$$\left\{\hat{a}^{(1)}_1(\alpha), \hat{a}^{(1)}_2(\alpha), \hat{a}^{(2)}_1(\alpha), \hat{a}^{(2)}_2(\alpha)\right\} \subset \mathfrak{G}^{\cup}(\alpha) \text{ for } \alpha \in [0,1].$$

In general, for $\alpha \in [0,1]$, we have

$$\begin{aligned}\mathfrak{G}^{\cap}(\alpha) &= \left\{\hat{a}(\alpha) : \hat{a} \in \tilde{A}^{(1)} \text{ and } \hat{a} \in \tilde{A}^{(2)}\right\} \\ &= \{\hat{a}(\alpha) : 1+\alpha \le \hat{a}(\alpha) \le 4-\alpha \text{ and } 2+\alpha \le \hat{a}(\alpha) \le 5-\alpha\} \\ &= \{\hat{a}(\alpha) : 2+\alpha \le \hat{a}(\alpha) \le 4-\alpha\} = [2+\alpha, 4-\alpha] = \tilde{A}^{(1)}_{\alpha} \cap \tilde{A}^{(2)}_{\alpha}\end{aligned}$$

and

$$\begin{aligned}\mathfrak{G}^{\cup}(\alpha) &= \left\{\hat{a}(\alpha) : \hat{a} \in \tilde{A}^{(1)} \text{ or } \hat{a} \in \tilde{A}^{(2)}\right\}\\ &= \{\hat{a}(\alpha) : 1+\alpha \le \hat{a}(\alpha) \le 4-\alpha \text{ or } 2+\alpha \le \hat{a}(\alpha) \le 5-\alpha\}\\ &= \{\hat{a}(\alpha) : 1+\alpha \le \hat{a}(\alpha) \le 5-\alpha\} = [1+\alpha, 5-\alpha] = \tilde{A}^{(1)}_{\alpha} \cup \tilde{A}^{(2)}_{\alpha}.\end{aligned}$$

According to (9.23), the membership functions of $\tilde{A}^{(1)} \cap \tilde{A}^{(2)}$ and $\tilde{A}^{(1)} \cup \tilde{A}^{(2)}$ are given by

$$\xi_{\tilde{A}^{(1)}\cap\tilde{A}^{(2)}}(x) = \sup_{\alpha\in[0,1]} \alpha \cdot \chi_{[2+\alpha,4-\alpha]}(x) \text{ and } \xi_{\tilde{A}^{(1)}\cup\tilde{A}^{(2)}}(x) = \sup_{\alpha\in[0,1]} \alpha \cdot \chi_{[1+\alpha,5-\alpha]}(x).$$

The α-level sets of $\tilde{A}^{(1)} \cap \tilde{A}^{(2)}$ and $\tilde{A}^{(1)} \cup \tilde{A}^{(2)}$ will be investigated in the subsequent discussion.

Proposition 9.4.3 *Let $\tilde{A}^{(1)}, \ldots, \tilde{A}^{(n)}$ be fuzzy sets in $\mathbb{R}^m$. Then, the following statements hold true.*

(i) *We have*

$$\mathfrak{G}^{\cup}(\alpha) = \tilde{A}^{(1)}_{\alpha} \cup \ldots \cup \tilde{A}^{(n)}_{\alpha} \text{ for all } \alpha \in [0,1]$$

and

$$\mathfrak{G}^{\cup}(\beta) \subseteq \mathfrak{G}^{\cup}(\alpha) \text{ for } \beta > \alpha.$$

(ii) *Suppose that $\tilde{A}^{(1)}_{\alpha} \cap \ldots \cap \tilde{A}^{(n)}_{\alpha} \neq \emptyset$ for all $\alpha \in [0,1]$. Then, we have*

$$\mathfrak{G}^{\cap}(\alpha) = \tilde{A}^{(1)}_{\alpha} \cap \ldots \cap \tilde{A}^{(n)}_{\alpha} \text{ for all } \alpha \in [0,1]$$

and

$$\mathfrak{G}^{\cap}(\beta) \subseteq \mathfrak{G}^{\cap}(\alpha) \text{ for } \beta > \alpha.$$

Proof. To prove part (i), for $\alpha \in [0,1]$, the inclusion

$$\mathfrak{G}^{\cup}(\alpha) \subseteq \tilde{A}^{(1)}_{\alpha} \cup \ldots \cup \tilde{A}^{(n)}_{\alpha}$$

follows from (9.22). Now, given any fixed $\alpha \in [0,1]$ and any $x \in \tilde{A}^{(1)}_{\alpha} \cup \ldots \cup \tilde{A}^{(n)}_{\alpha}$, we have $x \in \tilde{A}^{(i)}_{\alpha}$ for some $i = 1, \ldots, n$. Then, we define a function $\hat{a}$ on [0,1] by

$$\hat{a}(\beta) = \begin{cases} x, & \text{if } \beta = \alpha \\ y \text{ for some } y \in \tilde{A}^{(i)}_{\beta}, & \text{if } \beta \neq \alpha. \end{cases}$$

Then $\hat{a} \in \tilde{A}^{(i)}$. Therefore, we obtain $x = \hat{a}(\alpha) \in \tilde{A}^{(i)}_{\alpha}$ with $\hat{a} \in \tilde{A}^{(i)}$ for some $i = 1, \ldots, n$. This says that $x = \hat{a}(\alpha) \in \mathfrak{G}^{\cup}(\alpha)$. Therefore, we obtain the inclusion

$$\tilde{A}^{(1)}_{\alpha} \cup \ldots \cup \tilde{A}^{(n)}_{\alpha} \subseteq \mathfrak{G}^{\cup}(\alpha)$$

for all $\alpha \in [0,1]$. Using the nestedness of the α-level sets $\tilde{A}^{(i)}_{\alpha}$ for $i = 1, \ldots, n$, it is clear to see that the gradual set $\mathfrak{G}^{\cup}$ is also nested in the sense of $\mathfrak{G}^{\cup}(\beta) \subseteq \mathfrak{G}^{\cup}(\alpha)$ for $\beta > \alpha$.

To prove part (ii), for $\alpha \in [0,1]$, the inclusion

$$\mathfrak{G}^{\cap}(\alpha) \subseteq \tilde{A}^{(1)}_{\alpha} \cap \ldots \cap \tilde{A}^{(n)}_{\alpha}$$

follows from (9.21). Given any fixed $\alpha \in [0,1]$ and any $x \in \tilde{A}_\alpha^{(1)} \cap \ldots \cap \tilde{A}_\alpha^{(n)} \neq \emptyset$, we have $x \in \tilde{A}_\alpha^{(i)}$ for all $i = 1, \ldots, n$. Then, we define a function $\hat{a}$ on $[0,1]$ by

$$\hat{a}(\beta) = \begin{cases} x, & \text{if } \beta = \alpha \\ y \text{ for some } y \in \tilde{A}_\beta^{(1)} \cap \ldots \cap \tilde{A}_\beta^{(n)} \neq \emptyset, & \text{if } \beta \neq \alpha. \end{cases}$$

Then $\hat{a} \in \tilde{A}^{(i)}$ for all $i = 1, \ldots, n$. This says that $x = \hat{a}(\alpha) \in \mathfrak{G}^\cap(\alpha)$. Therefore, we obtain the inclusion

$$\tilde{A}_\alpha^{(1)} \cap \ldots \cap \tilde{A}_\alpha^{(n)} \subseteq \mathfrak{G}^\cap(\alpha)$$

for all $\alpha \in [0,1]$. This shows the desired equality. The nestedness of the gradual set $\mathfrak{G}^\cap$ can be similarly realized, and the proof is complete. ■

Now, we want to study the α-level sets of $\tilde{A}^{(1)} \cap \ldots \cap \tilde{A}^{(n)}$ and $\tilde{A}^{(1)} \cup \ldots \cup \tilde{A}^{(n)}$. Let $\tilde{A}$ be a fuzzy set in $\mathbb{R}^m$. Given any $\alpha \in (0,1]$ and any increasing convergent sequence $\{\alpha_m\}_{m=1}^\infty$ in $[0,1]$ with $\alpha_m > 0$ for all m and $\alpha_m \uparrow \alpha$, part (i) of Proposition 2.2.8 says that

$$\bigcap_{m=1}^{\infty} \tilde{A}_{\alpha_m} = \tilde{A}_\alpha, \tag{9.24}$$

which will be used for the subsequent discussion.

We say that two fuzzy sets $\tilde{A}$ and $\tilde{B}$ in $\mathbb{R}^m$ are identical, written by $\tilde{A} = \tilde{B}$, if and only if $\xi_{\tilde{A}} = \xi_{\tilde{B}}$, i.e. $\xi_{\tilde{A}}(x) = \xi_{\tilde{B}}(x)$ for all $x \in \mathbb{R}^m$.

Theorem 9.4.4 *Let $\tilde{A}^{(1)}, \ldots, \tilde{A}^{(n)}$ be fuzzy sets in $\mathbb{R}^m$ satisfying $\tilde{A}_\alpha^{(1)} \cap \ldots \cap \tilde{A}_\alpha^{(n)} \neq \emptyset$ for all $\alpha \in (0,1]$. Then, the α-level set of $\tilde{A}^\cap = \tilde{A}^{(1)} \cap \ldots \cap \tilde{A}^{(n)}$ is given by*

$$\tilde{A}_\alpha^\cap = \left(\tilde{A}^{(1)} \cap \ldots \cap \tilde{A}^{(n)}\right)_\alpha = \{x \in \mathbb{R}^m : \xi_{\tilde{A}^\cap}(x) \geq \alpha\} = \mathfrak{G}^\cap(\alpha) = \tilde{A}_\alpha^{(1)} \cap \ldots \cap \tilde{A}_\alpha^{(n)} \tag{9.25}$$

for every $\alpha \in (0,1]$, and

$$\begin{aligned} \tilde{A}_{0+}^\cap &= \left(\tilde{A}^{(1)} \cap \ldots \cap \tilde{A}^{(n)}\right)_{0+} = \bigcup_{0<\alpha\leq 1} \left(\tilde{A}^{(1)} \cap \ldots \cap \tilde{A}^{(n)}\right)_\alpha \\ &= \bigcup_{0<\alpha\leq 1} \mathfrak{G}^\cap(\alpha) = \bigcup_{0<\alpha\leq 1} \tilde{A}_\alpha^{(1)} \cap \ldots \cap \tilde{A}_\alpha^{(n)}. \end{aligned} \tag{9.26}$$

Proof. Given any $\alpha \in (0,1]$ and any increasing convergent sequence $\{\alpha_m\}_{m=1}^\infty$ in $(0,1]$ with $\alpha_m \uparrow \alpha$, using (9.24), we have

$$\bigcap_{m=1}^{\infty} \left(\tilde{A}_{\alpha_m}^{(1)} \cap \ldots \cap \tilde{A}_{\alpha_m}^{(n)}\right) = \left(\bigcap_{m=1}^{\infty} \tilde{A}_{\alpha_m}^{(1)}\right) \cap \ldots \cap \left(\bigcap_{m=1}^{\infty} \tilde{A}_{\alpha_m}^{(n)}\right) = \tilde{A}_\alpha^{(1)} \cap \ldots \cap \tilde{A}_\alpha^{(n)}. \tag{9.27}$$

From part (ii) of Proposition 9.4.3, we see that (9.27) is satisfied if and only if the following equality is satisfied

$$\bigcap_{m=1}^{\infty} \mathfrak{G}^\cap(\alpha_m) = \mathfrak{G}^\cap(\alpha). \tag{9.28}$$

We also have

$$\mathfrak{G}^{\cap}(\beta) \subseteq \mathfrak{G}^{\cap}(\alpha) \text{ for } \alpha, \beta \in [0,1] \text{ with } \beta > \alpha. \tag{9.29}$$

Given any fixed $x \in \mathbb{R}^m$, we define the following set

$$F_\rho = \left\{ \alpha \in [0,1] \,:\, \xi_{\tilde{A}^{\cap}}(x) = \alpha \cdot \chi_{\mathfrak{G}^{\cap}(\alpha)}(x) \geq \rho \right\}.$$

We are going to claim that the set F_ρ is closed for each $\rho \in \mathbb{R}$. Given any $\rho \in (0,1]$, for each $\alpha \in \text{cl}(F_\rho)$, the concept of closure says that there exists a sequence $\{\alpha_m\}_{m=1}^{\infty}$ in F_ρ satisfying $\alpha_m \to \alpha$. Therefore, we have $\alpha_m \in (0,1]$ with $\alpha_m \geq \rho > 0$ and $x \in \mathfrak{G}^{\cap}(\alpha_m)$ for all m, which also says that $\alpha > 0$, since

$$\alpha = \lim_{m\to\infty} \alpha_m \geq \rho > 0.$$

Therefore, there exists a subsequence $\{\alpha_{m_k}\}_{k=1}^{\infty}$ of $\{\alpha_m\}_{m=1}^{\infty}$ satisfying $\alpha_{m_k} \uparrow \alpha$ or $\alpha_{m_k} \downarrow \alpha$ as $k \to \infty$.

- Suppose that $\alpha_{m_k} \downarrow \alpha$. Then $\alpha_{m_k} \geq \alpha$ for all k. This says that $x \in \mathfrak{G}^{\cap}(\alpha_{m_k}) \subseteq \mathfrak{G}^{\cap}(\alpha)$ by (9.29).
- Suppose that $\alpha_{m_k} \uparrow \alpha$. Since $\alpha_{m_k} \in F_\rho$ for each k, it follows that $\alpha \in \text{cl}(F_\rho)$. Since $\text{cl}(F_\rho) \subseteq [0,1]$, it says that $\alpha \in [0,1]$. Since $x \in \mathfrak{G}^{\cap}(\alpha_{m_k})$ for all k, using (9.28), it follows that $x \in \mathfrak{G}^{\cap}(\alpha)$.

Therefore, we conclude that $x \in \mathfrak{G}^{\cap}(\alpha)$ for both cases. This also says that $\alpha \cdot \chi_{\mathfrak{G}^{\cap}(\alpha)}(x) \geq \rho$, i.e. $\alpha \in F_\rho$. Therefore, we obtain the inclusion $\text{cl}(F_\rho) \subseteq F_\rho$, which means that F_ρ is closed for each $\rho \in (0,1]$. Next, we are going to claim that, for any fixed $x \in \mathbb{R}^m$, the function

$$\eta_x(\alpha) = \alpha \cdot \chi_{\mathfrak{G}^{\cap}(\alpha)}(x)$$

is upper semi-continuous on $[0,1]$. Equivalently, it can be shown that the set F_ρ is closed for each $\rho \in \mathbb{R}$. We have shown that F_ρ is closed for each $\rho \in (0,1]$. If $\rho \notin [0,1]$ then the empty set $F_\rho = \emptyset$ is closed. If $\rho = 0$ then $F_\rho = [0,1]$ is also a closed set. This shows that the function η_x is indeed upper semi-continuous on $[0,1]$.

Given any fixed $\alpha \in (0,1]$, suppose that $x \in \tilde{A}^{\cap}_{\alpha}$. Now, we also assume that $x \notin \mathfrak{G}^{\cap}(\beta)$ for all $\beta \in [0,1]$ with $\alpha \leq \beta$. We want to lead to a contradiction. Under this assumption, we see that $\beta \cdot \chi_{\mathfrak{G}^{\cap}(\beta)}(x) < \alpha$ for all $\beta \in [0,1]$. Since η_x is upper semi-continuous on the closed and bounded set $[0,1]$, the supremum of function η_x is attained by Proposition 1.4.4. This says that

$$\xi_{\tilde{A}^{\cap}}(x) = \sup_{\beta\in[0,1]} \eta_x(\beta) = \sup_{\beta\in[0,1]} \beta \cdot \chi_{\mathfrak{G}^{\cap}(\beta)}(x) = \max_{\beta\in[0,1]} \beta \cdot \chi_{\mathfrak{G}^{\cap}(\beta)}(x) = \beta^* \cdot \chi_{\mathfrak{G}(\beta^*)}(x) < \alpha$$

for some $\beta^* \in [0,1]$, which violates $x \in \tilde{A}^{\cap}_{\alpha}$. Therefore, there exists $\beta_0 \in [0,1]$ satisfying

$$\beta_0 \geq \alpha \text{ and } x \in \mathfrak{G}^{\cap}(\beta_0) \subseteq \mathfrak{G}^{\cap}(\alpha)$$

by part (ii) of Proposition 9.4.3, which shows the inclusion $\tilde{A}^{\cap}_{\alpha} \subseteq \mathfrak{G}^{\cap}(\alpha)$. The following inclusion is obvious

$$\mathfrak{G}^{\cap}(\alpha) \subseteq \left\{ x \in \mathbb{R} \,:\, \sup_{\beta\in[0,1]} \beta \cdot \chi_{\mathfrak{G}^{\cap}(\beta)}(x) \geq \alpha \right\} = \{x \in \mathbb{R} \,:\, \xi_{\tilde{A}^{\cap}}(x) \geq \alpha\} = \tilde{A}^{\cap}_{\alpha}.$$

Therefore, we obtain the desired equalities (9.25) and (9.26). This completes the proof. ■

Example 9.4.5 Continued from Example 9.4.2, Theorem 9.4.4 says that the α-level sets of $\tilde{A}^{(1)} \cap \tilde{A}^{(2)}$ are given by

$$\left(\tilde{A}^{(1)} \cap \tilde{A}^{(2)}\right)_\alpha = \mathfrak{G}^{\cap}(\alpha) = \tilde{A}^{(1)}_\alpha \cap \tilde{A}^{(2)}_\alpha = [2+\alpha, 4-\alpha]$$

for $\alpha \in (0,1]$.

Theorem 9.4.6 *Let $\tilde{A}^{(1)}, \ldots, \tilde{A}^{(n)}$ be fuzzy sets in $\mathbb{R}^m$. Suppose that any one of the following conditions is satisfied.*

(a) *For any fixed $x \in \mathbb{R}^m$, the function*

$$\eta_x(\alpha) = \alpha \cdot \chi_{\mathfrak{G}^{\cup}(\alpha)}(x)$$

is upper semi-continuous on $[0,1]$.

(b) *Given any $\alpha \in (0,1]$ and any increasing convergent sequence $\{\alpha_m\}_{m=1}^{\infty}$ in $(0,1]$ with $\alpha_m \uparrow \alpha$, the following inclusion is satisfied*

$$\bigcap_{m=1}^{\infty} \left(\tilde{A}^{(1)}_{\alpha_m} \cup \ldots \cup \tilde{A}^{(n)}_{\alpha_m}\right) \subseteq \tilde{A}^{(1)}_\alpha \cup \ldots \cup \tilde{A}^{(n)}_\alpha. \tag{9.30}$$

(c) *Given any increasing sequence $\{\alpha_m\}_{m=1}^{\infty}$ in $(0,1]$, the following inclusion is satisfied*

$$\bigcap_{m=1}^{\infty} \left(\tilde{A}^{(1)}_{\alpha_m} \cup \ldots \cup \tilde{A}^{(n)}_{\alpha_m}\right) \subseteq \left(\bigcap_{m=1}^{\infty} \tilde{A}^{(1)}_{\alpha_m}\right) \cup \ldots \cup \left(\bigcap_{m=1}^{\infty} \tilde{A}^{(n)}_{\alpha_m}\right).$$

Then, the α-level set of $\tilde{A}^{\cup} = \tilde{A}^{(1)} \cup \ldots \cup \tilde{A}^{(n)}$ is given by

$$\tilde{A}^{\cup}_\alpha = \left(\tilde{A}^{(1)} \cup \ldots \cup \tilde{A}^{(n)}\right)_\alpha = \{x \in \mathbb{R}^m : \xi_{\tilde{A}^{\cup}}(x) \geq \alpha\} = \mathfrak{G}^{\cup}(\alpha) = \tilde{A}^{(1)}_\alpha \cup \ldots \cup \tilde{A}^{(n)}_\alpha \tag{9.31}$$

for every $\alpha \in (0,1]$, and

$$\begin{aligned} \tilde{A}^{\cup}_{0+} &= \left(\tilde{A}^{(1)} \cup \ldots \cup \tilde{A}^{(n)}\right)_{0+} = \bigcup_{0<\alpha\leq 1} \left(\tilde{A}^{(1)} \cup \ldots \cup \tilde{A}^{(n)}\right)_\alpha \\ &= \bigcup_{0<\alpha\leq 1} \mathfrak{G}^{\cup}(\alpha) = \bigcup_{0<\alpha\leq 1} \tilde{A}^{(1)}_\alpha \cup \ldots \cup \tilde{A}^{(n)}_\alpha. \end{aligned} \tag{9.32}$$

Proof. Suppose that condition (a) is satisfied. The arguments in the proof of Theorem 9.4.4 are still valid to obtain the desired equalities (9.31) and (9.32) by considering the upper semi-continuity of η_x.

Suppose that condition (b) is satisfied. From part (i) of Proposition 9.4.3, we see that the inclusion (9.30) is satisfied if and only if the following inclusion is satisfied

$$\bigcap_{m=1}^{\infty} \mathfrak{G}^{\cup}(\alpha_m) \subseteq \mathfrak{G}^{\cup}(\alpha). \tag{9.33}$$

We also have $\mathfrak{G}^{\cup}(\beta) \subseteq \mathfrak{G}^{\cup}(\alpha)$ for $\alpha, \beta \in [0,1]$. From the proof of Theorem 9.4.4, when the equality (9.28) is replaced by the following inclusion

$$\bigcap_{m=1}^{\infty} \mathfrak{G}^{\cup}(\alpha_m) \subseteq \mathfrak{G}^{\cup}(\alpha), \tag{9.34}$$

the same results can still be obtained. Therefore, by changing the role of (9.28) with (9.34) and using the arguments in the proof of Theorem 9.4.4, we can also show that the function η_x is upper semi-continuous on [0,1]. Therefore, using condition (a), the desired results can be obtained.

Suppose that condition (c) is satisfied. Then, from (9.24), it is clear to see that condition (b) is satisfied. Therefore, we also have the desired results, and the proof is complete. ∎

Example 9.4.7 Continued from Example 9.4.2, we are going to apply Theorem 9.4.6 to obtain the α-level sets of $\tilde{A}^{(1)} \cup \tilde{A}^{(2)}$. We shall check that the inclusion (9.30) in condition (b) will be satisfied. For $\alpha_m \uparrow \alpha$, since $\tilde{A}^{(1)}_\alpha \cup \tilde{A}^{(2)}_\alpha = [1+\alpha, 5-\alpha]$, we need to claim the inclusion

$$\bigcap_{m=1}^{\infty} \left[1+\alpha_m, 5-\alpha_m\right] \subseteq [1+\alpha, 5-\alpha].$$

Given x satisfying $1+\alpha_m \le x \le 5-\alpha_m$ for all $m = 1,2,\ldots$, by taking the limit as $m \to \infty$, we obtain $1+\alpha \le x \le 5-\alpha$, which proves the desired inclusion. Therefore, Theorem 9.4.6 says that

$$\left(\tilde{A}^{(1)} \cup \tilde{A}^{(2)}\right)_\alpha = \mathfrak{G}^{\cup}(\alpha) = \tilde{A}^{(1)}_\alpha \cup \tilde{A}^{(2)}_\alpha = [1+\alpha, 5-\alpha]$$

for $\alpha \in (0,1]$.

By referring to (9.30), we notice that the following inclusion

$$\left(\bigcap_{m=1}^{\infty} \tilde{A}^{(1)}_{\alpha_m}\right) \cup \ldots \cup \left(\bigcap_{m=1}^{\infty} \tilde{A}^{(n)}_{\alpha_m}\right) \subseteq \bigcap_{m=1}^{\infty} \left(\tilde{A}^{(1)}_{\alpha_m} \cup \ldots \cup \tilde{A}^{(n)}_{\alpha_m}\right).$$

is satisfied automatically, which is the reversed direction of inclusion.

9.4.3 Associativity

Now, we are going to study the associativity. Let $\tilde{A}, \tilde{B}, \tilde{C}$ be fuzzy sets in $\mathbb{R}^m$. In order to claim the following equalities

$$\left(\tilde{A} \cup \tilde{B}\right) \cup \tilde{C} = \tilde{A} \cup \left(\tilde{B} \cup \tilde{C}\right) = \tilde{A} \cup \tilde{B} \cup \tilde{C},$$

some sufficient conditions are needed.

Proposition 9.4.8 *Let $\tilde{A}, \tilde{B}, \tilde{C}$ be fuzzy sets in $\mathbb{R}^m$. Suppose that any one of the following conditions is satisfied.*

- *For any fixed $x \in \mathbb{R}^m$, the functions*

  $$\eta^{(1)}_x(\alpha) = \alpha \cdot \chi_{\tilde{A}_\alpha \cup \tilde{B}_\alpha}(x) \text{ and } \eta^{(2)}_x(\alpha) = \alpha \cdot \chi_{\tilde{B}_\alpha \cup \tilde{C}_\alpha}(x)$$

  are upper semi-continuous on [0,1].
- *Given any $\alpha \in (0,1]$ and any increasing convergent sequence $\{\alpha_m\}_{m=1}^{\infty}$ in* (0,1] *with $\alpha_m \uparrow \alpha$, the following inclusions*

  $$\bigcap_{m=1}^{\infty} \left(\tilde{A}_{\alpha_m} \cup \tilde{B}_{\alpha_m}\right) \subseteq \tilde{A}_\alpha \cup \tilde{B}_\alpha \text{ and } \bigcap_{m=1}^{\infty} \left(\tilde{B}_{\alpha_m} \cup \tilde{C}_{\alpha_m}\right) \subseteq \tilde{B}_\alpha \cup \tilde{C}_\alpha$$

  are satisfied.

- *Given any increasing sequence* $\{\alpha_m\}_{m=1}^{\infty}$ *in* (0,1], *the following inclusions*

$$\bigcap_{m=1}^{\infty}\left(\tilde{A}_{\alpha_m}\cup\tilde{B}_{\alpha_m}\right)\subseteq\left(\bigcap_{m=1}^{\infty}\tilde{A}_{\alpha_m}\right)\cup\left(\bigcap_{m=1}^{\infty}\tilde{B}_{\alpha_m}\right)$$

and

$$\bigcap_{m=1}^{\infty}\left(\tilde{B}_{\alpha_m}\cup\tilde{C}_{\alpha_m}\right)\subseteq\left(\bigcap_{m=1}^{\infty}\tilde{B}_{\alpha_m}\right)\cup\left(\bigcap_{m=1}^{\infty}\tilde{C}_{\alpha_m}\right)$$

are satisfied. Then, we have

$$(\tilde{A}\cup\tilde{B})\cup\tilde{C}=\tilde{A}\cup(\tilde{B}\cup\tilde{C})=\tilde{A}\cup\tilde{B}\cup\tilde{C}.$$

Proof. Let $\tilde{D}^{(1)}=\tilde{A}\cup\tilde{B}$ and $\tilde{D}^{(2)}=\tilde{B}\cup\tilde{C}$. Considering the function $\eta_x^{(1)}$ and using Theorem 9.4.6, we have

$$\tilde{D}_\alpha^{(1)}=\tilde{A}_\alpha\cup\tilde{B}_\alpha \text{ for all } \alpha\in(0,1]. \tag{9.35}$$

Similarly, considering the function $\eta_x^{(2)}$, we also have

$$\tilde{D}_\alpha^{(2)}=\tilde{B}_\alpha\cup\tilde{C}_\alpha \text{ for all } \alpha\in(0,1]. \tag{9.36}$$

Let $\tilde{E}^{(1)}=\tilde{D}^{(1)}\cup\tilde{C}$ and $\tilde{E}^{(2)}=\tilde{A}\cup\tilde{D}^{(2)}$. Then, the membership functions of $\tilde{E}^{(1)}$ and $\tilde{E}^{(2)}$ are given by

$$\begin{aligned}\xi_{\tilde{E}^{(1)}}(x)&=\sup_{\alpha\in(0,1]}\alpha\cdot\chi_{\tilde{D}_\alpha^{(1)}\cup\tilde{C}_\alpha}(x) \text{ (using (9.23) and part (i) of Proposition 9.4.3)}\\&=\sup_{\alpha\in(0,1]}\alpha\cdot\chi_{(\tilde{A}_\alpha\cup\tilde{B}_\alpha)\cup\tilde{C}_\alpha}(x) \text{ (using (9.35))}\\&=\sup_{\alpha\in(0,1]}\alpha\cdot\chi_{\tilde{A}_\alpha\cup\tilde{B}_\alpha\cup\tilde{C}_\alpha}(x)\end{aligned}$$

and

$$\begin{aligned}\xi_{\tilde{E}^{(2)}}(x)&=\sup_{\alpha\in(0,1]}\alpha\cdot\chi_{\tilde{A}_\alpha\cup\tilde{D}_\alpha^{(2)}}(x)=\sup_{\alpha\in(0,1]}\alpha\cdot\chi_{\tilde{A}_\alpha\cup(\tilde{B}_\alpha\cup\tilde{C}_\alpha)}(x) \text{ (using (9.36))}\\&=\sup_{\alpha\in(0,1]}\alpha\cdot\chi_{\tilde{A}_\alpha\cup\tilde{B}_\alpha\cup\tilde{C}_\alpha}(x).\end{aligned}$$

Let $\tilde{E}\equiv\tilde{A}\cup\tilde{B}\cup\tilde{C}$. Using (9.23) and part (i) of Proposition 9.4.3, we see that the membership function of $\tilde{E}$ is given by

$$\xi_{\tilde{E}}(x)=\sup_{\alpha\in(0,1]}\alpha\cdot\chi_{\tilde{A}_\alpha\cup\tilde{B}_\alpha\cup\tilde{C}_\alpha}(x).$$

Therefore, we obtain $\tilde{E}^{(1)}=\tilde{E}^{(2)}=\tilde{E}$. This completes the proof. ■

Example 9.4.9 Continued from Example 9.4.2, let $\tilde{A}^{(3)}$ be another fuzzy set with membership function given by

$$\xi_{\tilde{A}^{(3)}}(x)=\begin{cases}x-3 & \text{if } 3\leq x\leq 4\\ 1 & \text{if } 4<x<5\\ 6-x & \text{if } 5\leq x\leq 6\\ 0 & \text{otherwise.}\end{cases}$$

The α-level sets of $\tilde{A}^{(3)}$ are given by $\tilde{A}_\alpha^{(3)} = [3+\alpha, 6-\alpha]$. The second condition in Proposition 9.4.8 is satisfied, which can be realized from the similar argument in Example 9.4.7. Therefore, Proposition 9.4.8 says that

$$\left(\tilde{A}^{(1)} \cup \tilde{A}^{(2)}\right) \cup \tilde{A}^{(3)} = \tilde{A}^{(1)} \cup \left(\tilde{A}^{(2)} \cup \tilde{A}^{(3)}\right) = \tilde{A}^{(1)} \cup \tilde{A}^{(2)} \cup \tilde{A}^{(3)}.$$

The membership function is given by

$$\xi_{\tilde{A}^{(1)} \cup \tilde{A}^{(2)} \cup \tilde{A}^{(3)}}(x) = \sup_{\alpha \in (0,1]} \alpha \cdot \chi_{[1+\alpha, 6-\alpha]}(x).$$

For guaranteeing the following equalities

$$\left(\tilde{A} \cap \tilde{B}\right) \cap \tilde{C} = \tilde{A} \cap \left(\tilde{B} \cap \tilde{C}\right) = \tilde{A} \cap \tilde{B} \cap \tilde{C},$$

we do not need any extra sufficient conditions.

Proposition 9.4.10 *Let $\tilde{A}, \tilde{B}, \tilde{C}$ be fuzzy sets in $\mathbb{R}^m$ satisfying $\tilde{A}_\alpha \cap \tilde{B}_\alpha \cap \tilde{C}_\alpha \neq \emptyset$ for all $\alpha \in [0,1]$. Then, we have*

$$\left(\tilde{A} \cap \tilde{B}\right) \cap \tilde{C} = \tilde{A} \cap \left(\tilde{B} \cap \tilde{C}\right) = \tilde{A} \cap \tilde{B} \cap \tilde{C}. \tag{9.37}$$

Proof. Let $\tilde{D}^{(1)} = \tilde{A} \cap \tilde{B}$ and $\tilde{D}^{(2)} = \tilde{B} \cap \tilde{C}$. Using Theorem 9.4.4, we have

$$\tilde{D}_\alpha^{(1)} = \tilde{A}_\alpha \cap \tilde{B}_\alpha \text{ and } \tilde{D}_\alpha^{(2)} = \tilde{B}_\alpha \cap \tilde{C}_\alpha \text{ for all } \alpha \in (0,1].$$

Let $\tilde{E}^{(1)} = \tilde{D}^{(1)} \cap \tilde{C}$ and $\tilde{E}^{(2)} = \tilde{A} \cap \tilde{D}^{(2)}$. Then, the membership functions of $\tilde{E}^{(1)}$ and $\tilde{E}^{(2)}$ are given by

$$\xi_{\tilde{E}^{(1)}}(x) = \sup_{\alpha \in (0,1]} \alpha \cdot \chi_{\tilde{D}_\alpha^{(1)} \cap \tilde{C}_\alpha}(x) = \sup_{\alpha \in (0,1]} \alpha \cdot \chi_{(\tilde{A}_\alpha \cap \tilde{B}_\alpha) \cap \tilde{C}_\alpha}(x) = \sup_{\alpha \in (0,1]} \alpha \cdot \chi_{\tilde{A}_\alpha \cap \tilde{B}_\alpha \cap \tilde{C}_\alpha}(x)$$

and

$$\xi_{\tilde{E}^{(2)}}(x) = \sup_{\alpha \in (0,1]} \alpha \cdot \chi_{\tilde{A}_\alpha \cap \tilde{D}_\alpha^{(2)}}(x) = \sup_{\alpha \in (0,1]} \alpha \cdot \chi_{\tilde{A}_\alpha \cap (\tilde{B}_\alpha \cap \tilde{C}_\alpha)}(x) = \sup_{\alpha \in (0,1]} \alpha \cdot \chi_{\tilde{A}_\alpha \cap \tilde{B}_\alpha \cap \tilde{C}_\alpha}(x).$$

Let $\tilde{E} \equiv \tilde{A} \cap \tilde{B} \cap \tilde{C}$. Using (9.23) and part (ii) of Proposition 9.4.3, we see that the membership function of $\tilde{E}$ is given by

$$\xi_{\tilde{E}}(x) = \sup_{\alpha \in (0,1]} \alpha \cdot \chi_{\tilde{A}_\alpha \cap \tilde{B}_\alpha \cap \tilde{C}_\alpha}(x).$$

Therefore, we obtain $\tilde{E}^{(1)} = \tilde{E}^{(2)} = \tilde{E}$. This completes the proof. ■

Propositions 9.4.8 and 9.4.10 can be inductively extended by considering the fuzzy sets $\tilde{A}^{(1)}, \ldots, \tilde{A}^{(n)}$ in $\mathbb{R}^m$. Now, we want to consider the mixed set operations by including the parentheses. For example, we consider the following expression

$$\tilde{A} \equiv \left(\left(\tilde{A}^{(1)} \cap \tilde{A}^{(2)} \cap \tilde{A}^{(3)}\right) \cup \left(\tilde{A}^{(4)} \cap \tilde{A}^{(5)}\right)\right) \cap \left(\tilde{A}^{(6)} \cup \tilde{A}^{(7)} \cup \tilde{A}^{(8)} \cup \tilde{A}^{(9)}\right).$$

Let $\tilde{B}^{(1)} \equiv \tilde{A}^{(1)} \cap \tilde{A}^{(2)} \cap \tilde{A}^{(3)}$, $\tilde{B}^{(2)} \equiv \tilde{A}^{(4)} \cap \tilde{A}^{(5)}$, $\tilde{B}^{(3)} \equiv \tilde{A}^{(6)} \cup \tilde{A}^{(7)} \cup \tilde{A}^{(8)} \cup \tilde{A}^{(9)}$, and $\tilde{B}^{(4)} = \tilde{B}^{(1)} \cup \tilde{B}^{(2)}$. Now, we perform the following operations.

- Using (9.23) and part (ii) of Proposition 9.4.3, the membership function of $\tilde{B}^{(1)}$ is given by

  $$\xi_{\tilde{B}^{(1)}}(x) = \sup_{\alpha\in(0,1]} \alpha \cdot \chi_{\tilde{A}^{(1)}_\alpha \cap \tilde{A}^{(2)}_\alpha \cap \tilde{A}^{(3)}_\alpha}(x).$$

  Suppose that the conditions in Theorem 9.4.4 are satisfied. Then

  $$\tilde{B}^{(1)}_\alpha = \tilde{A}^{(1)}_\alpha \cap \tilde{A}^{(2)}_\alpha \cap \tilde{A}^{(3)}_\alpha \text{ for all } \alpha \in (0,1].$$

- Using (9.23) and part (ii) of Proposition 9.4.3, the membership function of $\tilde{B}^{(2)}$ is given by

  $$\xi_{\tilde{B}^{(2)}}(x) = \sup_{\alpha\in(0,1]} \alpha \cdot \chi_{\tilde{A}^{(4)}_\alpha \cap \tilde{A}^{(5)}_\alpha}(x).$$

  Suppose that the conditions in Theorem 9.4.4 are satisfied. Then

  $$\tilde{B}^{(2)}_\alpha = \tilde{A}^{(4)}_\alpha \cap \tilde{A}^{(5)}_\alpha \text{ for all } \alpha \in (0,1].$$

- Using (9.23) and part (i) of Proposition 9.4.3, the membership function of $\tilde{B}^{(3)}$ is given by

  $$\xi_{\tilde{B}^{(3)}}(x) = \sup_{\alpha\in(0,1]} \alpha \cdot \chi_{\tilde{A}^{(6)}_\alpha \cup \tilde{A}^{(7)}_\alpha \cup \tilde{A}^{8}_\alpha \cup \tilde{A}^{9}_\alpha}(x).$$

  Suppose that the conditions in Theorem 9.4.6 are satisfied. Then

  $$\tilde{B}^{(3)}_\alpha = \tilde{A}^{(6)}_\alpha \cup \tilde{A}^{(7)}_\alpha \cup \tilde{A}^{(8)}_\alpha \cup \tilde{A}^{(9)}_\alpha \text{ for all } \alpha \in (0,1].$$

- Using (9.23) and part (i) of Proposition 9.4.3, the membership function of $\tilde{B}^{(4)}$ is given by

  $$\xi_{\tilde{B}^{(4)}}(x) = \sup_{\alpha\in(0,1]} \alpha \cdot \chi_{\tilde{B}^{(1)}_\alpha \cup \tilde{B}^{(2)}_\alpha}(x).$$

  Suppose that the conditions regarding $\tilde{B}^{(1)}$ and $\tilde{B}^{(2)}$ in Theorem 9.4.6 are satisfied. Then

  $$\tilde{B}^{(4)}_\alpha = \tilde{B}^{(1)}_\alpha \cup \tilde{B}^{(2)}_\alpha \text{ for all } \alpha \in (0,1].$$

- Finally, using (9.23) and part (ii) of Proposition 9.4.3, the membership function of $\tilde{A}$ is given by

  $$\xi_{\tilde{A}}(x) = \sup_{\alpha\in(0,1]} \alpha \cdot \chi_{\tilde{B}^{(4)}_\alpha \cap \tilde{B}^{(5)}_\alpha}(x).$$

  Suppose that the conditions regarding $\tilde{B}^{(4)}$ and $\tilde{B}^{(5)}$ in Theorem 9.4.4 are satisfied. Then

  $$\tilde{A}_\alpha = \tilde{B}^{(4)}_\alpha \cap \tilde{B}^{(5)}_\alpha \text{ for all } \alpha \in (0,1].$$

  For $\alpha \in (0,1]$, we define

  $$\mathfrak{G}_\alpha \equiv \left(\left(\tilde{A}^{(1)}_\alpha \cap \tilde{A}^{(2)}_\alpha \cap \tilde{A}^{(3)}_\alpha\right) \cup \left(\tilde{A}^{(4)}_\alpha \cap \tilde{A}^{(5)}_\alpha\right)\right) \cap \left(\tilde{A}^{(6)}_\alpha \cup \tilde{A}^{(7)}_\alpha \cup \tilde{A}^{(8)}_\alpha \cup \tilde{A}^{(9)}_\alpha\right).$$

  Then, the membership function of $\tilde{A}$ is given by

  $$\xi_{\tilde{A}}(x) = \sup_{\alpha\in(0,1]} \alpha \cdot \chi_{\mathfrak{G}_\alpha}(x).$$

Now, we consider the following expression

$$\tilde{A}^\circ \equiv \left(\left(\tilde{A}^{(1)} \cap \tilde{A}^{(2)} \cap \tilde{A}^{(3)}\right) \cup \left(\tilde{A}^{(4)} \cap \tilde{A}^{(5)}\right)\right) \cap \left(\tilde{A}^{(6)} \cup \left(\tilde{A}^{(7)} \cup \tilde{A}^{(8)}\right) \cup \tilde{A}^{(9)}\right).$$

Then $\tilde{A} \neq \tilde{A}^\circ$ in general. However, if the similar conditions regarding $\tilde{A}^{(6)}, \tilde{A}^{(7)}, \tilde{A}^{(8)}, \tilde{A}^{(9)}$ in Proposition 9.4.8 are satisfied, then $\tilde{A} = \tilde{A}^\circ$.

9.4.4 Equivalence with the Conventional Situation

Now, we are going to present the equivalence with the conventional set operations of fuzzy sets in $\mathbb{R}^m$ by referring to Definition 3.2.9.

Theorem 9.4.11 **(Equivalence)** *Let $\tilde{A}^{(1)},\ldots,\tilde{A}^{(n)}$ be fuzzy sets in $\mathbb{R}^m$ satisfying $\tilde{A}_\alpha^{(1)} \cap \ldots \cap \tilde{A}_\alpha^{(n)} \neq \emptyset$ for all $\alpha \in [0,1]$. Suppose that the function $u_n^\cap : [0,1]^n \to [0,1]$ is both $\subseteq$-compatible and $\supseteq$-compatible with set intersection. Then, we have*

$$\tilde{A}^{(1)} \cap \ldots \cap \tilde{A}^{(n)} = \tilde{A}^{(1)} \sqcap \ldots \sqcap \tilde{A}^{(n)}. \tag{9.38}$$

Proof. From part (ii) of Proposition 9.4.3 and Definition 3.2.9, we have

$$\mathfrak{G}^\cap(\alpha) = \tilde{A}_\alpha^{(1)} \cap \ldots \cap \tilde{A}_\alpha^{(n)} = \left(\tilde{A}^{(1)} \sqcap \ldots \sqcap \tilde{A}^{(n)}\right)_\alpha \text{ for all } \alpha \in (0,1]. \tag{9.39}$$

From (9.25) in Theorem 9.4.4 and (9.39), we obtain

$$\left(\tilde{A}^{(1)} \sqcap \ldots \sqcap \tilde{A}^{(n)}\right)_\alpha = \mathfrak{G}^\cap(\alpha) = \left(\tilde{A}^{(1)} \cap \ldots \cap \tilde{A}^{(n)}\right)_\alpha \text{ for all } \alpha \in (0,1]. \tag{9.40}$$

Using (9.40) and Proposition 2.2.15, we obtain the equality (9.38). This completes the proof. ■

Theorem 9.4.12 **(Equivalence)** *Let $\tilde{A}^{(1)},\ldots,\tilde{A}^{(n)}$ be fuzzy sets in $\mathbb{R}^m$ satisfying $\tilde{A}_\alpha^{(1)} \cap \ldots \cap \tilde{A}_\alpha^{(n)} \neq \emptyset$ for all $\alpha \in [0,1]$. Then, we have*

$$\tilde{A}^{(1)} \cap \ldots \cap \tilde{A}^{(n)} = \tilde{A}^{(1)} \wedge \ldots \wedge \tilde{A}^{(n)}.$$

Proof. From part (ii) of Proposition 9.4.3 and Proposition 3.2.2, we have

$$\mathfrak{G}^\cap(\alpha) = \tilde{A}_\alpha^{(1)} \cap \ldots \cap \tilde{A}_\alpha^{(n)} = \left(\tilde{A}^{(1)} \wedge \ldots \wedge \tilde{A}^{(n)}\right)_\alpha \text{ for all } \alpha \in (0,1]. \tag{9.41}$$

From (9.25) and (9.41), we obtain

$$\left(\tilde{A}^{(1)} \wedge \ldots \wedge \tilde{A}^{(n)}\right)_\alpha = \mathfrak{G}^\cap(\alpha) = \left(\tilde{A}^{(1)} \cap \ldots \cap \tilde{A}^{(n)}\right)_\alpha \text{ for all } \alpha \in (0,1]. \tag{9.42}$$

Using (9.42) and Proposition 2.2.15, the proof is complete. ■

Theorem 9.4.13 **(Equivalence)** *Let $\tilde{A}^{(1)},\ldots,\tilde{A}^{(n)}$ be fuzzy sets in $\mathbb{R}^m$. Suppose that the following conditions are satisfied.*

- *The aggregation function $u_n^\cup : [0,1]^n \to [0,1]$ is both $\subseteq$-compatible and $\supseteq$-compatible with set union.*
- *Any one of the conditions* (a), (b), *and* (c) *in Theorem 9.4.6 is satisfied.*

Then, we have

$$\tilde{A}^{(1)} \cup \ldots \cup \tilde{A}^{(n)} = \tilde{A}^{(1)} \sqcup \ldots \sqcup \tilde{A}^{(n)}. \tag{9.43}$$

Proof. From part (i) of Proposition 9.4.3 and Definition 3.3.7, we have

$$\mathfrak{G}^\cup(\alpha) = \tilde{A}_\alpha^{(1)} \cup \ldots \cup \tilde{A}_\alpha^{(n)} = \left(\tilde{A}^{(1)} \sqcup \ldots \sqcup \tilde{A}^{(n)}\right)_\alpha \text{ for all } \alpha \in (0,1]. \tag{9.44}$$

From (9.31) in Theorem 9.4.6 and (9.44), we obtain

$$\left(\tilde{A}^{(1)} \sqcup \ldots \sqcup \tilde{A}^{(n)}\right)_\alpha = \mathfrak{G}^\cup(\alpha) = \left(\tilde{A}^{(1)} \cup \ldots \cup \tilde{A}^{(n)}\right)_\alpha \text{ for all } \alpha \in (0,1]. \tag{9.45}$$

Using (9.45) and Proposition 2.2.15, we obtain the equality (9.43). This completes the proof. ■

Theorem 9.4.14 **(Equivalence)** *Let* $\tilde{A}^{(1)}, \ldots, \tilde{A}^{(n)}$ *be fuzzy sets in* $\mathbb{R}^m$. *Suppose that any one of the conditions* (a), (b), *and* (c) *in Theorem 9.4.6 is satisfied. Then, we have*

$$\tilde{A}^{(1)} \cup \ldots \cup \tilde{A}^{(n)} = \tilde{A}^{(1)} \vee \ldots \vee \tilde{A}^{(n)}.$$

Proof. From part (i) of Proposition 9.4.3 and Proposition 3.3.2, we have

$$\mathfrak{G}^{\cup}(\alpha) = \tilde{A}^{(1)}_{\alpha} \cup \ldots \cup \tilde{A}^{(n)}_{\alpha} = \left(\tilde{A}^{(1)} \vee \ldots \vee \tilde{A}^{(n)}\right)_{\alpha} \text{ for all } \alpha \in (0,1]. \tag{9.46}$$

From (9.31) and (9.46), we obtain

$$\left(\tilde{A}^{(1)} \vee \ldots \vee \tilde{A}^{(n)}\right)_{\alpha} = \mathfrak{G}^{\cup}(\alpha) = \left(\tilde{A}^{(1)} \cup \ldots \cup \tilde{A}^{(n)}\right)_{\alpha} \text{ for all } \alpha \in (0,1]. \tag{9.47}$$

Using (9.47) and Proposition 2.2.15, the proof is complete. ■

9.5 Arithmetics Using Gradual Numbers

Let $\tilde{A}$ and $\tilde{B}$ be two fuzzy sets in $\mathbb{R}$ with membership function $\xi_{\tilde{A}}$ and $\xi_{\tilde{B}}$, respectively. According to Definition 9.1.6, for $\hat{a} \in \tilde{A}$ and $\hat{b} \in \tilde{B}$, the gradual numbers $\hat{a}$ and $\hat{b}$ are real-valued functions defined on [0,1] satisfying

$$\hat{a}(\alpha) \in \tilde{A}_{\alpha} \text{ and } \hat{b}(\alpha) \in \tilde{B}_{\alpha} \text{ for } \alpha \in [0,1].$$

The arithmetic $\hat{a}\circ\hat{b}$ is defined by

$$(\hat{a}\circ\hat{b})(\alpha) = \hat{a}(\alpha)\circ\hat{b}(\alpha)$$

for all $\alpha \in [0,1]$, where $\circ$ is an arithmetic operation $+, -, \times$ or $/$ for real numbers. When $\circ = /$ is taken, we need to assume that $\hat{b}(\alpha) \neq 0$ for each α. It is clear to see that $\hat{a}\circ\hat{b}$ is also a gradual number.

In the topic of set-valued analysis, given any two subsets A and B of $\mathbb{R}$, the arithmetic $A\circ B$ is defined by

$$A\circ B = \{a\circ b : a \in A \text{ and } b \in B\}, \tag{9.48}$$

where $\circ \in \{+, -, \times, /\}$. The division should avoid to divide by zero. We are going to follow this similar concept to define the arithmetic operation $\tilde{A} \circledcirc \tilde{B}$ of fuzzy sets $\tilde{A}$ and $\tilde{B}$ for $\circledcirc \in \{\oplus^{\circ}, \ominus^{\circ}, \otimes^{\circ}, \oslash^{\circ}\}$. Consider the following family

$$\left\{\hat{a}\circ\hat{b} : \hat{a} \in \tilde{A} \text{ and } \hat{b} \in \tilde{B}\right\}, \tag{9.49}$$

which consists of gradual numbers, where the arithmetic operation $\circ \in \{+, -, \times, /\}$ corresponds to the arithmetic operation $\circledcirc \in \{\oplus^{\circ}, \ominus^{\circ}, \otimes^{\circ}, \oslash^{\circ}\}$. Then, the above family (9.49) can generate a gradual set $\mathfrak{G}$ given by

$$\mathfrak{G}(\alpha) = \left\{\hat{a}(\alpha)\circ\hat{b}(\alpha) : \hat{a} \in \tilde{A} \text{ and } \hat{b} \in \tilde{B}\right\} \text{ for } \alpha \in [0,1]. \tag{9.50}$$

Using this gradual set and referring to (9.2), we can generate a fuzzy set $\tilde{C}^{\mathfrak{G}}$ in $\mathbb{R}$ with membership function given by

$$\xi_{\tilde{C}^{\mathfrak{G}}}(x) = \sup_{\alpha\in(0,1]} \alpha \cdot \chi_{\mathfrak{G}(\alpha)}(x). \tag{9.51}$$

In this case, we define $\tilde{A} \circledcirc \tilde{B} = \tilde{C}^{\mathfrak{G}}$.

Given any two fuzzy sets $\tilde{A}$ and $\tilde{B}$ in $\mathbb{R}$, by referring to (9.5), we can generate two gradual sets $\mathfrak{A}$ and $\mathfrak{B}$ corresponding to $\tilde{A}$ and $\tilde{B}$. More precisely, we have

$$\mathfrak{A}(\alpha) = \left\{\hat{a}(\alpha) : \hat{a} \in \tilde{A}\right\} \text{ and } \mathfrak{B}(\alpha) = \left\{\hat{b}(\alpha) : \hat{b} \in \tilde{B}\right\} \text{ for } \alpha \in [0,1]. \tag{9.52}$$

Part (i) of Proposition 9.1.9 says that

$$\mathfrak{A}(\alpha) = \tilde{A}_\alpha \text{ and } \mathfrak{B}(\alpha) = \tilde{B}_\alpha \text{ for } \alpha \in [0,1]. \tag{9.53}$$

According to (9.48), (9.50), (9.52), and (9.53), we see that

$$\mathfrak{G}(\alpha) = \mathfrak{A}(\alpha) \circ \mathfrak{B}(\alpha) = \tilde{A}_\alpha \circ \tilde{B}_\alpha \text{ for } \alpha \in [0,1], \tag{9.54}$$

From the basic properties of fuzzy sets, we have $\tilde{A}_\beta \subseteq \tilde{A}_\alpha$ and $\tilde{B}_\beta \subseteq \tilde{B}_\alpha$ for $\alpha, \beta \in [0,1]$ with $\alpha < \beta$. Using the equalities (9.54), we obtain

$$\mathfrak{G}(\beta) \subseteq \mathfrak{G}(\alpha) \text{ for } \alpha, \beta \in [0,1] \text{ with } \alpha < \beta. \tag{9.55}$$

This also says that the family $\{\mathfrak{G}(\alpha) : \alpha \in [0,1]\}$ is nested.

Example 9.5.1 Consider the fuzzy set $\tilde{A}$ in Example 9.1.3. Now, we consider another fuzzy set $\tilde{B}$ in $\mathbb{R}$ with membership function given by

$$\xi_{\tilde{B}}(r) = \begin{cases} (r - b^L)/(b_1 - b^L) & \text{if } b^L \le r \le b_1 \\ 1 & \text{if } b_1 < r \le b_2 \\ (b^U - r)/(b^U - b_2) & \text{if } b_2 < r \le b^U \\ 0 & \text{otherwise.} \end{cases}$$

According to (9.51), the membership functions of $\tilde{A} \oplus^\circ \tilde{B}$ is given by

$$\xi_{\tilde{A} \oplus^\circ \tilde{B}}(x) = \sup_{\alpha \in (0,1]} \alpha \cdot \chi_{\mathfrak{G}(\alpha)}(x),$$

where

$$\begin{aligned} \mathfrak{G}(\alpha) &= \tilde{A}_\alpha + \tilde{B}_\alpha \\ &= \left[(1-\alpha)a^L + \alpha a_1, (1-\alpha)a^U + \alpha a_2\right] + \left[(1-\alpha)b^L + \alpha b_1, (1-\alpha)b^U + \alpha b_2\right] \\ &= \left[(1-\alpha)\left(a^L + b^L\right) + \alpha\left(a_1 + b_1\right), (1-\alpha)\left(a^U + b^U\right) + \alpha\left(a_2 + b_2\right)\right] \end{aligned}$$

for $\alpha \in [0,1]$ by referring to (9.54). It is not hard to see that the α-level set of $\tilde{A} \oplus^\circ \tilde{B}$ is given by

$$\left(\tilde{A} \oplus^\circ \tilde{B}\right)_\alpha = \mathfrak{G}(\alpha) = \tilde{A}_\alpha + \tilde{B}_\alpha$$

for $\alpha \in [0,1]$.

Now, given any fuzzy set $\tilde{A}$ in $\mathbb{R}$, we want to study the meaning of $-\tilde{A}$ and $1/\tilde{A}$. Consider the following families

$$\left\{-\hat{a} : \hat{a} \in \tilde{A}\right\} \text{ and } \left\{1/\hat{a} : \hat{a} \in \tilde{A} \text{ with } \hat{a}(\alpha) \neq 0 \text{ for all } \alpha \in [0,1]\right\}, \tag{9.56}$$

which consist of gradual numbers. Then, the above two families (9.56) can generate two gradual sets $\mathfrak{G}^{(-)}$ and $\mathfrak{G}^{(/)}$ defined by

$$\mathfrak{G}^{(-)}(\alpha) = \left\{-\hat{a}(\alpha) : \hat{a} \in \tilde{A}\right\} \text{ and } \mathfrak{G}^{(/)}(\alpha) = \left\{1/\hat{a}(\alpha) : \hat{a} \in \tilde{A}\right\},$$

respectively. Using these two gradual sets and referring to (9.2), we can generate two fuzzy sets $\tilde{D}^{\mathfrak{G}^{(-)}}$ and $\tilde{D}^{\mathfrak{G}^{(/)}}$ in $\mathbb{R}$ with membership function given by

$$\xi_{\tilde{D}^{\mathfrak{G}^{(-)}}}(x) = \sup_{\alpha\in(0,1]} \alpha \cdot \chi_{\mathfrak{G}^{(-)}(\alpha)}(x) \text{ and } \xi_{\tilde{D}^{\mathfrak{G}^{(/)}}}(x) = \sup_{\alpha\in(0,1]} \alpha \cdot \chi_{\mathfrak{G}^{(/)}(\alpha)}(x),$$

respectively. In this case, we define

$$-\tilde{A} = \tilde{D}^{\mathfrak{G}^{(-)}} \text{ and } 1/\tilde{A} = \tilde{D}^{\mathfrak{G}^{(/)}}.$$

Since

$$-\tilde{A}_\alpha = \left\{-x : x \in \tilde{A}_\alpha\right\} \text{ and } 1/\tilde{A}_\alpha = \left\{1/x : x \in \tilde{A}_\alpha \text{ with } x \neq 0\right\},$$

we can similarly obtain

$$\mathfrak{G}^{(-)}(\alpha) = -\tilde{A}_\alpha \text{ and } \mathfrak{G}^{(/)}(\alpha) = 1/\tilde{A}_\alpha \text{ for } \alpha \in [0,1].$$

The following results can be obtained.

Proposition 9.5.2 *Given any two fuzzy sets $\tilde{A}$ and $\tilde{B}$ in $\mathbb{R}$, we have $\tilde{A} \ominus^\circ \tilde{B} = \tilde{A} \oplus^\circ (-\tilde{B})$. Suppose that $\xi_{\tilde{B}}(x) \neq 0$ for each $x \in \mathbb{R}$. Then, we have $\tilde{A} \oslash^\circ \tilde{B} = \tilde{A} \otimes^\circ (1/\tilde{B})$.*

By referring to (9.50), for $\circ \in \{+, \times\}$, since

$$\hat{a}(\alpha)\circ\hat{b}(\alpha) = \hat{b}(\alpha)\circ\hat{a}(\alpha) \text{ for all } \alpha \in [0,1],$$

it follows that the commutativity $\tilde{A} \circledcirc \tilde{B} = \tilde{B} \circledcirc \tilde{A}$ holds true for $\circledcirc \in \{\oplus^\circ, \otimes^\circ\}$. Using Proposition 9.5.2, we see that

$$\tilde{A} \ominus^\circ \tilde{B} = \tilde{A} \oplus^\circ (-\tilde{B}) = -\tilde{B} \oplus^\circ \tilde{A}.$$

If $\xi_{\tilde{B}}(x) \neq 0$ for each $x \in \mathbb{R}$, then

$$\tilde{A} \oslash^\circ \tilde{B} = \tilde{A} \otimes^\circ (1/\tilde{B}) = (1/\tilde{B})\otimes^\circ\tilde{A}.$$

Let $\tilde{C}$ be another fuzzy set in $\mathbb{R}$. For any $\hat{c} \in \tilde{C}$ and $\circ_1, \circ_2 \in \{+, -\}$, since the associativity

$$\left(\hat{a}(\alpha)\circ_1\hat{b}(\alpha)\right)\circ_2\hat{c}(\alpha) = \hat{a}(\alpha)\circ_1\left(\hat{b}(\alpha)\circ_2\hat{c}(\alpha)\right)$$

holds true for all $\alpha \in [0,1]$, it follows that the associativity

$$(\tilde{A}\circledcirc_1\tilde{B})\circledcirc_2\tilde{C} = \tilde{A}\circledcirc_1(\tilde{B}\circledcirc_2\tilde{C})$$

also holds true for $\circledcirc_1, \circledcirc_2 \in \{\oplus^\circ, \ominus^\circ\}$, where the arithmetic operations $\circledcirc_1$ and $\circledcirc_2$ correspond to the arithmetic operations $\circ_1$ and $\circ_2$. In this case, we can simply write

$$(\tilde{A}\circledcirc_1\tilde{B})\circledcirc_2\tilde{C} = \tilde{A}\circledcirc_1(\tilde{B}\circledcirc_2\tilde{C}) \equiv \tilde{A}\circledcirc_1\tilde{B}\circledcirc_2\tilde{C}.$$

The associativity for $\otimes^\circ$ can be similarly realized as follows

$$(\tilde{A} \otimes^\circ \tilde{B}) \otimes^\circ\tilde{C} = \tilde{A}\otimes^\circ (\tilde{B} \otimes^\circ \tilde{C}) \equiv \tilde{A} \otimes^\circ \tilde{B} \otimes^\circ \tilde{C}.$$

In general, let $\tilde{A}^{(1)}, \ldots, \tilde{A}^{(n)}$ be fuzzy sets in $\mathbb{R}$. Consider the arithmetic operation $\circledcirc_i \in \{\oplus^\circ, \ominus^\circ, \otimes^\circ, \oslash^\circ\}$ corresponding to the arithmetic operation $\circ_i \in \{+, -, \times, /\}$ for $i = 1, \ldots, n-1$. Then, we can generate a gradual set $\mathfrak{G}$ defined by

$$\begin{aligned}\mathfrak{G}(\alpha) &= \left\{\left(\hat{a}_1\circ_1\cdots\circ_{n-1}\hat{a}_n\right)(\alpha) : \hat{a}_i \in \tilde{A}^{(i)} \text{ for } i = 1, \ldots, n\right\}\\ &= \left\{\hat{a}_1(\alpha)\circ_1\cdots\circ_{n-1}\hat{a}_n(\alpha) : \hat{a}_i \in \tilde{A}^{(i)} \text{ for } i = 1, \ldots, n\right\}\end{aligned}$$

for $\alpha \in [0,1]$. Using this gradual set and referring to (9.2), we can generate a fuzzy set $\tilde{A}^{\mathfrak{G}}$ in $\mathbb{R}$ with membership function given by

$$\xi_{\tilde{A}^{\mathfrak{G}}}(x) = \sup_{\alpha \in (0,1]} \alpha \cdot \chi_{\mathfrak{G}(\alpha)}(x). \tag{9.57}$$

In this case, we define

$$\tilde{A}^{(1)} \circledcirc_1 \cdots \circledcirc_{n-1} \tilde{A}^{(n)} = \tilde{A}^{\mathfrak{G}},$$

where the operations $\otimes^\circ$ and $\oslash^\circ$ corresponding to the operations $\times$ and / have the highest priority for performing arithmetic calculations.

By referring to (9.54) and (9.55), we also see that

$$\mathfrak{G}(\alpha) = \tilde{A}^{(1)}_\alpha \circ_1 \cdots \circ_{n-1} \tilde{A}^{(n)}_\alpha \text{ for } \alpha \in [0,1], \tag{9.58}$$

and

$$\mathfrak{G}(\beta) \subseteq \mathfrak{G}(\alpha) \text{ for } \alpha, \beta \in [0,1] \text{ with } \alpha < \beta. \tag{9.59}$$

This also says that the family $\{\mathfrak{G}(\alpha) : \alpha \in [0,1]\}$ is nested.

Suppose that the parentheses are included in the expression $\tilde{A}^{(1)} \circledcirc_1 \cdots \circledcirc_{n-1} \tilde{A}^{(n)}$. Then, the terms within the parentheses have the highest priority for performing arithmetic calculations.

Example 9.5.3 Suppose that we want to calculate the following expression

$$\left(\tilde{A}^{(1)} \ominus^\circ \tilde{A}^{(2)}\right) \otimes^\circ \left(\tilde{A}^{(3)} \ominus^\circ \tilde{A}^{(4)} \oplus^\circ \tilde{A}^{(5)}\right) \ominus^\circ \left(\tilde{A}^{(6)} \otimes^\circ \left(\tilde{A}^{(7)} \oplus^\circ \tilde{A}^{(8)} \ominus^\circ \tilde{A}^{(9)} \ominus^\circ \tilde{A}^{(10)}\right)\right). \tag{9.60}$$

Then, we can generate a gradual set $\mathfrak{G}$ defined by

$$\begin{aligned}
\mathfrak{G}(\alpha) &= \Big\{ \left(\hat{a}_1 - \hat{a}_2\right) * \left(\hat{a}_3 - \hat{a}_4 + \hat{a}_5\right) - \left(\hat{a}_6 * \left(\hat{a}_7 + \hat{a}_8 - \hat{a}_9 - \hat{a}_{10}\right)\right)(\alpha) : \hat{a}_i \in \tilde{A}^{(i)} \\
&\qquad \text{for } i = 1, \dots, 10 \Big\} \\
&= \left(\tilde{A}^{(1)}_\alpha - \tilde{A}^{(2)}_\alpha\right) \times \left(\tilde{A}^{(3)}_\alpha \ominus^\circ \tilde{A}^{(4)}_\alpha \oplus^\circ \tilde{A}^{(5)}_\alpha\right) - \left(\tilde{A}^{(6)}_\alpha \times \left(\tilde{A}^{(7)}_\alpha + \tilde{A}^{(8)}_\alpha - \tilde{A}^{(9)}_\alpha - \tilde{A}^{(10)}_\alpha\right)\right)
\end{aligned}$$

for $\alpha \in [0,1]$. The membership function of expression (9.60) is given in the form of (9.57).

Now, we want to study the α-level sets of the arithmetic expression $\tilde{A}^{(1)} \circledcirc_1 \cdots \circledcirc_{n-1} \tilde{A}^{(n)}$.

Theorem 9.5.4 *Let $\tilde{A}^{(1)}, \dots, \tilde{A}^{(n)}$ be fuzzy sets in $\mathbb{R}$. Suppose that, for any fixed $x \in \mathbb{R}$, the function*

$$\eta_x(\alpha) = \alpha \cdot \chi_{\mathfrak{G}(\alpha)}(x)$$

is upper semi-continuous on $[0,1]$. *Then, the α-level set of*

$$\tilde{A}^{\mathfrak{G}} = \tilde{A}^{(1)} \circledcirc_1 \cdots \circledcirc_{n-1} \tilde{A}^{(n)}$$

is given by

$$\begin{aligned}
\tilde{A}^{\mathfrak{G}}_\alpha &= \left(\tilde{A}^{(1)} \circledcirc_1 \cdots \circledcirc_{n-1} \tilde{A}^{(n)}\right)_\alpha = \{x \in \mathbb{R} : \xi_{\tilde{A}^{\mathfrak{G}}}(x) \geq \alpha\} \\
&= \mathfrak{G}(\alpha) = \left\{ \hat{a}_1(\alpha) \circ_1 \cdots \circ_{n-1} \hat{a}_n(\alpha) : \hat{a}_i \in \tilde{A}^{(i)} \textit{ for } i = 1, \dots, n \right\} \\
&= \tilde{A}^{(1)}_\alpha \circ_1 \cdots \circ_{n-1} \tilde{A}^{(n)}_\alpha
\end{aligned}$$

for every $\alpha \in (0,1]$, *and*

$$\tilde{A}^{\mathfrak{G}}_{0+} = \left(\tilde{A}^{(1)} \circledcirc_1 \cdots \circledcirc_{n-1} \tilde{A}^{(n)}\right)_{0+} = \bigcup_{\alpha(0,1]} \left(\tilde{A}^{(1)} \circledcirc_1 \cdots \circledcirc_{n-1} \tilde{A}^{(n)}\right)_{\alpha} = \bigcup_{\alpha(0,1]} \mathfrak{G}(\alpha)$$
$$= \bigcup_{\alpha(0,1]} \left\{ \hat{a}_1(\alpha) \circ_1 \cdots \circ_{n-1} \hat{a}_n(\alpha) : \hat{a}_i \in \tilde{A}^{(i)} \textit{ for } i = 1, \ldots, n \right\}.$$

Proof. Given any fixed $0\alpha \in (0,1]$, suppose that $x \in \tilde{A}^{\mathfrak{G}}_{\alpha}$. Now, we also assume that $x \notin \mathfrak{G}(\beta)$ for all $\beta \in [0,1]$ with $\beta \geq \alpha$. We want to lead to a contradiction. Under this assumption, we see that $\beta \cdot \chi_{\mathfrak{G}(\beta)}(x) < \alpha$ for all $\beta \in [0,1]$. Since η_x is upper semi-continuous on the closed interval $[0,1]$ by the assumption, the supremum of function η_x is attained by Proposition 1.4.4. This says that

$$\xi_{\tilde{A}^{\mathfrak{G}}}(x) = \sup_{\beta \in [0,1]} \eta_x(\beta) = \sup_{\beta \in [0,1]} \beta \cdot \chi_{\mathfrak{G}(\beta)}(x) = \max_{\beta \in [0,1]} \beta \cdot \chi_{\mathfrak{G}(\beta)}(x) = \beta^* \cdot \chi_{\mathfrak{G}(\beta^*)}(x) < \alpha$$

for some $\beta^* \in [0,1]$, which violates $x \in \tilde{A}^{\mathfrak{G}}_{\alpha}$. Therefore, there exists $\beta_0 \in [0,1]$ with $\beta_0 \geq \alpha$ satisfying $x \in \mathfrak{G}(\beta_0) \subseteq \mathfrak{G}(\alpha)$ by (9.59), which shows the inclusion $\tilde{A}^{\mathfrak{G}}_{\alpha} \subseteq \mathfrak{G}(\alpha)$. The following inclusion is obvious

$$\mathfrak{G}(\alpha) \subseteq \left\{ x \in \mathbb{R} : \sup_{\beta \in [0,1]} \beta \cdot \chi_{\mathfrak{G}(\beta)}(x) \geq \alpha \right\} = \{x \in \mathbb{R} : \xi_{\tilde{A}^{\mathfrak{G}}}(x) \geq \alpha\} = \tilde{A}^{\mathfrak{G}}_{\alpha}.$$

Therefore, using (9.58), we obtain the desired equalities. This completes the proof. ■

Without assuming the upper semi-continuity, we also have the following interesting results.

Theorem 9.5.5 *Let* $\tilde{A}^{(1)}, \ldots, \tilde{A}^{(n)}$ *be fuzzy sets in* $\mathbb{R}$. *Suppose that given any* $\alpha \in (0,1]$ *and any increasing convergent sequence* $\{\alpha_m\}_{m=1}^{\infty}$ *in* $(0,1]$ *with* $\alpha_m \uparrow \alpha$, *the following inclusion is satisfied*

$$\bigcap_{m=1}^{\infty} \left(\tilde{A}^{(1)}_{\alpha_m} \circ_1 \cdots \circ_{n-1} \tilde{A}^{(n)}_{\alpha_m}\right) \subseteq \tilde{A}^{(1)}_{\alpha} \circ_1 \cdots \circ_{n-1} \tilde{A}^{(n)}_{\alpha}. \tag{9.61}$$

Then, the α*-level set of*

$$\tilde{A}^{\mathfrak{G}} = \tilde{A}^{(1)} \circledcirc_1 \cdots \circledcirc_{n-1} \tilde{A}^{(n)}$$

is given by

$$\tilde{A}^{\mathfrak{G}}_{\alpha} = \left(\tilde{A}^{(1)} \circledcirc_1 \cdots \circledcirc_{n-1} \tilde{A}^{(n)}\right)_{\alpha} = \mathfrak{G}(\alpha)$$
$$= \left\{ \hat{a}_1(\alpha) \circ_1 \cdots \circ_{n-1} \hat{a}_n(\alpha) : \hat{a}_i \in \tilde{A}^{(i)} \textit{ for } i = 1, \ldots, n \right\}$$
$$= \tilde{A}^{(1)}_{\alpha} \circ_1 \cdots \circ_{n-1} \tilde{A}^{(n)}_{\alpha}$$

for every $\alpha \in (0,1]$, *and*

$$\tilde{A}^{\mathfrak{G}}_{0+} = \left(\tilde{A}^{(1)} \circledcirc_1 \cdots \circledcirc_{n-1} \tilde{A}^{(n)}\right)_{0+} = \bigcup_{\alpha(0,1]} \mathfrak{G}(\alpha)$$
$$= \bigcup_{\alpha(0,1]} \left\{ \hat{a}_1(\alpha) \circ_1 \cdots \circ_{n-1} \hat{a}_n(\alpha) : \hat{a}_i \in \tilde{A}^{(i)} \textit{ for } i = 1, \ldots, n \right\}.$$

Proof. Using the equalities (9.58), we see that the inclusion (9.61) is satisfied if and only if the following inclusion is satisfied

$$\bigcap_{m=1}^{\infty} \mathfrak{G}(\alpha_m) \subseteq \mathfrak{G}(\alpha). \tag{9.62}$$

Given any fixed $x \in \mathbb{R}$, we define the following set

$$F_\rho = \left\{\alpha \in [0,1] \,:\, \alpha \cdot \chi_{\mathfrak{G}(\alpha)}(x) \geq \rho\right\}.$$

which also says that $F_\rho \neq \emptyset$ for $\rho \leq 1$ and $F_\rho = \emptyset$ for $\rho > 1$ by referring to (9.57). Next, we are going to prove $\mathrm{cl}(F_\rho) = F_\rho$ for each $\rho \in (0,1]$. Given any $\alpha \in \mathrm{cl}(F_\rho)$, there exists a sequence $\{\alpha_m\}_{m=1}^{\infty}$ in F_ρ satisfying $\alpha_m \to \alpha$. Therefore, we have $\alpha_m \geq \rho > 0$ and $x \in \mathfrak{G}(\alpha_m)$ for all m, which also says that $\alpha > 0$, since

$$\alpha = \lim_{m\to\infty} \alpha_m \geq \rho > 0.$$

Therefore, there exists a subsequence $\{\alpha_{m_k}\}_{k=1}^{\infty}$ of $\{\alpha_m\}_{m=1}^{\infty}$ satisfying $\alpha_{m_k} \uparrow \alpha$ or $\alpha_{m_k} \downarrow \alpha$ as $k \to \infty$.

- Suppose that $\alpha_{m_k} \downarrow \alpha$. Then $\alpha_{m_k} \geq \alpha$ for all k. This says that $x \in \mathfrak{G}(\alpha_{m_k}) \subseteq \mathfrak{G}(\alpha)$ by (9.59).
- Suppose that $\alpha_{m_k} \uparrow \alpha$. Since $\mathrm{cl}(F_\rho) \subseteq [0,1]$, it says that $\alpha \in [0,1]$. Since $x \in \mathfrak{G}(\alpha_{m_k})$ for all k, using (9.62), it follows that $x \in \mathfrak{G}(\alpha)$.

Therefore, we conclude that $x \in \mathfrak{G}(\alpha)$ for both cases. This also says that $\alpha \cdot \chi_{\mathfrak{G}(\alpha)}(x) \geq \rho$, i.e. $\alpha \in F_\rho$. Therefore, we obtain the inclusion $\mathrm{cl}(F_\rho) \subseteq F_\rho$, which means that F_ρ is a closed set for each $\rho \in (0,1]$. In order to apply Theorem 9.5.4, we need to show that the function η_x is upper semi-continuous on [0,1]. Equivalently, we claim that the set F_ρ is closed for each $\rho \in \mathbb{R}$. We consider the following cases.

- For each $\rho \in (0,1]$, the set F_ρ is closed as shown above.
- For $\rho > 1$, the set $F_\rho = \emptyset$ is empty, which is also closed.
- For $\rho \leq 0$, the set $F_\rho = I^\cap = [0,1]$ is closed.

We conclude that the function η_x is indeed upper semi-continuous on [0,1]. Therefore, the desired results follow from Theorem 9.5.4 immediately. ■

Suppose that, for any increasing sequence $\{\alpha_m\}_{m=1}^{\infty}$ in (0,1], the following inclusion is satisfied

$$\bigcap_{m=1}^{\infty} \left(\tilde{A}^{(1)}_{\alpha_m} \circ_1 \cdots \circ_{n-1} \tilde{A}^{(n)}_{\alpha_m}\right) \subseteq \left(\bigcap_{m=1}^{\infty} \tilde{A}^{(1)}_{\alpha_m}\right) \circ_1 \cdots \circ_{n-1} \left(\bigcap_{m=1}^{\infty} \tilde{A}^{(n)}_{\alpha_m}\right). \tag{9.63}$$

We are going to claim that the inclusion (9.61) is satisfied if and only if the inclusion (9.63) is satisfied. Indeed, according to the basic properties of fuzzy sets, given any $\alpha \in (0,1]$ and any increasing convergent sequence $\{\alpha_m\}_{m=1}^{\infty}$ in (0,1] with $\alpha_m \uparrow \alpha$, we have

$$\bigcap_{m=1}^{\infty} \tilde{A}^{(i)}_{\alpha_m} = \tilde{A}^{(i)}_{\alpha}, \tag{9.64}$$

which says that the inclusion (9.61) is satisfied if and only if the inclusion (9.63) is satisfied. Therefore, we can also obtain the interesting results from Theorem 9.5.5 when the inclusion (9.63) is considered.

Let $\tilde{A}^{(1)}, \ldots, \tilde{A}^{(n)}$ be fuzzy sets in $\mathbb{R}$. We consider the arithmetic

$$\tilde{A}^{\mathfrak{G}} = \tilde{A}^{(1)} \circledcirc_1 \cdots \circledcirc_{n-1} \tilde{A}^{(n)}$$

using gradual numbers. By referring to (7.62), we also consider the arithmetic

$$\tilde{A} = \tilde{A}^{(1)} \boxdot_1 \cdots \boxdot_{n-1} \tilde{A}^{(n)}$$

using the extension principle. Next, we are going to establish the equivalence between $\tilde{A}^{\mathfrak{G}}$ and $\tilde{A}$ under some suitable conditions.

Theorem 9.5.6 (Equivalent Arithmetic Operations) *Let $\tilde{A}^{(1)}, \ldots, \tilde{A}^{(n)}$ be fuzzy sets in $\mathbb{R}$ such that the membership functions of $\tilde{A}^{(1)}, \ldots, \tilde{A}^{(n)}$ are upper semi-continuous. Consider the arithmetic operations $\circledcirc_i \in \{\oplus^\circ, \ominus^\circ, \otimes^\circ\}$ and $\boxdot_i \in \{\boxplus, \boxminus, \boxtimes\}$ for $i = 1, \ldots, n-1$. Let*

$$\tilde{A}^{\mathfrak{G}} = \tilde{A}^{(1)} \circledcirc_1 \cdots \circledcirc_{n-1} \tilde{A}^{(n)} \text{ and } \tilde{A} = \tilde{A}^{(1)} \boxdot_1 \cdots \boxdot_{n-1} \tilde{A}^{(n)}.$$

Then $\tilde{A}^{\mathfrak{G}} = \tilde{A}$. Moreover, their α-level sets are given by

$$\begin{aligned}\tilde{A}^{\mathfrak{G}}_\alpha &= \tilde{A}_\alpha = \tilde{A}^{(1)}_\alpha \circ_1 \cdots \circ_{n-1} \tilde{A}^{(n)}_\alpha \\ &= \left\{ \hat{a}_1(\alpha) \circ_1 \cdots \circ_{n-1} \hat{a}_n(\alpha) : \hat{a}_i \in \tilde{A}^{(i)} \text{ for } i = 1, \ldots, n \right\}\end{aligned}$$

for every $\alpha \in [0,1]$.

Proof. Using part (i) of Theorem 7.5.4, we have

$$\tilde{A}_\alpha = \left(\tilde{A}^{(1)} \boxdot_1 \cdots \boxdot_{n-1} \tilde{A}^{(n)}\right)_\alpha = \tilde{A}^{(1)}_\alpha \circ_1 \cdots \circ_{n-1} \tilde{A}^{(n)}_\alpha \text{ for all } \alpha \in (0,1]. \tag{9.65}$$

In order to apply Theorem 9.5.5, we are going to check that the inclusion (9.61) is satisfied. Given any $\alpha \in (0,1]$ and any increasing convergent sequence $\{\alpha_m\}_{m=1}^\infty$ in (0,1] with $\alpha_m \uparrow \alpha$, according to the basic properties of fuzzy sets, we have $\bigcap_{m=1}^\infty \tilde{A}_{\alpha_m} = \tilde{A}_\alpha$. Then, using (9.65), we obtain

$$\begin{aligned}\bigcap_{m=1}^\infty \left(\tilde{A}^{(1)}_{\alpha_m} \circ_1 \cdots \circ_{n-1} \tilde{A}^{(n)}_{\alpha_m}\right) &= \bigcap_{m=1}^\infty \left(\tilde{A}^{(1)} \boxdot_1 \cdots \boxdot_{n-1} \tilde{A}^{(n)}\right)_{\alpha_m} = \bigcap_{m=1}^\infty \tilde{A}_{\alpha_m} = \tilde{A}_\alpha \\ &= \tilde{A}^{(1)}_\alpha \circ_1 \cdots \circ_{n-1} \tilde{A}^{(n)}_\alpha,\end{aligned}$$

which says that the inclusion (9.61) is satisfied. Therefore, using Theorem 9.5.5, we can obtain

$$\begin{aligned}\tilde{A}^{\mathfrak{G}}_\alpha &= \tilde{A}^{(1)}_\alpha \circ_1 \cdots \circ_{n-1} \tilde{A}^{(n)}_\alpha \\ &= \left\{ \hat{a}_1(\alpha) \circ_1 \cdots \circ_{n-1} \hat{a}_n(\alpha) : \hat{a}_i \in \tilde{A}^{(i)} \text{ for } i = 1, \ldots, n \right\}\end{aligned} \tag{9.66}$$

for every $\alpha \in (0,1]$. From (9.65) and (9.58), we have

$$\mathfrak{G}(\alpha) = \tilde{A}^{(1)}_\alpha \circ_1 \cdots \circ_{n-1} \tilde{A}^{(n)}_\alpha = \left(\tilde{A}^{(1)} \boxdot_1 \cdots \boxdot_{n-1} \tilde{A}^{(n)}\right)_\alpha = \tilde{A}_\alpha \text{ for all } \alpha \in (0,1]. \tag{9.67}$$

From (9.66) and (9.67), we obtain

$$\left(\tilde{A}^{(1)} \circledcirc_1 \cdots \circledcirc_{n-1} \tilde{A}^{(n)}\right)_\alpha = \mathfrak{G}(\alpha) = \tilde{A}_\alpha = \left(\tilde{A}^{(1)} \boxdot_1 \cdots \boxdot_{n-1} \tilde{A}^{(n)}\right)_\alpha \text{ for all } \alpha \in (0,1]$$

and

$$\left(\tilde{A}^{(1)}\circledcirc_1\cdots\circledcirc_{n-1}\tilde{A}^{(n)}\right)_0=\mathrm{cl}\left(\tilde{A}^{(1)}\circledcirc_1\cdots\circledcirc_{n-1}\tilde{A}^{(n)}\right)_{0+}=\mathrm{cl}\left(\bigcup_{0<\alpha\leq 1}\mathfrak{G}(\alpha)\right)$$
$$=\mathrm{cl}\left(\bigcup_{0<\alpha\leq 1}\tilde{A}_\alpha\right)=\mathrm{cl}\left(\{x:\xi_{\tilde{A}}(x)>0\}\right)=\mathrm{cl}\left(\tilde{A}_{0+}\right)$$
$$=\mathrm{cl}\left(\tilde{A}^{(1)}\boxdot_1\cdots\boxdot_{n-1}\tilde{A}^{(n)}\right)_{0+}=\left(\tilde{A}^{(1)}\boxdot_1\cdots\boxdot_{n-1}\tilde{A}^{(n)}\right)_0.$$

This completes the proof. ■

When the arithmetic operations are taken to be $\circledcirc_i=\oslash^\circ$ and $\odot_i=\oslash$ for some i, we need to assume that the denominator is never to be zero. This situation will complicate the proof of Theorem 9.5.6 for taking $\circledcirc_i=\oslash^\circ$ and $\odot_i=\oslash$. Although the case $\circledcirc_i=\oslash^\circ$ and $\odot_i=\oslash$ are not discussed here, it can still be similarly discussed by modifying the proof of Theorem 9.5.6, and it is left for the readers

10

Duality in Fuzzy Sets

The concept of α-level sets of fuzzy sets is an important tool to study the properties of fuzzy sets and their applications. In this chapter, the α-level sets will treated as the upper α-level sets. In other words, we are going to introduce a new concept of so-called lower α-level sets, which can also be regarded as the dual concept of upper α-level sets.

10.1 Lower and Upper Level Sets

Let $\tilde{A}$ be a fuzzy set in $\mathbb{R}^n$ with membership function $\xi_{\tilde{A}}$. For $\alpha \in (0,1]$, the α-level set $\tilde{A}_\alpha$ of $\tilde{A}$ that may be called the upper α-level set (or upper α-cut) of $\tilde{A}$ is defined by

$$\tilde{A}_\alpha = \left\{x \in \mathbb{R}^n : \xi_{\tilde{A}}(x) \geq \alpha\right\} = \left\{x \in \tilde{A}_0 : \xi_{\tilde{A}}(x) \geq \alpha\right\}.$$

Also, the strong α-level set (or upper strong α-level set) is defined by

$$\tilde{A}_{\alpha+} = \left\{x \in \mathbb{R}^n : \xi_{\tilde{A}}(x) > \alpha\right\} = \left\{x \in \tilde{A}_0 : \xi_{\tilde{A}}(x) > \alpha\right\}. \tag{10.1}$$

The 0-level set $\tilde{A}_0$ is defined by $\tilde{A}_0 = \mathrm{cl}(\tilde{A}_{0+})$. Now, the 0-level set $\tilde{A}_0$ is also called the **proper domain** of $\tilde{A}$, since the membership function $\xi_{\tilde{A}}$ vanishes outside $\tilde{A}_0$, i.e. $\xi_{\tilde{A}}(x) = 0$ for $x \notin \tilde{A}_0$. Based on the concept of proper domain, we shall consider the so-called lower α-level set (or lower α-cut) of $\tilde{A}$.

Definition 10.1.1 Let $\tilde{A}$ be a fuzzy set in a universal set $\mathbb{R}^n$ with proper domain $\tilde{A}_0$. For $\alpha \in [0,1]$, the following set

$${}_\alpha\tilde{A} = \left\{x \in \tilde{A}_0 : \xi_{\tilde{A}}(x) \leq \alpha\right\}$$

is called the **lower α-level set** of $\tilde{A}$. For $\alpha \in (0,1]$, we also define the **strong lower α-level set** of $\tilde{A}$ as

$${}_{\alpha-}\tilde{A} = \left\{x \in \tilde{A}_0 : \xi_{\tilde{A}}(x) < \alpha\right\}.$$

We notice that the lower α-level set ${}_\alpha\tilde{A}$ is considered in the proper domain $\tilde{A}_0$ rather than the whole universal set $\mathbb{R}^n$. In general, it is clear to see that

$${}_\alpha\tilde{A} = \left\{x \in \tilde{A}_0 : \xi_{\tilde{A}}(x) \leq \alpha\right\} \neq \left\{x \in \mathbb{R}^n : \xi_{\tilde{A}}(x) \leq \alpha\right\}.$$

Next, we present some interesting observations. We first recall that the notation $x \in A \backslash B$ means $x \in A$ and $x \notin B$.

Mathematical Foundations of Fuzzy Sets, First Edition. Hsien-Chung Wu.

Remark 10.1.2 Let $\tilde{A}$ be a fuzzy set in $\mathbb{R}^n$. Then, we have the following observations.

- For any $\alpha \in [0,1)$, using (10.1), we have

$$_{\alpha}\tilde{A} = \tilde{A}_0 \Big\backslash \left\{x \in \tilde{A}_0 : \xi_{\tilde{A}}(x) > \alpha\right\} = \tilde{A}_0 \Big\backslash \left\{x \in \mathbb{R}^n : \xi_{\tilde{A}}(x) > \alpha\right\} = \tilde{A}_0 \backslash \tilde{A}_{\alpha+}.$$

- Given $x \in {}_0\tilde{A}$, we have $x \in \tilde{A}_0$ and $\xi_{\tilde{A}}(x) = 0$.
- For any $\alpha, \beta \in [0,1]$ with $\alpha < \beta$, we have $_{\alpha}\tilde{A} \subseteq {}_{\beta}\tilde{A}$.
- For $\alpha \in (0,1]$, we have

$$\begin{aligned}\tilde{A}_\alpha &= \left\{x \in \mathbb{R}^n : \xi_{\tilde{A}}(x) \geq \alpha\right\} = \left\{x \in \tilde{A}_0 : \xi_{\tilde{A}}(x) \geq \alpha\right\} \\ &= \tilde{A}_0 \Big\backslash \left\{x \in \tilde{A}_0 : \xi_{\tilde{A}}(x) < \alpha\right\} = \tilde{A}_0 \backslash {}_{\alpha-}\tilde{A}.\end{aligned}$$

Proposition 10.1.3 *Let $\tilde{A}$ be a fuzzy set in $\mathbb{R}^n$. Then, we have the following properties.*

(i) *Suppose that $\alpha \in [0,1)$ and $\{\alpha_n\}_{n=1}^{\infty}$ is a decreasing sequence in $[0,1]$ satisfying $\alpha_n \downarrow \alpha$. Then, we have*

$$_{\alpha}\tilde{A} = \bigcap_{n=1}^{\infty} {}_{\alpha_n}\tilde{A}.$$

(ii) *For $\alpha \in (0,1]$, the following statements hold true.*
- *If $\alpha_n \uparrow \alpha$, then*

$$\bigcup_{n=1}^{\infty} {}_{\alpha_n}\tilde{A} \subseteq {}_{\alpha}\tilde{A} \text{ and } {}_{\alpha-}\tilde{A} \subseteq \bigcup_{n=1}^{\infty} {}_{\alpha_n-}\tilde{A} \subseteq \bigcup_{n=1}^{\infty} {}_{\alpha_n}\tilde{A}.$$

- *If $\alpha_n \uparrow \alpha$ with $\alpha_n < \alpha$ for all n, then*

$$_{\alpha-}\tilde{A} = \bigcup_{n=1}^{\infty} {}_{\alpha_n-}\tilde{A} = \bigcup_{n=1}^{\infty} {}_{\alpha_n}\tilde{A}.$$

(iii) *For $\alpha \in [0,1)$, we have*

$$_{\alpha}\tilde{A} = \bigcap_{\alpha<\beta\leq 1} {}_{\beta}\tilde{A} = \bigcap_{\alpha\leq\beta\leq 1} {}_{\beta}\tilde{A}.$$

(iv) *For $\alpha \in (0,1]$, we have*

$$_{\alpha-}\tilde{A} = \bigcup_{0\leq\beta<\alpha} {}_{\beta}\tilde{A}.$$

Proof. To prove part (i), since $\alpha_n \geq \alpha$ for all n, we have $_{\alpha}\tilde{A} \subseteq {}_{\alpha_n}\tilde{A}$ for all n, which implies

$$_{\alpha}\tilde{A} \subseteq \bigcap_{n=1}^{\infty} {}_{\alpha_n}\tilde{A}.$$

On the other hand, for $x \in \bigcap_{n=1}^{\infty} {}_{\alpha_n}\tilde{A}$, we have $\xi_{\tilde{A}}(x) \leq \alpha_n$ for all n, which implies

$$\xi_{\tilde{A}}(x) \leq \lim_{n\to\infty} \alpha_n = \alpha.$$

Therefore, we conclude that $x \in {}_{\alpha}\tilde{A}$.

To prove part (ii), given $x \in {}_{\alpha-}\tilde{A}$, we have $\xi_{\tilde{A}}(x) < \alpha$. Since $\alpha_n \uparrow \alpha$, given any $0 < \epsilon \leq \alpha - \xi_{\tilde{A}}(x)$, there exists an integer N satisfying $0 < \alpha - \alpha_N < \epsilon$, which says that $\xi_{\tilde{A}}(x) < \alpha_N$, i.e. $x \in \bigcup_{n=1}^{\infty} {}_{\alpha_n-}\tilde{A}$. Therefore, we obtain the inclusion

$$ {}_{\alpha-}\tilde{A} \subseteq \bigcup_{n=1}^{\infty} {}_{\alpha_n-}\tilde{A}. $$

On the other hand, since $\alpha_n \uparrow \alpha$, we have the following cases.

- If $\alpha_n \leq \alpha$ for all n, then ${}_{\alpha_n}\tilde{A} \subseteq {}_{\alpha}\tilde{A}$, which implies
$$ \bigcup_{n=1}^{\infty} {}_{\alpha_n}\tilde{A} \subseteq {}_{\alpha}\tilde{A}. $$
- If $\alpha_n < \alpha$ for all n, we see that $x \in {}_{\alpha_n}\tilde{A}$ implies $\xi_{\tilde{A}}(x) \leq \alpha_n < \alpha$, which says that ${}_{\alpha_n}\tilde{A} \subseteq {}_{\alpha-}\tilde{A}$ for all n. Therefore, we have
$$ \bigcup_{n=1}^{\infty} {}_{\alpha_n}\tilde{A} \subseteq {}_{\alpha-}\tilde{A}. $$

Then, we obtain the desired equalities and inclusions.

To prove part (iii), for $1 \geq \beta > \alpha \geq 0$, we have ${}_{\alpha}\tilde{A} \subseteq {}_{\beta}\tilde{A}$. Therefore, we have the inclusion

$$ {}_{\alpha}\tilde{A} \subseteq \bigcap_{\beta \in (\alpha,1]} {}_{\beta}\tilde{A}. $$

On the other hand, given any $\epsilon > 0$, for

$$ x \in \bigcap_{\beta \in (\alpha,1]} {}_{\beta}\tilde{A}, $$

we have $x \in {}_{\alpha+\epsilon}\tilde{A}$, since $\alpha + \epsilon > \alpha$. This says that $\xi_{\tilde{A}}(x) \leq \alpha + \epsilon$, which also implies $\xi_{\tilde{A}}(x) \leq \alpha$, since ϵ is an arbitrary positive number (i.e. we can take $\epsilon \to 0$). Therefore, we conclude that $x \in {}_{\alpha}\tilde{A}$.

To prove part (iv), for $\beta < \alpha$, if $x \in {}_{\beta}\tilde{A}$, i.e. $\xi_{\tilde{A}}(x) \leq \beta < \alpha$, then $x \in {}_{\alpha-}\tilde{A}$. Therefore, we obtain ${}_{\beta}\tilde{A} \subseteq {}_{\alpha-}\tilde{A}$, which implies the inclusion

$$ \bigcup_{0 \leq \beta < \alpha} {}_{\beta}\tilde{A} \subseteq {}_{\alpha-}\tilde{A}. $$

On the other hand, given any $x \in {}_{\alpha-}\tilde{A}$, we have $\xi_{\tilde{A}}(x) < \alpha$. By the denseness, there exists β_0 satisfying

$$ \xi_{\tilde{A}}(x) \leq \beta_0 < \alpha, \text{ i.e. } x \in \bigcup_{0 \leq \beta < \alpha} {}_{\beta}\tilde{A}. $$

Therefore, we obtain

$$ {}_{\alpha-}\tilde{A} \subseteq \bigcup_{0 \leq \beta < \alpha} {}_{\beta}\tilde{A}. $$

This shows the desired equality, and the proof is complete. ■

Example 10.1.4 Let $\tilde{A}$ be a fuzzy interval. Then, the upper α-level set $\tilde{A}_\alpha$ is a closed interval given by $\tilde{A}_\alpha = [\tilde{A}_\alpha^L, \tilde{A}_\alpha^U]$ for all $\alpha \in [0,1]$. From part (iv) of Proposition 2.2.8, for $\alpha \in [0,1)$, we have

$$\tilde{A}_{\alpha+} = \bigcup_{\alpha<\beta\leq 1} \tilde{A}_\beta = \bigcup_{\alpha<\beta\leq 1} \left[\tilde{A}_\beta^L, \tilde{A}_\beta^U\right].$$

Since $\tilde{A}_\alpha \subseteq \tilde{A}_\beta$ for $\beta < \alpha$, if we further assume that the endpoints $\tilde{A}_\alpha^L$ and $\tilde{A}_\alpha^U$ are continuous functions with respect to α on $[0,1]$, then $\tilde{A}_{\alpha+} = (\tilde{A}_\alpha^L, \tilde{A}_\alpha^U)$ is an open interval. In this case, Remark 10.1.2 says that the lower α-level set ${}_\alpha\tilde{A}$ is given by

$${}_\alpha\tilde{A} = \tilde{A}_0 \backslash \tilde{A}_{\alpha+} = \left[\tilde{A}_0^L, \tilde{A}_0^U\right] \backslash \left(\tilde{A}_\alpha^L, \tilde{A}_\alpha^U\right) = \left[\tilde{A}_0^L, \tilde{A}_\alpha^L\right] \cup \left[\tilde{A}_\alpha^U, \tilde{A}_0^U\right],$$

which is also a closed subset of $\mathbb{R}$. This also says that the membership function $\xi_{\tilde{A}}$ of $\tilde{A}$ is lower semi-continuous. We also see that the lower 1-level set is

$${}_1\tilde{A} = \tilde{A}_0 = \left[\tilde{A}_0^L, \tilde{A}_0^U\right].$$

Under the further assumptions, the membership function of $\tilde{A}$ is continuous.

10.2 Dual Fuzzy Sets

Let $\tilde{A}$ be a fuzzy set in $\mathbb{R}^n$ with membership function $\xi_{\tilde{A}}$. Recall that the membership function of complement fuzzy set of $\tilde{A}$ is denoted and defined by

$$\xi_{\tilde{A}^c} = 1 - \xi_{\tilde{A}},$$

which is also defined on $\mathbb{R}^n$. Since the 0-level set $\tilde{A}_0$ is treated as the proper domain of $\tilde{A}$ as described above, we consider the restrict function $\xi_{\tilde{A}^c}|_{\tilde{A}_0}$ of $\xi_{\tilde{A}^c}$ on $\tilde{A}_0$ and define a function $\xi_{\tilde{A}^\bullet} : \tilde{A}_0 \to [0,1]$ on $\tilde{A}_0$ by

$$\xi_{\tilde{A}^\bullet}(x) = \xi_{\tilde{A}^c}|_{\tilde{A}_0}(x) = 1 - \xi_{\tilde{A}}(x) \text{ for } x \in \tilde{A}_0. \tag{10.2}$$

Then, we use the notation $\tilde{A}^\bullet$ to denote the **dual fuzzy set** of $\tilde{A}$. The membership function of $\tilde{A}^\bullet$ is given in (10.2), which is also called a **dual membership function**. We also notice that $\xi_{\tilde{A}^\bullet} \neq \xi_{\tilde{A}^c}$, since their domains are different.

Remark 10.2.1 The relationship between the α-level sets of fuzzy set $\tilde{A}$ and dual fuzzy set $\tilde{A}^\bullet$ are given below

- We have

$$\tilde{A}_\alpha = \left\{x \in \mathbb{R}^n : \xi_{\tilde{A}}(x) \geq \alpha\right\} = \left\{x \in \mathbb{R}^n : \xi_{\tilde{A}^\bullet}(x) \leq 1 - \alpha\right\} = {}_{1-\alpha}\tilde{A}^\bullet.$$

- We have

$${}_\alpha\tilde{A} = \left\{x \in \mathbb{R}^n : \xi_{\tilde{A}}(x) \leq \alpha\right\} = \left\{x \in \mathbb{R}^n : \xi_{\tilde{A}^\bullet}(x) \geq 1 - \alpha\right\} = \tilde{A}^\bullet_{1-\alpha}.$$

- We have

$$\tilde{A}^\bullet_\alpha = \left\{x \in \mathbb{R}^n : \xi_{\tilde{A}^\bullet}(x) \geq \alpha\right\} = \left\{x \in \mathbb{R}^n : \xi_{\tilde{A}}(x) \leq 1 - \alpha\right\} = {}_{1-\alpha}\tilde{A}.$$

- We have

$${}_\alpha\tilde{A}^\bullet = \left\{x \in \mathbb{R}^n : \xi_{\tilde{A}^\bullet}(x) \leq \alpha\right\} = \left\{x \in \mathbb{R}^n : \xi_{\tilde{A}}(x) \geq 1 - \alpha\right\} = \tilde{A}_{1-\alpha}.$$

Let (U, τ_U) be a topological space, and let A be a set in $\mathbb{R}^n$. Then, the subset A can be endowed with a topology τ_A such that (A, τ_A) is a topological subspace of (U, τ_U). In other words, the subset C of A is a τ_A-closed subset of A if and only if $C = A \cap D$ for some τ_U-closed subset D of $\mathbb{R}^n$. In this case, we say that $f : (A, \tau_A) \to \mathbb{R}$ is upper semi-continuous on A if and only if $\{x \in A : f(x) \geq \lambda\}$ is a τ_A-closed subset of A for all $\lambda \in \mathbb{R}$. We also see that if f is upper semi-continuous on A then $-f$ is lower semi-continuous on A, and if f is lower semi-continuous on A then $-f$ is upper semi-continuous on A. We have the following observations.

- Suppose that $\tilde{A}$ is a fuzzy set in $\mathbb{R}^n$ such that its membership function $\xi_{\tilde{A}}$ is upper semi-continuous on $\mathbb{R}^n$. Then $\xi_{\tilde{A}}$ is also upper semi-continuous on the proper domain $\tilde{A}_0$. Indeed, the set
$$\left\{x \in \tilde{A}_0 : \xi_{\tilde{A}}(x) \geq \lambda\right\} = \tilde{A}_0 \cap \left\{x \in \mathbb{R}^n : \xi_{\tilde{A}}(x) \geq \lambda\right\}$$
is a $\tau_{\tilde{A}_0}$-closed subset of $\tilde{A}_0$.
- Suppose that $\tilde{A}$ is a fuzzy set in $\mathbb{R}^n$ such that its membership function $\xi_{\tilde{A}}$ is upper semi-continuous on the proper domain $\tilde{A}_0$. Then it is clear to see that the dual membership function $\xi_{\tilde{A}^\bullet} = 1 - \xi_{\tilde{A}}$ of $\tilde{A}^\bullet$ is lower semi-continuous on $\tilde{A}_0$.

Remark 10.2.2 Given $\tilde{A} \in \mathfrak{F}(\mathbb{R}^m)$ by referring to Definition 2.3.1, the membership function $\xi_{\tilde{A}}$ is upper semi-continuous and quasi-concave on $\tilde{A}_0$. It is also clear to see that the membership function $\xi_{\tilde{A}^\bullet}$ of dual fuzzy set $\tilde{A}^\bullet$ is lower semi-continuous and quasi-convex on $\tilde{A}_0$. This says that the lower α-level set ${}_\alpha\tilde{A}^\bullet$ is a closed and convex subset of $\tilde{A}_0$ for $\alpha \in (0,1]$.

Example 10.2.3 Let $\tilde{A}$ be a fuzzy interval with dual fuzzy set $\tilde{A}^\bullet$. Since the upper 0-level set $\tilde{A}_0$ is bounded and the lower α-level set ${}_\alpha\tilde{A}^\bullet$ is a closed and convex subset of $\mathbb{R}$ for $\alpha \in (0,1]$ by Remark 10.2.2, it follows that ${}_\alpha\tilde{A}^\bullet$ is also a bounded closed interval, and is denoted by
$${}_\alpha\tilde{A}^\bullet = \left[{}_\alpha\tilde{A}^{\bullet L}, {}_\alpha\tilde{A}^{\bullet U}\right].$$
Using Remark 10.2.1, we also have
$${}_\alpha\tilde{A}^\bullet = \tilde{A}_{1-\alpha} = \left[\tilde{A}_{1-\alpha}^L, \tilde{A}_{1-\alpha}^U\right],$$
which says that
$${}_\alpha\tilde{A}^{\bullet L} = \tilde{A}_{1-\alpha}^L \text{ and } {}_\alpha\tilde{A}^{\bullet U} = \tilde{A}_{1-\alpha}^U.$$
From part (iv) of Proposition 10.1.3, for $\alpha \in (0,1]$, we have
$${}_{\alpha-}\tilde{A}^\bullet = \bigcup_{0\leq\beta<\alpha} {}_\beta\tilde{A}^\bullet = \bigcup_{0\leq\beta<\alpha} \left[{}_\beta\tilde{A}^{\bullet L}, {}_\beta\tilde{A}^{\bullet U}\right].$$
We further assume that the endpoints ${}_\beta\tilde{A}^{\bullet L}$ and ${}_\beta\tilde{A}^{\bullet U}$ are continuous functions with respect to β on $[0,1]$. Since ${}_\beta\tilde{A}^\bullet \subseteq {}_\alpha\tilde{A}^\bullet$ for $\beta < \alpha$, we have
$${}_{\alpha-}\tilde{A}^\bullet = \left({}_\alpha\tilde{A}^{\bullet L}, {}_\alpha\tilde{A}^{\bullet U}\right)$$
is an open interval. In this case, from Remark 10.1.2, for $\alpha \in (0,1]$, the upper α-level set $\tilde{A}_\alpha^\bullet$ is given by
$$\tilde{A}_\alpha^\bullet = \tilde{A}_0^\bullet \setminus {}_{\alpha-}\tilde{A}^\bullet = \left[\tilde{A}_0^{\star L}, \tilde{A}_0^{\star U}\right] \setminus \left({}_\alpha\tilde{A}^{\bullet L}, {}_\alpha\tilde{A}^{\bullet U}\right) = \left[\tilde{A}_0^L, {}_\alpha\tilde{A}^{\bullet L}\right] \cup \left[{}_\alpha\tilde{A}^{\bullet U}, \tilde{A}_0^U\right],$$

which is a closed subset of $\mathbb{R}$. This also says that the membership function of $\tilde{A}^{\bullet}$ is upper semi-continuous. Therefore, under the further assumptions, we conclude that the membership functions of $\tilde{A}^{\bullet}$ and $\tilde{A}$ are continuous.

10.3 Dual Extension Principle

We consider the following onto crisp function

$$f : \mathbb{R}^{n_1} \times \cdots \times \mathbb{R}^{n_p} \to \mathbb{R}^n.$$

Let $\tilde{A}^{(i)}$ be a fuzzy set in $\mathbb{R}^{n_i}$ with dual fuzzy set $\tilde{A}^{(i\bullet)}$ for $i = 1, \ldots, p$. Inspired by (4.14), using the dual membership functions $\xi_{\tilde{A}^{(i\bullet)}}$ of $\tilde{A}^{(i\bullet)}$ for $i = 1, \ldots, p$, we define a new fuzzy set $\tilde{f}^{\bullet}(\tilde{A}^{(1)}, \tilde{A}^{(2)}, \ldots, \tilde{A}^{(n)})$ in $\mathbb{R}^n$ with membership function given by

$$\xi_{\tilde{f}^{\bullet}(\tilde{A}^{(1)},\tilde{A}^{(2)},\ldots,\tilde{A}^{(p)})}(y) = \inf_{\{\mathbf{x}\in\tilde{A}_0^{(1)}\times\cdots\times\tilde{A}_0^{(p)}\,:\,y=f(\mathbf{x})\}} \max\left\{\xi_{\tilde{A}^{(1\bullet)}}(x_1), \ldots, \xi_{\tilde{A}^{(n\bullet)}}(x_p)\right\}. \tag{10.3}$$

Then, we have

$$\begin{aligned}
&\xi_{\tilde{f}^{\bullet}(\tilde{A}^{(1)},\tilde{A}^{(2)},\ldots,\tilde{A}^{(p)})}(y) \\
&= \inf_{\{\mathbf{x}\in\tilde{A}_0^{(1)}\times\cdots\times\tilde{A}_0^{(p)}\,:\,y=f(\mathbf{x})\}} \max\left\{1-\xi_{\tilde{A}^{(1)}}(x_1), \cdots, 1-\xi_{\tilde{A}^{(p)}}(x_p)\right\} \\
&= \inf_{\{\mathbf{x}\in\tilde{A}_0^{(1)}\times\cdots\times\tilde{A}_0^{(p)}\,:\,y=f(\mathbf{x})\}} \left[1-\min\left\{\xi_{\tilde{A}^{(1)}}(x_1), \cdots, \xi_{\tilde{A}^{(p)}}(x_p)\right\}\right] \\
&= 1 - \sup_{\{\mathbf{x}\in\tilde{A}_0^{(1)}\times\cdots\times\tilde{A}_0^{(p)}\,:\,y=f(\mathbf{x})\}} \min\left\{\xi_{\tilde{A}^{(1)}}(x_1), \ldots, \xi_{\tilde{A}^{(p)}}(x_p)\right\} \\
&= 1 - \sup_{\{\mathbf{x}\in\mathbb{R}^{n_1}\times\cdots\times\mathbb{R}^{n_p}\,:\,y=f(\mathbf{x})\}} \min\left\{\xi_{\tilde{A}^{(1)}}(x_1), \ldots, \xi_{\tilde{A}^{(p)}}(x_p)\right\} \\
&= 1 - \xi_{\tilde{f}(\tilde{A}^{(1)},\tilde{A}^{(2)},\ldots,\tilde{A}^{(p)})}(y),
\end{aligned}$$

which implies

$$\xi_{\tilde{f}(\tilde{A}^{(1)},\tilde{A}^{(2)},\ldots,\tilde{A}^{(p)})}(y) + \xi_{\tilde{f}^{\bullet}(\tilde{A}^{(1)},\tilde{A}^{(2)},\ldots,\tilde{A}^{(p)})}(y) = 1. \tag{10.4}$$

The methodology for calculating formula (10.3) is then called the **dual extension principle**. This means that, instead of calculating formula (4.14), we can alternatively calculate formula (10.3) and use the duality (10.4) to recover formula (4.14). In this case, we also say that the fuzzy function $\tilde{f}^{\bullet}$ is **dual-fuzzified** from the crisp function f.

Let A be a set in $\mathbb{R}^n$. The characteristic function χ_A of A is defined to be

$$\chi_A(x) = \begin{cases} 1 & \text{if } x \in A \\ 0 & \text{if } x \notin A. \end{cases}$$

Now, we define the so-called **dual characteristic function** $\chi_A^{\bullet}$ of A as follows

$$\chi_A^{\bullet}(x) = \begin{cases} 0 & \text{if } x \in A \\ 1 & \text{if } x \notin A. \end{cases} \tag{10.5}$$

It is clear to see that

$$\chi_A(x) + \chi_A^{\bullet}(x) = 1 \text{ for each } x \in \mathbb{R}^n.$$

Let $\tilde{A}$ be a fuzzy set in $\mathbb{R}^n$. Recall that the (primal) decomposition theorem says that the membership function $\xi_{\tilde{A}}$ can be expressed as

$$\xi_{\tilde{A}}(x) = \sup_{\alpha\in[0,1]} \alpha \cdot \chi_{\tilde{A}_\alpha}(x) = \sup_{\alpha\in(0,1]} \alpha \cdot \chi_{\tilde{A}_\alpha}(x),$$

where $\chi_{\tilde{A}_\alpha}$ is the characteristic function of the α-level set $\tilde{A}_\alpha$. The (primal) decomposition theorem says that the membership function can be expressed in terms of upper α-level sets. In what follows, we are going to show that the membership function can also be expressed in terms of lower α-level sets as the following form

$$\xi_{\tilde{A}}(x) = \sup_{\alpha\in[0,1]} \alpha \cdot \chi^\bullet_{{}_\alpha\tilde{A}}(x) = \sup_{\alpha\in\mathcal{R}(\xi_{\tilde{A}})} \alpha \cdot \chi^\bullet_{{}_\alpha\tilde{A}}(x),$$

where $\chi^\bullet_{{}_\alpha\tilde{A}}$ is the dual characteristic function of lower α-level set ${}_\alpha\tilde{A}$.

Theorem 10.3.1 **(Dual Decomposition Theorem)** *Let $\tilde{A}$ be a fuzzy set in $\mathbb{R}^n$ with proper domain $\tilde{A}_0$. For $x \in \tilde{A}_0$, the membership degree $\xi_{\tilde{A}}(x)$ can be expressed in terms of lower α-level sets as follows*

$$\xi_{\tilde{A}}(x) = \sup_{\alpha\in(0,1]} \alpha \cdot \chi^\bullet_{{}_\alpha\tilde{A}}(x) = \max_{\alpha\in(0,1]} \alpha \cdot \chi^\bullet_{{}_\alpha\tilde{A}}(x) = \max_{\alpha\in(0,1]} \alpha \cdot \left[1 - \chi_{{}_\alpha\tilde{A}}(x)\right]. \tag{10.6}$$

Proof. Let $\alpha_0 = \xi_{\tilde{A}}(x)$. Suppose that $\alpha_0 = 0$. If $x \notin {}_\alpha\tilde{A}$ for some $\alpha \in (0,1]$, then $\xi_{\tilde{A}}(x) > \alpha \geq 0$, which contradicts $\xi_{\tilde{A}}(x) = \alpha_0 = 0$. Therefore, we have $x \in {}_\alpha\tilde{A}$, i.e. $\chi_{{}_\alpha\tilde{A}}(x) = 1$ for all $\alpha \in (0,1]$, which says that $\alpha \cdot \chi^\bullet_{{}_\alpha\tilde{A}}(x) = 0$ for all $\alpha \in (0,1]$. This shows that the equalities in (10.6) are satisfied.

Now, we assume $\alpha_0 > 0$. Then $x \in {}_{\alpha_0}\tilde{A}$. For $\alpha \in (0,1]$ with $\alpha < \alpha_0$, if $x \in {}_\alpha\tilde{A}$, then $\xi_{\tilde{A}}(x) \leq \alpha < \alpha_0$, which contradicts $\alpha_0 = \xi_{\tilde{A}}(x)$. Therefore, we have $x \notin {}_\alpha\tilde{A}$, i.e. $\chi^\bullet_{{}_\alpha\tilde{A}}(x) = 1$ for $\alpha \in (0,1]$ with $\alpha < \alpha_0$. If $\alpha \in (0,1]$ with $\alpha \geq \alpha_0$, then $x \in {}_{\alpha_0}\tilde{A} \subseteq {}_\alpha\tilde{A}$, which says that $x \in {}_\alpha\tilde{A}$, i.e. $\chi^\bullet_{{}_\alpha\tilde{A}}(x) = 0$ for $\alpha \geq \alpha_0$. Then, we obtain

$$\begin{aligned}\sup_{\alpha\in(0,1]} \alpha \cdot \chi^\bullet_{{}_\alpha\tilde{A}}(x) &= \max\left\{\sup_{\{\alpha\in(0,1]:\alpha<\alpha_0\}} \alpha \cdot \chi^\bullet_{{}_\alpha\tilde{A}}(x), \sup_{\{\alpha\in(0,1]:\alpha\geq\alpha_0\}} \alpha \cdot \chi^\bullet_{{}_\alpha\tilde{A}}(x)\right\}\\ &= \max\left\{\sup_{\{\alpha\in(0,1]:\alpha<\alpha_0\}} \alpha, 0\right\} = \max\{\alpha_0, 0\} = \alpha_0 = \xi_{\tilde{A}}(x).\end{aligned}$$

Since $\alpha_0 \in (0,1]$, the above supremum is attained. It means that

$$\xi_{\tilde{A}}(x) = \max_{\alpha\in(0,1]} \alpha \cdot \chi^\bullet_{{}_\alpha\tilde{A}}(x).$$

This completes the proof. ■

Remark 10.3.2 The decomposition theorem for dual fuzzy set $\tilde{A}^\bullet$ based on the upper α-level sets of $\tilde{A}^\bullet$ is given by

$$\xi_{\tilde{A}^\bullet}(x) = \sup_{\alpha\in(0,1]} \alpha \cdot \chi_{\tilde{A}^\bullet_\alpha}(x).$$

According to Theorem 10.3.1, the dual decomposition theorem for $\tilde{A}^\bullet$ based on the lower α-level sets of $\tilde{A}^\bullet$ is given by

$$\xi_{\tilde{A}^\bullet}(x) = \sup_{\alpha\in(0,1]} \alpha \cdot \chi^\bullet_{{}_\alpha\tilde{A}^\bullet}(x) = \max_{\alpha\in(0,1]} \alpha \cdot \chi^\bullet_{{}_\alpha\tilde{A}^\bullet}(x) = \max_{\alpha\in(0,1]} \alpha \cdot \left[1 - \chi_{{}_\alpha\tilde{A}^\bullet}(x)\right].$$

It is well known that the set of all rational numbers $\mathbb{Q}$ is dense in $\mathbb{R}$. This means that, given any $r \in \mathbb{R}$, there exist two sequences $\{p_n\}_{n=1}^{\infty}$ and $\{q_n\}_{n=1}^{\infty}$ in the countable set $\mathbb{Q}$ satisfying $p_n \uparrow r$ and $q_n \downarrow r$ as $n \to \infty$.

Theorem 10.3.3 **(Dual Decomposition Theorem)** *Let $\tilde{A}$ be a fuzzy set in $\mathbb{R}^n$ with proper domain $\tilde{A}_0$. For $x \in \tilde{A}_0$, the membership degree $\xi_{\tilde{A}}(x)$ can be expressed in terms of lower α-level sets as follows*

$$\xi_{\tilde{A}}(x) = \sup_{\alpha\in[0,1]} \alpha \cdot \chi^{\bullet}_{{}_\alpha\tilde{A}}(x) = \max_{\alpha\in[0,1]} \alpha \cdot \chi^{\bullet}_{{}_\alpha\tilde{A}}(x) = \max_{\alpha\in[0,1]} \alpha \cdot \left[1 - \chi_{{}_\alpha\tilde{A}}(x)\right] \tag{10.7}$$

$$= \sup_{\alpha\in\mathbb{Q}\cap(0,1]} \alpha \cdot \chi^{\bullet}_{{}_\alpha\tilde{A}}(x) = \max_{\alpha\in\mathbb{Q}\cap(0,1]} \alpha \cdot \chi^{\bullet}_{{}_\alpha\tilde{A}}(x) = \max_{\alpha\in\mathbb{Q}\cap(0,1]} \alpha \cdot \left[1 - \chi_{{}_\alpha\tilde{A}}(x)\right]. \tag{10.8}$$

Proof. Using Theorem 10.3.1, we can obtain the equalities (10.7). To prove the equalities (10.8), let $\xi_{\tilde{A}}(x) = \alpha_0$. We first assume that $\alpha_0 = 0$. From the proof of Theorem 10.3.1, we have $x \in {}_\alpha\tilde{A}$ for all $\alpha \in (0,1]$, which says that $\alpha \cdot \chi^{\bullet}_{{}_\alpha\tilde{A}}(x) = 0$ for all $\alpha \in \mathbb{Q} \cap (0,1]$. It follows that

$$\xi_{\tilde{A}}(x) = \alpha_0 = 0 = \sup_{\alpha\in\mathbb{Q}\cap(0,1]} \alpha \cdot \chi^{\bullet}_{{}_\alpha\tilde{A}}(x).$$

Now, we assume that $0 < \alpha_0$. Using (10.7), we have

$$0 < \xi_{\tilde{A}}(x) = \alpha_0 = \sup_{\alpha\in[0,1]} \alpha \cdot \chi^{\bullet}_{{}_\alpha\tilde{A}}(x) \geq \sup_{\alpha\in\mathbb{Q}\cap(0,1]} \alpha \cdot \chi^{\bullet}_{{}_\alpha\tilde{A}}(x). \tag{10.9}$$

Since $\alpha_0 > 0$, from the proof of Theorem 10.3.1 we have $x \notin {}_\alpha\tilde{A}$ for $\alpha \in (0,1]$ with $\alpha < \alpha_0$. The denseness also says that there exists a sequence $\{\alpha_n\}_{n=1}^{\infty}$ in $\mathbb{Q} \cap (0,1]$ satisfying $\alpha_n \uparrow \alpha_0$ with $\alpha_n < \alpha_0$. It follows that $x \notin {}_{\alpha_n}\tilde{A}$ for all n. Let

$$\Gamma = \{\alpha_n\}_{n=1}^{\infty} \subset \mathbb{Q} \cap (0,1].$$

Then, we have

$$\xi_{\tilde{A}}(x) = \alpha_0 = \lim_n \alpha_n = \sup_n \alpha_n = \sup_n \alpha_n \cdot \chi^{\bullet}_{{}_{\alpha_n}\tilde{A}}(x)$$

$$= \sup_{\alpha\in\Gamma} \alpha \cdot \chi^{\bullet}_{{}_\alpha\tilde{A}}(x) \leq \sup_{\alpha\in\mathbb{Q}\cap(0,1]} \alpha \cdot \chi^{\bullet}_{{}_\alpha\tilde{A}}(x). \tag{10.10}$$

Combining (10.9) and (10.10), we obtain the equality

$$\xi_{\tilde{A}}(x) = \sup_{\alpha\in\mathbb{Q}\cap(0,1]} \alpha \cdot \chi^{\bullet}_{{}_\alpha\tilde{A}}(x).$$

This completes the proof. ■

10.4 Dual Arithmetics of Fuzzy Sets

Let $\odot$ denote any one of the four basic arithmetic operations $\oplus, \ominus, \otimes, \oslash$ between fuzzy sets $\tilde{A}$ and $\tilde{B}$ in $\mathbb{R}$. The membership function of $\tilde{A} \odot \tilde{B}$ is defined by

$$\xi_{\tilde{A}\odot\tilde{B}}(z) = \sup_{\{(x,y)\in\mathbb{R}^2 : z = x\circ y\}} \min\left\{\xi_{\tilde{A}}(x), \xi_{\tilde{B}}(y)\right\}$$

for all $z \in \mathbb{R}$, where the operation $\circ \in \{+, -, \times, /\}$. Since the 0-level sets $\tilde{A}_0$ and $\tilde{B}_0$ are the proper domain of $\tilde{A}$ and $\tilde{B}$, respectively, i.e. $\xi_{\tilde{A}}(x) = 0$ for $x \notin \tilde{A}_0$ and $\xi_{\tilde{B}}(x) = 0$ for $x \notin \tilde{B}_0$, we have

$$\begin{aligned}\xi_{\tilde{A}\odot\tilde{B}}(z) &= \sup_{\{(x,y)\in\mathbb{R}^2 : z = x\circ y\}} \min\{\xi_{\tilde{A}}(x), \xi_{\tilde{B}}(y)\} \\ &= \sup_{\{(x,y)\in\tilde{A}_0\times\tilde{B}_0 : z = x\circ y\}} \min\{\xi_{\tilde{A}}(x), \xi_{\tilde{B}}(y)\}.\end{aligned} \tag{10.11}$$

Inspired by the above expression (10.11), we define a new operation between $\tilde{A}$ and $\tilde{B}$ using the dual membership functions as follows

$$\xi_{\tilde{A}\,\odot^\bullet\,\tilde{B}}(z) = \inf_{\{(x,y)\in\tilde{A}_0\times\tilde{B}_0 : z = x\circ y\}} \max\{\xi_{\tilde{A}^\bullet}(x), \xi_{\tilde{B}^\bullet}(y)\}. \tag{10.12}$$

We need to emphasize that

$$\inf_{\{(x,y)\in\tilde{A}_0\times\tilde{B}_0 : z = x\circ y\}} \max\{\xi_{\tilde{A}^\bullet}(x), \xi_{\tilde{B}^\bullet}(y)\} \neq \inf_{\{(x,y)\in\mathbb{R}^2 : z = x\circ y\}} \max\{\xi_{\tilde{A}^\bullet}(x), \xi_{\tilde{B}^\bullet}(y)\}.$$

However, this operation $\tilde{A} \odot^\bullet \tilde{B}$ is reasonable, since we consider the proper domains shown in (10.11). Then, we have

$$\begin{aligned}\xi_{\tilde{A}\,\odot^\bullet\,\tilde{B}}(z) &= \inf_{\{(x,y)\in\tilde{A}_0\times\tilde{B}_0 : z = x\circ y\}} \max\{1 - \xi_{\tilde{A}}(x), 1 - \xi_{\tilde{B}}(y)\} \\ &= 1 - \sup_{\{(x,y)\in\tilde{A}_0\times\tilde{B}_0 : z = x\circ y\}} \min\{\xi_{\tilde{A}}(x), \xi_{\tilde{B}}(y)\} \\ &= 1 - \sup_{\{(x,y)\in\mathbb{R}^2 : z = x\circ y\}} \min\{\xi_{\tilde{A}}(x), \xi_{\tilde{B}}(y)\} \\ &= 1 - \xi_{\tilde{A}\odot\tilde{B}}(z),\end{aligned}$$

which implies

$$\xi_{\tilde{A}\odot\tilde{B}}(z) + \xi_{\tilde{A}\,\odot^\bullet\tilde{B}}(z) = 1. \tag{10.13}$$

Therefore, we say that $\tilde{A} \odot^\bullet \tilde{B}$ is the **dual arithmetic** of $\tilde{A} \odot \tilde{B}$. This means that, instead of calculating $\tilde{A} \odot \tilde{B}$, we can alternatively calculate $\tilde{A} \odot^\bullet \tilde{B}$ and use the duality (10.13) to recover $\tilde{A} \odot \tilde{B}$. We are going to study the lower α-level sets of dual arithmetic $\tilde{A} \odot^\bullet \tilde{B}$ and establish the relationships between $\tilde{A} \odot \tilde{B}$ and $\tilde{A} \odot^\bullet \tilde{B}$.

Theorem 10.4.1 *Let $\tilde{A}$ and $\tilde{B}$ be two fuzzy sets in $\mathbb{R}$ with the dual fuzzy sets $\tilde{A}^\bullet$ and $\tilde{B}^\bullet$, respectively. Let the arithmetic operations $\odot \in \{\oplus, \ominus, \otimes\}$ correspond to the operations $\circ \in \{+, -, \times\}$. Then, the following statements hold true.*

(i) *We have the following inclusion*

$$(\tilde{A} \odot \tilde{B})_\alpha \supseteq \tilde{A}_\alpha \circ \tilde{B}_\alpha = {}_{1-\alpha}\tilde{A}^\bullet \circ_{1-\alpha} \tilde{B}^\bullet \text{ for all } \alpha \in [0,1].$$

(ii) *Suppose that the membership functions of $\tilde{A}$ and $\tilde{B}$ are upper semi-continuous. Then*

$$(\tilde{A} \odot \tilde{B})_\alpha = \tilde{A}_\alpha \circ \tilde{B}_\alpha = {}_{1-\alpha}\tilde{A}^\bullet \circ_{1-\alpha} \tilde{B}^\bullet \text{ for all } \alpha \in (0,1].$$

(iii) *Suppose that the membership functions of $\tilde{A}$ and $\tilde{B}$ are upper semi-continuous, and that the supports $\tilde{A}_{0+}$ and $\tilde{B}_{0+}$ are bounded. Then*

$$(\tilde{A} \odot \tilde{B})_\alpha = \tilde{A}_\alpha \circ \tilde{B}_\alpha = {}_{1-\alpha}\tilde{A}^\bullet \circ_{1-\alpha} \tilde{B}^\bullet \text{ for all } \alpha \in [0,1].$$

Proof. The results follow immediately from Theorem 7.1.1 and Remark 10.2.1. ■

We do not consider the operation $\oslash$ in Theorem 10.4.1. The reasons is that the case of zero denominator should be avoided. We also remark that the arguments in the proof of Theorem 10.4.1 are still available for the operation $\oslash$ by carefully excluding the zero denominator. In order not to complicate the argument of Theorem 10.4.1, we omit the case of operation $\oslash$.

Theorem 10.4.2 *Let $\tilde{A}$ and $\tilde{B}$ be two fuzzy sets in $\mathbb{R}$ with the dual fuzzy sets $\tilde{A}^\bullet$ and $\tilde{B}^\bullet$, respectively. Let the dual arithmetic operations $\odot^\bullet \in \{\oplus^\bullet, \ominus^\bullet, \otimes^\bullet\}$ correspond to the operations $\circ \in \{+, -, *\}$. Then, the following statements hold true.*

(i) *We have the inclusion*

$$_\alpha(\tilde{A} \odot^\bullet \tilde{B}) \supseteq {}_\alpha\tilde{A}^\bullet \circ {}_\alpha\tilde{B}^\bullet = \tilde{A}_{1-\alpha} \circ \tilde{B}_{1-\alpha} \text{ for all } \alpha \in [0,1].$$

(ii) *Suppose that the membership functions of $\tilde{A}$ and $\tilde{B}$ are upper semi-continuous on $\tilde{A}_0$ and $\tilde{B}_0$, respectively. Then*

$$_\alpha(\tilde{A} \odot^\bullet \tilde{B}) = {}_\alpha\tilde{A}^\bullet \circ {}_\alpha\tilde{B}^\bullet = \tilde{A}_{1-\alpha} \circ \tilde{B}_{1-\alpha} \text{ for all } \alpha \in [0,1].$$

Proof. To prove part (i), for $\alpha \in [0,1]$ and $z_\alpha \in {}_\alpha\tilde{A}^\bullet \circ {}_\alpha\tilde{B}$, there exist $x_\alpha \in {}_\alpha\tilde{A}^\bullet$ and $y_\alpha \in {}_\alpha\tilde{B}^\bullet$ satisfying $z_\alpha = x_\alpha \circ y_\alpha$ for $\circ \in \{+, -, \times\}$, where

$$\xi_{\tilde{A}^\bullet}(x_\alpha) \leq \alpha \text{ and } \xi_{\tilde{B}^\bullet}(y_\alpha) \leq \alpha.$$

Therefore, we have

$$\begin{aligned}\xi_{\tilde{A} \odot^\bullet \tilde{B}}(z_\alpha) &= \inf_{\{(x,y) \in \tilde{A}_0 \circ \tilde{B}_0 : z_\alpha = x \circ y\}} \max\{\xi_{\tilde{A}^\bullet}(x), \xi_{\tilde{B}^\bullet}(y)\} \\ &\leq \max\{\xi_{\tilde{A}^\bullet}(x_\alpha), \xi_{\tilde{B}^\bullet}(y_\alpha)\} \leq \alpha,\end{aligned}$$

which says that $z_\alpha \in {}_\alpha(\tilde{A} \odot^\bullet \tilde{B})$. This shows that the following inclusion

$$_\alpha\tilde{A}^\bullet \circ {}_\alpha\tilde{B}^\bullet \subseteq {}_\alpha(\tilde{A} \odot^\bullet \tilde{B}) \text{ for } \alpha \in [0,1].$$

To prove part (ii), in order to prove the other direction of inclusion, we further assume that the membership functions $\xi_{\tilde{A}}$ and $\xi_{\tilde{B}}$ of $\tilde{A}$ and $\tilde{B}$ are upper semi-continuous on $\tilde{A}_0$, which imply that the dual membership functions $\xi_{\tilde{A}^\bullet}$ and $\xi_{\tilde{B}^\bullet}$ are lower semi-continuous functions on $\tilde{A}_0$; that is, the nonempty lower α-level sets ${}_\alpha\tilde{A}^\bullet$ and ${}_\alpha\tilde{B}^\bullet$ are $\tau_{\tilde{A}_0}$-closed subsets of $\tilde{A}_0$ for all $\alpha \in [0,1]$. Given any $z_\alpha \in {}_\alpha(\tilde{A} \odot^\bullet \tilde{B})$, we have

$$\inf_{\{(x,y) \in \tilde{A}_0 \circ \tilde{B}_0 : z_\alpha = x \circ y\}} \max\{\xi_{\tilde{A}^\bullet}(x), \xi_{\tilde{B}^\bullet}(y)\} = \xi_{\tilde{A} \odot^\bullet \tilde{B}}(z_\alpha) \leq \alpha. \tag{10.14}$$

Since z_α is finite, it is clear to see that

$$F \equiv \{(x,y) \in \tilde{A}_0 \circ \tilde{B}_0 \; : \; z_\alpha = x \circ y\}$$

is a bounded subset of $\tilde{A}_0 \circ \tilde{B}_0$. We also see that the function $g(x,y) = x \circ y$ is continuous on $\tilde{A}_0 \circ \tilde{B}_0$. Since the singleton $\{z_\alpha\}$ is a closed subset of $\mathbb{R}$, it follows that the inverse image

$F = g^{-1}(\{z_\alpha\})$ of $\{z_\alpha\}$ is also a closed subset of $\tilde{A}_0 \circ \tilde{B}_0$. This says that F is a compact subset of $\tilde{A}_0 \circ \tilde{B}_0$. Now, we want to show that the function

$$f(x,y) = \max\{\xi_{\tilde{A}^\bullet}(x), \xi_{\tilde{B}^\bullet}(y)\}$$

is lower semi-continuous on $\tilde{A}_0 \circ \tilde{B}_0$, i.e. we want to show that

$$\{(x,y) \in \tilde{A}_0 \circ \tilde{B}_0 : f(x,y) \leq \alpha\}$$

is a closed subset of $\tilde{A}_0 \circ \tilde{B}_0$ for any $\alpha \in \mathbb{R}$.

- For $\alpha \in [0,1]$, we have
$$\{(x,y) \in \tilde{A}_0 \circ \tilde{B}_0 : f(x,y) \leq \alpha\} = \left\{(x,y) \in \tilde{A}_0 \circ \tilde{B}_0 : \xi_{\tilde{A}^\bullet}(x) \leq \alpha \text{ and } \xi_{\tilde{B}^\bullet}(y) \leq \alpha\right\}$$
$$= \left\{(x,y) \in \tilde{A}_0 \circ \tilde{B}_0 : x \in {}_\alpha\tilde{A}^\bullet \text{ and } y \in {}_\alpha\tilde{B}^\bullet\right\} = {}_\alpha\tilde{A}^\bullet \circ {}_\alpha\tilde{B}^\bullet,$$
which is a closed subset of $\tilde{A}_0 \circ \tilde{B}_0$, since ${}_\alpha\tilde{A}^\bullet$ and ${}_\alpha\tilde{B}^\bullet$ are closed subsets of $\tilde{A}_0$ and $\tilde{B}_0$, respectively.
- If $\alpha \notin [0,1]$, then $\{(x,y) \in \tilde{A}_0 \circ \tilde{B}_0 : f(x,y) \leq \alpha\} = \emptyset$ is a closed subset of $\tilde{A}_0 \circ \tilde{B}_0$.

Therefore, the function $f(x,y)$ is indeed lower semi-continuous on $\tilde{A}_0 \circ \tilde{B}_0$. By Proposition 1.4.4, the function f assumes its minimum on the closed and bounded subset F of $\tilde{A}_0 \circ \tilde{B}_0$; that is, from (10.14), we have

$$\min_{(x,y)\in F} f(x,y) = \min_{\{(x,y)\in \tilde{A}_0 \circ \tilde{B}_0 : z_\alpha = x \circ y\}} f(x,y) = \inf_{\{(x,y)\in \tilde{A}_0 \circ \tilde{B}_0 : z_\alpha = x \circ y\}} f(x,y) \leq \alpha.$$

In other words, there exists $(x_\alpha, y_\alpha) \in F$ satisfying $z_\alpha = x_\alpha \circ y_\alpha$ and

$$\max\{\xi_{\tilde{A}^\bullet}(x_\alpha), \xi_{\tilde{B}^\bullet}(y_\alpha)\} = f(x_\alpha, y_\alpha) = \min_{(x,y)\in F} f(x,y) \leq \alpha,$$

i.e. $\xi_{\tilde{A}^\bullet}(x_\alpha) \leq \alpha$ and $\xi_{\tilde{B}^\bullet}(y_\alpha) \leq \alpha$. Therefore, we obtain $x_\alpha \in {}_\alpha\tilde{A}^\bullet$ and $y_\alpha \in {}_\alpha\tilde{B}^\bullet$, which says that $z_\alpha \in {}_\alpha\tilde{A}^\bullet \circ {}_\alpha\tilde{B}^\bullet$. Therefore, we obtain the following inclusion

$${}_\alpha(\tilde{A} \odot^\bullet \tilde{B}) \subseteq {}_\alpha\tilde{A}^\bullet \circ {}_\alpha\tilde{B}^\bullet \text{ for all } \alpha \in [0,1].$$

This completes the proof. ∎

The related results regarding the mixed lower and upper α-level sets are presented below. Recall that the 0-level set $(\tilde{A} \odot \tilde{B})_0$ is the proper domain of the membership function $\xi_{\tilde{A}\odot\tilde{B}}$ of $\tilde{A} \odot \tilde{B}$.

Theorem 10.4.3 *Let $\tilde{A}$ and $\tilde{B}$ be two fuzzy sets in $\mathbb{R}$. Consider that the arithmetic operations $\odot \in \{\oplus, \ominus, \otimes\}$ correspond to the operations $\circ \in \{+, -, \times\}$. Suppose that the membership functions of $\tilde{A}$ and $\tilde{B}$ are upper semi-continuous. Then, we have*

$${}_1(\tilde{A} \odot \tilde{B}) = (\tilde{A} \odot \tilde{B})_0,$$

and, for $\alpha \in [0,1)$, we have

$$\begin{aligned} {}_\alpha(\tilde{A} \odot \tilde{B}) &= (\tilde{A} \odot \tilde{B})_0 \Big\backslash \bigcup_{\alpha<\beta\leq 1} (\tilde{A}_\beta \circ \tilde{B}_\beta) \\ &= \bigcup_{\alpha<\beta\leq 1} \left((\tilde{A} \odot \tilde{B})_0 \backslash (\tilde{A}_\beta \circ \tilde{B}_\beta)\right). \end{aligned} \tag{10.15}$$

We further assume that the supports $\tilde{A}_{0+}$ and $\tilde{B}_{0+}$ are bounded. Then, the above 0-level set $(\tilde{A} \odot \tilde{B})_0$ can be replaced by $\tilde{A}_0 \circ \tilde{B}_0$.

Proof. For $\alpha \in [0,1)$, we have

$$\begin{aligned}
{}_{\alpha}(\tilde{A} \odot \tilde{B}) &= (\tilde{A} \odot \tilde{B})_0 \Big\backslash (\tilde{A} \odot \tilde{B})_{\alpha+} \text{ (using Remark 10.1.2)}\\
&= (\tilde{A} \odot \tilde{B})_0 \Big\backslash \bigcup_{\alpha<\beta\leq 1} (\tilde{A} \odot \tilde{B})_{\beta} \text{ (using Proposition 2.2.8)}\\
&= (\tilde{A} \odot \tilde{B})_0 \Big\backslash \bigcup_{\alpha<\beta\leq 1} (\tilde{A}_{\beta} \circ \tilde{B}_{\beta}) \text{ (using Theorem 10.4.1).}
\end{aligned}$$

This completes the proof. ■

Theorem 10.4.4 *Let $\tilde{A}$ and $\tilde{B}$ be two fuzzy sets in $\mathbb{R}$ with the dual fuzzy set $\tilde{A}^{\bullet}$ and $\tilde{B}^{\bullet}$, respectively. Consider the dual arithmetic operations $\odot^{\bullet} \in \{\oplus^{\bullet}, \ominus^{\bullet}, \otimes^{\bullet}\}$ correspond to the operations $\circ \in \{+, -, \times\}$. Suppose that the membership functions of $\tilde{A}$ and $\tilde{B}$ are upper semi-continuous on $\tilde{A}_0$ and $\tilde{B}_0$, respectively. Then, for $\alpha \in (0,1]$, we have*

$$\begin{aligned}
(\tilde{A} \odot^{\bullet} \tilde{B})_{\alpha} &= (\tilde{A} \odot^{\bullet} \tilde{B})_0 \Big\backslash \bigcup_{0\leq\beta<\alpha} ({}_{\beta}\tilde{A}^{\bullet} \circ {}_{\beta}\tilde{B}^{\bullet})\\
&= \bigcup_{0\leq\beta<\alpha} \left((\tilde{A} \odot^{\bullet} \tilde{B})_0 \backslash ({}_{\beta}\tilde{A}^{\bullet} \circ {}_{\beta}\tilde{B}^{\bullet})\right). \qquad (10.16)
\end{aligned}$$

Proof. For $\alpha \in (0,1]$, we have

$$\begin{aligned}
(\tilde{A} \odot^{\bullet} \tilde{B})_{\alpha} &= (\tilde{A} \odot^{\bullet} \tilde{B})_0 \Big\backslash {}_{\alpha-}(\tilde{A} \odot^{\bullet} \tilde{B}) \text{ (using Remark 10.1.2)}\\
&= (\tilde{A} \odot^{\bullet} \tilde{B})_0 \Big\backslash \bigcup_{0\leq\beta<\alpha} {}_{\beta}(\tilde{A} \odot^{\bullet} \tilde{B}) \text{ (using Proposition 10.1.3)}\\
&= (\tilde{A} \odot^{\bullet} \tilde{B})_0 \Big\backslash \bigcup_{0\leq\beta<\alpha} ({}_{\beta}\tilde{A}^{\bullet} \circ {}_{\beta}\tilde{B}^{\bullet}) \text{ (using Theorem 10.4.2).}
\end{aligned}$$

This completes the proof. ■

Example 10.4.5 Let $\tilde{A}$ and $\tilde{B}$ be two fuzzy intervals. Using Theorem 10.4.1, for $\alpha \in [0,1]$, the upper α-level set of $\tilde{A} \odot \tilde{B}$ is given by

$$(\tilde{A} \odot \tilde{B})_{\alpha} = \tilde{A}_{\alpha} \circ \tilde{B}_{\alpha} = \left[\tilde{A}^L_{\alpha}, \tilde{A}^U_{\alpha}\right] \circ \left[\tilde{B}^L_{\alpha}, \tilde{B}^U_{\alpha}\right].$$

Now, we consider the lower α-level set of $\tilde{A} \odot \tilde{B}$. From Theorem 10.4.3, the lower 1-level set of $\tilde{A} \odot \tilde{B}$ is given by

$${}_1(\tilde{A} \odot \tilde{B}) = \tilde{A}_0 \odot \tilde{B}_0 = \left[\tilde{A}^L_0, \tilde{A}^U_0\right] \circ \left[\tilde{B}^L_0, \tilde{B}^U_0\right].$$

For $\alpha \in [0,1)$, we have

$${}_{\alpha}(\tilde{A} \odot \tilde{B}) = \tilde{A}_0 \odot \tilde{B}_0 \Big\backslash \bigcup_{\alpha<\beta\leq 1} (\tilde{A}_{\beta} \circ \tilde{B}_{\beta}) = \tilde{A}_0 \odot \tilde{B}_0 \Big\backslash \bigcup_{\alpha<\beta\leq 1} \left(\left[\tilde{A}^L_{\beta}, \tilde{A}^U_{\beta}\right] \circ \left[\tilde{B}^L_{\beta}, \tilde{B}^U_{\beta}\right]\right).$$

Suppose that we take $\circ = +$ and $\odot = \oplus$. Then, we have

$${}_1(\tilde{A} \oplus \tilde{B}) = \tilde{A}_0 + \tilde{B}_0 = \left[\tilde{A}^L_0, \tilde{A}^U_0\right] + \left[\tilde{B}^L_0, \tilde{B}^U_0\right] = \left[\tilde{A}^L_0 + \tilde{B}^L_0, \tilde{A}^U_0 + \tilde{B}^U_0\right].$$

For $\alpha \in [0,1)$, we have

$$_{\alpha}(\tilde{A} \oplus \tilde{B}) = \left[\tilde{A}_0^L + \tilde{B}_0^L, \tilde{A}_0^U + \tilde{B}_0^U\right] \setminus \bigcup_{\alpha<\beta\leq 1} \left[\tilde{A}_\beta^L + \tilde{B}_\beta^L, \tilde{A}_\beta^U + \tilde{B}_\beta^U\right].$$

In order to obtain a more simplified form of $_{\alpha}(\tilde{A} \oplus \tilde{B})$, we further assume that the endpoints $\tilde{A}_\alpha^L, \tilde{A}_\alpha^U, \tilde{B}_\alpha^L$ and $\tilde{B}_\alpha^U$ are continuous on $[0,1]$ with respect to α. Therefore, the endpoints $\tilde{A}_\beta^L + \tilde{B}_\beta^L$ and $\tilde{A}_\beta^U + \tilde{B}_\beta^U$ are continuous functions on $[0,1]$ with respect to β. It follows that

$$\bigcup_{\alpha<\beta\leq 1} \left[\tilde{A}_\beta^L + \tilde{B}_\beta^L, \tilde{A}_\beta^U + \tilde{B}_\beta^U\right] = \left(\tilde{A}_\alpha^L + \tilde{B}_\alpha^L, \tilde{A}_\alpha^U + \tilde{B}_\alpha^U\right)$$

is an open interval. Therefore, for $\alpha \in [0,1)$, we obtain

$$_{\alpha}(\tilde{A} \oplus \tilde{B}) = \left[\tilde{A}_0^L + \tilde{B}_0^L, \tilde{A}_\alpha^L + \tilde{B}_\alpha^L\right] \cup \left[\tilde{A}_\alpha^U + \tilde{B}_\alpha^U, \tilde{A}_0^U + \tilde{B}_0^U\right].$$

Example 10.4.6 Let $\tilde{A}$ and $\tilde{B}$ be two fuzzy intervals with dual fuzzy set $\tilde{A}^\bullet$ and $\tilde{B}^\bullet$ in $\mathbb{R}$, respectively. According to Example 10.2.3, the lower α-level sets of $\tilde{A}^\bullet$ and $\tilde{B}^\bullet$ are given by

$$_{\alpha}\tilde{A}^\bullet = \left[{}_{\alpha}\tilde{A}^{\bullet L}, {}_{\alpha}\tilde{A}^{\bullet U}\right] = \left[\tilde{A}_{1-\alpha}^L, \tilde{A}_{1-\alpha}^U\right] \text{ and } {}_{\alpha}\tilde{B}^\bullet = \left[{}_{\alpha}\tilde{B}^{\bullet L}, {}_{\alpha}\tilde{A}^{\bullet U}\right] = \left[\tilde{B}_{1-\alpha}^L, \tilde{B}_{1-\alpha}^U\right].$$

Using part (iii) of Theorem 10.4.2, the lower α-level set of the dual arithmetic $\tilde{A} \odot^\bullet \tilde{B}$ is given by

$$_{\alpha}(\tilde{A} \odot^\bullet \tilde{B}) = {}_{\alpha}\tilde{A}^\bullet \circ {}_{\alpha}\tilde{B}^\bullet = \left[{}_{\alpha}\tilde{A}^{\bullet L}, {}_{\alpha}\tilde{A}^{\bullet U}\right] \circ \left[{}_{\alpha}\tilde{B}^{\bullet L}, {}_{\alpha}\tilde{B}^{\bullet U}\right] = \left[\tilde{A}_{1-\alpha}^L, \tilde{A}_{1-\alpha}^U\right] \circ \left[\tilde{B}_{1-\alpha}^L, \tilde{B}_{1-\alpha}^U\right].$$

Using Theorem 10.4.4, for $0 < \alpha \leq 1$, we have

$$\begin{aligned}
(\tilde{A} \odot^\bullet \tilde{B})_\alpha &= (\tilde{A} \odot^\bullet \tilde{B})_0 \setminus \bigcup_{0\leq\beta<\alpha} ({}_{\beta}\tilde{A}^\bullet \circ {}_{\beta}\tilde{B}^\bullet) \\
&= (\tilde{A} \odot^\bullet \tilde{B})_0 \setminus \bigcup_{0\leq\beta<\alpha} \left(\left[{}_{\beta}\tilde{A}^{\bullet L}, {}_{\beta}\tilde{A}^{\bullet U}\right] \circ \left[{}_{\beta}\tilde{B}^{\bullet L}, {}_{\beta}\tilde{B}^{\bullet U}\right]\right) \\
&= (\tilde{A} \odot^\bullet \tilde{B})_0 \setminus \bigcup_{0\leq\beta<\alpha} \left(\left[\tilde{A}_{1-\beta}^L, \tilde{A}_{1-\beta}^U\right] \circ \left[\tilde{B}_{1-\beta}^L, \tilde{B}_{1-\beta}^U\right]\right).
\end{aligned}$$

For $0 < \alpha \leq 1$, we can similarly show that

$$\begin{aligned}
\bigcup_{0\leq\beta<\alpha} \left(\left[{}_{\beta}\tilde{A}^{\bullet L} + {}_{\beta}\tilde{B}^{\bullet L}, {}_{\beta}\tilde{A}^{\bullet U} + {}_{\beta}\tilde{B}^{\bullet U}\right]\right) &= \left({}_{\alpha}\tilde{A}^{\bullet L} + {}_{\alpha}\tilde{B}^{\bullet L}, {}_{\alpha}\tilde{A}^{\bullet U} + {}_{\alpha}\tilde{B}^{\bullet U}\right) \\
&= \left(\tilde{A}_{1-\alpha}^L + \tilde{B}_{1-\alpha}^L, \tilde{A}_{1-\alpha}^U + \tilde{B}_{1-\alpha}^U\right)
\end{aligned}$$

is an open interval. Therefore, for $0 < \alpha \leq 1$, we obtain

$$\begin{aligned}
(\tilde{A} \oplus^\bullet \tilde{B})_\alpha &= (\tilde{A} \oplus^\bullet \tilde{B})_0 \setminus \left({}_{\alpha}\tilde{A}^{\bullet L} + {}_{\alpha}\tilde{B}^{\bullet L}, {}_{\alpha}\tilde{A}^{\bullet U} + {}_{\alpha}\tilde{B}^{\bullet U}\right) \\
&= (\tilde{A} \oplus^\bullet \tilde{B})_0 \setminus \left(\tilde{A}_{1-\alpha}^L + \tilde{B}_{1-\alpha}^L, \tilde{A}_{1-\alpha}^U + \tilde{B}_{1-\alpha}^U\right).
\end{aligned}$$

10.5 Representation Theorem for Dual-Fuzzified Function

Let $\tilde{A}^{(i)}$ be fuzzy sets in U_i with the corresponding dual fuzzy sets $\tilde{A}^{(i\bullet)}$ for $i = 1, \ldots, p$. By referring to (10.3), the dual-fuzzified function $\tilde{f}^\bullet(\tilde{A}^{(1)}, \ldots, \tilde{A}^{(p)})$ is obtained from the dual extension principle.

According to the Dual Decomposition Theorem 10.3.1, the membership function of dual-fuzzified function $\tilde{A} \equiv \tilde{f}^{\bullet}(\tilde{A}^{(1)}, \ldots, \tilde{A}^{(p)})$ is given by

$$\xi_{\tilde{f}^{\bullet}(\tilde{A}^{(1)},\ldots,\tilde{A}^{(p)})}(y) = \xi_{\tilde{A}}(y)$$
$$= \sup_{\alpha\in(0,1]} \alpha \cdot \chi^{\bullet}_{{}_{\alpha}\tilde{A}}(y) = \max_{\alpha\in(0,1]} \alpha \cdot \chi^{\bullet}_{{}_{\alpha}\tilde{A}}(y) = \max_{\alpha\in(0,1]} \alpha \cdot \left[1 - \chi_{{}_{\alpha}\tilde{A}}(y)\right].$$

Next, we are going to obtain the representation for dual-fuzzified function as follows

$$\xi_{\tilde{f}^{\bullet}(\tilde{A}^{(1)},\ldots,\tilde{A}^{(p)})}(y) = \sup_{\alpha\in(0,1]} \alpha \cdot \chi^{\bullet}_{f({}_{\alpha}\tilde{A}^{(1\bullet)},\ldots,{}_{\alpha}\tilde{A}^{(p\bullet)})}(y).$$

Some useful lemmas are needed below.

Lemma 10.5.1 *Let $\tilde{A}$ be a fuzzy set in $\mathbb{R}^n$ with dual fuzzy set $\tilde{A}^{\bullet}$. For each fixed $x \in \tilde{A}_0^{\bullet}$, let*

$$\alpha_0 \equiv \sup_{\alpha\in[0,1]} \alpha \cdot \chi^{\bullet}_{{}_{\alpha}\tilde{A}^{\bullet}}(x). \tag{10.17}$$

Then, we have $x \in {}_{\alpha}\tilde{A}^{\bullet}$ for $\alpha \in [0,1]$ with $\alpha > \alpha_0$ and $x \notin {}_{\alpha}\tilde{A}^{\bullet}$ for $\alpha \in [0,1]$ with $\alpha < \alpha_0$.

Proof. Suppose that there exists $\alpha_1 \in [0,1]$ satisfying $\alpha_1 < \alpha_0$ and $x \in {}_{\alpha_1}\tilde{A}^{\bullet}$. Then $x \in {}_{\alpha}\tilde{A}^{\bullet}$ for all $\alpha \in [0,1]$ with $\alpha > \alpha_1$, since ${}_{\alpha}\tilde{A}^{\bullet} \subseteq {}_{\beta}\tilde{A}^{\bullet}$ for $\alpha, \beta \in [0,1]$ with $\alpha < \beta$. This says that

$$\sup_{\alpha\in[0,1]} \alpha \cdot \chi^{\bullet}_{{}_{\alpha}\tilde{A}^{\bullet}}(x) = \sup_{\alpha\in[0,1]} \alpha \cdot \left[1 - \chi_{{}_{\alpha}\tilde{A}^{\bullet}}(x)\right] \leq \alpha_1 < \alpha_0,$$

which is a contradiction. Therefore, we conclude $x \notin {}_{\alpha}\tilde{A}^{\bullet}$ for $\alpha \in [0,1]$ with $\alpha < \alpha_0$. On the other hand, if there exists $\alpha_2 \in [0,1]$ satisfying $\alpha_2 > \alpha_0$ and $x \notin {}_{\alpha_2}\tilde{A}^{\bullet}$, we have

$$\sup_{\alpha\in[0,1]} \alpha \cdot \chi^{\bullet}_{{}_{\alpha}\tilde{A}^{\bullet}}(x) = \sup_{\alpha\in[0,1]} \alpha \cdot \left[1 - \chi_{{}_{\alpha}\tilde{A}^{\bullet}}(x)\right] \geq \alpha_2 > \alpha_0,$$

which is also a contradiction. Therefore, we conclude $x \in {}_{\alpha}\tilde{A}^{\bullet}$ for $\alpha \in [0,1]$ with $\alpha > \alpha_0$. This completes the proof. ■

Lemma 10.5.2 *Let $\tilde{A}^{(i)}$ be fuzzy sets in $\mathbb{R}^{n_i}$ with dual fuzzy sets $\tilde{A}^{(i\bullet)}$ for $i = 1, \ldots, p$. We have*

$$\sup_{\alpha\in(0,1]} \max_{1\leq i\leq p} \left\{\alpha \cdot \chi^{\bullet}_{{}_{\alpha}\tilde{A}^{(i\bullet)}}(x_i)\right\} = \max_{1\leq i\leq p} \left\{\sup_{\alpha\in(0,1]} \alpha \cdot \chi^{\bullet}_{{}_{\alpha}\tilde{A}^{(i\bullet)}}(x_i)\right\}.$$

Proof. Let

$$\alpha_i = \sup_{\alpha\in(0,1]} \alpha \cdot \chi^{\bullet}_{{}_{\alpha}\tilde{A}^{(i\bullet)}}(x_i) \text{ and } \alpha_0 = \max_{1\leq i\leq p} \alpha_i \tag{10.18}$$

for $i = 1, \ldots, p$. Given any $\alpha \in [0,1]$ with $\alpha > \alpha_0$, we have $\alpha \in [0,1]$ with $\alpha > \alpha_i$ for all $i = 1, \ldots, p$. Lemma 10.5.1 says that $x_i \in {}_{\alpha}\tilde{A}^{(i\bullet)}$ for all $i = 1, \ldots, p$. Therefore, we obtain

$$\max_{1\leq i\leq p} \left\{\alpha \cdot \chi^{\bullet}_{{}_{\alpha}\tilde{A}^{(i\bullet)}}(x_i)\right\} = \max_{1\leq i\leq p} \left\{\alpha \cdot \left[1 - \chi_{{}_{\alpha}\tilde{A}^{(i\bullet)}}(x_i)\right]\right\} = 0.$$

Given any $\alpha \in [0,1]$ with $\alpha < \alpha_0$, we have $\alpha \in [0,1]$ with $\alpha < \alpha_i$ for some $i = 1, \ldots, p$. Lemma 10.5.1 says that $x_i \notin {}_{\alpha}\tilde{A}^{(i\bullet)}$ for some $i = 1, \ldots, p$. Therefore, we obtain

$$\max_{1\leq i\leq p} \left\{\alpha \cdot \chi^{\bullet}_{{}_{\alpha}\tilde{A}^{(i\bullet)}}(x_i)\right\} = \max_{1\leq i\leq p} \left\{\alpha \cdot \left[1 - \chi_{{}_{\alpha}\tilde{A}^{(i\bullet)}}(x_i)\right]\right\} = \alpha.$$

Then, we have

$$\sup_{\alpha\in(0,1]}\max_{1\le i\le p}\left\{\alpha\cdot\chi^{\bullet}_{{}_{\alpha}\tilde{A}^{(i\bullet)}}(x_i)\right\}$$

$$=\max\left\{\sup_{\{\alpha\in[0,1]:\alpha>\alpha_0\}}\max_{1\le i\le p}\left\{\alpha\cdot\chi^{\bullet}_{{}_{\alpha}\tilde{A}^{(i\bullet)}}(x_i)\right\},\sup_{\{\alpha\in[0,1]:\alpha<\alpha_0\}}\max_{1\le i\le p}\left\{\alpha\cdot\chi^{\bullet}_{{}_{\alpha}\tilde{A}^{(i\bullet)}}(x_i)\right\},\max_{1\le i\le p}\left\{\alpha_0\cdot\chi^{\bullet}_{\tilde{A}^{(i\bullet)}_{\alpha_0}}(x_i)\right\}\right\}$$

$$=\max\left\{0,\sup_{\{\alpha\in[0,1]:\alpha<\alpha_0\}}\alpha,\max_{1\le i\le p}\left\{\alpha_0\cdot\chi^{\bullet}_{\tilde{A}^{(i\bullet)}_{\alpha_0}}(x_i)\right\}\right\}$$

$$=\max\left\{\alpha_0,\max_{1\le i\le p}\left\{\alpha_0\cdot\chi^{\bullet}_{\tilde{A}^{(i\bullet)}_{\alpha_0}}(x_i)\right\}\right\}=\alpha_0,$$

which shows

$$\sup_{\alpha\in(0,1]}\max_{1\le i\le p}\left\{\alpha\cdot\chi^{\bullet}_{{}_{\alpha}\tilde{A}^{(i\bullet)}}(x_i)\right\}=\alpha_0=\max_{1\le i\le p}\left\{\sup_{\alpha\in(0,1]}\alpha\cdot\chi^{\bullet}_{{}_{\alpha}\tilde{A}^{(i\bullet)}}(x_i)\right\}\tag{10.19}$$

by using (10.18). From (10.17) and (10.18), we also have

$$\alpha_i=\sup_{\alpha\in\mathcal{R}(\xi_{\tilde{A}^{(i\bullet)}})}\alpha\cdot\chi^{\bullet}_{{}_{\alpha}\tilde{A}^{(i\bullet)}}(x_i)\text{ and }\alpha_0=\max_{1\le i\le p}\left\{\sup_{\alpha\in\mathcal{R}(\xi_{\tilde{A}^{(i\bullet)}})}\alpha\cdot\chi^{\bullet}_{{}_{\alpha}\tilde{A}^{(i\bullet)}}(x_i)\right\}.$$

Using (10.19), we obtain the desired equalities, and the proof is complete. ∎

Given any sets A_i in $\mathbb{R}^{n_i}$ for $i=1,\dots,p$, it is not hard to obtain the following equalities

$$\begin{aligned}1-\chi_{f(A_1,\dots,A_p)}(y)&=\chi^{\bullet}_{f(A_1,\dots,A_p)}(y)\\&=\inf_{\{(x_1,\dots,x_p):y=f(x_1,\dots,x_p)\}}\max\left\{\chi^{\bullet}_{A_1}(x_1),\dots,\chi^{\bullet}_{A_p}(x_p)\right\}\\&=\inf_{\{(x_1,\dots,x_p):y=f(x_1,\dots,x_p)\}}\max\left\{1-\chi_{A_1}(x_1),\dots,1-\chi_{A_p}(x_p)\right\}\\&=1-\sup_{\{(x_1,\dots,x_p):y=f(x_1,\dots,x_p)\}}\min\left\{\chi_{A_1}(x_1),\dots,\chi_{A_p}(x_p)\right\}.\end{aligned}\tag{10.20}$$

We are in a position to establish the representation theorem for dual-fuzzified function using the dual extension principle.

Theorem 10.5.3 **(Representation Theorem)** *Let $f:\mathbb{R}^{n_1}\times\cdots\times\mathbb{R}^{n_p}\to\mathbb{R}^n$ be an onto crisp function, and let $\tilde{f}^{\bullet}:\mathcal{F}(\mathbb{R}^{n_1})\times\cdots\mathcal{F}(\mathbb{R}^{n_p})\to\mathcal{F}(\mathbb{R}^n)$ be a fuzzy function extended from f via the dual extension principle defined in* (10.3). *Given any fuzzy sets $\tilde{A}^{(i)}$ in $\mathbb{R}^{n_i}$ with dual fuzzy sets $\tilde{A}^{(i\bullet)}$ for $i=1,\dots,p$, the membership function of $\tilde{f}^{\bullet}(\tilde{A}^{(1)},\tilde{A}^{(2)},\dots,\tilde{A}^{(p)})$ can be expressed as*

$$\xi_{\tilde{f}^{\bullet}(\tilde{A}^{(1)},\dots,\tilde{A}^{(p)})}(y)=\sup_{\alpha\in(0,1]}\alpha\cdot\chi^{\bullet}_{f({}_{\alpha}\tilde{A}^{(1\bullet)},\dots,{}_{\alpha}\tilde{A}^{(p\bullet)})}(y)=\sup_{\alpha\in(0,1]}\alpha\cdot\chi^{\bullet}_{f(\tilde{A}^{(1)}_{1-\alpha},\dots,\tilde{A}^{(p)}_{1-\alpha})}(y).$$

Proof. By the definition of dual extension principle, we have

$$\xi_{\tilde{f}^{\bullet}(\tilde{A}^{(1)},\ldots,\tilde{A}^{(p)})}(y) = \inf_{\{(x_1,\ldots,x_p)\in\tilde{A}_0^{(1)}\times\cdots\times\tilde{A}_0^{(p)}:y=f(x_1,\ldots,x_p)\}} \max_{1\leq i\leq p} \left\{\xi_{\tilde{A}^{(i\bullet)}}(x_i)\right\} \text{ (using (10.3))}$$

$$= \inf_{\{(x_1,\ldots,x_p)\in\tilde{A}_0^{(1)}\times\cdots\times\tilde{A}_0^{(p)}:y=f(x_1,\ldots,x_p)\}} \max_{1\leq i\leq p} \left\{ \sup_{\alpha\in(0,1]} \alpha\cdot\chi^{\bullet}_{{}_\alpha\tilde{A}^{(i\bullet)}}(x_i) \right\} \tag{10.21}$$

(using Remark 10.3.2).

Given any $\alpha \in [0,1]$, from 10.20, we have

$$\chi^{\bullet}_{f({}_\alpha\tilde{A}^{(1\bullet)},\ldots,{}_\alpha\tilde{A}^{(p\bullet)})}(y) = \inf_{\{(x_1,\ldots,x_p)\in\tilde{A}_0^{(1)}\times\cdots\times\tilde{A}_0^{(p)}:y=f(x_1,\ldots,x_p)\}} \max_{1\leq i\leq p} \left\{\chi^{\bullet}_{{}_\alpha\tilde{A}^{(i\bullet)}}(x_i)\right\}. \tag{10.22}$$

Therefore, we obtain

$$\begin{aligned}
&\sup_{\alpha\in(0,1]} \alpha\cdot\chi^{\bullet}_{f({}_\alpha\tilde{A}^{(1\bullet)},\ldots,{}_\alpha\tilde{A}^{(p\bullet)})}(y)\\
&= \sup_{\alpha\in(0,1]} \alpha\cdot\left(\inf_{\{(x_1,\ldots,x_p)\in\tilde{A}_0^{(1)}\times\cdots\times\tilde{A}_0^{(p)}:y=f(x_1,\ldots,x_p)\}} \max_{1\leq i\leq p} \left\{\chi^{\bullet}_{{}_\alpha\tilde{A}^{(i\bullet)}}(x_i)\right\}\right) \text{ (using (10.22))}\\
&= \sup_{\alpha\in(0,1]} \inf_{\{(x_1,\ldots,x_p)\in\tilde{A}_0^{(1)}\times\cdots\times\tilde{A}_0^{(p)}:y=f(x_1,\ldots,x_p)\}} \max_{1\leq i\leq p} \left\{\alpha\cdot\chi^{\bullet}_{{}_\alpha\tilde{A}^{(i\bullet)}}(x_i)\right\} \text{ (since } \alpha\geq 0)\\
&= \inf_{\{(x_1,\ldots,x_p)\in\tilde{A}_0^{(1)}\times\cdots\times\tilde{A}_0^{(p)}:y=f(x_1,\ldots,x_p)\}} \sup_{\alpha\in(0,1]} \max_{1\leq i\leq p} \left\{\alpha\cdot\chi^{\bullet}_{{}_\alpha\tilde{A}^{(i\bullet)}}(x_i)\right\}.
\end{aligned} \tag{10.23}$$

By applying Lemma 10.5.2 to (10.21) and (10.23), we obtain the desired equalities, where ${}_\alpha\tilde{A}^{(i\bullet)} = \tilde{A}^{(i)}_{1-\alpha}$ for $i = 1,\ldots,p$. This completes the proof. ∎

Bibliography

1 Abbasbandy, S. and Amirfakhrian, M., The Nearest Trapezoidal Form of a Generalized Left Right Fuzzy Number, *International Journal of Approximate Reasoning* 43 (2006) 166–178.

2 Abbasbandy, S. and Asady, B., The Nearest Trapezoidal Fuzzy Number to a Fuzzy Quantity, *Applied Mathematics and Computation* 156 (2004) 381–386.

3 Apostol, T.M., *Mathematical Analysis*, 2nd, Addison-Welesy, 1974.

4 Al-Qudah, Y. and Hassan, N., Operations On Complex Multi-Fuzzy Sets, *Journal of Intelligent and Fuzzy Systems* 33 (2017) 1527–1540.

5 Ban, A., Approximation of Fuzzy Numbers by the Trapezoidal Fuzzy Numbers Preserving the Expected Interval, *Fuzzy Sets and Systems* 159 (2008) 1327–1344.

6 Bán, J., Radon-Nikodym Theorem and Conditional Expectation of Fuzzy-Valued Measures and Variables, *Fuzzy Seta and Systems* 34 (1990) 383–392.

7 Banks, H.T. and Jacobs, M.Q., A Differential Calculus for Multifunctions, *Journal of Mathematical Analysis and Applications* 29 (1970) 246–272.

8 Bede, B. and Stefanini, L., Generalized Differentiability of Fuzzy-Valued Functions, *Fuzzy Sets and Systems* 230 (2013) 119–141.

9 Bobylev, V.N., Support Function of a Fuzzy Set and Its Characteristic Properties, *Math. Notes* 37 (1985) 281–285.

10 Bodjanova, S., Median Value And Median Interval of a Fuzzy Number, *Information Sciences* 172 (2005) 73–89.

11 Bose, R.K. and Sahani, D., Fuzzy Mappings and Fixed Point Theorems, *Fuzzy Sets and Systems* 21 (1987) 53–58.

12 Boukezzoula, R., Galichet, S., Foulloy, L., and Elmasry, M., Extended Gradual Interval Arithmetic and Its Application to Gradual Weighted Averages, *Fuzzy Sets and Systems* 257 (2014) 67–84.

13 Butnariu, D., Measurability Concepts for Fuzzy Mappings, *Fuzzy Sets and Systems* 31 (1989) 77–82.

14 Bzowski, A. and Urbański, M.K., A Note on Nguyen-Fuller-Keresztfalvi Theorem and Zadeh'S Extension Principle, *Fuzzy Sets and Systems* 213 (2013) 91–101.

15 Chakrabarty, K., Biswas, R., and Nanda, S., A Note on Fuzzy Union and Fuzzy Intersection, *Fuzzy Sets and Systems* 105 (1999) 499–502.

16 Chen, S.-H., Ranking Fuzzy Numbers with Maximizing Set wnd Minimizing Set, *Fuzzy Sets and Systems* 17 (1985) 113–129.

Mathematical Foundations of Fuzzy Sets, First Edition. Hsien-Chung Wu.

17 Colubi, A., Domínguez-Menchero, J.S., López-Díaz, M., and Ralescu, D.A., On the Formalization of Fuzzy Random Variables, *Information Sciences* 133 (2001) 3–6.

18 de Campos Ibánez, L.M. and González Munoz, A., A Subjective Approach for Ranking Fuzzy Numbers, *Fuzzy Sets and Systems* 29 (1989) 145–153.

19 Das, P., Fuzzy Vector Spaces under Triangular Norms, *Fuzzy Set and Systems* 25 (1988) 73–85.

20 Diamond, P. and Kloeden, P., Characterization of Compact Subsets of Fuzzy Sets, *Fuzzy Sets and System* 29 (1989) 341–348.

21 Diamond, P. and Kloeden, P., *Metric Spaces of Fuzzy Sets*, World Scientific, Singapore, 1994.

22 Diamond, P. and Pokrovskii, A., Chaos, Entropy and a Generalized Extension Principle, *Fuzzy Sets and Systems* 61 (1994) 277–283.

23 Driankov, A., Hellendoorn, H., and Reinfrank, M., *An Introduction to Fuzzy Control*, Springer-Verlag, 1993.

24 Dubois, D. and Prade, H., Additions of Interactive Fuzzy Numbers, *IEEE Trans. on Automatic Control* 26 (1981) 926–936.

25 Dubois, D. and Prade, H., Ranking Fuzzy Numbers in the Setting of Possibility Theory, *Information Sciences* 30 (1983) 183–224.

26 Dubois, D. and Prade, H., A Review of Fuzzy Set Aggregation Connectives, *Information Sciences* 36 (1985) 85–121.

27 Dubois, D. and Prade, H., The Mean Value of a Fuzzy Number, *Fuzzy Sets and Systems* 24 (1987) 279–300.

28 Dubois, D. and Prade, H., *Possibility Theory*, Springer-Verlag, NY, 1988.

29 Dubois, D. and Prade, H., Gradual Elements in a Fuzzy Set, *Soft Computing* 12 (2008) 165–175.

30 Dubois, D. and Prade, H., Gradualness, Uncertainty and Bipolarity: Making Sense of Fuzzy Sets, *Fuzzy Sets and Systems* 192 (2012) 3–24.

31 Facchinetti, G. and Ricci, R.G., A Characterization of a General Class of Ranking Functions on Triangular Fuzzy Numbers, *Fuzzy Sets and Systems* 146 (2004) 297–312.

32 Fard, O.S., Heidari, M., and Borzabadid, A.H., Fuzzy Taylor Formula: An Approach via Fuzzification of the Derivative and Integral Operators, *Fuzzy Sets and Systems* 358 (2019) 29–47.

33 Feng, Y., Convergence Theorems for Fuzzy Random Variables and Fuzzy Martingales, *Fuzzy Sets and Systems* 103 (1999) 435–441.

34 Feng, Y., Gaussian Fuzzy Random Variables, *Fuzzy Sets and Systems* 111 (2000) 325–330.

35 Filev, D.P. and Yager, R.R., An Adaptive Approach to Defuzzification Based on Level Sets, *Fuzzy Sets and Systems* 54 (1993) 355–360.

36 Fortemps, P. and Roubens, M., Ranking and Defuzzification Methods Based on Area Compensation, *Fuzzy Sets and Systems* 82 (1996) 319–330.

37 Fortin, J., Dubois, D., and Fargier, H., Gradual Numbers and Their Application to Fuzzy Interval Analysis, *IEEE Transactions on Fuzzy Systems* 16 (2008) 388–402.

38 Fullér, R. and Keresztfalvi, T., On Generalization of Nguyen's Theorem, *Fuzzy Sets and Systems* 41 (1990) 371–374.

39 Gebhardt, A., On Types of Fuzzy Numbers and Extension Principles. *Fuzzy Sets and Systems* 75 (1995) 311–318.

40 Gerla, G. and Scarpati, L., Extension Principles for Fuzzy Sets, *Information Sciences* 106 (1998) 49–69.

41 Goetschel, R.H., Representations with Fuzzy Darts, *Fuzzy Sets and Systems* 89 (1997) 77–105.

42 Gomes, L.T. and Barros, L.C., A Note on the Generalized Difference and the Generalized Differentiability, *Fuzzy Sets and Systems* 280 (2015) 142–145.

43 Grzegorzewski, P., Metrics and Orders in Space of Fuzzy Numbers, *Fuzzy Sets and Systems* 97 (1998) 83–94.

44 Grzegorzewski, P., Nearest Interval Approximation of a Fuzzy Number, *Fuzzy Sets and Systems* 130 (2002) 321–330.

45 Grzegorzewski, P., Trapezoidal Approximations of Fuzzy Numbers Preserving the Expected Interval - Algorithms and Properties, *Fuzzy Sets and Systems* 159 (2008) 1354–1364.

46 Heilpern, S., Fuzzy Mappings and Fixed Point Theorem, *Journal of Mathematical Analysis and Applications* 83 (1981) 566–569.

47 Herencia, J. and Lamata, M., A Total Order for the Graded Numbers Used in Decision Problems, *Int J. Uncertain Fuzziness Knowl-Based Syst* 7 (1999) 267–276.

48 Hu, B., Bi, L., Dai, S., and Li, S., Distances of Complex Fuzzy Sets and Continuity of Complex Fuzzy Operations, *Journal of Intelligent and Fuzzy Systems* 35 (2018) 2247–2255.

49 Jaballah, A. and Saidi, F.B., Uniqueness Results in the Representation of Families of Sets by Fuzzy Sets, *Fuzzy Sets and Systems* 157 (2006) 964–975.

50 Jung, J.S., Cho, Y.J., and Kim, J.K., Minimization Theorems for Fixed Point Theorems in Fuzzy Metric Spaces and Applications, *Fuzzy Sets and Systems* 61 (1994) 199–207.

51 Kaleva, O., Fuzzy Differential Equations, *Fuzzy Sets and Systems* 24 (1987) 301–317.

52 Kaleva, O., The Cauchy Problem for Fuzzy Differential Equation, *Fuzzy Sets and Systems* 35 (1990) 389–396.

53 Kaleva, O., The Calculus of Fuzzy Valued Functions, *Applied Mathematics Letters* 3 (1990) 55–59.

54 Katsaras, A.K. and Liu, B.D., Fuzzy Vector Spaces and Fuzzy Topological Vector Spaces, *Journal of Mathematical Analysis and Applications* 58 (1977) 135–146.

55 Kelley, J.L. and Namioka, I., *topological spaces*, Springer-Verlag, New York, 1961.

56 Kim, B.K. and Kim, J.H., Stochastic Integrals of Set-Valued Processes and Fuzzy Processes, *Journal of Mathematical Analysis and Applications* 236 (1999) 480–502.

57 Kim, K. and Park, S., Ranking Fuzzy Numbers with Index of Optimism, *Fuzzy Sets and Systems* 35 (1990) 143–150.

58 E.P. Klement, Operations on Fuzzy Sets - An Axiomatic Approach, *Fuzzy Sets and Systems* 27 (1982) 221–232.

59 Klement, E.P., Puri, M.L., and Ralescu, D.A., Limit Theorems for Fuzzy Random Variables, *Proc. Roy. Soc. Lond.* A407 (1986) 171–182.

60 Klir, G.J. and Yuan, B., *Fuzzy Sets and Fuzzy Logic: Theory and Applications* Prentice-Hall, NY, 1995.

61 Kloeden, P.E., Compact Supported Endographs and Fuzzy Sets, *Fuzzy Sets and Systems* 4 (1980) 193–201.

62 Krätschmer, V., A Unified Approach to Fuzzy Random Variables, *Fuzzy Sets and Systems* 123 (2001) 1–9.

63 Kupka, J., On Approximations of Zadeh's Extension Principle, *Fuzzy Sets and Systems* 283 (2016) 26–39.

64 Lakshmikantham, V. and Mohapatra, R.N., *Theory of Fuzzy Differential Equations and Inclusions*, Taylor & FRancis, 2003.

65 Lee, B.S., Lee, G.M., Cho, S.J., and Kim, D.S., A Common Fixed Point Theorem for a Pair of Fuzzy Mappings, *Fuzzy Sets and Systems* 98 (1998) 133–136.

66 Lee, J.-H. and Hyung, L.-K., Comparison of Fuzzy Values on a Continuous Domain, *Fuzzy Sets and Systems* 118 (2001) 419–428.

67 Lee-Kwang, H. and Lee, J.-H., A Method for Ranking Fuzzy Numbers and Its Application to Decision-Making, *IEEE Trans. on Fuzzy Systems* 7 (1999) 677–685.

68 Li, L., Random Fuzzy Sets and Fuzzy Martingales, *Fuzzy Sets and Systems* 69 (1995) 181–192.

69 Li, S. and Ogura, Y., Fuzzy Random Variables, Conditional Expectations and Fuzzy Valued Martingales, *Journal of Fuzzy Mathematics* 4 (1996) 905–927.

70 Li, S. and Ogura, Y., Convergence of Set-Valued and Fuzzy-Valued Martingales, *Fuzzy Sets and Systems* 101 (1999) 453–461.

71 Li, S., Ogura, Y., and Kreinovich, V., *Limit Theorems and Applications of Set-Valued and Fuzzy-Set-Valued Random Variables*, Kluwer Academic Piblishers, 2002.

72 Li, S., Ogura, Y., and Nguyen, H.T., Gaussian Processes and Martingales for Fuzzy Valued Random Variables with Continuous Parameter, *Information Sciences* 133 (2001) 7–21.

73 Liou, T.-S. and Wang, M.-J. J., Ranking Fuzzy Numbers with Integral Value, *Fuzzy Sets and Systems* 50 (1992) 247–255.

74 Lodwick, W.A. and Dubois, D., Interval Linear Systems as a Necessary Step in Fuzzy Linear Systems, *Fuzzy Sets and Systems* 281 (2015) 227–251.

75 López-Díaz, M. and Gil, M.A., Constructive Definitions of Fuzzy Random Variables, *Statistics and Probability Letters* 36 (1997) 135–143.

76 López-Díaz, M. and Gil, M.A., An Extension of Fubini's Theorem for Fuzzy Random Variables, *Information Sciences* 115 (1999) 29–41.

77 Ma, M., On Embedding Problems of Fuzzy Number Space: Part 5, *Fuzzy Sets and Systems* 55 (1993) 313–318.

78 Mabuchi, S., A Proposal for a Defuzzification Strategy by the Concept of Sensitivity Analysis, *Fuzzy Sets and Systems* 55 (1993) 1–14.

79 Medaglia, A.L., Fang, S.-C., Nuttle, H.L.W., and Wilson, J.R., An Efficient and Flexible Mechanism for Constructing Membership Functions, *European Journal of Operational Research* 139 (2002) 84–95.

80 Mesiar, R., Triangular-Norm-Based Addition of Fuzzy Intervals, *Fuzzy Sets and Systems* 91 (1997) 231–237.

81 Mizumoto, M. and Tanaka, K., Fuzzy Sets and Their Operations, *Information and Control* 48 (1981) 30–48.

82 Modarres, M. and Sadi-Nezhad, S., Ranking Fuzzy Numbers by Preference Ratio, *Fuzzy Sets and Systems* 118 (2001) 429–436.

83 Nasibov, E.N. and Peker, S., On the Nearest Parametric Approximation of a Fuzzy Number, *Fuzzy Sets and Systems* 159 (2008) 1365–1375.

84 Negoita, C.V. and Ralescu, D., *Applications of Fuzzy Sets to System Analysis*, Wiley, New York, 1975.

85 Nguyen, H.T., A Note on the Extension Principle for Fuzzy Sets, *Journal of Mathematical Analysis and Applications* 64 (1978) 369–380.

86 Nieto, J.J., The Cauchy Problem for COntinuous Fuzzy Differential Equations, *Fuzzy Sets and Systems* 102 (1999) 259–262.

87 Ogura, Y. and Li, S., Separability for Graph Convergence of Sequences of Fuzzy-Valued Random Variables, *Fuzzy Sets and Systems* 123 (2001) 19–27.

88 Park, J.Y. and Han, H.K., Existence and Uniqueness Theorem for a Solution of Fuzzy Differential Equations, *International Journal of Mathematics and Mathematical Sciences* 22 (1999) 271–279.

89 Puri, M.L. and Ralescu, D.A., Differentials of Fuzzy Function, *Journal of Mathematical Analysis and Applications* 91 (1983) 552–558.

90 Puri, M.L. and Ralescu, D.A., Convergence Theorem for Fuzzy Martingales, *Journal of Mathematical Analysis and Applications* 160 (1991) 107–122.

91 Puri, M.L. and Ralescu, D.A., The Concept of Normality for Fuzzy Random Variables, *The Annals of Probability Theory* 13 (1985) 1373–1379.

92 Puri, M.L. and Ralescu, D.A., Fuzzy Random Variables, *Journal of Mathematical Analysis and Applications* 114 (1986) 409–422.

93 Ralescu, D.A., A Generalization of the Representation Theorem, *Fuzzy Sets and Systems* 51 (1992) 309–311.

94 Ramík, J., Extension Principle in Fuzzy Optimization, *Fuzzy Sets and Systems* 19 (1986) 29–35.

95 Rojas-Medar, M., Bassanezi, R.C., and Román-Flores, H., A Generalization of the Minkowski Embedding Theorem and Applications, *Fuzzy Sets and Systems* 102 (1999) 263–269.

96 Rojas-Medar, M. and Román-Flores, H., On the Equivalence of Convergences of Fuzzy Sets, *Fuzzy Sets and Systems* 80 (1996) 217–224.

97 Román-Flores, H., The Compactness of $E(X)$, *Applied Mathematics Letters* 11 (1998) 13–17.

98 Román-Flores, H. and Rojas-Medar, M., Embedding of Level-Continuous Fuzzy Sets on Banach Spaces, *Information Sciences* 144 (2002) 227–247.

99 Román-Flores, H., Barros, L.C., and Bassanezi, R.C., A Note on Zadeh's Extensions, *Fuzzy Sets and Systems* 117 (2001) 327–331.

100 Royden, H.L., *Real Analysis*, 2nd, Macmillan, New York, 1968.

101 Sanchez, D., Delgado, M., Vila, M.A., and Chamorro-Martinez, J., On a Non-nested Level-Based Representation of Fuzziness, *Fuzzy Sets and Systems* 92 (2012) 159–175

102 Schaefer, H.H., *Topological Vector Spaces*, Springer-Verlag, 1966.

103 Seikkala, S., On the Fuzzy Initial Value Problem, *Fuzzy Sets and Systems* 24 (1987) 319–330.

104 Sengupta, A. and Pal, T.K., On Comapring Interval Numbers, *European Journal of Operational Research* 127 (2000) 28–43.

105 Šešelja, B., Stojić, D. and Tepavčević, A., On Existence of P-Valued Fuzzy Sets with a Given Collection of Cuts, *Fuzzy Sets and Systems* 161 (2010) 763–768.

106 Stojaković, M., Fuzzy Conditional Expectation, *Fuzzy Sets and Systems* 52 (1992) 53–60.

107 Stojaković, M., Fuzzy Random Variables, Expectation, and Martingales, *Journal of Mathematical Analysis and Applications* 184 (1994) 594–606.

108 Stojaković, M., Fuzzy Valued Measure, *Fuzzy Sets and Systems* 65 (1994) 95–104.

109 Tan, S.K., Wang, P.Z., and Lee, E.S., Fuzzy Set Operations Based on the Theory of Falling Shadows, *Journal of Mathematical Analysis and Applications* 174 (1993) 242–255.

110 Vitale, R.A., Approximation of Convex Set-Valued Functions, *Journal of Approximation Theory* 26 (1979) 301–316.

111 Voxman, W., Some Remarks on Distances between Fuzzy Numbers, *Fuzzy Sets and Systems* 100 (1998) 353–365.

112 Wang, L.-X., *A Course in Fuzzy Systems and Control*, Prentice-Hall.

113 Wang, X. and Kerre, E.E., Reasonable Properties for the Ordering of Fuzzy Quantities I, *Fuzzy Sets and Systems* 118 (2001) 375–385.

114 Wang, X. and Kerre, E.E., Reasonable Properties for the Ordering of Fuzzy Quantities II, *Fuzzy Sets and Systems* 118 (2001) 387–405.

115 Wang, Y.M., Yang, J.B., Xu, D.L., and Chin, K.S., 2006. On the centroids of fuzzy numbers. *Fuzzy Sets and Systems*, 157, 919–926.

116 Weber, S., A General Concept of Fuzzy Connectives, Negations and Implications Based on t-Norms and t-Conorms, *Fuzzy Sets and Systems* 11 (1983) 115–134.

117 Williamson, R.C., The Law of Large Numbers for Fuzzy Variables under a General Triangular Norm Extension Principle, *Fuzzy Sets and Systems* 41 (1991) 55–81.

118 Wong, C.K., Fuzzy Topology: Product and Quotient Theorems, *Journal of Mathematical Analysis and Applications* 43 (1973) 697–704.

119 Wu, C.-X. and Ma, M., Embedding Problem of Fuzzy Number Space: Part I, *Fuzzy Sets and Systems* 44 (1991) 33–38.

120 Wu, C.-X. and Ma, M., Embedding Problem of Fuzzy Number Space: Part II, *Fuzzy Sets and Systems* 45 (1992) 189–202.

121 Wu, C.X., Song, S., and Lee, E.S., Approximate Solutions, Existence, and Uniqueness of the Cauchy Problem of Fuzzy Differential Equations, *Journal of Mathematical Analysis and Applications* 202 (1996) 629–644.

122 Wu, H.-C., Fuzzy Reliability Analysis Based on Closed Fuzzy Numbers, *Information Sciences* 103 (1997) 135–159.

123 Wu, H.-C., The Improper Fuzzy Riemann Integral and Its Numerical Integration, *Information Sciences* 111 (1998) 109–137.

124 Wu, H.-C., The Fuzzy Riemann-Stieltjes Integral, *International Journal of Uncertainty, Fuzziness and Knowledge-Based Systems* 6 (1998) 51–67.

125 Wu, H.-C., The Fuzzy Riemann Integral and Its Numerical Integration, *Fuzzy Sets and Systems* 110 (2000) 1–25.

126 Wu, H.-C., Evaluate Fuzzy Riemann Integrals Using Monte Carlo Method, *Journal of Mathematical Analysis and Applications* 264 (2001) 324–343.

127 Wu, H.-C., Linear Regression Analysis for Fuzzy Input and Output Data Using the Extension Principle, *Computers and Mathematics with Applications* 45 (2003) 1849–1859.

128 Wu, H.-C., Pricing European Options Based on the Fuzzy Pattern of Black-Scholes Formula, *Computers and Operations Research* 31 (2004) 1069–1081.

129 Wu, H.-C., Fuzzy Reliability Estimation Using Bayesian Approach, *Computers and Industrial Engineering* 46 (2004) 467–493.

130 Wu, H.-C., Fuzzy Bayesian Estimation on Lifetime Data, *Computational Statistics* 19 (2004) 613–633.

131 Wu, H.-C., Bayesian System Reliability Assessment under Fuzzy Environments, *Reliability Engineering and System Safety* 83 (2004) 277–286.

132 Wu, H.-C., European Option Pricing under Fuzzy Environments, *International Journal of Intelligent Systems* 20 (2005) 89–102.

133 Wu, H.-C., Fuzzy Bayesian System Reliability Assessment Based on the Exponential Distribution, *Applied Mathematical Modelling* 30 (2006) 509–530.

134 Wu, H.-C., Using Fuzzy Sets Theory and Black-Scholes Formula to Generate Pricing Boundaries of European Options, *Applied Mathematics and Computation* 185 (2007) 136–146.

135 Wu, H.-C., Simulation for Queuing System under Fuzziness, *International Journal of Systems Science* 40 (2009) 587–600.

136 Wu, H.-C., Generalized Extension Principle, *Fuzzy Optimization and Decision Making* 9 (2010) 31–68.

137 Wu, H.-C., Hahn-Banach Extension Theorem over the Space of Fuzzy Elements, *Fuzzy Optimization and Decision Making* 9 (2010) 143–168.

138 Wu, H.-C., Solving the Fuzzy Earliness and Tardiness in Scheduling Problems by Using Genetic Algorithms, *Expert Systems with Applications* 37 (2010) 4860–4866.

139 Wu, H.-C. Decomposition and Construction of Fuzzy Sets and Their Applications to the Arithmetic Operations on Fuzzy Quantities. *Fuzzy Sets and Systems* 233 (2013) 1–25.

140 Wu, H.-C., Continuity of Fuzzified Functions Using the Generalized Extension Principle, *Symmetry* 9 (12), 299 (2017) (25 pages).

141 Wu, H.-C. Compatibility between Fuzzy Set Operations and Level Set Operations: Applications to Fuzzy Difference. *Fuzzy Sets and Systems* 353 (2018) 1–43.

142 Wu, H.-C. Convergence in Fuzzy Semi-Metric Spaces, *Mathematics* 2018, 6(9), 170 (39 pages).

143 Wu, H.-C. Fuzzy Semi-Metric Spaces, *Mathematics* 2018, 6(7), 106 (19 pages).

144 Wu, H.-C., Fuzzification of Real-Valued Functions Based on the Form of Decomposition Theorem: Applications to the Differentiation and Integrals of Fuzzy-Number-Valued Functions,*Soft Computing* 23 (2019) 6755–6775.

145 Wu, H.-C., Duality in Fuzzy Sets and Dual Arithmetics of Fuzzy Sets, *Mathematics* 2019, 7(1), 11 (24 pages).

146 Wu, H.-C., Intersection and Union of Non-Normal Fuzzy Sets Using Aggregation Functions, *Journal of Intelligent and Fuzzy Systems* 37 (2019) 4113–4132.

147 Wu, H.-C., Generalized Extension Principle for Non-Normal Fuzzy Sets, *Fuzzy Optimization and Decision Making* 18 (2019) 399–432.

148 Wu, H.-C., Arithmetic Operations of Non-Normal Fuzzy Sets Using Gradual Numbers, *Fuzzy Sets and Systems* 399 (2020) 1–19.

149 Wu, H.-C., Set Operations of Fuzzy Sets Using Gradual Elements, *Soft Computing* 24 (2020) 879–893.

150 Wu, H.-C., Arithmetics of Vectors of Fuzzy Sets, *Mathematics* 2020, 8(9), 1614 (42 pages).

151 Wu, H.-C., Generating Fuzzy Sets from the Families of Sets, *Journal of Intelligent and Fuzzy Systems* 41 (2021) 3061–3082.

152 Wu, J., An Embedding Theorem for Fuzzy Numbers on Banach Spaces and Its Applications, *Fuzzy Sets and Systems* 129 (2002) 57–63.

153 Wu, J. and Wu, C., The *w*-Derivatives of Fuzzy Mappings in Banach Spaces, *Fuzzy Sets and Systems* 119 (2001) 375–381.

154 Xue, X., Ha, M., and Ma, M., Random Fuzzy Number Integrals in Banach Spaces *Fuzzy Sets and Systems* 66 (1994) 97–111.

155 Yager, R.R., On a General Class of Fuzzy Connectives, *Fuzzy Sets and Systems* 4 (1980) 235–242.

156 Yager, R.R., A Characterization of the Extension Principle. *Fuzzy Sets and Systems* 18 (1986) 205–217.

157 Yager, R.R. and Filev, D.P., SLIDE: A Simple Adaptive Defuzzification Method, *IEEE Transactions on Fuzzy Systems* 1 (1993) 69–78.

158 Yager, R.R., Connectives and Quantifiers in Fuzzy Sets, *Fuzzy Sets and Systems* 40 (1991) 39–75.

159 Yager, R.R., Non-monotonic Set Theoretic Operations, *Fuzzy Sets and Systems* 42 (1991) 173–190.

160 Yeh, C.-T., A Note on Trapezoidal Approximations of Fuzzy Numbers, *Fuzzy Sets and Systems* 158 (2007) 747–754.

161 Yuan, Y., Criteria for Evaluation Fuzzy Ranking Methods, *Fuzzy Set and Systems* 43 (1991) 139–157.

162 Zadeh, L.A., Fuzzy Sets, *Information and Control* 8 (1965) 338–353.

163 Zadeh, L.A., The Concept of Linguistic Variable and Its Application to Approximate Reasoning I *Information Sciences* 8 (1975) 199–249.

164 Zadeh, L.A., The Concept of Linguistic Variable and Its Application to Approximate Reasoning II, *Information Sciences* 8 (1975) 301–357.

165 Zadeh, L.A., The Concept of Linguistic Variable and Its Application to Approximate Reasoning III, *Information Sciences* 9 (1975) 43–80.

Mathematical Notations

$\xi_{\tilde{A}}$: membership function of fuzzy set $\tilde{A}$
$\mathcal{R}(\xi_{\tilde{A}})$: range of membership function $\xi_{\tilde{A}}$
$\tilde{A}_\alpha = [\tilde{A}_\alpha^L, \tilde{A}_\alpha^U]$: α-level set is a closed interval
$\sup S$: supremum of S in Definition 1.1.2
$\inf S$: infimum of S in Definition 1.1.2
$\limsup_{n\to\infty} a_n$: limit superior of sequence $\{a_n\}_{n=1}^\infty$ in (1.2)
$\liminf_{n\to\infty} a_n$: limit inferior of sequence $\{a_n\}_{n=1}^\infty$ in (1.3)
$B(x;\epsilon)$: open ball (1.12)
$\mathrm{cl}(A)$: closure of A in Definition 1.2.5
χ_A: characteristic function of set A in (1.14)
$\tilde{A}_\alpha$: α-level set of fuzzy set $\tilde{A}$ in (2.1)
$\tilde{A}_{\alpha+}$: strong α-level set of fuzzy set $\tilde{A}$ in (2.7)
$\tilde{A}_{0+}$: support of set A in (2.3)
$\mathfrak{F}(\mathbb{R}^m)$: the set of fuzzy sets satisfying conditions in Definition 2.3.1
$\tilde{1}_{\{p\}}$: crisp vector with value p in (2.21)
$\tilde{A} = (a_1, a_2, a_3, a_4)_{LR}$: LR-fuzzy interval in Definition 2.3.8
$\mathfrak{F}_{LR}(\mathbb{R})$: set of all LR-fuzzy interval in Definition 2.3.8
$\tilde{A} \wedge \tilde{B}$: intersection of fuzzy sets in Definitions 3.2.1 and 3.2.5
$\tilde{A} \vee \tilde{B}$: union of fuzzy sets in Definitions 3.3.1 and 3.3.4
T_n: generalized t-norm in (3.9)
S_n: generalized s-norm in (3.27)
$\tilde{A}^{(1)} \sqcap \ldots \sqcap \tilde{A}^{(n)} = \sqcap_{i=1}^n \tilde{A}^{(i)}$: intersection of fuzzy sets in Definition 3.2.8
$\tilde{A}^{(1)} \sqcup \ldots \sqcup \tilde{A}^{(n)} = \sqcup_{i=1}^n \tilde{A}^{(i)}$: union of fuzzy sets in Definition 3.3.6
$\mathfrak{A}^\cap$: aggregation function for intersection in Definition 3.2.8
$\mathfrak{A}^\cup$: aggregation function for union in Definition 3.3.6
$\mathfrak{A}_n^L$ and $\mathfrak{A}_n^R$: aggregation functions in Definition 3.4.1
$\mathfrak{A}_n^L$: left-generated by $\mathfrak{A}_2$ in Definition 3.4.1
$\mathfrak{A}_n^R$: right-generated by $\mathfrak{A}_2$ in Definition 3.4.1
$S^\bullet = \{\alpha \in S : \lambda(\alpha) > 0\}$: in (5.36)
$A_\alpha^{(\eta)}$ and $A_{\alpha+}^{(\eta)}$: in (5.4)
$B_\alpha^{(\lambda)}$ and $B_{\alpha+}^{(\lambda)}$: in (5.40) and (5.41), respectively
$B_\alpha^{(\eta,\lambda)}$ and $B_{\alpha+}^{(\eta,\lambda)}$: in (5.62) and (5.63), respectively
$\tilde{A}^\star$ and $\tilde{A}^\circ$: generated by families in Theorem 5.3.1

Mathematical Foundations of Fuzzy Sets, First Edition. Hsien-Chung Wu.

$\tilde{A}^{\perp}$: usual case in (5.22)
$\tilde{A}^{\diamond}$: generated by using one function in (5.35)
$\tilde{A}^{\top}$: generated by using two functions in (5.58)
$\tilde{A} \overset{\kappa}{=} \tilde{B}$: permutably identical in Definition 5.6.1
$\mathcal{M} \overset{(\kappa,\eta)}{=} \tilde{A}$: permutably identical in Definition 5.6.5
$\xi_{\tilde{A}} \doteqdot \kappa \circ \xi_{\tilde{A}^{\star}}$: in Definition 5.6.11
$\tilde{\mathbf{f}}^{(EP)}$: fuzzification using the extension principle in (6.1)
$\tilde{\mathbf{f}}^{(DT)}$: fuzzification using the decomposition theorem in (6.8)
$\tilde{\mathbf{f}}^{(\diamond DT)}$: fuzzification in Subsection 6.2.1
$\tilde{\mathbf{f}}^{(\star DT)}$: fuzzification in Subsection 6.2.2
$\tilde{\mathbf{f}}^{(\dagger DT)}$: fuzzification in Subsection 6.2.3
$\tilde{\mathbf{A}} \oplus_{EP} \tilde{\mathbf{B}}$: addition in (6.2)
$\tilde{f}'(x^*)$: fuzzy derivative on open interval in Definition 6.4.1
$\widetilde{f'}^{(EP)}(\tilde{A})$: fuzzy derivative in Definition 6.4.3
$\widetilde{f'}^{(\diamond DT)}(\tilde{A})$: fuzzy derivative in Definition 6.4.5
$\widetilde{f'}^{(\star DT)}(\tilde{A})$: fuzzy derivative in Definition 6.4.7
$\widetilde{f'}^{(\dagger DT)}(\tilde{A})$: fuzzy derivative in Definition 6.4.9
$\int_E \tilde{f}(x)dv$: fuzzy Lebesgue integral in Definition 6.5.1
$(\diamond\mathrm{DT}) \int_{\tilde{A}}^{\tilde{B}} f(x)dx$: fuzzy Riemann integral in Definition 6.5.3
$(\star\mathrm{DT}) \int_{\tilde{A}}^{\tilde{B}} f(x)dx$: fuzzy Riemann integral in Definition 6.5.5
$(\dagger\mathrm{DT}) \int_{\tilde{A}}^{\tilde{B}} f(x)dx$: fuzzy Riemann integral in Definition 6.5.7
$(\mathrm{EP}) \int_{\tilde{A}}^{\tilde{B}} f(x)dx$: fuzzy Riemann integral in Definition 6.5.9
$(\star\mathrm{EP}) \int_{\tilde{A}}^{\tilde{B}} f(x)dx$: pseudo-fuzzy Riemann integral in Definition 6.5.10
$\tilde{A} \odot \tilde{B}$: arithmetics using min in (7.2)
$\tilde{A} \odot_{EP} \tilde{B}$: arithmetics using the extension principle in (7.14)
$\tilde{\mathbf{A}} \odot_{EP} \tilde{\mathbf{B}}$: arithmetics using the extension principle in (7.30)
$\tilde{A} \odot_{DT} \tilde{B}$: arithmetics using the decomposition theorem in (7.15)
$\tilde{A} \odot_{DT}^{\diamond} \tilde{B}$: arithmetics in (7.16)
$\tilde{A} \odot_{DT}^{\star} \tilde{B}$: arithmetics in (7.17)
$\tilde{A} \odot_{DT}^{\dagger} \tilde{B}$: arithmetics in (7.18)
$\tilde{\mathbf{A}} \ominus_{DT} \tilde{\mathbf{B}}$: difference using the decomposition theorem in (7.34)
$\tilde{\mathbf{A}} \ominus_{DT}^{\diamond} \tilde{\mathbf{B}}$: difference in (7.42)
$\tilde{\mathbf{A}} \ominus_{DT}^{\star} \tilde{\mathbf{B}}$: difference in (7.46)
$\tilde{\mathbf{A}} \ominus_{DT}^{\dagger} \tilde{\mathbf{B}}$: difference in (7.50)
$\tilde{\mathbf{A}} \oplus_{DT} \tilde{\mathbf{B}}$: addition using the decomposition theorem in (7.37)
$\tilde{A}^{(1)} \odot_1 \cdots \odot_{n-1} \tilde{A}^{(n)}$: arithmetics in (7.60)
$\tilde{A}^{(1)} \boxdot_1 \cdots \boxdot_{n-1} \tilde{A}^{(n)}$: arithmetics in (7.62)
$\tilde{A} \odot_{\mathfrak{g}} \tilde{B}$: binary operation in (7.93)
$\tilde{A} \odot_{\mathfrak{g}}^{*} \tilde{B}$: binary operation in (7.115)
$\tilde{A} \odot_{\mathfrak{G}} \tilde{B}$: binary operation in (7.118)
$\tilde{A} \oplus_{\mathfrak{g}} \tilde{B}$: addition in (7.94)
$\tilde{A} \ominus_{\mathfrak{g}} \tilde{B}$: difference in (7.95)
$\tilde{A} \ominus_{G_1} \tilde{B}$: type-I-generalized difference in (7.100)
$\tilde{A} \ominus_{G_2} \tilde{B}$: type-II-generalized difference in (7.103)

$\tilde{A} \ominus_H \tilde{B}$: Hausdorff difference in (7.97)
$\tilde{A} \ominus_{NH} \tilde{B}$: natural Hausdorff difference in (7.119)
$\tilde{A} \ominus_{FH} \tilde{B}$: fair Hausdorff difference in (7.144)
$\tilde{A} \ominus_{FH^*} \tilde{B}$: fair *-Hausdorff difference in (7.145)
$\tilde{A} \ominus_{GFH} \tilde{B}$: generalized fair Hausdorff difference in (7.146)
$\tilde{A} \ominus_{CH} \tilde{B}$: composite Hausdorff difference in (7.156)
$\tilde{A} \ominus^*_{CH} \tilde{B}$: composite *-Hausdorff difference in (7.157)
$\tilde{A} \ominus_{GCH} \tilde{B}$: generalized composite Hausdorff difference in (7.158)
$\tilde{A} \ominus_{CCH_1} \tilde{B}$: type-I-complete composite Hausdorff difference in (7.169)
$\tilde{A} \ominus_{CCH_2} \tilde{B}$: type-II-complete composite Hausdorff difference in (7.170)
$\tilde{A} \ominus^*_{CCH_1} \tilde{B}$: type-I-complete composite *-Hausdorff difference in (7.171)
$\tilde{A} \ominus^*_{CCH_2} \tilde{B}$: type-II-complete composite *-Hausdorff difference in (7.172)
$\tilde{A} \ominus_{GCCH_1} \tilde{B}$: generalized type-I-complete composite Hausdorff difference in (7.173)
$\tilde{A} \ominus_{GCCH_2} \tilde{B}$: generalized type-II-complete composite Hausdorff difference in (7.174)
$\tilde{\mathbf{A}} \circledast_{EP} \tilde{\mathbf{B}}$: inner product using the extension principle in (8.2)
$\tilde{\mathbf{A}} \circledast_{DT} \tilde{\mathbf{B}}$: inner product using the decomposition theorem in (8.8)
$\tilde{\mathbf{A}} \circledast^{\diamond}_{DT} \tilde{\mathbf{B}}$: inner product in in (8.9)
$\tilde{\mathbf{A}} \circledast^{\star}_{DT} \tilde{\mathbf{B}}$: inner product in in (8.11)
$\tilde{\mathbf{A}} \circledast^{\dagger}_{DT} \tilde{\mathbf{B}}$: inner product in in (8.13)
$\tilde{\mathbf{A}} \circledcirc_{EP} \tilde{\mathbf{B}}$: inner product using the extension principle in (8.32)
$\tilde{\mathbf{A}} \circledcirc_{DT} \tilde{\mathbf{B}}$: inner product using the decomposition theorem in (8.40)
$\tilde{A}^{\mathfrak{G}}$: fuzzy set generated by gradual set in (9.2)
$\mathfrak{G}_{\tilde{A}}$: gradual set generated by fuzzy set in (9.6)
$\mathfrak{G}^{\cap}$: gradual set for intersection in (9.19)
$\mathfrak{G}^{\cup}$: gradual set for union in (9.20)
${}_{\alpha}\tilde{A}$: lower α-level set in Definition 10.1.1
${}_{\alpha-}\tilde{A}$: strong lower α-level set in Definition 10.1.1
$\tilde{A}^{\bullet}$: dual fuzzy set in (10.2)
$\tilde{f}^{\bullet}(\tilde{A}^{(1)}, \tilde{A}^{(2)}, \ldots, \tilde{A}^{(n)})$: dual extension principle in (10.3)
$\chi^{\bullet}_{\tilde{A}}$: dual characteristic function in (10.5)
$\tilde{A} \odot^{\bullet} \tilde{B}$: dual arithmetics in (10.12)

Index

Mathematical Foundations of Fuzzy Sets, First Edition. Hsien-Chung Wu.

e

f

g

h